Contemporary Engineering Economics:
A Canadian Perspective

Chan S. Park
Auburn University

Kenneth C. Porteous
University of Alberta

Kenneth F. Sadler
Technical University of Nova Scotia

Ming J. Zuo
University of Alberta

Addison-Wesley Publishers Limited
Don Mills, Ontario • Reading, Massachusetts
Menlo Park, California • New York • Wokingham, England
Amsterdam • Bonn • Sydney • Tokyo • Madrid • San Juan

Executive Editor:	Joseph Gladstone
Managing Editor:	Linda Scott
Acquisitions Editor:	John Clelland
Editors:	Tom Shields/Barbara Brown

Much of the material in this publication was first published in **Contemporary Engineering Economics** by Chan S. Park (Addison-Wesley Publishing Company, Inc., 1993), with illustrations courtesy of The New York Times.

The photograph on the cover is reproduced courtesy of Eric S. Fordham.
The CAD drawing in the photograph is reproduced courtesy of Carroll Touch.

Main entry under title:
   Contemporary engineering economics: a Canadian perspective

1st Canadian ed.
Includes bibliographical references and index.
ISBN 0-201-42388-X

   1. Engineering economy. 2. Engineering economy—Canada. I. Park, Chan S.

TA177.4.C66 1994   658.15   C95-900076-3

The publishers will gladly receive information enabling them to rectify any errors in references or credits.

The authors and publisher have taken care in the preparation of this publication, but make no expressed or implied warranty of any kind and assume no responsibility for errors or omissions. No liability is assumed for accidental or consequential damages in connection with or arising out of the use of the information contained herein.

ISBN 0-201-42388-X

Printed in Canada

   B   C   D   E   MET 99 98 97 96

TRADEMARKS

Lotus and 1-2-3 are registered trademarks of Lotus Development Corporation.
WINDOWS is a registered trademark of the Microsoft Corporation.

Engineering economics is a required course or component in most engineering programs in North America. This textbook, *Contemporary Engineering Economics: A Canadian Perspective,* adopts a contemporary approach to the subject. Our philosophy on the teaching of engineering economics and the unique features of this textbook are discussed below relative to the following:

- What's "Contemporary" About Engineering Economics?
- Overview of the Text
- Content and Approach
- Addressing Educational Challenges
- Flexibility of Coverage
- Canadian Context and Content
- Supplements

What's "Contemporary" About Engineering Economics?

Decisions made during the engineering design phase of product development determine the majority (some say 85%) of the costs of manufacturing that product. As design and manufacturing processes become more complex, the engineer is making decisions that involve money more than ever before. Thus, the competent and successful engineer in the twenty-first century must have an improved understanding of the principles of science, engineering, and economics, coupled with relevant design experience. Increasingly, in the new world economy, successful businesses will rely on engineers with such expertise.

Economic and design issues are inextricably linked in the product/service life cycle. Therefore, one of our strongest motivations in writing this text was to bring the realities of economics and engineering design into the classroom and to help students integrate these issues when contemplating product development problems.

Another compelling motivation was to introduce the computer as a productivity tool for modeling and analyzing engineering decision problems once students have mastered fundamental concepts. Spreadsheets are currently the undisputed standard for automating complex engineering economic problems in industry and they are used increasingly in the classroom. This text introduces spreadsheets in dedicated sections at the ends of chapters.

In addition to spreadsheets, the end-of-chapter sections introduce the Windows-based software, **EzCash**, that is packaged with this text. Conventional engineering economic software has been less than completely successful for two reasons: (1) Programs often present knowledge in rigid ways with few possibilities for adapting to the needs of the individual student, and (2) the structure of the knowledge is usually hidden from the student. **EzCash** was developed to open *visually* the economic computing environment to the student's understanding. **EzCash** is an integrated package that includes the most frequently used economic analysis methods. It organizes information via graph-based structures that can be explored independently by a student.

Of course, our underlying motivation for writing this book was not simply to address contemporary needs, but to address as well the ageless goal of all educators: to help students to learn. Thus, thoroughness, clarity, and accuracy of presentation of essential engineering economics were our aims at every step in the development of the text.

Overview of the Text

Although it contains little advanced math and few truly difficult concepts, the introductory engineering economics course is often a curiously challenging one for the sophomores, juniors, and seniors who take it. There are several likely explanations for this difficulty:

- The course is the student's first analytical consideration of money (a resource with which he or she may have had little direct contact beyond paying for tuition, housing, food, and textbooks).
- An emphasis on theory—while critically important to forming the foundation of a student's understanding—may obscure for the student the fact that the course aims, among other things, to develop a very practical set of analytical tools for measuring project worth. This is unfortunate since, at one time or another, virtually every engineer—not to mention every individual—is responsible for the wise allocation of limited financial resources.
- The mixture of industrial, civil, mechanical, electrical, manufacturing, and other engineering undergraduates who take the course often fail to "see themselves" in the skills the course and text are intended to foster. This is perhaps less true for industrial engineering students, whom many texts take as their primary audience, but other disciplines are often motivationally shortchanged by a text's lack of applications that appeal directly to them.

Goals of the Text

This text aims not only to build a sound and comprehensive coverage of the *concepts* of engineering economics but also to address the student difficulties outlined above, all of which have their basis in an inattentiveness to the practical concerns of engineering economics. More specifically, this text has the following chief goals:

1. To build a thorough understanding of the theoretical and conceptual basis on which the practice of financial project analysis is built.
2. To satisfy the very practical need engineers have to make informed financial decisions when acting as a team member or project manager for an engineering project.
3. To incorporate all the critical decision-making tools—including the most contemporary, computer-oriented ones—that engineers bring to the task of making informed financial decisions.
4. To appeal to the full range of engineering disciplines for which this course is often required: industrial, civil, mechanical, electrical, computer, aerospace, chemical, and manufacturing engineering, as well as engineering technology.

Prerequisites

The text is intended for undergraduate engineering students at the sophomore level or above. The only mathematical background required is elementary calculus. (For Chapter 15, a first course in probability or statistics is helpful but not necessary, since the treatment of basic topics there is essentially self-contained.)

Content and Approach

Educators generally agree upon the proper contents and organization of an engineering economics text. A glance at the table of contents will show you that this text matches the standard embraced by most instructors and reflected in competing texts. However, one of our driving motivations was to supersede the standard in terms of depth of coverage and care with which difficult concepts are presented. Accordingly, the content and approach of *Contemporary Engineering Economics: A Canadian Perspective* reflect the following goals:

Thorough development of the concept of the time value of money

The notion of the time value of money and the interest formulas that model it form the foundation upon which all other topics in engineering economics are built. Because of their great importance, and because many students are being exposed to an analytical approach to money for the first time, interest topics are carefully and thoroughly developed in Chapters 2 and 3.

- Chapter 2 carefully examines the *conceptual* underpinnings of interest—the time value of money—including more "what-if" and graphical exploration than any other text.
- An understanding of the time value of money is extended via its real world complexities—effective interest, noncomparable payment and compounding periods, etc.—in Chapter 3.

Thorough, reasonably paced coverage of the major analysis methods

The equivalence methods—present worth, annual worth, and future worth—and rate of return analysis are the bedrock methods of project evaluation and comparison. This text carefully develops these topics in Chapters 4, 5, and 6, pacing them for maximum student comprehension of the subtleties, advantages, and disadvantages of each method.

- A separate, dedicated chapter (Chapter 5) on annual worth is presented to emphasize the circumstances in which that method of project analysis is preferred to other methods.
- The difficulties and exceptions associated with rate of return analysis are thoroughly covered in Chapter 6.
- Coverage of internal rate of return for nonsimple projects is included in Chapter 6 but marked as optional for those who wish to avoid this complication in an introductory course. The section can be omitted without disrupting the flow of topics.

More emphasis than competitive texts on developing after-tax project cash flows

Estimating and developing project cash flows is the first critical step in conducting an engineering economic analysis—further analysis, comparison of projects, and decision making all depend on intelligently developed project cash flows. A particularly important goal of this text is to build confidence in developing *after-tax* cash flows.

- Chapter 9 is a unique synthesis of previously developed topics (analysis methods, depreciation, and taxes) and is dedicated to building skill and confidence in developing after-tax cash flows for a series of fairly complex projects.
- The investment tax credit is covered briefly in text and examples in Chapter 9.

Complete coverage of the special topics that round out a comprehensive introduction to engineering economics

A number of special topics are important to a comprehensive understanding of introductory engineering economics. Chapters 10–15 cover the following topics, respectively:

- Inflation
- Project financing
- Replacement analysis
- Capital budgeting
- Public sector analysis
- Sensitivity and risk analyses

Recognizing that time availability and priorities vary from course to course and instructor to instructor, each one of these chapters is sufficiently self-contained that it may be skipped or covered out of sequence, as needed.

Addressing Educational Challenges

The features of *Contemporary Engineering Economics: A Canadian Perspective* were selected and shaped to address key educational challenges. It is the observation of both authors and publisher — based on many conversations with engineering educators — that certain challenges consistently frustrate both instructors and students across the engineering curriculum. Low student motivation and enthusiasm, student difficulty in developing problem-solving skills and intuition, challenges to integrate technology without shortchanging fundamental concepts and traditional methods, and student difficulty in prioritizing and remembering enormous amounts of information are among the key educational challenges that drove the features program in this textbook.

Building problem-solving skills and confidence

The examples in the text are formatted to maximize their usefulness as guides to problem solving. Further, they are intended to stimulate student curiosity to look beyond the mechanics of problem solving to "what-if" issues, alternative solution methods, and interpretation of solutions.

Example titles promote ease of student reference and review.

Example 2.19   A uniform series problem with "too much" information

Frequently, people share the cost of a lottery ticket. In one instance, 21 factory workers had agreed to pool $21 to play a lottery and split any winnings. Their winning ticket was worth $13,667,667, which would be distributed in 21 annual payments of $650,793. That meant each member of the pool would receive 21 annual payments of about $24,000, according to their lawyer. John Brown, one of the lucky workers, wanted to quit the factory and start his own business, which required him to secure a $250,000 bank loan. Brown offered to put up his future lottery earnings (as collateral) to secure the loan. If the bank's interest rate is 10% per year, how much can he borrow against his future lottery earnings?

Discussion sections at the beginning of complex examples help students begin organizing a problem-solving approach.

Discussion: We need to identify the relevant data in this problem, because some numbers ultimately have nothing to do with the solution method. Basically, Brown wants to borrow $250,000 from a bank, but there is no assurance from the bank that he will get the full amount. (Normally a lending officer determines the maximum amount that one can borrow based on the borrower's capability of repaying the loan.) If the bank views Brown's lottery earnings as his only source of future income for repaying the loan, the bank must find the equivalent present worth of his 21 annual receipts of $24,000 in order to set the maximum loan amount.

Solution

Given and **Find** heads help students identify critical data. This convention is employed in Chapters 2–10, then omitted in Chapters 11–15 after student confidence in setting up solution procedures has been established.

Given: $i = 10\%$ per year, $A = \$24,000$, and $N = 21$ years

Find: P

$$P = \$24,000(P/A, 10\%, 21) = \$24,000(8.6487) = \$207,569.$$

The bank would lend Brown a maximum of $207,569. He would have to borrow the remaining balance from other sources.

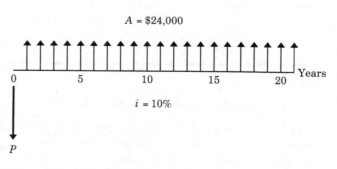

Figure 2.29   ■   Cash flow diagram (Example 2.19)

Comments: *Of course, the actual amount the bank is willing to lend could be far less than $207,569. Note that the critical data included the actual cash flow over the time rather than the total sum of the winnings because of the different time values of the payments.*

Capturing the student's imagination

Students want to know how the conceptual and theoretical knowledge they are acquiring will be put to use. This textbook incorporates real world applications and contexts in a number of ways to stimulate student enthusiasm and imagination:

Feature	Purpose	Sample References
Real-world, conceptual overview of engineering economics established in Chapter 1	To provide an engaging introduction to engineering economics via examples of its practical use	p. 4: GM's decision to build an electric car pp. 21–22:Motorola's investment in global portable phone network
Chapter-opening scenarios	To establish an interest and need to know chapter concepts within the context of a practical application	pp. 192–193, Chapter 4: Construction of a hydroelectric plant, pp. 470–471, Chapter 9: Flexible manufacturing of electronic circuit boards
An abundance of homework problems involving real engineering projects	To stimulate student interest and motivation with actual engineering investment projects, many taken from today's headlines	pp. 194–195, Prob. 3.79: Electronic "anti-noise" system for car interiors
A full range of engineering disciplines represented in problems, examples, chapter openers, and case studies	To illustrate the many disciplines that require engineering economics. Industrial, civil, mechanical, electrical, manufacturing, and other areas are all represented.	pp. 367–368, Prob. 6.50: Mechanical engineering pp. 114–115, Prob. 2.57: Electrical engineering p. 248, Prob. 4.22: Civil engineering p. 299, Prob. 5.40: Industrial engineering.

Harnessing the power of the computer

Three years ago, when Professor Chan Park began work on his U.S. edition of this textbook and Addison-Wesley began examining market desires and demands, the integration of the computer into the course and text was a sensitive issue, with most instructors' opinions varying from ambivalent to negative. A tremendous evolution appears to have swept the course since then: Students have greater access to and familiarity with the appropriate hardware and software tools, and instructors have greater inclination either to treat these topics explicitly in the course or to encourage students to experiment independently.

A remaining concern is that the computer will undermine true understanding of course concepts. This text does *not* promote the trivial or mindless use of computers as a replacement for genuine understanding of and skill in applying traditional solution methods. Rather, it focuses on the computer's productivity-enhancing benefits for complex project cash flow development and analysis. Specifically, this text includes a robust introduction to computer automation in the form of **Computer Notes**, which appear at the end of most chapters.

Spreadsheets are introduced via Lotus 1-2-3 examples. Where appropriate, conversion tables are included so that the built-in functions of both Excel and QuattroPro can be compared to Lotus 1-2-3 for ease in "translating" the example to another software.

EzCash, a Windows-based interactive analysis tool included free with this text, is demonstrated immediately after each spreadsheet example, using the same data and problem context. **EzCash** is a productivity tool intended to streamline complex and time-consuming analytical tasks within a user-friendly environment.

For both spreadsheet and **EzCash** coverage, the emphasis is on demonstrating a chapter concept that embodies some complexity that can be much more efficiently resolved by computer than by traditional longhand solutions. See, for example, Chapter 3 (pp. 177–180) in which the **Computer Notes** tackle the computational difficulty of generating a complete loan repayment schedule.

Synthesizing, reinforcing, and summarizing key ideas

Keeping "the forest and the trees" in perspective is always a challenge to engineering students. To facilitate the retention and understanding of important concepts, procedures, and reference items, this text utilizes a wide range of devices:

Feature	Purpose	Sample references
Point-by point chapter summaries	To highlight and reiterate key equations, terms, and concepts in an easy-to-digest format	pp. 287, Chapter 5: In addition to reviewing important chaper concepts, this feature includes a summary table comparing payback, equivalence, and rate of return methods.

Feature	Purpose	Sample references
Interpretive cash flow diagrams	To reinforce important concepts visually—a medium students are often more comfortable with than the written word.	p. 49, Fig. 2.8: Compounding effects on periodic interest payments p. 146, Fig. 3.17: Contrasting principal and interest portions of loan repayment cash flows
Summary tables	To capture and reiterate useful reference material and important concepts in a compact form	p. 87, Table 2.3: Discrete compounding formulas p. 419, Table 7.4: Book vs. tax depreciation
Flow charts, in-text lists	To reiterate important concepts in a brief, step-wise fashion	pp. 129–136, Section 3.2.2: Computational procedure for noncomparable compounding and payment periods p. 222, Fig. 4.11: Analysis period implied in comparing mutually exclusive alternatives

Flexibility of Coverage

For a typical three-credit-hour, one-semester course, the majority of topics in the text can be covered in the depth and breadth in which they are presented. For other arrangements—quarter terms or fewer credit-hours—Chapters 1–9 present the essential topics, with subsequent chapters presenting optional coverage. By varying the depth of coverage and supplementing the reading with some real engineering project case studies, there is enough material for a continuing, two-term engineering economics course.

Because the topics of the time value of money and interest relationships are so basic to the overall subject of engineering economics, they are treated in depth in Chapters 2 and 3. For those wishing a briefer coverage of these topics, we suggest covering all of Chapter 2, and Sections 3.1 and 3.2 of Chapter 3. Remaining topics in Chapter 3 may be omitted entirely or assigned as additional readings.

Canadian Context and Content

The "Canadianization" of the original U.S. edition has involved several types of changes. All tax-related explanations and calculations in the text and the **EzCash** software were changed to reflect current Revenue Canada tax regulations. Wherever possible, examples and problems have been modified to reflect a Canadian context and appropriate metric units. Miscellaneous modifications, additions, and deletions were made such that material was consistent with Canadian business practices and/or relatable to techniques presented in previous Canadian textbooks on engineering economy.

Some specific topics that have been addressed from the Canadian perspective are as follows:

- Commercial and personal loan practices, including mortgages, in Chapter 3.
- Capital Cost Allowance calculations in Chapter 7.
- Personal and corporate income tax calculations (Federal and Provincial) in Chapter 8.
- Tax effects arising from asset disposal (simplified and rigorous approaches) in Chapter 8.
- Application of capital tax factors for after-tax cash flow problem formulation in Chapter 9.

In some instances, the treatment involves considerable detail, but instructors may select sections which are consistent with the desired depth of coverage.

Supplement

An **Instructor's Manual** (ISBN #83036) is available to adopters of this text. In addition to complete solutions to all problems, it contains:

- Categorization of problems by concept tested.
- Transparency masters for key cash flow diagrams from the text.
- A selection of exam review questions and solutions which may be photo-copied and distributed to students.

Acknowledgments by Chan Park

I and the book team at Addison-Wesley are grateful to the many individuals who reviewed and contributed to the text. When we began to draw up a list of individuals who had a role in ensuring the high quality and accuracy of the text, we were astonished to discover that over 60 people had served in a range of capacities, including text reviewers, example reviewers, problem reviewers and contributors, software testers, spreadsheet advisors, and case study reviewers. We would like to thank each of them: Kamran Abedini, California Polytechnic-Pomona; James Alloway, Syracuse University; Mehar Arora, U. Wisconsin-

Stout; Joel Arthur, California State University-Chico; Robert Baker, University of Arizona; Robert Barrett, Cooper Union and Pratt Institute; Tom Barta, Iowa State University; Charles Bartholomew, Widener University; Bopaya Bidanda, University of Pittsburgh; James Buck, University of Iowa; Philip Cady, The Pennsylvania State University; Tom Carmichal, Southern College of Technology; Jeya Chandra, The Pennsylvania State University; Max C. Deibert, Montana State University; Stuart E. Dreyfus, University of California-Berkeley; W. J. Foley, RPI; Jane Fraser, Ohio State; Carl Haas, University of Texas-Austin; John Held, Kansas State University; T. Allen Henry, University of Alabama; R. C. Hodgson, University of Notre Dame; Philip Johnson, University of Minnesota; Harold Josephs, Lawrence Tech; Henry Kallsen, University of Alabama; W. J. Kennedy, Clemson University; Oh Keytack, University of Toledo; Stephen Kreta, California Maritime Academy; John Krogman, University of Wisconsin-Platteville; Dennis Kroll, Bradley University; Michael Kyte, University of Idaho; William Lesso, University of Texas-Austin; Martin Lipinski, Memphis State University; Robert Lundquist, Ohio State University; Richard Lyles, Michigan State University; Abu S. Masud, The Wichita State University; James Milligan, University of Idaho; Richard Minesinger, University of Massachusetts, Lowell; James S. Noble, University of Washington; Elizabeth Paté-Cornell, Stanford University; Cecil Peterson, GMI; J. K. Rao, California State University-Long Beach; Susan Richards, GMI; Mark Roberts, Michigan Tech; John Roth, Vanderbilt University; Bill Shaner, Colorado State University; Fred Sheets, California Polytechnic-Pomona; Dean Shupe, University of Cincinnati; Milton Smith, Texas Tech; Charles Stavridge, FAMU/FSU; Junius Storry, South Dakota State University; Frank E. Stratton, San Diego State University; Donna Summers, University of Dayton; Joe Tanchoco, Purdue; Deborah Thurston, University of Illinois-UC; Thomas Ward, University of Louisville.

In addition, we are especially grateful to: Richard Bernhard, North Carolina State, who provided detailed reviewing of the entire manuscript, as well as spreadsheet and **EzCash** reviews; Bruce Hartsough, University of California, Davis, who edited and contributed examples; Wayne Parker, Mississippi State University, who contributed several spreadsheet examples; Theo De Winter, Boston University, L. Jackson Turvaville, Tennessee Technological University, Wayne Knabach, South Dakota State University, and George Stukhart, Texas A&M, all of whom contributed original homework problems; Karen K. Renner, Chromalox Instruments and Controls Corp. (Case Study 4), George Prueitt, U.S. Army (Case Study 5), John Evans, Acustar, Inc. (Case Study 7), James Luxhoj, Rutgers University (Case Study 8) Frederick Davis, U.S. Air Force (Section 14.4.2), who provided original data and helped shape it into a case study; and Bushan Byragani, who assisted in developing the **EzCash** software.

Personally, I would like to express my deep gratitude to Donald A. Fowley, Executive Engineering Editor at Addison-Wesley. He oversaw the entire project with a great deal of interest. He has been a most knowledgeable, helpful, and creative editor. Faith Sherlock gracefully took over "mid-project" as Sponsoring Editor, and I wish to thank her for smoothing the transition period and for her diligent attention to marketing and sales issues without which no text can succeed in an intensely competitive market. I also wish to thank

Laurie McGuire, who served as development editor during the preparation of the manuscript. Her intuitive understanding and writing skill were of immense value for the successful completion of this manuscript. Thanks are also due Chi Jae Oh and Gyu-Tai Kim for their assistance in preparing the final draft.

I would like to thank Ed Unger, Head of Industrial Engineering at Auburn University, who provided me with resources and constant encouragement to complete the book.

Acknowledgments by the Canadian Authors

A number of individuals have contributed to our efforts in various ways. We would like to acknowledge the advice and assistance provided by the following: John R. McDougall, University of Alberta; Douglas A. Hackbarth, Stanley Engineering Consultants Ltd.; Ron G. Bryant, RBC Dominion Securities; Barry J. Walker, Peterson Walker Chartered Accountants; Carl F. Hunter, Dalcor Consultants Ltd.; Tony D'Andrea, Toronto Dominion Bank; Glen A. Mumey, University of Alberta; K.N. Gopalakrishnan, University of Alberta; Ron Pelot, Technical University of Nova Scotia; Peter Wilson, Technical University of Nova Scotia.

This project was conceived by the College Division of Addison-Wesley Canada. Joseph Gladstone, Executive Editor, John Clelland, Acquisitions Editor, Linda Scott, Managing Editor, and their editors developed an aggressive schedule and worked with us to make this book a reality in a timeframe of less than ten months. We would like to thank them for their help and patience.

Finally, the preparation of this manuscript would not have been possible without the support and understanding of our families.

Chan S. Park
Auburn University
Auburn, Alabama

Kenneth C. Porteous
University of Alberta
Edmonton, Alberta

Kenneth F. Sadler
Technical University of Nova Scotia
Halifax, Nova Scotia

Ming J. Zuo
University of Alberta
Edmonton, Alberta

CONTENTS

CHAPTER 1

Engineering Economic Decisions

S uppose as a college student you worked hard one summer and earned $2000. Assuming that your parents take care of your college expenses, what would you do with your money? There are many ways you could spend it—you might purchase stereo equipment, or a personal computer, for example. What if you already have those items? Surely keeping the money in your wallet is not the best way to manage it: You could at least deposit it in a savings account that earns interest. Alternately, you might buy a guaranteed investment certificate, which normally pays a higher rate of

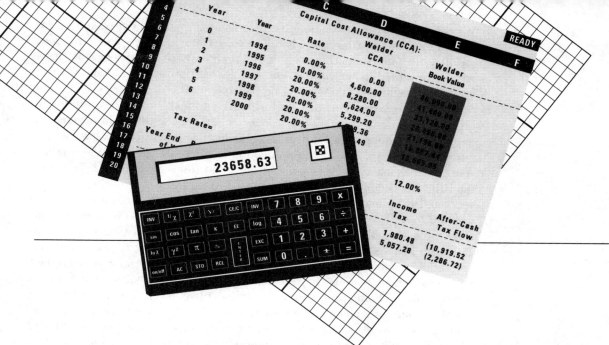

interest than a savings account. You might even take a chance and buy some bonds or stocks that could yield an even better return. If you think bonds and stocks are too passive an investment, you could develop a long-term business strategy of your own. You might purchase some painting equipment and start a small-scale painting business during your off-school hours. If your business prospers and brings in enough revenue, you will recover your expenses and earn a reasonable profit.

We can view all these situations as investment activities. They each have two factors in common: **time** and **risk**. The *time* factor can be seen in that some sacrifice (i.e., investment) is made in the *present*, with the expectation that it will be rewarded in the *future*. The *risk* is that the reward (i.e., profit) is of uncertain magnitude and may not be realized at all.[1]

The economic decisions that engineers have to make in business differ very little from those of the individual investor, except for the scale of concern. Suppose, for example, that a firm is producing a product with a lathe that was purchased 12 years ago. As the production engineer in charge of the product, you expect demand for it to continue into the foreseeable future. However, the lathe has begun to show its age: It has broken frequently during the last 2 years and has finally stopped operating altogether. Now you have to decide whether to replace or repair the damaged lathe. If you expect a more efficient lathe to be available in 1 or 2 years, you might repair it instead of replacing it. The main issue is whether you should make the considerable investment in a new lathe now or later. As an added complication, if demand for your product begins to decline, you may have to conduct an economic analysis to determine whether the declining profits from the project offset the cost of a new lathe.

Let us consider an engineering decision problem of much larger scale, taken from the real world. There is increasing public concern about poor air quality, which is caused primarily by gasoline-powered automobiles. To address this concern, General Motors Corporation has announced its intention to build an advanced electric car known as Impact. The biggest question remaining about the vehicle concerns its battery. With its current battery design, Impact's monthly operating cost would be roughly twice that of a conventional automobile. The primary advantage of the design, however, is that Impact does not emit any pollutants, a feature that could be very appealing at a time when government air-quality standards are becoming more rigorous and consumer interest in the environment is strong.

General Motors engineers have stated that the total market demand for Impact would have to be 100,000 cars to justify production. Despite General Motors management's decision to build the battery-powered electric car, their engineers were still uncertain whether the market demand for such a car would be sufficient to justify its production.

Obviously, this level of engineering decision is more complex and more significant to the company than the decision as to when a new lathe should be purchased. Projects of this nature involve large sums of money over long periods of time, and it is difficult to predict market demand accurately. An erroneous forecast of product demand can have serious consequences: With any

---

[1] W. F. Sharpe, *Investment,* Prentice-Hall, Englewood Cliffs, N.J., 1978.

overexpansion, you will have to pay unnecessary expenses for unused raw materials and finished product. In the case of Impact, if improved battery design never materializes, demand may remain insufficient to justify the project.

In this book we will consider many investment situations, personal as well as business. The focus, however, will be on evaluating engineering projects on the basis of economic desirability and on investment situations that face a typical firm.

1.1 The Role of Engineers in Business

Apple Computer, Microsoft Corporation, and Sun Microsystems all produce computer products and have a market value of several billion dollars. These companies were all started in the late 1970s or early 1980s by young college students with technical backgrounds. When they went into the computer business, the students initially organized their companies as proprietorships. As the businesses grew, they became partnerships and eventually converted to corporations. This chapter will introduce these three primary forms of business organization and briefly discuss the role of engineers in business.

1.1.1 Types of Business Organization

As an engineer, it is important to understand the nature of the business organization with which you are associated. This section will present some basic information about the type of organization you should choose if you decide to go into business for yourself.

There are three legal forms of business, each having certain advantages and disadvantages: proprietorship, partnership, and corporation.

Sole Proprietorship

A proprietorship is a business owned by one individual. This person is responsible for the firm's policies, owns all its assets, and is personally liable for its debts. A proprietorship has two major advantages: First, it can be formed easily and inexpensively. There are no legal and organizational requirements associated with setting up a proprietorship, and organizational costs are therefore virtually nil. Second, all the earnings of a proprietorship are taxed at the owner's personal tax rate, which is often lower than the rate at which corporate income is taxed. The major disadvantage of proprietorships is that they cannot issue stock and bonds, making it difficult for them to raise capital for any business expansion.

Partnership

A partnership is similar to a proprietorship except that it has more than one owner. Most partnerships are established by a written contract between the partners, which normally specifies salaries, contributions to capital, and the distribution of profits and losses. A partnership has many advantages, among which are its low cost and ease of formation. Because more than one person makes contributions, a partnership typically has a larger amount of capital available for business use. Since the personal assets of all the partners stand behind the business, a partnership can borrow money more easily from a bank. Each partner pays only personal income tax on his or her share of the partnership's taxable income.

On the negative side, under partnership law, each partner is liable for the business's debts. This means that the partners must risk all their personal assets, even those not invested in the business. And while each partner is responsible for his or her portion of the debts in the event of bankruptcy, if any partner cannot meet his or her pro rata claim, the remaining partners must take over the unresolved claims. Finally, the partnership has a limited life insofar as it must be dissolved and reorganized if one of the partners quits.

Corporation

A corporation is a legal entity under provincial or federal law. It is separate from its owners and managers. This separation gives the corporation four major advantages: (1) it can raise capital from a large number of investors by issuing stocks and bonds; (2) it permits easy transfer of ownership interest by trading shares of stock; (3) it allows limited liability—personal liability is limited to the amount of the individual's investment in the business; and (4) it is taxed differently than proprietorships and partnerships, and under certain conditions, the tax laws favor corporations. On the negative side, it is expensive to establish a corporation. Furthermore, the corporation is subject to numerous governmental requirements and regulations.

As a firm grows, it may need to change its legal form because this affects the extent to which it has control of its own operations and its ability to acquire funds. The legal form of organization also affects the risk borne by the owners in case of bankruptcy and the manner in which the firm is taxed. Apple Computer, for example, started out as a two-man garage operation. As the business grew, the owners felt constricted by the form of organization: It was difficult to raise capital for business expansion; they felt that the risk of bankrupcy was too high to bear; and as business income grew, their tax burden grew as well. Eventually, they found it necessary to convert the company into a corporation.

In Canada, the overwhelming majority of business firms are sole proprietorships, followed by corporations and partnerships. However, in terms of total business volume (sales dollars), the quantity of business done by sole proprietorships and partnerships is several times less than that of corporations. Since most business is conducted by companies of the corporation form, this text will generally address the economic decisions encountered in corporations.

1.1.2 The Role of Engineers in a Corporation

What role do engineers play within a firm? What specific tasks are assigned to the engineering staff, and what tools and techniques are available to it for improving the firm's profits? Engineers are called upon to participate in a variety of decision-making processes, ranging from manufacturing to marketing to financing decisions. We will restrict our focus, however, to various economic decisions related to engineering projects. We refer to these decisions as **engineering economic decisions.**

In manufacturing, engineering is involved in every detail of a product, from the conceptual design to the shipping. In fact, engineering decisions account for the majority (some say 85%) of the product costs. Engineers must consider the effective use of capital assets such as buildings and machinery. One of the engineer's primary tasks is to plan for the acquisition of equipment (capital expenditure) that will enable the firm to design and produce products economically.

Capital budgeting[2] refers to investment decisions involving fixed assets, and it therefore encompasses the entire process of analyzing projects and deciding if they should be included in the budget. The capital budgeting process includes several phases. First is the administration and organization of a capital expenditure program. This is followed by proposals and strategies for new investment opportunities. Next the future benefits and costs of investment opportunities must be estimated to determine whether they are worth undertaking. Finally, once the project is under way, its progress and contributions to the overall investment program must be reviewed regularly. Therefore, the entire capital budgeting process is of fundamental importance to the success or failure of the firm.

With the purchase of any fixed asset such as equipment, we need to estimate the profits (more precisely, cash flows) that the asset will generate during its service period. In other words, we have to make capital expenditure decisions based on predictions about the future. Suppose, for example, you are considering the purchase of a deburring machine to meet the anticipated demand for hubs and sleeves used in the production of gear couplings. You expect the machine to last 10 years. This purchase decision thus involves an implicit 10-year sales forecast for the gear couplings, which means that a long waiting period will be required before you will know whether the purchase was justified.

---

[2] This definition is given by Professor Eugene F. Brigham in his book, *Fundamentals of Financial Management*, Fifth Ed., The Dryden Press, 1990.

An inaccurate estimate of asset needs can have serious consequences. If you invest too much in assets, you incur unnecessarily heavy expenses. Spending too little on fixed assets also is harmful, for then the firm's equipment may be too obsolete to produce products competitively and without an adequate capacity you may lose a portion of your market share to rival firms. Regaining lost customers involves heavy marketing expenses and may even require price reductions and/or product improvements, all of which are costly.

■ 1.2 A Typical Engineering Project

In any product development, engineering is called upon to translate an idea into reality. A firm's growth and development largely depend upon a constant flow of ideas for new products, and for the firm to remain competitive, it has to make existing products better or produce them at a lower cost. Traditionally, the marketing department would propose a product and pass the recommendation to the engineering department. Engineering would work up a design and pass it on to manufacturing, which would make the product. With this type of product development cycle, a new product normally takes several months (or even years) to reach the market. A typical tool maker, for example, would take 3 years to develop and market a new machine tool. However, the Ingersoll-Rand Company, a leading tool maker, was able to cut down the normal development time by one third. How did the company do this? A group of engineers examined the current product development cycle to find out why things dragged on. They learned how to compress the crippling amount of time it took to bring products to life.[3] In the next section, we will present an example of how a design engineer's idea eventually turned into a popular consumer product.

1.2.1 How a Typical Project Idea Evolves

In many ways, the Gillette Sensor razor (Fig. 1.1), introduced in January 1990, is not a "typical" project. It represents the single most expensive project the Gillette Company has ever developed, and its success has been nothing short of phenomenal—it is currently the best-selling razor in the 39 countries in which it has been launched. Less than 2 years after its introduction, the 1 billionth Sensor cartridge was manufactured by Gillette.

A look at the evolution of the investment project that produced Sensor reveals a number of stages and events that are common to most engineering in-

---

[3] It took only a year to develop Ingersoll's new air grinder, a $225 flashlight-sized tool to finish and polish the pieces that become everything from barstools to jet planes.

Figure 1.1 ■ Gillette Corporation's new razor—Sensor (Courtesy of The Gillette Company)

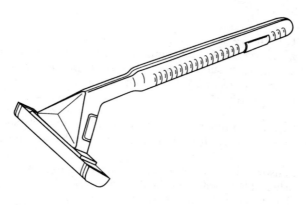

vestment projects. The following excerpted passages are from *Business Week*, January 29, 1990. Additional information has been provided by Gillette.

> There are 40 engineers, metallurgists, and physicists at Gillette Company's Reading (Britain) research facility who spend their days thinking about shaving and little else. In 1977, one of them had a bright idea. John Francis had already figured out how to create a thinner razor blade that would make Gillette's cartridges easier to clean. Then, the design engineer remembered a notion he had toyed with for years: He could set the thinner blades on springs so that they would follow the contours of a man's face. He built a simple proto-type, gave it a test, and thought: "This is pretty good." He passed the idea to his boss, then went on to the next project.[4]

The initial idea for an investment project may be born of inspiration or ne-cessity, and may entail an entirely new venture for a firm, or the expansion or improvement of current ventures. In the case of Sensor, John Francis's inspira-tion was for an improvement so profound that it caused consumers and com-petitors to view Sensor as the next generation in razors.

The case of Sensor also demonstrates that a long period of time may elapse between the moment an idea is articulated and the moment it is fully imple-mented as a completed, for-sale product. This time interval may vary widely from industry to industry. Computer manufacturers, for example, place a pre-mium on swift implementation of investment projects due to the rapidly evolv-ing technologies with which they work and compete.

---

[4] *Business Week,* January 29, 1990. Reprinted by permission of Business Week Magazine and The Gillette Company.

Selection of materials and processes are some of the most fundamental decisions affecting project costs. Gillette faced some particularly sticky issues with Sensor:

> Gillette used styrene plastic to mold blade cartridges for all its razors, because it's inexpensive and easy to work with. But a styrene spring, tests showed, lost some of its bounce over time. The engineers turned to a resin called Noryl, a stronger material that kept its bounce.[5]

In addition to innovating the material with which Sensor's blade cartridge would be constructed, Gillette made a remarkable production decision:

> Sensor's blades were to "float" on the springs independently of each other. That meant the blades had to be rigid enough to hold their shape—though each is no thicker than a sheet of paper. Engineers decided to attach each blade to a thicker steel support bar.
>
> The question was, how? For mass manufacturing, glue was too messy and too expensive. The answer was lasers. Engineers built a prototype laser that spot-welded each blade to a support without creating heat that would damage the blade edge, relying on a process more commonly used to make such things as heart pacemakers.[6]

As with any investment project, the materials and manufacturing processes that Sensor required had cost implications. Estimating costs is one of the critical tasks in any investment project:

> Sensor is the single most expensive project Gillette has ever taken on: By the time the razor hits stores, the company will have spent an estimated $200 million in research, engineering, and tooling.[7]

Engineers conducting economic analyses are used to considering research, material, tooling, and labor costs. They may not be used to factoring in such expenses as advertising, promotion, and public relations. In the case of Sensor, such costs added another $100 million to the project's overall cost. In making an accept or reject decision about a project, after estimating costs, the next concern is revenues:

> Gillette will need a huge win to justify its investment in Sensor. The new razor must add about four percentage points to Gillette's market share in North America and Europe just to recoup its ad budget.

---

[5] Ibid.

[6] Ibid.

[7] Ibid.

Those pressures also help explain why it took so long to get Sensor out the door. Gillette executives were reluctant to make the huge investment in manufacturing and marketing at a time when the company could still count on prodigious profits from its existing razors.[8]

Clearly, with a sizable investment, a firm needs a sizable return to justify undertaking an investment project. In the case of Sensor, the advertising budget alone was a significant expense; recovering that cost required that the company increase its market share by 4%. Note also that if existing ventures are already generating adequate income, a firm may be reluctant to undertake a risky new venture which could fail or which might jeopardize existing projects by taking resources and sales away from them. (In fact, in Sensor's case, the new product took sales away from other Gillette products, namely Trac II and Atra, but to a lesser extent than the company expected.)

Even with the most sophisticated market research to support it, no product is assured of success. Estimating project costs and revenues as exactly as possible is critical to a firm's decision to pursue a project. But finally the success or failure of a project cannot be judged until it is implemented and evaluated in comparison with predictions of its success. In the case of Sensor, success has been well documented, and the risk undertaken by Gillette has been justified.

Virtually every employee of a company is a potential source of ideas for a new product or the improvement of existing products, and many companies encourage their employees to present new ideas for evaluation. Many good ideas for product improvement come from the engineers actually involved in production or marketing. The process, of course, may take a long time. As our discussion illustrated, it took 13 years to realize John Francis's idea. Fortunately, not all engineering ideas take such a long time to develop into products.

1.2.2 Impact of an Engineering Project on the Firm's Financial Statements

Engineers must understand the business environment in which a company's major business decisions are made. It is important for an engineering project to generate profits, but it also must strengthen the firm's overall financial position. How do we measure The Gillette Company's success in the Sensor project? Will enough Sensor razors be sold, for example, to keep the blade business as Gillette's biggest source of profits? While the Sensor project will provide comfortable, reliable, low-cost shaving for the customers, the bottom line is its financial performance over the long run.

Regardless of business form, each company has to produce basic financial statements at the end of each operating cycle (typically a year). These financial statements provide the basis for future investment analysis. In practice, we seldom make investment decisions based solely on our estimate of a project's

---

[8] Ibid.

profitability because we must also consider its overall impact on the financial strength and position of the company.

Suppose that you were the CEO (Chief Executive Officer) of The Gillette Company. Let us further suppose that you even hold some shares in the company, which also makes you one of the company's many owners. What objectives would you set for the company? While all firms are in business in hopes of making a **profit**, what determines the market value of a company is not profits per se, but **cash flow**. It is, after all, the available cash that determines the future investments and growth of the firm. Therefore, one of your objectives should be to increase the company's value to its owners (including yourself) as much as possible. The **market price** of your company's stock to some extent represents the value of your company. Many factors affect your company's market value: present and expected future earnings, the timing and duration of these earnings, as well as the risk associated with them. Certainly, any successful investment decision will increase the company's market value. The stock price can be a good indicator of your company's financial health and may also reflect the market's attitude about how well your company is managed for the benefit of its owners.

In the case of The Gillette Company, the firm's financial position prior to and after introducing the Sensor to the market was as follows. (Appendix A provides a brief overview of the reporting format of financial statements.)

The Gillette Company at a Glance*
(Thousands of dollars, except per share amounts)

Year Ended (Dec. 31)	1990	1989
Net sales	$4,344,600	$3,818,500
Net income	$ 367,900	$ 284,700
Earnings per share	$3.20	$2.70
Total assets	$3,671,300	$3,114,000
Total liabilities	$2,805,900	$2,444,000
Net worth	$ 865,400	$ 670,000
Stock price (NYSE)	$63\frac{3}{4}$–$43\frac{1}{8}$	$49\frac{3}{4}$–$33\frac{1}{8}$

*Courtesy of The Gillette Company

Now let's examine how The Gillette Company's financial situation changed from 1989 to 1990. Gillette's 1990 sales were almost $4.34 billion. The $3.67 billion of assets (what it owns) and the $2.81 billion of liabilities (what it owes to creditors) were necessary to support these sales. The $0.86 billion of net worth indicates the portion of the company's assets that was provided by the investors (owners or stockholders). We can see that the company had earnings of $367.9 million, but only $310.6 million available to common stockholders (after paying out $57.3 million in cash dividends to its preferred stockholders). Gillette had about 96.06 million shares of common stock, so the company earned $3.20 per share of stock outstanding. (We calculate this earnings per share by dividing the net income available to common stockholders by the number of common stock shares outstanding.) This earnings per share indicates an increase of 19% compared with that of 1989.

Investors liked the new product, resulting in increased demand for the company's stock. This, in turn, caused stock prices, and hence shareholder wealth, to increase. In fact, this new, heavily promoted, high-tech Sensor razor turned out to be a smashing success and contributed to sending The Gillette Company's stock to an all-time high in early 1990. This caused Gillette's market value to increase about 20% during the first 6-month period. Any successful investment decision on Sensor's scale will tend to increase the firm's stock prices in the marketplace and promote long-term success. Thus, in making a large-scale engineering project decision, we must consider its possible effect on the firm's market value.

■ 1.3 Types of Engineering Economic Decisions

The story of how The Gillette Company successfully introduced a new product and regained the razor market share previously lost to competitors is typical: Someone has a good idea, executes it well, and obtains good results. Project ideas such as the Sensor can originate from many different levels in the organization. Since some ideas will be good while others will not, we need to establish procedures for screening projects. Many large companies have a specialized project analysis division that actively searches for new ideas, projects, and ventures. Once project ideas are identified, they are typically classified as (1) material and process selection, (2) equipment replacement, (3) new product and product expansion, (4) cost reduction, and (5) service improvement. This classification scheme allows management to address key questions. Can the existing plant, for example, be used to achieve the new production levels? Does the firm have the knowledge and skill to undertake this new investment? Does the new proposal warrant the recruitment of new technical personnel? The answers to these questions help firms screen out proposals that are not feasible given the company's resources.

Gillette's Sensor project represents a fairly complex engineering decision that required the approval of top executives and the board of directors. Virtually all big businesses at some time face investment decisions of this magnitude. In general, the larger the investment, the more detailed is the analysis required to support the expenditure. For example, expenditures to increase output of existing products, or to manufacture a new product, would invariably require a very detailed economic justification. Final decisions on new products and marketing decisions are generally made at a high level within the company. On the other hand, a decision to repair damaged equipment can be made at a lower level within the company. In this section, we will provide many real examples to illustrate each class of engineering economic decisions. At this point, our intention is not to provide the solution to each example but to describe the nature of the decision problems that a typical engineer would face in the real world.

1.3.1 Material and Process Selection

At the level of plant operations, engineers must make decisions involving materials, plant facilities, and the in-house capabilities of company personnel. Let us consider as an example the manufacture of food-processors. In terms of material selection, several of the parts could be made of plastic while others must be made of metal. Once materials have been chosen, engineers must consider the production methods, the shipping weight, and the method of packaging necessary to protect the different types of material. In terms of actual production, parts may be made in-house or purchased from an outside vendor. The decision as to which parts to produce in-house depends on the availability of machinery and labor. If the firm expects to produce the product for many years to come, it may be advantageous to purchase the required machinery and produce the product in-house.

What we have just described is a class of engineering decision problems that involve selecting the best course of action when there are several ways to meet the project's requirements. Which of several proposed items of equipment shall we purchase for a given purpose? The choice often turns on which item is expected to generate the largest savings (or return on the investment). We will provide two examples for this category.

Example 1.1

Engineers at General Motors want to investigate alternative materials and processes for the production of automotive exterior body panels. The engineers have identified two types of material: sheet metal and glass fiber reinforced polymer, known as plastic sheet molding compound (SMC), as shown in Fig. 1.2. Exterior body panels are traditionally made of sheet steel. With a low material cost, this sheet metal also lends itself to the stamping process, a very high-volume, proven manufacturing process. On the other hand, reinforced polymer easily meets the functional requirements of body panels (such as strength and resistance to corrosion). There is considerable debate among engineers as to the relative economic merits of steel as opposed to plastic panels. Much of the debate stems from the dramatically different cost structures of the two materials.

Description	Plastic SMC	Steel Sheet Stock
Material cost ($/lb)	$0.75	$ 0.35
Machinery investment	$2.1 million	$24.2 million
Tooling investment	$0.683 million	$ 4 million
Cycle time (min/part)	2.0	0.1

Figure 1.2 ■ Sheet molding compound process (Courtesy of Dow Plastics, a business group of The Dow Chemical Company)

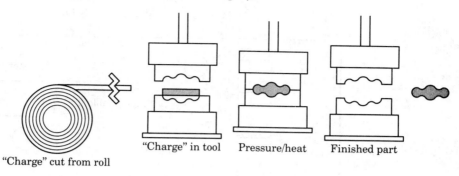

"Charge" cut from roll   "Charge" in tool   Pressure/heat   Finished part

Since plastic is petroleum based, it is inherently more expensive than steel, and because the plastic-forming process involves a chemical reaction, it has a slower cycle time. However, both machinery and tool costs for plastic are lower than steel due to the relatively low forming pressures, lack of tool abrasion, and single-stage pressing involved in handling. Thus, the plastic would require a lower initial investment but would incur higher material costs. Neither material is obviously superior economically. What, then, is the required annual production volume that would make the plastic material more economical?

Comments: *The choice of material will dictate the manufacturing process for the body panels. Many factors will affect the ultimate choice of the material, and engineers should consider all major cost elements, such as machinery and equipment, tooling, labor, and material. Other factors may include press and assembly, production and engineered scrap, the number of dies and tools, and the cycle times for various processes.*

Example 1.2

The Holley Automotive Division of Colt Industries supplies various automotive components to major North American automakers. One of these components is a solenoid assembly for transmission control, which controls the shift points (i.e., Park, Drive, Reverse, and so forth). Holley has a 3-year contract to supply solenoid assemblies to their customer. However, the contract terms specify that the total volume is to be split equally between Holley and another supplier. Moreover, at the end of the 3-year period, the volume for each supplier will be determined based on price. Currently, Holley's product cost is $2.00 higher than their competitors'. Consequently, Holley has undertaken a cost-reduction program in an effort to bring their cost in line with the competition. The engineers estimate that Holley can reduce the unit cost by up to $1.50 by reprocessing one component, the Insulator and Switch Assembly.

Figure 1.3 ■ Conversion from manual to automatic assembly process

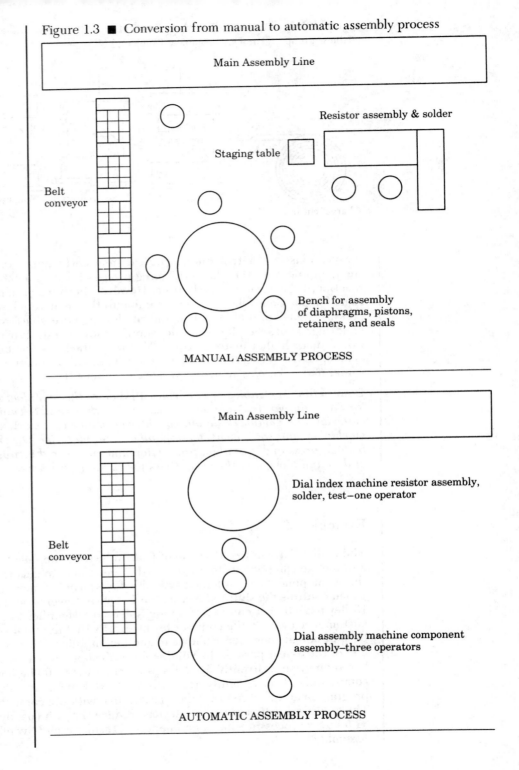

Main Assembly Line

Resistor assembly & solder

Staging table

Belt
conveyor

Bench for assembly
of diaphragms, pistons,
retainers, and seals

MANUAL ASSEMBLY PROCESS

Main Assembly Line

Dial index machine resistor assembly,
solder, test−one operator

Belt
conveyor

Dial assembly machine component
assembly–three operators

AUTOMATIC ASSEMBLY PROCESS

Currently, Holley purchases this component at $6.73 per unit from outside. If Holley decides to produce the component in-house, machinery for the deep drawing of metal will be required. To further reduce the unit component cost will also require an automatic assembly process that could result in a unit cost of about $3.80 (see Fig. 1.3.) The required machinery and new automatic assembly process would cost Holley about $120,000. Holley must reduce its costs to retain volume, but is not sure about making the component in-house at this time.

Comments: *Parts may be made in-house or purchased outside. The decision as to which parts to produce in-house depends on two factors, the machinery available in-house, and whether any additional machinery can also be used for future products. For this transmission solenoid, it may be cheaper to use the services of a supplier who has both the experience and the machinery for deep drawing of metal. If Holley is not skilled in deep drawing, it may require 6 months to get up to production. On the other hand, when Holley relies on a supplier, it can only assume that the supplier's labor problems will not halt production of Holley's parts.*

1.3.2 Equipment Replacement

This category of investment decisions involves considering the expenditure necessary to replace worn-out or obsolescent equipment. For example, a company may purchase ten large presses expecting to produce stamped metal parts for 10 years. After 5 years, however, it may become necessary to produce the parts in plastics, which would require retiring the presses early and purchasing plastic molding machines. Similarly, it may find that for competitive reasons, larger and more accurate parts are required, which will make the purchased machines obsolete earlier than expected. We will provide two real-world examples in this category.

Example 1.3

Moore Corporation Ltd. of Toronto is a worldwide producer of business forms and related paper products. The company presently has a number of plants at which raw paper is cut to size, printed on large printing presses, and finished in the finishing department. As a part of the finishing process, books of forms are bound by stapling and covered with heavy paper covers. One of the plants has four Stanley-Bostitch book machines which perform this operation.

Of the four book machines currently in use, only one is used every day. The company purchased this book machine in 1971, and it has been in service since that time. Over the years, this machine has been reliable, but during the past year, management has noticed a significant increase in machine downtime.

Moore is considering the purchase of a new book machine to replace the high-use machine. This $125,000 machine by Kidder/Schlumberger will not offer an increased production rate, but will result in a significant reduction in downtime. The training cost to operate this machine is very minimal.

Comments: *This example requires determining whether to replace existing equipment with more efficient equipment. The future expected cash inflows in this investment are the savings resulting from low operating costs, or the profits from additional volume produced by the new equipment, or both. Moore will eventually need to replace a deteriorating book machine in the near future, if not now. The proposed new machine is expensive, and Moore needs to know if the proposed investment is economically feasible.*

Example 1.4

Ampex Corporation produces a wide variety of tape cassettes for the commercial and government markets. The most popular commercial cassette type is the half-inch VHS cassette. Because of the increased competition, Ampex needs to price the product competitively in the market. To reduce the unit cost, the company could purchase the cassette shells more cheaply than it can currently produce them. The supplier has state-of-the-art equipment with which Ampex cannot

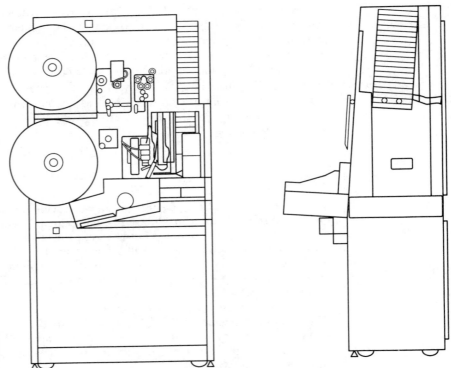

Figure 1.4 ■ A VHS cassette tape-loading machine, KING-2500 (Courtesy of King Instrument Corporation (a division of Otari Corporation))

compete. Presently, Ampex has 18 machines that load the cassette tape in the half-inch VHS cassette shells. Each loader requires one operator per shift. Ampex currently produces 25,000 half-inch tapes per week and operates 15 shifts per week, 50 weeks a year. However, the loaders Ampex currently has will not properly load the proposed shells.

One solution is to purchase eight KING-2500 VHS loaders at a cost of $40,000 each, shown in Fig. 1.4. For the machines to operate properly, Ampex will also have to purchase $20,827 worth of conveyer equipment. The new machines are much faster and will handle more than the current demand of 25,000 cassettes a week. The new loaders will require two people per machine, three shifts per day, 5 days a week. The new machines have an approximate lifespan of 8 years, and at the end of project life, Ampex expects the market value for each loader to be $3000. Using the purchased cassettes will result in a savings of $0.15 per cassette. The vendor has guaranteed a price of $0.77 per cassette for 3 years to get the new business.

The new loaders are simple to operate and, therefore, the training impact of the alternative is minimal. The cash inflows from the project will be the material savings per cassette of 15 cents and the labor savings of two employees per shift. This yields an annual savings in materials and labor costs of $187,500 and $122,065, respectively. If Ampex purchases the new loaders, it will ship the old machines to other plants for standby use.

Comments: *This example involves weighing an expenditure to replace serviceable but obsolete equipment. The purpose of this expenditure is to reduce the costs of labor, materials, or other items such as electricity.*

1.3.3 New Product and Product Expansion

Investments in this category are those that increase the revenues of the company if output is increased. There are two common types of expansion decision problems. The first category includes decisions about expenditures to increase the output of existing production or distribution facilities. In these situations, we are basically asking: Shall we build or otherwise acquire a new facility? The expected future cash inflows in this investment are the profits from the goods and services produced in the new facility. The second type of decision problem includes considering expenditures necessary to produce a new product or to expand into a new geographic area. These projects normally require large sums of money over long periods. We will provide two examples to illustrate the types of capacity expansion problems.

Example 1.5

DuPont, a leading chemical company, has set up a separate venture, Somos, to develop and market a rapid prototype system. This computer-driven technology allows design engineers to design and create plastic prototypes of complicated parts in just a few hours. An example of a rapid prototype system is shown in Fig. 1.5.

Rapid prototyping has allowed designers to complete projects that formerly took 6 months or more in less than 3 weeks. Assignments that took several weeks may now be done in a day or two. Rapid prototyping also saves tens of thousands of dollars per part in modeling costs compared to traditional methods.

The technology might also give birth to true just-in-time manufacturing for some businesses. An auto parts replacement shop, for example, might simply stock metal and plastic powders along with a library of computer programs that would allow it to build any part a customer needs on the spot. The venture will require an investment of $40 million on the part of DuPont and the prototype system once developed would be sold for $385,000.

Comments: *In this example, we are asking the basic questions involved in introducing a new product: Is it worth spending $40 million to market the*

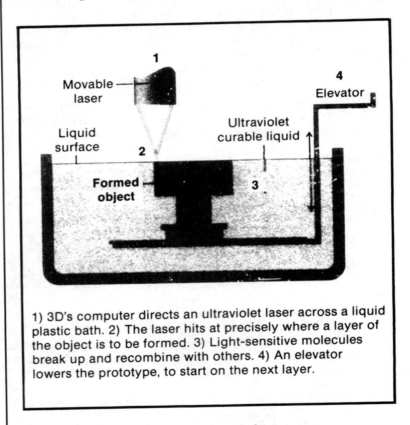

1) 3D's computer directs an ultraviolet laser across a liquid plastic bath. 2) The laser hits at precisely where a layer of the object is to be formed. 3) Light-sensitive molecules break up and recombine with others. 4) An elevator lowers the prototype, to start on the next layer.

Figure 1.5 ■ A rapid prototype system by Somos

rapid prototyping? How many sales are required to recover the investment? The future expected cash inflows in this example are the profits from the goods and services produced with the new product. Are these profits large enough to warrant the investment in equipment and the costs required to make and introduce the product?

Example 1.6

Motorola plans to build and operate a small and portable telephone that can be used anywhere on earth. The following details were taken from the *New York Times,* June 26, 1990.

> Motorola intends to charge less than $3500 for a 25-ounce handset that would fit in an overcoat pocket and could allow the user to make and receive calls from the North Pole to Antarctica. Motorola's system calls for the company to supplement antennas with a constellation of 77 satellites that would relay the calls, which would cost an estimated $1 to $3 a minute (Fig. 1.6). Potential users of the handsets are expected to include vacationers, business people, and engineers traveling in places where phone service is not available or where an international call can take hours to complete. Other users would include passengers aboard ships and planes and disaster relief crews working in places where all other communications had been disrupted.
>
> Motorola estimates that putting the system in service will cost $2.3 billion over 6 years and is seeking partners to help in the project. But the economics of a project like Motorola's are harder to forecast than the technical issues. Motorola says it will need at least 700,000 users to break even. After having consulted many international organizations about the proposed venture, company officials forecast that the system could attract as many as 5 million subscribers worldwide paying at least $100 a month by the year 2000.
>
> Besides the commercial challenges, Iridium [the project] will need the permission of many governments to receive calls from their soil. Regulatory decisions will be crucial in determining the number of subscribers that Iridium can reach. It will require certain frequencies and will need permission from a number of governments to receive— although not to send—calls from their soil.[9]

Comments: *This example illustrates an investment decision problem that requires expenditures necessary to expand into a geographic area not currently being served. These projects normally require a large sum of money over long periods of time. This type of project also requires strategic decisions that could change the fundamental nature of the business. Invariably, a very detailed analysis is required, and final decisions on new market ventures are generally made by the board of directors as a part of the strategic plan.*

---

[9] "Science Fiction Nears Reality: Pocket Phone for Global Calls," *New York Times,* June 26, 1990, pp. A1; C5. Copyright © 1990 by The New York Times Company. Reprinted by permission.

Figure 1.6 ■ Placing a call by satellite, Motorola Corporation

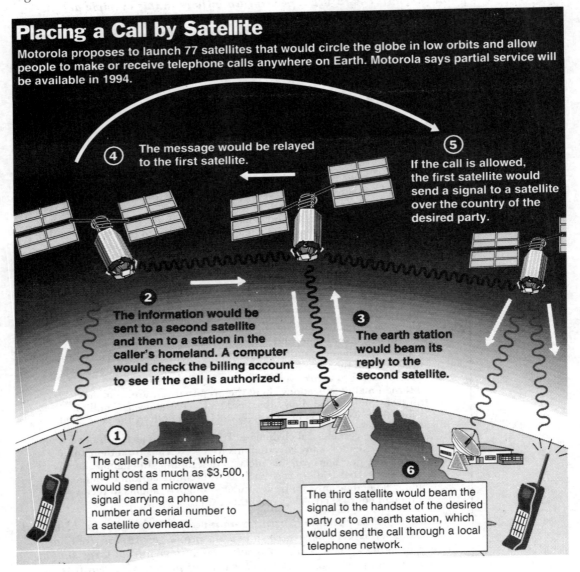

Placing a Call by Satellite

Motorola proposes to launch 77 satellites that would circle the globe in low orbits and allow people to make or receive telephone calls anywhere on Earth. Motorola says partial service will be available in 1994.

④ The message would be relayed to the first satellite.

⑤ If the call is allowed, the first satellite would send a signal to a satellite over the country of the desired party.

② The information would be sent to a second satellite and then to a station in the caller's homeland. A computer would check the billing account to see if the call is authorized.

③ The earth station would beam its reply to the second satellite.

① The caller's handset, which might cost as much as $3,500, would send a microwave signal carrying a phone number and serial number to a satellite overhead.

⑥ The third satellite would beam the signal to the handset of the desired party or to an earth station, which would send the call through a local telephone network.

1.3.4 Cost Reduction

A cost-reduction project is one that attempts to lower the firm's operating costs. Typically we need to consider whether we should buy equipment to perform an operation now done manually or spend money now in order to save money later. The expected future cash inflows on this investment are savings resulting from lower operating costs. We will give two examples for such cost-reduction projects.

Example 1.7

A major trucking company has begun installing communication systems on 2000 trucks (Fig. 1.7). Vehicles linked by such services carry a keyboard and display and transmission equipment about the size of a car battery. Drivers type in messages, which are bounced off a satellite and received by ground stations operated by the satellite company. The ground stations then relay the information by telephone line to the customer's home office. In tests last year, the trucking company concluded that the technology yielded direct gains to its bottom line.

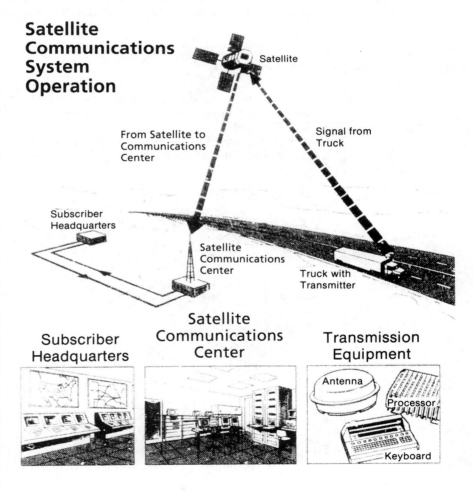

Figure 1.7 ■ Trucks linked by satellite

The company found that satellite messaging could cut 60% from its $5 million bill for long-distance communications with drivers. More important, the drivers reduced the number of "deadhead" kilometres—those driven without paying loads—by 0.5%. Applying all that improvement to all 400 million kilometres covered by its fleet each year would produce an extra $1.25 million of profits.

With innovations in transmission technology and miniaturization, the cost of satellite terminals has fallen in the last few years from about $20,000 to $4000.

Comments: *Equipping 2000 trucks with the satellite transmitters will cost $8 million (2000 × $4000). In addition, it is reasonable to suppose that construction of a message relaying system will cost about $2 million, for a total initial investment of $10 million. Expected annual savings are $4.25 million ($3 million in savings on long-distance communication plus $1.25 million on "deadhead" kilometres). The bottom line question is: Do expected savings justify the high initial cost of the investment?*

Example 1.8

An international manufacturer of prepared food items has made great progress in reducing energy costs by implementing several cogeneration projects. Cogeneration means that a large industrial firm produces its own electricity rather than purchasing it from the local utility. In most cases, the firm invests in the construction of a small power plant, usually

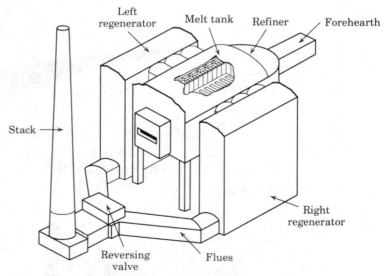

Figure 1.8 ■ A cogeneration unit

with a capacity of under 100 MW (megawatts), which is interconnected with existing distribution facilities owned by the local power utility.

One of the firm's chemical plants has a peak power usage of 90 million kWh (kilowatt hours) annually. On the average, however, the plant uses 80% of its peak usage, which would bring the average power usage to 72 million kWh annually. The local utility presently charges $0.09 per kWh, a rate considered high throughout the industry. Because the firm's power consumption is so large, its engineers are considering installing the cogeneration unit shown in Fig. 1.8. This cogeneration unit would allow the firm to generate its own power and avoid the annual $6,480,000 expense to the local utility.

The total initial investment for the cogeneration unit would be $10,500,000. The bulk of the investment—$10,000,000—is to purchase the power unit itself and pay for engineering, design, and site preparation. The remaining $500,000 includes the purchase of interconnection equipment, such as poles and distribution lines, which is required to interface the cogenerator with the existing utility facilities. Figure 1.8 illustrates the cogeneration design adopted by the company.

Comments: *Today, industrial power generation competes with utility power generation when an industrial company needs both electricity and process heat. We call this coincident production of both electricity and process heat cogeneration. The choice of a cogeneration system depends on many factors, including the project cost, hours per year of plant operation, quantity of steam demand, electrical and steam performance, fuel cost and availability, and electricity rates.*

1.3.5 Service Improvement

All the examples in this section were related to economic decisions in the manufacturing sector. The decision techniques we develop in this book are also applicable to various economic decisions involved in service industries. Example 1.9 will illustrate this.

Example 1.9

Bob Borg is an industrial engineering student working part-time as a bartender at Ryan's Tavern and Restaurant in Vancouver. The restaurant's menu ranges from various house salads, sandwiches, and burgers to fancy Mexican fajitas. Ryan's is separated into two major areas, the dining room (or floor) and the bar area, as shown in Fig. 1.9. On busy nights, 13 servers are responsible for the customers on the floor, while two servers are responsible for all customers not sitting at the bar but still in the bar area. Bob has often encountered a bottleneck at the service bar on busy nights. He is slowed most by making frozen drinks, which cost $3.75 each. Not only do frozen drinks take more time to make, they also do not have uniform consistency (ice clumps) and can vary in alcohol content, which is against the law.

Figure 1.9 ■ Ryan's Restaurant layout

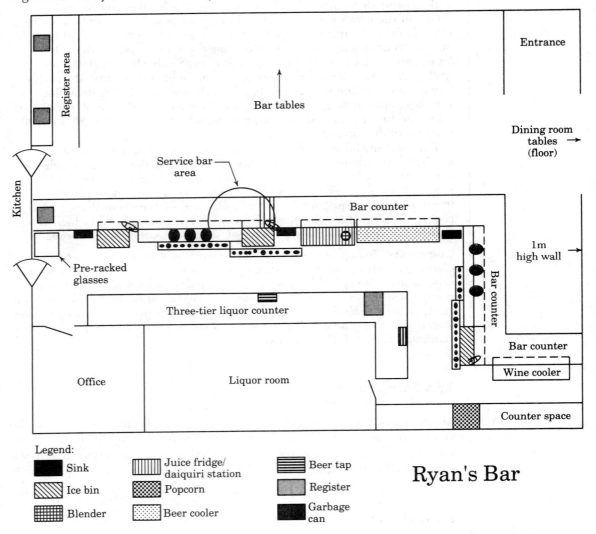

Legend:

■ Sink	▦ Juice fridge/ daiquiri station	▦ Beer tap			
▧ Ice bin	▨ Popcorn	▨ Register			
▤ Blender	░ Beer cooler	■ Garbage can			

Ryan's Bar

Once the server takes a drink order from a customer, he or she writes it on a ticket and rings it up at the bar. The server then goes to another table to take another order and returns to the bar to pick up the previous order. If the drink is not ready, the server is inclined to wait for the drink before returning to the floor area. There are two major reasons for this: First, if the server returns to his or her section without the drink, the customer who ordered it may suspect that the waiter forgot the drink and equate that with poor service. Second, if the server continues to the floor area without the drink, it means a special trip upon return from the floor area, or simply an extremely long wait for the customer if the server decides to consolidate as many moves as possible by bringing the drink out on the next trip. The key point, then, is to minimize both the server's waiting time at the bar and special trips. These extra trips will cause the server to fall behind in service to the customers and the situation thereby to snowball into a disastrous state called "in the weeds," in which all track of time is lost, along with short-term memory.

The majority of frozen drinks at Ryan's are made from five components: ice, alcohol, ice cream, limeade, and strawberries. When a frozen drink is made, some of these components are poured into a blender and mixed. Presently, Ryan's bar has one blender motor with two blender pitchers. The most common drink at Ryan's is the Strawberry Daiquiri, so one blender pitcher is used exclusively for Strawberry Daiquiris and the other is used for other types of daiquiris. Bob thinks that the addition of a frozen drink machine, namely the Taylor 450 C, which costs $4800, would alleviate the bottleneck at the bar and thereby improve customer service.

Comments: *Bob's main problem is to determine how much money Ryan's is losing due to server waiting time at the bar. How many more frozen drinks would Ryan's need to sell to justify the cost of an additional blender? This analysis should involve a comparison of the cost of operating the additional blender with the additional revenue generated by selling more frozen drinks.*

■ 1.4 Computers in Engineering Economic Analyses

There is no doubt that computer automation cannot take the place of a sound understanding of the principles of engineering economics—or any other course of study—for a variety of reasons. Computers are, after all, tools, and without mastering the principles of a subject, you are unlikely to be able to apply those tools knowledgeably or efficiently. Furthermore, in quizzes and exams throughout your academic career, you are unlikely to have the advantage of automation other than an engineering calculator. (Among other important tests, the Engineer-in-Training exam does not permit the use of computers!) And finally, some problems are simply more efficiently handled by traditional solution methods.

But as a busy student and, later, a practicing engineer, you will almost certainly want to put the efficiency of computers to work for you on some economic analysis problems. One of the fundamental lessons you will take away from this course is that "time is money." In an increasingly competitive business environment, the time- and effort-saving advantages of computers are often indispensable.

There are three broad categories of computer-aided engineering economics, as described in Table 1.1, and this text explicitly covers two of them: spreadsheets and an integrated analysis package, **EzCash**, which can be found on the back inside cover. A spreadsheet example example and a **EzCash** example are presented at the end of several chapters. It is assumed that you have at least passing familiarity with spreadsheets—enough to load the program, enter data, and request a calculation. Spreadsheets are demonstrated via Lotus 1-2-3 in

Table 1.1 Computer-Aided Engineering Economics Methods

Mode of Computer Automation	I/O–User Interface	Basic Functions	Advanced Functions
Programming	Code must be written in a programming language and debugged. Values to be manipulated must be read into the program.	Essentially, basic and advanced capabilities of programming are limited only by the talents of the programmer. However programming may be effort- and time-intensive. Prewritten subprograms may be available to streamline the programmer's task.	
Spreadsheets (e.g., Lotus® 1-2-3®, Excel, QuattroPro)	Visual interface is a screen consisting of columns, rows, and cells, in which user stores input data and equations for manipulating it. Equations may be built-in, or created by user.	Basic interest formulas, present, future, and annual equivalence and rate-of-return formulas. Some graphical abilities, including line graphs, pie charts, bar charts, etc., can be generated from spreadsheet contents.	With additional macro-programming by user, spreadsheets can be created and formatted to perform higher functions.
EzCash integrated analysis package	User interface is menu-driven. Calculation functions are all built-in.	Basic equivalence calculation, present, future, and annual worth functions, rate-of-return. Graphical abilities, including line graphs and bar graph.	External rate-of-return, benefit/cost ratio, project balance, spreadsheet calculations, capital cost allowance calculations, after-tax analysis, loan analysis, replacement analysis, and sensitivity analysis functions are built-in.

this text; however, "conversion tables" for Excel and QuattroPro are included in later chapters. Because the keystrokes for built-in functions are similar for these three packages, you shuld be able to translate easily from the Lotus example presented if you use one of the other two packages. Some fundamental differences between the three packages are shown in Table 1.2.

Despite the advantages of spreadsheets, you are likely to find the **EzCash** software as effective for some problems and more effective for many problems. **EzCash** is almost entirely menu-driven, so little or no preparation or practice

Table 1.2 Comparison of Basic Functions of Spreadsheets

Function	Lotus 1-2-3	Excel	QuattroPro
To enter data (numerical or textual)	With cursor on desired cell, type data and press "return."	With cursor on desired cell, type data and press "return."	With cursor on desired cell, type data and press "return."
To indicate a range of cells	A . . B	A:B	A . . B
To enter negative numbers	Precede number with a negative sign.	Precede number with a negative sign.	Precede number with a negative sign.
To calculate	Automatic, but can be set to manual.	Automatic, but can be set to manual.	Automatic, but can be set to manual.

Figure 1.10 ■ Hierarchy of efficiency and user-friendliness of methods of computer-aided engineering economics

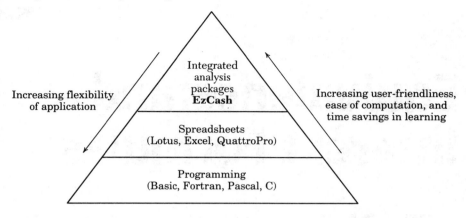

will be required for you to explore the program. Unlike spreadsheets, no macro-programming or formatting is required, even for advanced applications. **EzCash** also has a superior graphical environment (including visualization of cash flow diagrams).

After discovering some of the powers of spreadsheets and **EzCash** in the examples presented, you should feel encouraged to experiment with them on other, traditional examples and homework problems. With a little practice, you will quickly discern which tools are better for which problems as well as which problems are more troublesome to solve by computer than not (see Fig. 1.10).

Summary

- The factors **time** and **risk** are the defining aspects of investments.
- The three types of business organization are
  1. Proprietorship
  2. Partnership
  3. Corporation

  Although corporations make up only a small portion of the businesses in Canada, they conduct most of the business volume in sales dollars. Thus, our focus in future chapters will be primarily on the role of engineering economic decisions within the corporation.
- **Capital budgeting** refers to all the investment decisions regarding fixed assets. The facet of capital budgeting that is of most interest from an engineering economics point of view is the evaluation of costs and benefits associated with making a capital investment.
- The five main types of engineering economic decisions are
  1. Material and process selection
  2. Equipment replacement
  3. New product and product expansion
  4. Cost reduction
  5. Service improvement
- The efficiency of computers in streamlining some complex economic analysis represents a savings in time for the analyst and money for the firm. Spreadsheets and an analytical package called **EzCash** will be used throughout this text to demonstrate the benefits of computers.

CHAPTER 2

Equivalence and Interest Formulas

You may have already won $2 million!!! Just peel the game piece off the Instant Winner Sweepstakes ticket and mail it to us along with your orders for subscriptions to your two favorite magazines. As a Grand Prize Winner you may choose between a $1 million cash prize paid immediately or $100,000 per year for 20 years — that's $2 million!!!

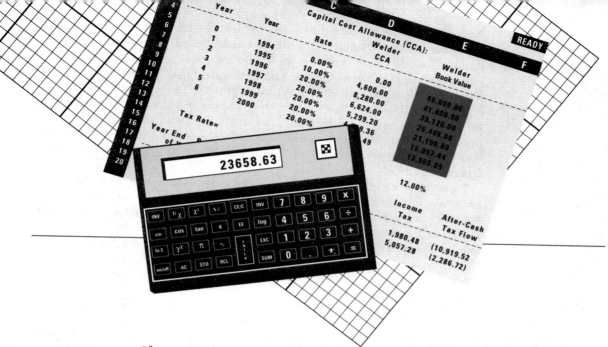

If you were the winner of the jackpot described above, you might well wonder why the value of one prize, $1 million paid immediately, is so much lower than the $2 million paid in 20 installments. Isn't receiving the $2 million overall a lot better than receiving $1 million now? The answer to your question involves the principles we will discuss in this chapter, namely, the operation of interest and the time value of money.

Now suppose that you acted on your first impulse and selected as your prize the annual payments totaling $2 million. You may be surprised by how your decision stands up under economic analysis. First, most people familiar with investments would tell you that receiving $1 million today is likely to prove a far better deal than taking $100,000 a year for 20 years. In fact, based on the principles you will learn in this chapter, the real present value of your earnings—the value that you could receive today in the financial marketplace for the promise of $100,000 a year for the next 20 years—can be shown to be worth considerably less than $1 million. And that is before we even consider the effects of inflation!

We can also use the techniques presented in this chapter to show that if you save your winnings for the first 6 years and then spend every cent of your winnings in the remaining 14 years, you are likely to come out wealthier than if you do the reverse and spend for 6 years and then save for 14! (Both examples assume, of course, that the economy will remain stable.) The reason for this surprising result is the **time value of money;** that is, the earlier a sum of money is received, the more it is worth, because over time money can earn more money, or interest.

In engineering economic analysis, we regard the principles discussed in this chapter as the underpinning for nearly all project investment analysis because we always need to account for the effect of interest operating on sums of cash over time. Interest formulas allow us to place different cash flows received at different times in the same time frame and thus to compare them. As will become apparent, almost our entire study of engineering economic analysis is built on the principles introduced in this chapter.

■ 2.1 Interest: The Cost of Money

Most of us are familiar in a general way with the concept of interest. We know that money left in a savings account earns interest so that the balance over time is greater than the sum of the deposits, and that borrowing to buy a car means repaying an amount over time that includes interest and is therefore greater than the amount borrowed. What may be unfamiliar to us is the idea that, in the financial world, money itself is a commodity, and like other goods that are bought and sold, money costs money.

The cost of money is established and measured by an **interest rate,** a percentage that is periodically applied and added to an amount (or varying amounts) of money over a specified length of time. When money is borrowed, the interest paid is the charge to the borrower for the use of the lender's property; when money is loaned or invested, the interest earned is the lender's gain from providing a good to another. **Interest,** then, may be defined as the cost of having money available for use. In this section, we will examine how interest operates in a free-market economy and establish a basis for understanding the more complex interest relationships that follow later in the chapter.

2.1.1 Money Has a Time Value

The operation of interest reflects the fact that money has a time value. This is why amounts of interest depend on lengths of time; interest rates, for example, are typically given in terms of a percentage per year. This principle of the time value of money can be formally defined as follows: The economic value of a sum depends on when it is received. Because money has **earning power** over time (it can be put to work, earning more money for its owner), a dollar received today has a greater value than a dollar received at some future time.

The changes in the value of a sum of money over time can become extremely significant when we deal with large amounts of money, long periods of time, or high interest rates. For example, at a current annual interest rate of 10%, $1 million will earn $100,000 in interest in a year; thus, waiting a year to receive $1 million clearly involves a significant sacrifice. In deciding among alternative proposals, then, we must take into account the operation of interest and the time value of money to make valid comparisons of different amounts at various times.

It is important to differentiate between the time value of money as we use it in this chapter and the effects of inflation, which we study in Chapter 10. The notion that a sum is worth more the earlier it is received can refer to its earning potential over time, to decreases in its value due to inflation over time, or to both. Since the earning power of money and its loss of value due to inflation represent different analytical techniques, we will consider these issues separately.

2.1.2 The Elements of Transactions Involving Interest

Many types of transactions involve interest—for example, borrowing or investing money, purchasing machinery on credit—but certain elements are common to all of them:

1. Some initial amount of money, called the **principal** in transactions of debt or investment.
2. The **interest rate,** which measures the cost or price of money, expressed as a percentage per period of time.
3. A period of time, called the **interest period,** that determines how frequently interest is calculated. (Note that even though the interest period is often some other length of time, interest rates are frequently quoted in terms of an annual percentage rate. We will discuss this potentially confusing aspect of interest in Chapter 3.)
4. The specified length of time that marks the duration of the transaction and thereby establishes a certain **number of interest periods.**
5. A **plan for receipts or disbursements** that yields a particular cash flow pattern over the length of time. (For example, we might have a series of equal monthly payments that repay a loan.)
6. A **future amount of money** that results from the cumulative effects of the interest rate over a number of interest periods.

For the purposes of calculation, these elements are represented by the following variables:

- $A_n =$ Discrete payment or receipt occurring at the end of some interest period.
- $i =$ The interest rate per interest period.
- $N =$ The total number of interest periods.
- $P =$ A sum of money at a time chosen for purposes of analysis as time zero, sometimes referred to as the **present value** or **present worth.**
- $F =$ A future sum of money at the end of the analysis period. This may be specified as F_N, the sum at the end of N interest periods.
- $A =$ An end-of-period payment or receipt in a uniform series, continuing for N periods. This is a special situation where $A_1 = A_2 = \cdots = A_N$.
- $V_n =$ An equivalent sum of money at the end of a specified period n considering the effect of the time value of money. Note that $V_0 = P$ and $V_N = F$.

Because we make frequent use of these symbols in this text, it is important to become familiar with them. Note, for example, the distinction between A, A_n, and A_N. A_n refers to a specific payment or receipt, at the end of period n, in any series of payments or receipts. A_N is the final payment or receipt in such a series because N refers to the total number of interest periods. A refers to any series of cash flows where all payments or receipts are equal.

An Example Using Elements in an Interest Transaction

As an example of how the elements we have just defined appear in a particular situation, suppose that a construction company buys a machine for $5000 and borrows the money from a bank at an 8% annual interest rate. In addition, the company pays a $100 loan origination fee when the loan commences. The bank offers two repayment plans, one with equal payments made at the end of every year for the next 5 years, and the other with a single payment made after the loan period of 5 years. These payment plans are summarized in Table 2.1.

Table 2.1 Repayment Plans for Example Given in Text (for $N = 5$ years and $i = 8\%$)

	Receipts	Payments	
		Plan 1	Plan 2
Year 0 ($n = 0$)	$P = \$5000.00$	$ 100.00	$ 100.00
Year 1 ($n = 1$)		$A_1 = \$1252.50$	0
Year 2 ($n = 2$)		$A_2 = $ 1252.50	0
Year 3 ($n = 3$)		$A_3 = $ 1252.50	0
Year 4 ($n = 4$)		$A_4 = $ 1252.50	0
Year 5 ($n = 5$)		$A_5 = $ 1252.50	$F = \$7345.50$

In Plan 1 the principal amount, P, is $5000, and the interest rate, i, is 8%. The interest period is 1 year, and the duration of the transaction is 5 years, which means that there are 5 interest periods $(N = 5)$. It bears repeating that while 1 year is a common interest period, interest is frequently calculated at other intervals: monthly, quarterly, or semiannually, for instance. For this reason, we use the term **period** rather than **year** in defining the preceding list of variables. The receipts and disbursements planned over the duration of this transaction yield a cash flow pattern of 5 equal payments, A, of $1252.50 each, paid at year-end during years 1–5. (You'll have to accept these amounts on faith for now—the following section presents the formula used to arrive at the amount of the equal payments, given the other elements of the problem.)

For Plan 2, we have most of the elements of Plan 1, except that instead of 5 equal repayments we have a single future repayment, F, of $7345.50.

Cash Flow Diagrams

It is convenient to represent problems involving the time value of money in graphic form with a **cash flow diagram** (see Fig. 2.1), which represents time by a horizontal line marked off with the number of interest periods specified. The cash flows over time are represented by arrows at the relevant periods: upward arrows for positive flows (receipts) and downward arrows for negative flows (disbursements). Note, too, that the arrows actually represent **net cash flows:** Two or more receipts or disbursements made at the same time are summed and shown as a single arrow. For example, $1000 received during the same period as a $250 payment would be recorded as an upward arrow of $750. The lengths of the arrows can suggest the relative values of particular cash flows.

Cash flow diagrams function in a manner similar to the free body diagrams or circuit diagrams that most engineers use frequently: They provide a convenient summary of all the important elements of a problem as well as a check point for determining whether we've properly converted a problem statement into its appropriate parameters. The text makes frequent use of this graphic tool, and you are strongly encouraged to develop the habit of using well-labeled cash flow diagrams as a means of identifying and summarizing the pertinent information in a cash flow problem. Similarly, a table such as the one shown in Table 2.1 can help you organize information in another summary format.

End-of-period Convention

In practice, cash flows can occur at the beginning or in the middle of an interest period, or at practically any point in time. One of the simplifying assumptions we make in engineering economic analysis is the end-of-period convention, which is the practice of placing all cash flow transactions at the end of an interest period. This assumption relieves us of the responsibility of dealing with

Figure 2.1 ■ Cash flow diagrams: (a) the generalized cash flow diagram, (b) the cash flow diagram for Plan 1 of the loan repayment summarized in Table 2.1.

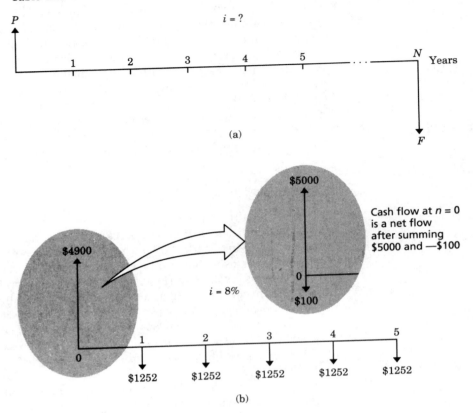

(a)

(b)

the effects of interest *within* an interest period, which would greatly complicate our calculations.

It is important to be aware of the fact that, like many of the simplifying assumptions and estimates we make in our modeling of engineering economic problems, the end-of-period convention inevitably leads to small discrepancies between our model and real-world results.

Suppose, for example, that $100,000 is deposited on the second day of the year in an account with an interest period of 1 year and an interest rate of 10% per year. In such a case, the difference of 2 days would cause an interest income loss of $10,000. This example gives you a sense of why financial institutions choose interest periods that are less than 1 year even though they usually quote their rate in terms of annual percentage.

Armed with an understanding of the basic elements involved in interest problems, we can begin to look at the details of calculating interest.

2.1.3 Methods of Calculating Interest

There are many ways that money can be loaned and repaid, and there are an equal number of ways that money can earn interest. Usually, at the end of each interest period, the interest earned on the principal amount is calculated according to a specified interest rate. The two computational schemes for calculating this earned interest are said to yield either **simple interest** or **compound interest.** Engineering economic analysis uses the compound interest scheme almost exclusively.

Simple Interest

The first approach considers the interest to be earned on only the **principal** amount during each interest period. In other words, under simple interest, the interest earned during each interest period does not earn additional interest in the remaining periods, *even though you do not withdraw it.*

In general, for a deposit of P dollars at a simple interest rate of i for N periods, the total earned interest I would be

$$I = (iP)N \tag{2.1}$$

The total amount available at the end of N periods, F, thus would be

$$F = P + I = P(1 + iN). \tag{2.2}$$

Simple interest is commonly used with bonds, which we will soon review in Chapter 3.

Compound Interest

Under the compound interest scheme, the interest each period is based on the total amount owed at the end of the previous period. This total amount includes the original principal plus the accumulated interest that has been left in the account. In this case, you are in effect increasing the deposit amount by the amount of interest earned. In general, if you deposited (invested) P dollars at interest rate i, you would have $P + iP = P(1 + i)$ dollars at the end of one period. With the entire amount (principal and interest) reinvested at the same rate i for another period, you would have, at the end of the second period,

$$P(1 + i) + i[P(1 + i)] = P(1 + i)(1 + i)$$
$$= P(1 + i)^2.$$

Continuing, we see that the balance after period 3 is

$$P(1 + i)^2 + i[P(1 + i)^2] = P(1 + i)^3.$$

This interest-earning process repeats, and after N periods, the total accumulated value (balance) F will grow to

$$F = P(1 + i)^N. \qquad (2.3)$$

Example 2.1   Compound interest

Suppose you deposit $2000 in a bank savings account that pays interest at a rate of 10% compounded annually. Assume that you don't withdraw the interest earned at the end of each period (year), but let it accumulate. How much could you withdraw at the end of year 3?

Solution

Given: $P = \$2000$, $N = 3$ years, and $i = 10\%$ per year

Find: F

Applying Eq. (2.3) to our 3-year, 10% case, we obtain

$$F = \$2000(1 + 0.10)^3 = \$2662.$$

The total interest earned is $662, which is $62 more than would be accumulated with simple interest. We can keep track of the process of interest accumulation more precisely as follows:

Period	Amount at Beginning of Interest Period	Interest Earned for Period	Amount at End of Interest Period
1	$2000	$2000(0.10)	$2200
2	2200	2200(0.10)	2420
3	2420	2420(0.10)	2662

Comments: *At the end of the first year, you would have $2000 plus $200 in interest, or a total of $2200. In effect, at the beginning of the second year, you would be depositing $2200, rather than $2000. Thus, at the end of the second year, the interest earned would be 0.10($2200) = $220, and the balance would be $2200 + $220 = $2420. This is the amount you would be depositing at the beginning of the third year, and the interest earned for that period would be 0.10($2420) = $242. With the beginning* *principal amount of $2420 plus the $242 interest, the total balance would be $2662 at the end of year 3. If the total balance were then withdrawn, the net cash flow in this case would appear as*

Year	Cash Flow
0	−$2000
1	0
2	0
3	2662

which is the same as the value worked out previously. (See Fig. 2.2.)

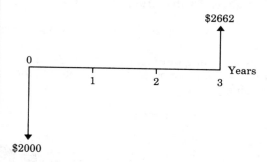

Figure 2.2 ■ Cash flow diagram for Example 2.1

2.1.4 Simple Interest Versus Compound Interest

From Eq.(2.3), the total interest earned over N periods is

$$I = F - P = P[(1 + i)^N - 1]. \qquad (2.4)$$

When compared with simple interest, the additional interest earned with compound interest is

$$\Delta I = P[(1 + i)^N - 1] - (iP)N \qquad (2.5)$$

$$= P[(1 + i)^N - (1 + iN)]. \qquad (2.6)$$

As either i or N becomes large, the difference in interest earnings also becomes large, so the effect of compounding is further pronounced. Note that when $N = 1$, compound interest is the same as simple interest.

Example 2.2 Comparing simple and compound interest

Compare the simple and compound interest earned by depositing $1000 for 5 years at 12% interest.

Solution

Given: $P = \$1000$, $i = 12\%$ per year, $N = 5$ years
Find: I

- Simple interest calculation:

$$I = \$1000(0.12)\,(5) = \$600$$

- Compound interest calculation:

$$I = \$1000(1 + 0.12)^5 - \$1000 = \$762.34$$

The difference in earned interest is $162.34. Figure 2.3 shows how the difference between compound and simple interest increases over the five compounding periods.

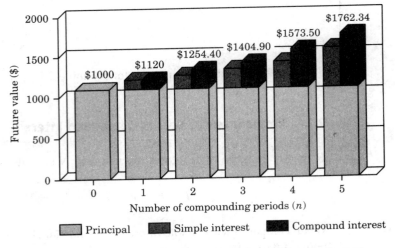

Figure 2.3 ■ Simple versus compound interest (Example 2.2)

■ 2.2 Economic Equivalence

The observation that money has a time value leads us to an important question: If receiving $100 today is not the same as receiving $100 at any future point, how do we measure and compare various cash flows? How do we know, for example, whether we should prefer to have $20,000 today, $50,000

Figure 2.4 ■ Which cash flow is worth more? We cannot compare these cash flows unless we know what interest rate operates on them.

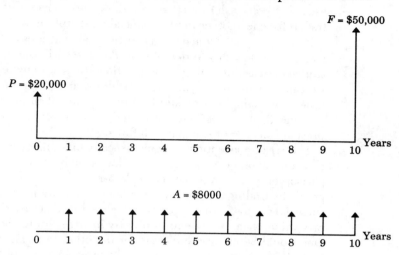

10 years from now, or $8000 each year for the next 10 years (Fig. 2.4)? In this section, we will describe the basic analytical techniques for making these comparisons. Then, in Section 2.3, we will use these techniques to develop a series of formulas that can greatly simplify our calculations.

2.2.1 Definition and Simple Calculations

The central question in deciding among alternative cash flows involves comparing their economic worth. This would be a simple matter if we did not need to consider the time value of money in our comparison: In such a case, we could simply add the individual cash flows, treating receipts as positive cash flows and payments (disbursements) as negative cash flows. The fact that money has a time value makes our calculations more complicated. We need to know more than just the size of a cash flow in order to determine completely its economic effect. In fact, as we will discuss in this section, we have to know several things:

- Its magnitude.
- Its direction—is it a receipt or a disbursement?
- Its timing—when does the transaction occur?
- The interest rate in operation during the time period under consideration.

It follows that to assess the economic impact of a *series* of cash flows, we must consider the impact of each individual cash flow.

Calculations for determining the economic effects of one or more cash flows are based on the concept of economic equivalence. **Economic equivalence** exists between cash flows that have the same economic effect and thus could be traded for one another in the financial marketplace, which we assume to exist.

Economic equivalence refers to the fact that a cash flow—whether it is a single cash flow or a series of cash flows—can be said to be converted to an *equivalent* cash flow at any point in time; thus, for any sequence of cash flows, we can find an equivalent single cash flow at a given interest rate and a given time. For example, we could find the equivalent future value, F, of a present amount, P, at interest rate i at period n; or we could determine the equivalent present value, P, of N equal cash flows, A.

The strict conception of equivalence, which limits us to converting a cash flow into another equivalent cash flow, may be extended to include the comparison of alternatives. For example, we could compare the value of two proposals by finding the equivalent value of each at any common point in time. If financial proposals that appear to be quite different turn out to have same monetary value, then we are **economically indifferent** to choosing between them: One would be an even exchange for the other in terms of economic effect, so there is no reason to prefer one to the other in terms of their economic value.

One way to see the concepts of equivalence and economic indifference at work in the real world is to note the variety of payment plans offered by lending institutions for consumer loans. Table 2.2 extends the example we developed earlier to include three different repayment plans for a loan of $5000 for 5 years at 8% interest. You will notice, perhaps to your surprise, that the three plans require significantly different repayment patterns and even different total amounts of repayment. However, because money has time value, these plans are equivalent, and the bank is economically indifferent to a consumer's choice of plan. We will now discuss how such equivalence relationships are established.

Table 2.2 Typical Payment Plans for Bank Loan of $5000 (for N = 5 years and i = 8%)

	Repayments		
	Plan 1	Plan 2	Plan 3
Year 1 (n = 1)	$1252.50	0	$ 400.00
Year 2 (n = 2)	1252.50	0	400.00
Year 3 (n = 3)	1252.50	0	400.00
Year 4 (n = 4)	1252.50	0	400.00
Year 5 (n = 5)	1252.50	$7345.50	5400.00
Total of payments[1]	$6262.50	$7345.50	$7000.00
Total interest paid[2]	$1262.50	$2345.50	$2000.00

Plan 1: Equal annual payments
Plan 2: End-of-loan-period repayment of P and I
Plan 3: Annual payment of I and end-of-loan repayment of P

---

[1] Ignores timing effects

[2] Total interest paid = total of payments − $5000

Equivalence Calculations: A Simple Example

Equivalence calculations can be viewed as an application of the compound interest relationships we developed in Section 2.1. Suppose, for example, that we invest $1000 at 12% annual interest for 5 years. The formula developed for calculating compound interest, $F = P(1 + i)^N$ (Eq. 2.3), expresses the equivalence between some present amount, P, and a future amount, F, for a given interest rate, i, and a number of interest periods, N. Therefore, at the end of the investment period, our sums grow to

$$\$1000(1 + 0.12)^5 = \$1762.64.$$

Thus we can say that at 12% interest, $1000 received now is *equivalent to* $1762.64 received in 5 years and that we could trade $1000 now for the promise of receiving $1762.64 in 5 years. Example 2.3 further demonstrates the application of this basic technique.

Example 2.3 Equivalence

Suppose you are offered the alternative of receiving either $3000 at the end of 5 years or P dollars today. There is no question that the $3000 will be paid in full (no risk). Having no current need for the money, you would deposit the P dollars in an account that pays 8% interest. What value of P would make you indifferent in your choice between P dollars today and the promise of $3000 at the end of 5 years from now?

Discussion: Our job is to determine the present amount that is economically equivalent to $3000 in 5 years, given the investment potential of 8% per year. Note that the problem statement assumes that you would exercise your option of using the earning power of your money by depositing it. The "indifference" ascribed to you refers to *economic* indifference; that is, within a marketplace where 8% is the applicable interest rate, you could trade one cash flow for the other.

Solution

Given: $F = \$3000$, $N = 5$ years, $i = 8\%$ per year
Find: P
Equation: Eq. (2.3), $F = P(1 + i)^N$

Rearranging to solve for P,

$$P = F/(1 + i)^N.$$

Substituting,

$$P = \$3000/(1 + 0.08)^5 = \$2042.$$

Figure 2.5 ■ Cash flow 1 is economically equivalent to cash flow 2 at an interest rate of 8%.

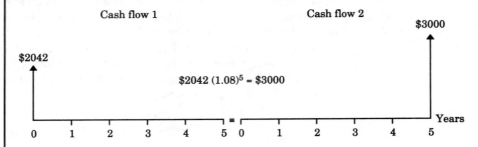

We can summarize the problem graphically in Fig. 2.5.

Comments: *In this example, it is clear that if P is anything less than $2042, you would prefer the promise of $3000 in 5 years to P dollars today; if P were greater than $2042, you would prefer P. It is less obvious that at a lower interest rate, P must be higher to be equivalent to the future amount. For example, at i = 4%, P = $2466.*

2.2.2 General Observations about Equivalence Calculations

In spite of their numerical simplicity, the examples we have developed reflect several important general principles, which we will explore.

   1. **Equivalence calculations that compare alternatives require a common time basis.** Just as we must convert fractions to common denominators to add them together, we must convert cash flows to a common basis to compare their value. One aspect of this basis is the choice of a single point in time at which to make our calculations. In Example 2.3, if we had been given the magnitude of each cash flow and had been asked to determine whether they are equivalent, we could have chosen *any* reference point and used the compound interest formula to find the value of each cash flow at that point. As you can readily tell, the choice of $n = 0$ or $n = 5$ would make our problem simpler because we would need to make only one set of calculations: At 8% interest, either convert $2042 at time 0 to its equivalent value at time 5, or convert $3000 at time 5 to its equivalent value at time 0. (To see that we can choose a different reference point, take a look at Example 2.4.)

   When selecting a point in time at which to compare the value of alternative cash flows, we commonly use either the present time, which yields what is called the **present worth** of the cash flows, or some point in the future, which yields their **future worth.** The choice of point in time often depends on the circumstances surrounding a particular decision, or it may be chosen for convenience. If the present worth is known, for instance, for two of three alternatives, all three may be compared by simply calculating the present worth of the third.

Example 2.4 Equivalent cash flows are equivalent at any common point in time

In Example 2.3, we determined that, given an interest rate of 8% per year, receiving $2042 today is equivalent to receiving $3000 in 5 years. Are these cash flows also equivalent at the end of year 3?

Discussion: This problem is summarized in Fig. 2.6. The solution consists of solving two equivalence problems: (1) What is the future value of $2042 after 3 years at 8% interest? (Part a of the solution); and (2) Given the sum of $3000 after 5 years and an interest rate of 8%, what is the equivalent sum after 3 years? (Part b).

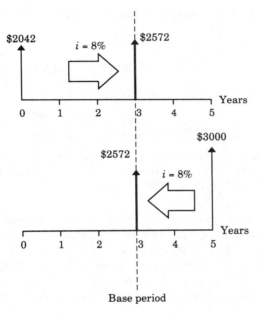

Figure 2.6 ■ Selection of a base period for equivalence calculation (Example 2.4)

Solution

Given:

(a) $P = \$2042$, $i = 8\%$ per year, $N = 3$ years
(b) $F = \$3000$, $i = 8\%$ per year, $N = 5 - 3 = 2$ years

Find: (a) V_3 for Part (a); (b) V_3 for Part (b); (c) Are these two values equivalent?

Equation:

(a) $F = P(1 + i)^N$
(b) $P = F/(1 + i)^N$

Notation: The usual terminology of F and P is confusing in this example, since the cash flow at $n = 3$ is considered a future sum in Part (a) of the solutions and a past cash flow in Part (b) of the solution. To simplify matters, we are free to arbitrarily designate a reference point, $n = 3$, and understand that it need not be *now* or *the present*. Therefore, we assign a single variable, V_3, for the *equivalent* cash flow at $n = 3$.

(a) The equivalent worth of $2042 after three years is

$$V_3 = \$2042(1 + 0.08)^3$$

$$= \$2572.$$

(b) The equivalent worth of the sum $3000, two years earlier is

$$V_3 = F/(1 + i)^N$$

$$= \$3000/(1 + 0.08)^2$$

$$= \$2572.$$

(Note that $N = 2$ because that is the number of periods during which discounting is calculated in order to arrive back at year 3.)

Comments: *While our solution doesn't strictly prove that the two cash flows are equivalent at any time, they will be equivalent at any time as long as we use an interest rate of 8%.*

2. Equivalence depends on interest rate. The equivalence between two cash flows is a function of both the cash flow pattern and the interest rate or rates that operate on those cash flows. This is easy to grasp in relation to our simple example: $1000 received now is equivalent to $1762.64 received five years from now *only* at a 12% interest rate. Any change in the interest rate will destroy the equivalence between these two sums, as we demonstrate in Example 2.5.

Example 2.5 Changing the interest rate destroys equivalence

In Example 2.3, we determined that, given an interest rate of 8% per year, receiving $2042 today is equivalent to receiving $3000 in five years. Are these cash flows equivalent at an interest rate of 10%?

Discussion: We can test for equivalence by finding the equivalent value of $2042 at $N = 5$.

Solution

Given: $P = \$2042$, $i = 10\%$ per year, $N = 5$ years
Find: F: Is it equal to $3000?

Equation: $F = P(1 + i)^N$

$$F = \$2042(1 + 0.10)^5$$
$$= \$2042(1.6015)$$
$$= \$3270$$

Therefore, the change in interest rate destroys the equivalence between the two cash flows.

3. Equivalence calculations may require the conversion of multiple payment cash flows to a single cash flow. In all the examples presented thus far, we have limited ourselves to the simplest case of converting a single payment at one time to an equivalent single payment at another time. In comparing alternative cash flows, part of the task of converting the flows to a common basis is to reduce them to a single cash flow at a single time. We perform such a calculation in Example 2.6.

Example 2.6 Equivalence calculations with multiple payments

Suppose that you borrow $1000 from a bank for 3 years at 10% annual interest. The bank offers you two options: (1) repaying the loan all at once at the end of 3 years, or (2) repaying the interest charges for each year at the end of that year. The repayment schedules for the two options are as follows:

Options	Year 1	Year 2	Year 3	Total[1]
Option 1: End-of-year repayment of interest, principal repaid at end of loan	$100	$100	$1100	$1300
Option 2: One end-of-loan repayment of both principal and interest	0	0	1331	1331

[1] Ignores timing effects

Determine whether these options are equivalent, assuming that the appropriate interest rate for our comparison is 10%.

Discussion: Since we pay the principal after three years in either plan, we can remove the repayment of principal from our analysis. This is an important point: We can ignore common elements of alternatives being compared. Now we focus entirely on comparing the interest payments. It is easy to notice that under Option 1 we will pay a total of $300 interest, while under Option 2 we will pay a total of $331. Before we conclude that we prefer Option 1, we should remember that a comparison of the two cash flows is based on a *combination of payment amounts and timing of those payments.* To make our comparison, we must compare the equivalent value of each option at a single point in time. Since Option 2 is already a single payment at $n = 3$ years, it is simplest to convert the Option 1 cash flow pattern to a single value at $n = 3$. To do this, we must convert the three disbursements of Option 1 to their respective equivalent values at $n = 3$. At that point, since they are at a common time, we can simply sum them in order to compare them to the $331 sum in Option 2.

Solution

Given: The cash flow diagrams in Fig. 2.7, $i = 10\%$ per year

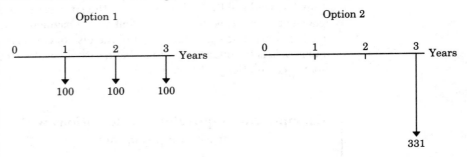

Figure 2.7 ■ Equivalent cash flow diagram for Option 1 and Option 2 (excluding the common principal payment at the end of year 3)

Find: A single future value, F, of the flows in Option 1

Equation: $F = P(1 + i)^N$, applied to each disbursement in the cash flow diagram

N in Eq. (2.3) is the number of interest periods upon which interest is in effect; n is the period number (i.e., for year 1, $n = 1$). Therefore, we determine its value by finding the interest period for each payment. Thus, for each payment in the series, N can be calculated by subtracting n from the total number of years of the loan (3). That is, $N = 3 - n$. Once we've found the value of each payment, we sum them:

$$F_3 \text{ for } \$100 \text{ at } n = 1: \quad \$100(1 + 0.10)^{3-1} = \$121$$

$$F_3 \text{ for } \$100 \text{ at } n = 2: \quad \$100(1 + 0.10)^{3-2} = \$110$$

$$F_3 \text{ for } \$100 \text{ at } n = 3: \quad \$100(1 + 0.10)^{3-3} = \underline{\$100}$$

$$\text{Total} = \$331$$

By converting the cash flow in Option 1 to a single future payment at year 3, we can compare it to Option 2 and see that the two interest payments are equivalent. Thus, we would be economically indifferent in choosing between the two plans. Note that the final interest payment in Option 1 does not accrue any compound interest.

Figure 2.8 ■ Options 1 and 2 are equivalent.

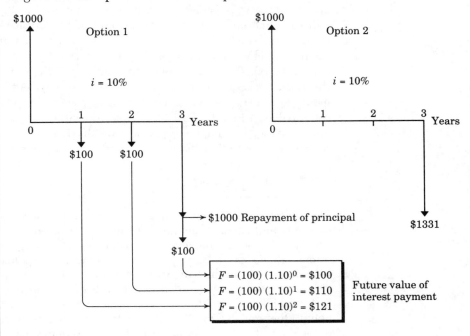

Comments: *If it is difficult to grasp the fact that Options 1 and 2 are equivalent even though $31 more of interest is paid under Option 2, focus again on the concept of the time value of money and note that Option 1 excludes the opportunity provided by Option 2 of earning interest on the deferred interest payments (see Fig. 2.8). Under Option 1, the interest must be paid at the end of each interest period. Because of the time value of money, the $100 paid at the end of period 1 has a future worth of $121 from the perspective in time of the end of period 3. The person under Option 2 can actually realize that future worth by depositing $100 in an account at 10% interest. Of course, the person may not be able to get 10% interest, or possibly could earn a higher rate. This brings up the question, what is the appropriate interest rate to use in equivalence calculations? We will address this topic in Chapter 13. If, for example, the person could only earn 8%, the two options would not be equivalent.*

4. **Equivalence is maintained regardless of point of view.** As long as we use the same interest rate in equivalence calculations, equivalence can be maintained regardless of point of view. In Example 2.6, the two options were equivalent at an interest rate of 10% from the banker's point of view. What about from a borrower's point of view? Suppose you borrow $1000 from a bank and deposit it in another bank that pays 10% interest annually. Then, you make the future loan repayments out of this savings account. With Option 1, your savings account at the end of year 1 will show a balance of $1100 after crediting the interest earned during the first period. Now you withdraw $100 from this savings account (the exact amount required to pay the loan interest during

the first year) and make the first year interest payment to the bank. This leaves only $1000 in your savings account. At the end of year 2, your savings account will earn another interest in the amount of $1000(0.10) = $100, making the end-of-year balance a total of $1100. Now you withdraw another $100 to make the required loan interest payment. After this payment, your remaining balance will be $1000. This balance will grow again at 10%, so you will have $1100 at the end of year 3. After making the last loan payment ($1100), you have no money left in your accounts. For Option 2, you can keep track of the yearly account balances in a similar fashion, and you will find that no money will be left after making the lump sum payment of $1331. If the borrower had used the same interest rate that the bank used, the two options would be equivalent.

2.2.3 Looking Ahead

The preceding examples are intended to give you some insight into the basic concepts and calculations involved in economic equivalence. Obviously, the variety of financial arrangements possible for borrowing and investing money is extensive, as is the variety of time-related factors (e.g., maintenance costs over time, increased productivity over time) in alternative proposals for various engineering projects. It is important to recognize that even the most complex relationships incorporate the basic principles we have introduced in this section. In the remainder of this chapter, we will represent most cash flow diagrams in the context of an initial deposit with a subsequent pattern of withdrawals, or an initial borrowed amount with a subsequent pattern of repayments.

A cash flow diagram representation of a more complicated equivalence calculation appears in Fig. 2.9. The cash flow diagrams summarize contracts offered to two well-known professional quarterbacks, Vinny Testaverde and Troy Aikman. If we were limited to the methods developed in this section, a comparison between the two contracts would involve a large number of calculations. Fortunately, in the analysis of many transactions, certain cash flow patterns emerge that may be categorized, and for many of these we can derive formulas that can be used to simplify our work. In Section 2.3, we develop these formulas.

■ 2.3 Development of Interest Formulas

Now that we have established some working assumptions and notations and have developed a preliminary understanding of the concept of equivalence, we will develop a series of interest formulas for use in more complex comparisons of cash flows.

As we begin to compare *series* of cash flows instead of single payments, the analysis required becomes more complicated. However, whenever we can identify patterns in the cash flow transactions, we can take advantage of them by developing concise expressions for computing either the present or future worth of the series. We will classify five major categories of cash flow transactions, develop interest formulas for them, and present several working examples of

Figure 2.9 ■ Which quarterback is higher paid?

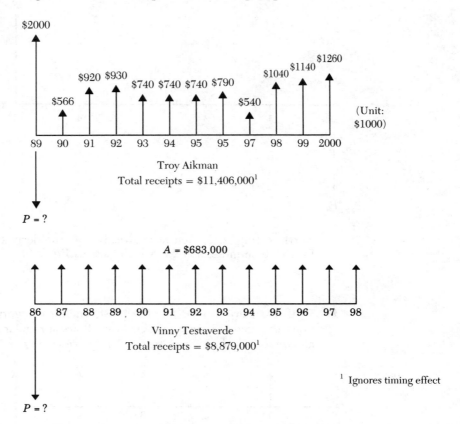

Troy Aikman
Total receipts = $11,406,000[1]

$P = ?$

$A = \$683,000$

Vinny Testaverde
Total receipts = $8,879,000[1]

[1] Ignores timing effect

$P = ?$

each type. Before we present the details, however, we will describe them briefly in the following section.

Before beginning this section, it is recommended that you review the notation in Section 2.1.2. These variables will be used consistently throughout the development of interest formulas.

2.3.1 The Five Types of Cash Flows

Whenever we can identify patterns in cash flow transactions, we may use them in developing concise expressions for computing either the present or future worth of the series. For this purpose, we will classify cash flow transactions into five categories: (1) single cash flow, (2) equal-cash-flow series, (3) linear gradient series, (4) geometric gradient series, and (5) irregular series. To simplify the description of various interest formulas, we will use the following notation:

1. **Single Cash Flow:** The simplest case involves the equivalence of a single present amount and its future worth. Thus, the single-cash-flow formulas deal with only two amounts: a single present amount, P, and its future

Figure 2.10 ■ Single cash flow

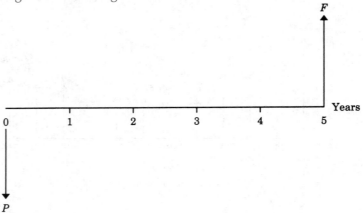

worth, F (Fig. 2.10). You have already seen the derivation of one formula for this case in Section 2.2, which gave us Eq. (2.3):

$$F = P(1 + i)^N.$$

2. **Uniform Series:** Probably the most familiar category includes transactions arranged as a series of equal cash flows at regular intervals, known as an *equal-cash-flow series* (or *uniform series*) (Fig. 2.11). This describes

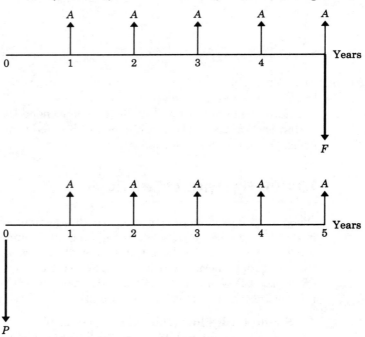

Figure 2.11 ■ Uniform series

Figure 2.12 ■ Linear gradient series

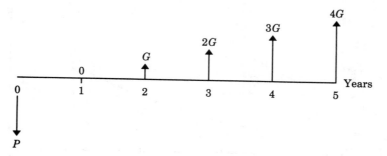

the cash flows, for example, of the common installment loan contract, which arranges for the repayment of a loan in equal periodic installments. The equal-cash-flow formulas deal with the equivalence relations to P, F, and A, the constant amount of the cash flows in the series.

3. **Linear Gradient Series:** While many transactions involve series of cash flows, the amounts are not always uniform; however, they may vary in some regular way. One common pattern of variation occurs when each cash flow in a series increases (or decreases) by a fixed amount (Fig. 2.12). A 5-year loan-repayment plan might specify, for example, a series of annual payments that increased by $500 each year. We call such a cash flow pattern a *linear gradient series* because its cash flow diagram produces an ascending (or descending) straight line, as you will see in Section 2.3.5. In addition to P, F, and A, the formulas used in such problems involve the constant amount, G, of the change in each cash flow.

4. **Geometric Gradient Series:** Another kind of gradient series is formed when the increase in cash flow is determined, not by some fixed amount like $500, but by some fixed *rate*, expressed as a percentage. For example in a 5-year financial plan for a project, the cost of a particular raw material might be budgeted to increase at a rate of 4% per year. The curving gradient in the diagram of such a series suggests its name: a *geometric gradient series* (Fig. 2.13). In the formulas dealing with such series, the rate of change is represented by a lowercase g.

5. **Irregular Series:** Finally, a series of cash flows may be irregular, exhibiting no regular overall pattern. Even in such a series, however, one or more of the patterns already identified may appear over portions of the total series. The cash flows may be equal, for example, for five consecutive periods in a ten-period series. When such patterns appear, the formulas for dealing with them may be applied and the results included in calculating an equivalent value for the entire series.

Interest Tables

Interest formulas like the one developed in Eq. (2.3), $F = P(1 + i)^N$, allow us to substitute the known values of a particular situation into the equation and solve for the unknown. Before the development of the hand calculator, solving these equations was very tedious. With a large value of N, for example, one might need to solve an equation such as $F = \$20,000(1 + 0.12)^{15}$, and the more complex formulas require even more involved calculations. To simplify the

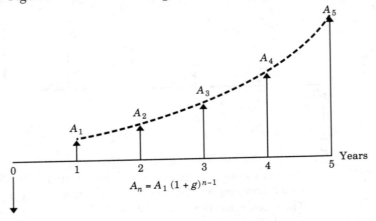

Figure 2.13 ■ Geometric gradient series

$$A_n = A_1 (1 + g)^{n-1}$$

process, tables of compound interest factors were developed that allow us to find the appropriate factor for a given interest rate and number of interest periods. Even with calculators available, it is still convenient to use such tables, and they are included in this text in Appendix C. Take some time now to become familiar with the arrangement of these tables and, if you can, locate the compound interest factor for the example just presented, where we know P, and to find F we need to know the factor by which to multiply \$20,000 when the interest rate, i, is 12% and the number of periods is 15.

Factor Notation

As we develop interest formulas in the rest of this chapter, we will express the resulting compound interest factors in a conventional notation that can be substituted in a formula to indicate precisely which table factor to use in solving an equation. In the preceding example, for instance, the formula derived as Eq. (2.3) is $F = P(1 + i)^N$. In common language, this tells us that to determine what future amount, F, is equivalent to a present amount, P, we need to multiply P by a factor expressed as 1 plus the interest rate, raised to the power given by the number of interest periods. To specify how the interest tables are to be used, we may also express that factor in a functional notation as follows: $(F/P,i,n)$, which is read as "Find F, given P, i, and N." This is known as the **single-cash-flow compound amount factor.** When we incorporate the table factor in the formula, it is expressed as follows:

$$F = P(1 + i)^N = P(F/P, i, N).$$

Thus, in the preceding example, where we had $F = \$20,000(1.12)^{15}$, we can write $F = \$20,000(F/P, 12\%, 15)$. The table factor tells us to use the 12% interest table and find the factor in the F/P column for $N = 15$. Because using the interest tables is often the easiest way to solve an equation, this factor notation is included for each of the formulas derived in the following sections.

2.3.2 Single-Cash-Flow Formulas

We begin our coverage of interest formulas by considering the simplest of cash flows: single cash flows.

Compound Amount Factor

Given a present sum P invested for N interest periods at interest rate i, what sum will have accumulated at the end of the N periods? You probably noticed quickly that this description matches the case we first encountered in describing compound interest. To solve for F (the future sum) we use Eq. (2.3):

$$F = P(1 + i)^N = P(F/P, i, N).$$

Because of its origin in compound interest calculation, the factor $(F/P,i,N)$ is known as the *single-cash-flow compound amount factor*. Like the concept of equivalence, this factor is one of the foundations of engineering economic analysis. Given this factor, all the other important interest formulas can be derived.

This process of finding F is often called the **compounding process.** The cash flow transaction is illustrated in Fig. 2.14. (Note the time scale convention. The first period begins at $n = 0$ and ends at $n = 1$.) If a calculator is handy, it is easy enough to calculate $(1 + i)^N$ directly. However, the appropriate interest table can also be used, and a wide range of i and N values can be found there.

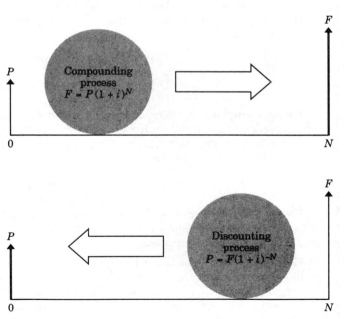

Figure 2.14 ■ Equivalence relationship between F and P

Example 2.7 Single amounts: find *F*, given *i*, *N*, *P*

If you had $2000 now and invested it at 10%, how much would it be worth in 8 years?

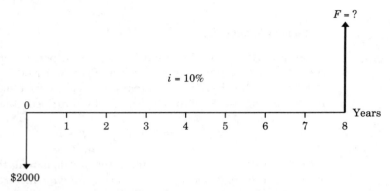

Figure 2.15 ■ Cash flow diagram (Example 2.7)

Solution

Given: $P = \$2000$, $i = 10\%$ per year, and $N = 8$ years
Find: F

We can solve this problem in any of three ways:

- **Using a calculator:** You can simply use a calculator to evaluate the $(1 + i)^n$ term. (Financial calculators are preprogrammed to solve most future-value problems.)

$$F = \$2000(1 + 0.10)^8$$

$$= \$4287.18$$

- **Using compound interest tables:** Using the interest tables, locate the compound amount factor for $i = 10\%$ and $N = 8$, and substitute that number into the equation. Compound interest tables appear in Appendix C of this book.

$$F = \$2000(F/P, 10\%, 8) = \$2000(2.1436) = \$4287.20$$

This is essentially identical to the value obtained by the direct evaluation of the single-cash-flow compound amount factor. This slight deviation is due to rounding errors.

- **Using a computer:** Many financial software programs for solving interest problems are available for microcomputers. Spreadsheet programs such as Lotus 1-2-3 or Excel also provide financial functions for evaluating various interest formulas. The software **EzCash**, which accompanies this book, also provides an easy way to calculate the *F* value in graphical form.

Example 2.8 Single amounts: find N, given P, F, i

You have just purchased 100 shares of Interprovincial Pipe Line stock at $33 per share. You will sell the stock when its market price has doubled. If you expect that the stock price will increase 15% per year, how long should you wait before selling the stock?

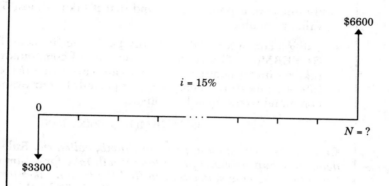

Figure 2.16 ■ Cash flow diagram (Example 2.8)

Solution

Given: $P = \$3300$, $F = \$6600$, $i = 15\%$ per year
Find: N (years)

Using the single-cash-flow compound amount factor, we write

$$F = P(1 + i)^N = P(F/P, i, N)$$

$$\$6600 = \$3300(1 + 0.15)^N = \$3300(F/P, 15\%, N)$$

$$2 = (1.15)^N = (F/P, 15\%, N)$$

Again, we could use a calculator, the interest tables (using trial and error), or a computer to find N.

• Using a calculator: Solving for N gives

$$\log 2 = N \log 1.15$$

$$N = \frac{\log 2}{\log 1.15}$$

$$= 4.96 \approx 5 \text{ years}$$

- Using compound interest tables: In the table for $i = 15\%$, locate the single-payment compound amount factors closest to 2.0 and note the corresponding values of N:

$$(F/P, 15\%, 4) = 1.749$$

$$(F/P, 15\%, 5) = 2.011$$

By linear interpolation, we find that it takes almost 5 years for the value to double.

- Using a computer: Within Lotus 1-2-3, the financial function @CTERM(i, F, P) computes the number of compounding periods it will take an investment (P) to grow to a future value (F), earning a fixed interest rate (i) per compounding period. In our example, the Lotus command would look like this:

@CTERM(15%, 6600, 3300)

Comments: *A very handy rule of thumb, called the* Rule of 72, *can determine approximately how long it will take for a sum of money to "double." The rule states that to find the time it takes for the present sum of money to grow by a factor of 2, we divide 72 by the interest rate. For our example, the interest rate is 15%. Therefore, the Rule of 72 indicates 72/15 = 4.8, or roughly 5 years for a sum to double. This is, in fact, very close to our exact solution.*

Present Worth Factor

Finding the present worth of a future sum is simply the reverse of compounding and is known as **discounting process.** In Eq. (2.3), we can see that if we were to find a present sum P, given a future sum F, we simply solve for P.

$$P = F\left[\frac{1}{(1 + i)^N}\right] = F(P/F, i, N) \qquad (2.7)$$

The factor $1/(1 + i)^N$ is known as the **single-cash-flow present worth factor,** and is designated $(P/F,i,N)$. Tables have been constructed for the P/F factors for various values of i and N. The interest rate i and the P/F factor are referred to as **discount rate** and **discounting factor,** respectively.

Example 2.9 Single amounts: find P, given F, i, N

Suppose that $1000 is to be received in 6 years. At an annual interest rate of 9%, what is the present worth of this amount?

Solution

Given: $F = \$1000$, $i = 9\%$ per year, and $N = 6$ years

Find: P

$$P = \$1000(1 + 0.09)^{-6} = \$1000(0.5963) = \$596.30$$

Figure 2.17 ■ Cash flow diagram (Example 2.9)

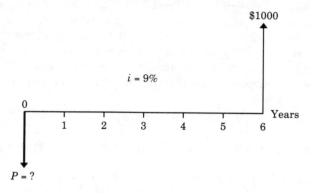

A calculator may be the best way to make this simple calculation. To have $1000 in the savings account at the end of 6 years, you must deposit $596.30 now.

We can also use the interest tables to find:

$$P = \$1000(P/F, 9\%, 6) = \$1000(0.5963) = \$596.30.$$

Again, you could use a financial calculator or computer to find the present worth.

Graphic Views of the Compounding and Discounting Processes

Figure 2.18 illustrates the characteristics of the F/P and P/F factors with variations in i and N. Figure 2.18(a) shows how $1 (or any other sum) grows over

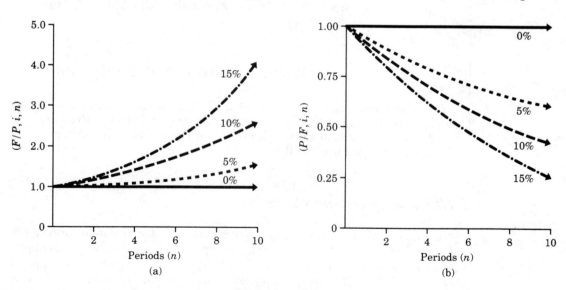

(a) (b)

Figure 2.18 ■ The characteristics of the F/P and P/F factors with variations in i and n

time at various rates of interest. Note the rapid increases in future value F with increases in either N or i. The higher the rate of interest, the faster the rate of growth. The interest rate is, in fact, a *growth rate*.

Figure 2.18(b) shows how interest factors for discounting decrease as the discounting period increases. The curves in the figure show that the present worth of a sum to be received at some future time (F) decreases (1) as the payment period extends further into the future and (2) as the discount rate increases.

Example 2.10   Effect of varying discounting rates

What is the present worth of a lump sum of $1 million received 50 years from now if your interest rate is (a) 6%, (b) 10%, and (c) 20%?

Solution

Given: F = $1 million; N = 50 years; i = 6%, 10%, or 20% per year
Find: P at each of the interest rates

(a)   at i = 6%:

$$P = \$1,000,000(P/F, 6\%, 50) = \$1,000,000(0.054288) = \$54,288$$

(b)   at i = 10%:

$$P = \$1,000,000(P/F, 10\%, 50) = \$1,000,000(0.008519) = \$8519$$

(c)   at i = 20%:

$$P = \$1,000,000(P/F, 20\%, 50) = \$1,000,000(0.000110) = \$110$$

As seen in our calculation, when relatively high discount rates are applied, funds due in the future are worth very little today. Even at relatively low discount rates, the present worth of funds due in the distant future is quite small.

2.3.3 Uneven-Cash-Flow Series

A common cash flow transaction involves a series of disbursements or receipts. Familiar arrangements such as car loans and insurance payments are examples of a cash flow series. These transactions typically involve identical sums paid at regular intervals. However, when there is no clear pattern to the amounts in the series, we call the transaction an **uneven-cash-flow series.** Recall, for example, the contractual salary payments shown for quarterback Troy Aikman in Fig. 2.9.

We can find the present worth of Aikman's contract—or any other uneven stream of cash flows—by calculating the present value of each individual cash flow and summing the results. Once the present worth is found, we can make other equivalence calculations, such as calculating future worth by using the interest factors developed in the previous section.

Example 2.11 Present value of an uneven series by decomposition into single payments

A growing electronics firm wishes to set aside money now to invest over the next 4 years in automating their customer service department. They can earn 10% on a lump sum deposited now, and they wish to withdraw the money in the following increments:

Year 1: $25,000 to purchase a computer and database software designed for customer service use.

Year 2: $3000 to purchase additional computer memory to accommodate anticipated growth in use of the system.

Year 3: no expenses.

Year 4: $5000 to purchase software upgrades.

How much money must be deposited now to cover the anticipated payments over the next 4 years?

Discussion: This problem is equivalent to asking what value of P would make you indifferent in your choice between P dollars today and the future expense stream of ($25,000, $3000, $0, $5000). One way to deal with a series of cash flows is to calculate the equivalent present value of each single payment and sum the present values to find P. In other words, the cash flow is broken into three parts as shown in Fig. 2.19.

Solution

Given: Uneven cash flow shown in Fig. 2.19, $i = 10\%$ per year
Find: P

$$P = \$25,000(P/F, 10\%, 1) + \$3000(P/F, 10\%, 2) + \$5000(P/F, 10\%, 4)$$

$$= \$28,622$$

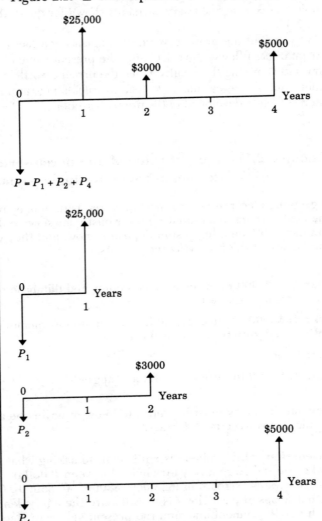

Figure 2.19 ■ Decomposition of uneven cash flow series

Example 2.12 Present worth of a uniform series by decomposition

Consider the magazine sweepstakes problem introduced at the beginning of this chapter. If you win you have the choice of either accepting $1 million now or taking the 20 annual installments ($100,000 each year, a total of $2 million). Suppose that you will earn 8% interest from your deposit in a bank. Which option would be more desirable?

Solution

Given: Cash flows shown below for the two options, $i = 8\%$ per year
Find: P for each option

Period	Option 1	Option 2
0	$1,000,000	
1		$100,000
2		100,000
3		100,000
$\vdots$		$\vdots$
20		100,000

The equivalent present worth for Option 1 is already known, $P_{\text{Option 1}} = \$1,000,000$. Let's compute the equivalent present worth for Option 2. Although the cash flow is not an uneven stream, we can solve for the present value of each payment just as we did for each of the uneven payments in Example 2.11. This requires using 20 different $(P/F, 8\%, n)$ factors.

$$P_{\text{Option 2}} = \$100,000(P/F, 8\%, 1) + \$100,000(P/F, 8\%, 2) + \cdots$$

$$+ \ \$100,000(P/F, 8\%, 20)$$

$$= \$100,000[0.9259 + 0.8573 + \cdots + 0.2145]$$

$$= \$100,000(9.818)$$

$$= \$981,800$$

Since $P_{\text{Option 1}} > P_{\text{Option 2}}$, Option 1 is a better choice. Even though Option 2 would pay a total of $2 million over 20 years, the equivalent worth at 8% interest is less than receiving one lump sum of $1 million now.

Comments: *This "brute force" approach of breaking cash flows into single amounts will always work, but it is slow and subject to error because of the many factors that must be included in the calculation. We will develop more efficient methods in the next sections for cash flows with certain patterns.*

2.3.4 Equal-Cash-Flow Series

As we learned in Example 2.11, the present worth of a stream of future cash flows can always be found by summing the present worth figures of each individual cash flow. However, cash flow regularities within the stream, such as we saw in Example 2.12, may allow the use of short-cuts, such as finding the

Figure 2.20 ■ Equal-payment series

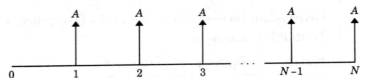

present worth of a uniform series. We often encounter transactions in which there is a uniform series of payments. Rental payments, bond interest payments, and commercial installment plans are based on a uniform payment series. (Consider also Vinny Testaverde's contractual salary payment shown in Fig. 2.9.) The cash flow diagram for an equal-payment series is shown in Fig. 2.20, where A represents an end-of-period cash flow in a uniform series continuing for N periods.

Compound Amount Factor—Find *F*, Given *A*, *i*, *N*

Suppose we are interested in the future amount F of a fund to which we contribute A dollars each period and on which we earn interest at a rate of i per period. The contributions are made at the end of each of the next N periods. We graphically illustrate these transactions in Fig. 2.21. Looking at this diagram, we see that, if an amount A is invested at the end of each period for N periods, the total amount F which can be withdrawn at the end of N periods will be the sum of the compound amounts of the individual deposits.

As shown in Fig. 2.22, the A dollars we put into the fund at the end of the first period will be worth $A(1 + i)^{N-1}$ at the end of N periods. The A dollars we put into the fund at the end of the second period will be worth $A(1 + i)^{N-2}$, and so forth. Finally, the last A dollars that we contribute at the end of the Nth period will be worth exactly A dollars at that time. This means there exists a series in the form

$$F = A(1 + i)^{N-1} + A(1 + i)^{N-2} + \cdots + A(1 + i) + A$$

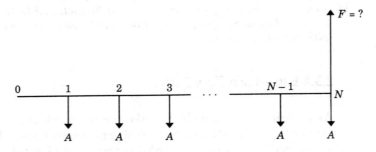

Figure 2.21 ■ Cash flow diagram of the relationship between A and F

Figure 2.22 ■ The future worth of a cash flow series can be found by summing the future worth figures of each individual flow.

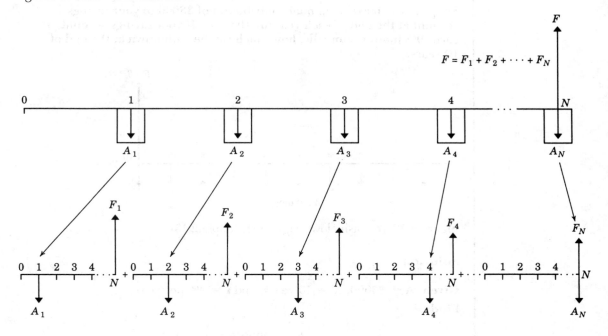

or, expressed alternatively,

$$F = A + A(1 + i) + A(1 + i)^2 + \cdots + A(1 + i)^{N-1}. \tag{2.8}$$

Multiplying Eq. (2.8) by $(1 + i)$ results in

$$(1 + i)F = A(1 + i) + A(1 + i)^2 + \cdots + A(1 + i)^N. \tag{2.9}$$

Subtracting the Eq. (2.8) from Eq. (2.9) to eliminate common terms gives us

$$F(1 + i) - F = -A + A(1 + i)^N.$$

Solving for F yields

$$F = A \left[\frac{(1 + i)^N - 1}{i} \right] = A(F/A, i, N). \tag{2.10}$$

The bracketed term is called the **equal-cash-flow-series compound amount factor**, or **uniform-series compound amount factor;** its factor notation is $(F/A,i,N)$. This interest factor has been calculated for various combinations of i and N in the interest tables.

Example 2.13 Uniform series: find *F*, given *i*, *A*, *N*

Suppose you make an annual contribution of $3000 to your savings account at the end of each year for 10 years. If your savings account earns 9% interest annually, how much can be withdrawn at the end of 10 years?

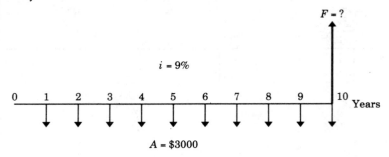

Figure 2.23 ■ Cash flow diagram (Example 2.13)

Solution

Given: A = $3000, N = 10 years, and i = 9% per year
Find: F

$$F = \$3000(F/A, 9\%, 10)$$
$$= \$3000(15.1929)$$
$$= \$45,578.70$$

Example 2.14 Handling time shifts in a uniform series

In Example 2.13, the first deposit of the ten-deposit series was made at the end of period 1 and the remaining nine deposits were made at the end of each following period. Suppose that all deposits were made at the beginning of each period instead. How would you compute the balance at the end of period 10?

Solution

Given: Cash flow as shown in Fig. 2.24, i = 9% per year
Find: F_{10}

In terms of Fig. 2.24, each payment has been shifted one year earlier, compared to Example 2.13; thus each payment would be compounded for one extra year. Note that with the end-of-year deposit, the ending balance (F) was $45,578.70. With the beginning-of-year deposit, the same

Figure 2.24 ■ Example 2.14: Find *F*, when the equal-payment series shifts by one period.

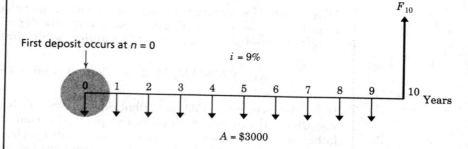

First deposit occurs at *n* = 0

i = 9%

A = $3000

balance accumulates by the end of period 9. This balance can earn interest for one additional year. Therefore, we can easily calculate the resulting balance by

$$F_{10} = \$45{,}578.70(1.09) = \$49{,}680.78.$$

Comments: *Another way to determine the ending balance is to compare the two cash flow patterns. By adding the $3000 deposit at period 0 to the original cash flow and subtracting the $3000 deposit at the end of period 10, we obtain the second cash flow. Therefore, the ending balance can be found by making adjustment to the $45,578.70:*

$$F_{10} = \$45{,}578.70 + \$3000(F/P, 9\%, 10) - \$3000 = \$49{,}680.70.$$

Sinking-fund Factor—Find *A*, Given *F, i, N*

If we solve Eq. (2.10) for *A*, we obtain

$$A = F\left[\frac{i}{(1 + i)^N - 1}\right] = F(A/F, i, N). \tag{2.11}$$

The term within the brackets is called the **equal-cash-flow-series sinking-fund factor,** or simply **sinking-fund factor,** and is referred to by the notation $(A/F,i,N)$. A sinking fund is an interest-bearing account into which a fixed sum is deposited each interest period; it is established for the purpose of replacing fixed assets.

Example 2.15　Combination of a uniform series and single present and future amount

To help you reach your $5000 goal 5 years from now, your father offers to give you $500 now. You plan to get a part-time job and make 5 additional deposits at the end of each year. (The first deposit is made at the end of first year.) If all of your money is deposited in a bank that pays 7% interest, how large must your annual deposits be?

Discussion: If your father reneges on his offer, the calculation of the required annual deposit is easy because your five deposits fit the standard end-of-period pattern for a uniform series. Just evaluate

$$A = \$5000(A/F, 7\%, 5) = \$5000(0.1739) = \$869.50.$$

If you do receive the $500 contribution from your father at $n = 0$, you may divide the deposit series into two parts: one contributed by your father at $n = 0$ and the five equal annual deposit series by yourself. Then you can use the F/P factor to find how much your father's contribution will be worth at the end of year 5 at a 7% interest rate. Let's call this amount F_c. The future value of your five annual deposits must then make up the difference, $\$5000 - F_c$.

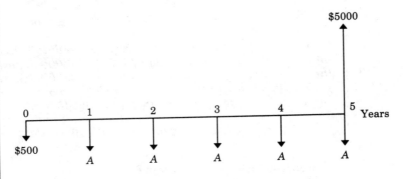

Figure 2.25 ■ Cash flow diagram (Example 2.15)

Solution

Given: Cash flow as shown in Fig. 2.25, $i = 7\%$ per year, and $N = 5$ years

Find: A

$$
\begin{aligned}
A &= (\$5000 - F_c)\,(A/F, 7\%, 5) \\
&= [\$5000 - 500(F/P, 7\%, 5)]\,(A/F, 7\%, 5) \\
&= [\$5000 - 500(1.4025)]\,(0.1739) \\
&= \$747.55
\end{aligned}
$$

Capital Recovery Factor (Annuity Factor)—Find *A*, Given *P, i, N*

We can determine the amount of a periodic cash flow A if we know P, i, and N. Figure 2.26 illustrates the situation. To relate P to A, recall the relationship

Figure 2.26 ■ Cash diagram of the relationship between P and A

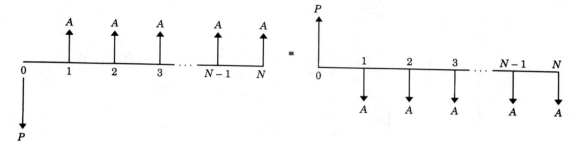

between P and F in Eq. (2.3), $F = P(1 + i)^N$. replacing F in Eq. (2.11) by $P(1 + i)^N$ gives us

$$A = P(1 + i)^N \left[\frac{i}{(1 + i)^N - 1} \right]$$

or

$$A = P \left[\frac{i(1 + i)^N}{(1 + i)^N - 1} \right] = P(A/P, i, N). \qquad (2.12)$$

Now we have an equation for determining the value of the series of end-of-period cash flows A when the present sum P is known. The portion within the brackets is called the **equal-cash-flow-series capital recovery factor,** or simply **capital recovery factor,** and is designated $(A/P, i, N)$. In finance, this A/P factor is referred to as an **annuity factor,** indicating a series of payments of a fixed, or constant, amount for a specified number of periods.

Example 2.16   Uniform series: find A, given P, i, N

Suppose that a small biotechnology firm has borrowed $100,000 to purchase laboratory equipment. The loan carries an interest rate of 8% per year and is to be repaid in equal installments over the next 5 years. Compute the amount of this annual installment.

Solution

Given: $P = \$100,000$, $i = 8\%$ per year, $N = 5$ years
Find: A

$$A = \$100,000(A/P, 8\%, 5)$$

$$= \$100,000(0.25046)$$

$$= \$25,046$$

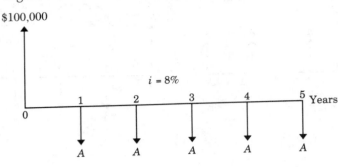

Figure 2.27 ■ Cash flow diagram (Example 2.16)

Present Worth Factor—Find P, Given A, i, N

What would you have to invest now in order to withdraw A dollars at the end of each of the next N periods? The situation we face is just the opposite of the equal-cash-flow capital recovery factor—A is known, but P has to be determined. With the capital recovery factor given in Eq. (2.12), solving for P gives us

$$P = A\left[\frac{(1 + i)^N - 1}{i(1 + i)^N}\right] = A(P/A, i, N). \qquad (2.13)$$

The term in brackets is called the **equal-cash-flow-series present worth factor**; its factor rotation is $(P/A,i,N)$.

Example 2.17   Uniform series: find P, given A, i, N

In Example 2.12, we used the P/F factors to find the present worth of an equal-payment series. Compute the present worth of the 20 annual installments of $100,000 for the sweepstake receipts at an 8% rate of interest.

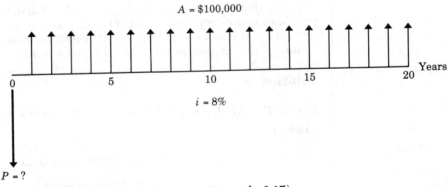

Figure 2.28 ■ Cash flow diagram (Example 2.17)

Solution

Given: $A = \$100,000$, $i = 8\%$ per year, $N = 20$ years
Find: P

$$P = \$100,000(P/A,8\%,20)$$

$$= \$100,000(9.818)$$

$$= \$981,800,$$

which is the same value obtained in Example 2.12.

Example 2.18 Comparing two options: find P, given A, i, N

Suppose you were offered the following two options:

- A 4-year annuity with payments of $2000 at the end of each year.
- A lump-sum payment (P) now.

You have no need for the money during the next 4 years. Therefore, if you accept either the annuity or the lump-sum payment, you would simply deposit the payments in a savings account that pays 6% interest. How large must the lump-sum payment (P) be to make it equivalent to the annuity?

Period	Option 1	Option 2
0	P	
1		$2000
2		2000
3		2000
4		2000

Solution

Given: $A = \$2000$, $i = 6\%$ per year, $N = 4$ years
Find: P

$$P = \$2000(P/A, 6\%, 4) = \$2000(3.4651) = \$6930.21$$

Example 2.19 A uniform series problem with "too much" information

Frequently, people share the cost of a lottery ticket. In one instance, 21 factory workers had agreed to pool $21 to play a lottery and split any winnings. Their winning ticket was worth $13,667,667, which would be distributed in 21 annual payments of $650,793. That meant each member of the pool would receive 21 annual payments of about $24,000, according to their lawyer. John Brown, one of the lucky workers, wanted to quit the factory and start his own business, which required him to secure a $250,000 bank loan. Brown offered to put up his future lottery earnings (as collateral) to secure the loan. If the bank's interest rate is 10% per year, how much can he borrow against his future lottery earnings?

Discussion: We need to identify the relevant data in this problem, because some numbers ultimately have nothing to do with the solution method. Basically, Brown wants to borrow $250,000 from a bank, but there is no assurance from the bank that he will get the full amount. (Normally a lending officer determines the maximum amount that one can borrow based on the borrower's capability of repaying the loan.) If the bank views Brown's lottery earnings as his only source of future income for repaying the loan, the bank must find the equivalent present worth of his 21 annual receipts of $24,000 in order to set the maximum loan amount.

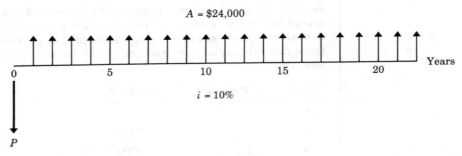

Figure 2.29 ■ Cash flow diagram (Example 2.19)

Solution

Given: $i = 10\%$ per year, $A = \$24,000$, and $N = 21$ years

Find: P

$$P = \$24,000(P/A, 10\%, 21) = \$24,000(8.6487) = \$207,569.$$

The bank would lend Brown a maximum of $207,569. He would have to borrow the remaining balance from other sources.

Comments: *Of course, the actual amount the bank is willing to lend could be far less than $207,569. Note that the critical data included the actual cash flow over the time rather than the total sum of the winnings because of the different time values of the payments.*

Graphical View of Equal-Payment-Series Interest Factors

Figure 2.30 illustrates the responses of the F/A, A/F, A/P, and P/A factors to variations in i and N. The plot of the F/A factor is quite similar to that of the F/P factor in Fig. 2.18, but the F/A factor provides more rapid increases in F with increases in either N or i. On the other hand, the plot of the P/A factor indicates that, to provide the same amount of P value for a fixed N, the A value has to be increased for an increased value of i. Relationships similar to the P/A factor can be observed for other factors. Understanding these relationships provides insights into how these factors reflect the time value of money.

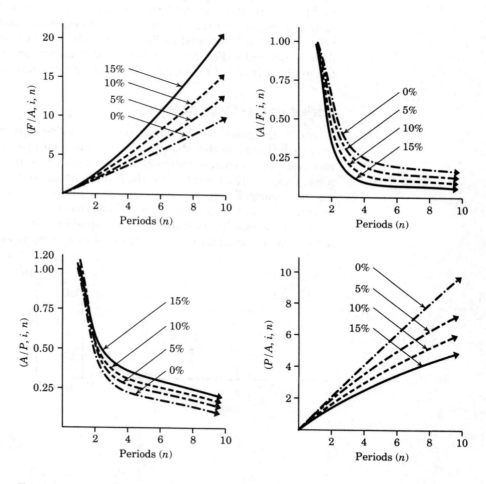

Figure 2.30 ■ The characteristics of equivalent series factors with variations in i and n

2.3.5 Linear Gradient Series

Engineers frequently encounter situations involving periodic cash flows that increase or decrease by a constant amount (G) form period to period. This situation occurs often enough to warrant the use of special equivalence factors relating the arithmetic gradient to other cash flows. Figure 2.31 illustrates such a gradient series, $A_n = (n - 1)G$. The gradient G can be either positive (an *increasing* gradient series) or negative (a *decreasing* gradient series). Note that the origin of the gradient series is at the end of the first period with a zero value.

Unfortunately, the generalized form of the increasing or decreasing gradient series does not correspond to the form taken by most engineering economic problems. A typical problem involving a linear gradient includes an initial payment during period 1 that increases by G during some number of interest periods, a situation illustrated in Fig. 2.32. This contrasts with the generalized form illustrated in Fig. 2.31 in which there is no payment during period 1 and the gradient is added to the previous payment beginning in period 2.

Gradient Series as Composite Series

To utilize the generalized gradient series in solving typical problems, we must view cash flows like that shown in Fig. 2.32 as a **composite series,** or a set of two cash flows, each corresponding to a form we can recognize and easily solve. Figure 2.33 illustrates that the form in which we find a typical cash flow can be separated into two components: a uniform series of N payments of amount A_1, and the gradient series of increments of constant amount G. The need to view cash flows involving linear gradient series as composites of two series is very important for the solution of problems, as we shall now see.

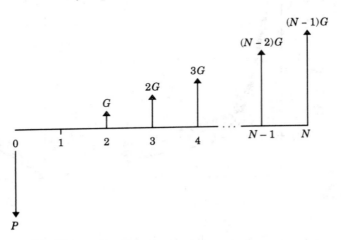

Figure 2.31 ■ Cash flow diagram of a gradient series

Figure 2.32 ■ Cash flow diagram for a typical problem involving linear gradient series. (Compare to Fig. 2.31.)

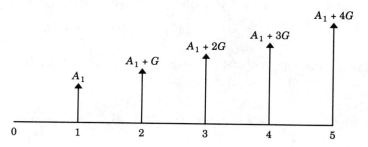

Present Worth Factor—Linear Gradient: Find *P*, Given *A, G, N, i*

How much would you have to deposit now to withdraw the gradient amounts specified in Fig. 2.31? To find an expression for the present amount *P*, we apply the single-payment present worth factor to each term of the series and obtain

$$P = 0 + G/(1 + i)^2 + 2G/(1 + i)^3 + \cdots + (N - 1)G/(1 + i)^N$$

or

$$P = \sum_{n=1}^{N} (n - 1)G(1 + i)^{-n}. \tag{2.14}$$

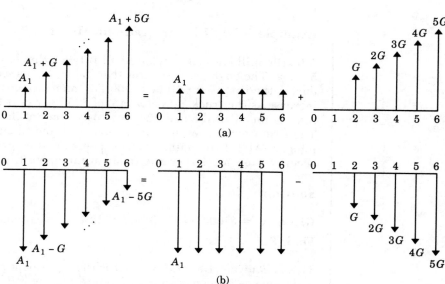

Figure 2.33 ■ Linear gradient series as composites

Letting $G = a$ and $1/(1 + i) = x$ yields

$$P = 0 + ax^2 + 2ax^3 + \cdots + (N - 1)ax^N$$
$$= ax[0 + x + 2x^2 + \cdots + (N - 1)x^{N-1}]. \qquad (2.15)$$

Since an arithmetic-geometric series $\{0, x, 2x^2, \ldots, (N - 1)x^{N-1}\}$ has the finite sum of

$$0 + x + 2x^2 + \cdots + (N - 1)x^{N-1} = \frac{1 - Nx^{N-1} + (N - 1)x^N}{(1 - x)^2},$$

we can rewrite Eq. (2.15) as

$$P = ax\left[\frac{1 - Nx^{N-1} + (N - 1)x^N}{(1 - x)^2}\right]. \qquad (2.16)$$

Replacing the original values for a and x, we obtain

$$P = G\left[\frac{(1 + i)^N - iN - 1}{i^2(1 + i)^N}\right] = G(P/G, i, N) \qquad (2.17)$$

The factor in brackets is known as the **gradient-series present worth factor** and is represented by the notation $(P/G, i, N)$.

Example 2.20   Linear gradient: find P, given A_1, G, i, N

A textile mill has just purchased a lift truck that has a useful life of 5 years. The engineer estimates that the maintenance costs for the truck during the first year will be $1000. Maintenance costs are expected to increase as the truck ages at a rate of $250 per year over the remaining life. Assume that the maintenance costs occur at the end of each year. The firm wants to set up a maintenance account that earns 12% annual interest. All future maintenance expenses will be paid out from this account. How much does the firm have to deposit in the account now?

Solution

Given: $A_1 = \$1000$, $G = \$250$, $i = 12\%$ per year, and $N = 5$ years
Find: P

This is equivalent to asking, what is the equivalent present worth for this maintenance expenditure if 12% interest is used? The cash flow may be broken into its two components, as shown in Fig. 2.34. The first

Figure 2.34 ■ Cash diagram (Example 2.20)

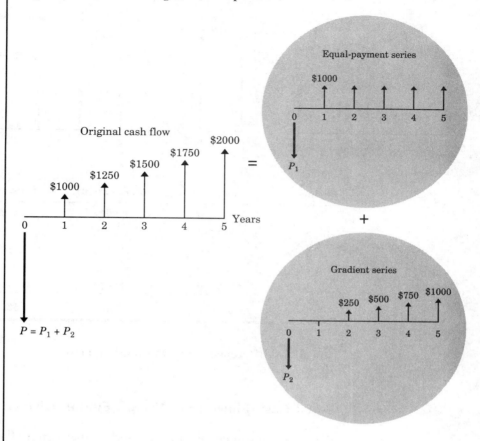

component is an equal-payment series (A_1), and the second is a linear gradient series (G).

$$P = P_1 + P_2$$

$$P = A_1(P/A, 12\%, 5) + G(P/G, 12\%, 5)$$

$$= \$1000(3.6048) + \$250(6.397)$$

$$= \$5204$$

Note that the value of N in the gradient factor is 5, not 4, because by definition of the series the first gradient value begins at period 2.

Comments: *As a check, we can compute the present worth of the cash flow by using the $(P/F, 12\%, n)$ factors.*

Figure 2.35 ■ Converting a gradient series into an equivalent uniform series

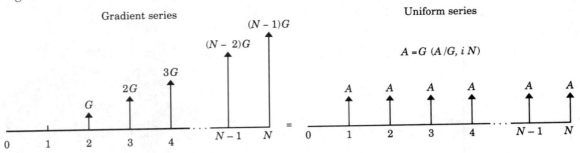

Gradient series   Uniform series

Period (n)	Cash Flow	($P/F, 12\%, n$)	Present Worth
1	$1000	0.8929	$ 892.90
2	1250	0.7972	996.50
3	1500	0.7118	1067.70
4	1750	0.6355	1112.13
5	2000	0.5674	1134.80
		Total	$5204.03

The slight difference is due to rounding error.

Equal-Cash-Flow-Series Factor: Find A, Given A_1, G, i, N

We can obtain an equal-cash-flow series equivalent to the gradient series, as depicted in Fig. 2.35, by substituting Eq. (2.17) into Eq. (2.12) for P to obtain

$$A = G\left[\frac{(1 + i)^N - iN - 1}{i[(1 + i)^N - 1]}\right] = G(A/G, i, N). \qquad (2.18)$$

The bracketed factor is called the **gradient-to-equal cash-flow series conversion factor**; its factor notation is (A/G,i,N).

Example 2.21   Linear gradient: find A, given A_1, G, i, N

John and Barbara just opened two savings accounts at their credit union. The accounts earn 10% annual interest. John wants to deposit $1000 in his account at the end of the first year and increase this amount by $300

Figure 2.36 ■ Cash flow diagram of a combination of uniform and gradient series (Example 2.21)

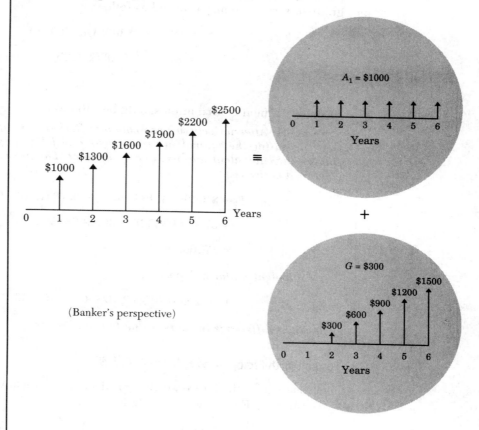

(Banker's perspective)

for each of the following 5 years. Barbara wants to deposit an equal amount each year for next 6 years. What should be the size of the Barbara's annual deposit so that the two accounts would have equal balances at the end of 6 years?

Solution

Given: $A_1 = \$1000$, $G = \$300$, $i = 10\%$, and $N = 6$
Find: A

Since we use the end-of-period convention unless otherwise stated, this series begins at the end of the first year, and the last contribution occurs at the end of the sixth year. We can separate the constant portion of $1000 from the series, leaving the strictly gradient series of $0, 0, 300, 600, \ldots, 1500$.

To find the equal-payment series beginning at the end of year 1 and ending at year 6 that would have the same present worth as that of the gradient series, we may proceed as follows:

$$A = \$1000 + \$300(A/G, 10\%, 6)$$

$$= \$1000 + \$300(2.2235)$$

$$= \$1667.05$$

Barbara's annual contribution should be $1667.05.

Comments: *Alternatively, we can compute Barbara's annual deposit by first computing the equivalent present worth of John's deposits and then finding the equivalent uniform annual amount. The present worth of this combined series is*

$$P = \$1000(P/A, 10\%, 6) + \$300(P/G, 10\%, 6)$$

$$= \$1000(4.3553) + 300(9.6842)$$

$$= \$7260.20.$$

The equivalent uniform deposit is

$$A = \$7260.20(A/P, 10\%, 6) = \$1667.01.$$

(The slight difference in cents is due to rounding error.)

Future Worth Factor—Find F, Given A, G, i, N

The equation for the future worth equivalent of a gradient series can be found by substituting Eq. (2.18) into Eq. (2.10) for A:

$$F = \frac{G}{i}\left[\frac{(1 + i)^N - 1}{i} - N\right] = G(F/G, i, N). \qquad (2.19)$$

Example 2.22 Declining linear gradient: find F, given A_1, G, i, N

Suppose that you make a series of annual deposits into a bank account that pays 10% interest. The initial deposit at the end of the first year is $1200. The deposit amounts decline by $200 in each of the next 4 years. How much would you have right after the fifth deposit?

Solution

Given: Cash flow shown in Fig. 2.37, $i = 10\%$ per year, $N = 5$ years

Find: F

Figure 2.37 ■ Cash flow diagram (Example 2.22)

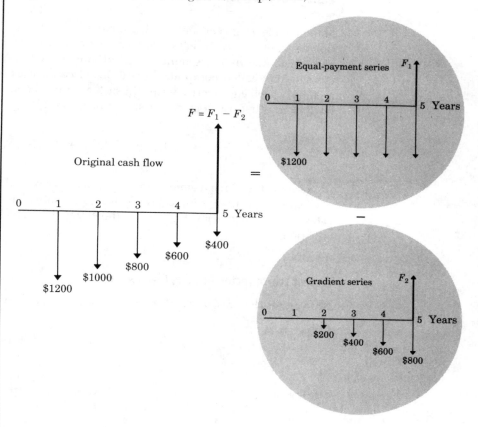

The cash flow includes a decreasing gradient series. Recall that we derived the linear gradient factors for an increasing gradient series. For a decreasing gradient series, the solution is most easily obtained by separating the flow into two components: a uniform series and an increasing gradient, which is *subtracted* from the uniform series (see Fig. 2.37). The future value is

$$F = F_1 - F_2$$

$$= A_1(F/A, 10\%, 5) - \$200(F/G, 10\%, 5)$$

$$= A_1(F/A, 10\%, 5) - \$200(P/G, 10\%, 5)\,(F/P, 10\%, 5)$$

$$= \$1200(6.105) - \$200(6.862)(1.611)$$

$$= \$5115.$$

2.3.6 Geometric Gradient Series

Many engineering economy problems, particularly those relating to construction costs, involve cash flows that increase or decrease over time by a constant percentage known as a **geometric gradient**. (Growth by geometric gradient is sometimes called **compound growth**.) Price changes due to inflation can exemplify such a geometric series. In such a series, the nth payment, A_n, is related to the first payment, A_1, as shown in Eq. (2.20):

$$A_n = A_1(1 + g)^{n-1}, \qquad n = 1, 2, \ldots, N, \qquad (2.20)$$

where g = the percentage change in payment per period.

Clearly, a positive g indicates an increasing series, and a negative g indicates a decreasing series. Figure 2.38 illustrates the cash flow diagram for this situation.

Present Worth Factor—Find P, Given A_1, g, i, N

Notice that the present worth P_n of any cash flow A_n at interest rate i is

$$P_n = A_n(1 + i)^{-n} = A_1(1 + g)^{n-1}(1 + i)^{-n}$$

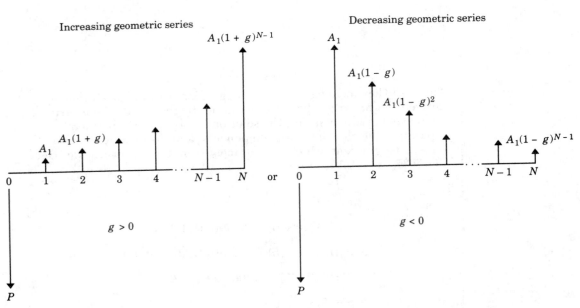

Figure 2.38 ■ A geometrically increasing or decreasing gradient series at a constant rate

We can find an equation for the present amount, P, by applying the *single-cash-flow present worth factor* to each term of the series.

$$P = \sum_{n=1}^{N} A_1(1 + g)^{n-1}(1 + i)^{-n} \tag{2.21}$$

Bringing the constant term $A_1(1 + g)^{-1}$ outside the summation yields

$$P = \frac{A_1}{(1 + g)} \sum_{n=1}^{N} \left[\frac{1 + g}{1 + i} \right]^n . \tag{2.22}$$

Let $a = A_1/(1 + g)$ and $x = (1 + g)/(1 + i)$. Then, we rewrite Eq. (2.22) as

$$P = a(x + x^2 + x^3 + \cdots + x^N) . \tag{2.23}$$

Since the summation in Eq. (2.23) represents the first N terms of a geometric series, we may obtain the closed-form expression as follows. First, multiply Eq. (2.23) by x:

$$xP = a(x^2 + x^3 + x^4 + \cdots + x^{N+1}) . \tag{2.24}$$

Then, subtract Eq. (2.24) from Eq. (2.23):

$$P - xP = a(x - x^{N+1})$$

$$P(1 - x) = a(x - x^{N+1})$$

$$P = \frac{a(x - x^{N+1})}{1 - x}, \text{ where } x \neq 1 . \tag{2.25}$$

Replacing the original values for a and x, we obtain:

$$P = \begin{cases} A_1 \left[\dfrac{1 - (1 + g)^N(1 + i)^{-N}}{i - g} \right] & \text{if } i \neq g \\ \dfrac{NA_1}{1 + i}, & \text{if } i = g \end{cases} \tag{2.26}$$

or

$$P = A_1(P/A_1, g, i, N) .$$

The factor within brackets is called the **geometric-gradient-series present worth factor** and is designated $(P/A_1, g, i, N)$. In the special case where $i = g$, Eq. (2.22) becomes $P = [A_1/(1 + i)]N$.

Example 2.23 Geometric gradient: find P, given A_1, g, i, N

A municipal power plant is expected to generate a net revenue of $500,000 at the end of its first year, and this annual amount will increase by 8% per year for the next 5 years. To finance a new construction project, the municipal government wants to issue a bond that pays 6% annual interest. (A **bond** is a long-term promissory note issued by a business or governmental unit.) The amount of bond financing will depend on the equivalent present worth of the expected future earnings from this power plant, which will be used to pay off the bonds. What would be the maximum amount of bond financing that could be secured?

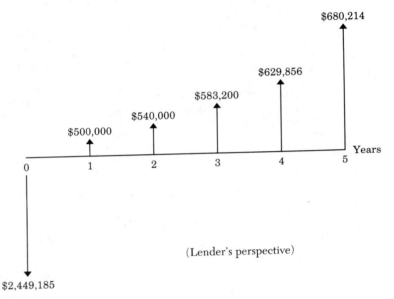

(Lender's perspective)

Figure 2.39 ■ Cash flow for a municipal power plant project (Example 2.23)

Solution

Given: $A_1 =$ $500,000 per year, $g =$ 8% per year, $N =$ 5 years, and $i =$ 6% per year

Find: P

Assuming that all the net revenues would occur at the end of each year, we determine the equivalent present worth by substituting these values back into Eq. (2.26):

$$P = A_1(P/A_1, g, i, N)$$

$$= \$500,000(P/A_1, 8\%, 6\%, 5)$$

$$= \$500,000\left[\frac{1 - (1 + 0.08)^5(1 + 0.06)^{-5}}{0.06 - 0.08}\right]$$

$$= \$2,449,185.$$

Future Worth Factor—Find *F*, Given *A₁*, *g*, *N*, *i*

The future worth equivalent of the geometric series can be obtained by multiplying Eq. (2.26) by the *F/P* factor, $(1 + i)^N$.

$$F = \begin{cases} A_1 \left[\dfrac{(1 + i)^N - (1 + g)^N}{i - g} \right], & \text{if } i \neq g \\[2ex] NA_1(1 + i)^{N-1}, & \text{if } i = g \end{cases} \tag{2.27}$$

or

$$F = A_1(F/A_1, g, i, N).$$

Example 2.24   Geometric gradient: find *A₁*, given *F*, *g*, *i*, *N*

A self-employed individual, Jimmy Carpenter, is opening a retirement account at a bank. His goal is to accumulate $1,000,000 in the account by the time he retires from work in 20 years. A local bank is willing to open such a retirement account that pays 8% interest compounded annually, throughout the 20 years. Jimmy expects his annual income will increase at a 6% annual rate during his working career. He wishes to start with a deposit at the end of year 1 (A_1) and increase the deposit at a rate of 6% each year thereafter. What should be the size of his first deposit (A_1)? The first deposit will occur at the end of year 1, and the subsequent deposits will be made at the end of each year. The last deposit will be made at the end of year 20.

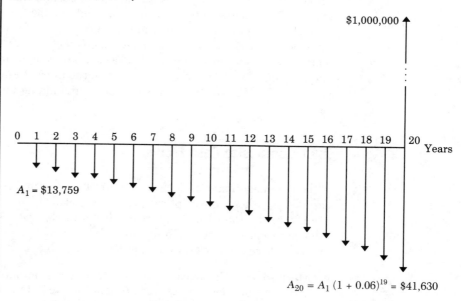

Figure 2.40 ■ Cash flow diagram (Example 2.24)

Solution

Given: $F = \$1,000,000$, $g = 6\%$ per year, $i = 8\%$ per year, and $N = 20$ years

Find: A_1

$$F = A_1(F/A_1, g, i, N)$$
$$= A_1(F/A, 6\%, 8\%, 20)$$
$$= A_1(72.6780)$$

Solving for A_1 yields

$$A_1 = \$1,000,000/72.6780 = \$13,759.32$$

Table 2.3 summarizes the interest formulas developed in this section and the cash flow situations in which they should be used. Recall that all these interest formulas are applicable only when the interest (compounding) period is the same as the payment period (e.g., annual compounding with annual payment). Also, in this table we present some useful interest factor relationships.

■ 2.4 Unconventional Equivalence Calculations

Throughout the preceding section, we occasionally applied two or more methods of attacking example problems, even though we had standard interest factor equations by which to solve them. It is important that you become adept at examining problems from unusual angles and seeking out unconventional solution methods, because not all cash flow problems conform to the neat patterns for which we have discovered and developed equations. Two categories of problems that demand unconventional treatment are composite (mixed) cash flows and problems in which we must determine the interest rate implicit in a financial contract. We will begin this section by examining instances of composite cash flows.

2.4.1 Composite Cash Flows

Although many financial decisions do involve constant or systematic change in cash flows, others consist of cash flows exhibiting no overall pattern. To illustrate, consider the cash flow stream shown in Fig. 2.41. We want to compute the equivalent present worth for this mixed payment series at an interest rate of 15%. There are two ways to approach the problem.

- **Method 1:** Multiply each payment by the appropriate $(P/F, 15\%, n)$ factors and then sum these products to obtain the present worth of the cash flows,

Table 2.3 Summary of Discrete Compounding Formulas with Discrete Payments

Flow Type	Factor Notation	Formula	Cash Flow Diagram	Factor Relationship
S I N G L E	Compound amount $(F/P,i,N)$	$F = P(1+i)^N$	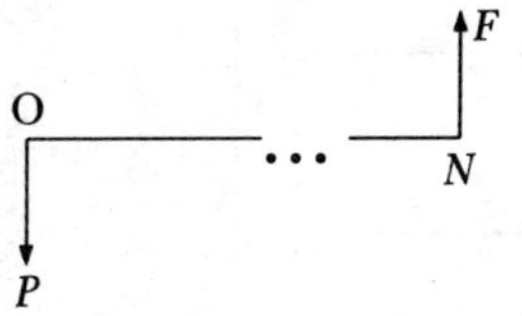	$(F/P,i,N) = i(F/A,i,N) + 1$
	Present worth $(P/F,i,N)$	$P = F(1+i)^{-N}$		$(P/F,i,N) = 1 - (P/A,i,N)i$
E Q U A L	Compound amount $(F/A,i,N)$	$F = A\left[\dfrac{(1+i)^N - 1}{i}\right]$	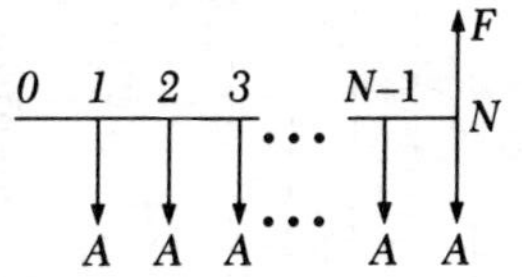	
	Sinking fund $(A/F,i,N)$	$A = F\left[\dfrac{i}{(1+i)^N - 1}\right]$		$(A/F,i,N) = (A/P,i,N) - i$
F L O W	Present worth $(P/A,i,N)$	$P = A\left[\dfrac{(1+i)^N - 1}{i(1+i)^N}\right]$	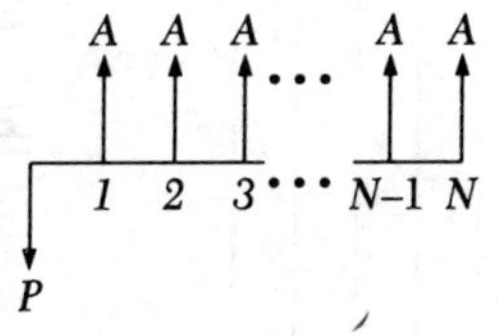	
S E R I E S	Capital recovery $(A/P,i,N)$	$A = P\left[\dfrac{i(1+i)^N}{(1+i)^N - 1}\right]$		$(A/P,i,N) = \dfrac{i}{1 - (P/F,i,N)}$
G R A D I E N T	Uniform gradient Present worth $(P/G,i,N)$	$P = G\left[\dfrac{(1+i)^N - iN - 1}{i^2(1+i)^N}\right]$	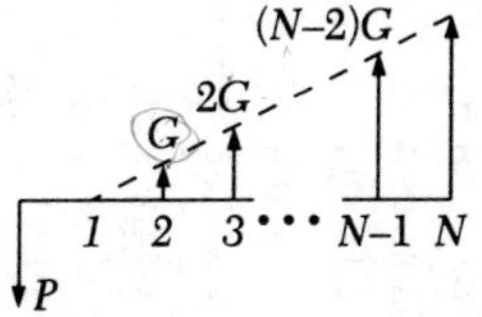	$(F/G,i,N) = (P/G,i,N)(F/P,i,N)$ $(A/G,i,N) = (P/G,i,N)(A/P,i,N)$
S E R I E S	Geometric gradient Present worth $(P/A_1,g,i,N)$	$P = \begin{cases} A_1\left[\dfrac{1 - (1+g)^N(1+i)^{-N}}{i-g}\right] \\ \dfrac{NA_1}{1+i} \quad (\text{if } i = g) \end{cases}$	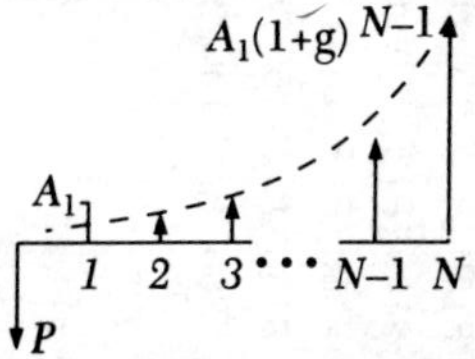	$(F/A_1,g,i,N) = (P/A_1,g)$

Figure 2

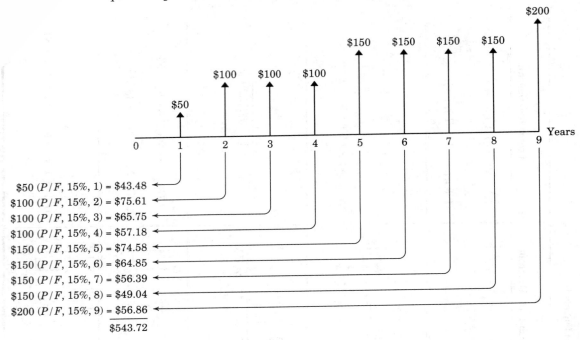

$50\ (P/F,\ 15\%,\ 1) = \$43.48$
$100\ (P/F,\ 15\%,\ 2) = \$75.61$
$100\ (P/F,\ 15\%,\ 3) = \$65.75$
$100\ (P/F,\ 15\%,\ 4) = \$57.18$
$150\ (P/F,\ 15\%,\ 5) = \$74.58$
$150\ (P/F,\ 15\%,\ 6) = \$64.85$
$150\ (P/F,\ 15\%,\ 7) = \$56.39$
$150\ (P/F,\ 15\%,\ 8) = \$49.04$
$200\ (P/F,\ 15\%,\ 9) = \$56.86$
$\overline{}$
$\$543.72$

$543.72. Recall that this is the same procedure we used to solve uneven cash flow series problems, described in Section 2.3.3. Figure 2.41 illustrates this computational process.

- **Method 2:** Group the cash flow components according to the type of cash flow pattern into which they fit, such as the single cash flow, equal-cash-flow series and so forth, as shown in Fig. 2.42. Then, the solution procedure involves the following steps:

Step 1: Find the present worth of $50 due in year 1:

$$\$50(P/F, 15\%, 1) = \$43.48$$

Step 2: Recognize that a $100 equal-cash-flow series will be received during years 2 through 4. Thus, we could determine the value of a 4-year annuity, subtract from it the value of a 1-year annuity, and have remaining the value of a 4-year annuity whose first payment is due in year 2. This result is achieved by subtracting the $(P/A,15\%,1)$ for a 1 year, 15% annuity from that for a 4-year annuity and then multiplying the difference by $100:

$$\$100[(P/A, 15\%, 4) - (P/A, 15\%, 1)] = \$100(2.8550 - 0.8696)$$

$$= \$198.54$$

Figure 2.42 ■ Equivalent present worth calculation for an uneven cash-flow series using *P/F* and *P/A* factors

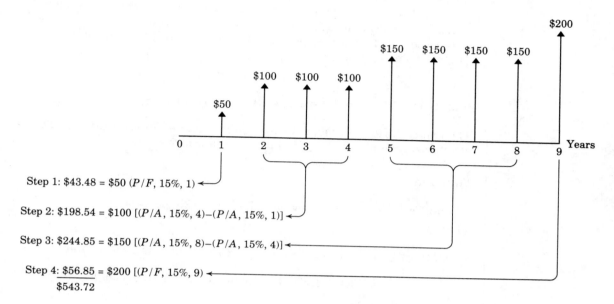

Step 1: $43.48 = $50 (*P/F*, 15%, 1)

Step 2: $198.54 = $100 [(*P/A*, 15%, 4)–(*P/A*, 15%, 1)]

Step 3: $244.85 = $150 [(*P/A*, 15%, 8)–(*P/A*, 15%, 4)]

Step 4: $56.85 = $200 [(*P/F*, 15%, 9)
$543.72

or alternately, $100 (*P/A*, 15%, 3) (*P/F*, 15%, 1) = $198.54.
Thus, the equivalent present worth of the annuity component of the uneven stream is $198.54.

Step 3: We have another equal-payment series, which starts in year 5 and ends in year 8.

$$\$150[(P/A,15\%,8) - (P/A,15\%,4)] = \$150(4.4873 - 2.8550)$$

$$= \$244.85$$

or alternately,
$150 (*P/A*, 15%, 4) (*P/F*, 15%, 4) = $244.85.

Step 4: Find the equivalent present worth of the $200 due in year 9:

$$\$200(P/F, 15\%, 9) = \$56.85$$

Step 5: Sum the components:

$$P = \$43.48 + \$198.54 + \$244.85 + \$56.85 = \$543.72$$

Either the uneven payment series method in Fig. 2.41 or the method utilizing both (*P/A*, i, n) and (*P/F*, i, n) factors can be used to solve problems of this type. If the annuity component runs for many years, however, the second method is much easier. For example, the second method would be clearly superior for finding the equivalent present worth of a stream consisting of $50 in year 1, $200 in years 2 through 19, and $500 in year 20.

Also, note that in some instances we may want to find the equivalent value of a stream of payments at some point other than the present (year 0). In this case, we proceed as before but compound and discount to some other points in time, say year 2, rather than year 0.

Example 2.25 Cash flows with subpatterns

The two cash flows in Fig. 2.43 are equivalent at an interest rate of 12% compounded annually. Determine the unknown value X.

Solution

Given: Cash flows as shown in Fig. 2.43, $i = 12\%$ per year

Find: X

- **Method 1:** Compute the present worth of each cash flow at time 0.

$$P_1 = \$100(P/A, 12\%, 2) + \$300(P/A, 12\%, 3)\,(P/F, 12\%, 2)$$

$$= \$743.42$$

$$P_2 = X(P/A, 12\%, 5) - X(P/F, 12\%, 3)$$

$$= 2.8930X$$

Since the two flows are equivalent, $P_1 = P_2$.

$$743.42 = 2.8930X$$

Solving for X, we obtain $X = \$256.97$.

- **Method 2:** We may select a time point other than 0 for comparison. The best choice of a base period is largely determined by the patterns of cash flows. Obviously, we want to select a base period that requires the minimum number of interest factors for the equivalence calculation. Cash flow 1 represents a combined series of two

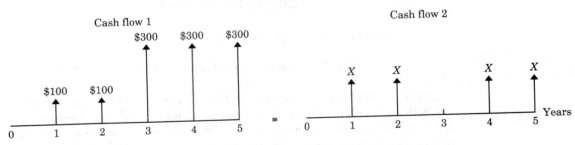

Figure 2.43 ■ Equivalence calculation (Example 2.25)

equal-payment cash flows, whereas cash flow 2 can be viewed as an equal-payment series with the third payment missing. For cash flow 1, computing the equivalent worth at period 3 will require only two interest factors.

$$V_{3,1} = \$100(F/A, 12\%, 3) + \$200 + \$300(P/A, 12\%, 2)$$

$$= \$1044.46$$

For cash flow 2, computing the equivalent worth of the equal payment series at period 3 will also require two interest factors.

$$V_{3,2} = X(F/A, 12\%, 3) + X(P/A, 12\%, 2) - X$$

$$= 4.0645X$$

Therefore, the equivalence would be obtained by letting $V_{3,1} = V_{3,2}$:

$$1044.46 = 4.0645X$$

Solving for X yields $X = \$256.97$, which is the same result obtained from Method 1. In this example, the alternative solution by shifting the time point of comparison will require only four interest factors, whereas Method 1 would require five interest factors.

Example 2.26 Combination of linear and geometric gradients

Referring to Example 2.23, suppose that the government actually borrowed \$2 million via the sale of bonds to finance the power plant project. The bondholders were promised payments of \$120,000 in interest at the end of each year for 5 years and repayment of the \$2 million principal (known as **face value** of the bonds) at the end of year 5. The government will meet the bond obligations with part of the revenues from the power plant and will deposit any left-over revenues each year in a special account that earns 6% annual interest. After paying off all the bond interest and principal, how much money will be left in this special account?

Discussion: Let us compute the annual revenues generated by the power plant over the next 5 years. The revenue during the first year is \$500,000, and the subsequent annual revenue increases at a rate of 8%. We can estimate the future revenues as follows:

Period		Revenue
1	\$500,000	= \$500,000
2	$\$500,000(1 + 0.08) = \$500,000(1 + 0.08)^1$	= \$540,000
3	$\$540,000(1 + 0.08) = \$500,000(1 + 0.08)^2$	= \$583,200
4	$\$583,200(1 + 0.08) = \$500,000(1 + 0.08)^3$	= \$629,856
5	$\$629,856(1 + 0.08) = \$500,000(1 + 0.08)^4$	= \$680,244

Figure 2.44 ■ Cash flow diagram (Example 2.26)

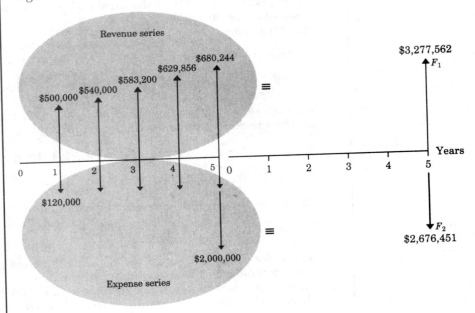

In Fig. 2.44, these annual revenues are shown as cash receipts.
    Second, we need to compute the annual bond interest payment. With the principal of $2 million at an annual interest rate of 6%, the annual bond interest due is $120,000. When the bond matures at the end of year 5, the principal amount of $2 million will be paid to the bondholders. In Fig. 2.44, the bond interest payments are shown as disbursements.

Solution

Given: Revenue and expense flows shown in Fig. 2.44, $i = 6\%$ per year

Find: Net future worth of the two flows at the end of 5 years

We may approach the problem by computing the equivalent of the revenue series at the end of 5 years (F_1) and then subtracting the equivalent of all the bond obligations at that time if they were deferred (F_2). The differential amount represents the left-over funds in the account.

$$F_1 = A_1(F/A_1, g, i, N) = \$500,000(F/A_1, 8\%, 6\%, 5)$$

$$= \$3,277,562$$

$$F_2 = \$120,000(F/A, 6\%, 5) + \$2,000,000$$

$$= \$2,676,451$$

$$F_1 - F_2 = \$3{,}277{,}562 - \$2{,}676{,}451$$

$$= \$601{,}111$$

To verify the ending balance in the special account, we can calculate the actual year-by-year transactions:

n	Beginning Balance	Interest Earned	Revenue Received	Bond Payment	Ending Balance
1	$ 0	$ 0	$500,000	−$120,000	$ 380,000
2	380,000	380,000(0.06)	540,000	−120,000	822,800
3	822,800	822,800(0.06)	583,200	−120,000	1,335,368
4	1,335,368	1,335,368(0.06)	629,856	−120,000	1,925,346
5	1,925,346	1,925,346(0.06)	680,244	−2,120,000	601,111

The government will have $601,111 in the special account at the end of 5 years. This figure agrees with the one calculated above.

Comments: *The first calculation did not require the actual determination of the individual revenue flows. Only the value of the initial flow and growth rate are necessary.*

2.4.2 Determining Unknown Interest Rates

Thus far, we have assumed that a typical interest rate is given in the equivalence calculations. Now we can use the same interest formulas developed earlier to determine the interest rates explicit in the equivalence problems. The rates of interest are specified in the contracts for most commercial loans, so there often may be no need to find these rates. However, when you invest in financial assets such as stocks, you may want to know at what rate of growth (or rate of return) your asset is appreciating over the years. (This kind of calculation is the basis of rate-of-return analysis, covered in Chapter 6.) Although we can use interest tables to find the interest rate implicit in single payments and annuities, it is more difficult to find the interest rate implicit in an uneven series of payments. A trial-and-error procedure or computer software may be used to calculate the interest rate in such cases. To illustrate, we will show two examples.

Example 2.27 Calculating an unknown interest rate with a single factor

Your credit union is willing to lend you $20,000 for your home remodeling project. You must sign a loan contract calling for payments of $3116.40 at the end of each of the next 10 years. What interest rate is the credit union offering you?

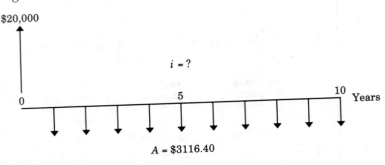

Figure 2.45 ■ Loan cash flow diagram (Example 2.27)

$20,000

$i = ?$

0   5   10 Years

A = $3116.40

Solution

Given: $P = \$20{,}000$, $N = 10$ years, $A = \$3116.40$

Find: i

The problem may be solved using either the equal-payment-series present worth formula,

$$P = A(P/A, i, N),$$

or the capital recovery factor (annuity factor),

$$A = P(A/P, i, N).$$

Using the P/A factor,

$$P = A(P/A, i, N)$$

$$\$20{,}000 = \$3116.40(P/A, i, 10)$$

$$(P/A, i, 10) = \frac{\$20{,}000}{\$3116.40} = 6.4177.$$

Thus, we want to find a P/A factor that equals or is close to 6.4177 when $n = 10$. If we simply begin at the first interest table and skim through the pertinent listing in each, we eventually see that at $i = 9\%$ the factor value matches the 6.4177 we derived in our calculations. Thus, the interest rate on this loan contract is 9%.

Example 2.28 Calculating an unknown interest rate with multiple factors

Consider again the sweepstakes problem introduced at the beginning of this chapter. Suppose that you had decided to accept the 20 annual installments of $100,000 instead of receiving one lump sum of $1 million.

If you are like most jackpot winners, you will be tempted to spend your winnings to improve your lifestyle during the first several years. Only after you've gotten the spending "out of your system," will you save the later sums for investment purposes. Suppose that you are considering the following two options:

- **Option 1:** You save your winnings for the first 6 years and then spend every cent of the winnings in the remaining 14 years.

- **Option 2:** You do the reverse and spend for 6 years and then save for 14 years.

What interest rate on the savings will make these two options equivalent? (The cash flows into savings for the two options are shown in Fig. 2.46.)

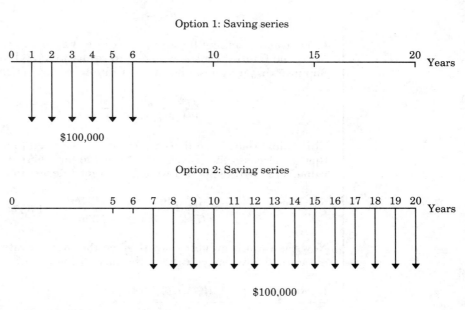

Figure 2.46 ■ Equivalence calculation (Example 2.28)

Solution

Given: Cash flows shown in Fig. 2.46
Find: i at which the two flows are equivalent

To compare the alternatives, we may compute the present worth for each option at period 0. By selecting time period 6, however, we can establish the same economic equivalence with fewer interest factors. That is, we calculate the equivalent value for each option at the end of period 6 (V_6), remembering that the end of period 6 is also the beginning of period 7.

(Recall from Example 2.4 that the choice of the point in time at which to compare two cash flows for equivalence is arbitrary.)

- For Option 1:

$$V_{6,1} = \$100,000(F/A, i, 6).$$

- For Option 2:

$$V_{6,2} = \$100,000(P/A, i, 14).$$

To be equivalent, these values must be the same.

$$\$100,000(F/A, i, 6) = \$100,000(P/A, i, 14)$$

$$\frac{(F/A,\ i,\ 6)}{(P/A,\ i,\ 14)} = 1$$

Here, we are looking for an interest rate that makes the ratio 1. When using the interest tables, we need to resort to a trial-and-error method. Suppose that we guess the interest rate to be 6%. Then

$$\frac{(F/A, 6\%, 6)}{(P/A, 6\%, 14)} = \frac{6.97532}{9.29498} = 0.7504$$

This is less than 1. To increase the ratio, we need to use an i value such that it increases the $(F/A, i, 6)$ factor value but decreases the $(P/A, i, 14)$ value. This will happen if we use a larger interest rate. Let's try $i = 10\%$.

$$\frac{(F/A, 10\%, 6)}{(P/A, 10\%, 14)} = \frac{7.71561}{7.36668} = 1.0474$$

Now the ratio is greater than 1. Since the interest rate appears to be very close to 10%, we may evaluate the ratio using $i = 9\%$. As a result, we find

Interest Rate	$(F/A, i, 6)/(P/A, i, 14)$
6%	0.7504
9%	0.9662
10%	1.0474

The interest rate is between 9% and 10%, and may be approximated by linear interpolation as shown in Fig. 2.47.

$$i = 9\% + (10\% - 9\%)\left[\frac{1 - 0.9662}{1.0474 - 0.9662}\right]$$

$$= 9\% + 1\%\left[\frac{0.0338}{0.0812}\right]$$

$$= 9.42\%$$

Figure 2.47 ■ Linear interpolation to find unknown interest rate (Example 2.28)

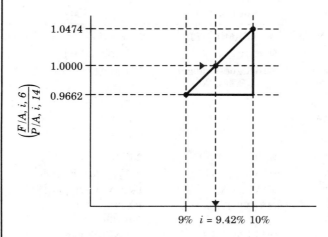

At 9.42% interest, the options are equivalent and you may decide to indulge your desire to spend like crazy for the first 6 years. However, if you could obtain a higher interest rate, you would be wiser to save for 6 years, and spend for the next 14.

Comments: *The last two examples demonstrate that finding an interest rate is an iterative process, which is more complicated and generally less precise than the problem of finding an equivalent worth at a known interest rate. Since computers and financial calculators can speed the process of finding unknown interest rates, these tools are highly recommended for these types of problem solving.*

 ## 2.5 Computer Notes

With the advent of personal computers (and workstations), we can easily access a great deal of computing power at a fractional cost. In particular, electronic spreadsheets such as Lotus 1-2-3, Excel, and QuattroPro provide many useful financial functions that can expedite equivalence calculations.

Most of the equivalence equations we've studied in Chapter 2 are treated as built-in functions in electronic spreadsheets. However, despite their power, one drawback of these spreadsheets is the macro-programming required to perform some higher economic functions. The **EzCash** software which accompanies this text does not require any macro-programming, an efficiency that should enhance its usefulness to you significantly as you study subsequent chapters. An installation guide for **EzCash** is found in Appendix B.

Table 2.4 summarizes the built-in equivalence functions for Lotus 1-2-3, Excel, and QuattroPro. As you will quickly see, the keystrokes and choices are similar for all three. Thus, although we begin with a Lotus 1-2-3 example, you should easily be able to translate the procedure described if you use a different spreadsheet package.

Table 2.4 Built-in Equivalence Functions

Function Description	Lotus 1-2-3	Excel[1]	QuattroPro
Given $i, A, P, F, type$[2]; find N	@CTERM(i, F, P)	NPER ($i, A, P, F, type$)	@CTERM (i, F, P)
Given $i, A, N, P, type$; find F	@FV(A, i, N)	FV($i, N, A, P, type$)	@FV (A, i, N)
Given cash flow series; find i (You must enter a reasonable guess for i and indicate the range of cells that contain the cash flow. The first cash flow in the series must be negative.)	@IRR(guess, range)	IRR(range, guess)	@IRR(guess, range)
Given i, range, $type$; find P	@NPV(i, range)	NPV(i, range)	@NPV(i, range, $type$)
Given $P, i, N, F, type$; find A	@PMT(P, i, N)	PMT ($i, N, P, F, type$)	@PMT(P, i, N)
Given $A, i, N, F, type$; find P	@PV(A, i, N)	PV($i, N, A, F, type$)	@PV(A, i, N)
Given $P, F, N, A, type$ and guess; find i	@RATE(F, P, N)	RATE($N, A, P, F, type$, guess)	@RATE(F, P, N)
Given A, F, i; find N	@TERM(A, i, F)	see NPER above	@TERM (A, i, F)

[1] Excel functions are shown without equal signs; remember to type an equal sign at the beginning of every formula but not before functions in nested formulas.

[2] *"Type"* indicates end- or beginning-of-period convention: 0 = end of period; 1 = beginning of period.

We will use Lotus 1-2-3 and **EzCash** to calculate the present worth of Troy Aikman's contract, shown in Fig. 2.9. You are encouraged to explore both methods.

2.5.1 Equivalence Calculation Using Lotus 1-2-3

For the salary payments shown in Fig. 2.9, we want to compute the value of Troy Aikman's contract at the time of signing. We assume a 6% interest rate.

Figure 2.48 shows a reasonable format for the spreadsheet. We enter the known cash flows, interest rate, and number of periods as shown. In cells F14, F16, and F18, the present, future, and annual worth functions have been entered. Then Lotus 1-2-3 automatically calculates and presents the numerical value.

One of the most useful features of any spreadsheet is to allow you to do relatively quick "what if" analyses. For example: What if Troy Aikman can earn 10% instead of 6% interest on his cash flow? Enter this new interest amount in cell F11, and you will see that Lotus automatically recalculates the present, future, and annual worth amounts. (You should obtain the following values: $P = \$7335.73$, $F = \$20,929.70$, $A = \$1129.43$.)

2.5.2 Equivalence Calculation Using EzCash

As just demonstrated, the spreadsheet can be a convenient tool in solving many types of equivalence problems. It even allows us to answer some "what-if"

Figure 2.48 ■ Equivalence calculation using Lotus 1-2-3 (Troy Aikman's contract shown in Fig. 2.9)

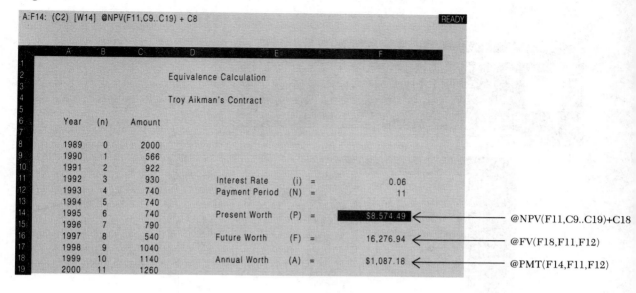

questions by changing a single cell value. In some situations, however, you may need a series of adjustments in cell values to address such questions. Suppose that Aikman's contract runs 12 years instead of 11. Certainly, you need to add the twelfth year data into the worksheet. Consequently, you must also update the cell range values in the @NPV function. Manually, this can be a tedious task. **Macros** (built-in programming functions) can help you automate a Lotus 1-2-3 task that you perform manually. Unfortunately, macro programming can be slow and may require a steep learning curve.

As an alternative, **EzCash** is a very powerful computer-aided economic analysis tool that does not require any programming on the part of users. As you will see in the examples to follow, its input or output can be presented in graphical form, so that you can have a better visual aid in term of user-interface. We will re-examine Aikman's contract using the **EzCash** software.

Pull-down Menu Operation

As in Lotus, the first step is to enter the cash flows. Move the graphic cursor to the top of the screen and click **Edit**, followed by **Build** (Fig. 2.49a).

Now the software automatically opens a **Cash Flow Editor** window on the left side of the screen. As shown in Fig. 2.9, there is no pattern to the cash flows for the first four, so enter the individual cash amounts given for the first four years. (To check the amount of a particular cash flow once it has been entered, scan the **Cash Flow** window box on the right side of the screen, as shown in Fig. 2.49b.)

Figure 2.49 ■ Entering cash flows using the **EzCash** program: (a) opening screen to edit (or enter) the cash flow, (b) entering individual cash flows, (c) entering cash flow patterns

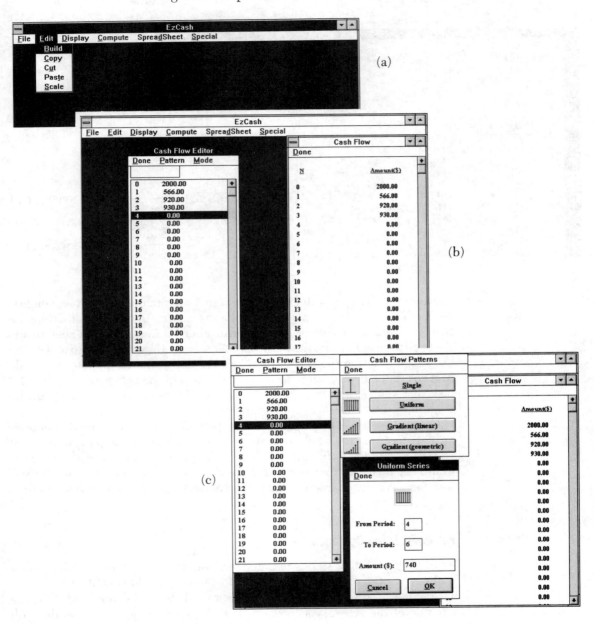

For years 4 through 6, there is an equal cash flow series, so click the **Pattern** menu in the **Cash Flow Editor** window and select **Uniform** from the **Cash Flow Patterns** window (Fig. 2.49c). You will be prompted to specify the number of the beginning and ending cash flow period and the uniform amount of the series in the **Uniform Series** window. Clicking **OK** will complete entering the uniform cash flow series. Now return to the **Cash Flow Editor** by clicking the **Done** option in the **Uniform Series** and the **Cash Flow Patterns** window, and enter the rest of the cash flows.

Displaying Cash Flow Plot and Present Worth Table

When the cash flows are all entered, exit the **Cash Flow Editor** by clicking the **Done** option and select **Display** from the main menu (Fig. 2.50a). Now click **Cash Flow Plot** and a diagram of the cash flows just entered will be displayed on screen (Fig. 2.50b). You may print the cash flow diagram to a printer by clicking the **Print** option. Clicking the **Done** option will bring you back to the main menu.

By clicking **Display** in the main menu followed by **Present Worth Table**, you will be prompted for **Lower Bound**, **Upper Bound**, and the **Increment** that **EzCash** needs in order to generate a present worth table (Fig. 2.50c). The present values of the cash flows in Troy Aikman's contract are shown in Fig. 2.50c when the interest rate changes from 1% to 10%, with a step size of 1%. This is a very convenient way of doing sensitivity analysis on the interest rate.

Equivalent Calculations

Selecting **Compute** from the main menu followed by **Equivalent Worth**, you will be prompted with three choices of equivalence calculations: **Present Worth**, **Future Worth**, and **Annual Worth** (Fig. 2.51a). By clicking **Present Worth**, you will be prompted to specify the interest rate. Note that the interest rate should be entered as a percentage, as shown in Fig. 2.51b. Once the interest rate is entered and the **OK** option is selected, **EzCash** automatically calculates the present worth (Fig. 2.51b). If you have selected **Future Worth**, you will be prompted to specify the interest rate and the **Base Period**, which is the period at which you wish to calculate equivalence. **EzCash** will then automatically calculate the equivalent worth (Fig. 2.51c). If **Annual Worth** is selected and the interest rate is specified, **EzCash** will automatically calculate the annual worth (Fig. 2.51d).

What if we want to change Aikman's rate of interest in **EzCash**? Return to the **Interest Rate** window, and follow the same steps as before, this time indicating 10% as the interest rate (Fig. 2.52a). This simple sensitivity analysis is as easily done in **EzCash** as in Lotus 1-2-3. Suppose we want to add a twelfth year to Aikman's contract? As mentioned in the Lotus 1-2-3 presentation, doing so in a spreadsheet requires manually changing some of the built-in formulas. In **EzCash**, you simply return to **Edit** mode in the main menu, select **Build**, and add the additional cash flow. Another efficiency of the **EzCash** software is that it allows you to choose a base period other than 0 for calculating equivalent worth, once again without manually changing any built-in functions (Fig. 2.52b).

Figure 2.50 ■ (a) The choices under the **Display** mode, (b) displaying **Cash Flow Plot**, (c) displaying **Present Worth Table**

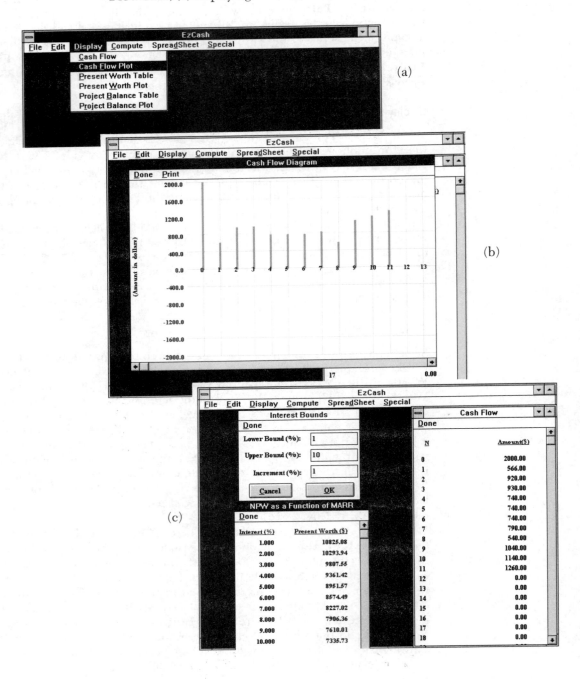

(a)

(b)

(c)

Figure 2.51 ■ Equivalence Calculations: (a) choices in **Compute** mode and under **Equivalent Worth**, (b) **Present Worth** calculation, (c) **Future Worth** calculation, (d) **Annual Worth** calculation

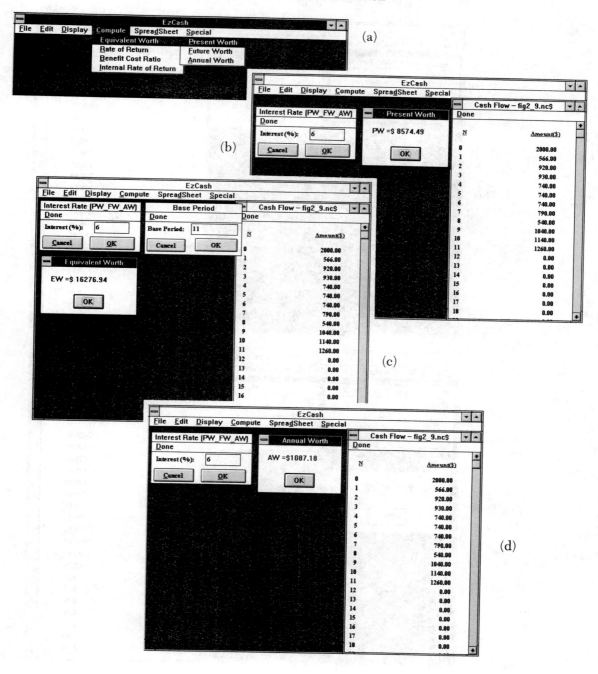

Figure 2.52 ■ You can answer many "what-if" questions without re-entering the cash flows: (a) by specifying a new interest rate, (b) by selecting a base period other than zero for equivalent worth calculation

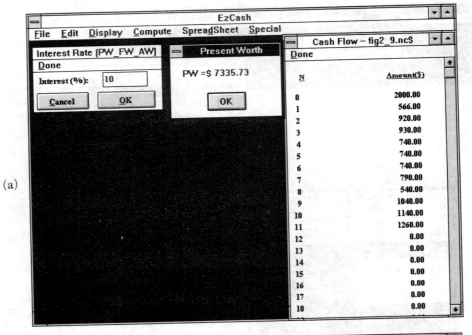

(a)

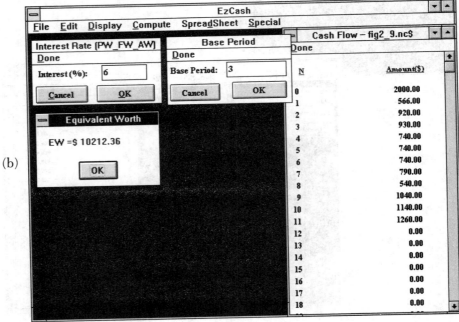

(b)

Summary

- Money has a time value because it can earn more money over time. A number of terms involving the time value of money were introduced in this chapter:

  Interest is the cost of money. More specifically, it is a cost to the borrower, and an earning to the lender, above and beyond the initial sum borrowed or loaned.

  Interest rate is a percentage periodically applied to a sum of money to determine the amount of interest to be added to that sum.

  Simple interest is the practice of charging an interest rate only to an initial sum.

  Compound interest is the practice of charging an interest rate to an initial sum *and* to any previously accumulated interest that has not been withdrawn from the initial sum. Compound interest is by far the most commonly used system in the real world.

  Economic equivalence exists between individual cash flows and/or patterns of cash flows that have the same value. Even though the amounts and timing of the cash flows may differ, the appropriate interest rate makes them equal.

- The compound interest formula is perhaps the single most important equation in this text:

$$F = P(1 + i)^N,$$

  where P is a present sum, i is the interest rate, N is the number of periods for which interest is compounded, and F is the resulting future sum. All the other important interest formulas are derived from this one.

- **Cash flow diagrams** are visual representations of cash inflows and outflows along a timeline. They are particularly useful for helping us detect which of the five patterns of cash flows is represented by a particular problem.

- The five patterns of cash flows are:
  1. Single cash flow: A single present or future cash flow.
  2. Uniform series: A series of flows of equal amounts at regular intervals.
  3. Linear gradient series: A series of flows increasing or decreasing by a fixed amount at regular intervals.
  4. Geometric gradient series: A series of flows increasing or decreasing by a fixed percentage at regular intervals.
  5. Irregular series: A series of flows exhibiting no overall pattern. However, patterns might be detected for portions of the series.

- Cash flow patterns are significant because they allow us to develop **interest formulas,** which streamline the solution of equivalence problems. Table 2.3 summarizes the important interest formulas that form the foundation for all other analyses you will conduct in engineering economic analysis.

Problems

2.1 You deposit $1000 in a savings account that earns a 12% simple interest per year. To double your balance, you wait at least ( ) years. But if you deposit the $1000 in another savings account that earns a 10% interest compounded yearly, it will take ( ) years to double your balance.

2.2 Compare the interest earned by $500 for 10 years at 8% simple interest with that earned by the same amount for 10 years at 8% compounded annually.

2.3 You are considering investing $2000 at an interest rate of 6% compounded annually for 3 years or investing the

$2000 at 7% per year simple interest rate for 3 years. Which option is better?

2.4 You are going to borrow $2000 from a bank at an interest rate of 13% compounded annually. You are required to make three equal annual repayments in the amount of $847.04 per year, with the first repayment occurring at the end of year 1. In each year, show the interest payment and principal payment.

2.5 Suppose you have the alternative of receiving either $3000 at the end of 3 years or P dollars today. Having no current need for the money, you would deposit the P dollars in a bank that pays 6% interest. What value of P would make you indifferent in your choice between P dollars today and the promise of $3000 at the end of 3 years?

2.6 Suppose that you are obtaining a personal loan from your uncle in the amount of $5000 for two years to cover your college expenses. If your uncle always earns 9% interest (annually) on his money invested in various sources, what minimum lump-sum payment two years from now would make your uncle happy?

2.7 What will be the amount accumulated by each of these present investments?

(a) $7200 in 8 years at 9% compounded annually
(b) $675 in 11 years at 4% compounded annually
(c) $3500 in 31 years at 7% compounded annually
(d) $11,000 in 7 years at 8% compounded annually

2.8 What is the present worth of these future payments?

(a) $4300—6 years from now at 9% compounded annually
(b) $6200—15 years from now at 12% compounded annually

(c) $10,000—5 years from now at 6% compounded annually
(d) $20,000—10 years from now at 8% compounded annually

2.9 For an interest rate of 9% compounded annually, find

(a) How much can be loaned now if $3000 will be repaid at the end of 5 years?
(b) How much will be required in 4 years to repay a $2000 loan now?

2.10 How many years will it take an investment to triple itself if the interest rate is 8% compounded annually?

2.11 You bought 100 shares of Northern Telecom at $5200 on December 31, 1993. Your intention is to keep the stock until it doubles in value. If you expect 12% annual growth for Northern Telecom, how many years do you expect to hold the stock? Compare this solution with that obtained using the Rule of 72 discussed in Example 2.8.

2.12 From the interest tables in the text, determine the following value of the factors by interpolation:

(a) The single-payment compound amount factor for 38 periods at 6.5% interest.
(b) The single-payment present worth factor for 57 periods at 8% interest.

2.13 If you desire to withdraw the following amounts over the next 5 years from a savings account that earns a 6% interest compounded annually, how much do you need to deposit now?

n	Amount
2	$2000
3	3000
4	5000
5	3000

2.14 If $500 is invested now, $700 two years from now, and $900 four years from now at an interest rate of 5% compounded annually, what will be the total amount in 10 years?

2.15 The *Hockey Times* headline of April 1993, blared: "Jim Smith Signed for $5 Million." A reading of the article revealed that on April 1, 1991, Jim Smith, a junior hockey scoring sensation, signed a $5 million package with the Calgary Flames. The terms of the contract were $500,000 immediately, $400,000 per year for 5 years (first payment after 1 year), and $500,000 per year for 5 years (first payment at year 6). If Jim's interest rate is 8% per year, what would be his contract worth at the time of contract signing?

2.16 How much invested now at 5% would be just sufficient to provide three withdrawals with the first withdrawal in the amount of $1000 occurring 2 years hence, $2000 5 years hence, and $3000 7 years hence?

2.17 What is the future worth of a series of equal year-end deposits of $1500 for 10 years in a savings account that earns 9% annual interest?

2.18 What is the future worth of the following series of payments?

  (a) $2000 at the end of each year for 10 years at 7% compounded annually
  (b) $1000 at the end of each year for 5 years at 8.25% compounded annually
  (c) $500 at the end of each year for 20 years at 10% compounded annually
  (d) $300 at the end of each year for 15 years at 12.25% compounded annually

2.19 What equal annual series of payments must be paid into a sinking fund to accumulate the following amount?

  (a) $15,000 in 18 years at 6% compounded annually
  (b) $20,000 in 15 years at 8% compounded annually
  (c) $5000 in 20 years at 9% compounded annually
  (d) $10,000 in 5 years at 8% compounded annually

2.20 Part of the income that a machine generates is put into a sinking fund to replace the machine when it wears out. If $500 is deposited annually at 6% interest, how many years must the machine be kept before a new machine costing $10,000 can be purchased?

2.21 A no-load (commission free) mutual fund has grown at a rate of 15% compounded annually since its beginning. If it is anticipated that it will continue to grow at this rate, how much must be invested every year so that $10,000 will be accumulated at the end of 5 years?

2.22 What equal annual payment series is required to repay the following present amounts?

  (a) $10,000 in 4 years at 10% interest compounded annually
  (b) $5000 in 3 years at 12% interest compounded annually
  (c) $6000 in 5 years at 8% interest compounded annually
  (d) $80,000 in 30 years at 9% interest compounded annually

2.23 You have borrowed $10,000 at an interest rate of 15%. Equal payments will be made over a 3-year period (first payment at the end of the first year). The annual payment will be ( ) and the interest payment for the second year will be ( ).

2.24 What is the present worth of the following series of payments?

  (a) $700 at the end of each year for 12 years at 5% compounded annually
  (b) $500 at the end of each year for 10 years at 10% compounded annually

(c) $1000 at the end of each year for 5 years at 8.25% compounded annually

(d) $2000 at the end of each year for 8 years at 9.75% compounded annually

2.25 From the interest tables in Appendix C determine the following value of the factors by interpolation.

(a) The capital recovery factor for 36 periods at 6.25% interest.

(b) The equal-cash-flow-series present worth factor for 125 periods at 9.25% interest.

2.26 An individual deposits an annual bonus into a savings account that pays 6% interest compounded annually. The size of the bonus increases by $100 each year and the initial bonus amount was $300. Determine how much will be in the account immediately after the fifth deposit.

2.27 Five annual deposits in the amounts of ($800, $700, $600, $500, and $400) are made into a fund that pays interest at a rate of 8% compounded annually. Determine the amount in the fund immediately after the fifth deposit.

2.28 Compute the value of P in the cash flow in Fig. 2.53. Assume $i = 10\%$.

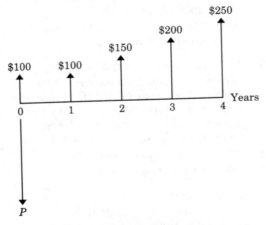

Figure 2.53 ■ Cash flow diagram (Problem 2.28)

2.29 What is the equal-payment series for 15 years that is equivalent to a payment series of $10,000 at the end of the first year decreasing by $500 each year over 15 years. Interest is 7% compounded annually.

2.30 Suppose that an oil well is expected to produce 10,000 barrels of oil during its first production year. However, its subsequent production (yield) is expected to decrease by 10% over the previous year's production. The oil well has a proven reserve of 100,000 barrels.

(a) Suppose that the price of oil is expected to be $30 per barrel for next several years. What would be the present worth of the anticipated revenue stream at an interest rate of 15% compounded annually over next 7 years?

(b) Suppose that the price of oil is expected to start at $30 per barrel during the first year, but to increase at the rate of 5% over the previous year's price, what would be the present worth of the anticipated revenue stream at an interest rate of 15% compounded annually over next 7 years?

(c) Reconsider (b) above. After 3-years' production, you decide to sell the oil well. What would be the fair price for the oil well?

2.31 An engineer has estimated the annual toll revenues from a proposed toll highway over 20 years as follows:

$$A_n = (500,000)n(1.05)^{n-1},$$

$$n = 1,2,\ldots,20.$$

During an assessment of this project, the engineer was asked to present the estimated total present value of toll revenue at an interest rate of 8%. Assuming annual compounding, find the present value of the estimated toll revenue using **EzCash** software.

2.32 What is the amount of 10 equal annual deposits made at the beginning of each year that can provide 5 annual borrowings, when the first borrowing of $1000 is made at the end of year 1, and subsequent borrowings increase at the rate of 6% per year over the previous year's, if

(a) the interest rate is 8% compounded annually?
(b) the interest rate is 6% compounded annually?

2.33 By using only those factors given in the interest tables, find the values of the following factors which are not given in your tables. Show the relationship between the factors using the factor notation and then calculate the value of the factor. Then compare the solution using the factor formulas to directly calculate the factor values.

Example:
$$(F/P, 8\%, 10) = (F/P, 8\%, 4)(F/P, 8\%, 6)$$
$$= 2.159$$

(a) $(P/F, 8\%, 48)$
(b) $(A/P, 8\%, 36)$
(c) $(P/A, 8\%, 125)$

2.34 Prove the following relationships among interest factors:

(a) $(F/P, i, N) = i(F/A, i, N) + 1$
(b) $(P/F, i, N) = 1 - (P/A, i, N)i$
(c) $(A/F, i, N) = (A/P, i, N) - i$
(d) $(A/P, i, N) = i/[1 - (P/F, i, N)]$

2.35 Find the present worth of the cash receipts in Fig. 2.54 where $i = 10\%$ compounded annually, with only three interest factors.

2.36 Find the equivalent present worth of the cash receipts in Fig. 2.55 where $i = 10\%$. In other words, how much do you have to deposit now so that you will be able to withdraw $100 at the end of the first year, $100 at the end of the second year, and so forth, where the bank pays you a 10% annual interest on your balance?

2.37 What value of A makes the two annual cash flows in Fig. 2.56 equivalent at 10% interest compounded annually?

2.38 The two cash flow transactions in Fig. 2.57 are said to be equivalent at 10% interest compounded annually. Find the unknown X value that satisfies the equivalence.

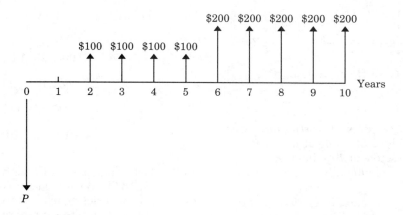

Figure 2.54 ■ Cash flow diagram (Problem 2.35)

Figure 2.55 ■ Cash flow diagram
(Problem 2.36)

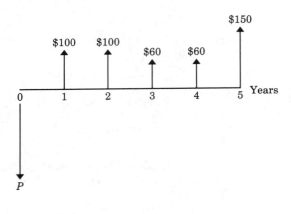

Figure 2.57 ■ Cash flow diagram
(Problem 2.38)

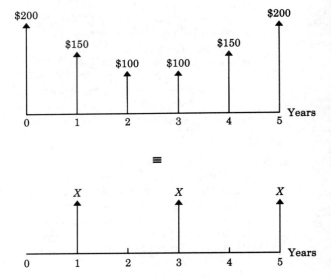

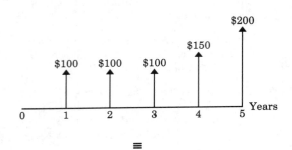

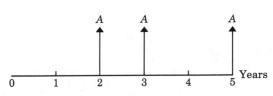

Figure 2.56 ■ Cash flow diagram
(Problem 2.37)

compounded annually—that is, how much can you withdraw at the end of year 4 when you make the series of deposits and the bank pays a 10% annual interest on your deposit balance? Use only three interest factors.

2.41 The following equation describes the conversion of a cash flow into an equivalent equal-cash-flow series with $N = 10$. Given the equation, reconstruct the original cash flow diagram.

$$A = [(800 + 20(A/G, 6\%, 7)]$$
$$\times (P/A, 6\%, 7)\,(A/P, 6\%, 10)$$
$$+ [300(F/A, 6\%, 3) - 500]$$
$$\times (A/F, 6\%, 10)$$

2.42 Consider the cash flow given in Fig. 2.60. What value of C makes the inflow series equivalent to the outflow series at an interest rate of 12%?

2.39 Solve for the present worth of this cash flow in Fig. 2.58 using at most three interest factors at 10% interest compounded annually.

2.40 Find the equivalent future worth of the cash flows in Fig. 2.59 at 10% interest

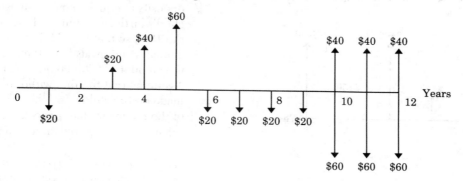

Figure 2.58 ■ Cash flow diagram (Problem 2.39)

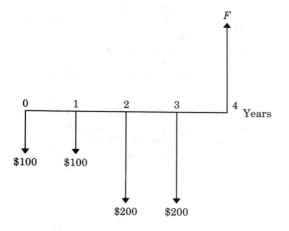

Figure 2.59 ■ Cash flow diagram
(Problem 2.40)

2.43 Find the value of X in Fig. 2.61 so that the two cash flows in the figure are equivalent for an interest rate of 10%.

2.44 What single amount at the end of the fourth year is equivalent to a uniform annual series of $5000 per year for 10 years, if the interest rate is 9% compounded annually?

2.45 In computing either the equivalent present worth (P) or future worth (F) for the cash flow given in Fig. 2.62, at $i = 10\%$, identify all the correct equations from the list below to compute them.

(1) $P = A(P/A, 10\%, 6)$
(2) $P = A + A(P/A, 10\%, 5)$

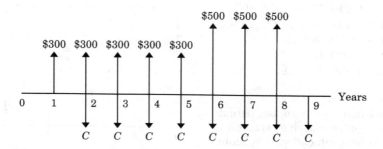

Figure 2.60 ■ Cash flow diagram (Problem 2.42)

Figure 2.61 ■ Cash flow diagram
(Problem 2.43)

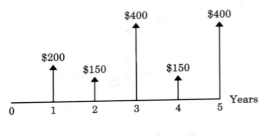

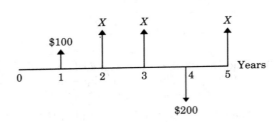

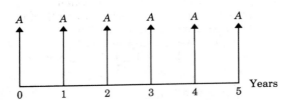

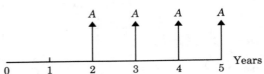

Figure 2.62 ■ Cash flow diagram
(Problem 2.45)

(3) $P = A(P/F, 10\%, 5) + A(P/A, 10\%, 5)$
(4) $F = A(F/A, 10\%, 5) + A(F/P, 10\%, 5)$
(5) $F = A + A(F/A, 10\%, 5)$
(6) $F = A(F/A, 10\%, 6)$
(7) $F = A(F/A, 10\%, 6) - A$

2.46 On the day a baby was born, the parents decided to establish a savings account for the child's college education. Any money that is put into the account will earn an interest rate of 8% compounded annually.

They will make a series of annual deposits in equal amounts on each of the child's birthdays from the first through the 18th, so that the child can make four annual withdrawals from the account in the amount of $20,000 on each birthday. Assuming that the first withdrawal will be made on the child's 18th birthday, which of the following statements are correct to calculate the required annual deposit?

(1) $A = (\$20,000 \times 4)/18$
(2) $A = \$20,000(F/A, 8\%, 4)\,(P/F, 8\%, 21) \times (A/P, 8\%, 18)$
(3) $A = \$20,000(P/A, 8\%, 18)\,(F/P, 8\%, 21) \times (A/F, 8\%, 4)$
(4) $A = [\$20,000(P/A, 8\%, 3) + \$20,000] \times (A/F, 8\%, 18)$
(5) $A = \$20,000[(P/F, 8\%, 18) + (P/F, 8\%, 19) + (P/F, 8\%, 20) + (P/F, 8\%, 21)]\,(A/P, 8\%, 18)$

2.47 Find the equivalent equal-cash-flow series (A) using an A/G factor, such that the two cash flows in Fig. 2.63 are equivalent at 12% compounded annually.

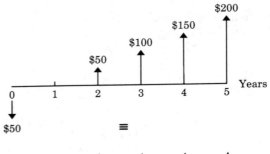

Figure 2.63 ■ Cash flow diagram
(Problem 2.47)

2.48 Consider the following cash flow:

Year End	Amount
0	$ 500
1	1000
2	1000
3	1000
4	1000
5	1000

In computing F at the end of year 5 at an interest rate of 12%, which of the following statements is incorrect?

(a) $F = 1000(F/A, 12\%, 5)$
    $- 500(F/P, 12\%, 5)$

(b) $F = 500(F/A, 12\%, 6)$
    $+ 500(F/A, 12\%, 5)$

(c) $F = [500 + 1000(P/A, 12\%, 5)]$
    $\times (F/P, 12\%, 5)$

(d) $F = [500(A/P, 12\%, 5) + 1000]$
    $\times (F/A, 12\%, 5)$

2.49 At what rate of interest compounded annually will an investment triple itself in 10 years?

2.50 Determine the interest rate (i) that makes the pairs of cash flows in Fig. 2.64 economically equivalent.

2.51 You have $5000 available for investment in stock. You are looking for a growth stock whose value can grow to $23,000 over 5 years. What kind of growth rate are you looking for?

2.52 Read the following letter from a magazine publisher:

Dear Parent:

Currently your *Growing Child/Growing Parent* subscription will expire with your 24-month issue. To renew on an annual basis until your child reaches 72 months would cost you a total of $63.84 ($15.96 per year). We feel it is so important for you to continue

Figure 2.64 ■ Cash flow diagram (Problem 2.50)

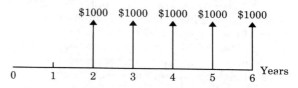

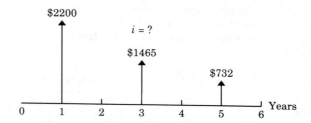

receiving this material until the 72nd month that we offer you an opportunity to renew now for $57.12. Not only is this a savings of 10% over the regular rate, it is an excellent inflation hedge for you against increasing rates in the future. Please act now by sending $57.12.

(a) If your money is worth 6% per year, determine whether this offer can be of any value.

(b) What rate of interest would cause you to be indifferent between the two renewal options?

2.53 A major lottery corporation sold a total of 36.1 million tickets at $1 each during the first week of January 1994. As prize money, a total of $41 million is to be distributed ($1,952,381 at the *beginning* of each year) over the next 21 years. The distribution of the first-year prize money is distributed immediately, and the remaining lottery proceeds will be put into the province's educational reserve fund, which earns an interest at the rate of 6% compounded annually. After making the last prize distribution (at the beginning of year 21), how much would be left over in the reserve account?

2.54 *The Sporting News* carried the following story on June 19, 1989. "Dallas Cowboys quarterback Troy Aikman, the number one pick in the National Football League draft, will earn either $11,406,000 over 12 years or $8,600,000 over 6 years. Aikman, represented by Leigh Steinberg, must declare which plan he prefers. The $11 million package is deferred through the year 2000, while the nondeferred arrangement ends after the 1994 season. Regardless which plan is chosen, Aikman will be playing through the 1994 season."[2]

Deferred Plan		Nondeferred Plan	
1989	$ 2,000,000	1989	$2,000,000
1990	566,000	1990	900,000
1991	920,000	1991	1,000,000
1992	930,000	1992	1,225,000
1993	740,000	1993	1,500,000
1994	740,000	1994	1,975,000
1995	740,000		
1996	790,000		
1997	540,000		
1998	1,040,000		
1999	1,140,000		
2000	1,260,000		
Total	$11,406,000	Total	$8,600,000

(a) If Aikman's interest rate is 6%, which plan is the better choice?
(b) What interest rate would make the two plans economically equivalent?

2.55 Fairmont Textile has a plant whose employees have been having trouble with Carpal Tunnel Syndrome (CTS) (inflammation of the nerves that pass

through the carpal tunnel, a tight space at the base of the palm), resulting from long-term repetitive activities, such as years of sewing operation. It seems that 15 of the employees working in this facility have developed signs of CTS over the last 5 years. Avon Mutual, their insurance firm, has been steadily increasing Fairmont's liability insurance because of this problem. Avon Mutual is willing to lower the insurance premiums to $16,000 a year (from the current $30,000 a year) for the next 5 years if Fairmont implements an acceptable CTS-prevention program that includes making the employees aware of CTS and how to reduce chances of developing it. What would be the maximum amount that Fairmont should invest in the CTS-prevention program to make this program worthwhile? The firm's interest rate is 12% compounded annually.

2.56 Kersey Manufacturing Company, a small plastic fabricator, needs to purchase an extrusion molding machine for $120,000. Kersey will borrow money from a bank at an interest rate of 9% over 5 years. Kersey expects its product sales to be slow during the first year but to increase subsequently at an annual rate of 10%. Kersey therefore arranges with the bank to pay off the loan on a "balloon scale," which results in the lowest payment at the end of first year, each subsequent payment to be just 10% over the previous one. Determine these five annual payments.

2.57 The R&D department of Boswell Electronics Company has developed a voice-recognition system that could lead to greater acceptance of personal computers among the Japanese. Currently, the complicated and voluminous set of Japanese characters required for a keyboard make it unwieldy and bulky in comparison to the trim, easily portable keyboards Westerners are accustomed to. Boswell

---

[2] Reprinted by permission of *The Sporting News*.

has used voice-recognition systems in more traditional settings, such as medical reporting in the English language. However, adapting this technology to the Japanese language could result in a breakthrough new level of acceptance of personal computers in Japan. The investment required to develop a full-scale commercial version would cost BEC $10 million, which will be financed at an interest rate of 12%. The system will sell for about $4000 (or net cash profit of $2000) and run on high-powered PCs. The product will have a 5-year market life. Assuming that the annual demand for the product remains constant over the market life, how many units does BEC have to sell each year just to pay off the initial investment and interest?

CHAPTER 3

Extending Equivalence to Real World Transactions

In the 1890s, McDougall & Secord, a private company based in Edmonton, Alberta, purchased a commercial property at the corner of 101 Street and Jasper Avenue in Edmonton. The property had an area of approximately 1000 square metres and was originally purchased for a sum of $50. During the subsequent 100 years, this location has become the key commercial intersection of the financial district of Edmonton. This land is now valued at an estimated $4 million.

In the 1970s, the same company purchased 4,800 shares of stock in a publicly traded company for $12,504. The stock was recently sold for $167,837.

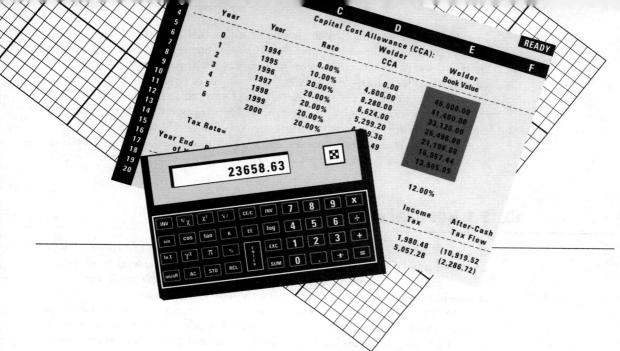

Year	Year	Rate	Capital Cost Allowance (CCA): Welder CCA	Welder Book Value
0	1994	0.00%	0.00	46,000.00
1	1995	10.00%	4,600.00	41,400.00
2	1996	20.00%	8,280.00	33,120.00
3	1997	20.00%	6,624.00	26,496.00
4	1998	20.00%	5,299.20	21,196.80
5	1999	20.00%	9.36	16,957.44
6	2000	20.00%	49	13,565.95

Tax Rate=

12.00%

Income Tax	After-Cash Tax Flow
1,980.48	(10,919.52)
5,057.28	(2,286.72)

`23658.63`

Although the land has increased much more in absolute value, the stock investment actually produced a higher interest rate or rate of return. The land investment has been held for over 100 years, whereas the stock investment was held for about 20 years. Both these investments represent superior investment decisions.

Acknowledgement: John R. McDougall, McDougall & Secord, Limited, Edmonton, Alberta.

In this chapter, we will look at several concepts that are crucial to managing money. In Chapter 2, we examined the value of money at different points in time and developed various interest formulas for this purpose. Using these basic interest formulas, we will extend the concept of equivalence to determine the interest rates implicit in many financial contracts. To this end, we will introduce several examples in the area of loan transactions. Many commercial loans require interest to compound more frequently than once a year—monthly or quarterly, for example. To consider more frequent compounding, we must begin with the concepts of nominal and effective interest.

■ 3.1 Nominal and Effective Interest Rates

In all our examples in Chapter 2, we assumed that payments are received once a year, or annually. However, some of the most familiar financial transactions in both personal financial matters and engineering economic analysis involve nonannual payments; for example, monthly mortgage payments and quarterly earnings on savings accounts. Thus, if we are to compare different cash flows with different compounding periods, we need to address them on a common basis. The need to do this has led to the development of the concepts of **nominal interest rate** and **effective interest rate.**

3.1.1 Nominal Interest Rate

Even if a financial institution uses a unit of time other than a year—a month or quarter, for instance—in calculating interest payments, it usually quotes the interest rate on an annual basis. It is commonly stated as

$$r\% \text{ compounded } M\text{-ly},$$

where

r = the nominal interest rate per year,

M = the compounding frequency, or the number of compounding periods per year,

r/M = the interest rate per compounding period.

Many banks, for example, state the interest arrangement for credit cards in this way

"18% compounded monthly."

We say 18% is the *nominal interest rate* or *annual percentage rate* (APR), and the compounding frequency is monthly (12). To obtain the interest rate per compounding period, we divide 18% by 12 to obtain 1.5% per month (see Fig. 3.1).

Although the annual percentage rate, or APR, is commonly used by financial institutions and is familiar to customers, when compounding takes place more frequently than annually, the APR does not explain precisely the amount of interest that will accumulate in a year. To explain the true effect of more frequent compounding on annual interest amounts, we need to introduce the term effective interest rate.

3.1.2 Effective Interest Rate

The *effective interest rate* is the one that truly represents the interest earned in a year or some other time period. For instance, in our credit card example, the 1.5% rate represents an effective monthly interest rate—on a monthly basis, it is the rate that predicts the actual interest payment on your outstanding credit card balance.

Suppose you obtain a loan from a bank at an interest rate of 12% compounded monthly. Here, 12% represents the nominal interest rate. The interest rate per month is 1% (12/12). The total annual interest payment (assuming no withdrawals) for a principal amount of $1 may be computed using the formula given in Eq. (2.3). With $P = \$1$, $i = 12\%/12$, and $N = 12$, we obtain

$$F = P(1 + i)^N$$

$$= \$1(1 + 0.01)^{12}$$

$$= \$1.1268.$$

This implies that for each dollar borrowed for one year, you owe $1.1268 at the end of the year, including the principal and interest. You can easily obtain the annual interest payment by subtracting the principal amount from the F value.

$$I = F - P$$

$$= 1.1268 - 1$$

$$= 12.68 \text{ cents}$$

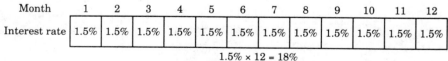

Month	1	2	3	4	5	6	7	8	9	10	11	12
Interest rate	1.5%	1.5%	1.5%	1.5%	1.5%	1.5%	1.5%	1.5%	1.5%	1.5%	1.5%	1.5%

1.5% × 12 = 18%
Nominal interest rate

Figure 3.1 ■ The nominal interest rate is found by summing the individual interest rates per period.

For each dollar borrowed, you pay an equivalent annual interest of 12.68 cents. In terms of an effective annual interest rate (i_a), we can rewrite the interest payment as a percentage of the principal amount:

$$i_a = (1 + 0.01)^{12} - 1 = 0.1268, \text{ or } 12.68\%.$$

Thus, the effective annual interest rate is 12.68%. We may relate the nominal to the annual effective rate as shown in Fig. 3.2.

For the same loan made at an interest rate of 12% compounded quarterly, the interest rate per compounding period would be 3% (12/4) for each of the 3-month periods during the year. Since there are 4 quarters in a year, we find

$$i_a = (1 + 0.03)^4 - 1 = 0.1255, \text{ or } 12.55\%.$$

Table 3.1 shows the effective interest rates at other compounding intervals for 12% APR and several other frequently encountered nominal rates. As you can see, depending on the frequency of compounding, the effective interest earned or paid by you can differ significantly from the APR. Therefore, truth-in-lending laws require that financial institutions quote both nominal interest rate and compounding frequency when you deposit or borrow money.

Clearly, compounding more frequently increases the amount of interest paid for the year at the same nominal interest rate. We can generalize the result for

Month	1	2	3	4	5	6	7	8	9	10	11	12
$r/12$	1%	1%	1%	1%	1%	1%	1%	1%	1%	1%	1%	1%

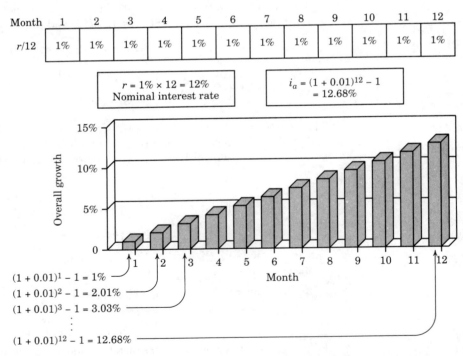

$$r = 1\% \times 12 = 12\%$$
Nominal interest rate

$$i_a = (1 + 0.01)^{12} - 1 = 12.68\%$$

$(1 + 0.01)^1 - 1 = 1\%$
$(1 + 0.01)^2 - 1 = 2.01\%$
$(1 + 0.01)^3 - 1 = 3.03\%$
$\vdots$
$(1 + 0.01)^{12} - 1 = 12.68\%$

Figure 3.2 ■ Relationship between the nominal interest rate and the annual effective interest rate per period

Table 3.1 Nominal and Effective Interest Rates with Different Compounding Periods

Nominal Rate	Compounding Annually	Compounding Semiannually	Effective Rate Compounding Quarterly	Compounding Monthly	Compounding Daily
4%	4.00%	4.04%	4.06%	4.07%	4.08%
5	5.00	5.06	5.09	5.12	5.13
6	6.00	6.09	6.14	6.17	6.18
7	7.00	7.12	7.19	7.23	7.25
8	8.00	8.16	8.24	8.30	8.33
9	9.00	9.20	9.31	9.38	9.42
10	10.00	10.25	10.38	10.47	10.52
11	11.00	11.30	11.46	11.57	11.62
12	12.00	12.36	12.55	12.68	12.74

any arbitrary compounding scheme by providing a formula for effective interest rate computation. Assuming that the nominal interest rate is r and there are M compounding periods in the year, then the effective annual interest rate (i_a) can be calculated as follows:

$$i_a = (1 + r/M)^M - 1. \tag{3.1}$$

When $M = 1$, we have the special case of annual compounding. Substituting $M = 1$ into Eq. (3.1), we find it reduces to $i_a = r$. That is, when compounding takes place once annually, effective interest is equal to nominal interest. Thus, in all our Chapter 2 examples, where we considered only annual interest, we were by definition using effective interest rates.

Example 3.1 Determining the compounding period

Consider the following bank advertisement appearing in a local newspaper: "Open a Term Deposit (TD) account at CIBC and you get a guaranteed rate of return on as little as $1000. It's a smart way to manage your money for months."

TD Type	6-month	1-year	2-year	5-year
Rate (nominal)	8.00%	8.50%	9.20%	9.60%
Yield (effective)	8.16%	8.87%	9.64%	10.07%
Minimum deposit	$1000	$1000	$10,000	$20,000
Compounding period	—	—	—	—

Find the compounding period for each TD.

Solution

Given: $r = 8\%$ per year, i_a
Find: M

First we will consider the 6-month TD. The nominal interest rate is 8% per year, and the effective annual interest rate (yield) is 8.16%. Using Eq. (3.1), we obtain

$$0.0816 = (1 + 0.08/M)^M - 1$$

or

$$1.0816 = (1 + 0.08/M)^M.$$

By trial and error, we find $M = 2$, indicating semiannual compounding. Thus, the 6-month TD earns 8% interest compounded semiannually. Similarly, we find the compounding periods for the other TDs: For the 1-year TD, we note that the difference between nominal and effective interest is greater than in the case of the 6-month TD, where the compounding period was semiannual. Greater difference between nominal and effective rates suggests a greater number of compounding periods, so let us guess at the other end of the spectrum, that is, daily compounding. In fact, this guess proves to be correct, as: $i_a = (1 + 0.085/365)^{365} - 1 = 8.87\%$. The same reasonable guess will prove true for the 2-year and 5-year TD—both are compounded daily.

Example 3.2 Future value of a certificate of deposit

Suppose you purchase the 6-month TD in Example 3.1. How much would you have when the TD matures at the end of 6 months?

Solution

Given: $r = 8\%$ per year, $M = 2$ periods per year
Find: F

If you purchase the 6-month TD now, it will earn an 8% interest compounded semiannually. This means that your TD will earn 4% interest during the first 6 months.

$$F = P(1 + r/M)^N$$
$$= \$1000(1 + 0.04)^1$$
$$= \$1040$$

Comments: *If you do not cash the TD when it matures, it will normally be renewed automatically at the original interest rate. For example, if you leave the TD in the bank for another 6 months, your TD will earn 4% interest on $1040.*

$$TD \text{ value after 1-year deposit} = \$1040(1 + 0.04)$$
$$= \$1081.60$$

Note that this is equivalent to earning 8.16% interest on $1000 for 1 year.

3.1.3 Effective Interest Rate per Payment Period

We can generalize the result of Eq. (3.1) to compute the effective interest rate for any time duration. As you will see later, we normally compute the effective interest rate based on payment (transaction) period. For example, if cash flow transactions occur quarterly but interest is compounded monthly, we may wish to calculate the effective interest rate per quarter. To consider this, we may re-define Eq. (3.1) as

$$i = (1 + r/M)^C - 1, \tag{3.2}$$

or, $\quad i = [1 + r/(CK)]^C - 1, \tag{3.3}$

where

C = the number of compounding periods per payment period,

K = the number of payment periods per year,

M = the number of compounding periods per year.

Note that $M = CK$ in Eq. (3.3).

Example 3.3   Effective rate per payment period

Suppose that we make quarterly deposits to a savings account that earns 9% interest compounded monthly. Compute the effective interest rate per payment period.

Solution

Given: $r = 9\%$, $C = 3$ compounding periods per quarter, $K = 4$ quarterly payments per year, and $M = 12$ componding periods per year

Find: i

Using Eq. (3.2) we compute the effective interest rate per quarter as

$$i = (1 + 0.009/12)^3 - 1$$
$$= 2.27\%.$$

Comments: *For the special case of annual payments with annual compounding, we obtain* i = i_a *with C = M.* *Figure 3.3 illustrates the relationship between the nominal and effective interest rates.*

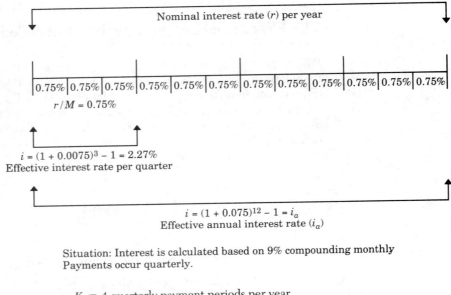

Situation: Interest is calculated based on 9% compounding monthly
Payments occur quarterly.

K = 4 quarterly payment periods per year
C = 3 compounding periods per quarter
r = 9%
M = 12 monthly compounding periods per year
r/M = 0.75%, the interest rate per month (compounding period)

Figure 3.3 ■ Functional relationships among r, i, and i_a for monthly compounding with quarterly payments (Source: *Advanced Engineering Economics*, Chan S. Park and Gunter P. Sharp-Bette. Copyright © 1990, John Wiley & Sons, Inc. Reprinted by permission of John Wiley & Sons, Inc.)

3.1.4 Continuous Compounding

To be competitive on the financial market or to entice potential depositors, some financial institutions offer very frequent compounding. As the number of compounding periods (M) becomes very large, the interest rate per compounding period (r/M) becomes very small. As M approaches infinity and r/M approaches zero, we approximate the situation of **continuous compounding**.

By taking limits on both sides of Eq.(3.3), we obtain the effective interest rate per payment period as

$$i = \lim_{CK \to \infty} [(1 + r/CK)^C - 1]$$
$$= \lim_{CK \to \infty} (1 + r/CK)^C - 1$$
$$= \lim_{CK \to \infty} [(1 + r/CK)^{CK}]^{1/K} - 1$$
$$= (e^r)^{1/K} - 1$$

Therefore,

$$i = e^{r/K} - 1. \tag{3.4}$$

To calculate the effective annual interest rate for continuous compounding, we set K equal to 1, resulting in:

$$i_a = e^r - 1. \tag{3.5}$$

As an example, the effective annual interest rate for a nominal interest rate of 12% compounded continuously is $i_a = e^{0.12} - 1 = 12.7497\%$.

Example 3.4   Calculating an effective interest rate

Assuming a $1000 initial deposit and 8% APR, find the effective interest rate per quarter, at a nominal rate of 8% compounded (a) weekly, (b) daily, and (c) continuously. Also find the final balance at the end of 3 years under each compounding scheme (see Fig. 3.4).

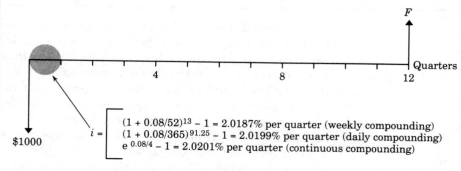

$$i = \begin{bmatrix} (1 + 0.08/52)^{13} - 1 = 2.0187\% \text{ per quarter (weekly compounding)} \\ (1 + 0.08/365)^{91.25} - 1 = 2.0199\% \text{ per quarter (daily compounding)} \\ e^{0.08/4} - 1 = 2.0201\% \text{ per quarter (continuous compounding)} \end{bmatrix}$$

Figure 3.4 ■ Calculation of effective interest rate per quarter (Example 3.4)

Solution

Given: $r = 8\%$, $M, C, K = 4$ quarterly payments per year, $P = \$1000$, $N = 12$ quarters
Find: i, F

(a) Weekly compounding: $r = 8\%$, $M = 52$, $C = 13$ weeks per quarter, $K = 4$ payments per year;

$$i = (1 + 0.08/52)^{13} - 1 = 2.0187\%.$$

With $P = \$1000$, $i = 2.0187\%$, and $N = 12$ (quarters) in Eq. (2.3),

$$F = \$1000(1 + 0.020187)^{12} = \$1271.03.$$

(b) Daily compounding: $r = 8\%$, $M = 365$, $C = 91.25$ days per quarter, $K = 4$;

$$i = (1 + 0.08/365)^{91.25} - 1 = 2.0199\%.$$

$$F = \$1000(1 + 0.020199)^{12} = \$1271.21.$$

(c) Continuous compounding: $r = 8\%$, $M \to \infty$, $C \to \infty$, $K = 4$;

$$i = e^{0.08/4} - 1 = 2.0201\%.$$

$$F = \$1000(1 + 0.020201)^{12} = \$1271.23.$$

Comments: *Note that the difference between daily compounding and continuous compounding is rather negligible. Many banks offer a continuous compounding to entice deposit customers, but the extra benefits are small.*

■ 3.2 Equivalence Analysis Using Effective Interest

As we have just seen, a real-world complication to our development of interest formulas in Chapter 2 is the notion of effective interest—the effect that interest has when it is compounded at intervals more frequent than annually. (Note that rarely if ever in the real world does compounding take place *less* frequently than annually; therefore, this text will not explicitly cover such a situation.)

An additional complication, hinted at when we developed the equation for effective interest rate *per payment period* is the situation in which compounding and payments do not occur at the same intervals. As we said in Section 3.1.3, in such cases we need to determine the effective interest rate per payment period. In the sections that follow, we will explore this complication and how to deal with it more thoroughly.

To begin, two situations can arise with regard to frequency of compounding and frequency of payments in a real-world financial transaction:

1. *Comparable payment and compounding periods*—payments and interest compounding take place at the same time intervals (e.g., annual payments with annual compounding).

2. *Noncomparable payment and compounding periods*—payments and interest compounding occur at different time intervals (e.g., monthly payments with quarterly compounding).

3.2.1 When Payment Periods and Compounding Periods Coincide

All of the examples in Chapter 2 assumed annual payments and annual compounding. Whenever we have a situation where the compounding and payment periods are equal ($M = K$), whether annual or some other interval, we can follow this solution method:

1. Identify the number of compounding periods (M) per year.
2. Compute the effective interest rate per payment period to use

$$i = r/M.$$

3. Determine the number of payment or compounding periods to use

$$N = (M)(\text{number of years}).$$

Example 3.5 Calculating auto loan payments

Suppose you want to buy a car. You have surveyed the dealers' newspaper advertisements, and the following ad has caught your attention:

> 7.9% Annual Percentage Rate! 48-month financing on all V-6 Cougars in stock. 80 to choose from. ALL READY FOR DELIVERY! Prices Starting as Low as $10,599. You just add GST and 1% of dealer's freight. We will pay the tag, title, and license.

$$\text{GST at 7\%} = 10{,}599(0.07) = \$741.93$$

$$\text{Dealer's freight} = 10{,}599(0.01) = \$105.99$$

$$\text{Total purchase price} = \$11{,}446.92$$

You can afford to make a down payment of $1446.92, so the net amount to be financed is $10,000. What would be the monthly payment? (See Fig. 3.5.)

Solution

The ad does not specify a compounding period, but in automobile financing the interest and the payment periods are almost always both monthly. Thus, the 7.9% APR means 7.9% compounded monthly.

Figure 3.5 ■ A car loan cash flow transaction (Example 3.5)

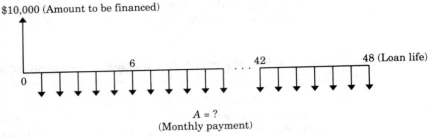

$10,000 (Amount to be financed)

A = ?
(Monthly payment)

Given: $P = \$10,000$, $r = 7.9\%$ per year, $K = 12$ payments per year, $N = 48$ months, $M = 12$

Find: A

In this situation, we can easily compute the monthly payment using Eq. (2.12):

$$i = 0.079/12 = 0.6583\% \text{ per month}$$

$$N = (12)(4) = 48 \text{ months}$$

$$A = \$10,000(A/P, 0.6583\%, 48) = \$243.66$$

Example 3.6 Term loan and remaining balance

Mr. Smith has just financed a term loan of $100,000 for his business. The term loan is for 15 years at an interest rate of 12% compounded monthly. The agreement is to make equal monthly payments.

(a) What is Mr. Smith's monthly payment?
(b) After making the 100th monthly payment, Mr. Smith would like to pay off the remainder of the loan in a lump sum. What is the required lump sum?

Solution

(a) **Given:** $P = \$100,000$, $r = 12\%$, $K = 12$ payment periods per year, $N = 180$ months

Find: A

We first compute the effective interest rate per month and the total number of payment periods.

$$i = 12\%/12 = 1\% \text{ per month}$$

$$N = (12)(15) = 180 \text{ months}$$

Then, we determine the monthly payments.

$$A = \$100,000(A/P, 1\%, 180)$$

$$= \$1200.17$$

(b) The lower half of Fig. 3.6 shows the cash flow diagram that applies to this part of the problem. We can compute the amount that Mr. Smith owes after making the 100th payment by calculating the equivalent worth of the remaining 80 payments at the end of 100th month.

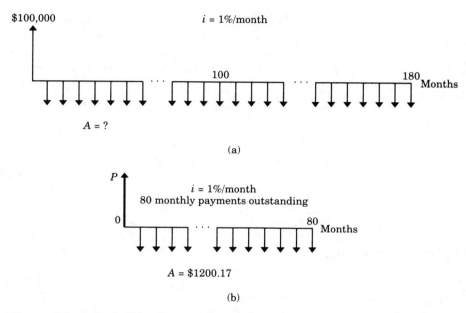

Figure 3.6 ■ Cash flow diagram (Example 3.6)

Given: $A = \$1200.17$, $i = 1\%$ per month, $N = 80$ months
Find: Remaining balance after 100 months (P)

$$P = \$1200.17(P/A, 1\%, 80)$$

$$= \$65,875.07$$

If Mr. Smith desires to pay off the remainder of the loan at the end of 100th payment, he must come up with $65,875.07.

3.2.2 When Payment Periods Differ from Compounding Periods

A number of situations involve cash flows occurring at intervals that are not the same as the compounding intervals. Whenever payment and compounding periods differ from each other, *we must transform one or the other so that they both conform to the same unit of time.* For example, if payments occur quarterly and compounding occurs monthly, we can most logically proceed by calculating the effective interest rate per quarter. On the other hand, if payments occur monthly and compounding occurs quarterly, we may find the equivalent monthly interest rate. (We may find it more convenient to convert our monthly payments into an equivalent quarterly payment for the sake of quarterly compounding calculations.) The bottom line is that the compounding and payment periods must be the same in order to proceed with equivalency analysis.

The specific computational procedure for noncomparable compounding and payment periods is:

1. Identify the number of compounding periods per year (M), the number of payment periods per year (K), and the number of compounding periods per payment period (C).
2. Compute the effective interest rate per payment period.

   - For discrete compounding, compute

   $$i = (1 + r/M^C) - 1.$$

   - For continuous compounding, compute

   $$i = e^{r/K} - 1.$$

3. Find the total number of payment periods:

   $$N = (K) \text{ (number of years)}.$$

4. Use i and N in the appropriate formulas in Table 2.3.[1]

Example 3.7   Compounding more frequent than payments

Suppose you make equal quarterly deposits of $1000 into a fund that pays interest at a rate of 12% compounded monthly. Find the balance at the end of year 1 (Fig. 3.7).

Solution

Given: A = $1000 per quarter, r = 12% per year, M = 12 compounding periods per year, N = 4 quarters

Find: F

---

[1] *Advanced Engineering Economics*, Chan S. Park and Gunter P. Sharp-Bette. Copyright © 1990, John Wiley & Sons, Inc. Reprinted by permission of John Wiley & Sons, Inc.

Figure 3.7 ■ Cash flow diagram (Example 3.7)

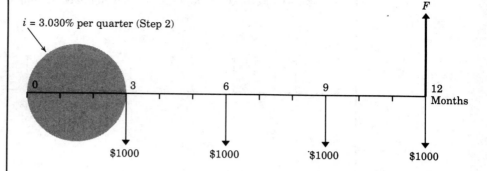

i = 3.030% per quarter (Step 2)

We follow the procedure for noncomparable compounding and payment periods described above.

1. Identify the parameter values for *M*, *K*, and *C*.

$$M = 12 \text{ compounding periods per year}$$

$$K = 4 \text{ payment periods per year}$$

$$C = 3 \text{ compounding periods per payment period}$$

2. Use Eq. (3.2) to compute effective interest.

$$i = (1 + 0.12/12)^3 - 1$$

$$= 3.030\% \text{ per quarter}$$

3. Find the total number of payment periods, *N*.

$$N = (K) \text{ (number of years)} = (4)(1) = 4$$

4. Use *i* and *N* in the appropriate equivalence formulas.

$$F = \$1000(F/A, 3.030\%, 4) = \$4185.50$$

Example 3.8   Equivalent worth of a quarterly series with continuous compounding

A series of equal quarterly receipts of $500 extends over a period of 5 years. What is the present worth of this quarterly payment series at 8% interest compounded continuously? (See Fig. 3.8.)

Figure 3.8 ■ A present-worth calculation for an equal-payment series with an interest rate of 8% compounded continuously (Example 3.8)

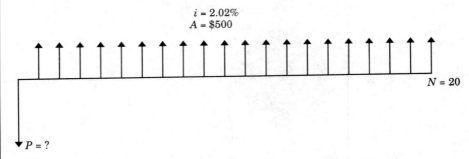

$i = 2.02\%$
$A = \$500$

$N = 20$

$P = ?$

Discussion: Since the payments are quarterly, we need to compute i per quarter for our equivalence calculations. The required steps are as follows.

$$i = e^{r/K} - 1$$

$$= e^{0.08/4} - 1$$

$$= 0.0202 = 2.02\% \text{ per quarter}$$

$$N = (4 \text{ payment periods per year}) (5 \text{ years})$$

$$= 20 \text{ periods}$$

Solution

Given: $i = 2.02\%$ per quarter, $N = 20$ quarters, and $A = \$500$ per quarter
Find: P

Using the $(P/A, i, N)$ factor with $i = 2.02\%$ and $N = 20$, we find that

$$P = A(P/A, 2.02\%, 20)$$

$$= 500(16.3199)$$

$$= \$8159.96.$$

Compounding Less Frequent Than Payments

The accounting methods used by most firms record cash transactions that occur within a compounding period as if they have occurred at the end of that period. For example, when cash flows occur daily, but the compounding is performed monthly, the cash flows within each month are summed (ignoring interest) and treated as a single payment on which interest is calculated.

The two following examples contain identical parameters for savings situations in which compounding occurs less frequently than payments. However, two different underlying assumptions are used regarding how interest is calculated. In Example 3.9, we assume that whenever a deposit is made, it starts to earn interest. In Example 3.10, we assume that the deposits made within a quarter do not earn interest until the end of that quarter. As a result, in Example 3.9, we transform the compounding period to conform to the payment period. In Example 3.10, we lump several payments to match the compounding period. In the real world, which assumption is applicable depends on the transactions and the financial institutions involved. In this textbook, we assume that whenever the timing of a cash flow is specified, one cannot move it to another time point without considering the time value of the money, i.e., the practice illustrated in Example 3.9 should be followed.

Example 3.9 Compounding less frequent than payments: effective rate per payment period

Suppose you make $1000 monthly deposits to a registered retirement savings plan that pays interest at a rate of 10% compounded quarterly. Compute the balance at the end of 10 years.

Solution

Given: r = 10% per year, M = 4 quarterly compounding periods per year, K = 12 payment periods per year, A = $1000 per month, N = 120 months, interest is accrued on flow during the compounding period
Find: i, F

As in the case of Example 3.6, we follow the procedure for noncomparable compounding and payment periods:

1. The parameter values for M, K, and C are:

    M = 4 compounding periods per year

    K = 12 payment periods per year

    C = 1/3 compounding periods per payment period.

2. As shown in Fig. 3.9, the effective interest rate per payment period is calculated using Eq. (3.2):

$$i = (1 + 0.10/4)^{1/3} - 1$$

$$= 0.826\% \text{ per month.}$$

3. Find N:

$$N = (12)(10) = 120 \text{ payment periods.}$$

Figure 3.9 ■ Calculation of an equivalent monthly interest rate when the quarterly interest rate is specified (Example 3.9)

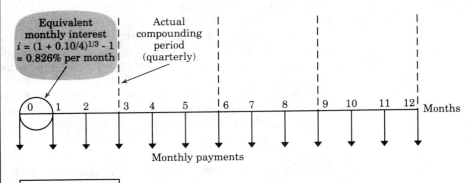

4. Use i and N in the appropriate equivalence formulas (see Fig. 3.10):

$$F = \$1000(F/A, 0.826\%, 120)$$

$$= \$203,815.78.$$

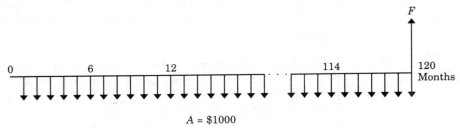

$A = \$1000$

Figure 3.10 ■ Cash flow diagram (Example 3.9)

Example 3.10 Compounding less frequent than payment: summing cash flows to the end of compounding period

Consider Example 3.9. Assume that money deposited during a quarter (the compounding period) will not earn any interest. Compute F at the end of 10 years.

Solution

Given: Same as for Example 3.8, no interest on flow within the compounding period

Find: F

Figure 3.11 ■ Equivalent cash flow diagram (Example 3.10)

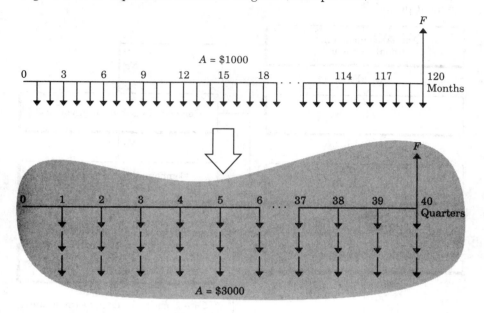

In this case, the three monthly deposits during each quarter period are placed at the end of each quarter. Then the payment period coincides with the interest period.

$$i = 0.10/4 = 2.5\% \text{ per quarter}$$

$$A = 3(\$1000) = \$3000 \text{ per quarter}$$

$$N = 4(10) = 40 \text{ payment periods}$$

$$F = \$3000(F/A, 2.5\%, 40) = \$202{,}207.66$$

Comments: *There will be $1608.12 less in the balance than with the compounding situation in Example 3.9 (Fig.3.11), a fact that is consistent with our understanding that increasing frequency of compounding increases future value. In the real world, many financial institutions use the practice illustrated in this example. As an investor in this situation, you should reasonably ask yourself whether it makes sense to make deposits in an interest-bearing account more frequently than interest is paid. In the interim between interest compounding, you may be tying up your funds prematurely and foregoing other opportunities to earn interest.*

Figure 3.12 illustrates a decision chart to sum up how you can proceed to find **effective interest rate per payment period** given the various possible compounding/interest arrangements.

Figure 3.12 ■ A decision flow chart to compute effective interest rate per payment period (i)

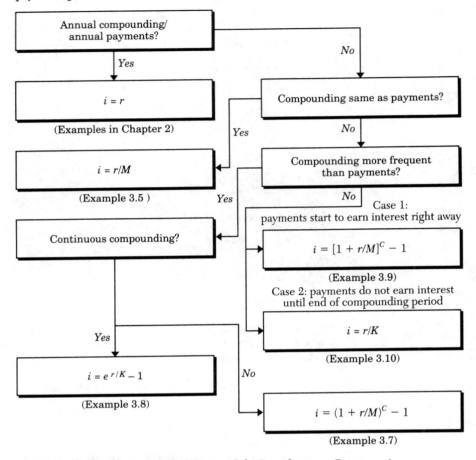

3.2.3 Equivalence Calculations with Continuous Payments

As we have seen so far, interest can be compounded annually, semiannually, monthly, or even continuously. Discrete compounding is appropriate for many financial transactions such as mortgages, bonds, and installment loans, which require payments or receipts at discrete times. In most businesses, however, transactions occur continuously throughout the year. In these circumstances, we may describe the financial transactions by a continuous flow of money, for which continuous compounding and discounting are more realistic. This section illustrates how we establish the economic equivalence between cash flows under continuous compounding.

The continuous cash flows represent situations where money flows at a known rate and continuously throughout a given time period. Many daily cash flow transactions in business can be viewed as continuous cash flows. An advantage of the continuous flow approach is that it more closely models the realities of business transactions. Costs for labor, carrying inventory, and operating and maintenance equipment are typical examples. Others include capital improvement projects that conserve energy, water, or process steam, whose savings can occur continuously.

Single-Payment Transactions

We will first illustrate how the single-payment formulas for continuous compounding and discounting are derived. Suppose that you invested P dollars at a nominal rate of $r\%$ interest for N years. If interest is compounded continuously, the effective annual interest is $i = e^r - 1$. As shown in Example 3.7, we obtain the future value of the investment at the end of N years with the F/P factor by substituting $e^r - 1$ for i.

$$
\begin{aligned}
F &= P(1 + i)^N \\
&= P(1 + e^r - 1)^N \\
&= Pe^{rN} \cdot
\end{aligned}
$$

This implies that \$1 invested now at an interest rate of $r\%$, compounded continuously, accumulates to e^{rN} dollars at the end of N years.

Correspondingly, the present value of F due N years from now and discounted continuously at an interest rate of $r\%$ is equal to

$$
P = Fe^{-rN}.
$$

We can say that the present value of \$1 due N years from now, discounted continuously at an annual interest rate of $r\%$, is equal to e^{-rN} dollars.

Continuous Funds—Flow

Suppose that an investment's future cash flow per unit of time (e.g., per year) can be expressed by a continuous function $(f(t))$, which can take any shape. Assume also that the investment promises to generate cash of $f(t)\Delta t$ dollars between t and $t + \Delta t$, where t is a point in the time interval $0 \le t \le N$ (see Fig. 3.13). If the nominal interest rate is constant r during this time interval, the present value of this cash stream is given approximately by the expression

$$
\sum (f(t)\Delta t)e^{-rt},
$$

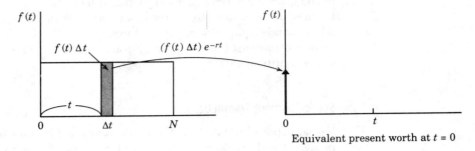

Figure 3.13 ■ Discounting a flow function $f(t)$ at a nominal rate of $r\%$

Equivalent present worth at $t = 0$

where e^{-rt} is the discounting factor that converts future dollars into present dollars. With the project's life extending from 0 to N, we take the summation over all subperiods in the interval from 0 to N. As the division of the interval becomes smaller and smaller, that is, as Δt approaches zero, we obtain the present value expression by the integral

$$P = \int_0^N f(t)e^{-rt}\,dt. \qquad (3.6)$$

Similarly, the future value expression of the cash flow stream is given by the expression

$$F = Pe^{rN} = \int_0^N f(t)e^{r(N-t)}\,dt, \qquad (3.7)$$

where $e^{r(N-t)}$ is the compounding factor that converts present dollars into future dollars. It is important to observe that the time unit is year, because the effective interest rate is expressed in terms of year. Therefore, all time units in equivalence calculations must be converted into years. Table 3.2 summarizes some typical continuous cash functions that can facilitate equivalence calculations.

Example 3.11   Comparison of daily flows and daily compounding with continuous flows with continuous compounding

Consider the situation where money flows daily. Suppose you own a retail shop and generate $200 cash each day. You establish a special business account and deposit these daily cash flows in an account for 15 months

Table 3.2 Summary of Interest Factors for Continuous Cash Flows with Continuous Compounding

Type of Cash Flow	Cash Flow Function	Parameters To Find	Parameters Given	Algebraic Notation	Factor Notation
Uniform (step)	$f(t) = \overline{A}$	P	$\overline{A}$	$\overline{A}\left[\dfrac{e^{rN} - 1}{re^{rN}}\right]$	$(P/\overline{A}, r, N)$
		$\overline{A}$	P	$P\left[\dfrac{re^{rN}}{e^{rN} - 1}\right]$	$(\overline{A}/P, r, N)$
		F	$\overline{A}$	$\overline{A}\left[\dfrac{e^{rN} - 1}{r}\right]$	$(F/\overline{A}, r, N)$
		$\overline{A}$	F	$F\left[\dfrac{r}{e^{rN} - 1}\right]$	$(\overline{A}/P, r, N)$
Gradient (ramp)	$f(t) = Gt$	P	G	$\dfrac{G}{r^2}(1 - e^{-rN}) - \dfrac{G}{r}(Ne^{-rN})$	
Decay	$f(t) = ce^{-jt}$, jt = decay rate with time	P	c, j	$\dfrac{c}{r + j}(1 - e^{-(r+j)N})$	

Source: *Advanced Engineering Economics*, Chan S. Park and Gunter P. Sharp-Bette. Copyright © 1990, John Wiley & Sons, Inc. Reprinted by permission of John Wiley & Sons, Inc.

that earns an interest rate of 6%. Compare the accumulated cash values at the end of 15 months, assuming

(a) daily compounding
(b) continuous compounding, respectively.

Solution

(a) With daily compounding:

Given: $A = \$200$ per day, $r = 6\%$ per year, $M = 365$ compoundings per year, $N = 455$ days
Find: F

Assuming there are 455 days in the 15-month period, we find

$$i = 6\%/365$$
$$= 0.01644\% \text{ per day}$$
$$N = 455 \text{ days.}$$

The balance at the end of 15 months will be

$$F = 200(F/A, 0.01644\%, 455)$$
$$= 200(472.4095)$$
$$= \$94{,}482.$$

(b) With continuous compounding:

Given: $\overline{A}$ = \$7300 per year, r = 6% per year compounded continuously, $N = 1.25$ years

Find: F

Now we approximate this discrete cash flow series by a uniform continuous cash flow function shown in Fig. 3.14. Here we have the situation where an amount flows at the rate of $\overline{A}$ per year for N years. Note that our time unit is a year. Thus, a 15-month period is 1.25 years. Then, the cash flow function is expressed as

$$f(t) = \overline{A}, 0 \le t \le 1.25$$
$$= 200(365)$$
$$= \$73{,}000 \text{ per year.}$$

Substituting these values back into Eq. (3.7) yields

$$F = \int_0^{1.25} 73{,}000 e^{0.06(1.25-t)} \, dt$$
$$= \$73{,}000 \left[\frac{e^{0.075} - 1}{0.06} \right]$$
$$= \$94{,}759.$$

The factor in the bracket is known as the **funds flow compound-amount factor** and is designated $(F/\overline{A}, r, N)$, as shown in Table 3.2. Also notice that the difference between the two methods is only \$277 (less than 0.3%).

Figure 3.14 ■ Comparison between daily transaction and continuous fund flow transaction (Example 3.11)

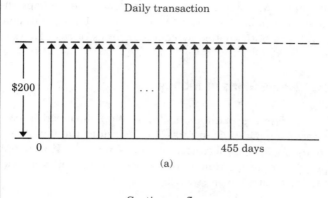

Daily transaction

$200

0   · · ·   455 days

(a)

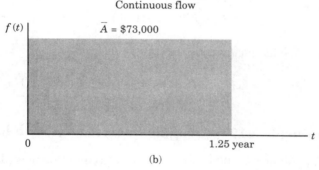

Continuous flow

$f(t)$ $\bar{A}$ = $73,000

0   1.25 year t

(b)

Comments: *As shown in this example, the differences between discrete (daily) and continuous compounding have no practical significance in most cases. Consequently, as a mathematical convenience, instead of assuming that money flows in discrete increments at the end of each day, we could assume that money flows continuously during the time period at a uniform rate.*

■ 3.3 Changing Interest Rates

Up to this point, we have assumed a constant interest rate in our equivalence calculations. When an equivalence calculation extends over several years, more than one interest rate may be applicable to properly account for the time value of money. This is to say, over time, interest rates available in the financial marketplace fluctuate and a financial institution committed to a long-term loan may find itself in the position of losing the opportunity to earn higher interest because some of its holdings are tied up in a lower interest loan. A financial institution may attempt to protect itself from such lost earning opportunities by building gradually increasing interest rates into a long-term loan at the outset. Variable home mortgage loans are perhaps the most common example of variable interest rates. In this section we will consider variable interest rates in both single-payment and series of cash flows.

3.3.1 Single Sums of Money

To illustrate the mathematical operations involved in computing equivalence under changing interest rates, first consider the investment of a single sum of money, P, in a savings account for N periods. If i_n denotes the interest rate appropriate during time period n, the future worth equivalent for a single sum of money can be expressed as

$$F = P(1 + i_1)(1 + i_2)\ldots(1 + i_{N-1})(1 + i_N), \tag{3.8}$$

and solving for P yields the inverse relation

$$P = F[(1 + i_1)(1 + i_2)\ldots(1 + i_{N-1})(1 + i_N)]^{-1}. \tag{3.9}$$

Example 3.12   Changing interest rates with lump-sum amount

You deposit $2000 in a registered retirement savings plan (RRSP) that pays interest at 12% compounded quarterly for the first 2 years and 9% compounded quarterly for the next 3 years. Determine the balance at the end of 5 years (see Fig. 3.15).

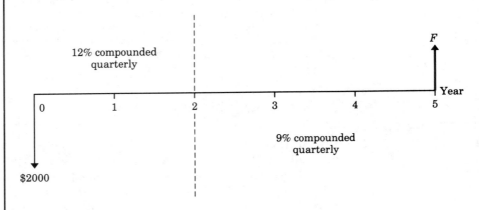

Figure 3.15 ■ Changing interest rates (Example 3.12)

Solution

Given: $P = \$2000$, $r = 12\%$ per year for first 2 years, 9% per year for last 3 years, $M = 4$ compounding periods per year, $N = 20$ quarters
Find: F

We will compute the value of F in two steps. First we will compute the balance at the end of 2 years, B_2. With 12% compounded quarterly,

$$i = 12\%/4 = 3\%$$

$$N = 4(2) = 8 \text{ (quarters)}$$

$$B_2 = \$2000(F/P, 3\%, 8)$$

$$= \$2000(1.2668)$$

$$= \$2533.60.$$

Since the fund is not withdrawn but reinvested at 9% compounded quarterly, as our second step, we compute the final balance as follows.

$$i = 9\%/4 = 2.25\%$$

$$N = 4(3) = 12 \text{ (quarters)}$$

$$F = B_2(F/P, 2.25\%, 12)$$

$$= \$2533.60(1.3060)$$

$$= \$3309.$$

3.3.2 Series of Cash Flows

We can easily extend the consideration of changing interest rates to a series of cash flows. In this case, the present worth of a series of cash flows can be represented as

$$P = A_1(1 + i_1)^{-1} + A_2[(1 + i_1)^{-1}(1 + i_2)^{-1}] + \cdots$$
$$+ A_N[(1 + i_1)^{-1}(1 + i_2)^{-1}\cdots(1 + i_N)^{-1}]. \qquad (3.10)$$

The future worth of a series of cash flows is given by the inverse of Eq. (3.10):

$$F = A_1[(1 + i_2)(1 + i_3)\cdots(1 + i_N)]$$
$$+ A_2[(1 + i_3)(1 + i_4)\cdots(1 + i_N)] + \cdots + A_N. \qquad (3.11)$$

The uniform-series equivalent is obtained in two steps. First, find the present worth equivalent of the series using Eq. (3.10). Then, solve for A after establishing the following equivalence equation:

$$P = A(1 + i_1)^{-1} + A[(1 + i_1)^{-1}(1 + i_2)^{-1}] + \cdots$$
$$+ A[(1 + i_1)^{-1}(1 + i_2)^{-1} \cdots (1 + i_N)^{-1}]. \tag{3.12}$$

Example 3.13   Changing interest rates with uneven cash flow series

Consider the cash flow in Fig. 3.16 with the interest rates indicated. Determine the uniform series equivalent for the cash flow series.

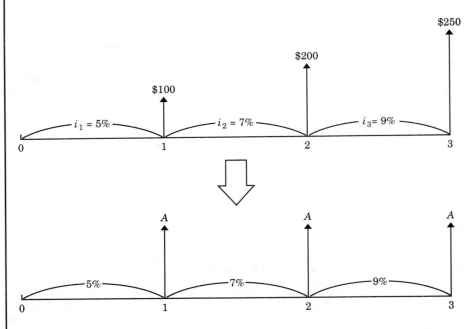

Figure 3.16 ■ Equivalence calculation with changing interest rates (Example 3.13)

Discussion: In this problem and many others, the easiest approach involves collapsing the original flow into a single equivalent amount, for example, at time 0, and then converting the single amount into the final desired form.

Solution

Given: Cash flows and interest rates shown in Fig. 3.16, $N = 3$

Find: A

Using Eq. (3.10), we find the present worth:

$$P = 100(P/F, 5\%, 1) + 200(P/F, 5\%, 1)(P/F, 7\%, 1)$$

$$+ 250(P/F, 5\%, 1)(P/F, 7\%, 1)(P/F, 9\%, 1)$$

$$= \$477.41.$$

Then, we obtain the uniform-series equivalent as follows:

$$\$477.41 = A(P/F, 5\%, 1) + A(P/F, 5\%, 1)(P/F, 7\%, 1)$$

$$+ A(P/F, 5\%, 1)(P/F, 7\%, 1)(P/F, 9\%, 1)$$

$$= 2.6591A$$

$$A = \$179.54.$$

■ 3.4 Commercial Loan Transactions

Commercial loans are among the most significant financial transactions that involve interest. While there are many types of loans, we will focus on those most frequently used by individuals and in business—**amortized loans**. **Add-on loans** will also be covered to illustrate possible variations from the standard **amortized loans**. Moreover, since it is often important to know how much interest is represented in the loan payments, we will examine several methods for determining interest and principal amounts, including the **Rule of 78ths.**

3.4.1 Amortized Loans

One of the most important applications of compound interest involves loans that are to be paid off in **installments** over time. If the loan is to be repaid in *equal* periodic amounts (weekly, monthly, quarterly, or annually) it is said to be an amortized loan. Examples of amortized loans include automobile loans, loans for appliances, home mortgage loans, and most business debt other than very short-term loans. Mortgage loans normally have interest compounded semi-annually (see Section 3.5). Most other commercial loans have interest compounded monthly.

So far in this text we have considered many instances of amortized loans in which we calculated present or future values of the loans, or the amounts of the installment payments. An additional aspect of amortized loans which will be of great interest to us is calculating the amount of interest versus principal that is paid off in each installment. As we shall explore more fully in Chapter 9, the interest paid on a loan is an important element in calculating taxable income and thus has repercussions for both personal and business loan transactions. For now, we will focus on several methods of calculating interest and principal paid at any point in the life of a loan.

For a typical amortized loan, the amount of interest owed for a specified period is calculated based on the remaining balance of the loan at the beginning of the period. We can develop a set of formulas for computing the remaining loan balance, interest payment, and principal payment for a specified period. Suppose we borrow an amount P at an interest rate i and agree to repay this principal sum P including interest in equal payments A over N periods. The payment size (A) is $A = P(A/P, i, N)$. Figure 3.17 illustrates the cash flow for

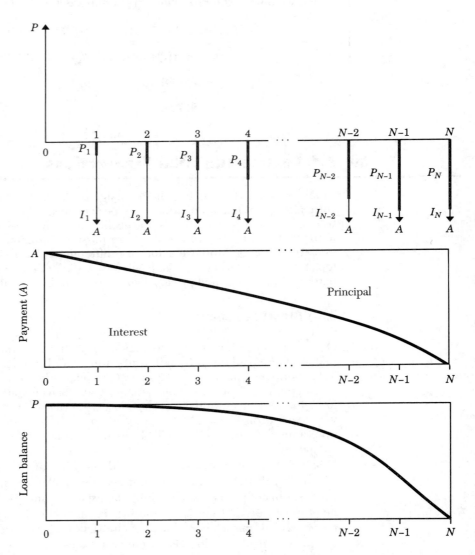

Figure 3.17 ■ Components of an amortized loan (interest and principal payments)

such a fixed loan. In this figure, each payment A is divided into an amount that is interest and a remaining amount for the principal payment. Let

- B_n = remaining balance at the end of period n, with $B_0 = P$
- I_n = interest payment in period n, where $I_n = B_{n-1}i$
- P_n = principal payment in period n.

Then, we define each payment as

$$A = P_n + I_n. \tag{3.13}$$

There are several ways to determine the interest and principal payments for an amortized loan; three methods will be presented here. There are no clear-cut reasons why one method is better than another. Method 1 may be easier to adopt in automating the computational process through a spreadsheet application, whereas methods 2 and 3 may be suitable for obtaining a quick solution when a time period is specified. You should therefore become comfortable with at least one of these methods for use in future problem solving. Pick the one that comes most naturally to you.

Method 1—Tabular Method

The first method is a tabular method by which we compute progressively the interest charge for a given period based on the balance remaining at the beginning of that period. The interest due at the end of the first period will be

$$I_1 = B_0 i = Pi.$$

Thus the principal payment at that time will be

$$P_1 = A - Pi,$$

and the balance remaining after the first payment will be

$$B_1 = B_0 - P_1 = P - P_1.$$

At the end of the second period, we will have

$$I_2 = B_1 i = (P - P_1)i$$

$$P_2 = A - (P - P_1)i = (A - Pi) + P_1 i = P_1(1 + i)$$

$$B_2 = B_1 - P_2 = P - (P_1 + P_2).$$

By continuing, we can show that, with the nth payment,

$$B_n = P - (P_1 + P_2 + \cdots + P_n)$$
$$= P - [P_1 + P_1(1 + i) + \cdots + P_1(1 + i)^{n-1}]$$
$$= P - P_1(F/A, i, n)$$
$$= P - (A - Pi)(F/A, i, n).$$

Then, the interest payment during the nth payment period is expressed as

$$I_n = (B_{n-1})i. \qquad (3.14)$$

The portion of payment A at period n that is used to reduce the remaining balance is

$$P_n = A - I_n. \qquad (3.15)$$

Example 3.14　Loan balance, principal, and interest—tabular method

Suppose you secure a car loan in the amount of $4620 from a local bank. The loan officer computes your monthly payment as follows:

$$\text{Contract amount} = \$4620$$
$$\text{Contract period} = 36 \text{ months}$$
$$\text{Annual percentage rate} = 12\%$$
$$\text{Compounding frequency} = \text{monthly}$$
$$\text{Monthly installments} = \$153.45.$$

Figure 3.18 is the cash flow diagram. Construct the loan payment schedule by showing the remaining balance, interest payment, and principal payment at the end of each period over the life of the loan.

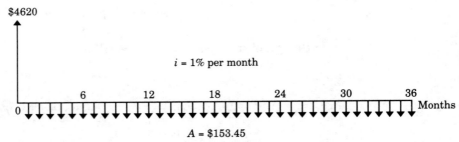

Figure 3.18 ■ A cash flow transaction of an installment loan (Example 3.14)

Table 3.3 Loan Payment Schedule for Repayment of Principal Amount of $4620

Payment No.	Payment Size	Reduction in Principal	Interest Payment	Balance
1	$153.45	$107.25	$46.20	$4512.75
2	153.45	108.32	45.13	4404.43
3	153.45	109.41	44.04	4295.02
4	153.45	110.50	42.95	4184.52
5	153.45	111.60	41.85	4072.92
6	153.45	112.72	40.73	3960.20
7	153.45	113.85	39.60	3846.35
8	153.45	114.99	38.46	3731.36
9	153.45	116.14	37.31	3615.23
10	153.45	117.30	36.15	3497.93
11	153.45	118.47	34.98	3379.46
12	153.45	119.66	33.79	3259.80
13	153.45	120.85	32.60	3138.95
14	153.45	122.06	31.39	3016.89
15	153.45	123.28	30.17	2893.61
16	153.45	124.51	28.94	2769.09
17	153.45	125.76	27.69	2643.33
18	153.45	127.02	26.43	2516.32
19	153.45	128.29	25.16	2388.03
20	153.45	129.57	23.88	2258.46
21	153.45	130.87	22.58	2127.60
22	153.45	132.17	21.28	1995.42
23	153.45	133.50	19.95	1861.93
24	153.45	134.83	18.62	1727.10
25	153.45	136.18	17.27	1590.92
26	153.45	137.54	15.91	1453.38
27	153.45	138.92	14.53	1314.46
28	153.45	140.31	13.14	1174.15
29	153.45	141.71	11.74	1032.45
30	153.45	143.13	10.32	889.32
31	153.45	144.56	8.89	744.76
32	153.45	146.00	7.45	598.76
33	153.45	147.56	5.99	451.30
34	153.45	148.94	4.51	302.36
35	153.45	150.43	3.02	151.94
36	153.45	151.93	1.52	0

Solution

Given: $P = \$4620$, $A = \$153.45$ per month, $r = 12\%$ per year, $M = 12$ compounding periods per year, $N = 36$ months

Find: B_n, I_n for $n = 1$ to 36

First we can easily see how the bank calculated the monthly payment of $153.45. Since the effective interest rate per payment period on this loan transaction is 1% per month, we establish the following equivalence relationship:

$$\$153.45(P/A, 1\%, 36) = \$153.45(30.1075) = \$4620.$$

Then, the loan payment schedule can be constructed as in Table 3.3. The interest due at $n = 1$ is \$46.20, 1% of the \$4620 outstanding during the first month. The \$107.25 left over is applied to the principal, reducing the amount outstanding in the second month to \$4512.75. The interest due in the second month is 1% of \$4512.75, or \$45.13, leaving \$108.32 for repayment of the principal. At $n = 36$, the last \$153.45 payment is just sufficient to pay the interest on the outstanding loan principal and to repay the outstanding principal.

Comments: *Certainly, generation of a loan repayment schedule such as Table 3.3 can be a tedious and time-consuming process without the computer. An electronic spreadsheet can solve this type of problem more effectively. As you will see in Section 3.7, EzCash software has a built-in function to analyze common loan problems and create such a loan payment schedule very quickly. We will demonstrate this loan analysis feature of EzCash software in Section 3.7.*

Method 2—Remaining Balance Method

Alternatively, we can derive B_n by computing the equivalent payments remaining after the nth payment. Thus, the balance with $N - n$ payments remaining is

$$B_n = A(P/A, i, N - n) \tag{3.16}$$

and the interest payment during period n is

$$I_n = (B_{n-1})i = A(P/A, i, N - n + 1)i, \tag{3.17}$$

where $A(P/A, i, N - n + 1)i$ is the balance remaining at the end of period $n - 1$, and

$$P_n = A - I_n = A - A(P/A, i, N - n + 1)i$$
$$= A[1 - (P/A, i, N - n + 1)i].$$

Figure 3.19 ■ Calculating the remaining loan balance based on method 2

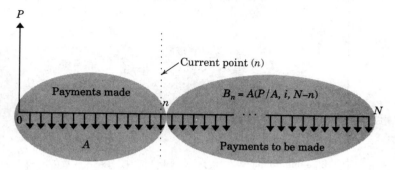

Knowing the interest factor relationship from Table 2.3, $(P/F, i, n) = 1 - (P/A, i, n)i$, we obtain

$$P_n = A(P/F, i, N - n + 1). \qquad (3.18)$$

As we can see in Fig. 3.19, this latter method provides more concise expressions for computing the loan balance, interest payment, and principal payment.

Example 3.15   Loan balances, principal, and interest— remaining balance method

Repeat Example 3.14 using method 2 by computing the outstanding loan balance after making the 6th payment in Table 3.3. For the 6th payment, compute both the interest and principal payments.

Solution

Given: (as for Example 3.14)
Find: B_6, I_6, and P_6

At the end of the 6th month, you would still owe 30 payments. As shown in Fig. 3.20, we can compute the equivalent value of these payments at time $n = 6$ with the time scale shifted by 6:

$$B_6 = \$153.45(P/A, 1\%, 30) = \$3960.20.$$

Using Eqs. (3.16) and (3.17), we compute I_6 as follows:

$$I_6 = \$153.45(P/A, 1\%, 31)(0.01)$$

$$= (\$4072.91)(0.01)$$

$$= \$40.73$$

$$P_6 = \$153.45(P/F, 1\%, 31) = \$112.72$$

Figure 3.20 ■ Cash flow diagram (Example 3.15)

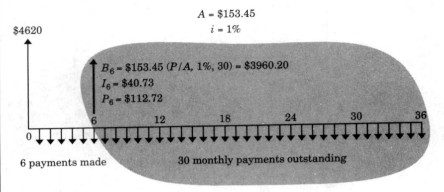

$A = \$153.45$
$i = 1\%$

$\$4620$

$B_6 = \$153.45 \ (P/A, 1\%, 30) = \$3960.20$
$I_6 = \$40.73$
$P_6 = \$112.72$

6 payments made   30 monthly payments outstanding

or simply subtracting the interest payment from the monthly payment:

$$P_6 = \$153.45 - \$40.73 = \$112.72.$$

To verify our results, compare the answer to the value given in Table 3.3.

Method 3—Computing Equivalent Worth at *N*th Payment

The third method is to compute the remaining balance by the following equivalence relationship:

$$B_n = P(F/P, i, n) - A(F/A, i, n). \tag{3.19}$$

The first term $P(F/P,i,n)$ indicates the equivalent amount of P at the end of period n, whereas the second term indicates the equivalent lump sum amount of the n payments at the end of period n. The difference between the sums is the remaining balance of the loan. Calculation of the interest payment is the same as in Eq. (3.17).

Example 3.16   Loan balance, principal, and interest— equivalent worth of *n* payments method

Repeat Example 3.15 using method 3 just described.

Solution

Given: (same as for Example 3.15)

Figure 3.21 ■ Loan balance calculation based on method 3 (Example 3.16)

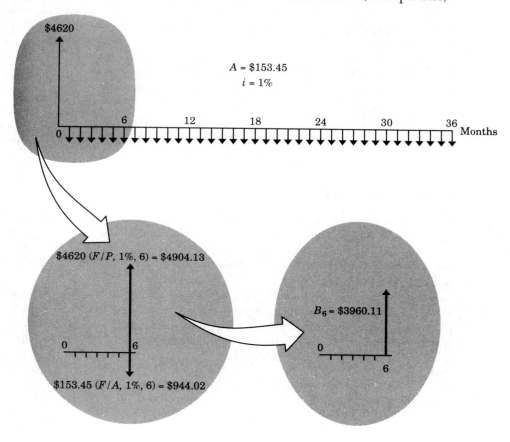

Find: $B_6 = ?$

$$B_6 = \$4620(F/P, 1\%, 6) - \$153.45(F/A, 1\%, 6)$$

$$= \$4620(1.0615) - \$153.45(6.1520)$$

$$= \$3960.11$$

The slight deviation is due to rounding error (see Fig. 3.21).

3.4.2 Add-on Loans

The **add-on loan** is totally different from the popular amortized loan. In this type of loan, the total interest to be paid is precalculated and added to the principal. The principal plus this precalculated interest amount is then paid in equal installments. In such a case, the interest rate quoted is not the effective interest rate, but what is known as **add-on interest**. If we borrow $5000 for a year at an add-on rate of 12% with equal payments due at the end of each month, a typical financial institution might compute the installment payments as follows.

Total add-on interest = (0.12) ($5000) (1) = $600

Principal plus add-on interest = $5000 + $600 = $5600

Monthly installments = $5600/12 = $466.67

Notice that the add-on interest is simple interest. The net cash flow from the borrower's viewpoint should look like Fig. 3.22.

Effective Interest Loan Rate

Once the monthly payment is determined, the financial institution will compute the APR based on this payment, and you will be told what this value will be. Even though the add-on interest is specified along with the APR value, many ill-informed borrowers would think that they are actually paying 12% interest for this installment loan.

Putting yourself in the lender's position, compute the APR value of the loan just described. Since you are making monthly payments with monthly compounding, you need to find the effective interest rate that makes the present $5000 sum equivalent to 12 future monthly payments of $466.67. In this situation, we are solving for i in the equation

$$\$466.67 = \$5000(A/P, i, 12)$$

or

$$0.0933 = (A/P, i, 12).$$

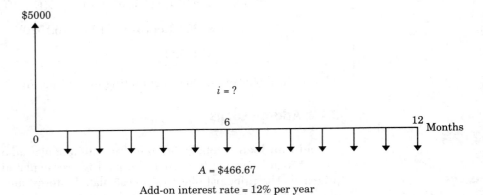

Figure 3.22 ■ An example of an add-on loan

You know the value of the A/P factor, but you do not know the interest rate i. As a result, you need to look through several interest tables and determine i by interpolation as shown in Fig. 3.23. The interest rate is between 1.75% and 2%, and may be computed by a **linear interpolation:**

A/P Factor	i
0.0931	1.75%
0.0933	?
0.0946	2.00%

$$i = 1.75\% + 0.25\%\left[\frac{0.0933 - 0.0931}{0.0946 - 0.0931}\right] = 1.7833\%$$

The nominal interest rate for this add-on loan is $1.7833 \times 12 = 21.40\%$, and the effective annual interest rate is $(1 + 0.017833)^{12} - 1 = 23.63\%$ rather than the 12% quoted add-on interest. When you take a loan, you should not confuse the add-on interest rate stated by the lender with the actual interest cost of the loan.

As an alternative to the linear interpolation based on interest tables, you can also calculate the effective interest rate for the add-on loan by many financial calculators. If you solved the equation on either a financial calculator or a computer, the exact value would be 1.7879%. In practice, the small error introduced by this type of interpolation is rarely of significance.

In the real world, truth-in-lending laws require that APR information always be provided in mortgage and other loan situations—you would not have to calculate nominal interest as a prospective home buyer (although you might be

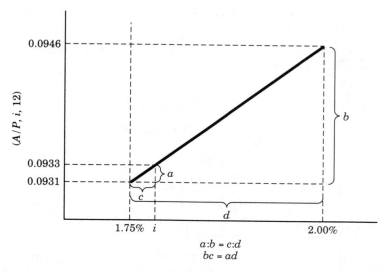

Figure 3.23 ■ Linear interpolation of unknown interest rate

interested in calculating actual or effective interest). However, in later engineering economic analyses, you will discover that solving for implicit interest rates or rates of return on investment is regularly part of the solution procedure. Our purpose is to periodically give you some practice with this type of problem, even though the example/problem scenario does not exactly model the real-world information you would be given.

Example 3.17   Effective interest rate for an add-on loan

For our add-on loan example above, with an add-on rate of 12% but $N = 3$ years, compute the effective annual rate if the principal plus interest is paid in equal monthly payments.

Solution

Given: Add-on rate = 12% per year, loan amount $(P) = \$5000$, $N = 3$ years

Find: i_a

First we determine the amount of add-on interest:

$$iPN = (0.12)(\$5000)(3) = \$1800.$$

Then we add this simple interest amount to the principal and divide the total amount by 36 months to obtain A:

$$A = (\$5000 + \$1800)/36 = \$188.89.$$

The equivalence relation between A and P must satisfy the following equation:

$$A = P(A/P, i, 36)$$

$$\$188.89 = \$5000(A/P, i, 36)$$

$$(A/P, i, 36) = 0.0378.$$

We may use the interest factor tables of 1.75% and 2.0%, respectively:

i	$(A/P, i, 36)$
1.75%	0.03768
?	0.03778
2.00%	0.03923

$$i = 1.75\% + 0.25\% \left[\frac{0.03778 - 0.03768}{0.03923 - 0.03768} \right] = 1.7661\%$$

The effective annual interest rate is

$$i_a = (1 + 0.01766)^{12} - 1 = 23.38\%.$$

Comments: *Notice that as the number of installment periods increases for a fixed add-on rate, the annual effective rate decreases (for 12 months, $i_a = 23.63\%$). For either of the two time periods, we find the effective annual rate to be nearly double the add-on rate.*

3.4.3 The Rule of 78ths

To determine the outstanding loan balances in Example 3.14, we used Eq. (3.15). Recall that the outstanding balances were computed at the end of each payment period (that is, month). You may notice that the interest charge per period is a constant fraction of the principal at the beginning of the period. The installment payment first goes toward paying the interest charge, with the remainder being used to reduce the principal. This payment scheme is called the **conventional method.** There is another repayment scheme, called the **Rule of 78ths**, that is used by some financial institutions (such as General Motors Acceptance Corporation of Canada). The rule uses uniform periodic repayments, but differs from the conventional repayment method for amortized loans in that the interest rate is not constant, but rather is greater than average initially and then declines throughout the life of the loan. In effect, the lender is laying early claim to interest payments, and consequently the unpaid principal is larger. Since interest is in effect the *profit* the lender makes and that profit has a time value, you can understand why the lender might want to collect it earlier rather than later.

According to the Rule of 78ths, the interest charged during a given month is figured by applying a changing fraction to the total interest over the loan period. The numerator of the fraction is the number of remaining months of the loan period. The denominator is the sum of the numbers representing the months of the loan. For our example, in the case of a 1-year loan, the fraction used in figuring the interest charge for the first period would be 12/78, 12 being the number of remaining months of the loan, and 78 being the sum of $1 + 2 + \cdots + 11 + 12$. For the second month, the fraction would be 11/78, and so on. To illustrate how the Rule of 78ths works, let's consider Example 3.18.

Example 3.18 Loan balance, principal, and interest—Rule of 78ths method

Suppose you borrow $1000 from a bank for 1 year, payable in 12 equal monthly payments. The bank computes the monthly payment at an add-on interest rate of 15.6% per year. To initiate the loan, the bank also charges an acquisition fee of $25, which for calculation purposes is added to the principal.

$$\text{Contract amount} = \$1000$$

$$\text{Contract period} = 12 \text{ months}$$

$$\text{Add-on interest} = \$1000(0.156)\,(1) = \$156$$

$$\text{Acquisition fee} = \$25$$

$$\text{Monthly payment} = (\$1000 + \$156 + \$25)/12 = \$98.42$$

(a) Determine the remaining balance after making two monthly payments by the Rule of 78ths.
(b) Determine the remaining balance after making two monthly payments by the conventional method.

Solution

Given: $P = \$1025$, $A = \$98.42$ per month, $N = 12$ months
Find: B_2

(a) **The Rule of 78ths:** In this example, if you pay back the loan immediately, you will owe

$$\$1025 + \$156(12/78) = \$1049.$$

After 1 month, you have paid $98.42, of which $156(12/78) + $156(11/78) = $46 goes to pay interest. The remainder reduces the principal amount by $98.42 − $46. Therefore, the total outstanding balance after 1 month is

$$\$1025 - (\$98.42 - \$46) = \$972.58.$$

After 2 months, you have paid $196.84, of which $156(12/78) + $156(11/78) + $156(10/78) = $66 goes to pay interest. The remainder reduces the principal by $196.84 − $66. Thus, the total outstanding balance after 2 months is

$$\$1025 - [2(\$98.42) - \$66] = \$894.16$$

After n months, you have paid $98.42(n)$, of which

$$\$156\{[12 + 11 + \cdots + (12 - n)]/78\}$$

goes to pay interest due, and the remainder reduces the principal amount. Thus, the amount due after n payments is

$$\$1025 - \$98.42(n) + \$156[12 + 11 + \cdots + (12 - n)]/78.$$

Table 3.4 The Balances Due After n Payments Under the Rule of 78ths and Conventional Method

Period	Monthly Payment	Rule of 78ths			Conventional Method		
		Principal	Interest	Balance	Principal	Interest	Balance
0	$ 0	$+25 °	$24	$1049	$25 °	$ 0	$1025
1	98.42	76.42	22	972.58	75.35	23.07	949.65
2	98.42	78.42	20	894.16	77.05	21.37	872.60
3	98.42	80.42	18	813.74	78.78	19.64	793.81
4	98.42	82.42	16	731.32	80.56	17.86	713.26
5	98.42	84.42	14	646.90	82.37	16.05	630.89
6	98.42	86.42	12	560.48	84.22	14.20	546.67
7	98.42	88.42	10	472.06	86.12	12.30	460.55
8	98.42	90.42	8	381.64	88.06	10.36	372.49
9	98.42	92.42	6	289.22	90.04	8.38	282.45
10	98.42	94.42	4	194.80	92.06	6.36	190.39
11	98.42	96.42	2	98.38	94.14	4.28	96.25
12	98.42	98.42	0	0	96.25	2.17	0
		Total = $156			Total = $156		

°Acquisition fee that must be added on to the loan principal

(b) **Conventional Method:** First, find the effective interest rate for the loan:

$$P = A(P/A, i, 12)$$

$$\$1025 = \$98.42(P/A, i, 12).$$

Solving for i yields 2.2504% per month. Now, using Eq. (3.16), the remaining balance right after making the second payment is

$$B_2 = \$98.42(P/A, 2.2504\%, 10)$$

$$= \$872.60$$

Table 3.4 summarizes the amount due during each period over the life of the loan.

Comments: *The real difference between the two methods may be observed by inspecting the balance columns of Table 3.4. The total interest payments over the life of the loan are exactly the same for both methods. Suppose, however, that after 6 months the borrower wishes to retire the current loan. The borrower who financed by the Rule of 78ths would be penalized $13.81.*

In Example 3.18, we assumed the loan life to be 1 year. If the loan life extends several years, the same Rule of 78ths can be applied. For example, if the loan life extends 24 months instead of 12 months, the fraction during the first month is 24/300, because there are 24 remaining payment periods and the sum of the loan periods is $300 = 1 + 2 + \cdots + 24$. (In this case, the rule could be called the Rule of 300ths. However, we always follow the convention by referring to the procedure as the Rule of 78ths, irrespective of the value of N.) Thus, we can generalize the results for a loan life of N periods by

$$B_n = P - A(n) + (I/Sum)[N + (N - 1) + \cdots(N - n)]$$

or

$$B_n = P - A(n) + (I/Sum)[(n + 1)(N) - n(n + 1)/2], \qquad (3.20)$$

where

$$P = \text{loan amount, including acquisition fee}$$

$$A = \text{installment amount}$$

$$I = \text{total add-on interest}$$

$$N = \text{loan life}$$

$$Sum = \text{sum of the loan periods.}$$

■ 3.5 Mortgages

A mortgage is generally a long-term amortized loan used primarily for the purpose of purchasing a piece of property such as a home. The mortgage itself is a legal document in which the borrower agrees to give the lender certain rights to the property being purchased as security for the loan. The borrower is referred to as the mortgagor and the lender as the mortgagee. The mortgage document specifies the rights that the lender has to the property in the event of default by the borrower on the terms of the mortgage. In the sections that follow we will explain some concepts related to mortgages and give examples on how payment schedule, regular payment amount, and the interest charges are calculated. The amount of the loan—the cash that you actually borrow—is called the **principal**. The difference between the price of the property and the amount that one owes on the mortgage is called the purchaser's **equity**.

3.5.1 Types of Mortgages

Canadian banks, credit unions, trust companies, mortgage companies, private lenders, and others offer four main types of mortgages. These four types of mortgages are briefly described below.

1. **National Housing Act (NHA) mortgage**: These mortgages are loans granted under the provisions of the National Housing Act of 1954. Lenders are insured against loss by the Canada Mortgage and Housing Corporation (CMHC). Borrowers must pay an application fee which usually includes the property appraisal fee to CMHC and an insurance fee. The insurance fee is usually added to the principal amount of the mortgage, although it may be paid in cash.

2. **Conventional mortgage**: The conventional mortgage is the most common type of financing for principal residence or residential investment property. In this type of mortgage, the loan amount generally does not exceed 75% of the appraised value or the purchase price of the property, whichever is lower. The purchaser is responsible for raising the other 25% as a down payment.

3. **High-ratio mortgage**: If a potential buyer is unable to raise the necessary 25% funding to complete the purchase of the property, then he or she may obtain a high-ratio mortgage. Essentially, these are conventional mortgages that exceed the 75% referred to above. These mortgages must, by law, be insured. High-ratio mortgages are available for up to 95% of the purchase price or of the appraisal, whichever is lower.

4. **Collateral mortgage**: A collateral mortgage provides backup protection of a loan that is filed against a property. It is secondary to a main form of security taken by the lender for the loan, for example, a promissory note. When the promissory note is paid off, the collateral mortgage is automatically discharged. The money borrowed may be used for the purchase of the property itself or for other purposes, such as home improvements and other investments.

3.5.2 Terms and Conditions of Mortgages

To make the best mortgage decision, one has to consider many factors. The key factors are amortization, term of the mortgage, whether the mortgage is open or closed, interest rate, payment schedule, prepayment privilege, and portability. A brief explanation of each of these concepts is provided below:

1. **Amortization**: Amortization refers to the number of years it would take to repay a mortgage loan in full for a given interest rate and payment schedule. The usual amortization period is 25 years, although there is a wide range of choices. The longer the amortization period, the smaller the regular payment (usually monthly) and the larger the total interest payments.

2. **Term of the mortgage:** Term refers to the number of months or years which the mortgage—the legal document—covers. Terms may vary from 6 months to 10 years. At the end of a term, the unpaid principal is due and payable. One has the option to renew the mortgage with the same bank or refinance it through a different lending institution.

3. **Interest rate**: By law, mortgages must contain a statement showing, among other things, the rate of interest calculated annually or semi-annually. Mortgage interest has traditionally been quoted as a nominal annual rate based on semi-annual compounding. A fixed-rate mortgage is one where the rate of interest is set for the whole term. A variable-rate mortgage is one where the rate of interest varies according to the premium interest rate set by the lender every month. For both a fixed-rate mortgage and a variable-rate mortgage, the required regular payment amount does not change within the term. However, the interest portion of the regular payment amount for a payment period is dependent on the outstanding balance at the beginning of the period and the interest rate applicable for the period.

4. **Open or closed mortgage**: An open mortgage provides the borrower with the flexibility to repay the loan more quickly. One can pay off the mortgage in full at any time before the term is over without any penalty or extra charges. Because of this flexibility, the interest rate for an open mortgage is higher than a closed mortgage, when other conditions remain the same. A closed mortgage does not allow the borrower to repay the loan more quickly than agreed. Payments must be made as specified in the agreement. If the borrower wants to pay off the mortgage before the term is over, a penalty charge is applied. The penalty charge is often equal to the greater of (1) three months' interest on the amount of the prepayment; and (2) the interest rate differential. The "interest rate differential" refers to the amount, if any, by which the existing interest exceeds the interest at the rate at which the lender would lend to the same borrower for a term commencing on the prepayment date and expiring at the existing term date. However, even for a closed mortgage, there may be some prepayment privileges such that no penalty will be levied.

5. **Payment schedule**: Most mortgage loans are amortized. Constant regular payments are made to pay off the principal. Monthly payments are the most common, although some mortgages may be paid weekly, bi-weekly, semi-monthly, quarterly, semi-annually, or annually.

6. **Prepayment privileges**: Many financial institutions offer closed mortgages with some prepayment privileges. For example, the borrower taking the closed mortgage may be allowed to make a prepayment of up to 10% of the original principal every calendar year or every anniversary. Another privilege allows the borrower to increase the regular payment by up to 100% once per year. If you are able to take advantage of these prepayment privileges, you may dramatically decrease the actual amortization period.

7. **Portability**: Some lenders offer mortgages with a feature called portability. This feature means that a borrower can sell one home and buy another during the term of the mortgage. The mortgage can be transferred from one property to the other without penalty.

3.5.3. An Example of Mortgage Calculations

Throughout this book, we will assume that the quoted mortgage interest rates are based on semi-annual compounding.

Example 3.19 Closed mortgage with prepayment privileges

John Montgomery is considering buying a $125,000 home with a $25,000 down payment. He can get a conventional mortgage in the amount of $100,000 with a 3-year term at 8% per annum from the Toronto Dominion Bank. He has selected an amortization period of 25 years. The mortgage is a closed mortgage with the following prepayment privileges:

- Once each calendar year, on any regular payment date, John can prepay on account of principal a sum not more than 10% of the original borrowed amount, without notice or charge. If this privilege is not exercised in a certain year, it cannot be carried forward to the following years.

- Once each calendar year, on any regular payment date, on written notice, John, without charge can increase the amount of the regular installment of principal and interest. The total of such increases cannot exceed 100% of the installment of principal and interest set out in the mortgage. If the regular installment has been increased, the mortgagor may decrease the installments to an amount not less than the installment of principal and interest set out in the mortgage, on written notice, without charge.

- On any regular payment date, John can prepay the whole or any part of the principal amount then outstanding on payment of an amount equal to the greater of

  (1) 3 months' interest, at the rate specified in the mortgage, on the amount prepaid or
  (2) the amount, if any, by which interest at the rate specified in the mortgage exceeds interest at the prevailing rate, calculated on the amount of principal prepayment, for a term commencing on the date of prepayment and expiring on the maturity date of the mortgage.

Answer the following questions regarding the mortgage in consideration:

(a) What is the amount of his regular payment if he chooses to pay weekly, semi-monthly, or monthly?
(b) What would the balance be at the end of the term for each of the three payment frequencies if the calculated regular payment amount is followed exactly?
(c) Assume that John has selected the option of monthly payment because he receives only one salary payment per month. In year 2, he

increases his monthly payment by 50%. In year 3, he doubles his calculated monthly payment in (a). What is the balance of the mortgage at the time of renewal?

(d) In addition to (c), if he makes lump sum payments of $8000 and $10,000 at the first and the second anniversaries, respectively, what is the balance of the mortgage at the time of renewal?

(e) After John has made only the calculated monthly payments for 1 year, the interest rate for a 2-year term has dropped to 6%, what would be the total penalty charge if he chooses to payoff his mortgage completely? What if the prevailing rate for a 2-year mortgage has increased to 9%?

Solution

Given: $P = \$100,000$, $r = 8\%$ per year, $M = 2$ compounding periods per year, amortization = 25 years, term = 3 years

Find:

(a) The regular payment amounts on the following payment schedules: weekly, semi-monthly, and monthly.

Figure 3.24(a) illustrates the cash flow diagram for the computation of regular payment

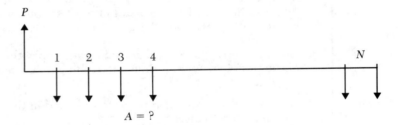

Figure 3.24(a) ■ Cash flow diagram for the term loan (Example 3.19(a)).

where

N is the number of payment periods in the amortization period.

- For weekly payment: $N = (52)(25) = 1300$ weeks

$$i_{wk} = (1 + r/M)^C - 1 = (1 + 0.08/2)^{1/26} - 1 = 0.1510\%$$

$$A_{wk} = \$100,000 \, (A/P, 0.1510\%, 1300) = \$175.68$$

- For semi-monthly payment: $N = (25)(24) = 600$ half-months

$$i_{1/2\ mon} = (1 + 0.08/2\ )^{1/12} - 1 = 0.3274\%$$

$$A_{1/2\ mon} = \$100{,}000\ (A/P,\ 0.3274\%,\ 600) = \$380.98$$

- For monthly payment: $N = (25)(12) = 300$ months

$$i_{mon} = (1 + 0.08/2)^{1/6} - 1 = 0.6558\%$$

$$A_{mon} = \$100{,}000\ (A/P,\ 0.6558\%,\ 300) = \$763.20$$

Comments: *The more frequent the payments, the smaller the total amount paid per month (i.e., $A_{mon} > 2 \times A_{1/2\ mon} > 4.33 \times A_{wk}$)*

(b) What are the end-of-term balances for weekly, semi-monthly, and monthly payments?

Figure 3.24(b) illustrates the cash flow diagram for the end-of-term balance calculation

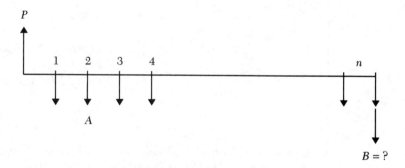

Figure 3.24(b) ■ Equivalent cash flow diagram (Example 3.19(b))

where

n is the number of payment periods in a term,

A is the calculated regular payment amount, and

B is the end-of-term balance to be computed.

- For weekly payment: $n = (3)(52) = 156$ weeks

$$i_{wk} = 0.1510\%$$

$$B_{wk} = P(F/P, i_{wk}, n) - A_{wk}(F/A, i_{wk}, n)$$

$$= \$100,000 \ (F/P, \ 0.1510\%, \ 156) - \$175.68 \ (F/A, \ 0.1510\%, \ 156)$$

$$= \$95,655.93$$

- For semi-monthly payment: $n = (3)(24) = 72$ half-months

$$i_{1/2 \, mon} = 0.3274\%$$

$$B_{1/2 \, mon} = \$100,000 \ (F/P, \ 0.3274\%, \ 72)$$

$$- \ \$380.98 \ (F/A, \ 0.3274\%, \ 72)$$

$$= \$95,655.54$$

- For monthly payment: $n = (3)(12) = 36$ months

$$i_{mon} = 0.6558\%$$

$$B_{mon} = \$100,000 \ (F/P, \ 0.6558\%, \ 36)$$

$$- \ \$763.20 \ (F/A, \ 0.6558\%, \ 36)$$

$$= \$95,655.54$$

Comments: *Note that the end-of-term balances are the same for all three payment options. This is because all three options will pay off the mortgage loan in exactly 25 years, if the calculated regular payment amounts are made for 25 years.*

(c) What is the end-of-term balance, when monthly payments and some prepayment privileges are used?

$$B_{(c)} = \$100,000 \ (F/P, \ 0.6558\%, \ 36)$$

$$- \ \$763.20 \ (F/A, \ 0.6558\%, \ 36)$$

$$- \ \$381.60 \ (F/A, \ 0.6558\%, \ 24)$$

$$- \ \$381.60 \ (F/A, \ 0.6558\%, \ 12)$$

$$= \$81,023.51$$

Figure 3.25(a) shows the payments for the whole term.

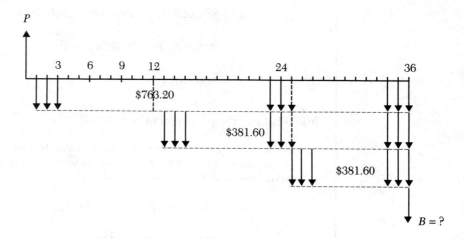

Figure 3.25(a) ■ Equivalent cash flow diagram (Example 3.19(c))

(d) **What is the end-of-term balance with some additional lump sum payments?**

Notice that the lump sum payments are within the prepayment privilege limits. We can utilize the result calculated in (c), as shown in Figure 3.25(b).

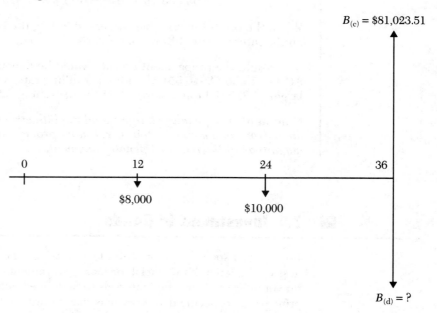

Figure 3.25(b) ■ Equivalent cash flow diagram (Example 3.19(d))

$$B_{(d)} = \$81,023.51 - \$8000\ (F/P,\ 0.6558\%,\ 24)$$

$$- \$10,000\ (F/P,\ 0.6558\%,\ 12)$$

$$= \$60,848.71$$

(e) **What are the prepayment penalties?**

To find the prepayment penalty, we need to calculate the total prepayment amount first (i.e., the balance of the loan after monthly payments have been made for a year).

$$B = \$100,000\ (F/P,\ 0.6558\%,\ 12)$$

$$- \$763.20\ (F/A,\ 0.6558\%,\ 12)$$

$$= \$98,663.79$$

Three months simple interest $= \$98,663.79 \times 0.6558\% \times 3 = \$1941.11$

When the prevailing rate has dropped to 6%, the interest rate differential can be estimated as

$$\$98,663.79 \times (8\% - 6\%) \times 2 = \$3946.55.$$

When the prevailing rate has increased to 9%, the penalty charged based on the interest rate differential is zero.

As a result, the prepayment penalty would be $3946.55 (i.e., the larger of $1941.11 and $3946.55) when the prevailing rate is 6% and $1941.11 (the larger of $1941.11 and zero) when the prevailing rate is 9%.

Comment: *The penalty charge calculation illustrated in this case is only an approximate method. However, it does provide an indication of the magnitude of the required penalty payment.*

■ 3.6 Investment in Bonds

Bonds are a specialized form of a loan in which the creditor—usually a business or the federal, provincial, or local government—promises to pay a stated rate of interest at specified intervals for a defined period and then to repay the principal at a specific date known as the maturity date of the bond. Bonds are an important financial instrument by which the business world may raise funds to finance engineering projects.

In the case of bonds the lenders are investors (known as bondholders) who may be individuals or other businesses. Bonds can be a significant investment opportunity. In addition to the interest they earn, once purchased by the initial bondholder, they can be sold again for amounts other than their stated face value and thus enhance the bondholder's opportunity to increase the return on his or her initial investment. Given these complications, the concept of economic equivalence can be important in determining the worth of bonds, and the following sections will illustrate some typical bond investment problems in the context of economic equivalence.

3.6.1 Bond Terminology

We will first look at how a typical bond may be issued on the financial market. We will consider a government bond issued by Ontario Hydro. Shown in Figure 3.26(a) and Figure 3.26(b) is a reproduction of a $1000 bond issued by Ontario Hydro and known as "Ontario Hydro 9 1/4s of 2004". The certificate is the document one receives as evidence of an investment. Before explaining how bond values are determined, some of the terms associated with a typical bond will be defined.

Debentures (versus mortgage bonds): Traditionally, a debenture is a debt security or an unsecured promise to pay. That is, no property or assets are pledged as security for the loan. Most government bonds fall in this category. The Ontario Hydro bond is precisely this. You will note that this bond is guaranteed as to principal and interest by the province of Ontario. Hence the issue is known as a *Provincial Guarantee*. If a company or government agency issues bonds and backs them with specific pieces of property such as buildings, the bonds are called *mortgage bonds*.

Par value: Individual bonds are normally issued in even denominations, such as $1000 or multiples of $1000. The stated face value on the individual bond is termed the *par value*. For this type of bond, it is usually set at $1000, which is the case for the Ontario Hydro bond.

Maturity date: Bonds generally have a specified *maturity date* on which the par value is to be repaid. The Ontario Hydro bonds, which were issued on March 10, 1977, will mature on January 6, 2004; thus, they had a 26.17-year maturity at time of issue.

Coupon rate: The interest paid on the par value of a bond is called the annual *coupon rate*. For example, the Ontario Hydro bonds have a $1000 par value and pay $92.50 in interest (9.25%) each year ($46.25 every 6 months). The bond's coupon interest is $92.50 so its annual coupon rate is 9.25%. Even though the coupon rate is stated as an annual rate, Ontario Hydro will pay 4.625% interest semi-annually on the face value (i.e., $46.25) to the bond holders. Shown in Figure 3.26(c) is a sample of one of the 54 coupons originally attached to the Ontario Hydro bond. Each coupon has a date on which, or after which, it may be cashed.

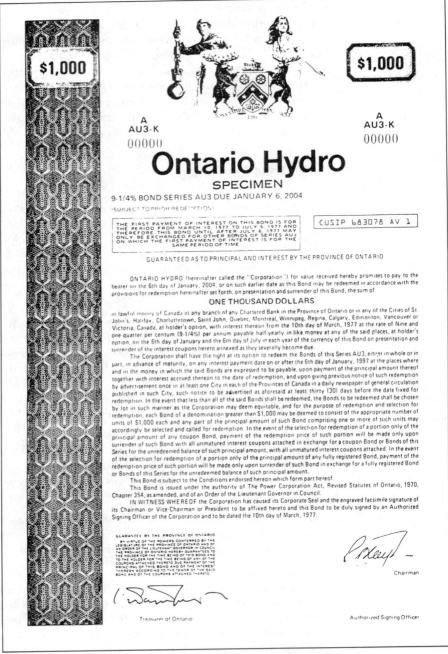

Figure 3.26(a) ■ The face side of the Ontario Hydro bond certificate
(Source: *The Canadian Securities Course.* Prepared and published by the
Canadian Securities Institute, Toronto, Ontario. Copyright 1990.)

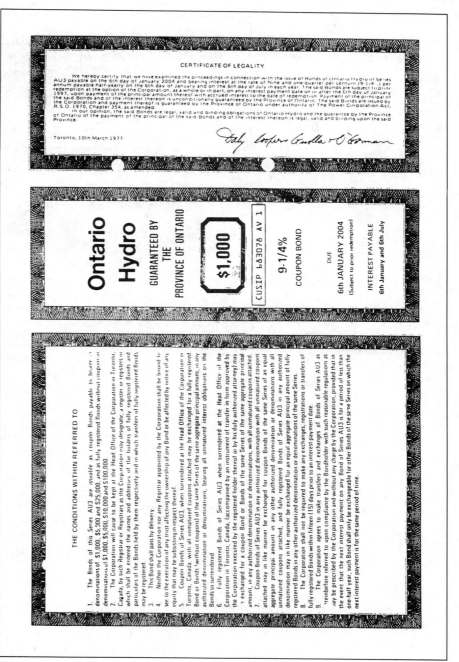

Figure 3.26(b) ■ The back side of the Ontario Hydro bond certificate (Source: *The Canadian Securities Course.* Prepared and published by the Canadian Securities Institute, Toronto, Ontario. Copyright 1990.)

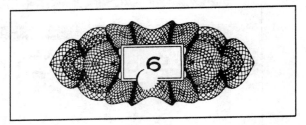

Sample of one of the 54 coupons originally attached
to the $1,000 Ontario Hydro 9¼s of January 6, 2004.
Top illustration: face of the sixth coupon which paid
$46.25 on January 6, 1980. Bottom illustration: back
of same coupon is numbered "6". Other coupons
originally attached to the certificate were numbered
up to 54 and payable up to January 6, 2004.

Figure 3.26(c) ■ Sample of one of the 54 coupons originally attached to the
Ontario Hydro bond certificate (Source: *The Canadian Securities Course*.
Prepared and published by the Canadian Securities Institute, Toronto, Ontario.
Copyright 1990.)

Discount or premium bond: A bond that sells below its par value is called a
discount bond. When a bond sells above its par value, it is called a *premium
bond*. A bond that is purchased for its par value is said to have been bought *at
par*. The Ontario Hydro bonds were offered *at par*. The price one has to pay
to purchase a bond is called the bond's *market value*.

3.6.2 Bond Valuation

Bonds can be traded just like stocks on the market. Once purchased, a bond
may be kept for its maturity or for a variable number of interest periods before
being sold. Bonds may be purchased or sold at prices other than face value,
depending on the economic environment. Furthermore, bond prices change
over time as a result of various factors – the risk of nonpayment of interest or
the par value at maturity, supply and demand, market interest rates, and the

future outlook for economic conditions. These factors affect the **yield to maturity** (or **return on investment**). The yield to maturity represents the actual interest earned from a bond over the holding period. We will explain these values with numerical examples in the following section.

Yield to Maturity

The yield to maturity on a bond is the interest rate that establishes the equivalence between all future interest/face value receipts and the market price of the bond. To illustrate the point, let us consider buying a $1000 denomination of the Ontario Hydro bond on January 7, 1994. Recall that the par value is $1000, but assume the market price is $1088. Since the interest will be paid semi-annually, the interest rate per payment period will be simply 4.625%, and there will be 20 interest payments over 10 years. The resulting cash flow to the investor is shown in Fig. 3.27.

The yield to maturity is found by determining the interest rate that makes the present worth of the receipts equal to the market price of the bond:

$$\$1088 = \$46.25(P/A, i, 20) + \$1000(P/F, i, 20).$$

The value of i that makes the present worth of the receipts equal to $1088 lies between 3% and 4%. Solving for i by interpolation yields $i = 3.98\%$.

Present Worth of Receipts	i
$1084.94	4%
1088.00	?
1241.76	3%

$$i = 3\% + 1\% \left[\frac{\$1241.76 - \$1088}{\$1241.76 - \$1084.94} \right] = 3.98\%$$

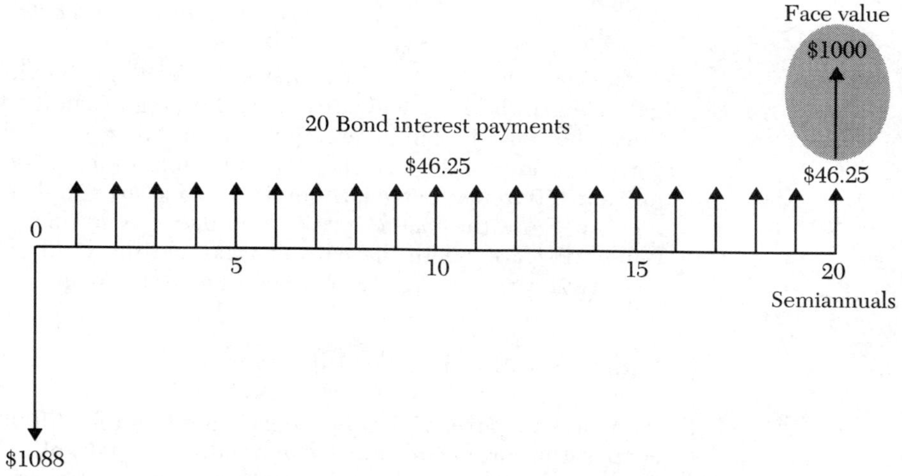

Figure 3.27 ■ A typical cash flow transaction associated with an investment in the Ontario Hydro bond

Note that this is 3.98% yield on maturity per semiannual period. The nominal (annual) yield is $2 \times 3.98\% = 7.96\%$ compounded semiannually. When compared with the coupon rate of $9\frac{1}{4}\%$ (or 9.25%), purchasing the bond at a price higher than the face value brings about a lower yield (the difference is 1.29%). The effective annual interest rate is then

$$i_a = (1 + 0.0398)^2 - 1 = 8.12\%.$$

This 8.12% represents the **effective annual yield** to maturity on the bond. Notice that when a bond is purchased at a par value, the yield to maturity will be exactly the *same* as the coupon rate of the bond, provided that they are both expressed as effective rates for the same length of time.

Until now, we have observed differences in nominal and effective interest rates as a result of frequency of compounding. In the case of bonds, the reason is very different: The stated (par) value of the bond and the actual price paid for it are different. The nominal interest is stated as a percentage of par value, but when the bond is sold at a premium, you earn the same nominal interest on a larger initial investment; hence, your effective interest earnings are smaller than the stated nominal rate.

Current Yield

The **current yield** of a bond is the annual interest earned as a percentage of the current market price. This current yield provides an indication of the annual return realized from the bond investment. For our example of the Ontario Hydro Bond, the current yield is computed as follows:

$$\$46.25/1088 = 4.25\% \text{ semiannually}$$

$$2 \times 4.25\% = 8.5\% \text{ per year (nominal current yield)}$$

$$i_a = (1 + 0.0425)^2 - 1 = 8.68\%.$$

This effective current yield is 0.56% larger than the yield to maturity computed above (8.12%). The current yield is larger than the yield to maturity because the bond is sold at premium. If the bond were sold at a discount, the current yield would be smaller than the yield to maturity. There can be a significant difference between the yield to maturity and the current yield of a bond, because the market price of a bond may be more or less than its face value. Moreover, both the current yield and the yield to maturity may differ considerably from the stated coupon value of the bond.

Example 3.20   Bond yields

John Brewer purchased a new corporate bond for $1000. The issuing corporation promised to pay the bondholder $45 interest on the $1000 face (par) value of the bond every 6 months, and to repay the $1000 at the end of 10 years. After 2 years John sold the bond to Kimberly Crane for $900.

(a) What was the yield on John's investment?
(b) If Kimberly keeps the bond for its remaining 8-year life, what is the yield to maturity on her bond investment?
(c) What was the current yield when Kimberly purchased the bond?

Solution

Given: Initial purchase price = par value = $1000, coupon rate = 9% per year paid semiannually, 10 years maturity, sold after 2 years for $900

Find: (a) Yield to initial owner, (b) yield to maturity for second owner, (c) current yield at 2 years

(a) Since John has received $45 every 6 months for 2 years, the cash flow diagram for this situation would look like Fig. 3.28. John should find the yield on bond investment by determining the interest rate that makes the expenditure of $1000 equivalent to the present worth of the semiannual receipts and selling price:

$$\$1000 = \$45(P/A, i, 4) + \$900(P/F, i, 4).$$

Try $i = 3\%$:

$$\$45(P/A, 3\%, 4) + \$900(P/F, 3\%, 4) = \$966.91.$$

The present worth of the annual receipts is too low. Try a lower interest rate, say, $i = 2\%$:

$$\$45(P/A, 2\%, 4) + \$900(P/F, 2\%, 4) = \$1002.81.$$

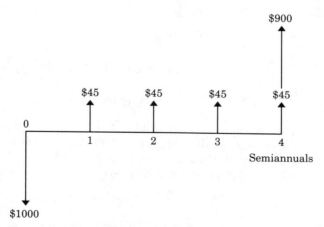

Figure 3.28 ■ John Brewer's cash flow transaction associated with the bond investment (Example 3.20)

Figure 3.29 ■ Kimberly Crane's bond investment (Example 3.20)

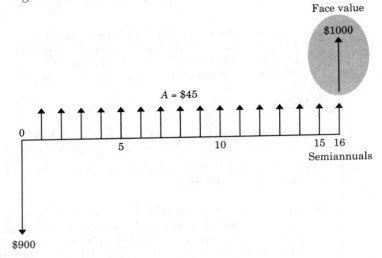

The yield on bond investment per 6 months is very close to 2% (to be exact, 2.08%). This means the annual nominal yield is $2 \times 2\% = 4\%$. (The effective annual yield is $(1.02)^2 - 1 = 4.04\%$.)

(b) Kimberly's bond investment would look like Fig. 3.29.

$$\$900 = \$45(P/A, i, 16) + \$1000(P/F, i, 16)$$

Try $i = 5.4\%$:

$$\$45(P/A, 5.4\%, 16) + \$1000(P/F, 5.4\%, 16) = \$905.17.$$

Try $i = 5.5\%$:

$$\$45(P/A, 5.5\%, 16)$$

$$+ \ \$1000(P/F, 5.5\%, 16) = \$895.38.$$

$$i = 5.4\% + (0.10\%) \left[\frac{\$905.17 - \$900}{\$905.17 - \$895.38} \right]$$

$$= 5.453\%$$

The nominal (annual) yield is $2 \times 5.453\% = 10.906\%$, and the effective annual yield is $(1 + 0.05453)^2 - 1 = 11.20\%$. Compare this interest with the coupon rate of 9%.

(c) The current yield when Kimberly purchased the bond is

$$\$45/\$900 = 5\% \text{ per semiannual}$$

$$2 \times 5\% = 10\% \text{ per nominal}$$

$$i_a = (1 + 0.05)^2 - 1 = 10.25\%.$$

The Value of a Bond over Time

Reconsider the Ontario Hydro bond investment introduced earlier. If the nominal (annual) yield to maturity remains constant at 7.96%, what will the value of the bond be 1 year after it was purchased? We can find this value using the same valuation procedure, but now the term to maturity is only 9 years.

$$\$46.25(P/A, 3.98\%, 18) + \$1000(P/F, 3.98\%, 18) = \$1081.79$$

The value of the bond has decreased because there is a smaller number of interest payments to be received.

Now suppose interest rates in the economy have risen since the Ontario Hydro bonds were issued, and as a result, the going rate of interest is 10%. Both the coupon interest payments and the maturity value would remain constant, but now 10% values would have to be used in calculating the value of the bond. The value of the bond at the end of the first year would be

$$\$46.25(P/A,5\%,18) + \$1000(P/F, 5\%, 18) = \$956.16.$$

Thus, the bond would sell at a discount under its par value.

The arithmetic of the bond price increase should be clear, but what is the logic behind it? We can explain the reason for the decrease as follows: The fact that the going bond market interest has risen to 10% means that, if we had $1000 to invest, we would buy new bonds rather than Ontario Hydro bonds, since the new bonds will pay $100 of interest every year instead of $92.50. If we are interested in buying Ontario Hydro Bonds, we would be willing to pay less than face value because they would have lower interest payments. All investors would recognize these facts, and as a result, Ontario Hydro bonds would be discounted to a lower price of $956.16, at which point, they would provide the same yield to maturity (rate of return) to a potential investor as the new bonds, 10%.

3.7 Computer Notes

As mentioned in Section 3.4, the generation of a loan payment schedule can be tedious. The power of an electronic spreadsheet or the **EzCash** software that accompanies this text can greatly facilitate the generation of such a schedule.

3.7.1 Application of Lotus 1-2-3

Using Fig. 3.30 as a model, set up your Lotus 1-2-3 screen and specify the contract amount, contract period, and the interest rate. Lotus 1-2-3 calculates the monthly payment, with the interest as well as the principal itemized for each period. By separating the interest payment from the periodic payment, you can generate the remaining loan balance for each period.

Figure 3.30 ■ Formula inputs and resulting loan repayment schedule generated by Lotus 1-2-3

A:F8: (C2) [W13] @PMT($F$5,$F$7/1200,$F$6)   READY

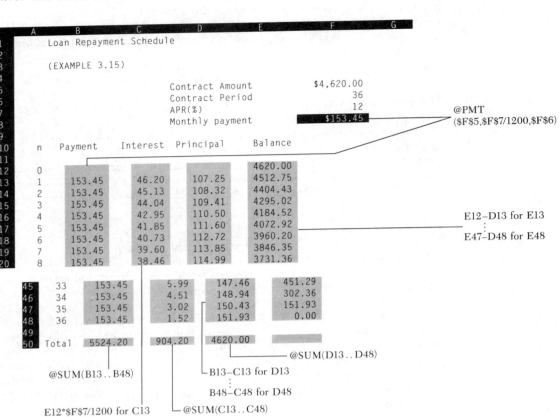

Figure 3.30 is annotated to show the cell formulas and values used. Note the difference between F7 and $F$7. The latter notation indicates the cell location that is absolute. Whenever a cell formula contains this absolute cell, Lotus must use the data stored physically in that cell. Note also that the given APR interest rate must be divided by 12 for monthly compounding to obtain the effective monthly interest rate. To convert the interest rate to decimal, you also need to divide the effective interest rate per payment period by 100.

Table 3.5 compares the built-in principal and interest balance functions for Lotus 1-2-3, Excel, and QuattroPro.

3.7.2 Application of EzCash

Even using the Copy function of Lotus 1-2-3, whereby a piece of data can be copied into many cells at once, you can see that a great deal of manipulation

Table 3.5 Built-in Loan Functions

Functional Description	Lotus 1-2-3	Excel	QuattroPro
Loan repayment size: Given i, N, P and *type*; Find A.	@PMT(P,i,N)	PMT($i,N,P,type$)	@PMT(P,i,N)
Interest payment: Given i, n, N, P or F, *type*; Find portion of A that is interest for a given period n.		IPMT($i,n,N,P,F,type$)	@IPAYMT($i,n,N,P,F,type$)
Principal payment: Given i, n, N, P or F, *type*; Find portion of A that is principal for a given period n.		PPMT($i,n,N,P,F,type$)	@PPAYMT($i,n,N,P,F,type$)

i = loan interest rate $\quad N$ = loan term $\qquad F$ = future amount $\qquad$ *type* = timing of payment:
n = period $\qquad\qquad P$ = loan amount $\quad A$ = installment size $\qquad\qquad\qquad$ 0 for end of period
$\qquad\qquad\qquad\qquad\qquad\qquad\qquad\qquad\qquad\qquad\qquad\qquad\qquad\qquad$ 1 for beginning of period

If no value is entered for F or type, each is assumed to be zero.

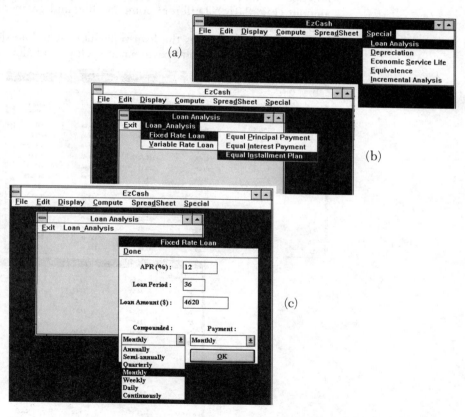

Figure 3.31 ■ (a) Select **Special** mode in the main menu and then **Loan Analysis**. (b) Select **Loan Analysis**, **Fixed Rate Loan**, and **Equal Installment Plan** in sequence. (c) Enter the APR, loan period, and loan amount, and select the compounding and payment frequencies.

was required to generate the Lotus 1-2-3 schedule. **EzCash** can generate the loan repayment schedule in a much more efficient way.

From the main menu, select the **Special** mode and then **Loan Analysis** (Fig. 3.31a). Clicking **Loan Analysis** again will give you two choices: **Fixed Rate Loan** and **Variable Rate Loan**. If you choose **Fixed Rate Loan**, the next pull-down menu you see allows you to select the type of loan repayment method (Fig. 3.31b):

Equal Principal Payment uses equal principal payments, decreasing interest payments and, thus, decreasing periodic payments over the life of the loan.

Equal Interest Payment uses constant interest payments over the life of the loan, with principal repaid as a lump sum at the final period.

Equal Installment Plan uses increasing principal payments, decreasing interest payments, and equal periodic payments over the life of the loan.

In Example 3.19(a), an equal installment plan is used, so enter this option and specify the loan rate (APR), number of loan payment periods, and loan amount (Fig. 3.31c). Different compounding and payment frequencies should also be selected.

EzCash next generates the loan repayment schedule showing principal and interest payments and total payment in each period along with the cumulative

Loan Analysis (a)

N	Principal Payment	Interest Payment	Cumulative Interest	Total Payment	Remaining Balance
1	107.25	46.20	46.20	153.45	4512.75
2	108.32	45.13	91.33	153.45	4404.43
3	109.41	44.04	135.37	153.45	4295.02
4	110.50	42.95	178.32	153.45	4184.52
5	111.60	41.85	220.17	153.45	4072.92
6	112.72	40.73	260.90	153.45	3960.20
7	113.85	39.60	300.50	153.45	3846.35
8	114.99	38.46	338.96	153.45	3731.36
9	116.14	37.31	376.28	153.45	3615.22
10	117.30	36.15	412.43	153.45	3497.93
11	118.47	34.98	447.41	153.45	3379.46
12	119.66	33.79	481.20	153.45	3259.80
13	120.85	32.60	513.80	153.45	3138.95

Loan Analysis (b)

N	Principal Payment	Interest Payment	Cumulative Interest	Total Payment	Remaining Balance
23	133.50	19.95	771.28	153.45	1861.92
24	134.83	18.62	789.90	153.45	1727.09
25	136.18	17.27	807.17	153.45	1590.91
26	137.54	15.91	823.08	153.45	1453.37
27	138.92	14.53	837.61	153.45	1314.46
28	140.31	13.14	850.75	153.45	1174.15
29	141.71	11.74	862.50	153.45	1032.44
30	143.13	10.32	872.82	153.45	889.32
31	144.56	8.89	881.71	153.45	744.76
32	146.00	7.45	889.16	153.45	598.76
33	147.46	5.99	895.15	153.45	451.29
34	148.94	4.51	899.66	153.45	302.36
35	150.43	3.02	902.68	153.45	151.93
36	151.93	1.52	904.20	153.45	0.00
	Total Interest		904.20		

Figure 3.32 ■ Monthly balances and terminal loan balance

interest and the remaining loan balance after each payment (Fig. 3.32). It also calculates the total interest paid over the life of the loan. Several data pages are needed to show the complete loan repayment schedule. You can scroll these data pages by using the scroll bar provided by the Windows operating system.

Summary

- Interest is most frequently quoted by financial institutions as an **APR** or **annual percentage rate**. However, compounding occurs more often than annually, and the APR does not account for the effect of this more frequent compounding. This situation leads to the distinction between nominal and effective interest:

  Nominal interest is a stated rate of interest for a given period (usually a year).

  Effective interest is the actual rate of interest, which accounts for the interest amount accumulated over a given period. The effective rate is related to the APR by the following equation:

  $$i = (1 + r/M)^M - 1, \qquad (3.21)$$

  where r = APR, M = number of compounding periods, and i = effective interest rate.

- In any equivalence problem, the interest rate to use is the **effective interest rate per payment period**:

  $$i = (1 + r/M)^C - 1, \qquad (3.22)$$

  where C = number of compounding periods per payment period. Figure 3.12 outlines the possible relationships between compounding and payment periods and indicates what version of the effective interest formula to use.

- The equations for effective interest of **continuous compounding** is

  $$i = e^{r/K} - 1, \qquad (3.23)$$

  where K is the number of payment periods per year.

  The difference in accumulated interest between continuous compounding and very frequent compounding ($M > 50$) is minimal.

- Cash flows, as well as compounding, can be continuous. Table 3.2 shows the interest factors to use for **continuous cash flows** with continuous compounding.

- Nominal (and hence effective) interest rates may fluctuate over the life of a cash flow series. Some forms of home mortgages and bond yields are typical examples.

Problems

3.1 A loan company offers money at 2% per month compounded monthly.

  (a) What is the nominal interest rate?
  (b) What is the effective annual interest rate?
  (c) How many years will it take an investment to triple itself if interest is compounded monthly?

  (d) How many years will it take an investment to triple itself if the nominal rate is compounded continuously?

3.2 A department store has offered you a credit card that charges interest at 1.5% per month compounded monthly. What is the nominal interest (annual percentage) rate for this credit card? What is the the effective annual interest rate?

3.3 The Toronto Dominion Bank advertised the following information: interest 7.55%—effective annual yield 7.842%. There is no mention of the compounding period in the advertisement. Can you figure out the compounding scheme used by the bank?

3.4 War Eagle Financial Sources, which makes small loans to college students, offers to lend $400 with the borrower required to pay $26.61 at the end of each week for 16 weeks. Find the interest rate per week. What is the nominal interest rate per year? What is the effective interest rate per year?

3.5 A financial institution is willing to lend you $40 if you will repay $45 at the end of one week.

(a) What is the nominal interest rate?
(b) What is the effective annual interest rate?

3.6 Easy-Loan Company, which engages in making small loans to college students, offers to lend $200 with the borrower required to pay $10.75 at the end of each week for 25 weeks to pay off the debt. Find the approximate interest rate per week. What is the nominal rate per year? What is the effective annual interest rate?

3.7 As a typical middle-class consumer, you are making monthly payments on your home mortgage (9% annual interest rate), car loan (12%), home improvement loan (14%), and past-due charge accounts (18%). Immediately after getting a $100 monthly raise, your friendly mutual broker tries to sell you some investment funds, with a guaranteed return of 10% per year. Assuming that your only other investment alternative is savings accounts, should you buy?

3.8 What will be the amount accumulated by each of these present investments?

(a) $675 in 11 years at 4% compounded semiannually
(b) $3500 in 31 years at 7% compounded quarterly
(c) $11,000 in 7 years at 8% compounded monthly

3.9 How many years will it take an investment to triple itself if the interest rate is 8% compounded

(a) quarterly
(b) monthly
(c) continuously

3.10 A series of equal quarterly payments of $5000 for 25 years is equivalent to what present amount at an interest rate of 8% compounded

(a) quarterly
(b) monthly
(c) continuously

3.11 What is the future worth of an equal payment series of $1000 each for 5 years if the interest rate is 9% compounded continuously?

3.12 Suppose that $1000 is placed in a bank account at the end of each quarter over the next 10 years. What is the future worth at the end of 10 years when the interest rate is 8% compounded

(a) quarterly
(b) monthly
(c) continuously

3.13 A series of equal quarterly deposits of $1000 extends over a period of 3 years. What is the future worth of this quarterly deposit series at 12% compounded monthly?

(a) $F = 4(\$1000) \, (F/A, 12\%, 3)$
(b) $F = \$1000(F/A, 3\%, 12)$

(c) $F = \$1000(F/A, 1\%, 12)$

(d) $F = \$1000(F/A, 3.03\%, 12)$

Which of the above equations is correct?

3.14 If the interest rate is 8.8% compounded continuously, the required quarterly payment to repay a loan of $10,000 in 3 years is ( ).

3.15 What is the future worth of a series of equal monthly payments of $1000 that extends over a period of 5 years at 12% interest compounded

(a) quarterly

(b) monthly

(c) continuously

3.16 What will be the required quarterly payment to repay a loan of $10,000 in 5 years if the interest rate is 6% compounded continuously?

3.17 A series of equal quarterly payments of $1000 extends over a period of 5 years. What is the present worth of this quarterly time series at 8% interest compounded continuously?

3.18 Suppose you deposit $500 at the end of each quarter for 5 years at an interest rate of 8% compounded continuously. What equal end-of-year deposit over 5 years would accumulate the same amount at the end of 5 years under the same interest compounding?

(a) $A = [\$500(F/A, 2\%, 20)](A/F, 8\%, 5)$

(b) $A = \$500(F/A, e^{0.02} - 1, 4)$

(c) $A = \$500(F/A, e^{0.02} - 1, 20)$
   $\times (A/F, 8\%, 5)$

(d) None of the above

Which of the above equations is correct?

3.19 A series of equal quarterly payments of $1000 for 25 years is equivalent to what future amount at the end of 10 years at an interest rate of 8% compounded continuously?

3.20 What is the future worth of the following series of payments?

(a) $1000 at the end of each 6-month period for 10 years at 8% compounded semiannually

(b) $500 at the end of each quarter for 20 years at 10% compounded quarterly

(c) $300 at the end of each month for 15 years at 12% compounded monthly

3.21 What equal series of payments must be paid into a sinking fund to accumulate the following amount?

(a) $20,000 in 15 years at 8% compounded semiannually when payments are semiannual

(b) $5000 in 20 years at 9% compounded quarterly when payments are quarterly

(c) $10,000 in 5 years at 8% compounded monthly when payments are monthly

3.22 James Hogan is purchasing a $12,000 automobile, which is to be paid for in 36 monthly installments of $398.56. What effective annual interest is being paid for this financing arrangement?

3.23 A loan of $6000 is to be financed to assist in buying an automobile. Based upon monthly compounding for 30 months, the end-of-the-month equal payment is quoted as $217.19. What nominal interest rate in percentage is being charged?

3.24 You are purchasing a $4000 used automobile, which is to be paid for in 36 monthly installments of $150. What nominal interest rate is being paid on this financing arrangement?

3.25 Suppose a young newlywed couple are planning to buy a home 2 years from now. To save the down payment required at the time of purchasing a home worth

$180,000 (let's assume 10% of the sales price, $18,000), they have decided to set aside some money from their salaries at the end of each month. If they can earn a 6% interest (compounded monthly) on their savings, determine the equal amount this couple must deposit each month until they reach a point to buy the home.

3.26 What is the present worth of the following series of payments?

(a) $500 at the end of each 6-month period for 10 years at 10% compounded semiannually

(b) $1000 at the end of each quarter for 5 years at 8% compounded quarterly

(c) $2000 at the end of each month for 8 years at 8% compounded monthly

3.27 What is the amount of the quarterly deposits A such that you will be able to withdraw the amounts shown in Fig. 3.33, if the interest rate is 8% compounded quarterly?

3.28 Georgi Rostov deposits $4000 in a savings account that pays 6% interest compounded quarterly. Three years later he deposits $4500. Two years after the $4500 deposit, he makes another deposit in the amount of $2500. Four years after the $2500 deposit, half of the accumulated funds is transferred to a fund that pays 7% interest compounded

monthly. How much money will be in each account 6 years after the transfer?

3.29 A man is planning to retire in 30 years. He wishes to deposit a regular amount every 3 months until he retires so that beginning 1 year following his retirement he will receive annual payments of $30,000 for the next 10 years. How much must he deposit if the interest rate is 8% compounded quarterly?

3.30 A building is priced at $75,000. If a down payment of $25,000 is made and a payment of $500 every month thereafter is required, how many years will it take to pay for the building? Interest is charged at a rate of 9% compounded monthly.

3.31 You are considering buying a new car worth $15,000. You can finance the car by either withdrawing cash from your savings accounts which earns 8% interest or borrowing $15,000 from your dealer for 4 years at 11%. You could earn interest of $5635 from your savings account for 4 years if you left the money in the account. If you borrow $15,000 from your dealer, you only pay $3609 in interest over 4 years. *So it makes sense to borrow for your new car and keep your cash in your savings account.* Do you agree or disagree with the statement above? Justify your reasoning with numerical calculation.

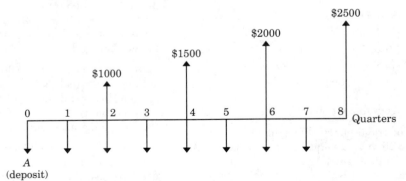

Figure 3.33 ■ A cash flow diagram (Problem 3.27)

3.32 The *Engineering Economist* (a professional journal) offers 3 types of subscriptions payable in advance: 1 year at $14, 2 years at $27, and 3 years at $38.50. If money can earn 6% interest compounded quarterly, which subscription should you take? (Assume that you plan to subscribe the journal over the next 3 years.)

3.33 A couple is planning to finance their 3-year-old child's college education. Money can be deposited at 8% compounded quarterly. What quarterly deposit must be made from the child's 3rd birthday to his/her 18th birthday to provide $20,000 on each birthday from the 18th to the 21st? (Note that the last deposit is made on the date of the first withdrawal.)

3.34 Sam Salvetti is planning to retire in 10 years. Money can be deposited by 8% compounded quarterly. What quarterly deposit must be made at the end of each quarter until he retires so that he can make a withdrawal of $15,000 semiannually over 5 years after his retirement? Assume that his first withdrawal occurs at the end of 6 months after his retirement.

3.35 Emily Lacy received $200,000 from an insurance company after her husband's death. Emily wants to deposit this amount in a savings account which earns interest at a rate of 8% compounded monthly. Then, she would like to make 60 equal monthly withdrawals over the 5-year deposit period, such that, when she makes the last withdrawal, the savings accounts will have a balance of zero. How much can she withdraw each month?

3.36 Anita Tahani, who owns a travel agency, bought an old house to use as her business office. She found that the ceiling was poorly insulated and the heat loss could be cut significantly if 6 inches of foam insulation were installed. She estimated that with the insulation she could cut the heating bill by $20 per month and the air conditioning cost by $15 per month. Assuming that the summer season is 3 months (June, July, August) of the year and that the winter season is another 3 months (December, January, and February) of the year, how much can she spend on insulation if she expects to keep the property for 3 years? Assume that there would be neither heating nor air conditioning during the fall and spring seasons. If she decides to install the insulation, it will be done at the beginning of May. Anita's interest rate is 1% compounded monthly.

3.37 A local newspaper is offering the following arrangement. You may pay for 11 months' subscription in a lump sum now and make no other payments until the following year, or you may pay the regular amount at the end of each month for the next 12 months. If the interest rate is 9% compounded monthly, which option is the better choice?[2]

3.38 Income from a project is projected to decline at a constant rate from an initial value of $500,000 at time 0 to a final value of $40,000 at the end of year 3. If interest is compounded continuously at a nominal annual rate 9%, determine the present value of this continuous cash flow.

3.39 A sum of $100,000 will be received uniformly over a 5-year period beginning 2 years from today. What is the present value of this deferred funds flow if interest is compounded continuously at a nominal rate of 10%?

3.40 A small chemical company, a producer of an epoxy resin, expects its production volume to decay exponentially according to the relationship

$$\nu_t = 3e^{-0.2t},$$

where ν_t is the production rate at time t. Simultaneously, the unit price is

---

From *Engineering Economy,* 7th Ed., Problem 5.41, J. Thuesen and W. J. Fabrycky, Prentice-Hall, 1989.

Problems **185**

expected to increase linearly over time at the rate of

$$u_t = \$50(1 + 0.08t).$$

What is the expression for the present worth of sales revenues from $t = 0$ to $t = 20$ at 15% interest compounded continuously?

3.41 Consider the cash flow diagram in Fig. 3.34 that represents three different interest rates applicable over the 5-year time span.

(a) Calculate the equivalent amount P at the present.
(b) Calculate the single-payment equivalent to F at $n = 5$.
(c) Calculate the equal-payment-series cash flow A that runs from $n = 1$ to $n = 5$.

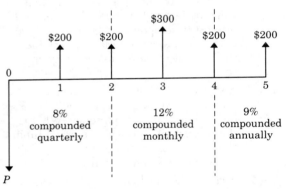

Figure 3.34 ■ Cash flow diagram (Problem 3.41)

3.42 Consider the cash flow transactions depicted in Fig. 3.35, with the changing interest rates specified.

(a) What is the equivalent present worth? (In other words, how much do you have to deposit now so that you can withdraw \$300 at the end of

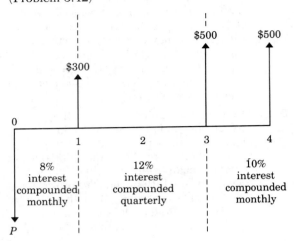

Figure 3.35 ■ Cash flow diagram (Problem 3.42)

year 1, \$500 at the end of year 3, and \$500 at the end of year 4?)

(b) What is the single effective annual interest rate over 4 years?

3.43 Compute the present worth for the cash flows in Fig. 3.36 with different interest rates.

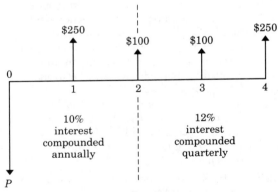

Figure 3.36 ■ Cash flow diagram (Problem 3.43)

3.44 An automobile loan of \$12,000 at a nominal rate of 9% compounded monthly for 60 months requires equal end-of-month payments of \$249.10.

Complete the following table for the first 6 payments, as you would expect a bank to calculate the values:

End of Month (n)	Interest Payment	Repayment of Principal	Remaining Loan Balance
1			$11,840.90
2			
3		$161.50	
4	$86.39		
5	$85.17		
6			$11,027.32

3.45 Mr. Smith wants to buy a new car which costs $12,500. He will make a down payment in the amount of $2500. He would like to borrow the remainder from a bank at an interest rate of 9% compounded monthly. He agrees to pay the loan payment monthly for a period of 2 years. Select the correct answer for the following questions.

(a) What is the amount of the monthly payment (A)?

$A = \$10{,}000(A/P, 0.75\%, 24)$

$A = \$10{,}000(A/P, 9\%, 2)/12$

$A = \$10{,}000(A/F, 0.75\%, 24)$

$A = \$12{,}500(A/F, 9\%, 2)/12$

(b) Mr. Smith has made 12 payments and wants to figure out the remaining balance right after 12th payment. What is the remaining balance?

$B_{12} = 12A$

$B_{12} = A(P/A, 9\%, 1)/12$

$B_{12} = A(P/A, 0.75\%, 12)$

$B_{12} = 10{,}000 - 12A$

3.46 Talhi Hafid is considering the purchase of a used automobile. The price including the title and taxes is $6240, but Talhi is able to make a $1240 down payment. The balance, $5000, will be borrowed from his credit union at an interest rate of 9% compounded quarterly. The loan should be paid in 48 equal monthly payments. Compute the monthly payment. What is the total amount of interest he has to pay over the life of the loan?

3.47 Suppose you are in the market for a new car worth $15,000. You are offered to make a $1000 down payment now and to pay the balance in equal end-of-month payments, $389.64, over a 48-month period. Consider the following situations:

(a) Instead of going through the dealer's financing, you want to make a down payment of $1000 and take out an auto loan from a bank at 12% compounded monthly. What would be your monthly payment to pay off the loan in 4 years?

(b) If you are going to accept the dealer's offer, what would be the effective rate of interest per month the dealer charges on your financing?

3.48 Bob Pearson borrowed $10,000 from a bank at an interest rate of 12% compounded monthly. This loan will be repaid in 60 equal monthly installments over 5 years. Right after his 30th payment he desires to pay the remainder of the loan in a single payment. Compute the total amount he must pay.

3.49 David Kapamagian borrowed money from a bank to finance a small fishing boat. The bank's loan terms allowed him to defer payments for 6 months and to make 36 equal end-of-month payments thereafter. The original bank note was for $3000 with an interest rate of 12% compounded monthly. After 16 monthly payments David found himself in a financial bind and went to a loan company for assistance. Fortunately, the loan company has offered to pay his debts in one lump sum if he will pay them $73.69 per month for the next 40 months. What monthly rate of interest is the loan company charging on this transaction?

3.50 You are buying a home for $170,000. If you make a down payment of $30,000 and take out a mortgage on the rest at 12% per annum, what is your monthly payment to retire the mortgage in 15 years?

3.51 With a $150,000 home mortgage loan with a 20-year amortization at 9% APR, compute the total payments on principal and interest over first 5 years of ownership.

3.52 A lender requires that monthly mortgage payments be no more than 25% of gross monthly income with a maximum amortization of 30 years. If you can make only a 10% down payment, what is the minimum monthly income needed to purchase a $180,000 house when the interest rate is 9% per annum?

3.53 If to buy a $150,000 house, you take out a 9% (APR) mortgage for $100,000 and 5 years later you sell the house for $200,000 (after all other selling expenses), what equity would you realize with a 30-year amortization and a 5-year term?

3.54 Just before their 13th monthly payment:

- Family A had a balance of $30,000 on a 9%, 30-year mortgage.
- Family B had a balance of $30,000 on a 9%, 15-year mortgage.
- Family C had a balance of $30,000 on a 9%, 23-year mortgage.

How much interest did each family pay for the 13th payment?

3.55 Jason Wang acquired a mortgage in the amount of $90,000 on May 1, 1990. The mortgage had a term of 5 years, an interest rate of 9% per annum, and an amortization period of 20 years. Jason had selected monthly payments and his first

payment occurred on June 1, 1990.

(a) What was his monthly payment on the mortgage?
(b) He took advantage of the prepayment privileges of the mortgage by making lump sum payments of $8000, $5000, $9000, and $7000 on December 1, 1991, October 1, 1992, January 1, 1993, and April 1, 1994. What was the balance of the mortgage on May 1, 1994?
(c) Starting June 1, 1994, Jason doubled his monthly payment amount because there was no penalty charge for such an increase. What would be the balance of the mortgage at the end of the 5-year term?

3.56 A restaurant is considering purchasing a lot adjacent to its business to provide adequate parking space for its customers. The restaurant needs to borrow $20,000 to secure the lot. A deal has been made between a local bank and the restaurant, so that the restaurant would pay the loan back over a 5-year period with the following payment terms: a 15%, 20%, 25%, 30%, and 35% of the initial loan at the end of first, second, third, fourth, and fifth years, respectively.

(a) What rate of interest is the bank earning from this loan transaction?
(b) What would be the total interest paid by the restaurant over the 5-year period?

3.57 Don Harrison's current salary is $30,000 per year, and he is planning to retire 30 years from now. He anticipates that his annual salary will increase by $1500 each year (first year, $30,000; second year, $31,500; and so forth), and he plans to deposit 10% of his yearly salary into a retirement fund that earns 7% interest compounded quarterly. What will be the amount accumulated at the time of his retirement?

3.58 Katerina Unger wants to purchase a set of furniture worth $5000, and plans to finance the furniture for 2 years from the furniture store. She is told the interest rate is only 1% per month, and her monthly payment is computed as follows:

- Installment period = 24 months
- Interest = $24(0.01)($5000) = 1200
- loan processing fee = $25
- Total amount owed = $5000 + $1200 + $25 = 6225
- Monthly payment = $6225/24 = $259.38 per month

(a) What is the annual effective interest rate i Katerina is paying for her loan transaction? What is the nominal interest (annual percentage rate) for this loan?

(b) Katerina bought the furniture and made 12 such monthly payments. Now she wants to pay off the remaining installments with one lump-sum payment (at the end of 12 months). How much does she owe the furniture store?

3.59 You purchase a piece of furniture worth $5000 on credit through a local furniture store. You are told that your monthly payment will be $146.35 at a 10% add-on interest rate over 48 months, which includes an acquisition fee of $25. After making 15 payments, you have decided to pay off the balance.

(a) Compute the remaining balance based on the conventional amortized loan.

(b) If the store applies the Rule of 78ths, what will the payoff amount be?

3.60 Paula Wu bought a new car for $11,600. A dealer's financing was available through a local bank at an interest rate of 17.5% compounded monthly. The dealer financing required a 10% down payment and 48 equal monthly payments. Because the interest rate was rather high, she checked her credit union for possible financing. The loan officer at the credit union quoted an interest rate of 13.5% for a new car loan and 15.5% for a used car loan. But to be eligible for the loan, Paula had to be a member of the union at least for 6 months. Since she joined the union 2 months ago, she had to wait 4 more months to apply for the loan. Now she decided to go ahead with the dealer's financing, and 4 months later refinanced the balance through the credit union at an interest rate of 15.5%.

(a) Compute the monthly payment to the dealer.

(b) Compute the monthly payment to the union.

(c) What is the total interest payment for each loan transaction?

3.61 A house can be purchased for $85,000, and you have $17,000 cash for a down payment. You are considering the following financing options:

- **Option 1:** Getting a new standard mortgage with a 10% (APR) interest, 30-year amortization and 5-year term.

- **Option 2:** Assuming seller's old mortgage, which has an interest rate of 8.5% (APR), a remaining term of 5 years (original term of 10 years and amortization of 30 years), and a remaining balance of $35,394, you can obtain a second mortgage for the remaining balance ($32,606) at 12% (APR) with a 10-year amortization and a 3-year term.

(a) What is the effective interest rate of the combined mortgage?

(b) Compute the monthly payments for each option over the mortgage life.

(c) Compute the total interest payment for each option.

(d) What homeowner's interest rate makes the two financing options equivalent?

3.62 A loan of $6000 is to be financed over a period of 30 months. The agency quotes a nominal rate of 6% for the first 18 months and a nominal rate of 9% for any remaining unpaid balance after 18 months, with monthly compounding being used. Based on these rates, what equal end-of-the-month payment for 30 months would be required to repay the loan with interest?

3.63 Robert Carré financed his office furniture from a furniture dealer. The dealer's terms allowed him to defer payments for 6 months and to make 36 equal end-of-month payments thereafter. The original note was for $12,000 with interest at 12% compounded monthly. After 26 monthly payments Robert found himself in a financial bind and went to a loan company for assistance. The loan company offered to pay his debts in one lump sum if he would pay them $204 per month for the next 30 months.

(a) Determine the original monthly payment made to the furniture store.
(b) Determine the lump-sum payoff amount the loan company will make.
(c) What monthly rate of interest is the loan company charging on this loan?

3.64 If you borrow $60,000 with a 30-year amortization, 9%(APR) variable rate and the interest rate can be changed every 5 years:

(a) What is the initial monthly payment?
(b) If at the end of 5 years the lender's interest rate is 10% (APR), what will the new monthly payment be?

3.65 The Jimmy Corporation issued a new series of bonds on January 1, 1981. The bonds were sold at par ($1000), have a 12% coupon, and mature in 30 years, on December 31, 2010. Coupon interest payments are made semiannually (on June 30 and December 31).

(a) What was the yield to maturity (YTM) of the bond on January 1, 1981?

(b) What was the price of the bond on January 1, 1989, 8 years later, assuming that the interest rate had fallen to 9%?

(c) On July 1, 1989, the bonds sold for $922.38. What was the YTM at that date? What was the current yield at that date?

3.66 A $1000, 10% semiannual bond is purchased for $1020. If the bond is sold at the end of 3 years and 6 interest payments, what should the selling price be to yield a 10% return on your investment?

3.67 Mr. Gonzalez wishes to sell a bond that has a face value of $2000. The bond bears an interest rate of 8% with bond interests payable semiannually. Four years ago, $1800 was paid for the bond. At least a 9% return (yield) in investment is desired. What must be the minimum selling price?

3.68 Candi Yamaguchi is considering purchasing a 6% bond with a face value of $1000 and interest paid semiannually. She desires to earn 9% annual return on her investment. Assume that the bond will mature to its face value 5 years hence. What is the required purchasing price of the bond?

(a) $P = \$60(P/A, 6\%, 5) + \$1000(P/F, 6\%, 5)$
(b) $P = \$90(P/A, 9\%, 5) + \$1000(P/F, 9\%, 5)$
(c) $P = \$30(P/A, 4.5\%, 10) + \$1000(P/F, 4.5\%, 10)$
(d) $P = \$30(P/A, 3\%, 10) + \$1000(P/F, 3\%, 10)$

Which of the above equations is the correct one?

3.69 Suppose you have the choice of investing in (1) a zero coupon bond that costs $513.60 today, pays nothing during its life, and then pays $1000 after 5 years or (2) a bond that costs $1000 today, pays $113 in interest semiannually, and matures at the end of 5 years. Which bond would provide the higher yield?

3.70 Suppose you were offered a 12-year, 15% coupon, $1000 par value bond at a price of $1298.68. What rate of interest (yield to maturity) would you earn if you bought the bond and held it to maturity? (Semiannual interest)

3.71 The Diversified Products Company has two bond issues outstanding. Both bonds pay $100 semiannual interest plus $1000 at maturity. Bond A has a remaining maturity of 15 years and Bond B a maturity of 1 year. What will be the value of each of these bonds now when the going rate of interest is 9%?

3.72 The AirJet Service Company's bonds have 4 years remaining to maturity. Interest is paid annually, the bonds have a $1000 par value, and the coupon interest rate is 8%.

  (a) What is the yield to maturity at a current market price of $1115?
  (b) Would you pay $930 for one of these bonds if you thought that the market rate of interest was 9.5%?

3.73 Suppose Ford sold an issue of bonds with a 15-year maturity, a $1000 par value, a 12% coupon rate, and semiannual interest payments.

  (a) Two years after the bonds were issued, the going rate of interest on bonds such as these fell to 9%. At what price would the bonds sell?

  (b) Suppose that 2 years after the issue the going interest rate had risen to 13%. At what price would the bonds sell?

  (c) Today, the closing price of this bond is $783.58. What is the current yield?

3.74 Jim Norton, a engineering junior, has received in the mail guaranteed line of credit applications from two different banks. Each bank offers a different annual fee and finance charge.

Terms	Bank A	Bank B
Annual fee	$20	$30
Finance charge	1.55% monthly interest rate	16.5% annual percentage rate

Jim expects his average monthly balance after payment to be $400 and plans to keep the cards for only 24 months. (After graduation he would apply for a new card.) Jim's interest rate (on his savings account) is 6% compounded quarterly.

  (a) Compute the effective annual interest rate for each card.
  (b) Which bank's credit card should Jim choose?

3.75 Ms Kennedy borrowed $4909 from a bank to finance a car at an add-on interest rate of 6.105%. The bank calculated the monthly payments as follows:

  • Contract amount = $4909
  • Contract period = 42 months
  • Add-on interest at 6.105% = $4909(0.06105)(3.5) = $1048.90
  • Acquisition fee = $25
  • Loan charge = $1048.90 + $25 = $1073.90

- Total of payments = $4909 + $1073.90 = $5982.90
- Monthly installment = $5982.90/42 = $142.45

After making the 7th payment, Ms Kennedy wants to pay off the remaining balance. The following is the letter from the bank explaining the net balance Ms Kennedy owes:

Dear Ms Kennedy:

The following is an explanation of how we arrived at the payoff amount on your loan account.

Original note amount	$5982.90
Less: 7 payments made @ $142.45 each	997.15
	$4985.75
Loan charge (interest)	$1073.90
Less: Acquisition fee	25.00
	$1048.90

Rebate factor from Rule of 78ths chart is 0.6589 (Loan ran 8 months on a 42-month term) $1048.90 multiplied by 0.6589 = $691.12 $691.12 represents the unearned interest rebate. Therefore:

Balance	$4985.75
Less unearned and interest rebate	691.12
Payoff amount	$4294.63

If you have any further questions concerning these matters, please contact us.

Sincerely,

S. Govia
Vice President

(a) Compute the effective annual interest rate for this loan.

(b) Compute the annual percentage rate (APR) for this loan.

(c) Show how would you derive the rebate factor (0.6589).

(d) Verify the payoff amount using the Rule of 78ths formula.

(e) Compute the payoff amount using the interest factor $(P/A, i, N)$.

3.76 The following is the promotional pamphlet prepared by Trust Company:

"Lower your monthly car payments as much as 48%." Now you can buy the car you want and keep the monthly payments as much as 48% lower than they would be if you financed with a conventional auto loan. Trust Company's *Alternative Auto Loan* (AAL)^{SM} makes the difference. It combines the lower monthly payment advantages of leasing with tax and ownership of a conventional loan. And if you have your monthly payment deducted automatically from your Trust Company checking account, you will save $\frac{1}{2}$% on your loan interest rate. Your monthly payments can be spread over 24, 36, or 48 months.

		Monthly Payment	
Amount Financed	Financing Period (Months)	Alternative Auto Loan	Conventional Auto Loan
	24	$249	$477
$10,000	36	211	339
	48	191	270
	24	498	955
$20,000	36	422	678
	48	382	541

The amount of the final payment will be based on the residual value of the car at the end of the loan. Your monthly payments are kept low because you make principal payments on only a portion of the loan and not on the residual value of the car. Interest is computed on the full amount of the loan. At the end of the loan period you may:

1. Make the final payment and keep the car.
2. Sell the car yourself, repay the note (remaining balance), and keep any profit you make.
3. Refinance the car.

4. Return the car to Trust Company in good working condition and pay only a return fee.

So, if you've been wanting a special car but not the high monthly payments that could go with it, consider the *Alternative Auto Loan.* For details, ask at any Trust Company branch.

Note 1: The chart above is based on the following assumptions: Conventional auto loan 13.4% Annual Percentage Rate. *Alternative Auto Loan* 13.4% Annual Percentage Rate.

Note 2: It is assumed that residual value is 50% of sticker price for 24 months; 45% for 36 months. The amount financed is 80% of sticker price.

Note 3: Monthly payments are based on principal payments equal to the depreciation amount on the car and interest in the amount of the loan.

Note 4: The residual value of the automobile is determined by a published residual value guide in effect at the time your Trust Company's *Alternative Auto Loan* is originated.

Note 5: The minimum loan amount is $10,000 (Trust Company will lend up to 80% of the sticker price). The annual household income requirement is $25,000.

Note 6: Trust Company reserves the right of final approval based on the customer's credit history.

(a) Show how the monthly payments were computed for the *Alternative Auto Loan* by the bank.

(b) Suppose that you have decided to finance a new car for 36 months from Trust Company. Assume also that your are interested in ownership of the car (not leasing). If you decided to go with the *Alternative Auto Loan*, you will make the final payment and keep the car at the end of 36 months. Assume that your opportunity cost rate (personal interest rate) is an interest rate of 8% compounded monthly. (You may view this opportunity cost rate as an interest rate at which you can invest your money in some financial instrument such as a savings account.) Compare this alternative option with the conventional option and determine your choice.

3.77 Registered Education Savings Plan (RESP): Parents often look for ways to set aside money for their children to help supplement the costs of a college or university education. One way of achieving this goal is to invest in an RESP. Although the funds invested in an RESP are not tax-deductible, the interests accumulated in such funds are tax deferred until withdrawn, at which time they are taxable to the child attending a post-secondary school. The following are some additional rules governing an RESP:

- Maximum annual contributions: $1500 per child.
- Maximum life time contributions: $31,500 per child.
- Maximum age of child to withdraw funds: 21 years.
- Maximum number of years to receive full amount of funds: 4 years.
- Maximum age to change beneficiary of the plan: 13 or 16 (varies by plan).

Based on past statistics, the cost of post secondary education in Canada will reach at least $25,000 per year by year 2008, and it will continue to rise at a rate of at least 7% per year. If the funds in the RESP that a parent is considering will earn an interest rate of 15% per year, what is the annual amount that he or she has to invest in the RESP in order to accumulate enough funds for a child born in 1990? What if the interest rate earned will be 12% per year? Ignore the tax payments at time of withdrawals.

3.78 Canadian mortgages are quite standardized; they are amortized loans quoted with nominal interest rates based on semiannual compounding. Most banks and trust companies offer competitive interest rates. However, as a consumer, you have to decide the length of the term of your mortgage and what features you would like your mortgage to have. The special features include prepayment privileges, how penalty charges are calculated, assumability, and portability. The interest rates are different for different terms. The following lists the residential mortgage rates of a few lending institutions as of March 30, 1994.

Based on the data below, assume you are considering a mortgage in the amount of $100,000. Calculate the monthly payment if you have selected the Toronto Dominion Bank, an amortization of 24 years, and a term of 2 years. Also answer the following questions:

1. What factors affect your choice of a longer term or a shorter term?
2. What factors affect the choice of a closed or an open mortgage?
3. What factors affect your choice of a lender?

3.79 Researchers at Lotus Engineering in Britain have developed a way to cut interior (car) noise with microphones and microchips, a technique dubbed "antinoise." As shown in Fig. 3.37 they mount microphones, which monitor noise, in the car's passenger compartment and feed the information to a small computer. The computer analyzes the sound and produces a signal, through the car's stereo speakers, that is identical in pitch and volume but reversed, so that the peaks of its sound waves coincide with the troughs of the original noise, effectively canceling most of what is known as the resonant boom.

Suppose that it will cost Lotus $5 million to commercialize the system and that it is considering securing these funds from a bank. The entire sum can be financed at 12% annual interest over 10 years. Lotus pays only the interest due at the end of each year and the entire principal at the end of the life of the loan. Alternatively, Lotus may issue corporate bonds to finance the investment. The lowest rate that Lotus can get on the financial market is 9.5% compounded semiannually. Lotus thinks this rate is competitive with other bonds issued by other companies in the same category. It will cost $300,000 to issue

Lenders	6 months	1 year open	1 year closed	2 years	3 years	4 years	5 years	7 years	10 years
Bank of Montreal	6.5	7.5	7.0	7.75	8.25	8.625	8.95	9.25	9.5
Bank of Nova Scotia	6.25	7.5	7.0	7.75	8.25	8.625	8.95	9.25	
CIBC	6.5	7.5	7.0	7.75	8.25	8.625	8.95	9.25	
Hong Kong Bank of Canada	6.5	7.5	7.0	7.75	8.25	8.625	8.95	9.25	
Toronto Dominion	6.5	7.5	7.0	7.75	8.25	8.625	8.95		9.5
Investors Group	6.25	7.75	6.75	7.5	8.0	8.375	8.75	9.25	9.5
Capital City Savings	6.5	7.5	7.0	7.75	8.25	8.625	8.95		
Royal LePage Mortgage Services	6.5	7.75	7.0	7.75	8.25	8.625	8.75	9.25	9.5

Figure 3.37 ■ Quieting a car electronically (Problem 3.79)

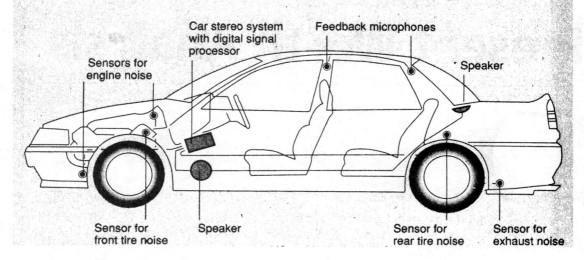

Quieting a Car Electronically

In Lotus Engineering's adaptive noise control system, microphones and sensors mounted throughout the car monitor noise. The impulses from the microphones are fed into a specialized computer chip, known as a digital signal processor. The processor analyzes the content of the noise and produces an acoustic mirror image of it, or 'anti-noise,' which is played back through the car's stereo speakers. The peaks of the anti-noise's soundwaves match the troughs of the original noise, effectively nullifying it.

Car stereo system with digital signal processor

Feedback microphones

Sensors for engine noise

Speaker

Sensor for front tire noise

Speaker

Sensor for rear tire noise

Sensor for exhaust noise

the bond. To promote the bond for sale, Lotus must discount the market price of the bond. This means that Lotus must raise $5,570,000 to net the $5,000,000 investment. (The extra amount of $570,000 is to cover the issuing costs and bond discounting.) The bond's interest will be payable annually and it will mature at the end of 10 years. If the bond is further discounted, Lotus can also set a lower interest rate on the bond. It has been estimated that whenever the bond is discounted a full percentage point, Lotus will be able to lower the bond interest rate by 0.2%. What would be the minimum selling price so that this bond financing is economically preferred over the conventional bank financing?

CHAPTER 4

Present Worth Analysis

O

ver the past 30 years, a British Columbia entrepreneur amassed a small fortune developing real estate. He sold more than 700 hectares of timber and farmland to raise $800,000 to build a small hydroelectric plant, known as Edgemont Hydro, which has been a decade in the making. The design for the plant, which the entrepreneur developed using his military training as a civil engineer, is relatively simple. A 7-metre-deep canal, blasted out of solid rock, just above the higher of two dams on his property, carries water 350 metres along the river to a "trash rack," where leaves and other debris are caught. A 2-metre-wide pipeline capable of holding 1.5 million kilograms of liquid then funnels the water into the powerhouse at 3.5 metres per second, creating the thrust to run the turbines.

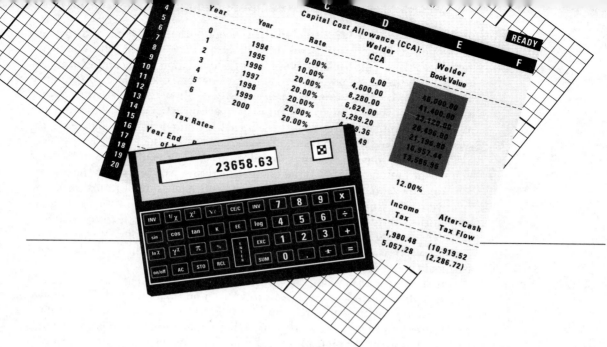

Government regulations encourage private power development, and any electricity generated must be purchased by the power company that owns the provincial power grid. The plant can generate 6 million kilowatt hours per year. Suppose that after paying income taxes and operating expenses, the annual income from the hydroelectric plant will be $120,000. With normal maintenance, the plant is expected to provide service for at least 50 years. Was the $800,000 investment a wise one? How long does this entrepreneur have to wait to recover his initial investment, and will he ever make a profit?

In Chapters 2 and 3, we presented the concept of the time value of money and developed techniques for establishing cash flow equivalence with compound interest factors. This background provides a foundation for accepting or rejecting a capital investment—the economic evaluation of a project's desirability. Forthcoming coverage of investment worth will allow us to go a step beyond accepting or rejecting an investment to comparisons of alternative investments. We will determine how to compare alternatives on equal basis and select the wisest alternative from an economic standpoint.

The three common measures based on cash flow equivalence are (1) the equivalent present worth, (2) the equivalent future worth, and (3) the equivalent annual worth. The present worth represents a measure of future cash flow relative to the time point "now" with provisions that account for earning opportunities. Future worth is a measure of the cash flow at some future planning horizon, considering the earning opportunities of the intermediate cash flows. Annual worth is a measure of the cash flow in terms of the equivalent equal payments on an annual basis.

Our treatment of measures of investment worth is divided into three chapters. Chapter 4 introduces two measures based on the basic cash flow equivalence techniques: present worth and future worth analysis. Because the annual worth approach has many useful engineering applications in estimating the unit cost, Chapter 5 is devoted to annual cash flow analysis. Chapter 6 presents measures of investment worth based on yield, known as rate of return analysis.

We must also recognize that an important part of the capital budgeting process is the estimation of relevant cash flows. For all examples in this chapter and those in Chapters 5 and 6, net cash flows can be viewed as before-tax values or as after-tax values for which the tax effects have been precalculated. Since some organizations (e.g., governments, Crown corporations, nonprofit associations) are not taxable, the before-tax situation can be a valid base for that type of economic evaluation. The procedures for determining after-tax net cash flows in taxable situations are developed in Chapter 9.

■ 4.1 Describing Project Cash Flows

In Section 1.3, we described many engineering economic decision problems without providing suggestions on how actually to solve them. What do Examples 1.1 through 1.8 have in common? Note that all these problems involve two dissimilar types of amounts. First, there is the investment, which is usually made in a lump sum at the beginning of the project. Although not literally made "today," it is made at a specific point in time that for analytical purposes is called today, or time 0. Second, there is a stream of cash benefits that are expected to result from this investment over a period of future years.

Such an investment made in a fixed asset is similar to that made by a bank when it lends money. The essential characteristic of both transactions is that funds are committed today in the expectation of earning a return in the future.

In the case of the bank loan, the future return takes the form of interest plus repayment of the principal. This is known as the loan cash flows. In the case of the fixed asset, the future return takes the form of profits generated by productive use of the asset. The representation of these future earnings along with the capital expenditures and annual expenses (such as wages, raw materials, operating costs, maintenance costs, and income taxes) is the **project cash flows.** Example 4.1 illustrates a typical procedure for obtaining a project's cash flows.

Example 4.1   Identifying project cash flows

Business at your design engineering firm has been brisk. To keep up with the increasing workload, you are considering the purchase of a new CAD/CAM system costing $300,000, which would provide 5000 hours of productive time per year. Your firm puts a lot of effort into drawing new product designs. At present, this is all done manually by design engineers. If you purchase the system, 40% of its productive time will be devoted to drawing (CAD) and the remainder to CAM. While drawing, the system is expected to out-produce design engineers by a factor of 3:1. You estimate that the annual out-of-pocket cost of maintaining the CAD/CAM system will be $175,000, including any tax effects. The expected useful life of the system is 8 years, after which the equipment will have no residual value. As an alternative, you could hire more design engineers. Each normally works 2000 hours per year, and 60% of this time is productive. The total cost for a design engineer is $25 per hour. Identify the net cash flows (benefits and costs) associated with the drawing activities if the CAD/CAM system is purchased instead of hiring more design engineers.

Discussion: As in real-world engineering economic analysis, this problem contains a great deal of data from which we must extract and interpret the critical cash flows. You would be wise to begin organizing your solution with a table consisting of the following categories.

Year (n)	Cash Inflows (Benefits)	Cash Outflows (Costs)	Net Cash Flows
0			
1			
⋮			
8			

Solution

Given: Cost and benefit information as stated above

Find: Net cash flow in each year over the life of the new system

Although the figures could be interpreted differently, let's allocate all of the purchase price and annual costs of the system to the CAD activity. (You could logically argue that only 40% of these costs should be included.) At time 0, the only cash flow is a cost of $300,000. The gross benefits are the avoided costs of the additional design engineers. The system would provide the equivalent of $5000 \times 40\% \times 3 = 6000$ productive hours of design engineers' time, or $6000/60\% = 10,000$ hours of paid time per year. The avoided cost is 10,000 hours per year $\times$ $25 per hour = $250,000 per year. The net benefits in each of the 8 years are the gross benefits less the maintenance costs ($250,000 − $175,000) = $75,000 per year. Now we are ready to summarize a cash flow table as follows:

Year	Cash Inflows	Cash Outflows	Net Cash Flows
0	$ 0	$300,000	−$300,000
1	250,000	175,000	75,000
2	250,000	175,000	75,000
⋮	⋮	⋮	⋮
8	250,000	175,000	75,000

Comments: *In this example, we considered only the benefits associated with the CAD portion of the system. We could quantify some benefits attributable to CAM as well. Suppose that the CAD benefits alone justify the acquisition. Then, it is obvious that had other benefits such as rapid response to customer demand and improvement in product quality been considered as well, acquisition would have been even more clearly justified.*

In Example 4.1, if the company purchases the CAD/CAM system for $300,000 now, it expects an annual savings of $75,000 for 8 years. (Note that these savings occur in discrete lumps at the ends of years.)

Year (n)	Net Cash Flow (A_n)
0	−$300,000
1	75,000
2	75,000
3	75,000
4	75,000
5	75,000
6	75,000
7	75,000
8	75,000

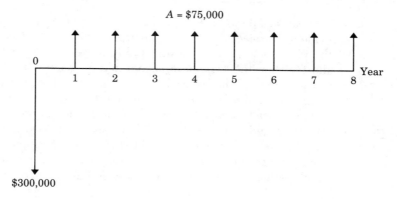

Figure 4.1 ■ Cash flow diagram for a CAD project (Example 4.1)

$A = \$75,000$

0

1 2 3 4 5 6 7 8   Year

$300,000

We can draw a cash flow diagram of this situation (Fig. 4.1). Assuming these cost savings and cash flow estimates are correct, should management give the go-ahead for installation of the system? If the management has decided not to purchase the CAD/CAM system, what do they do with the $300,000 (assuming they have it in the first place)? The company could buy $300,000 of Canadian government Treasury Bills. Or it could invest the amount in other cost saving projects. How would the company compare cash flows that differ both in timing and amount for the alternatives it is considering? This is an extremely important question because virtually every engineering investment decision involves a comparison of alternatives. These are the types of questions this chapter is designed to help you answer.

■ 4.2 Initial Project Screening Method

Before studying the present worth analysis, we will review a screening method commonly used in evaluating capital investments. One of the primary concerns of most businesspeople is whether and when the money invested in a project can be recovered. The **payback method** screens projects on the basis of how long it takes for net receipts to equal investment outlays.

A common standard used in determining whether or not to pursue a project is that no project may be considered unless its payback period is shorter than some specified period of time. (The choice of this time limit is largely determined by the management policy. For example, high-tech firms such as computer manufacturers would set a short time limit for any new investment because their products rapidly become obsolete.) If the payback period is within the acceptable range, a formal project evaluation (such as the present worth analysis) may begin. It is important to remember that payback screening is *not* an end itself, but rather a method of screening out certain unacceptable investment alternatives before progressing to an analysis of potentially acceptable ones.

4.2.1 Payback Period

The payback period is the number of years required to recover the investment made in a project. If the company makes investment decisions *solely* based on the payback period, it considers only those projects with a payback period shorter than the maximum acceptable payback period. (However, due to shortcomings of the payback screening method, which we will discuss, it is rarely used as the *only* decision criterion.)

What does the payback period tell us? One consequence of insisting that each proposed investment has a short payback period is that investors can assure themselves of being restored to their initial position within a short span of time. By restoring their initial position, investors can take advantage of additional, perhaps better, investment possibilities that may come along.

Example 4.2 Payback period—uniform annual benefit

Consider the cash flows given in Example 4.1. Determine the payback period for this CAD/CAM project.

Solution

Given: Initial cost = $300,000, annual net benefits = $75,000
Find: Payback period

Given a uniform stream of receipts, we can easily calculate the payback period by dividing the initial cash outlay by the annual receipt.

$$\text{Payback period} = \frac{\text{Initial cost}}{\text{Uniform annual benefit}} = \frac{\$300,000}{\$75,000}$$

$$= 4 \text{ years}$$

If the company's policy is to consider only projects with a payback period of 5 years or less, this CAD/CAM project passes the initial screening.

In Example 4.2, dividing the initial payment by annual receipts to determine payback period is a simplification we can do because annual receipts are uniform. Wherever the expected cash flows vary from year to year, however, the payback period must be determined by adding the expected proceeds for each year until the sum is equal to or greater than zero. The significance of this procedure can be easily explained. The cumulative cash flow equals zero at the point that cash inflows exactly match or pay back the cash outflows; thus, the project has reached the payback point. Similarly, if the cumulative cash flows *exceed* zero, the cash inflows exceed the cash outflows, and the project has begun to generate a profit, thus exceeding its payback point. To illustrate, consider Example 4.3.

Example 4.3   Payback period—uneven cash flow series

A company plans to invest funds to purchase a new machine. The projected after-tax cash flows are as follows:

Period	Cash Flow	Cumulative Cash Flow
0	−$1000	−$1000
1	−500	−1500
2	500	−1000
3	700	−300
4	1000	+700
5	1500	+2200
6	500	+2700

Solution

Given: Cash flow as in Fig. 4.2

Find: Payback period

As we see from the cumulative cash flow in Fig.4.2, the total investment is recovered during year 4. If the firm's stated maximum payback period is 3 years, the project would be rejected.

Comments: *In Example 4.2, we assumed that cash flows only occur in discrete lumps at ends of years. If cash flows occur continuously throughout the year, the payback period calculation needs adjustment. A negative balance of $300 remains at the start of year 4. If the $1000 is expected to be received as a more or less continuous flow during the 4th year, the total investment will be recovered at three-tenths ($300/$1000) of the way through the 4th year. In this situation, the payback period is thus 3.3 years.*

4.2.2 Benefits and Flaws of Payback Screening

The simplicity of the payback method is one of its most appealing qualities. Initial project screening by the payback method reduces the information search by focusing on that time at which the firm expects to recover the initial investment. It may also eliminate some alternatives, thus reducing the firm's further analysis efforts.

Too Much Emphasis on Project Liquidity

The principal objection to the payback method is its failure to measure profitability. Simply measuring how long it will take to recover the initial investment outlay contributes little to gauging the earning power of a project. Because

Figure 4.2 ■ Illustration of conventional payback period (Example 4.3)

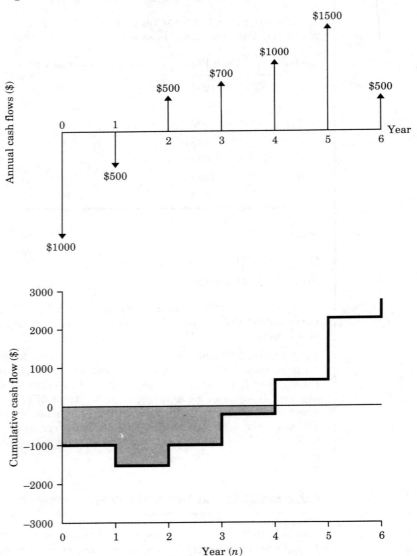

payback period analysis ignores differences in the timing of cash flows, it fails to recognize the difference between the present and future value of money. Because it also ignores all proceeds after the payback period, it does not allow for the possible advantages of a project with a longer economic life.

By way of illustration, consider two investment projects. Each requires an initial investment outlay of $90,000. Project 1, with expected annual cash proceeds of $30,000 for the first three years for the duration of its 6-year economic life, has a payback period of 3 years. Project 2 is expected to generate annual cash proceeds of $25,000 for 6 years; hence, its payback period is 4 years. If the company's maximum payback period is set to 3 years, then project 1 would pass the initial project screening, whereas project 2 would fail even though it is clearly the more profitable investment.

n	Project 1	Project 2
0	−$90,000	−$90,000
1	30,000	25,000
2	30,000	25,000
3	30,000	25,000
4	1,000	25,000
5	1,000	25,000
6	1,000	25,000
	$ 3,000	$60,000

In summary, using payback period as the *sole* criterion may well lead to an undue emphasis on liquidity at the expense of profitability.

Discounted Payback Period

To remedy some shortcomings of the payback period, we may modify the procedure to consider the cost of funds (interest) used to support the project. This modified payback period is often referred to as the *discounted payback period*. In other words, we may define the discounted payback period as the number of years required to recover the investment from *discounted* cash flows.

For the project in Example 4.3, suppose the company requires a rate of return of 15%. To determine the period necessary to recover both the capital investment and the cost of funds required to support the investment, we may construct Table 4.1 showing cash flows and costs of funds to be recovered over

Table 4.1 Payback Period Calculation Considering the Cost of Funds (Example 4.3)

Period	Cash Flow	Cost of Funds (15%)°	Cumulative Cash Flow
0	−$1000	0	−$1000
1	−500	−$1000(0.15) = −150	−1650
2	500	−1650(0.15) = −248	−1398
3	700	−1398(0.15) = −210	−909
4	1000	−908(0.15) = −136	−44
5	1500	−44(0.15) = −7	1449
6	500	1449(0.15) = 217	2166

°Cost of funds = Unrecovered beginning balance × interest rate

Figure 4.3 ■ Illustration of discounted payback period

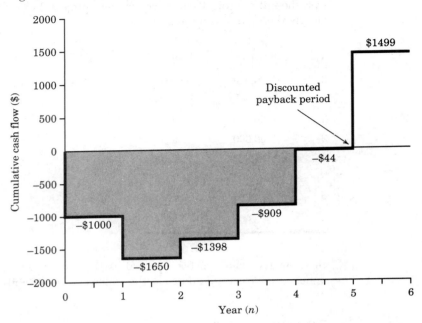

the project life. To illustrate, let's consider the cost of funds during the first year: With $1000 committed at the beginning of the year, the interest would be $150. With no receipt from the project during the first year, the total commitment grows to $1150. During year 2, with $500 additional investment at the beginning of year 2, the total commitment in the project reaches $1650. Then, the cost of funds during the second year would be $248 (1650 × 0.15). But with $500 receipt from the project, the net commitment reduces to $1398. When this process repeats for the remaining project years, we find that the net commitment to project is over during year 5. In other words, assuming that all cash flows are year-end flows, the project must remain in use 5 years in order for the company to cover its cost of capital and also recover the funds invested in the project. Certainly, this modified measure is an improved one, but it does not show the complete picture of the project profitability, either. Figure 4.3 illustrates this relationship.

■ 4.3 Present Worth Analysis

Until 1950s, the payback method was widely used as a means of making investment decisions. As flaws in this method were recognized, however, businesspeople began to search for methods to improve project evaluations. This led to the development of discounted cash flow techniques (DCF), which take into account the time value of money. One of the DCFs is the **net present worth method** (NPW). A capital investment problem is essentially one of determining whether the anticipated cash inflows from a proposed project are sufficiently attractive to invest funds in the project. In developing the NPW criterion, we will use the concept of cash flow equivalence discussed in Chapter 2. As we observed in Chapter 2, the most convenient point at which to calculate the equivalent values is often time 0. Under the NPW criterion, the present worth of all cash inflows is compared against the present worth of all cash outflows that are associated with an investment project. The difference between the present worth of these cash flows, called the net present worth (NPW), determines whether or not the project is an acceptable investment. When two or more projects are under consideration, NPW analysis further allows us to select the best project by comparing their NPW figures.

4.3.1 The Net Present Worth Criterion

We will first summarize the basic procedure for applying the present worth criterion to a typical investment project.

- Determine the interest rate that the firm wishes to earn on its investments. This represents an interest rate at which the firm can always invest the money in its **investment pool**. We often refer to this interest rate as either a **required rate of return** or a **minimum attractive rate of return** (MARR). Usually this selection will be a policy decision by top management. It is possible for the MARR to change over the life of a project, as we have seen in Section 3.3, but for now we will use a single rate of interest in calculating NPW.

- Estimate the service life of the project.
- Estimate the cash inflow for each period over the service life.
- Estimate the cash outflow for each period over the service life.
- Determine the net cash flows (net cash flow = cash inflow − cash outflow).
- Find the present worth of each net cash flow at the MARR. Add up these present worth figures; their sum is defined as the project's NPW:

$$PW(i) = \frac{A_0}{(1 + i)^0} + \frac{A_1}{(1 + i)^1} + \frac{A_2}{(1 + i)^2} + \cdots + \frac{A_N}{(1 + i)^N}$$

$$= \sum_{n=0}^{N} \frac{A_n}{(1 + i)^n}$$

$$= \sum_{n=0}^{N} A_n(P/F, i, n) \tag{4.1}$$

where

$$PW(i) = \text{NPW calculated at } i,$$

$$A_n = \text{net cash flow at end of period } n,$$

$$i = \text{MARR (or cost of capital)},$$

$$N = \text{service life of the project}$$

A_n will be positive if the corresponding period has a net cash inflow or negative if there is a net cash outflow.

- Here, a positive NPW means the equivalent worth of inflows are greater than the equivalent worth of outflows, so project makes a profit. Therefore, if the $PW(i)$ is positive for a single project, the project should be accepted; if negative, it should be rejected. The decision rule is

If $PW(i) > 0$, accept the investment.

If $PW(i) = 0$, remain indifferent.

If $PW(i) < 0$, reject the investment.

Note that the decision rule is for a single project evaluation where you can estimate the revenues as well as costs associated with the project. As you will find in Section 4.5, when you are comparing alternatives with the *same* revenues, you can compare them based on the *cost only*. In this situation (because you are minimizing costs, rather than maximizing profits), you should accept the project that results in the largest, i.e., *least* negative, NPW.

Example 4.4 Net present worth—uniform flows

Consider the investment cash flows associated with the CAD/CAM project in Example 4.1. If the firm's MARR is 15%, compute the NPW of this project. Is this project acceptable?

Solution

Given: Cash flows in Fig. 4.1, MARR = 15% per year
Find: NPW

Since the CAD/CAM project requires an initial investment of $300,000 at $n = 0$ followed by the eight equal annual savings of $75,000, we can easily determine the NPW as follows:

$$PW(15\%)_{\text{outflow}} = \$300,000$$

$$PW(15\%)_{\text{inflow}} = \$75,000(P/A, 15\%, 8)$$

$$= \$336,549.$$

Then, the NPW of the project is

$$PW(15\%) = PW(15\%)_{\text{inflow}} - PW(15\%)_{\text{outflow}}$$

$$= \$336{,}549 - \$300{,}000$$

$$= \$36{,}549$$

or, using Eq. (4.1),

$$PW(15\%) = -\$300{,}000 + \$75{,}000(P/A, 15\%, 8)$$

$$= \$36{,}549.$$

Since $PW(15\%) > 0$, the project would be acceptable.

Now let's consider an example where the investment cash flows are not uniform over the service life of the project.

Example 4.5 Net present worth—uneven flows

Tiger Machine Tool Company is considering the proposed acquisition of a new metal-cutting machine. The required initial investment of $75,000 and the projected cash benefits[1] over the project's 3-year life are as follows.

End of Year	Cash Flow
0	−$75,000
1	24,400
2	27,340
3	55,760

You have been asked by the president of the company to evaluate the economic merit of the acquisition. The firm's MARR is known to be 15%.

Solution

Given: Cash flows as tabulated, MARR = 15% per year
Find: NPW

---

[1] As stated in the beginning of this chapter, we are treating the net cash flows as before tax values or as having their tax effects precalculated. Explaining the process of obtaining the after tax cash flows requires an understanding of income taxes and the role of depreciation, which is developed in Chapters 8 and 9.

Bringing each flow to its equivalent at time 0 we find

$$PW(15\%) = -\$75{,}000 + \$24{,}400(P/F, 15\%, 1) + \$27{,}340(P/F, 15\%, 2)$$

$$+ \$55{,}760(P/F, 15\%, 3)$$

$$= \$3553$$

Since the project results in a surplus of $3553, the project would be acceptable.

In Example 4.5, we just computed the NPW of the project at a fixed interest rate of 15%. When we compute the NPW at varying interest rates, we obtain Table 4.2. Plotting the NPW as a function of interest rate gives Fig. 4.4, the **present worth profile.** (As you will see in Section 4.6, you may use a spreadsheet program such as Lotus to generate Table 4.2 or Fig. 4.4. CASH software also has a built-in command to generate both the present worth table and the present worth profile by specifying the lower and upper bounds of the MARR.)

Figure 4.4 indicates the investment project has a positive NPW if the interest rate is below 17.45% and a negative NPW if the interest rate is above 17.45%. As we will see in Chapter 6, this break-even interest rate is known as the *internal rate of return*. If the firm's MARR is 15%, the project has a NPW of $3553 and so may be accepted. The figure of $3553 measures the immediate gain in present worth to the firm following the acceptance of the project. On the other hand, at $i = 20\%$, $PW(20\%) = -\$3412$, the firm should reject the project. (Note that either accepting or rejecting an investment is influenced by the choice of a MARR. So it is crucial to estimate MARR correctly. We will

Table 4.2 Present Worth Amounts at Varying Interest Rates (Example 4.5)

i(%)	PW(i)	i(%)	PW(i)
0	$32,500	20	−$ 3,412
2	27,743	22	− 5,924
4	23,309	24	− 8,296
6	19,169	26	− 10,539
8	15,296	28	− 12,662
10	11,670	30	− 14,673
12	8,270	32	− 16,580
14	5,077	34	− 18,360
16	2,076	36	− 20,110
17.46°	0	38	− 21,745
18	−750	40	− 23,302

°Break-even interest rate (also known as the rate of return)

Figure 4.4 ■ Present worth profile (Example 4.5)

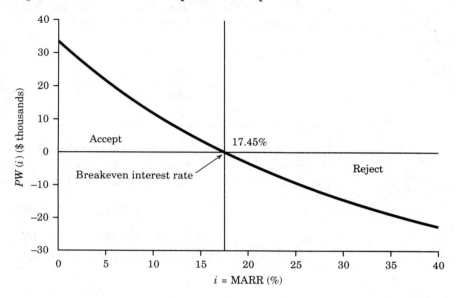

defer this important issue until Chapter 13, so for now we assume that the firm has such an accurate MARR estimate available for use in investment analysis.)

4.3.2 The Meaning of Net Present Worth

In present worth analysis, we assume that all funds in a firm's treasury can be placed in investments yielding a return equal to the MARR. We may view these funds as an **investment pool**. Alternatively, if no funds are available for investment, we assume the firm can borrow them at the MARR from the capital market. In this section, we will examine these two views in explaining the meaning of MARR in NPW calculation.

Investment Pool Concept

An investment pool is equivalent to a firm's treasury where all fund transactions are administered and managed by the firm's comptroller. The firm may withdraw funds from this investment pool for other investment purposes, but if left in the pool, the funds will earn at the MARR. Thus, in investment analysis, net cash flows will be net cash flows relative to this investment pool. To illustrate the investment pool concept, we consider again the project in Example 4.5 that required an investment of $75,000.

If the firm did not invest in the project and left $75,000 in the investment pool for 3 years, these funds would grow

$$\$75,000(F/P, 15\%, 3) = \$114,066.$$

Suppose the company decided instead to invest $75,000 in the project described in Example 4.5. Then the firm would receive a stream of cash inflows during its project life of 3 years in the amounts of

Period (n)	Net Cash Flow (A_n)
1	$24,400
2	27,340
3	55,760

Since the funds that return to the investment pool earn interest at the rate of 15%, it would be of interest to see how much the firm would benefit from this investment. For this alternative, the returns after reinvestment are

$$\$24,400(F/P, 15\%, 2) = \$32,269$$

$$\$27,340(F/P, 15\%, 1) = \$31,441$$

$$\$55,760(F/P, 15\%, 0) = \underline{\$55,760}$$

$$\text{Total} \quad \$119,470$$

These returns total $119,470. The additional cash accumulation at the end of 3 years from investing in the project is

$$\$119,470 - \$114,066 = \$5404.$$

If we compute the equivalent present worth of this net cash surplus at time 0, we obtain

$$\$5404(P/F, 15\%, 3) = \$3553,$$

which is exactly the same as the NPW of the project computed by Eq. (4.1). Clearly, on the basis of its positive NPW, the alternative of purchasing a new machine should be preferred to simply leaving the funds in the investment pool at the MARR. Thus, in PW analysis, any investment is assumed to be returned at the MARR. If there is a surplus at the end of the project, then PW(MARR) > 0. Figure 4.5 illustrates the reinvestment concept related to the firm's investment pool.

Borrowed Funds Concept

Suppose that the firm does not have $75,000 at the outset. In fact, the firm doesn't have to have an investment pool at all. Let's further assume that the firm obtains all its capital by borrowing from a bank at the interest rate of 15%, invests in the project, and uses the proceeds from the investment to pay off the principal and interest on the bank loan. How much is left over for the firm at the end of the project period?

Figure 4.5 ■ The concept of investment pool with a company as a lender and the project as a borrower

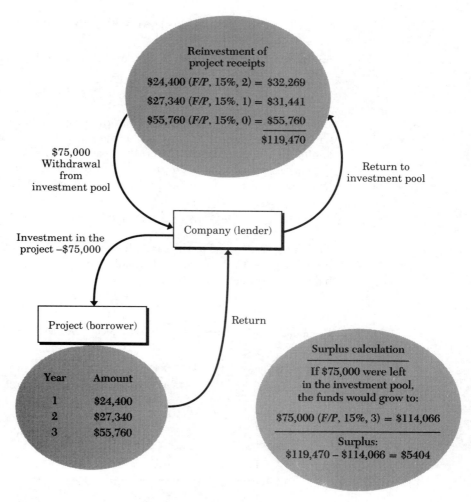

At the end of first year, the interest on the project use of the bank loan would be $75,000(0.15) = $11,250$. Therefore, the total loan balance grows to $75,000(1 + 0.15) = $86,250$. Then, the firm receives $24,400 from the project and applies the entire amount to repay the portion of the loan. This repayment leaves a balance due of

$$75,000(1 + 0.15) - \$24,400 = \$61,850.$$

This amount becomes the net amount the project is borrowing at the beginning of year 2. (This net amount is known as **project balance.**) At the end of period 2, the bank debt grows to $61,850(1.15) = $71,128$, but with the receipt of $27,340, the project balance reduces to

$$\$61,850(1.15) - \$27,340 = \$43,788.$$

Similarly, at the end of year 3, the project balance becomes

$$\$43,788(1.15) = \$50,356.$$

But, with the receipt of $55,760 from the project, the firm should be able to pay off the remaining balance and come out with a surplus in the amount of $5404. In other words, the firm fully repays its initial bank loan and interest at the end of period 3, with a resulting profit of $5404. If we compute the equivalent present worth of this net profit at time 0, we obtain

$$PW(15\%) = \$5404(P/F, 15\%, 3) = \$3553.$$

The result is identical to the case where we directly computed the NPW of the project at $i = 15\%$, shown in Example 4.5. Figure 4.6 illustrates the project balance as a function of time.

Period (n)	0	1	2	3
Beginning project balance	$ 0	−$75,000	−$61,850	−$43,788
Interest	0	−11,250	−9,278	−6,568
Cash receipt (payment)	−75,000	+24,400	+27,340	+55,760
Project balance	−$75,000	−$61,850	−$43,788	+$5,404

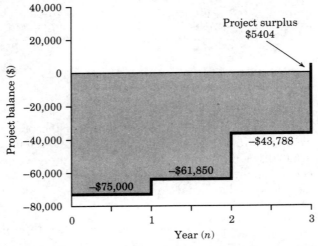

Figure 4.6 ■ Project balance diagram as a function of time (a negative project balance indicates the amount of remaining loan that needs to be paid off)

■ 4.4 Variations to PW Analysis

As variations to the present worth analysis, we will consider two additional measures of investment worth. They are **future worth analysis** and **capitalized equivalent worth analysis.** (The equivalent annual worth is another variation of the present worth measure, but we present it in Chapter 5.) Future worth analysis calculates the future worth of an investment undertaken. Capitalized equivalent worth analysis calculates the present worth of a project with a perpetual life span.

4.4.1 The Future Worth Analysis

The NPW measures the surplus in an investment project at time 0. On the other hand, the net future worth (NFW) measures this surplus at time period other than 0. This NFW analysis is particularly useful for an investment situation where we need to compute the equivalent worth of a project at the end of its investment period, rather than its beginning. As an example, it may take seven to ten years to build a new nuclear power plant due to complexities of engineering design and the many time-consuming regulatory procedures that must be satisfied to ensure public safety. In this situation, it is more common to measure the worth of the investment at the time of commercialization, that is, to conduct a NFW analysis at the end of the investment period.

Net Future Worth Criterion and Calculations

When A_n represents the cash flow at time n for $n = 0, 1, 2, \ldots, N$ for a typical investment project that extends over N periods, the net future worth (NFW) expression at the end of period N is

$$FW(i) = A_0(1 + i)^N + A_1(1 + i)^{N-1} + A_2(1 + i)^{N-2} + \cdots + A_N$$

$$= \sum_{n=0}^{N} A_n(1 + i)^{N-n}$$

$$= \sum_{n=0}^{N} A_n(F/P, i, N - n) \qquad (4.2)$$

As you might expect, the decision rule for the NFW criterion is the same as for the NPW: For a single project evaluation,

If $FW(i) > 0$, accept the investment.

If $FW(i) = 0$, remain indifferent.

If $FW(i) < 0$, reject the investment.

Example 4.6 Net future worth—at the end of project

Consider the project's cash flows in Example 4.5. Compute the NFW at the end of year 3 at $i = 15\%$.

Solution

Given: Cash flows in Example 4.5, MARR $= 15\%$ per year
Find: NFW

As seen in Fig. 4.7, the NFW of this project at an interest rate of 15% would be

$$FW(15\%) = -\$75,000(F/P, 15\%, 3) + \$24,400(F/P, 15\%, 2)$$
$$+ \$27,340(F/P, 15\%, 1) + \$55,760$$
$$= \$5404.$$

Since $FW(15\%) > 0$, the project is acceptable. Note that we reached the same conclusion under the present worth analysis.

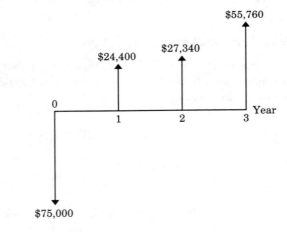

$$
\begin{array}{lr}
-\$75,000\ (F/P, 15\%, 3) \dotfill & -\$114,066 \\
\$24,400\ (F/P, 15\%, 2) \dotfill & \$32,269 \\
\$27,340\ (F/P, 15\%, 1) \dotfill & \$31,441 \\
\$55,760 \dotfill & \underline{\$55,760} \\
& \text{NFW} = \$5,404
\end{array}
$$

Figure 4.7 ■ Future worth calculation (Example 4.6)

Example 4.7 Net future worth—at intermediate time

Horton Corporation (HC), a Montreal-based robot-manufacturing company, has developed a new advanced-technology robot that incorporates such advanced technology as vision systems, tactile sensing, and voice recognition. These features allow a robot to roam the corridors of a hospital or office building without following a predetermined track or bumping into objects. HC's marketing department plans to target sales of the robot toward major hospitals to ease the nurses' workload by performing such low-level duties as delivering medicines and meals.

The firm would need a new plant to manufacture the robots; this plant could be built and made ready for production in 2 years. The plant would require a 10-hectare site, which can be purchased for $1.5 million in year 0. Building construction would begin in early year 1 and continue through year 2. The building would cost an estimated $10 million, involving a $4 million payment due to the contractor at the end of year 1 and another $6 million payable at the end of year 2. The necessary manufacturing equipment would be installed late in year 2 and would be paid for at the end of year 2. The equipment would cost $13 million, including transportation and installation. When the project terminates, the land is expected to have an after-tax market value of $2 million, the building an after-tax value of $3 million, and the equipment an after-tax value of $3 million.

For capital budgeting purposes, assume that cash flows occur at the end of each year. Because the plant would begin operations at the beginning of year 3, the first operating cash flows would occur at the end of year 3. The plant's estimated economic life is 6 years after completion, with the following expected after-tax operating cash flows in millions:

Calendar Year	94	95	96	97	98	99	2000	01	02
Year after Completion			*0*	*1*	*2*	*3*	*4*	*5*	*6*
After-tax cash flows:									
A. Operating revenue:				6	8	13	18	14	8
B. Investment:									
Land	−1.5								+2
Building		−4	−6						+3
Equipment			−13						+3
Net cash flow	−1.5	−4	−19	6	8	13	18	14	16

Compute the equivalent worth of this investment at the start of operation. Assume that HC's MARR is 15%.

Solution

Given: Cash flows above, MARR = 15% per year
Find: NFW at the end of calender year 2

One solution method involves calculating the present worth, then transforming this to the equivalent worth at year 2. First, we compute $PW(15\%)$ at time 0 of this project.

$$PW(15\%) = -\$1.5 - \$4(P/F, 15\%, 1) - \$19(P/F, 15\%, 2)$$
$$+ \$6(P/F, 15\%, 3) + \$8(P/F, 15\%, 4) + \$13(P/F, 15\%, 5)$$
$$+ \$18(P/F, 15\%, 6) + \$14(P/F, 15\%, 7) + \$16(P/F, 15\%, 8)$$
$$= \$13.91 \text{ million.}$$

Then,

$$FW(15\%) = PW(15\%)\,(F/P, 15\%, 2)$$
$$= \$18.40 \text{ million.}$$

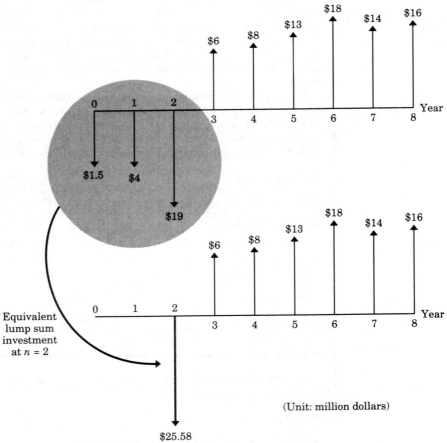

Figure 4.8 ■ Cash diagram for the robot project (Example 4.7) (unit: million dollars)

A second method brings all flows prior to year 2 up to that point, and discounts future flows back to year 2. The equivalent worth of the earlier investment when the plant begins full operation is

$$-\$1.5(F/P, 15\%, 2) - \$4(F/P, 15\%, 1) - \$19 = -\$25.58 \text{ million,}$$

producing an equivalent flow as shown in Fig. 4.8. Discounting the future flows to the start of operation, we obtain

$$FW(15\%) = -\$25.58 + \$6(P/F, 15\%, 1) + \$8(P/F, 15\%, 2) + \cdots$$

$$+ \$16(P/F, 15\%, 6)$$

$$= \$18.40 \text{ million.}$$

Comments: *If another company is willing to purchase the plant and the right to manufacture the robots when the plant is just completed (year 2), HC would set the price of the plant at $43.98 million ($18.40 + $25.58) at a minimum.*

4.4.2 Capitalized Equivalent Method

Another special case of the PW criterion is useful when the life of a proposed project is *perpetual* or the planning horizon is extremely long. Many public projects such as bridges, waterway constructions, irrigation systems, and hydro dams for electricity are expected to generate benefits over an extended period of time (or forever). In this section, we will examine the **capitalized equivalent (CE(i))** method for evaluating such projects.

Perpetual Service Life

Consider the cash flow series shown in Fig. 4.9. How do we determine the PW for an infinite (or almost infinite) uniform series of cash flows or repeated

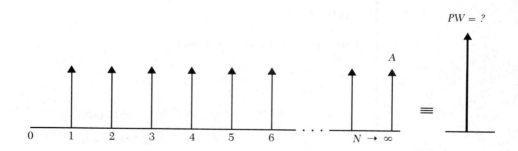

Figure 4.9 ■ Equivalent present worth of an infinite cash flow series

cycle of cash flows? The process of computing the PW cost for this infinite series is referred to as the **capitalization** of project cost. The cost is known as the **capitalized cost.** It represents the amount of money that must be invested today to yield a certain return A at the end of each and every period *forever*, assuming an interest rate of i. Observe the limit of the uniform series present worth factor as N approaches infinity:

$$\lim_{N \to \infty} (P/A, i, N) = \lim_{N \to \infty} \left[\frac{(1 + i)^N - 1}{i(1 + i)^N} \right] = \frac{1}{i}.$$

Thus, it follows that

$$PW(i) = A(P/A, i, N \to \infty) = \frac{A}{i}. \tag{4.3}$$

Another way of looking at this, $PW(i)$ dollars today, is to ask what constant income stream could be generated by this in perpetuity. Clearly, the answer is $A = iPW(i)$. If withdrawals were greater than A, they would be eating into the principal, which would eventually reduce to 0.

Example 4.8   Capitalized equivalent cost

An engineering school has just completed a new engineering complex worth $50 million dollars. A campaign, targeting alumni, is planned to raise funds for future maintenance costs, estimated at $2 million per year. Any unforeseen costs above $2 million per year would be obtained by raising tuition. Assuming that the school can create a trust fund that earns 8% interest annually, how much has to be raised now to cover the perpetual string of $2 million annual costs?

Solution

Given: A = $2 million, i = 8% per year, $N = \infty$
Find: $CE(8\%)$

The capitalized cost equation is

$$CE(i) = \frac{A}{i}$$

$$CE(8\%) = \$2,000,000/0.08$$

$$= \$25,000,000.$$

Comments: *It is easy to see that this lump-sum amount should be sufficient to pay the maintenance expense for the school forever. Suppose*

the school deposited $25 million in a bank that paid 8% interest annually.
At the end of the first year, the $25 million would earn 8%($25 million) =
$2 million interest. If this interest were withdrawn, the $25 million would
remain in the account. At the end of the second year, the $25 million
balance would again earn 8%($25 million) = $2 million. The annual
withdrawal could be continued forever, and the endowment (gift funds)
would always remain at $25 million.

When a Project's Service Life Is Extremely Long

The benefits of typical civil engineering projects, such as bridge and highway construction, although not perpetual, can last for many years. In this section, we will examine the use of the $CE(i)$ criterion to approximate the NPW of such long-lived engineering projects.

Example 4.9 Comparison of present worth for long life and infinite life

Consider the hydroelectric power plant project introduced at the beginning of this chapter. One of the main questions was whether the plant would be a profitable investment. Figure 4.10 illustrates when and in what quantities the entrepreneur spent his $800,000 (not considering

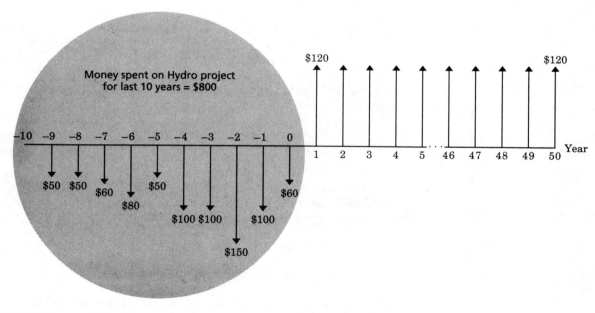

(Unit: thousand dollars)

Figure 4.10 ■ Net cash flow diagram for the Edgemont Hydro project (Example 4.9)

the time value of money) during the last 10 years. We will examine the situation by computing the project worth at varying interest rates. (a) If the entrepreneur's interest rate is 8%, compute the NPW (at time 0 in Fig. 4.10) of this project with a 50-year service life and infinite service, respectively. (b) Repeat part (a) assuming an interest rate of 12%.

Solution

Given: Cash flow in Fig. 4.10 (to 50 years or ∞), $i = 8\%$ or 12%
Find: NPW at time 0

Now we will compute the equivalent total investment and the equivalent worth of receiving the future power revenues at the start of generation, that is, at time 0.

(a) At $i = 8\%$ and with a service life of 50 years:
We can make use of two uniform series elements in the invested cash flow to help us find the equivalent total investment at the start of power generation. Using "K" to indicate thousand,

$$F_1 = -\$50K(F/A, 8\%, 10) - \$10K(F/P, 8\%, 7) - \$30K(F/P, 8\%, 6)$$
$$- \$50K(F/A, 8\%, 4)\,(F/P, 8\%, 1) - \$50K(F/P, 8\%, 2) - \$10K$$
$$= - \$1101K.$$

The equivalent total benefits at the start of generation is

$$F_2 = \$120K(P/A, 8\%, 50) = \$1468.$$

Summing, we find the net equivalent worth at the start of power generation:

$$F_1 + F_2 = -\$1101K + \$1468K$$
$$= \$367K.$$

With the infinite cash flows, the net equivalent worth is called the capitalized equivalent worth. The investment portion prior to time 0 is identical, so the capitalized equivalent worth is

$$CE(8\%) = -\$1101K + \$120K/(0.08)$$
$$= \$399K.$$

Note that the difference between the perpetual service life and a planning horizon of 50 years is only $32,000.

(b) At $i = 12\%$ and with a service life of 50 years:
Proceeding as for (a), the equivalent total investment at the start of generation is

$$F_1 = -\$50K(F\!/\!A, 12\%, 10) - \$10K(F\!/\!P, 12\%, 7) - \$30K(F\!/\!P, 12\%, 6)$$

$$\quad - \$50K(F\!/\!A, 12\%, 4)\,(F\!/\!P, 12\%, 1) - \$50K(F\!/\!P, 12\%, 2) - \$10K$$

$$= -\$1299K.$$

Equivalent total benefits at the start of generation:

$$F_2 = \$120K(P\!/\!A, 12\%, 50) = \$997K.$$

Net equivalent worth at the start of power generation:

$$F_1 + F_2 = -\$1299K + \$997K$$

$$= -\$302K.$$

With the infinite cash flows, the capitalized equivalent worth at the current time is

$$CE(12\%) = -\$1299K + \$120K/(0.12)$$

$$= -\$299K.$$

Note that the difference between the perpetual service life and a planning horizon of 50 years is merely $3000. This demonstrates that we may closely approximate the present worth of long cash flows (that is, 50 years or more) by using the capitalized equivalent value. The accuracy of the approximation improves as the interest rate increases.

Comments: *At $i = 12\%$, this investment is not a profitable one, but at 8% it is. This indicates the importance of using the appropriate i in investment analysis. Once again, the issue of selecting an appropriate i is presented in Chapter 13.*

■ 4.5 Comparing Mutually Exclusive Projects

Until now, we have considered situations in which only one project was under consideration, and we were determining whether to pursue it, based on whether its present worth or future worth met our MARR requirements. We were making a reject or accept decision about a *single* project.

In the real world of engineering practice, however, it is more typical for us to have two or more choices of projects for accomplishing a business objective. (As we shall see, even when it appears we have only one project to consider, the implicit "do nothing" alternative must be factored into our decision making.) As such, the greatest use of NPW and NFW analysis from our point of view is in selecting between alternative projects. In this section, we will begin to analyze multiple projects via NPW and NFW analysis. But first, some background about mutually exclusive projects and their useful lives is in order.

4.5.1 The Meaning of Mutually Exclusive and the "Do-Nothing" Alternative

Mutually exclusive means that any one of several alternatives will fulfill the same need and that selecting one alternative means that the others will be excluded. Take, for example, buying versus leasing an automobile for business use—when one alternative is accepted, the other is excluded.

We will use the terms *alternative* and *project* interchangeably to mean a *decision option*. In many situations, undertaking even a single alternative or project entails making a decision between two alternatives. Why? Because implicitly we have the "do-nothing" alternative. That is, some process or system is already in place to accomplish our business objectives and, just as for any new project under consideration, we can calculate its cash flows. If the current system begins to fail or is expected to begin to fail in meeting all of our objectives, we may propose a new project to replace or augment it. Nevertheless, in such a situation we *usually* retain the alternative of doing nothing, which means continuing to rely on the current system.

4.5.2 Revenue Projects Versus Service Projects

In comparing mutually exclusive alternatives, we need to classify investment projects into either service or revenue projects. **Service projects** are those whose revenues do not depend on the choice of project. For example, an electric utility is considering building a new power plant to meet the peak-load demand during either hot summer or cold winter days. There are two alternative service projects that could meet this peak-load demand: a combustion turbine plant or a fuel-cell power plant. No matter which plant is selected, the firm will generate the same amount of revenue from the customers. The only difference is how much it will cost to generate electricity from each plant. If we were to compare these service projects, we would be interested in knowing which plant can provide the cheapest power (production cost). Further, if we were to use the NPW criterion to compare these alternatives, we would choose the alternative with the largest (i.e., *least negative*) present value over the service life.

On the other hand, **revenue projects** are those whose revenues depend on the choice of alternative. For example, a TV manufacturer is considering marketing two types of high-resolution monitor. With its present production capacity, the firm can market only one of them. Because each model requires dramatically different production processes, each model can also have a different market price, production costs, and other capital requirements. In this situation, if we were to use the NPW criterion, we would select the model that promises to bring in a higher NPW.

4.5.3 Analysis Period

The **analysis period** is the time span over which the economic effects of an investment will be evaluated. The analysis period may also be called the **study period** or **planning horizon.** The length of the analysis period may be determined in several ways: It may be a predetermined amount of time set by company policy, or it may be either implied or explicit in the need the company is trying to fulfill—for example, a diaper manufacturer determines the need to dramatically increase production over a 10-year period in response to an anticipated "baby boom." In either of these situations, we consider the analysis period to be a *required service period.* When no required service period is stated at the outset, the analyst must choose an appropriate analysis period over which to study the alternative investment projects. In such a case, one convenient choice of analysis period is the period of useful life of the investment project.

When useful life of the investment project does not match the analysis or required service period, we must make adjustments in our analysis. A further complication, when we are considering two or more mutually exclusive projects, is that the investments themselves may have differing useful lives. We *must* compare projects with different useful lives over an *equal time span*, which may require further adjustments in our analysis. (Figure 4.11 is a flow chart showing the possible combinations of the analysis period and the useful life of the investment.) In the sections that follow, we will explore in more detail how to handle situations in which project lives differ from the analysis period and from each other. But we begin with the most straightforward situation, when the project lives and analysis period coincide.

4.5.4 Analysis Period Equals Project Lives

Let's begin our analysis with the simplest situation where the project lives equal the analysis period. In this case, we compute the NPW for each project and select the one with highest NPW for revenue projects or least negative NPW for service projects. Example 4.10 will illustrate this point.

Figure 4.11 ■ Analysis period implied in comparing mutually exclusive alternatives

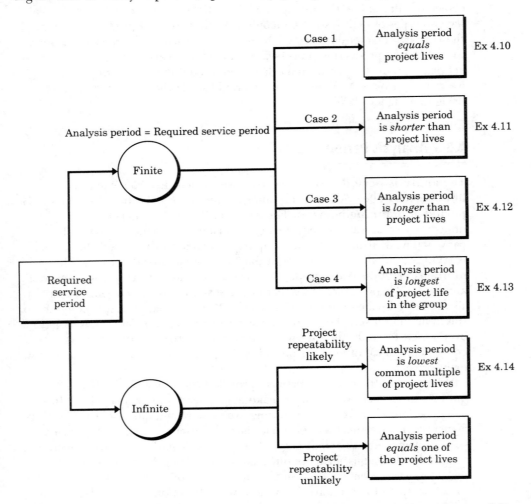

Example 4.10   Present worth comparison—equal lives

Bullard Company (BC) operates two machining centers with three operators to manufacture machine tables, saddles, machine bases, and other similar parts. The existing machine centers have been slow and have finally broken down, and would require a considerable expenditure to become operational. BC is considering upgrading the machining center by replacing broken equipment and introducing new operating procedures. Three methods of machining-center use are considered along with Bullard's existing methods:

- **Option 1:** Operating the machining center without an automatic pallet changer.

- **Option 2:** Operating the machining center with an automatic pallet changer and one operator.

- **Option 3:** Operating the machining center with an automatic pallet changer and two operators.

Different methods of use will affect the parts and their corresponding cycle times due to changes in handling, operator fatigue, and delay allowances. The time allowed for loading and unloading parts is reduced in the pallet-changer/one-operator case, because most parts will be loaded and unloaded while another part on the second pallet is being machined. In the pallet-changer/two-operator method, one operator can cover for the second to a degree that set-up as well as part unit times can be shared. BC has estimated the incremental investment costs as well as the labor and set-up savings for the alternative machining-center methods compared with the existing methods as follows:

Machining-Center Methods

	Option 1	Option 2	Option 3
Investment	$319,000	$369,800	$369,800
Proceeds from sales of broken machines	50,000	50,000	50,000
Net investment cost	$269,000	$319,800	$319,800
Annual Net Savings:			
Direct labor	$ 61,800	$ 68,800	$ 92,800
Set-up	19,700	19,700	25,500
Total savings	$ 81,500	$ 88,500	$118,300
Service life	5 years	5 years	5 years

All savings figures represent any tax effects. Once a method is chosen, BC expects to operate the machining centers over the next 5 years. Which option would be selected based on the use of the NPW measure at $i = 12\%$?

Solution

Given: Cash flows for three projects, $i = 12\%$ per year

Find: The NPW of each project, which to select

These are revenue projects with equal service lives. Since the required service period is 5 years, we should select the analysis period of 5 years. Since the analysis period coincides with the project lives, we simply compute the NPW value for each option. The equivalent NPW figures at $i = 12\%$ would be as follows:

- For Option 1:

$$PW(12\%)_{\text{Option 1}} = -\$269,000 + \$81,500(P/A, 12\%, 5)$$

$$= \$24,789$$

- For Option 2:

$$PW(12\%)_{\text{Option 2}} = -\$319,800 + \$88,500(P/A, 12\%, 5)$$

$$= -\$777$$

- For Option 3:

$$PW(12\%)_{\text{Option 3}} = -\$319,800 + \$118,300(P/A, 12\%, 5)$$

$$= \$106,645$$

Clearly, option 3 is the most economical option. We know that option 2 is inferior to option 3 even before calculating the NPW values. This is simply because option 2 costs the same as option 3 but generates less savings. Given the nature of BC parts and shop orders, the management has decided that the best way to operate would be with an automatic pallet changer and two operators.

4.5.5 Project Lives Differ from a Specified Analysis Period

In Example 4.10, we assumed the simplest scenario possible in analyzing mutually exclusive projects: The projects had useful lives equal to each other *and* to the required service period. In practice, this is seldom the case. Often project lives do not match the required analysis period and/or do not match each other. For example, two machines may perform exactly the same function, but one lasts longer than the other and both of them last longer than the analysis period for which they are being considered. In the following sections and examples, we will develop the techniques for dealing with these complications.

Project Lives Longer than the Analysis Period

Project lives rarely conveniently coincide with a firm's predetermined required analysis period; they are often too long or too short. The case of project lives that are too long is the easier one to address.

Consider the case of a firm that undertakes a 5-year production project when all of the alternative equipment choices have useful lives of 7 years. In such a case, we analyze each project for only as long as the required service period (in this case, 5 years). We are then left with some unused portion of the equipment (in this case, 2 years' worth), which we include as **salvage value** in our analysis. Salvage value is the amount of money for which the equipment could be sold after its service to the project has been rendered or the dollar measure of its remaining usefulness. (Later in this text we will present methods for determining salvage value; for now, they will be given to you as a precalculated aspect of examples and problems.)

A common instance of project lives longer than the analysis period occurs in the construction industry, where a building project may have a relatively short completion time, but the equipment purchased has a much longer useful life.

Example 4.11 Present worth comparison—project lives longer than the analysis period

Waste Management Company (WMC) has won a contract that requires the firm to remove radioactive material from government-owned property and transport it to a designated dumping site. This task requires a specially made ripper-bulldozer to dig and load the material onto a transportation vehicle. Approximately 400,000 tonnes of waste must be moved in a period of 2 years. Model A costs $150,000 and has a life of 6000 hours before requiring any major overhaul. Two units of model A would be required to remove the material within 2 years, and the operating cost for each unit would run to $40,000 per year for 2000 hours of operation. At this operational rate, the model would be operable for 3 years, and at the end of that time, it is estimated that the salvage value will be $25,000 each.

The larger model B costs $480,000, has a life of 12,000 hours without any major overhaul, and costs $45,000 to operate for 2000 hours/year to complete the job within 2 years. The estimated salvage value of model B at the end of 6 years is $60,000.

Since the lifetime of either model exceeds the required service period of 2 years (as shown in Fig. 4.12), WMC has to assume something about the used equipment at the end of that time. Therefore, the engineers at WMC estimate that after 2 years the model A units could be sold for $45,000 each and the model B units for $250,000 each. After considering all income taxes, WMC has summarized the resulting cash flows for each project as follows:

Period	Model A		Model B	
0	−$300		−$480	
1	−80		−45	
2	−80	+ 90	−45	+ 250
3	−80	+50	−45	
4			−45	
5			−45	
6			−45	+60

Figure 4.12 ■ (a) Cash flow for model A. (b) Cash flow for model B.
(c) Comparison of unequal-lived projects when the required service period is
shorter than the individual project life (Example 4.11)

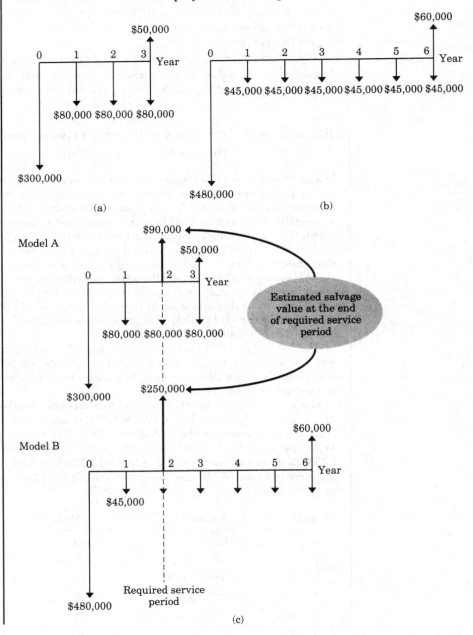

Here, the figures in the boxes represent the estimated salvage values at the end of the analysis period (end of year 2). Assuming that the firm's MARR is 15%, which option would be acceptable?

Solution

Given: Cash flows for two alternatives as shown in Fig. 4.12, $i = 15\%$ per year
Find: The NPW for each alternative, which is preferred

First, note that these are service projects. Since the firm explicitly estimated the market values of the assets at the end of the analysis period (2 years), we can compare the two models directly. Since the benefits (removal of the wastes) are equal, we can concentrate on the costs:

$$PW(15\%)_A = -\$300 - \$80(P/F, 15\%, 1) + \$10(P/F, 15\%, 2)$$

$$= -\$362$$

$$PW(15\%)_B = -\$480 - \$45(P/F, 15\%, 1) + \$205(P/F, 15\%, 2)$$

$$= -\$364$$

Model A has the greater PW (least negative) and thus would be preferred.

Project Lives Shorter than the Analysis Period

When project lives are shorter than the required service period, we must consider how, at the end of the project lives, we will satisfy the rest of the required service period. **Replacement projects**—additional projects to be implemented when the initial project has reached the limits of its useful life—are needed in such a case. Sufficient replacement projects must be analyzed to match or exceed the required service period.

To simplify our analysis, we sometimes assume that the replacement project will be exactly the same as the initial project, with the same corresponding costs and benefits. However, this assumption is not necessary. For example, depending on our forecasting skills, we may decide that a different kind of technology—in the form of equipment, materials, or processes—is a preferable replacement. Whether we select exactly the same alternative or a new technology as the replacement project, we are ultimately likely to have some unused portion of the equipment to consider as salvage value, just as in the case when project lives are longer than the analysis period. On the other hand, we may decide to lease the necessary equipment or subcontract the remaining work for the duration of the analysis period. In this case, we can probably exactly match our analysis period and not worry about salvage values.

In any event, we must make some initial guess concerning the method of completing the analysis period at its outset. Later, when the initial project life is closer to its expiration, we may revise our analysis with a different replacement project. This is only reasonable, since economic analysis is an ongoing activity in the life of a company and an investment project, and we should always use the most reliable, up-to-date data we can reasonably acquire.

Example 4.12 Present worth comparison—project lives shorter than analysis period

The Smith Novelty Company, a mail-order firm, wants to install an automatic mailing system to handle product announcements and invoices. The firm has a choice between two different types of machines. The two machines are designed differently but have identical capacities and do exactly the same job. The $12,500 semiautomatic model A will last 3 years, while the fully automatic model B will cost $15,000 and last 4 years. The expected cash flows for the two machines including maintenance, salvage value, and tax effects are as follows:

n	Model A	Model B
0	−$12,500	−$15,000
1	−5,000	−4,000
2	−5,000	−4,000
3	−3,000	−4,000
4		−2,500
5		

As business grows to a certain level, neither of the models can handle the expanded volume at the end of year 5. If that happens, a fully computerized mail-order system will need to be installed to handle the increased business volume. With this scenario, which model should the firm select at MARR = 15%?

Solution

Given: Cash flows for two alternatives as shown in Fig. 4.13, analysis period of 5 years, $i = 15\%$

Find: The NPW of each alternative, which to select

Since both models have a shorter life than the required service period (5 years), we need to make an explicit assumption of how the service

Figure 4.13 ■ Comparison for unequal-lived projects when the required service period is longer than the individual project life (Example 4.12)

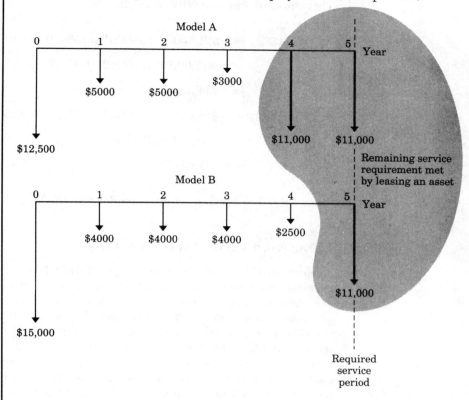

requirement is to be met. Suppose that the company considers leasing comparable equipment that has an annual lease payment of $11,000 (after taxes) for the remaining required service period. In this case, the cash flow would look like Fig. 4.13.

n	Model A	Model B
0	−$12,500	−$15,000
1	−5,000	−4,000
2	−5,000	−4,000
3	−3,000	−4,000
4	−11,000	−2,500
5	−11,000	−11,000

Here, the boxed figures represent the annual lease payments. Note that both alternatives now have the same analysis period of 5 years. Therefore, we can use NPW analysis.

$$PW(15\%)_A = -\$12,500 - \$5000(P/A, 15\%, 2) - \$3000(P/F, 15\%, 3)$$

$$- \$11,000(P/A, 15\%, 2)\,(P/F, 15\%, 3)$$

$$= -\$34,359,$$

$$PW(15\%)_B = -\$15,000 - \$4000(P/A, 15\%, 3) - \$2500(P/F, 15\%, 4)$$

$$- \$11,000(P/F, 15\%, 5)$$

$$= -\$31,031.$$

Since these are service projects, model B is the better choice.

Analysis Period Is Dictated by Choice of a Project

As seen in the preceding pages, equal future time periods are generally necessary to achieve comparability of alternatives. In some situations, however, revenue projects with different lives can be comparable *if they require only a one-time investment* because the task or need within the firm is a one-time task or need. We may find an example of this situation in the extraction of a fixed quantity of a natural resource such as oil or coal. Consider two mutually exclusive processes, one that requires 10 years to recover some coal and a second that can accomplish the task in only 8 years. There is no need to continue the project if the short-lived process is used and all the coal has been retrieved. In this example, the two processes can be compared even though the cash flows cover differing time spans. The revenues must be included in the analysis even if the price of coal is constant, because of the time value of money. While the total revenue is equal for either process, that for the faster process has a larger present worth. Therefore, the two projects should be compared using the NPW of each over its own life. Note that in this case the analysis period is determined by and coincides with the life of the selected project. (Here we are still, in effect, assuming an analysis period of 10 years.)

Example 4.13 Present worth comparison—a case where different analysis periods can be used

The family-owned Foothills Ranching Company owns the mineral rights for land used for growing grain and grazing cattle. Recently, oil was discovered on this property. The family has decided to extract the oil, sell the land, and retire. The company can either lease the necessary equipment and extract and sell the oil itself, or it can lease the land to an oil-production company. If the company chooses the former, the net annual cash flow after taxes from operations will be $500,000 at the end of each year for the next 5 years, and the company can sell the land for a net cash flow of $1,000,000 in 5 years when the oil is depleted. If the company chooses the latter, the production company can extract all the oil in only 3 years, and the company can sell the land for a net cash flow of $800,000 in 3 years. (The difference in resale value of the land is due to the increasing rate of land appreciation anticipated for this property.) The net cash flow from the lease payments to the company will be $630,000 at the *beginning* of each of the next 3 years. All benefits and costs associated with the two alternatives have been accounted for in the figures listed above. Which option should the firm select at $i = 15\%$?

Solution

Given: Cash flows shown in Fig. 4.14, $i = 15\%$ per year

Find: The NPW of each alternative, which to select

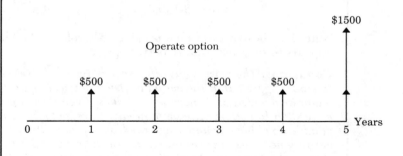

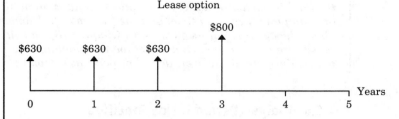

Figure 4.14 ■ Comparison of unequal-lived revenue projects where the analysis period coincides with the longest-lived project in the mutually exclusive group (Example 4.13). In our example, the analysis period is 5 years assuming zero cash flows in years 4 and 5 for the lease option.

As illustrated in Fig. 4.14, the cash flows associated with each option would look like this.

n	Operate	Lease
0		$630,000
1	$ 500,000	630,000
2	500,000	630,000
3	500,000	800,000
4	500,000	
5	1,500,000	

After depletion of the oil, the project will terminate.

$$P/W(15\%)_{\text{operate}} = \$500,000(P/A, 15\%, 4) + \$1,500,000(P/F, 15\%, 5)$$

$$= \$2,173,250$$

$$P/W(15\%)_{\text{lease}} = \$630,000 + \$630,000(P/A, 15\%, 2)$$

$$+ \$800,000(P/F, 15\%, 3)$$

$$= \$2,180,013$$

Note that these are revenue projects. Therefore, the leasing option appears to be a marginally better option.

Comments: *The very slight difference between the two NPW amounts ($6763) suggests that the actual decision between operating and leasing might be decided on noneconomic issues. Even if the operating option were slightly better, the company might prefer to forgo the small amount of additional income and select the lease option rather than undertake an entirely new business venture and run their own operation. A variable that might also have a critical effect on this decision is the sales value of the land in each alternative. The value of land is often difficult to forecast over any long period of time, and the firm may feel some uncertainty about the accuracy of its guesses. In Chapter 15, we will discuss sensitivity analysis, which is a method by which we can factor uncertainty about the accuracy of project cash flows into our analysis.*

4.5.6 Analysis Period Is Not Specified

Our coverage so far has focused on situations in which an analysis period is known. When an analysis period is not specified either explicitly by company policy or practice or implicitly by the projected life of the investment project, it is up to the analyst to choose an appropriate one. In such a case, the most convenient procedure is to choose one based on the useful lives of the alternatives. When the alternatives have equal lives, this is an easy selection. When alternatives' lives differ, we must select an analysis period that allows us to compare different-lived projects on an equal time basis, that is, a **common service period.**

Lowest Common Multiple of Project Lives

A required service period that may be assumed is infinity. If we anticipate an investment project will be ongoing at roughly the same level of production for some indefinite period, we may choose to consider an analysis period of infinity. It is certainly possible to do so mathematically, though the analysis is likely to be complicated and tedious. Therefore, in the case of an indefinitely ongoing investment project, we typically select a finite analysis period by using the lowest common multiple of project lives. For example, if alternative A has a 3-year useful life and alternative B has a 4-year life, we may select 12 years as the analysis period (or the common service period). We would consider alternative A through 4 life cycles and alternative B through 3 life cycles; in each case, we would use the alternatives completely. We then accept the finite model's results as a good prediction of what will be the economically wisest course of action for the foreseeable future. The following example is a case in which we conveniently use the lowest common multiple of project lives as our analysis period.

Example 4.14   Present worth comparison—unequal lives, lowest common multiple method

Consider Example 4.12. Suppose that both models A and B can handle the increased future volume and that the system is not going to be phased out at the end of 5 years. Instead, the current mode of operation is expected to continue for an indefinite period of time. We also assume that these two models will be available in the future without significant changes in price and operating costs. At MARR = 15%, which model should the firm select?

Solution

Given: Cash flows for two alternatives as shown in Fig. 4.15, $i = 15\%$ per year, indefinite period of need

Find: The NPW of each alternative, which to select

Recall that the two mutually exclusive alternatives have different lives, but provide identical annual benefits. In such a case, we ignore the common benefits and can make the decision based solely on costs, as long as a common analysis period is used for both alternatives.

Figure 4.15 ■ Comparison of unequal-lived projects when the required
service period is infinite and project repeatability likely (Example 4.14)

Model A

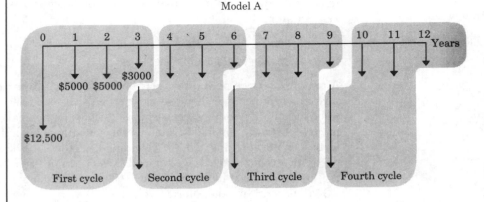

Model B

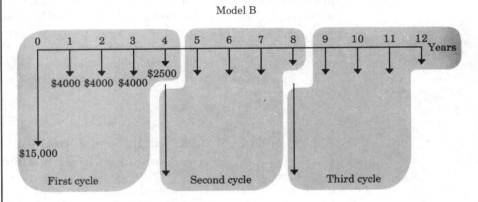

To make the two projects comparable, let's assume that after either the
3- or 4-year period the system would be reinstalled repeatedly using the
same model and that the same costs would apply. The lowest common
multiple of 3 and 4 is 12, so we will use 12 years as the *common analysis
period*. Note that any cash flow difference between the alternatives will
be revealed during the first 12 years. After that, the same cash flow
pattern repeats every 12 years for an indefinite period. The replacement
cycles and cash flows are shown in Fig. 4.15.

- Model A: There are 4 replacements in a 12-year period. The PW for
  the first investment cycle is

$$\text{PW}(15\%) = -\$12{,}500 - \$5000(P/A, 15\%, 2)$$

$$- \$3000(P/F, 15\%, 3)$$

$$= -\$22{,}601.$$

With 4 replacement cycles, the total PW is

$$PW(15\%) = -\$22,601[1 + (P/F, 15\%, 3)$$
$$+ (P/F, 15\%, 6) + (P/F, 15\%, 9)]$$
$$= -\$53,657.$$

- Model B: The PW for the first investment cycle is

$$PW(15\%) = -\$15,000 - \$4000(P/A, 15\%, 3)$$
$$- \$2500(P/F, 15\%, 4)$$
$$= -\$25,562.$$

With 3 replacement cycles in 12 years, the total PW is

$$PW(15\%) = -\$25,562[1 + (P/F, 15\%, 4) + (P/F, 15\%, 8)]$$
$$= -\$48,534.$$

The PW of model B is less costly, so that would be the model of choice.

Comments: *In Example 4.14, an analysis period of 12 years seems reasonable. The number of actual reinvestment cycles needed with each type of system will depend on the technology of the future system, so we may or may not actually need the 4 (model A) or 3 (model B) reinvestment cycles we used in our analysis. The validity of the analysis also depends on the costs of system and labor remaining constant. If we assume constant dollar prices, this analysis would provide us with a reasonable result. (As you will see in Example 5.9, the annual worth approach will make it mathematically easier to solve this type of comparison.)*

Other Common Analysis Periods

In some cases, the lowest common multiple of project lives is an unwieldy analysis period to consider. Suppose, for example, that you were considering alternatives with lives of 7 and 12 years, respectively. Besides making for tedious calculations, an 84-year analysis period may lead to inaccuracies, since over a long period of time we can be less and less confident about the ability to install identical replacement projects with identical costs and benefits. In a case like this, it would be reasonable to use the useful life of one of the alternatives by either factoring in a replacement project or salvaging the remaining useful life as the case may be. The important rule is to compare both projects on the same time basis.

In this section, we will demonstrate how to create a NPW table and NPW plot as a function of interest rate by rate by using Lotus 1-2-3 and **EzCash** programs. We will use the cash flow data described in Example 4.5.

4.6.1 Creating a NPW Table and Graph by Lotus 1-2-3

Using Fig. 4.16 as a model, enter the known time periods and their cash flows (figures in parentheses represent negative amounts). Enter the range of interest rates in column E. To calculate the NPW of the cash flows at a specified interest rate, you can use the @NPV(i, range) function. The @NPV(i, range) function assumes that cash flows occur at the *ends* of periods. To find the NPW of an investment where you make an initial cash outflow *immediately*, or at period 0, and follow it by a series of future cash flows, you must factor the initial flow separately because it is not affected by the interest.

If A_0 is your initial flow (at period 0), $A_1 \ldots A_N$ is a range of future cash flows, and i is the periodic interest rate, the total NPW is calculated by

$$+A_0 + @NPV(i, \text{range}).$$

In our example, to compute the NPW at $i = 0$, the cell formula for F5 is

$$\$C\$8 + @NPV(E5/100, \$C\$9..\$C\$11).$$

```
A:F5: (C0) [W13] +$C$8+@NPV(E5/100,$C$9..$C$11)                    READY
```

				Interest Rate (%)	Present Worth	
		Example 4.5				
				0	$32,500	
	n	Cash Flow		2	$27,744	$C$8+
				4	$23,309	@NPV(E7/100,
	0	($75,000)		6	$19,169	$C$9..$C$11)
	1	$24,400		8	$15,296	
	2	$27,340		10	$11,670	
	3	$55,760		12	$8,270	
				14	$5,077	
				16	$2,076	
				18	($750)	
				20	($3,412)	
				22	($5,924)	$C$8+
				24	($8,296)	@NPV(E17/100,
				26	($10,539)	$C$9..$C$11)
				28	($12,662)	
				30	($14,673)	

Figure 4.16 ■ Creating a present worth table using Lotus 1-2-3

The $ sign in front of each cell number indicates that Lotus 1-2-3 must use the data stored in these absolute locations. Figure 4.16 shows that the NPW of the investment will be zero when the interest rate is somewhere between 16% and 18%.

You can now easily build a NPW graph using the "Graph" command in Lotus 1-2-3, which appears in the main menu as shown in Fig. 4.17. To create a line graph of this data, begin by defining the type of graph as follows:

/**G**raph **T**ype **L**ine

Now you are prompted to specify the data ranges. The X data range will display the interest rate as the x-axis label. Highlight E5 through E20 to specify the X data range (Fig. 4.18a). The data range for the numeric data to be graphed are the NPW figures obtained in column F: highlighted cells F5 through F20

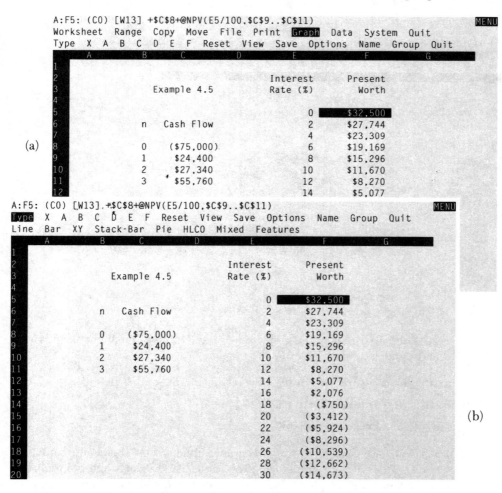

(a)

(b)

Figure 4.17 ■ (a) Select the "Graph" mode to initiate the plotting. (b) Specify type of graph by selecting "Line."

Figure 4.18 ■ (a) Enter x-data by highlighting the interest rate column; (b) Enter y-data by highlighting the present worth data column.

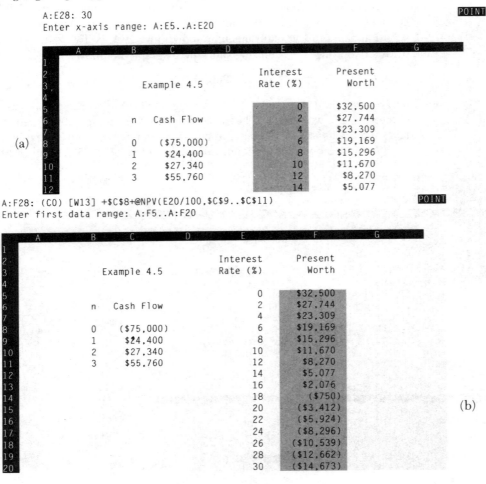

A:E28: 30   POINT
Enter x-axis range: A:E5..A:E20

A:F28: (C0) [W13] +$C$8+@NPV(E20/100,$C$9..$C$11)   POINT
Enter first data range: A:F5..A:F20

(Fig. 4.18b). To complete the graph, select the "Options" command to add titles, legends, and secondary notes to the graph.

First title   **Present Worth as a Function of Interest Rate**

X-axis title   **Interest Rate (%)**

Y-axis title   **Present Worth ($)**

When you have finished, view the graph. Your screen should be similar to Fig. 4.19. To print your graph using the Lotus 1-2-3 PrintGraph program, you must first save the graph settings in a special file. This is called a graph file and has a file extension of .PIC. Then, you leave the worksheet and enter the PrintGraph program to obtain a hard copy.

Figure 4.19 ■ Present worth curve for Example 4.5 generated by Lotus 1-2-3

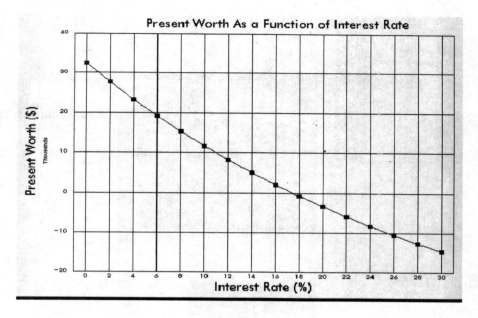

4.6.2 Creating an NPW Table and Graph by EzCash

To create an NPW table and graph, access the **Cash Flow Editor** by selecting **Edit** followed by **EzCash** and **Build**. Since the periodic cash flows have no regular pattern, they are entered individually (Fig. 4.20a). Click **Done** and select the **Display** mode followed by **Cash Flow Plot** (Fig. 4.20b). The software displays the cash flow diagram (Fig. 4.20c).

Select **Present Worth Plot** from the **Display** menu options. Now you are prompted to set the upper and lower bounds of the interest range (Fig. 4.21a). After these data are entered, click **Ok**. **EzCash** calculates the net present worth for various interest rates in the specified range and displays the results in the form of a graph (Fig. 4.21b). The break-even interest rate (rate of return) is displayed in the upper right-hand corner of the screen. In our example, this break-even interest rate is 17.46%. When there is no break-even interest rate within the specified bounds, **EzCash** will display the message "No ROR within specified limits". The interest rate limits can be changed by clicking on the **Done** button which returns you to the Interest Bounds window. A hard copy of the graph is obtained by selecting **Print**.

To obtain net present worth data in tabular form select **Present Worth Table** from the menu choices under **Display**. Again, an Interest Bounds window opens (Fig. 4.21c). After entering the required data, click **Ok** and the table of net present worth values is displayed (Fig. 4.21d). You can scroll up and down this table to find the desired interest rate.

Figure 4.20 ■ Generating a NPW table and plot with **EzCash**—Part I

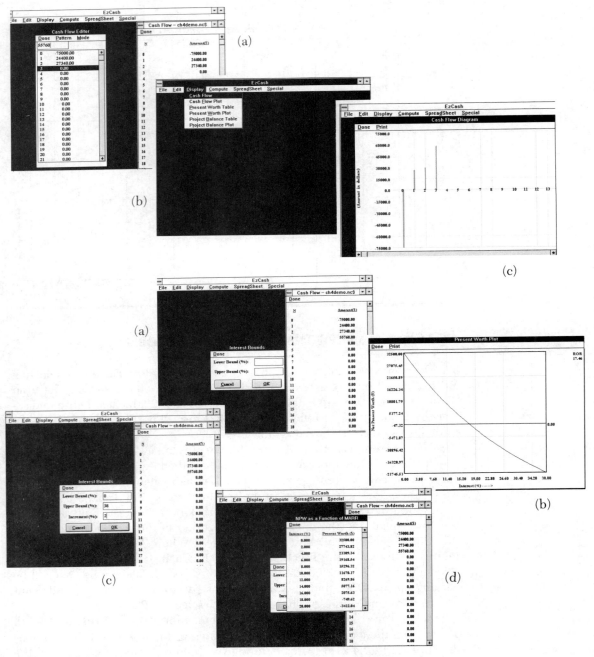

Figure 4.21 ■ Steps required to obtain a NPW table and plot by using **EzCash**—Part II

Summary

In this chapter, we presented the concept of present worth analysis based on cash flow equivalence along with the payback period. We observed the following important results.

- **Present worth** is an equivalence method of analysis in which a project's cash flows are discounted to a single present value. It is perhaps the most efficient analysis method we can use for determining project acceptability on an economic basis. Other analysis methods that we will study in Chapters 5 and 6 are built on a sound understanding of present worth. Table 4.3 summarizes present worth and its related equivalence methods.

- The **MARR** or **minimum attractive rate of return** is an interest rate at which the firm can always earn or borrow money. It is generally dictated by management and is the rate at which NPW analysis should be conducted.

- **Revenue projects** are those for which the income generated depends on the choice of project. **Service projects** are those for which income remains the same, regardless of which project is selected.

- **Mutually exclusive** means that when one of several alternatives that meet the same need is selected, the others will be rejected.

- When it is not specified by management or company policy, the **analysis period** to use in comparing mutually exclusive projects may be chosen by the individual analyst. Several efficiencies to selecting an analysis period can be applied; they are highlighted in Figure 4.11.

Table 4.3 Summary of Project Analysis Methods

Analysis Method	Equation Form(s)	Description	Comments
Payback period	—	A method for determining *when* in a project's history it breaks even (i.e., revenues repay costs).	Should be used as a screening only. Reflects liquidity, not profitability of project.
Discounted payback period	—	A variation of payback period which factors in the time value of money.	Reflects a truer measure of payback point than regular payback period.
Net present worth (NPW)	$PW(i) = \sum_{n=0}^{N} A_n(P/F, i, n)$	An *equivalence* method which translates a project's cash flows into a net present worth value.	For investments involving disbursements *and* receipts, the decision rule is: If $PW(i) > 0$, accept If $PW(i) = 0$, remain indifferent If $PW(i) < 0$, reject
			When comparing multiple alternatives, select the one with the greatest NPW value. For investments involving only disbursements (service projects), this is the project with the least negative NPW.

Table 4.3 Summary of Project Analysis Methods (continued)

Analysis Method	Equation Form(s)	Description	Comments
Net future worth (NFW)	$FW(i) = PW(i)(F/P, i, N)$	An *equivalence* method variation of NPW; a project's cash flows are translated into a net future worth value.	Most common real-world application occurs when we wish to determine a project's value at commercialization (a future date) *not* its value when we begin investing (the "present").
Capitalized equivalent (CE)	$PW(i) = \dfrac{A}{i}$	An *equivalence* method variation of NPW; calculates the NPW of a perpetual or very long-lived project that generates a constant annual net cash flow.	Most common real-world applications are civil engineering projects with lengthy service lives (i.e., $N > 50$) and *equal annual* costs (or incomes).
Annual equivalence (AE)	$AE(i) = PW(i)(A/P, i, N)$	An *equivalence* method and variation of NPW; a project's cash flows are translated into an annual equivalent sum.	AE analysis facilitates the comparison of unit costs and profits. In certain circumstances, AE may also avoid the complication of lowest common multiple of project lives which may arise in NPW analysis.
Rate of return (ROR) or internal rate of return (IRR)	$PW(i) = 0$	A *relative* percentage method which measures the yield as a percentage of investment over the life of a project.	Rate of return is stated in a form more intuitively understandable to most laypersons and financial administrators. Unlike equivalence methods, which are absolutes, IRR when calculated must be compared to a predetermined acceptable rate of return known as MARR. The decision rule is: If IRR > MARR, accept If IRR = MARR, remain indifferent If IRR < MARR, reject

Problems

Note: Unless otherwise stated, all cash flows given in the problems represent after-tax cash flows. The interest rate (MARR) is also given on after-tax basis.

4.1 Camptown Togs, Inc., a children's clothing manufacturer, has always found payroll processing to be costly because it must be done by a clerk to verify the piece-goods coupons collected for each employee and to calculate the types of tasks performed by each employee. Recently an industrial engineer designed a system that partially automates the process with a scanner by reading the piece-goods coupons. Management is enthusiastic about this because it utilizes some of their personal computer systems that were purchased recently. It is expected that this new automated system will save $30,000 per year in labor. The new system will cost about $25,000 to build and test prior to operation. It is expected that operating costs including income taxes will be about $5000 per year. The system will have a 5-year useful life. The expected net salvage value of the system is estimated to be $3000.

(a) Identify the cash inflows over the life of the project.
(b) Identify the cash outflows over the life of the project.
(c) Determine the net cash flows over the life of the project.

4.2 In Problem 4.1:

(a) How long does it take to recover the investment?
(b) If the firm's interest rate is 15% after tax, what would be the discounted payback period for this project?

4.3 For each of the following cash flows:

Project's Cash Flow ($)

n	A	B	C	D
0	−$2000	−$6000	−$5000	−$8000
1	300	2000	2000	5000
2	300	1500	2000	3000
3	300	1500	2000	−2000
4	300	500	5000	1000
5	300	500	5000	1000
6	300			2000
7	300			3000
8	300			

(a) Calculate the payback period for each project.
(b) Is it meaningful to calculate a payback period for project D?
(c) Assuming $i = 10\%$, calculate the discounted payback period for each project.

4.4 Consider the following sets of investment projects. All projects have a 3-year investment life:

Period	Project's Cash Flow			
n	A	B	C	D
0	−$2000	−$2000	−$2000	−$2000
1	0	500	1000	800
2	0	800	800	800
3	2500	1000	500	800

(a) Compute the net present worth of each project at $i = 10\%$.
(b) Plot the present worth as function of interest rate (from 0% to 40%) for project B.

4.5   It is desired to know if the building of a new warehouse is justified under the following conditions:

The proposal is for a warehouse costing $100,000 with an expected useful life of 35 years and a net salvage value (net proceeds from sale after tax adjustments) of $25,000. Annual receipts of $17,000 are expected, annual maintenance and administrative costs will be $4000/year, and annual income taxes are $2000.

Then which of the following statements are correct?

(a) The proposal is justified for a MARR of 9%.

(b) The proposal has a net present worth of $62,730.50 when 6% is used as the interest rate.

(c) The proposal is acceptable as long as MARR $\leq$ 10.77%.

(d) All of the above are correct.

4.6   Consider the following sets of investment projects. All projects have a 3-year investment life.

Period	Project's Cash Flow			
n	A	B	C	D
0	-$2,000	0	$2,000	-$2,000
1	5,400	-3,000	-8,000	-500
2	14,400	1,000	2,000	7,500
3	7,200	2,000	4,000	6,500

(a) Compute the net present worth of each project at $i = 12\%$.

(b) Compute the net future worth of each project at $i = 12\%$. Which project(s) are acceptable?

4.7   Consider the project balances for a typical investment project with a service life of 4 years.

n	A_n	Project Balance
0	-$1000	-$1000
1	( )	-1100
2	( )	-800
3	460	-500
4	( )	0

(a) Construct the original cash flows of the project.

(b) What is the interest rate used in computing the project balance?

(c) At $i = 15\%$, would this project be acceptable?

4.8   Consider the project balance diagram in Fig. 4.22 for a typical investment project with a service life of 5 years. The numbers in the figure indicate the beginning project balances.

(a) From the project balance diagram, construct the original cash flows of the project.

(b) What is the conventional payback period (without interest) of this project?

4.9   Consider the following cash flows and present worth profile in Fig. 4.23.

	Net Cash Flows ($)	
Year	Project 1	Project 2
0	-100	-100
1	40	30
2	80	Y
3	X	80

Figure 4.22 ■ Project balance diagram (Problem 4.8)

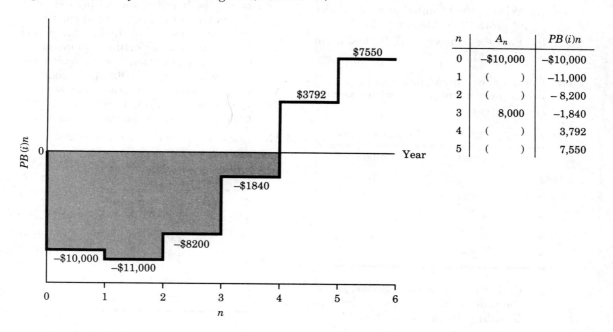

n	A_n	$PB(i)_n$
0	−$10,000	−$10,000
1	( )	−11,000
2	( )	− 8,200
3	8,000	−1,840
4	( )	3,792
5	( )	7,550

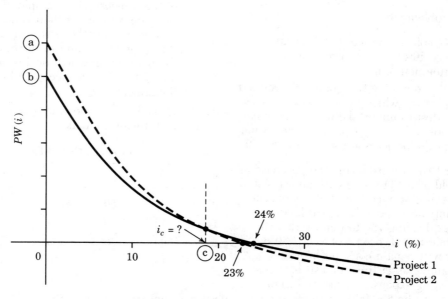

Figure 4.23 ■ Plot of NPW as a function of i (Problem 4.9)

(a) Determine the values for X and Y.

(b) What is the terminal project balance of project 1 at MARR $= 24\%$?

(c) Find the values for (a), (b) and (c) in Fig. 4.23.

4.10 Consider the project balances for a typical investment project with a service life of 5 years.

n	A_n	Project Balance
0	$-\$1000$	$-\$1000$
1	( )	-900
2	490	-500
3	( )	0
4	( )	-100
5	200	( )

(a) Construct the original cash flows of the project and the terminal balance by filling the blanks in the table.

(b) Determine the interest rate used in the project balance calculation and compute the present worth of this project at the computed interest rate.

4.11 In Problem 4.3:

(a) Graph the project balances (at $i = 10\%$) of each project as a function of n.

(b) By examining the graphical results in part (a), which project appears to be safest to undertake if there is some possibility of premature termination of the projects at the end of year 2?

4.12 Your firm is considering the purchase of an old office building with an estimated remaining service life of 50 years. The tenants have recently signed long-term leases, leading you to believe that the current rental income of $250,000 per year will remain constant over the remaining asset life. You estimate that operating expenses including income taxes will be $60,000 for the first year, and that they will increase by $3000

each year thereafter. You estimate that razing the building and selling the lot it stands on will realize a net amount of $20,000 at the end of the 50-year period. If you have the opportunity to invest your money elsewhere and thereby earn interest at the rate of 10% per annum, what would be the maximum amount you would be willing to pay for the building and lot at the present time?

4.13 Consider the following investment project.

n	A_n	i
0	$-\$1000$	10%
1	400	12
2	400	14
3	500	15
4	500	13
5	300	10

Suppose the company's reinvestment opportunities are changing over the life of the project as shown above (that is, the firm's MARR changes over the life of the project). For example, the company can invest their funds available now at 10% for the first year, 12% for the second year, and so forth. Calculate the net present worth of this investment and determine the acceptability of the investment.

4.14 Consider the following sets of investment projects:

Project's Cash Flow

n	A	B	C	D	E
0	$-\$100$	$-\$50$	$-\$100$	$-\$100$	$-\$5$
1	10	100	0	10	10
2	90	-30	0	20	30
3	10		30	30	-40
4	-10		70	40	
5	-40		130	125	

(a) Compute the future worth at the end of life for each project at $i = 15\%$.

(b) Determine the acceptability of each project.

4.15 Refer to Problem 4.14.

(a) Plot the future worth for each project as a function of interest rate (0%–50%).

(b) Compute the project balance of each project at $i = 15\%$.

(c) Compare the terminal project balances calculated in (b) with the results obtained in Problem 4.14(a).

(d) Without using the interest factor tables, compute the future worth based on the project balance concept.

4.16 In Problem 4.3, compute the future worth for each project at $i = 12\%$.

4.17 In Problem 4.4, compute the future worth for each project at $i = 10\%$.

4.18 The maintenance money for a new building has been sought. Mr. Kendall would like to make a donation to cover all future expected maintenance cost for the building. The maintenance costs are expected to be $20,000 each year for the first 5 years, $30,000 for each year 6 through 10, and $40,000 each year after that. (The building has an indefinite service life.)

(a) If the money is placed in an account that will pay 11% interest compounded annually, how large should the gift be?

(b) What is the equivalent annual maintenance cost over the infinite service life?

4.19 Consider the investment project whose cash flow pattern repeats itself every 5 years forever as shown in Fig. 4.24. At an interest rate of 10%, compute the capitalized equivalent amount for this project.

4.20 A group of concerned citizens has established a trust fund at a bank that pays 6% interest, compounded quarterly, to preserve a historical building by providing annual maintenance funds of $5000 forever. Compute the capitalized equivalent amount for these building maintenance expenses.

4.21 A bridge just constructed costs $1,000,000. It is estimated that the same bridge should be renovated every 50 years at a cost of $800,000. Annual repairs and maintenance are estimated to be $30,000 per year.

(a) If interest rate is 5%, determine the capitalized cost of the bridge.

(b) Suppose that the bridge should be renovated every 30 years instead of 50 years in (a). What is the capitalized cost of the bridge?

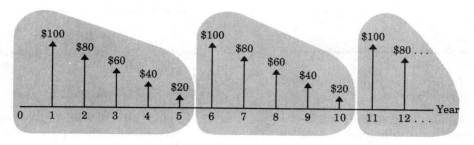

Figure 4.24 ■ Cash flow diagram (Problem 4.19)

(c) Repeat (a) and (b) with the interest rate of 10%. What can you say about the effect of interest on the results?

4.22 To decrease costs of operating a lock in a large river, a new system of operation is proposed. It will cost $450,000 to design and build. It is estimated that it will have to be reworked every 10 years at a cost of $50,000. In addition, there will be an expenditure of $40,000 at the end of the fifth year for a new type of gear that will not be available until then. The annual operating costs are expected to be $30,000 for the first 15 years and $25,000 a year thereafter. Compute the capitalized cost of perpetual service at $i = 10\%$.

4.23 Consider the following cash flow data for two competing investment projects.

Cash Flow Data
(unit: $ thousand)

n	Project A	Project B
0	−$200	−$1520
1	−1300	−165
2	−235	420
3	575	420
4	575	580
5	575	580
6	575	580
7	575	580
8	475	480
9	375	280
10	660	440

At $i = 10\%$, which project would be a better choice?

4.24 Consider the cash flows for the following investment projects. Assume MARR = 12%.

Project's Cash Flow

n	A	B	C	D	E
0	−$1000	−$1000	−$2000	$1000	−$1200
1	900	600	900	−300	400
2	500	500	X	−300	400
3	100	500	900	−300	400
4	50	100	X	−300	400

(a) Suppose A and B are mutually exclusive projects. Which project would be selected based on the NPW criterion?

(b) Repeat (a) using the NFW criterion.

(c) Find the minimum value of X that makes the project C acceptable?

(d) Would you accept D at $i = 20\%$?

(e) Assume that projects D and E are mutually exclusive. Which project would you select based on the NPW criterion?

4.25 Consider the following two mutually exclusive investment projects. Assume MARR = 15%.

Project's Cash Flow

n	A	B
0	−$4000	−$2000
1	2410	1310
2	2930	1720

(a) Which alternative would be selected using the NPW criterion?

(b) Which alternative would be selected using the NFW criterion?

4.26 Consider the following two mutually exclusive investment projects. Assume MARR = 15%.

Project's Cash Flow

n	A	B
0	−$300	−$800
1	0	1150
2	690	40

(a) Using the NPW criterion, which project would be selected?

(b) Sketch the $PW(i)$ function for each alternative on the same chart between 0% and 100%. For what range of i would you prefer project B?

4.27 Two methods of carrying away surface run-off water from a new subdivision are being evaluated:

- **Method A:** Dig a ditch where the first cost is $20,000 and $5500 of redigging and shaping are required at 5-year intervals forever.
- **Method B:** Lay concrete pipe where the first cost is $45,000 with a replacement required at 50-year intervals at a net cost of $65,000 forever.

At $i = 12\%$, which method is better?

4.28 Jones Construction Company needs a temporary office building at a construction site. Two types of heating schemes are considered. The first method is to use "bottled gas" for floor type furnaces. The second method is to install electric radiant panels in the walls and ceiling. This temporary building will be used for 5 years before being dismantled.

	Bottled Gas	Electric Panels
Investment cost	$6000	$8500
Service life	5 years	5 years
Salvage value	0	$1000
Annual O&M cost	$2000	$1000
Estimated extra expenses for income taxes		$220

(a) Compare the alternatives based on the present worth criterion at $i = 10\%$.

(b) Compare the alternatives based on the equivalent future worth method at $i = 10\%$.

4.29 Two alternative machines are being considered for a certain manufacturing process. Machine A has a first cost of $60,000 and its salvage value at the end of 6 years of estimated service life is $22,000. Operating costs of this machine are estimated to be $5000 per year. Extra income taxes with this machine are estimated to be $1400 per year. Machine B has a first cost of $35,000 and its estimated salvage value at the end of 6 years' service is estimated to be negligible. The annual operating costs will be $8000. Compare these two alternatives by the present worth method at $i = 10\%$

4.30 A certain motor is rated at 10 horsepower (hp) and costs $800. Its full load efficiency is specified to be 85%. A new design, high-efficiency motor of the same size has an efficiency of 90% but costs $1200. It is estimated that the motors will operate at rated 10 hp output for 1000 hours a year, and the cost of energy will be $0.07 per kilowatt-hour. Each motor is expected to have a 15-year life. The first motor will have a salvage value of $200, the second motor will have a salvage value of $300, both at the end of 15 years. Consider the MARR to be 8%. (*Note:* 1 hp = 0.7457 kW.)

(a) Determine which motor should be installed based on the NPW criterion.

(b) In (a), what if the motors operated 2000 hours a year instead of 1000 hours a year? Would the same motor in (a) be the choice?

4.31 Consider the following two mutually exclusive investment projects.

Project's Cash Flow

n	A	B
0	-$100	-$200
1	60	120
2	50	150
3	50	

Which project would be selected under infinite planning horizon with project repeatability likely based on the NPW criterion? Assume $i = 10\%$.

4.32 Consider the following two mutually exclusive investment projects with unequal service lives.

Project's Cash Flow

n	A1	A2
0	-$900	-$1800
1	-400	-300
2	-400	-300
3	-400 + 200	-300
4		-300
5		-300
6		-300
7		-300
8		-300 + 500

(a) What assumption(s) do you need to compare a set of mutually exclusive investments with unequal service lives?

(b) With the assumption(s) defined in (a) and using $i = 10\%$, determine which project should be selected.

(c) If your analysis period (study period) will be just 3 years, what should be the salvage value of project A2 at the end of year 3 to make the two alternatives economically indifferent?

4.33 Consider the following two mutually exclusive projects B1 and B2.

Project's Cash Flow and Salvage Value

n	B1 Cash Flow	B1 Salvage Value	B2 Cash Flow	B2 Salvage Value
0	-$12,000		-$10,000	
1	-2,000	$6,000	-2,100	$6,000
2	-2,000	4,000	-2,100	3,000
3	-2,000	3,000	-2,100	1,000
4	-2,000	2,000		
5	-2,000	2,000		

Salvage values represent the net proceeds (after tax) from disposal of the assets if they are sold at the end of each year. Both B1 and B2 will be available (or can be repeated) with the same costs and salvage values for an indefinite period.

(a) With the infinite planning horizon assumption, which project is a better choice at MARR = 12%?

(b) With a 10-year planning horizon, which project is a better choice at MARR = 12%?

4.34 Consider the following cash flows for two types of models.

Project's Cash Flow

n	Model A	Model B
0	-$2000	-$3000
1	1000	2000
2	1000	2000
3	1000	

Both models have no salvage value upon their disposal (at the end of its respective service life). The firm's MARR is known to be 15%.

(a) Notice that both models have a different service life. However, model B will be available in the future with the same cash flows. If your firm uses the present worth as a decision criterion, which model should be selected assuming that your firm will need either model for an indefinite period?

(b) Suppose that your firm will need either model for only two years. Determine the salvage value of model A at the end of year 2 that makes both models indifferent (equally likely).

4.35 An electric utility is taking bids on the purchase, installation, and operation of microwave towers.

	Cost per Tower	
	Bid A	Bid B
Equipment cost	$35,000	$30,000
Installation cost	$15,000	$30,000
Annual maintenance and inspection fee	$500	$750
Annual extra income taxes		$200
Life	50 years	40 years
Salvage value	$0	$0

Which is the most economical bid, if the interest rate is considered to be 8%? Either tower will have no salvage value after 20 years' use.

4.36 Consider the two different payment plans for Troy Aikman's contract in Problem 2.54. Compare these two

mutually exclusive plans based on the NPW method.

4.37 Consider the following two investment alternatives.

	Project's Cash Flow	
n	A1	A2
0	−$10,000	−$20,000
1	5,500	0
2	5,500	0
3	5,500	X
PW(15%)	?	6,300

The firm's MARR (minimum attractive rate of return) is known to be 15%.

(a) Compute the PW(15%) for A1.
(b) Compute the unknown cash flow X in year 3 for A2.
(c) Compute the project balance (at 15%) of A1 at the end of period 3.
(d) If these two projects are mutually exclusive alternatives, which project would you select?

4.38 For each of the following after-tax cash flows:

	Project's Cash Flow			
n	A	B	C	D
0	−$2500	−$7000	−$5000	−$5000
1	350	2500	−2000	−500
2	350	−2000	2000	−500
3	350	1500	2000	1000
4	300	500	2000	1000
5	300	500	5000	1000
6	300	500	5000	2000
7	300		2000	3000
8	300			

(a) Compute the project balance of each project as a function of project year at $i = 10\%$.

(b) Compute the net future worth of each project at $i = 10\%$.

(c) If projects C and D are mutually exclusive, which project would you select at $i = 10\%$?

(d) Suppose that projects B and C are mutually exclusive. Assume also that the required service period is 8 years, and the company considers leasing comparable equipment that has an annual lease expense of $1000 for the remaining required service period. Which project is a better choice?

4.39 Consider the following set of independent investment projects:

Project Cash Flows

n	A	B	C
0	−$100	−$100	$100
1	50	40	−40
2	50	40	20
3	50	40	−80
4	−500	10	
5	400	10	
6	400		

Assume MARR = 10% for the following questions.

(a) Compute the net present worth for each project and determine the acceptability of each project.

(b) Compute the net future worth of each project at the end of each project period and determine the acceptability of each project.

(c) Compute the project worth of each project at the end of 6 years with variable MARRs: 10% for $n = 0$ to $n = 3$ and 15% for $n = 4$ to $n = 6$.

4.40 Consider the following project balance profiles for proposed investment projects.

Project Balances

n	A	B	C
0	−$1000	−$1000	−$1000
1	−1000	−650	−1200
2	−900	−348	−1440
3	−690	−100	−1328
4	−359	85	−1194
5	105	198	−1000
Interest rate used	10%	?	20%
Net present worth	?	$79.57	?

Project balance figures are rounded to nearest dollars.

(a) Compute the net present worth of projects A and C, respectively.

(b) Determine the cash flows for project A.

(c) Identify the net future worth of project C.

(d) What would be the interest rate used for the project balance calculations for project B?

4.41 Consider the following project balance profiles for proposed investment projects. $A_2 = \$500$ for C.

Project Balances

n	A	B	C
0	−$1000	−$1000	−$1000
1	−800	−680	−530
2	−600	−302	X
3	−400	−57	−211
4	−200	233	−89
5	0	575	0
Interest rate used	0%	18%	12%

Project balance figures are rounded to nearest dollars.

(a) Compute the net present worth of each investment.
(b) Determine the project balance at the end of period 2 for project C?
(c) Determine the cash flows for each project.
(d) Identify the net future worth of each project.

4.42 In the construction of a bi-level mall, it is planned to install only 9 escalators at the start, although the ultimate design calls for 16. The question arises in the design whether to provide necessary facilities (stair supports, wiring conduits, motor foundations, etc.) to permit the installation of the additional escalators when needed merely by the purchase and installation of the escalators themselves, or to defer the investment in these facilities until the escalators need to be installed.

- **Option 1:** Provide these facilities now for all 7 future escalators at $200,000.
- **Option 2:** Defer the facility investment as needed. It is planned to install two more escalators in 2 years, three more

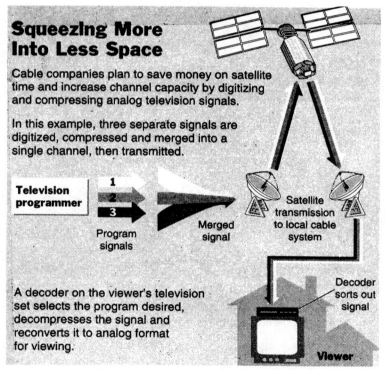

Figure 4.25 ■ Digital compression technique used to transmit more programs through cable channels

in 5 years, and the last two in 8 years. The installation of these facilities at the time they are required is estimated to cost $100,000 in year 2, $160,000 in year 5, and $140,000 in 8 years.

Additional annual expenses are estimated at $3000 for each escalator facility installed. At an interest rate of 12%, compare the net present worth of each option over 8 years.

4.43 Cable television companies and their equipment suppliers are on the verge of installing new technology that will pack many more channels into cable networks, creating a potential programming revolution with implications for broadcasters, telephone companies, and the consumer electronics industry.

Digital compression uses computer techniques to squeeze 3 to 10 programs into a single channel (Fig. 4.25). A cable system fully using digital compression technology would be able to offer well over 100 channels, compared with about 35 for the average cable television system now. Combined with increased use of optical fibers it might be possible to offer as many as 300 channels.

A cable company is considering installing this new technology to increase subscription sales and save on satellite time. The company estimates that the installation will take place over 2 years. The system is expected to have an 8-year service life with the following savings and expenditures:

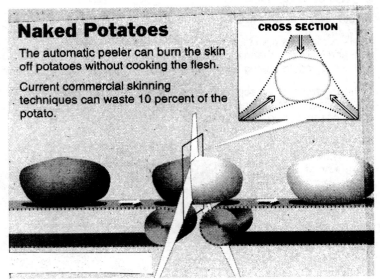

Figure 4.26 ■ Peeling potatoes by laser beams

	Digital Compression
Investment:	
First year	$3,200,000
Second year	4,000,000
Annual savings in satellite time	2,000,000
Incremental annual revenues due to new subscriptions	4,000,000
Incremental annual expenses	1,500,000
Incremental annual income taxes	1,300,000
Economic service life	8 years
Net salvage values	1,200,000

Note that the project has a 2-year investment period followed by an 8-year service life (a total 10-year life project). This implies that the first annual savings will occur at the end of year 3 and the last savings will occur at the end of 10 years. If the firm's MARR is 15%, justify the economic worth of the project based on the NPW method.

4.44 A large food-processing corporation is considering using laser technology to speed up and eliminate waste in the potato-peeling process (see Fig. 4.26). To implement the system, the company anticipates needing $3 million to purchase the industrial strength lasers. The system will save $1,200,000 per year in labor and materials. However, it will require an additional operating and maintenance cost of $250,000. There is also an increase in annual income taxes of $150,000. The system is expected to have a 10-year service life and will have a salvage value of about $200,000. If the company's MARR is 18%, justify the economics of the project based on the NPW method.

CHAPTER 5

Annual Equivalent Worth Analysis

Y ou see two refrigerators in a department store. Brand A sells for $999.99 and uses $85 worth of electricity a year. Brand B costs $100 more to purchase but only $65 a year to run. Given that either refrigerator should last at least ten years without repair, which one would you buy? Given a choice between these refrigerators, which differ only in price and energy efficiency, a careful consumer would buy brand B. It costs more to buy but less to operate. How much will the more expensive refrigerator save its owner over ten years? How would you calculate the annual operating cost per kilowatt-hour per year for each brand?

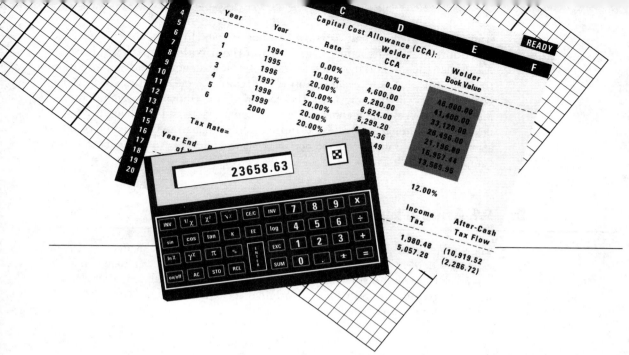

Suppose you are considering buying a new car. If you expect to drive 18 000 km per year, can you figure out how much it costs per kilometre? You would have good reason to want to know that figure if you were being reimbursed by your employer on a per-kilometre basis for the business use of your car. Or consider a real estate developer planning to build a shopping center of 50,000 m². What would be the minimum annual rental fee per square metre required to recover the initial investment?

Annual equivalence analysis is the method by which we calculate these and other unit costs. As its name suggests, **annual equivalent worth analysis,** or AE, is also the method by which we can determine an equivalent *annual* rather than overall (that is, present or future) worth of a project. AE analysis, along with present worth analysis, is the second major equivalence technique for placing alternatives on a common basis of comparison. In this chapter, we will develop the annual equivalent criterion and demonstrate a number of situations in which annual equivalence analysis is preferable to other methods of comparison.

■ 5.1 Annual Equivalent Criterion

The annual equivalent worth (AE) criterion is a basis for measuring investment worth by determining equal payments on an annual basis. Knowing that we can convert any lump-sum cash amount into a series of equal annual payments, we may first find the NPW for the original series and then multiply the NPW by the capital recovery factor:

$$AE(i) = PW(i)(A/P, i, N). \tag{5.1}$$

The accept/reject decision rule for a single *revenue* project is

If $AE(i) > 0$, accept the investment.

If $AE(i) = 0$, remain indifferent.

If $AE(i) < 0$, reject the investment.

Notice that the factor $(A/P, i, N)$ in Eq. (5.1) is positive for $-1 < i < \infty$. This indicates that the $AE(i)$ value will be positive if and only if $PW(i)$ is positive. In other words, accepting a project that has a positive $AE(i)$ value is equivalent to accepting a project that has a positive $PW(i)$ value. Therefore, the AE criterion should provide a basis for evaluating a project that is consistent with the NPW criterion.

As with the present worth analysis, when you are comparing mutually exclusive *service* projects whose revenues are the same, you may compare them based on *cost* only. In this situation, you will select the alternative with the largest (least negative) annual equivalent worth.

Example 5.1   Annual equivalent worth by conversion from PW

The Sky Communications Company (SCC) is planning to develop satellite-based systems that will enable airline passengers to make phone calls or send facsimiles from a plane flying virtually anywhere in the world. The systems will use a network of satellites to bounce signals from a plane to ground stations, which are linked with normal telephone networks. Because the systems will use digital technology, calls will be crisper than those made on today's airborne telephones, which use conventional radio technology. But the systems will not allow people on the ground to call airborne passengers, and one stumbling block remains: The airlines have yet to install the digital cockpit electronics needed. Five international airlines have agreed to offer the inflight telephone service on their flights if SCC develops a satellite-based telephone system. Having the system installed for 120 airplanes from the five airlines, SCC has estimated the projected cash flows (in million dollars) as follows:

n	$A_n(\$)$
0	-15
1	-3.5
2	5
3	9
4	12
5	10
6	8

In the first place, SCC wants to determine whether this project can be justified at MARR = 15%. Then the company wants to know what would be the annual benefit (or loss) that can be generated from the airplanes.

Discussion: When a cash flow has no special pattern, it is easiest to find AE in two steps: (1) Find the NPW (or NFW) of the flow; (2) find the AE of the NPW (or NFW). This is the method presented below. You might try any other method with this cash flow to prove how difficult that might be.

Solution

Given: Cash flow in Fig. 5.1, $i = 15\%$
Find: AE

We first compute the NPW at $i = 15\%$.

$$PW(15\%) = -\$15 - \$3.5(P/F, 15\%, 1) + \$5(P/F, 15\%, 2) + \cdots$$
$$+ \$10(P/F, 15\%, 5) + \$8(P/F, 15\%, 6)$$
$$= \$6.946 \text{ million.}$$

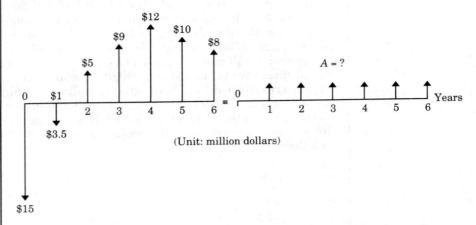

Figure 5.1 ■ Cash flow diagram (Example 5.1)

$12

$9 $10

$5 $8   A = ?

0    $1   0   Years
         2   3   4   5   6   =   1   2   3   4   5   6

    $3.5

                              (Unit: million dollars)

$15

Since $PW(15\%) > 0$, the project would be acceptable under the NPW analysis. Now, spreading the NPW over the project life gives

$$AE(15\%) = \$6.946(A/P, 15\%, 6) = \$1.835 \text{ million.}$$

Since $AE(15\%) > 0$, the project is worth undertaking. This positive AE value indicates that the project is expected to bring in a net annual benefit of $1.835 million over the project life.

There are some situations where we may observe a cyclic cash flow pattern over the project life. Unlike in Example 5.1, where we first computed the NPW of the entire cash flow and then calculated AE value from this NPW, we can compute the AE by just examining the first cash flow cycle. Then compute the NPW for the first cash flow cycle and derive the AE over the first cash flow cycle. This short-cut method provides the same solution as calculating the NPW of the entire project and then computing the AE from this NPW.

Example 5.2   Annual equivalent worth—repeating cash flow cycles

SOLEX Company is producing electricity directly from solar energy by using a large array of solar cells and selling the power to the local utility company. SOLEX has decided to use amorphous silicon cells because of their low initial cost, but the cells degrade over time, resulting in lower conversion efficiency and power output. The cells must be replaced every 4 years, resulting in a particular cash flow pattern that repeats itself as shown in Fig. 5.2. Determine the annual equivalent cash flows at $i = 12\%$.

Figure 5.2 ■ Conversion of repeating cash flow cycles into an equivalent annual payment series (Example 5.2)

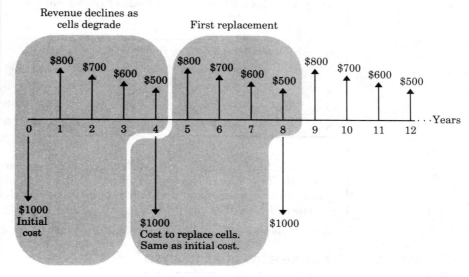

(Unit: thousand dollars)

Solution

Given: Cash flows in Fig. 5.2, $i = 12\%$

Find: Annual equivalent benefit

To calculate the AE, we need only consider one cycle over its 4-year period. For $i = 12\%$, we first obtain the NPW for the first cycle

$$PW(12\%) = -\$1,000,000$$
$$+ [\$800,000 - \$100,000(A/G, 12\%, 4)] (P/A, 12\%, 4)$$
$$= -\$1,000,000 + \$2,017,150$$
$$= \$1,017,150.$$

Then, we calculate the AE value over the 4-year life cycle:

$$AE(12\%) = \$1,017,150(A/P, 12\%, 4)$$
$$= \$334,880.$$

Then we can say that the two cash flow series are equivalent:

Original Cash Flows		Annual Equivalent Flows	
n	A_n	n	A_n
0	−$1,000,000 ≡	0	$ 0
1	800,000	1	334,880
2	700,000	2	334,880
3	600,000	3	334,880
4	500,000	4	334,880

We can extend this cash flow equivalency over the remaining cycles of the cash flow. The reasoning, of course, is that each similar set of five values (one disbursement and four receipts) is equivalent to four annual receipts of $334,880 each.

5.1.1 Benefits of AE Analysis

Example 5.1 should look familiar to you: It is exactly the situation we encountered in Chapter 2 when we converted a mixed cash flow into a single present value and then into a series of equivalent cash flows. In the case of Example 5.1, you may wonder why we bothered to convert NPW to AE since we already know the project is acceptable from NPW analysis. In fact, the example was mainly an exercise to familiarize you with the AE calculation.

However, in the real world there are a number of situations in which AE analysis is preferred or demanded over NPW analysis. Consider, for example, the fact that corporations issue annual reports and develop yearly budgets in which they find it more useful to present the annual cost or benefit of an ongoing project, rather than its overall cost or benefit. Some additional situations in which AE analysis is preferred include:

1. **Consistency of report formats.** Financial managers more commonly work with yearly rather than overall costs in any number of internal and external reports. Engineering managers may be required to submit project analyses in annual form for consistency and ease of use by other members of the corporation and stockholders.
2. **Need for unit costs.** In many situations, projects must be broken into unit costs for ease of comparison with alternatives. *Make-or-buy* and *reimbursement* analyses are key examples and will be discussed further in this chapter.
3. **Unequal project lives.** As we saw in Chapter 4, comparing projects with unequal service lives is complicated by the need to determine the lowest common multiple life. For the special situation of indefinite service period and replacement with identical projects, we can avoid this complication by

using AE analysis. This situation will also be discussed in more detail in this chapter.

5.1.2 Capital Costs Versus Operating Costs

The AE method is sometimes called the **annual equivalent cost** method when only costs are involved. In this case, there are two kinds of costs that revenues must cover: *operating costs* and *capital costs*. Operating costs are incurred by the operation of physical plant or equipment to provide service; they include such items as labor and raw materials. Capital costs are incurred by purchasing the assets used in production and service. Normally, capital costs are nonrecurring (that is, one-time costs), whereas operating costs recur as long as the asset is owned.

Because operating costs recur over the life of a project, they tend to be estimated on an annual basis anyway, so for the purposes of an annual equivalent cost analysis, they require no special calculation on our part. However, because capital costs tend to be one-time costs, in conducting an annual equivalent cost analysis, we must translate this one-time cost into its annual equivalent over the life of the project. The annual equivalent of a capital cost is given a special name: **capital recovery cost,** designated $CR(i)$.

There are two general monetary transactions associated with the purchase and eventual retirement of a capital asset, its initial cost (P) and its salvage value (S). Taking into account these sums, we calculate the capital recovery factor as:

$$CR(i) = P(A/P,i,N) - S(A/F,i,N). \qquad (5.2)$$

Recalling the algebraic relationships between factors in Table 2.3, notice that the $(A/P,i,N)$ factor can be expressed as

$$(A/P,i,N) = (A/F,i,N) + i.$$

Then, we may rewrite the $CR(i)$ as

$$\begin{aligned} CR(i) &= P(A/P,i,N) - S[(A/P,i,N) - i] \\ &= (P - S)(A/P,i,N) + iS. \qquad (5.3) \end{aligned}$$

We may interpret this situation thus: To obtain the machine, one borrows a total of P dollars, S dollars of which are returned at the end of the Nth year. The first term $(P - S)(A/P,i,N)$ implies that the balance $(P - S)$ be paid back in equal installments over the N-year period at a rate of i, and the second term iS implies that interest in the amount iS is paid on S until it is repaid. Thus, the amount to be financed is $P - S(P/F,i,N)$, and the installments of this loan over the N-period are

$$
\begin{aligned}
AE(i) &= -[P - S(P/F,i,N)](A/P,i,N) \\
&= -P(A/P,i,N) + S(P/F,i,N)(A/P,i,N) \\
&= -P(A/P,i,N) + S(A/F,i,N) \\
&= -CR(i).
\end{aligned} \tag{5.4}
$$

The minus sign in front of the capital recovery factor indicates that these installments are payments or cash outflows. Therefore, the $CR(i)$ tells us what the bank would charge each year for such an arrangement. Many auto leases are based on this arrangement in that most require a guarantee of S dollars in salvage. From an industry viewpoint, $CR(i)$ is the annual cost to the firm of owning the asset.

With this information, we can determine the amount of annual savings required to recover the capital and operating costs associated with a project. As an illustration, we consider Example 5.3.

Example 5.3 Annual equivalent worth—capital recovery cost

Consider a machine that costs $5000 and has a 5-year useful life. At the end of the 5 years, it can be sold for $1000 after tax adjustment. If the firm could earn an after-tax revenue of $1100 per year with this machine, should it be purchased at an interest rate of 10%? (All benefits and costs associated with the machine are accounted for in these figures.)

Solution

Given: $I = \$5000$, $S = \$1000$, $A = \$1100$, $N = 5$ years, $i = 10\%$ per year
Find: AE, and determine whether to purchase

We will compute the capital costs in two different ways:

- **Method 1:** We will first compute the NPW of the cash flows and then compute the AE from the calculated NPW:

$$
\begin{aligned}
PW(10\%) &= -\$5000 + \$1100(P/A, 10\%, 5) \\
&\quad + \$1000(P/F, 10\%, 5) \\
&= -\$5000 + \$1100(3.7908) + \$1000(0.6209) \\
&= -\$209.22 \\
AE(10\%) &= -\$209.22(A/P, 10\%, 5) = -\$55.19.
\end{aligned}
$$

This negative AE value indicates that the machine does not generate sufficient revenue to recover the original investment, so we may reject the project. In fact, there will be an equivalent loss of $55.19 per year over the machine's life (see Fig. 5.3a).

Figure 5.3 ■ Alternative ways of computing capital recovery cost for an investment (Example 5.3)

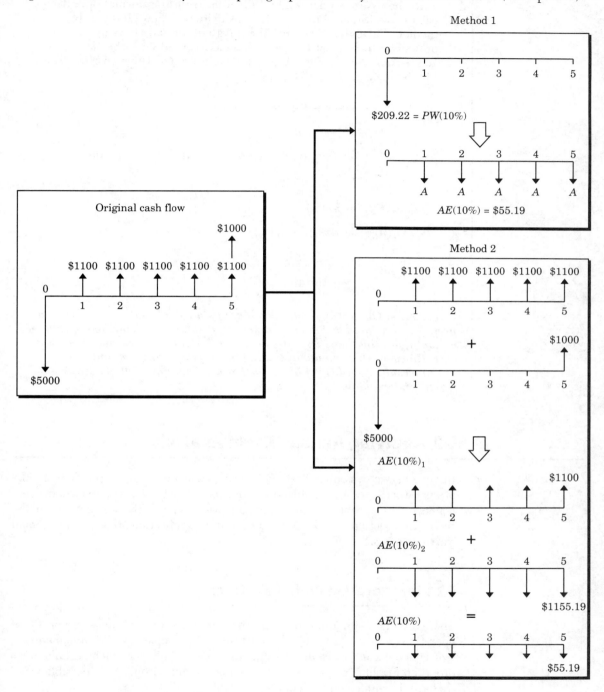

- **Method 2:** The second method is to separate the cash flows associated with the asset acquisition and disposal from the normal operating cash flows. Since the operating cash flows—the $1100 yearly income—are already given in equivalent annual flows $(AE(i)_2)$, we only need to convert the cash flows associated with the asset acquisition and disposal into equivalent annual flows $(AE(i)_1)$ (see Fig. 5.3b). Using Eq. (5.3),

$$CR(i) = (P - S)(A/P,i,N) + iS$$

$$AE(i)_1 = -CR(i)$$

$$= -[(\$5000 - \$1000)\,(A/P, 10\%, 5) + \$1000(0.10)]$$

$$= -\$1155.19$$

$$AE(i)_2 = \$1100$$

$$AE(10\%) = AE(i)_1 + AE(i)_2$$

$$= -\$1155.19 + \$1100$$

$$= -\$55.19.$$

Comments: *Obviously, method 2 saves us a calculation step, so we may prefer it to method 1. We may interpret method 2 as determining that the annual operating benefits must be at least $1155.19 to recover the asset cost. However, the annual operating benefits actually amount to only $1100, resulting in loss of $55.19 per year. Therefore, the project is not worth undertaking.*

■ 5.2 Applying Annual Worth Analysis

In general, we can solve most engineering economic analysis problems by the present worth methods that were introduced in Chapter 4. However, some economic analysis problems can be solved more efficiently by annual worth analysis. In this section, we will introduce several applications that call for annual worth analysis.

5.2.1 Unit Cost/Unit Profit Calculation

There are many situations in which we want to know the unit cost (or profit) of operating an asset. As we briefly illustrated in Example 5.4, the annual equivalent concept can be useful in estimating the savings per machine hour for a proposed machine acquisition. This unit cost comparison is only possible when we have the annual equivalent cost available.

Example 5.4   Equivalent worth per unit of time

Consider the investment in the metal-cutting machine in Example 4.5. Recall that this three-year investment was expected to generate a NPW of $3553. Suppose that the machine will be operated for 2000 hours per year. Compute the equivalent savings per machine hour at $i = 15\%$.

Solution

Given: NPW = $3553, N = 3 years, i = 15% per year, 2000 machine hours per year

Find: Equivalent savings per machine hour

We first compute the annual equivalent savings from the use of the machine. Since we already know the NPW of the project, we obtain the AE by

$$AE(15\%) = \$3553(A/P, 15\%, 3) = \$1556.$$

With an annual usage of 2000 hours, the equivalent savings per machine hour would be

$$\text{Savings per machine hour} = \$1556/2000 = \$0.78/\text{hr}.$$

Note that we cannot simply divide the NPW amount ($3553) by the total number of machine hours over the three-year period (6000 hours), or $0.59/hr. This $0.59 figure represents the immediate savings in present worth for each hourly use of the equipment, but does not consider the time over which the savings occur. Once we have the annual equivalent worth, we can divide by the desired time unit if the compounding period is one year. If the compounding period is shorter, then the equivalent worth should be calculated for the compounding period.

Example 5.5 illustrates the use of the annual equivalent cost concept in estimating the printing cost per page for a proposed in-house printing project.

Example 5.5   Equivalent worth per unit of production

National Engineering Service is considering printing two publications in its print shop that presently are contracted to outside vendors. The two publications are *CENTS* magazine and *INSIDE*, an internal newsletter for stockholders and employees.

CENTS magazine is a quarterly publication, 13 pages in length. Fifty thousand (50,000) copies are needed per quarter. *INSIDE* is printed monthly, is three pages in length, and 4500 copies are needed per issue. National Engineering has itemized the cost data for decision making as follows:

- Annual printing cost by outside vendor:

CENTS (50,000 copies × 4 quarters)	$137,444
INSIDE (4500 copies × 12 months)	67,728
Total annual costs	$205,172

CENTS (13 pages × 50,000 × 4)	2,600,000 pages
INSIDE (3 pages × 4500 × 12)	162,000 pages
Total pages	2,762,000 pages

The printing cost (outside contract) per page:

$$\$205{,}172/2{,}762{,}000 = 7.43 \text{ cents per page.}$$

- Required expenditures if printed internally:

    1. Capital costs:

Two-color printer	$136,000
Binding machine	90,200
Site preparation	45,000
Total	$271,200

    The printing equipment has a 6-year service life with $20,000 salvage value.

    2. Annual operating costs:

Annual wages and salaries	$120,840
Annual supplies (paper, ink, etc.)	$ 45,200
Annual operating and maintenance cost	$ 15,000
Annual incremental income taxes	$ 24,000
Total	$205,040

    Determine the in-house printing cost per page if the firm's MARR is 12%.

Solution

Given: $P = \$271{,}200$, $S = \$20{,}000$, $N = 6$ years, $i = 12\%$ per year, 2,762,000 pages per year

Find: Equivalent cost per page

First, we will compute the capital costs associated with the in-house printing option. Using Eq. (5.3), we obtain

$$CR(12\%) = (\$271,200 - \$20,000)\,(A/P, 12\%, 6) + \$20,000(0.12)$$
$$= \$63,498.$$

Then, as shown in Fig. 5.4, the total annual equivalent worth becomes:

$$\text{annual equivalent worth} = -\text{ capital costs} - \text{operating costs}$$
$$AE(12\%) = -\$63,498 - \$205,040$$
$$= -\$268,538.$$

The negative sign indicates that AE is a cash outflow or cost. In comparing the options, we are interested in the magnitude of this quantity and can ignore the negative sign.

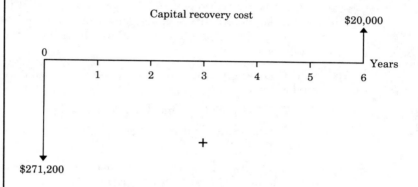

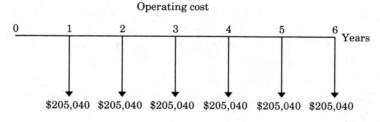

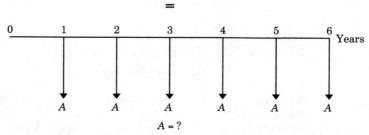

Figure 5.4 ■ Annual equivalent costs associated with the in-house printing option (Example 5.5)

We obtain the printing cost (in-house) per page

$$\$268,538/2,762,000 = 9.72 \text{ cents per page.}$$

The in-house printing option is not advisable at the current volume because it costs 2.29 cents more per page than does the outside contract.

5.2.2 Make-or-Buy Decision

Make-or-buy problems are among the most common types of business decisions (in fact, Example 5.5 was a special case of make-or-buy decision). At any given time, a firm has the option of either buying an item or producing it. If either the "make" or the "buy" alternative requires the acquisition of machinery and equipment, then it is an investment decision. Since the cost of an outside service (the "buy" alternative) is usually quoted in terms of dollars per unit, it is easier to compare the two alternatives if the differential costs of the "make" alternative are also given in dollars per unit. This unit cost comparison requires the use of annual worth analysis. The specific procedure is as follows:

- Step 1: Determine the time span (planning horizon) for which the part (or product) will be needed.
- Step 2: Determine the annual quantity of the part (or product).
- Step 3: Obtain the unit cost of purchasing the part (or product) from the outside firm.
- Step 4: Determine the equipment, manpower, and all other resources required to make the part (or product).
- Step 5: Estimate the net cash flows associated with the "make" option over the planning horizon.
- Step 6: Compute the annual equivalent worth of producing the part (or product).
- Step 7: Compute the unit cost of making the part (or product) by dividing the magnitude of the annual equivalent worth by the required annual volume.
- Step 8: Choose the option with the minimum unit cost.

Example 5.6 Equivalent worth—make or buy (embellishment of Example 1.4)

Ampex Corporation currently produces both videocassette cases and metal particle magnetic tape for commercial use. Due to the projected increased demand for metal particle tapes, Ampex is deciding between the increased internal production of empty cassette cases and magnetic tape or purchasing empty cassette cases from an outside vendor. If it purchases the cases from a vendor, Ampex must also buy a specialized equipment to load magnetic tapes, since their current loading machine is not compatible with the cassette cases produced by the vendor. The projected production rate of cassettes is 79,815 units per week for 48 weeks of operation per year. The planning horizon is 7 years. The accounting department has itemized the annual costs associated with each option as follows:

- Make option (annual costs):

Labor	$1,445,633
Materials	$2,048,511
Incremental overhead	$1,088,110
Total annual cost	$4,582,254

- Buy option:

  Capital costs:

Acquisition of a new loading machine	$405,000
Salvage value at end of 7 years	$ 45,000

  Annual operating costs:

Labor	$ 251,956
Purchasing empty cassette ($0.85/unit)	$3,256,452
Incremental overhead	$ 822,719
Total annual operating costs	$4,331,127

After considering the effects of income taxes, the accounting department has summarized the net cash flows for each option as follows:

Net Cash Cost

Year	Make Option	Buy Option
0		$ 405,000
1	$2,749,352	2,575,533
2	2,749,352	2,559,003
3	2,749,352	2,570,338
4	2,749,352	2,578,435
5	2,749,352	2,584,218
6	2,749,352	2,584,218
7	2,749,352	2,549,989

(Note the conventional assumption that the cash flows occur in discrete lumps at ends of years, as shown in Fig. 5.5.) Assuming that the Ampex's MARR is 14%, calculate the unit cost under each option.

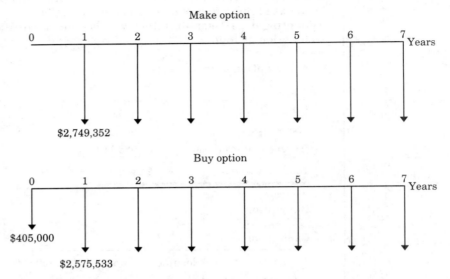

Figure 5.5 ■ Make-or-buy analysis (Example 5.6)

Solution

Given: Cash flows for two options, $i = 14\%$
Find: AE cost for the options, and which one is preferred

The required production volume is

$$79{,}815 \text{ units/week} \times 48 \text{ weeks} = 3{,}831{,}120 \text{ units per year.}$$

Now we need to calculate the annual equivalent worth under each option. Since the "make option" is already given in annual basis, we need only calculate the annual equivalent cost for the "buy option."

$$AE(14\%)_{\text{make}} = -\$2{,}749{,}352$$

$$AE(14\%)_{\text{buy}} = -[\$405{,}000 + \$2{,}575{,}533(P/F,14\%,1)$$

$$+ \ \$2{,}559{,}033(P/F,14\%,2) + \dots$$

$$+ \ \$2{,}549{,}989(P/F,14\%,7)](A/P,14\%,7)$$

$$= -\$2{,}666{,}187.$$

Obviously, this annual equivalent calculation indicates that Ampex would be better off by buying the cassette cases from outside. However, Ampex wants to know the unit costs in order to set a price for the product. In this situation, we need to calculate the unit cost of producing the cassette tapes under each option. This is done by dividing the magnitude of the annual equivalent worth for each option by the annual quantity required:

- Make option:

$$\text{unit cost} = \$2{,}749{,}352/3{,}831{,}120 = 71.67 \text{ cents/unit.}$$

- Buy option:

$$\text{unit cost} = \$2{,}666{,}187/3{,}831{,}120 = 69.69 \text{ cents/unit.}$$

Buying the empty cassette cases from the outside vendor and loading the tape in-house will save Ampex 1.98 cents per cassette.

Comments: *There are two important noneconomic factors that should also be considered. The first is the question of whether the quality of the supplier's part is better than (or equal to) what the firm is presently manufacturing or worse. The second is the reliability of the supplier in terms of providing the needed quantities of the part on a timely basis. A reduction in quality or reliability should virtually rule out a switch from making to buying.*

5.2.3 Cost Reimbursement

Companies often have the need to calculate the cost of equipment that corresponds to a unit of use of that equipment. A familiar example is an employer's reimbursement of costs for the use of an employee's personal car for business purposes. If an employee's job is dependent on obtaining and using a personal vehicle in the employer's behalf, reimbursement on the basis of the employee's overall costs per kilometre seems fair. Although many car owners think of costs in terms of outlays for gasoline, oil, tires, and tolls, a careful examination shows that in addition to these operating costs, which are directly related to the use of the car, there are also *ownership costs* which occur whether or not the vehicle is driven.

Ownership costs include depreciation, insurance, finance charges, registration fees, scheduled maintenance, accessory costs, and storage. Even if the vehicle is permanently stored, a portion of each of these costs occurs. **Depreciation** is the loss in value of the vehicle during the time it is owned due to (1) the passage of time, (2) its mechanical and physical condition, and (3) the number of kilometres it is driven. (This type of depreciation, known as *economic depreciation* (as opposed to accounting depreciation), is discussed in Chapter 7.)

Operating costs include nonscheduled repairs and maintenance, gasoline, oil, tires, parking and tolls, and taxes on gasoline and oil. Certainly, the more a car is used the greater these costs become.

Once the cost of owning and operating a personal vehicle is determined, you may wonder what the minimum reimbursement rate per kilometre should be so that you can break even. Here we are looking at the reimbursement cost equation that is solely determined by a decision variable, reimbursement rate. The reimbursement rate that is just enough to equal the cost of owning and operating is known as the **break-even point**. Example 5.7 illustrates the process of obtaining the cost of reimbursement for the use of an employee's vehicle for business use.

Example 5.7 Break-even point—per unit of equipment use

Sam Tucker is a sales engineer at Buford Chemical Engineering Company. Sam owns two vehicles, and one of them is entirely dedicated to his business use. His business car is a 1994 compact-size automobile purchased with personal savings for $11,000. Based on his own records and an analysis published in the *Financial Post*, Sam has estimated the costs of owning and operating his business vehicle for the first three years as follows:

	First Year	Second Year	Third Year
Expected kilometres driven	14,500	13,000	11,500
Depreciation	2,879	1,776	1,545
Scheduled maintenance	100	153	220
Insurance	635	635	635
Registration and taxes	78	57	50
Total ownership cost	$ 3,692	$ 2,621	$ 2,450
Nonscheduled repairs	35	85	200
Replacement tires	35	30	27
Accessories	15	13	12
Gasoline and taxes	688	650	522
Oil	80	100	100
Parking and tolls	135	125	110
Total operating costs	$ 988	$ 1,003	$ 971
Total of all costs	$ 4,680	$ 3,624	$ 3,421

Sam expects to drive 14,500, 13,000, and 11,500 business kilometres respectively for the next three years. If his interest rate is 6%, what should be the reimbursement rate per kilometre so that Sam can break even?

Discussion: You may wonder why the initial cost of the vehicle ($11,000) is not explicitly considered in the costs Sam has estimated. The answer is in the depreciation listings. In true economic analysis, capital costs are not considered all at once in the year in which they are incurred. Rather, the costs are spread out over the useful life of the asset, based on how much of the cost of the asset is used up each year. Thus, the depreciation amount of $2879 during the first year represents the fact that, if the subcompact car were bought for $11,000 and then sold at the end of the first year when it had been driven 14,500 kilometres, Sam would expect the sale price to be $2879 less than the original purchase price. In Chapter 7, we will explore depreciation in more detail and learn the conventions for its calculation. For this example, depreciation amounts are given to you to more closely represent a true economic analysis of the problem.

Solution

Given: Yearly costs and kilometres, $i = 6\%$ per year

Find: Equivalent cost per kilometre

Suppose Buford pays Sam $X per kilometre for his personal car. Assuming that Sam expects to travel 14,500 kilometres the first year, 13,000 kilometres the second year, and 11,500 kilometres the third year, his annual reimbursements would be

Year	Total Kilometres Driven	Reimbursement ($)
1	14,500	$(X)(14,500) = 14,500X$
2	13,000	$(X)(13,000) = 13,000X$
3	11,500	$(X)(11,500) = 11,500X$

As depicted in Fig. 5.6, the annual equivalent reimbursement would be

$$[14,500X(P/F, 6\%, 1) + 13,000X(P/F, 6\%, 2)$$
$$+ 11,500X(P/F, 6\%, 3)] (A/P, 6\%, 3)$$
$$= 13,058X.$$

The annual equivalent costs of owning and operating would be

$$[\$4680(P/F, 6\%, 1) + \$3624(P/F, 6\%, 2) + \$3421(P/F, 6\%, 3)](A/P, 6\%, 3)$$
$$= \$3933.$$

Then, the minimum reimbursement rate should be

$$13,058X = \$3933$$
$$X = 30.12 \text{ cents per kilometre.}$$

If Buford pays him 30.12 cents per kilometre or more, Sam's decision to use his car for business makes sense economically.

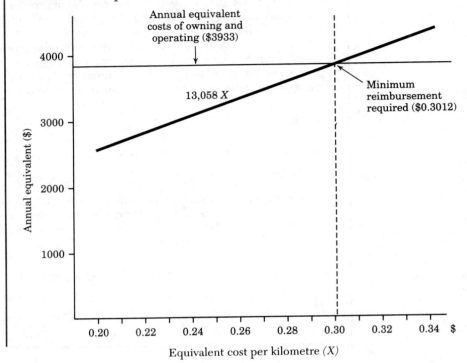

Figure 5.6 ■ Annual equivalent reimbursement as a function of cost per kilometre (Example 5.7)

5.2.4 Mutually Exclusive Alternatives

In this section, we will consider the situation where we need to compare two or more mutually exclusive alternatives based on the annual equivalent worth. In Section 4.5, we discussed the general principles that should be applied when we compare mutually exclusive alternatives with unequal service lives. The same general principles should be applied in comparing mutually exclusive alternatives based on the annual equivalent worth—we must compare mutually exclusive alternatives in *equal time spans*. Therefore, we must give a careful consideration of the time period covered by the analysis, the *analysis period*. We will consider two situations: (1) The analysis period equals project lives, and (2) the analysis period is different from project lives.

Analysis Period Equals Project Lives

Let's begin our comparison with the simplest situation where the project lives equal the analysis period. In this situation, we compute the AE value for each project and select the one that has the largest AE. For service projects this is the least negative AE.

In many situations, we need to compare a set of different design alternatives where each design would produce the same number of units (constant revenues), but require different amounts of investment and operating costs (because of different degrees of mechanization). Example 5.8 will illustrate this situation.

Example 5.8 Annual worth comparison—equal project lives

Hamilton Couplings Company is considering two types of manufacturing system to produce its shaft couplings: (1) cellular manufacturing system (CMS) and (2) flexible manufacturing system (FMS). Operating cost, initial investment, and salvage value for each alternative are estimated as follows:

Items	CMS	FMS
Average number of pieces produced/year	544,000	544,000
Project life	6 years	6 years
Annual labor costs	$1,169,600	$ 707,200
Annual material costs	$ 832,320	$ 598,400
Annual incremental overhead costs	$3,150,000	$ 1,950,000
Annual tooling costs	$ 470,000	$ 300,000
Annual inventory costs	$ 141,000	$ 31,500
Annual income taxes	$1,650,000	$ 1,917,000
Total annual operating costs	$7,412,920	$ 5,504,100
Investment	$5,000,000	$12,500,000
Net salvage value	$ 500,000	$ 1,000,000

Figure 5.7 illustrates the cash flows associated with each alternative. The firm's MARR is 15%. Which alternative would be a better choice based on the annual equivalent cost? Determine also the unit production cost under each system.

Solution

Given: Cash flows shown in Fig. 5.7, $i = 15\%$ per year

Find: Unit production cost for each alternative, and select the better alternative

Since we can assume that both manufacturing systems would provide the same level of revenues over the analysis period, we can compare these alternatives based on cost only. (These are service projects.) As we

Figure 5.7 ■ Comparison of mutually exclusive alternatives with equal project lives (Example 5.8)

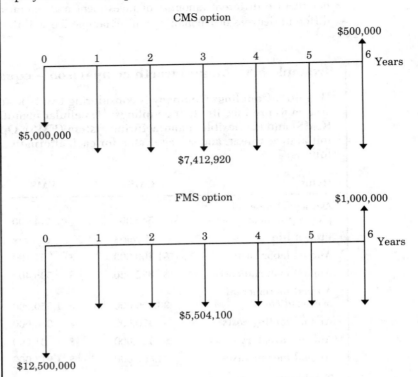

discussed in Section 5.1.2, we can separate the capital costs and the operating costs.

CMS Option	Annual Equivalent Worth ($)
Capital costs − ($5,000,000 − $500,000)(A/P,15%,6) − $500,000(0.15)	− $1,264,066
Operating costs	− 7,412,920
Total	− $8,676,986
Unit production cost ($8,676,986/544,000)	$15.95 per part

FMS Option	
Capital costs − ($12,500,000 − $1,000,000)(A/P,15%,6) − $1,000,000(0.15)	− $3,188,720
Operating costs	− 5,504,100
Total	− $8,692,820
Unit production cost ($8,692,820/544,000)	$15.98 per part

Since $AE(15\%)_{CMS} > AE(15\%)_{FMS}$, we would select CMS. Although the FMS would provide an incremental annual savings of \$1,908,820 in operating costs, the savings do not justify the incremental investment of \$7,500,000. In terms of unit production cost, each part produced under FMS would cost 3 cents more than under CMS.

Comments: *Note that the CMS option was marginally preferred to the FMS option. However, there are dangers in relying solely on the easily quantified savings in input factors—such as labor, energy, and materials—from FMS and in not considering gains from improved manufacturing performance that are more difficult and subjective to quantify. Factors such as improved product quality, increased manufacturing flexibility (rapid response to customer demand), reduced inventory levels, and increased capacity for product innovation are frequently ignored because we have inadequate means for quantifying their benefits. If these intangible benefits were considered, however, the FMS option could come out better than the CMS option.*

Analysis Period Different from Project Lives

In Section 4.5.3, we learned that in present worth analysis there must be a *common analysis period* (least common multiple periods) when we compare mutually exclusive alternatives. Annual worth analysis also requires establishing such a common analysis period, but it offers some computational advantages as opposed to the present worth analysis, provided the following criteria are met:

1. There is a continuing requirement for the service of the selected alternative.
2. Each alternative will be replaced by an identical asset that has the same costs and performance.

When these two criteria are in effect, we may solve for the AE of each project based on its initial life span, rather than the lowest common multiple of the projects' lives.

Example 5.9 Annual equivalent worth comparison—unequal project lives

Consider Example 4.14, where a mail-order firm was considering two different mailing systems (models A and B) for product announcement and invoice handling. In that example, we assumed that both models would be available in the future without significant changes in price and operating costs. Determine which project should be selected based on the annual worth analysis.

Figure 5.8 ■ Comparison of unequal-lived projects with an indefinite analysis period using the equivalent annual worth criterion (Example 5.9)

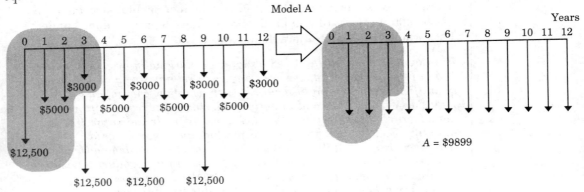

Model A

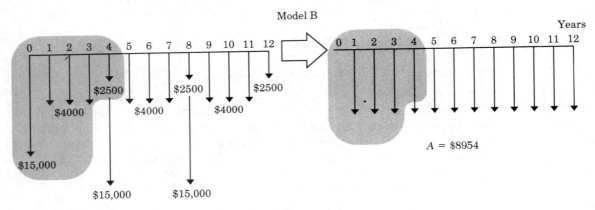

Model B

Solution

Given: Cost cash flows shown in Fig. 5.8, $i = 15\%$ per year

Find: AE, and which alternative is preferred

An alternative procedure for solving Example 4.14 is to compute the annual equivalent worth of an outlay of $12,500 for model A every 3 years, and the annual equivalent cost of an outlay of $15,000 for model B every 4 years. Notice that the AE of each 12-year cash flow is the same as that of the corresponding 3- or 4-year cash flow, as shown in Fig. 5.8. From Example 4.14, we calculate

- Model A:
  For 3-year life:

$$PW(15\%) = -\$22,601$$

$$AE(15\%) = -\$22,601(A/P,15\%,3)$$

$$= -\$9899.$$

For the 12-year period (computed for the complete analysis period):

$$PW(15\%) = -\$53,657$$

$$AE(15\%) = -\$53,657(A/P, 15\%, 12)$$

$$= -\$9899.$$

- Model B:
  For 4-year life:

$$PW(15\%) = -\$25,562$$

$$AE(15\%) = -\$25,562(A/P, 15\%, 4)$$

$$= -\$8,954.$$

For the 12-year period (computed for the complete analysis period):

$$PW(15\%) = -\$48,534$$

$$AE(15\%) = -\$48,534(A/P, 15\%, 12)$$

$$= -\$8,954.$$

Notice that the annual equivalent values that were calculated based on the common service period are the same as those that were obtained over their initial life spans. Thus, for alternatives with unequal lives, we will obtain the same selection by comparing NPW over a common service period using repeated projects or comparing AE for initial lives.

■ 5.3 Minimum Cost Analysis

Another valuable extension of AE analysis is minimum-cost analysis. This method is useful when we have two or more cost components that are affected differently by the same design element. That is, for a single design variable, some costs increase while others decrease. When the equivalent annual total cost of a design variable is a function of increasing and decreasing cost components, we usually can find the optimal value that will minimize its cost.

$$AE(i) = a + bx + \frac{c}{x}, \tag{5.5}$$

where x is a common design variable, and a, b, and c are constants.

To find the value of the common design variable that minimizes the $AE(i)$, we need to take the first derivative, equate the result to zero, and solve for x:

$$\frac{dAE(i)}{dx} = b - \frac{c}{x^2}$$

$$= 0$$

$$x^\circ = \sqrt{\frac{c}{b}} \qquad (5.6)$$

The value x° is the minimum cost point for the design alternative.

Example 5.10 Minimum-cost analysis

A constant electric current of 5000 amperes is to be transmitted a distance of 1970 metres from a power plant to a substation for 24 hours a day, 365 days a year. Copper conductors can be installed for $16.50 per kilogram and will have an estimated life of 25 years and a salvage value of $1.65 per kilogram. Power loss from conductors is inversely proportional to the cross-sectional area (A) of the conductor. It is known that the resistance of a conductor is 0.00098 Ω for one square centimetre cross section. The cost of energy is $0.0375 per kilowatt-hour, the interest rate is 9%, and the density of copper is 8894 kilograms per cubic metre. For the data given, calculate the optimum cross-sectional area (A) of the conductor.

Discussion: This is a classical minimum-cost example to design the cross-sectional area of an electrical conductor that involves the increasing and decreasing cost components. Since resistance is inversely proportional to the size of the conductor, the energy loss will decrease with the increased conductor size. More specifically, the energy loss in kilowatt-hours in a conductor due to resistance is equal to

$$\text{energy loss in kilowatt-hour} = \frac{I^2 R}{1000A} T$$

$$= \frac{(5000^2)\,(0.00098)}{1000A}\,(24 \times 365)$$

$$= \frac{214{,}620}{A}\ \text{kWh,}$$

where I = the current flow in amperes, R = the resistance in the conductor in ohms, and T = the number of hours.

Since the electrical resistance is inversely proportional to the area of the cross section (A), the total energy loss in dollars per year for a specified conductor material is

$$\text{annual energy loss cost} = \frac{214{,}620}{A} \, (L)$$

$$= \frac{214{,}620}{A} \, (\$0.0375)$$

$$= \frac{\$8048}{A},$$

where L = lost energy in dollars per kilowatt-hour.

As we increase the size of the conductor, however, it will cost more to build. First, we need to calculate the total amount of conductor material in kilograms. Since the cross-sectional area is given in square centimetres, we need to convert it to square metres before finding the material mass.

$$\text{material mass in kilograms} = \frac{(1970)(8894)A}{100^2}$$

$$= 1752(A)$$

$$\text{total material cost} = 1752(A)(\$16.50)$$

$$= \$28{,}908(A)$$

Here, we are looking for the trade-off between the cost of installation and the cost of energy loss.

Solution

Given: Cost components as a function of cross-sectional area (A), $N = 25$ years, $i = 9\%$

Find: Optimal A value

Since the copper material will be salvaged at the rate of $1.65 per kilogram at the end of 25 years, we can compute the capital recovery cost as follows:

$$CR(9\%) = [\$28{,}908A - \$1.65(1752A)] \, (A/P, \, 9\%, \, 25) + \$1.65(1752A) \, (0.09)$$

$$= 2648A + 260A$$

$$= 2908A.$$

Using Eq. (5.5), we express the total annual equivalent worth as a function of a design variable (A) as follows:

$$AE(9\%) = \overbrace{-2908A}^{\text{capital cost}} - \underbrace{\frac{8048}{A}}_{\text{operating cost}}$$

To find the annual equivalent cost of minimum magnitude, we use the result of Eq. (5.6).

$$\frac{dAE(9\%)}{dA} = -2908 + \frac{8048}{A^2} = 0$$

$$A^\circ = \sqrt{\frac{8048}{2908}}$$

$$= 1.664 \text{ cm}^2$$

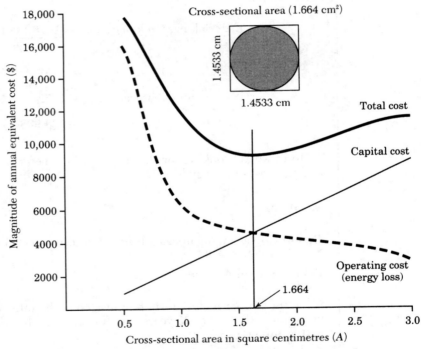

Figure 5.9 ■ Optimal cross-sectional area for a copper conductor (Example 5.10). Note that the minimum point coincides with the crossing point of the capital cost and operating cost lines. In general, this is not always true. Since the cost components can have a variety of cost patterns, the minimum point usually does not occur at the crossing point.

The minimum annual equivalent total cost is

$$AE(9\%) = -2908(1.664) - \frac{8048}{1.664}$$

$$= -\$9674.$$

Figure 5.9 illustrates the nature of this design trade-off problem.

 ## 5.4 Computer Notes

As indicated in Section 5.3, minimum cost analysis is useful for an engineering design decision with two more cost components that are affected by a common design variable. This section illustrates this concept using Lotus 1-2-3 as our tool. The data given in Example 5.10 indicated that

Electric current (A)	5000
Transmission distance (m)	1970
Operating hours per year (hrs.)	8760
Service life (years)	25
Material costs ($/kg)	16.5
Scrap value ($/kg)	1.65
Electrical resistance (Ω)	0.00098
Cost of energy ($/kWh)	0.0375
Material density (kg/m^3)	8894
MARR (%)	9

These data have been used to set up the spreadsheet shown in Fig. 5.10. In addition, a cross-sectional area (A) is specified as a user input.

Illustrating the power of the spreadsheet approach, the cross-sectional area (A) is shown as a range of values from 1 square centimetre to 2 square centimetres, with the increment of 0.01 square centimetre. Note that the minimum cost occurs with A value in between 1.6 square centimetres and 1.7 square centimetres. You can search the optimal value (minimum cost) by varying the value of A within this bound. At A=1.664, the annual equivalent cost is $9677, the same number (a slight difference is due to rounding errors) as previously calculated in the text discussion of this example.

Figure 5.10 ■ Lotus 1-2-3 output: Determining the optimal cross-sectional area for a copper transmission line (Example 5.10)

G19: (.0) +E19+F19

	A	B	C	D	E	F	G
1							
2		Cross-Sectional Area (cm^2)			1,664		
3							
4		Electric Current (amps)			5000		
5		Transmission Distance (m)			1970		
6		Operating Hours per Year (hrs.)			8760		
7		Service Life (years)			25		
8		Material Cost ($/kg)			16.50		
9		Scrap Value ($/kg)			1.65		
10		Electrical Resistance (ohms)			0.00098		
11		Cost of Energy ($/kwh)			0.0375		
12		Material Density (kg/m^3)			8894		
13		MARR (%)			9		
14							
15	Design	Energy	Material	Total	Annual Equivalent Worth of:		
16	Area (A)	Loss	Weight	Material	Capital	Energy	Total
17	(cm^2)	(kWh)	(kg)	Cost	Recovery	Loss	
18							
19	1.664	128,620[a]	6,413[b]	48,100[c]	−4,840[d]	−4,837[e]	−9,677[f]
20							
21	1,000	214,620	3,854	28,906	−2,909	−8,048	−10,957
22	1,100	195,109	4,240	31,797	−3,200	−7,317	−10,516
23	1,200	178,850	4,625	34,688	−3,490	−6,707	−10,197
24	1,300	165,092	5,010	37,578	−3,781	−6,191	−9,972
25	1,400	153,300	5,396	40,469	−4,072	−5,749	−9,821
26	1,500	143,080	5,781	43,359	−4,363	−5,366	−9,729
27	1,600	134,138	6,167	46,250	−4,654	−5,030	−9,684
28	1,700	126,247	6,552	49,141	−4,945	−4,734	−9,679
29	1,800	119,233	6,938	52,031	−5,236	−4,471	−9,707
30	1,900	112,958	7,323	54,922	−5,527	−4,236	−9,762
31	2,000	107,310	7,708	57,813	−5,817	−4,024	−9,842
32							

Cell Formulas
a: $E$4^2°$E$10°$E$6/(1000°A19)
b: $E$5°$E$12°A19/(100^2)
c: C19°$E$8
d: −@PMT((D19-C19°$E$9), $E$13/100, $E$7) + $E$9°C19°$E$13/100)
e: −B19°$E$11
f: E19 + F19

The spreadsheet formulas are written as functions of the data in the upper section of the spreadsheet. To allow maximum flexibility in calculations, these input data are expressed as absolute locations in all formulas. A plot of these total costs against design area (A) is easily obtained, as explained in Chapter 4. This plot is shown as Fig. 5.11.

The spreadsheet can now be used to explore a wide variety of scenarios by merely changing input data values. This exploration is the basic concept of sensitivity analysis, and is considerably eased by spreadsheet analysis.

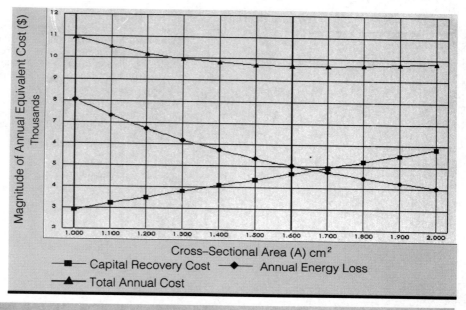

Figure 5.11 ■ Lotus 1-2-3 output for sizing optimal cross-sectional area

Summary

- **Annual equivalent worth analysis,** or **AE**, is—along with present worth analysis—one of two main analysis techniques based on the concept of equivalence. The equation for AE is $AE(i) = PW(i)(A/P,i,N)$. AE analysis will yield the same decision result as PW analysis. (See Table 15.1)

- The **capital recovery cost factor,** or $CR(i)$, is one of the most important applications of AE analysis. It allows managers to calculate an annual equivalent cost of capital for ease of itemization with annual operating costs. The equation for $CR(i)$ is:

$$CR(i) = (P - S)(A/p,i,N) + iS,$$

where P = initial cost and S = salvage value.

- AE analysis is recommended over NPW analysis in many key real-world situations for the following reasons:
  1. For inclusion in many financial reports, an annual equivalent value is preferred to a present worth value.
  2. Calculation of unit costs is often required in order to determine reasonable pricing for sale items.
  3. Calculation of cost per unit of use is required to reimburse employees for business use of personal cars.
  4. Make-or-buy decisions usually require developing unit costs for the various alternatives.
  5. Minimum cost analysis is easily done based on annual equivalent worth.

- AE analysis requires less calculation than NPW analysis for comparing projects with different useful lives under the following conditions:
  1. When the required service period for the projects is ongoing infinitely.
  2. When the projects can be replaced by identical projects with identical costs.

In this special case, we can calculate AE for each project's initial life instead of the lowest common multiple of lives.

Table 5.1 Summary of Project Analysis Methods

Analysis Method	Equation Form(s)	Description	Comments
Payback period	—	A method for determining *when* in a project's history it breaks even (i.e., revenues repay costs).	Should be used for screening only. Reflects liquidity, not profitability of project.
Discounted payback period	—	A variation of payback period which factors in the time value of money.	Reflects a truer measure of payback point than regular payback period.
Net present worth (NPW)	$PW(i) = \sum_{n=0}^{N} A_n(P/F, i, n)$	An *equivalence* method which translates a project's cash flows into a net present worth value.	For investments involving disbursements *and* receipts, the decision rule is: If $PW(i) > 0$, accept. If $PW(i) = 0$, remain indifferent. If $PW(i) < 0$, reject. When comparing multiple alternatives, select the one with the greatest NPW value. For investments involving only disbursements (service projects), this is the project with the least negative NPW.
Net future worth (NFW)	$FW(i) = PW(i)(F/P, i, N)$	An *equivalence* method variation of NPW; a project's cash flows are translated into a net future worth value.	Most common real-world application occurs when we wish to determine a project's value at commercialization (a future date) *not* its value when we begin investing (the "present").
Capitalized equivalent (CE)	$PW(i) = \dfrac{A}{i}$	An *equivalence* method variation of NPW; calculates the NPW of a perpetual or very long-lived project that generates a constant annual net cash flow.	Most common real-world applications are civil engineering projects with lengthy service lives (i.e., $N > 50$) and *equal annual* costs (or incomes).
Annual equivalence (AE)	$AE(i) = PW(i)(A/P, i, N)$	An *equivalence* method and variation of NPW; a project's cash flows are translated into an annual equivalent sum.	AE analysis facilitates the comparison of unit costs and profits. In certain circumstances, AE may also avoid the complication of lowest common multiple of project lives which may arise in NPW analysis.

Table 5.1 Summary of Project Analysis Methods (continued)

Analysis Method	Equation Form(s)	Description	Comments
Rate of return (ROR) or internal rate of return (IRR)	$PW(i) = 0$	A *relative* percentage method which measures the yield as a percentage of investment over the life of a project.	Rate of return is stated in a form more intuitively understandable to most laypersons and financial administrators. Unlike equivalence methods, which are absolutes, IRR when calculated must be compared to a predetermined acceptable rate of return known as MARR. The decision rule is: If IRR > MARR, accept; If IRR = MARR, remain indifferent; If IRR < MARR, reject

Problems

Note: Unless otherwise stated, all cash flows given in the problems represent after-tax cash flows. The interest rate (MARR) is also given on an after-tax basis.

5.1 Consider the following cash flows and compute the equivalent annual worth at $i = 12\%$.

	A_n	
n	Investment	Revenue
0	−$10,000	
1		$500
2		500
3		500
4		500
5		500
6	+10,000	500

5.2 Consider the cash flow diagram in Fig. 5.12. Compute the equivalent annual worth at $i = 10\%$.

Figure 5.12 ■ Cash flow diagram (Problem 5.2)

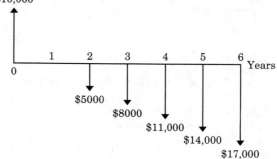

5.3 Consider the cash flow diagram in Fig. 5.13. Compute the equivalent annual worth at $i = 8\%$.

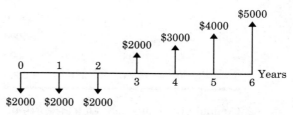

Figure 5.13 ■ Cash flow diagram (Problem 5.3)

5.4 Consider the cash flow diagram in Fig. 5.14. Compute the equivalent annual worth at $i = 15\%$.

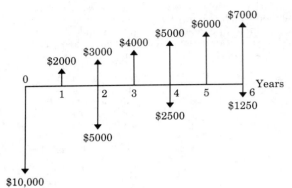

Figure 5.14 ■ Cash flow diagram (Problem 5.4)

5.5 Consider the cash flow diagram in Fig. 5.15. Compute the equivalent annual worth at $i = 9\%$.

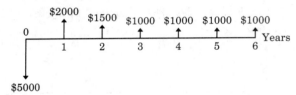

Figure 5.15 ■ Cash flow diagram (Problem 5.5)

5.6 Consider the following sets of investment projects:

Project's Cash Flow

n	A	B	C	D
0	−$2000	−$6000	−$4000	−$8000
1	300	2000	−2000	3000
2	400	2000	1000	3000
3	500	1500	2000	3000
4	600	1500	1000	3000
5	700	1500	1000	2000

Compute the equivalent annual worth of each project at $i = 12\%$ and determine the acceptability of each project.

5.7 In Problem 5.6, plot the equivalent annual worth of each project as a function of interest rate (0%–50%).

5.8 Consider the following sets of investment projects:

Period	Project's Cash Flow			
n	A	B	C	D
0	−$4000	−$4000	−$4000	−$4000
1	0	1500	3000	1800
2	0	1800	2000	1800
3	5500	2100	1000	1800

Compute the equivalent annual worth of each project at $i = 13\%$ and determine the acceptability of each project.

5.9 Consider the following project's cash flows:

	Net Cash Flow	
n	Investment	Operating Income
0	−$1000	
1		$400
2		400
3	−1000	500
4		400
5		400
6	−1000	500
7		400
8		400
9		500

Find the equivalent annual worth for this project at $i = 10\%$, and determine the acceptability of the project.

5.10 The owner of a business is considering investing $35,000 for new equipment. He estimates that the net cash flows during the first year will be $2000, but these will increase by $2500 per year the next year and each year thereafter. The equipment is estimated to have a 10-year service life and a net salvage value at this time of $3000. The firm's interest rate is 15%.

(a) Determine the annual capital cost for the equipment.
(b) Determine the equivalent annual savings (revenues).
(c) Is this a wise investment?

5.11 Nelson Electronics Company just purchased a soldering machine to be used in its assembly cell for flexible disk drives. This soldering machine cost $200,000. Because of the specialized function it performs, its useful life is estimated to be 5 years. At that time its salvage value is also estimated to be $50,000. What is the capital cost recovery for this investment if the firm's interest rate is 20%?

5.12 Beginning next year a foundation will support an annual seminar on campus by the earnings of a $50,000 gift it received this year. It is felt that 10% interest will be realized for the first 10 years, but that plans should be made to anticipate an interest rate of 6% after that time. What amount should be added to the foundation now to fund the seminar at the $5000 level into infinity?

5.13 The present price (year 0) of heating oil is $0.22 per litre and its cost is expected to increase by $0.02 per year. (Heating oil at the end of year 1 will cost $0.24 per litre.) Mr. Graves uses about 4000 litres of heating oil during a winter season for space heating. He has an opportunity to buy a storage tank for $300, and at the end of 4 years he can sell the storage tank for $150. The tank has a capacity to supply 4 years of Mr. Graves' heating needs, so he can buy 4 years of heating oil at its present price ($0.22). He can invest his money elsewhere at 8%. Should he purchase the storage tank? Assume that heating oil purchased on a pay-as-you-go basis is paid for at the end of the year. (However, heating oil purchased for the storage tank is purchased now.)

5.14 Consider the following advertisement that appeared in a local paper.

Pools-Spas-Hot Tubs–Pure Water without Toxic Chemicals: The IONETICS water purification system has proven highly effective in killing algae and bacteria in pools and spas. Here is how it works: The "Ion Chamber" installed in the return water line contains copper/silver electrodes. A safe, low-voltage current is sent through those electrodes from a "Computerized Controller." Copper and silver ions enter the water stream and the pool or spa where they attack and kill the algae and bacteria. These charged, dead microorganisms mutually attract, forming larger particles easily removed by the existing filtration system. The IONETICS system can make your pool water pure enough to drink without the use of chlorine or other toxic chemicals. The ion level need only be tested about once per week and ion output is easily adjusted. The comparative costs between the conventional chemical system (chlorine) and the IONETICS systems are as follows:

Item	Conventional	IONETICS
Annual chemical cost	$471	
Annual Ionetics		$85
Annual pump costs ($0.667/kilowatt hour)	$576	$100
Capital investment		$1200

Note that the IONETICS system pays for itself less than two years!

Assume that the IONETICS system has a 12-year service life and your interest rate is 6%. What is the equivalent monthly cost of operating the IONETICS system?

5.15 A construction firm is considering establishing an engineering computing center. This center will be equipped with three engineering workstations that would cost $50,000 each, and each has a service life of 5 years. The expected salvage value of each work-station is $5000. The annual operating and maintenance cost would be $12,000 for each workstation. At a MARR of 20%, determine the equivalent annual cost for operating the engineering center.

5.16 Consider the cash flows for the following investment projects:

Project's Cash Flow

n	A	B
0	−$4000	$5000
1	1000	−1300
2	X	−1300
3	1000	−1300
4	1000	−1300

(a) For project A, find the value of X that makes the equivalent annual receipts equal the equivalent annual disbursement at $i = 12\%$.

(b) Would you accept B at $i = 20\%$ based on AE criterion?

5.17 An industrial firm can purchase a special machine for $20,000. A down payment of $2000 is required, and the balance can be paid in 5 equal year-end installments at 7% interest on the unpaid balance. As an alternative, the machine can be purchased for $18,000 in cash. If the firm's MARR is 10%, deter-mine which alternative should be ac-cepted, based on the annual equivalent method.

5.18 An industrial firm is considering purchasing several programmable con-trollers and automating their manufac-turing operations. It is estimated that the equipment will initially cost $100,000 and the labor to install it will cost $35,000. A service contract to maintain the equipment will cost $5000 per year. Trained service personnel will have to

be hired at annual salary of $30,000. It is also estimated that there will be about $10,000 annual income-tax savings (cash inflow). How much will this investment in equipment and services have to increase the annual revenues after taxes to just break even? The equipment is estimated to have an operating life of 10 years with no salvage value because of obsolescence. The firm's MARR is 10%.

5.19 The Engineering department of a large firm is overly crowded. In many cases, several engineers occupy the same office. It is evident that the distractions caused by these crowded conditions consider-ably reduce the productive capacity of the engineers. Management is consider-ing the possibility of new facilities for the department, which could result in fewer engineers per office and a private office for some. For an office presently occupied by 5 engineers, what minimum increase in company revenue per individual is required to justify the assignment of only 3 engineers to an office if the following data apply?

• The office size is 4 × 8 metres.

• The average annual salary of each engineer is $60,000.

• The cost of building per square metre is $500.

• The estimated life of the building is 25 years.

• The estimated salvage value of the building is 10% of the first cost.

• Annual taxes, insurance, and maintenance are 6% of the first cost.

• The cost of janitor service, heating and illumination, etc., is $30.00 per square metre per year.

• The interest rate is 10%.

5.20 Two 150-hp (horsepower) motors are being considered for installation at a municipal sewage-treatment plant. The first costs $4500 and has an operating efficiency of 83%. The second costs $3600 and has an efficiency of 80%. Both motors are projected to have zero salvage value after a life of 10 years. If all the charges such as insurance, maintenance, etc., amount to a total of 15% of the original cost of each motor, and if power costs are a flat 5 cents per kilowatt-hour, how many minimum hours of full-load operation per year are necessary to justify purchase of the more expensive motor at $i = 6\%$? (Note that 1 hp = 746 W = 0.746 kW.)

5.21 Danford Company, a farm-equipment manufacturer, currently produces 20,000 units of gas filters for use in its lawn-mower production annually. The following costs are reported based on the previous year's production:

Direct materials	$ 60,000
Direct labor	180,000
Variable overhead (power and water)	135,000
Fixed overhead (light and heat)	70,000
Total cost	$445,000

It is anticipated that the gas-filter production will last 5 years. If the company continues to produce the product in-house, the annual direct material costs will increase at the rate of 5%. (For example, the annual material costs during the first production year will be $63,000.) The direct labor will also increase at the rate of 6% per year. However, the variable overhead costs would increase at the rate of 3%, but the fixed overhead would remain at the current level over the next 5 years. Tompkins Company has offered to sell Danford 20,000 units of gas filters for $25 per unit. If Danford accepts the offer, some of the manufacturing

facilities presently used to manufacture the gas filter could be rented to a third party at annual rental of $35,000. Additionally, $3.50 per unit of the fixed overheard applied to gas-filter production would be totally eliminated. The firm's interest rate is known to be 15%. What is the unit cost of buying the gas filter from outside? Should Danford accept Tompkins' offer, and why?

5.22 Sentech Environmental Consulting (SEC), Inc., designs plans and specifications for asbestos abatement (removal) projects involving public, private, and governmental buildings. Currently, SEC must also conduct an air test before allowing the reoccupancy of a building from which asbestos has been removed. SEC sends the air-test sample to a subcontracted laboratory for analysis by a transmission electron microscope (TEM). In subcontracting the TEM analysis, SEC charges its client $100 above the subcontractor's analysis fee. The only expenses in this system are the costs of shipping the air-test samples to the subcontractor and the labor involved in this shipping. As business grows, SEC needs to consider either continuing to subcontract the TEM analysis to outside companies or developing its own TEM laboratory. With recent government regulations requiring the removal of asbestos from buildings, SEC expects about 1000 air-sample testings per year over 8 years. The firm's MARR is known to be 15%.

- **Subcontract option:** The client is charged $400 per sample, which is $100 above the subcontracting fee of $300. Labor expenses are $1500 per year, and the shipping expenses are estimated to be $0.50 per sample.
- **TEM purchase option:** The purchase and installation cost for the TEM is $415,000. The equipment would last for 8 years with no salvage value. The design and renovation cost is estimated to be $9500. The client is charged $300

per sample based on the current market price. One full-time manager and two part-time technicians are expected to be needed to operate the laboratory. Their combined annual salaries are $50,000. Material required to operate the lab includes carbon rods, copper grids, filter equipment, and acetone. These annual material costs are estimated at $6000. Utility costs, operating and maintenance costs, and indirect labor needed to maintain the lab are estimated to be $18,000 per year. The extra income tax expenses would be $20,000.

(a) Determine the cost of air-sample test by the TEM (in-house).

(b) What is the required number of air samples per year to make two options equivalent?

5.23 A company is currently paying an employee $0.25 per kilometre to drive his or her car for company business. The company is considering supplying the employee with a car, which would involve the following: purchase of the car for $15,000 with an estimated 3-year life, a net salvage value of $5000, taxes and insurance costing $500 per year, and operating and maintenance expenses of $0.10 per kilometre. If the interest rate is 10% and they anticipate the employee's travels to be 12,000 kilometres annually, what is the equivalent cost per kilometre (without considering any income tax)?

5.24 An electrical automobile can be purchased for $25,000. It is estimated to have a life of 15 years with annual travel of 20,000 kilometres. A new set of batteries will have to be purchased every 3 years at a cost of $3000. Annual maintenance to the vehicle is estimated to be $350 per year. The cost of recharging the batteries is estimated to be $0.015 per kilometre. The salvage value of the batteries and the vehicle at the end of 15 years is estimated to be $500. Consider the MARR to be 7%.

What is the cost per kilometre to own and operate this vehicle, based on the above estimates? The $3000 cost of the batteries is a net value with the old batteries traded in for the new ones.

5.25 A 40-kilowatt generator is estimated to cost $30,000 completely installed and ready to operate. The annual maintenance for this machine is estimated to be $500. The annual energy generated is estimated to be 100,000 kilowatt-hours. If the value of the energy generated is considered to be $0.08 per kilowatt-hour, how many hours each year will it take to generate sufficient energy to equal the purchase price? Consider the MARR to be 9%. Also consider the salvage value of the machine to be $2000 at the end of its estimated life of 15 years. What is the annual worth of this machine? How long will it take before this machine becomes profitable?

5.26 A large university facing severe parking problems on its campus is considering constructing parking decks off campus. Then, using a shuttle service, students could be picked up at the off-campus parking deck and quickly transported to various locations on campus. The university would charge a small fee for each shuttle ride, and the students could quickly and economically travel to their classes. The funds raised by the shuttle would be used to pay for the trolleys, which cost about $150,000 each. The trolley has a 12-year service life with an estimated salvage value of $3000. To operate each trolley, the following additional expenses must be considered:

Item	Annual Expenses
Driver	$25,000
Maintenance	7,000
Insurance	2,000

If students pay 10 cents for each ride, determine the annual ridership per trolley (number of shuttle rides per year) required to justify the shuttle project, assuming an interest rate of 6%.

5.27 The following cash flows represent the potential annual savings associated with two different types of production processes, each requiring an investment of $10,000.

n	Process A	Process B
0	−$10,000	−$10,000
1	9,120	6,350
2	6,840	6,350
3	4,560	6,350
4	2,280	6,350

Assuming an interest rate of 18%,

(a) Determine the equivalent annual savings for each process.
(b) Determine the hourly savings for each process if there were 2000 hours of operation per year.
(c) Which process should be selected?

5.28 A certain factory building has an old lighting system, and the lighting costs for this building average $20,000 a year. A lighting consultant tells the factory supervisor that the monthly lighting bill can be reduced to $8000 a year by investing $50,000 in relighting the factory building. There will be additional maintenance costs of $3000 per year if the new lighting system is installed. If the old lighting system has zero salvage value and the new lighting system is estimated to have a life of 20 years, what is the net annual benefit for this investment in new lighting? Consider the MARR to be 12%. Also consider that the new lighting system has zero salvage value at the end of its life.

5.29 Travis Wenzel has $2000 to invest. Normally, he would deposit the money in his savings account, which earns 6% interest compounded monthly. However, he is considering the three alternative investment opportunities:

- **Option 1:** Purchasing a bond for $2000 that has a face value of $2000 and pays $100 every 6 months for 3 years. The bond will mature 3 years from now.
- **Option 2:** Buying and holding a growth stock that grows 11% per year for 3 years.
- **Option 3:** Making a personal loan of $2000 to a friend and receiving $450 per year for 3 years.

Determine the equivalent annual cash flows for each option, and select the best option.

5.30 Consider the cash flows for the following investment projects (assume MARR = 12%):

	Project's Cash Flow		
n	A	B	C
0	−$2000	−$3000	−$6000
1	1000	1600	1800
2	800	1500	1800
3	600	1500	2000
4	400	1000	2000

(a) Suppose that projects A and B are mutually exclusive. Which project would you select based on the AE criterion?
(b) Assume that projects B and C are mutually exclusive. Which project would you select based on the AE criterion?

5.31 An airline is considering two types of engine systems for use in its planes.

Each has the same life and the same maintenance and repair record.

- System A costs $100,000 and uses 200,000 litres per 1000 hr. of operation at the average load encountered in passenger service.
- System B costs $200,000 and uses 160,000 litres per 1000 hr. of operation at the same level.

Both engine systems have 3-year lives before any major overhaul of the engine systems, with 10% salvage values of their initial investment. If jet fuel costs $0.25 a litre currently and its price is expected to increase at the rate of 6% due to degrading engine efficiency (each year), which engine system should the firm install, assuming 2000 hr. of operation per year? Assume a MARR of 10% using the AE criterion. What is the equivalent operating cost per hour for each engine?

5.32 Norton Auto-Parts, Inc., is considering one of two forklift trucks for their assembly plant. Truck A costs $15,000, requires $3000 annually in operating expenses, and will have a $5000 salvage value at the end of its 3-year service life. Truck B costs $20,000, but requires only $2000 annually in operating expenses. Its service life is 4 years at which time its expected salvage value is $8000. The firm's MARR is 12%. Assuming that the need for trucks is 12 years and that no significant changes are expected in the future price and functional capacity for both trucks, select the most economical truck based on AE analysis.

5.33 A small manufacturing firm is considering the purchase of a new machine to modernize one of its current production lines. Two types of machines are available on the market. The lives of machine A and machine B are 4 years and 6 years, respectively, but the firm does not expect to need the service of either machine for more than 5 years. The machines have the following expected receipts and disbursements.

Item	Machine A	Machine B
First cost	$6000	$8500
Service life	4 years	6 years
Estimated salvage value	$500	$1000
Annual O&M costs	$700	$520
Change oil filter every other year	$100	None
Engine overhaul	$200 (every 3 years)	$280 (every 4 years)

The firm can always lease a machine at $3000/year, fully maintained by the leasing company. After 4 years' use, the salvage value for machine B will remain constant at $1000.

(a) How many different decision alternatives are there?

(b) Which decision appears to be the best at $i = 10\%$?

5.34 A continuous electric current of 5000 amperes is to be transmitted from a power plant to a substation located 10,000 metres away. Copper conductors can be installed for $13.00 per kilogram, will have an estimated life of 25 years, and can be salvaged for $2.17 per kilogram. Power loss from each conductor will be inversely proportional to the cross-sectional area of the conductors and may be expressed as $2008/A$ kilowatt, where A is given in square centimetres. The cost of energy is $0.0825 per kilowatt-hour, the interest rate is 11%, and the density of copper is 8894 kg/m^3.

(a) Calculate the optimum cross-sectional area of the conductor.

(b) Calculate the annual equivalent total cost for the value obtained in part (a).

(c) Graph the two individual cost factors and the total cost as a function of cross-sectional area A and discuss the impact of increasing energy cost on the optimum obtained in part (a).

5.35 As a result of the conflict in the Persian Gulf, Kuwait is studying the feasibility of running a steel pipeline across the

Arabian Peninsula to the Red Sea. The pipeline will be designed to handle 3 million barrels (1 barrel = 0.159m³) or 477,000 cubic metres of crude oil per day at optimum conditions. The length of the line will be 600 miles or 966 kilometres. Calculate the optimum pipeline diameter that will be used for 20 years for the following data at $i = 10\%$:

- Pumping power = $1.785 \times 10^{-11} Q\Delta P$ kilowatts

- Q = volume flow rate, cubic m/hr
- $\Delta P = \frac{65.1Q\mu L}{1000D^4}$, pressure drop, Pa
- L = pipe length, m
- D = pipe diameter, m
- $t = 0.01 D$, pipeline wall thickness, m
- $\mu = 12,657$ kg/m·hr oil viscosity
- Power cost, $0.020 per kilowatt hour
- Oil cost, $18 per barrel
- Pipeline cost, $2.20 per kilogram of steel
- Pump and motor costs, $262/kW
- Density of steel, 7849 kg/m³

5.36 A corporate executive jet with a seating capacity of 20 has the following cost factors:

Initial cost	$12,000,000
Service life	15 years
Salvage value	$2,000,000
Crew costs per year	$225,000
Fuel cost per kilometre	$0.40
Landing fee	$250
Maintenance per year	$237,500
Insurance cost per year	$166,000
Catering per passenger trip	$75

The company flies three round trips from Boston to London per week, a distance of 5494 kilometres one way. How many passengers must be carried on an average trip in order to justify the use of the jet if the first-class round-trip fare is $3400? The firm's MARR is 15%. (Ignore any income-tax consequences.)

5.37 A plastic manufacturing company owns and operates a polypropylene production facility which converts the propylene

from one of its cracking facilities to polypropylene plastics for outside sale. The polypropylene production facility is currently forced to operate at less than capacity due to lack of enough propylene production capacity in its hydrocarbon cracking facility. The chemical engineers are considering alternatives for supplying additional propylene to the polypropylene production facility. Some of the feasible alternatives are: (Option 1) Build a pipeline to the nearest outside supply source, (Option 2) Provide additional propylene by truck from an outside source. The engineers also gathered the following projected cost estimates:

- Future costs for purchased propylene excluding delivery: $0.43 per kilogram
- Cost of pipeline construction: $120,000 per pipeline kilometre
- Estimated length of pipeline: 300 km
- Transportation costs by tank truck: $0.10 per kilogram utilizing common carrier
- Pipeline operating costs: $0.010 per kilogram, excluding capital costs
- Projected additional propylene needs: 90 million kilograms per year
- Projected project life: 20 years
- Estimated disposal cost of the pipeline: 8% of the installed costs

Determine the propylene cost per kilogram under each option if the firm's MARR is 18%. Which option is more economical?

5.38 Food Preservation Inc., is considering an investment of $7 million to construct a food irradiation plant. (See Fig. 5.16.) This technology destroys organisms that cause spoilage and disease, thus extending the shelf life of fresh foods and the distances over which it can be shipped. The plant can handle about 200,000 kilograms of produce in an hour, and it will be operated for 3600 hours a year. The net expected operating and maintenance costs (considering any income-tax effects) would be $4 million per year. The plant is expected to have a useful life of 15 years with a net salvage value of $700,000. The firm's interest rate is 15%.

Figure 5.16 ■ How food is irradiated (Problem 5.38)

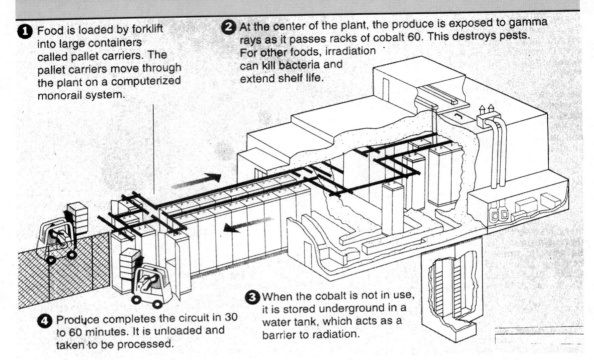

How Food Is Irradiated

The proposed food irradiation plant will be used to kill bacteria in fruit and vegetables. It can handle about 200,000 kg of produce in an hour, passing it counterclockwise around a rack of cobalt 60.

1 Food is loaded by forklift into large containers called pallet carriers. The pallet carriers move through the plant on a computerized monorail system.

2 At the center of the plant, the produce is exposed to gamma rays as it passes racks of cobalt 60. This destroys pests. For other foods, irradiation can kill bacteria and extend shelf life.

3 When the cobalt is not in use, it is stored underground in a water tank, which acts as a barrier to radiation.

4 Produce completes the circuit in 30 to 60 minutes. It is unloaded and taken to be processed.

(a) If the investors of the company want to recover the plant investment within 6 years of operation (rather than 15 years), what would be the equivalent after-tax annual revenues that must be generated?

(b) To generate such annual revenues in part (a), what minimum processing fee per kilogram should the company charge to their producers?

5.39 An electrical utility is experiencing a sharp power demand which continues to grow at a high rate in a certain local area. Two alternatives are under consideration. Each alterative is designed to provide enough capacity during the next 25 years. Both alternatives will consume the same amount of fuel, so the fuel cost is not considered in the analysis.

- **Alternative A:** Increase the generating capacity now so that the ultimate demand can be met without additional expenditures later. It would require an initial investment of $30 million and it is estimated that this plant facility would be in service for 25 years with a salvage value of $0.85 million. The annual operating and maintenance costs (including income taxes) would be $0.4 million.

- **Alternative B:** Spend $10 million now and this will be followed by future additions during the 10th year and the 15th year costing $18 million and $12 million, respectively. This facility would be also sold 25 years from now with a salvage value of $1.5 million. The annual operating and maintenance costs (including income taxes) initially will be $250,000, increasing by $0.35 million after the second addition (from 11th year to 15th year) and by $0.45 million during the final 10 years. (Assume that these costs begin a year subsequent to the actual addition.)

If the firm uses 15% as a MARR, which alternative should be undertaken based on the annual equivalent criterion?

5.40 A large refinery-petrochemical complex is planning to manufacture caustic soda, which will use feed water of 50,000 litres per day. Two types of feeder-water storage installation are being considered over 40 years of useful life.

- **Option 1:** Build a 100,000-litre tank on a tower. The cost of installing the tank and tower is estimated to be $164,000. The salvage value is estimated to be negligible.

- **Option 2:** Place a tank of equal capacity on a hill, which is 150 metres away from the refinery. The cost of installing the tank on the hill including the extra length of service lines is estimated to be $120,000 with negligible salvage value. Because of its hill location, it will require an additional investment of $12,000 in pumping equipment. The pumping equipment is expected to have a service life of 20 years with a salvage value of $1000 at the end of that time. The annual operating and maintenance cost (including any income tax effects) for the pumping operation is estimated at $1000.

If the firm's MARR is known to be 12%, which option is better on the basis of equivalent annual cost?

5.41 The government of Bahrain, an island nation off the east coast of Saudia Arabia, is completing plans to build a desalting plant to help ease a critical drought on the island. The drought and new construction on this island have combined to leave it with an urgent need for a new water source. A modern desalting plant could produce fresh water from seawater for $1000 an acre foot. An acre foot is 1,233,900 litres, or enough to supply two households for a year. On Bahrain, the cost from natural sources is about the same as for desalting. The $3 million plant, with a daily desalting capacity of 0.4 acre foot, can produce 493,600 litres of fresh water a day (enough to supply 295 households daily), more than a quarter of the island's total needs. The desalting plant has an estimated service life of 20 years with no appreciable salvage value. The annual operating and maintenance costs would be about $250,000. Assuming an interest rate of 10%, what should be the minimum monthly water bill for each household?

5.42 A chemical company is considering two types of incinerator to burn solid waste generated by a chemical operation. Both incinerators have a burning capacity of 20 tonnes per day. The following data have been compiled for comparison:

	Incinerator A	Incinerator B
Installed cost	$1,000,000	$650,000
Annual O&M costs	$40,000	$75,000
Service life	20 years	10 years
Salvage value	$50,000	$30,000
Income taxes	$30,000	$20,000

If the firm's MARR is known to be 13%, determine the processing cost per tonne of solid waste by each incinerator. Assume that incinerator B will be available in the future at the same cost.

5.43 The City of Halifax wishes to compare two plans for supplying water to a newly developed subdivision.

- Plan A will take care of requirements for the next 15 years; at the end of that period, the first cost of $400,000 will have to be duplicated to meet subsequent years' requirements. The facilities installed at dates 0 and 15 may be considered permanent; however, certain supporting equipment will have to be replaced every 30 years from the installation dates at a cost of $75,000. Operating costs are $31,000 a year for the first 15 years and $62,000 thereafter, although they are expected to increase by $1000 a year beginning in the 21st year.

- Plan B will supply all requirements for water indefinitely into the future, although it will only be operated at half capacity for the first 15 years. Annual costs over this period will be $35,000 and will increase to $55,000 beginning in the 16th year. The initial cost of Plan B is $550,000; the facilities can be considered permanent, although it will be necessary to replace $150,000 of

equipment every 30 years after initial installation.

The city will charge the subdivision the use of water based on the equivalent annual cost. At an interest rate of 10%, determine the equivalent annual cost for each plan, and make a recommendation to the city.

5.44 A utility firm is considering building a 50-megawatt geothermal plant that generates electricity from naturally occurring hot underground water (see Fig. 5.17). The binary geothermal system will cost $85 million to build and $6 million (including any income-tax

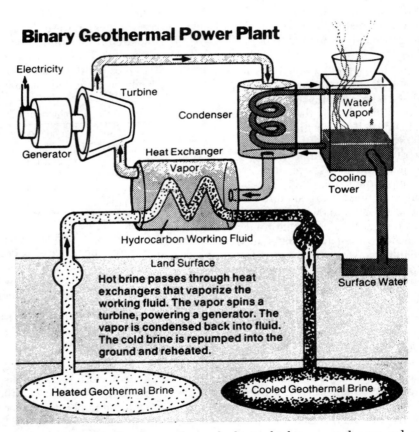

Figure 5.17 ∎ Binary geothermal plant which uses underground water for power (Problem 5.44)

effect) to operate per year. (There is virtually no fuel cost compared with a conventional fossil fuel plant.) The geothermal plant is to last for 25 years. At that time, the expected salvage value will be about the same as the plant removal cost. The plant will be in operation for 70% (plant utilization factor) of the year (or 70% of 8760 hours per year). If the firm's MARR is 14% per year, determine the cost of generating electricity per kilowatt-hour.

5.45 A Veterans Affairs Canada hospital is to decide which type of boiler fuel system will most efficiently provide the required steam energy output for heating, laundry, and sterilization purposes. The present boilers were installed in the early 1930s and are now obsolete. Much of the auxiliary equipment is also old and in need of repair. Because of these general conditions an engineering recommendation was made to replace the entire plant with a new boiler plant building housing modern equipment. The cost of demolishing the old boiler plant would be almost a complete loss as the salvage value of the scrap steel and used brick was estimated to be only about $1000. The hospital's engineer finally selected two alternative proposals as being worthy of more intensive analysis. The hospital's annual energy requirement, measured in terms of steam output, is approximately 65,830,000 kilograms of steam. As a rule of thumb for analysis, one kilogram of steam is approximately 2.3 megajoules (MJ), and one cubic metre of natural gas is approximately 37.2 MJ. The two alternatives were as follows:

• **Alternative 1:** A new, coal-fired boiler plant. This boiler plant would cost $1,770,300. To meet the requirements for particulate emissions from the chimney as set by Environment Canada, this coal-fired boiler, even burning low sulphur coal, would need an electrostatic precipitator which would cost approximately $100,000. This plant would last for 20 years. One kilogram of dry coal yields about 33.2 MJ. To convert the 65,830,000 kilograms of steam energy to the common denominator of MJs, it is necessary to multiply by 2.3. To find MJ input requirements, it is necessary to divide by the relative boiler efficiency for type of fuel. The boiler efficiency for coal is 0.75. The coal price is estimated to be $36.00 per tonne.

• **Alternative 2:** A gas-fired boiler plant with No. 2 fuel oil as standby. This system would cost $889,200 with an expected service life of 20 years. Since small household or commercial gas users who are entirely dependent on gas have priority, large plants must have oil switchover capability. It has been estimated that 6% of 65,830,000 kilograms of steam energy (or 3,950,000 kilograms) would come from the oil switch. The boiler efficiency under each fuel would be 0.78 for gas and 0.81 for oil, respectively. The heat value of natural gas is approximately 37,200 MJ/ MCM (thousand cubic metres), and for No. 2 fuel oil it is 38.9 MJ/L. The estimated gas price is $88.20/MCM, and the fuel oil (No. 2) price is $0.22 per litre.

(a) Calculate the annual fuel costs for each alternative.

(b) Determine the unit cost per steam kilogram for each alternative. Assume $i = 10\%$.

(c) Which alternative is more economical?

CHAPTER 6

Rate of Return Analysis

One of the highest prices ever paid for a painting at public auction was $53.9 million for Vincent Van Gogh's *Irises,* at Sotheby's on November 11, 1987. The seller was art collector John Whitney Payson, who had purchased the painting for $80,000 in 1947. Mr. Payson's investment in art brought him a rate of return of 17.68%.

Laidlaw Inc. is a Canadian transportation and waste management company. If you had invested $10,000 in Laidlaw stock on July 31, 1980, a decade later you would have stock worth $454,000 without taking into consideration any dividend payments. This represents a whopping 46.45% rate of return on investment.

Johnson Controls spent more than $2.5 million retrofitting a government complex and installing a computerized building management system. As a result, the energy bill for the complex dropped from an average of $6 million a year to $3.5 million. Moreover, both parties will benefit from the 10-year life contract. Johnson recovers half the money saved in reduced utility costs (about $1.2 million a year over 10 years); the government has the other half to spend elsewhere. It is estimated that Johnson Controls will receive a 46.98% rate of return on their investment in this energy-control system.

What do all these rate of return figures really represent? How do we compute them? And once computed, how do we use them in evaluating investment projects? Our consideration of the concept of rate of return in this chapter will answer these and other questions.

Along with the NPW and AE, the third primary measure of investment worth is **rate of return.** As shown in Chapter 4, the NPW measure is easy to calculate and apply. Nevertheless, many engineers and financial managers prefer rate of return analysis to the NPW method because they find it intuitively more appealing to analyze investments in terms of percentage rates of return than in dollars of NPW. Consider the following statements regarding an investment's profitability:

- This project will bring in a 15% rate of return on investment.
- This project will result in a net surplus of $10,000 in NPW.

The rate of return figure is somewhat easier to understand because many of us are so familiar with savings and loan interest rates, which are in fact rates of return.

In this chapter, we will examine five aspects of rate of return analysis: (1) the concept of return on investment, (2) investment classification, (3) calculation of a rate of return, (4) development of an internal rate of return criterion, and (5) the comparison of mutually exclusive alternatives based on the rate of return.

■ 6.1 Rate of Return

Many different terms refer to rate of return, including yield (that is, the yield to maturity, commonly used in bond valuation), internal rate of return, and marginal efficiency of capital. We will first review three common definitions for rate of return. Then we will use the definition of the internal rate of return as a measure of profitability for a single investment project.

6.1.1 Return on Investment

There are several ways of defining the concept of rate of return on investment. The first is based on a typical loan transaction, and the second on the mathematical expression of the present worth function.

Definition 1 *Rate of return is defined as the interest rate earned on the unpaid balance of an amortized loan.*

Suppose that a bank lends $10,000 and receives interest payments of $1000 at the end of each year for 3 years, with the $10,000 loan being repaid at the end of the third year. In this situation, the bank is said to earn a return of 10% on its investment of $10,000.

Year	Unpaid Balance at Beginning Year	Return on Unpaid Balance (10%)	Payment Received	Unpaid Balance at End of Year
0	−$10,000	$ 0	$ 0	−$10,000
1	−10,000	−1,000	+1,000	−10,000
2	−10,000	−1,000	+1,000	−10,000
3	−10,000	−1,000	+11,000	0

Here a negative balance indicates an unpaid balance.

If, however, a bank lends $10,000 and is repaid $4021 at the end of each year for 3 years, determining the rate of return becomes more complicated. In this situation, only part of the $4021 annual payment represents interest, and the remainder goes toward repaying principal. By the calculation method you will learn in Section 6.3, we can show that this loan also has a return of 10%, in the same sense as the loan described in the previous loan repayment schedule: Namely, the annual payments will repay the loan itself and additionally provide a return of 10% on the *amount still outstanding each year*. Note that for both repayment schedules shown, the 10% interest is calculated only for each year's outstanding balance. When we make the last payment, the outstanding principal is eventually reduced to zero. We say the bank received a 10% rate of return on this loan transaction.

$$A = \$10,000(A/P, 10\%, 3) = \$4021$$

Year	Unrecovered Balance at Beginning Year	Return on Unrecovered Balance (10%)	Payment Received	Unrecovered Balance at End of Year
0	−$10,000	$ 0	$ 0	−$10,000
1	−10,000	−1,000	+4,021	−6,979
2	−6,979	−698	+4,021	−3,656
3	−3,656	−366	+4,021	0

Note that the ending balance is zero. This indicates that the bank can break even at a 10% rate of interest. The rate of return then becomes the rate of discount that equates the present value of future cash receipts to the cost of the project. This observation prompts the second definition on rate of return.

Definition 2 *Rate of return is the break-even interest rate, i°, which equates the present worth of a project's cash outflows to the present worth of its cash inflows, or*

$$PW(i^\circ) = PW_{\text{cash inflows}} - PW_{\text{cash outflows}}$$

$$= 0.$$

Note that the NPW expression is equivalent to

$$PW(i^\circ) = \frac{A_0}{(1 + i^\circ)^0} + \frac{A_1}{(1 + i^\circ)^1} + \cdots + \frac{A_N}{(1 + i^\circ)^N} = 0. \qquad (6.1)$$

Here we know the value of A_n for all n, but not the value of i°. Since it is the only unknown, we can solve for i°. (There will inevitably be N values of i° that satisfy this equation. In most project cash flows, you would be able to find a unique positive i° that satisfies Eq. (6.1). However, you may encounter some cash flow that cannot be solved for a single rate of return greater than -100%. By the nature of the NPW function in Eq. (6.1), it is certainly possible to have more than one rate of return for a certain type of cash flow.[1] For some cash flows, we may not find any rate of return at all.)

Note that the i° formula in Eq. (6.1) is simply the NPW formula, Eq. (4.1), solved for the particular interest rate (i°) at which $PW(i)$ is equal to zero. Multiplying both sides of Eq. (6.1) by $(1 + i^\circ)^N$, we obtain

$$PW(i^\circ)(1 + i^\circ)^N = FW(i^\circ) = 0.$$

If we multiply both sides of Eq. (6.1) by the capital recovery factor, $(A/P, i^\circ, N)$, we obtain the relationship $AE(i^\circ) = 0$. Therefore, the i° of a project may be defined as the rate of interest that equates the present worth, future worth, and annual equivalent worth of the entire series of cash flows to zero.

6.1.2 The Internal Rate of Return

We will introduce the rate of return based on the return on invested capital in terms of a project investment. This project's return is referred to as the internal rate of return (IRR) or the *true interest yield promised by an investment project over its useful life.*

Return on Invested Capital

Definition 3 *Internal rate of return is the interest rate earned on the unrecovered project balance of investment such that, when the project terminates, the unrecovered project balance will be zero.*

---

[1] You will always have N of them. The issue is whether they are real or imaginary. If they are real, are they in the $(-100\%, \infty)$ interval or not?

We can view an investment project as analogous to a bank loan. Suppose a lender invests $10,000 in a computer with a 3-year useful life and equivalent annual labor savings of $4021. Here, we may view the investing firm as the lender and the project as the borrower. The cash flow transaction between them would be identical to the second amortized loan transaction in Definition 1.

n	Beginning Project Balance	Return on Invested Capital	Cash Payment	Ending Project Balance
0	$ 0	$ 0	−10,000	−10,000
1	−10,000	−1,000	4,021	−6,979
2	−6,979	−697	4,021	−3,656
3	−3,656	−365	4,021	0

In our project balance calculation, we see that 10% is earned on $10,000 during year 1, 10% is earned on $6979 during year 2, and 10% is earned on $3656 during year 3. This indicates that the firm earns a 10% rate of return on funds that remain *internally* invested in the project. Since it is a return *internal* to the project, we refer to it as the **internal rate of return,** or IRR. Notice also that there is only one cash outflow occurring at time 0, and the present worth of this outflow will simply be $10,000. There are three equal receipts, and the present worth of these inflows is $4021(P/A, 10\%, 3) = \$10,000$. Since the NPW = $PW_{inflow} - PW_{outflow} = \$10,000 - \$10,000 = 0$, 10% also satisfies Definition 2 for rate of return.

Even though our simple example above implies that i° coincides with IRR, only Definitions 1 and 3 correctly describe the true meaning of internal rate of return. As we will see later, if the cash expenditures of an investment are not restricted to the initial period, there may exist several break-even interest rates (i°s) that satisfy Eq. (6.1), but there may not be a rate of return concept *internal* to the project.

■ 6.2 Investment Classification

As hinted in the previous section, for certain series of project cash flows, we may uncover the complication of multiple i° values that satisfy Eq. (6.1). By analyzing and classifying cash flows, we may anticipate this difficulty and adjust our analysis approach.

6.2.1 Simple Versus Nonsimple Investments

We can classify an investment project by counting the number of sign changes in its net cash flow sequence. A change from either "+" to "−" or "−" to "+" is counted as one sign change. (We ignore a zero cash flow.) Then, a **simple**

investment is defined as one in which the initial cash flows are negative and there is only one sign change in the net cash flow. A **nonsimple project** is one for which there is more than one sign change in the cash flow series. Multiple i°s, as we will see later, occur only in nonsimple projects. The different investment possibilities may be illustrated as follows:

Investment	Cash Flow Sign at Period					
Type	0	1	2	3	⋯	N
Simple	−	+	+	+	⋯	+
Simple	−	−	+	+	⋯	+
Nonsimple	−	+	−	+	⋯	−
Nonsimple	−	+	+	−	⋯	+

An investment with the cash flow sequence of either (−$1000, $250, $800, $1200) or (−$1000, −$500, $300, $1000, $2000) represents a simple investment variety. These types of investment would reveal the NPW profile shown in Fig. 6.1. There is only one i° crossing the i-axis. As an example of nonsimple investment, a project with the cash flow sequence (−$1000, $3900, −$5030, $2145) represents a nonsimple investment. The NPW profile for this investment will have the shape shown in Fig. 6.2, with crossings of the i axis at 10%, 30%, and 50%. Not all nonsimple investments will have a net present value with multiple crossings of the i-axis. In the next section, we will illustrate when to expect such multiple crossings by examining the types of cash flows.

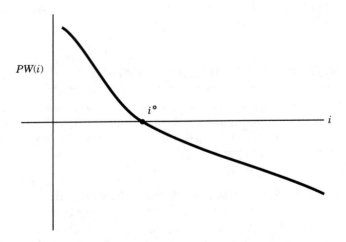

Figure 6.1 ■ NPW profile for a simple investment

Figure 6.2 ■ NPW profile for a typical nonsimple investment

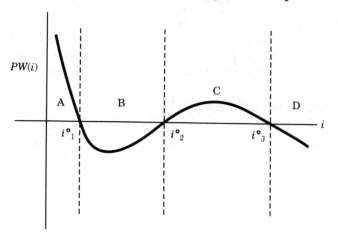

6.2.2 Predicting Multiple i*s

Here we will focus on the initial problem of whether we can predict a unique i^* for a project by examining its cash flow pattern. There are two useful rules that allow us to do so by focusing on sign changes in: (1) net cash flows and (2) accounting net profit (accumulated net cash flows).

Net Cash Flow Rule of Signs

There are several ways to predict the number of i^*s. One useful method for predicting an upper limit on the number of positive i^*s of a cash flow stream is to apply the rule of signs: *The number of real* i*s *that are greater than* −100% *for a project with* N *periods is never greater than the number of sign changes in the sequence of the* A_n.

An example would be

Period	Net Cash Flow	Sign Change
0	−$100	
1	−20	
2	50	☐1
3	0	
4	60	
5	−30	☐1
6	100	☐1

There are 3 sign changes in the cash flow sequence, so there are 3 or fewer real positive i^*s.

It must be emphasized that the rule of signs provides an indication only of the possibility of multiple rates of return: The rule only predicts the *maximum* number of possible i^*s. Many projects have multiple sign changes in their cash flow sequence but still possess a unique real i^* in the $(-100\%, +\infty)$.

Accumulated Cash Flow Sign Test

The accumulated cash flow is the sum of the net cash flows up to and including a given time. If the rule of cash flow signs indicates multiple i^*s, we should proceed to the *accumulated* cash flow sign test to possibly eliminate some alternatives.

If we let A_n represent the net cash flow in period n and S_n represent the accumulated cash flow up to period n, we have the following:

Period (n)	Cash Flow (A_n)	Accumulated Cash Flow (S_n)
0	A_0	$S_0 = A_0$
1	A_1	$S_1 = A_0 + A_1$
2	A_2	$S_2 = A_0 + A_1 + A_2$
$\vdots$	$\vdots$	$\vdots$
N	A_N	$S_N = A_0 + A_1 + A_2 + \cdots + A_N$

We then examine the sequence of accumulated cash flows $(S_0, S_1, S_2, S_3, \dots, S_N)$ to determine the number of sign changes.

Rule of Cumulative Cash Flow Sign

If the series S_n starts negatively and changes sign only once, there exists a unique positive i^*. This cumulative cash flow sign rule is a more discriminating test for identifying the uniqueness of i^* than the previous method described.

Example 6.1 Predicting the number of i^*s

Predict the number of real positive rate(s) of return for each cash flow series:

Period	A	B	C	D
0	−$100	−$100	$ 0	−$100
1	−200	+50	−50	+50
2	+200	−100	+115	0
3	+200	+60	−66	+200
4	+200	−100		−50

Solution

Given: Four cash flow series and cumulative flow series

Find: The upper limit on number of i^* for each series

The cash flow rule of signs indicates the following possibilities for the positive values of i^*:

Project	Number of Sign Changes in Net Cash Flows	Possible Number of Positive Values of i^*
A	1	1 or 0
B	4	4, 3, 2, 1 or 0
C	2	2, 1, or 0
D	2	2, 1, or 0

For cash flows B, C, and D, we would like to apply the more discriminating cumulative cash flow test to see if we can specify a smaller number of possible values of i^*:

Project B		Project C		Project D	
Net Cash Flow	Cumulative Cash Flow	Net Cash Flow	Cumulative Cash Flow	Net Cash Flow	Cumulative Cash Flow
−$100	−$100	$ 0	$ 0	−$100	−$100
+50	−50	−50	−50	+50	−50
−100	−150	+115	+65	0	−50
+60	−90	−66	−1	+200	+150
−100	−190			−50	+100

Recall the test: If the series starts negatively and changes sign only once, there exists a unique positive i^*. Only project D begins negatively and passes the test; we may predict a unique i^* value, rather than 2, 1, or 0 as predicted by the cash flow rule of signs. Projects B and C fail the test and we cannot eliminate the possibility of multiple i^*s. (If projects do not begin negatively, they are borrowing projects rather than investment projects.)

■ 6.3 Computational Methods for Determining i^*

We may find the i^* by several procedures, each of which has advantages and disadvantages. We will discuss some of the most practical methods here. They are

- the direct-solution method,
- the trial-and-error method, and
- the graphic method.

6.3.1 Direct-Solution Method

For the very special cases of a project with only a two-flow transaction (an investment followed by a single future payment) or a project with a service life of 2 years of return, we can seek a direct mathematical solution for rate. These two cases will be examined in Example 6.2.

Example 6.2 i° by direct solution: two flows and two period cases

Consider two investment projects with the following cash flow transactions. Compute the rate of return for each project.

n	Project 1	Project 2
0	$-\$1000$	$-\$2000$
1	0	1300
2	0	1500
3	0	
4	$+1500$	

Solution

Given: Cash flows for two projects
Find: i° for each

- **For Project 1:** Solving for i° in $PW(i^\circ) = 0$ is identical to solving $FW(i^\circ) = 0$ because FW equals PW times a constant. We could do either here, but we will set $FW(i^\circ) = 0$ to demonstrate the latter. Using the single-payment future worth relationship, we obtain

$$FW(i^\circ) = -\$1000(F/P, i^\circ, 4) + \$1500$$

$$= 0$$

$$\$1500 = \$1000(F/P, i^\circ, 4) = \$1000(1 + i^\circ)^4$$

$$1.5 = (1 + i^\circ)^4.$$

Taking a natural log on both sides, we obtain

$$ln(1.5) = 4ln(1 + i^\circ)$$

$$ln(1 + i^\circ) = 0.1014.$$

Solving for i^* yields

$$i^* = e^{0.1014} - 1$$

$$= 0.1067, \text{ or } 10.67\%.$$

- **For Project 2:** We may write the NPW expression for this project as follows:

$$PW(i) = -\$2000 + \frac{\$1300}{(1+i)} + \frac{\$1500}{(1+i)^2} = 0.$$

Let $X = \frac{1}{(1+i)}$. We may then rewrite the $PW(i)$ as a function of X.

$$PW(i) = -\$2000 + \$1300X + \$1500X^2 = 0$$

This is a quadratic equation having the following solution[2]

$$X = \frac{-\$1300 \pm \sqrt{\$1300^2 - 4(\$1500)(-\$2000)}}{2(\$1500)}$$

$$= \frac{-\$1300 \pm \$3700}{\$3000}$$

$$= 0.8 \text{ or, } -1.667.$$

Replacing X-values and solving for i gives us

$$0.8 = \frac{1}{(1+i)}, \quad i = 25\%$$

$$-1.667 = \frac{1}{(1+i)}, \quad i = -160\%$$

Since an interest rate less than -100% has no economic significance, we find that the project's i^* is 25%.

Comments: *These projects had very simple cash flows. When cash flows are more complex, we generally must use a trial-and-error method or a computer to find i^*.*

---

[2] Given $aX^2 + bX + c = 0$, the solution of the quadratic equation is

$$X = \frac{-b \pm \sqrt{b^2 - 4ac}}{2a}.$$

$$X = \frac{-b \pm \sqrt{b^2 - 4ac}}{2a}$$

6.3.2 Trial-and-Error Method for Simple Investment

The first step in the trial-and-error method is to make an estimated guess[3] at the value of i^*. For a simple investment, we compute the present worth of net cash flows using the guessed interest rate and observe whether it is positive, negative, or zero. Suppose the $PW(i)$ is negative. Since we are aiming for a value of i that makes $PW(i) = 0$, we must raise the present worth of the cash flow. To do this we lower the interest rate and go through the process again. On the other hand, if $PW(i)$ is positive, we raise the interest rate in order to lower $PW(i)$. We continue until $PW(i)$ is approximately equal to zero. Whenever we reach the point where $PW(i)$ is bounded by one negative and one positive value, we use **linear interpolation** to approximate the i^*. This is a somewhat tedious and inefficient process. (In general, this trial-and-error method does not work for nonsimple investments.)

Example 6.3 i^* by trial and error

Agdist Corporation distributes agricultural equipment. The board of directors is considering a proposal to establish a facility to manufacture an electronically controlled "intelligent" crop sprayer invented by a professor at a local university. This independent project would require an investment of $10 million in assets and would produce an annual after-tax net benefit of $1.8 million over a service life of 8 years. All costs and benefits are included in these figures. When the project terminates, the net proceeds from the sale of the assets would be $1 million (see Fig. 6.3). Compute the rate of return of this project.

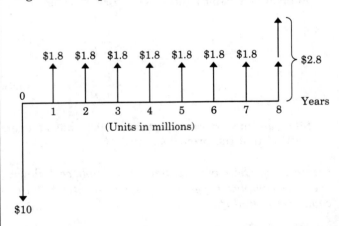

Figure 6.3 ■ Cash flow diagram for a simple investment (Example 6.3)

---

[3] As we see later in this chapter, the ultimate objective of finding the i^* is to compare it against the MARR. Therefore, it is a good idea to use the MARR as the initial guess value.

Solution

Given: Initial investment $(I) = \$10$ million, $A = \$1.8$ million, $S = 1$ million, $N = 8$ years
Find: i°

We start with a guessed interest rate of 8%. The present worth of the cash flows in millions of dollars is

$$PW(8\%) = -\$10 + \$1.8(P/A, 8\%, 8) + \$1(P/F, 8\%, 8) = \$0.88.$$

Since this present worth is positive, we must raise the interest rate to bring this value toward zero. When we use an interest rate of 12%, we find that

$$PW(12\%) = -\$10 + \$1.8(P/A, 12\%, 8) + \$1(P/F, 12\%, 8) = -\$0.65.$$

We have bracketed the solution: $PW(i)$ will be zero at i somewhere between 8% and 12%. Using straight-line interpolation, we approximate

$$i^\circ \cong 8\% + (12\% - 8\%)\left[\frac{0.88 - 0}{0.88 - (-0.65)}\right]$$

$$= 8\% + 4\%(0.5752)$$

$$= 10.30\%.$$

Now we will check to see how close this value is to the precise value of i°. If we compute the present worth at this interpolated value, we obtain

$$PW(10.30\%) = -\$10 + \$1.8(P/A, 10.30\%, 8) + \$1(P/F, 10.30\%, 8)$$

$$= -\$0.045.$$

As this is not zero, we may recompute the i° at a lower interest rate, say 10%.

$$PW(10\%) = -\$10 + \$1.8(P/A, 10\%, 8) + \$1(P/F, 10\%, 8) = \$0.069$$

With another round of linear interpolation, we approximate

$$i^\circ \cong 10\% + (10.30\% - 10\%)\left[\frac{0.069 - 0}{0.069 - (-0.045)}\right]$$

$$= 10\% + 0.30\%(0.6053)$$

$$= 10.18\%.$$

At this interest rate,

$$PW(10.18\%) = -\$10 + \$1.8(P/A, 10.18\%, 8) + \$1(P/F, 10.18\%, 8)$$

$$= \$0.0007,$$

which is practically zero, so we may stop here. In fact, there is no need to be more precise about these interpolations because the final result can be more accurate than the basic data, which ordinarily are only rough estimates. Computing the i° for this problem on computer, incidentally, gives us 10.1819%.

6.3.3 Graphic Solution

The most easily generated and understandable graphic method of solving for i° is the **NPW profile.** In this graph, the horizontal axis indicates interest rate, and the vertical axis indicates NPW. For a given project's cash flows, the NPW is calculated at an interest rate of zero (which gives the vertical axis intercept) and several other interest rates. These points are plotted and a curve sketched. Since i° is defined as the interest rate at which $PW(i^\circ) = 0$, the point at which the curve crosses the horizontal axis closely approximates the i°. We will consider the method of graphic solution by type of investment.

- **Simple investment:** As an example, we can generate the NPW profile for the cash flow in Fig. 6.4:

  1. We first use $i = 0$ in this equation, to obtain $PW(0\%) = \$636$, which is the vertical axis intercept.
  2. Substituting in several other interest rates—5%, 10%, 20%, and 30%—we plot these values of $PW(i)$ as well.

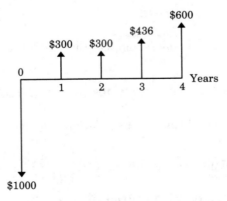

Figure 6.4 ■ Cash flow diagram

Figure 6.5 ■ NPW profile for a simple investment

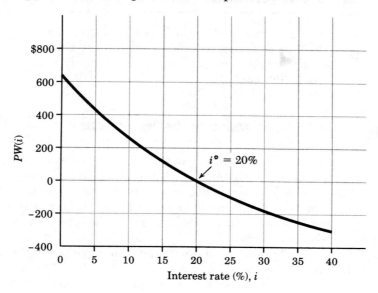

The result is Fig. 6.5, which shows the curve crossing the horizontal axis at roughly 20%. This value can be verified by other methods if we desire substitution. Note that in addition to establishing the interest rate that makes $PW(20\%) = 0$, the NPW profile indicates where positive and negative NPW values fall, thus giving us a broad picture of the interest rates for which the project is acceptable or unacceptable.

- **Nonsimple investment:** As mentioned in the previous section, it is possible to have more than one value of i^* that makes $PW(i^*) = 0$; that is, more than one i^* for a project. In such a case, the NPW profile would cross the horizontal axis more than once. We discussed methods for predicting the number of i^* values by looking at cash flows in Section 6.2.1. However, generating a NPW profile to discover multiple i^*s is as useful and informative as any other method.

We don't need laborious manual calculations or plots to find i^*. Many financial calculators have built-in functions for calculating i^*. It is worth noting here that many spreadsheet packages have i^* functions, which solve Eq. (6.1) very rapidly. This is normally done by entering the cash flows through a computer keyboard or by reading a cash flow data file. A better method of solving i^* problems is to use a computer-aided economic analysis program, **EzCash** that finds i^* visually by specifying the lower and upper bounds of the interest search limit, and generates NPW profiles when given a cash flow series. In addition to the savings of calculation time, the advantage of the computer-generated profiles is their precision.

■ 6.4 Internal Rate of Return Criterion

Now we have classified investment projects and learned methods to determine the i° value for a given project's cash flows. Our objective is to develop an accept/reject decision rule that gives us results consistent with those obtained from NPW analysis.

6.4.1 Relationship to the PW Analysis

As we already observed in Chapter 4, NPW analysis is dependent on the rate of interest used for the NPW computation. A different rate may change a project from acceptable or it may change the ranking of projects. Consider again the NPW profile as drawn for a simple project in Fig. 6.1. For interest rates below i° this project should be accepted; for interest rates above i° it should be rejected.

On the other hand, for certain nonsimple projects, the NPW may look like the one shown in Fig. 6.2. The use of the NPW analysis would lead to the acceptance of the projects in regions A and C, but rejection in regions B and D. Of course, this result goes against one's intuition—a higher interest rate would change an unacceptable project into an acceptable one. The situation graphed in Fig. 6.2 is one of the cases of multiple i°s, mentioned in Definition 2. Therefore, for a simple investment situation in Fig. 6.1, the i° can serve as an appropriate index to either accept or reject the investment. However, for a nonsimple investment in Fig. 6.2, it is not clear which i° to use in making an accept/reject decision. Therefore, the i° value fails to provide an appropriate measure of profitability for an investment project with multiple rates of return.

6.4.2 Accept/Reject Decision Rules for Simple Investments

Why are we interested in finding the particular interest rate that equates a project's cost with the present worth of its receipts? Again, we may easily answer this by examining Fig. 6.1. In this figure, we notice two important characteristics of the NPW profile. First, as we compute the project's $PW(i)$ at a varying interest rate (i), we see that the NPW becomes positive for $i < i^\circ$, indicating that the project would be acceptable under the PW analysis for those values of i. Second, the NPW becomes negative for $i > i^\circ$, indicating that the project is unacceptable for those values of i. Therefore, the i° serves as a *break-even* interest rate. By knowing this break-even rate, we will be able to make an accept/reject decision that is consistent with the NPW analysis.

Note that, for a simple investment, i° is indeed IRR of the investment (see Section 6.1.2). Merely knowing the i° is not enough to apply this method, however. Because firms typically wish to do better than break even (recall that at

NPW = 0 we were indifferent to the project), a minimum acceptable rate of return (MARR) is indicated by company policy, management, or the project decision maker. At the MARR the company will more than break even. Thus, the IRR becomes a useful gauge against which to judge project acceptability, and the decision rule for a simple project is

If IRR > MARR, accept the project.

If IRR = MARR, remain indifferent.

If IRR < MARR, reject the project.

Note that this decision rule is designed to be applied for a single project evaluation. When we have to compare mutually exclusive investment projects, we need to apply the incremental analysis, as we shall see in Section 6.5.1.

6.4.3 Accept/Reject Decision Rules for Nonsimple Investment

When applied to simple projects, the i° provides an unambiguous criterion for measuring profitability. However, when multiple rates of return occur, none of them is an accurate portrayal of project acceptability or profitability. Clearly, then, we should place a high priority on discovering this situation early in our analysis of a project's cash flows. The quickest way to predict multiple i°s is to generate a NPW profile and check to see if it crosses the horizontal axis more than once.

In addition to the NPW profile, there are good—although somewhat more complex—analytical methods for predicting multiple i°s. Perhaps more importantly, there is a good method, which uses an **external rate of return,** of refining our analysis when we do discover multiple i°s. An external rate of return allows us to calculate a single accurate rate of return; it is covered in optional Section 6.4.4.

If you choose to avoid these more complex applications of rate of return techniques, you must at a minimum be able to predict multiple i°s via the NPW profile and, when they occur, select an alternative method such as NPW or AE analysis for determining project acceptability.

6.4.4 Computing IRR for Nonsimple Investments[4]

To comprehend the nature of multiple i°s, we need to understand the investment situation represented by any cash flow. The net investment test will indi-

---

[4] This optional advanced topic may be omitted at the instructor's discretion.

cate whether the i° computed represents the true rate of return earned on the money invested in the project while it is actually in the project. As we shall see, the phenomenon of multiple i°s occurs only when the net investment test fails. When multiple positive rates of return for a cash flow are found, in general none is suitable as a measure of project profitability, and we must proceed to the next analysis step: introducing an external rate of return.

Net Investment Test

A project is said to be a **net investment** when the project balances computed at the project's i°, $PB(i^\circ)_n$, are either less than or equal to zero throughout the life of the investment with $A_0 < 0$. The investment is *net* in the sense that the firm does not overdraw on its return at any point and hence is not *indebted* to the project. This type of project is called a **pure investment.** (On the other hand, **pure borrowing** is defined as the situation where $PB(i^\circ)_n$ are positive or zero throughout the life of the loan with $A_0 > 0$.) Simple investments will always be pure investments. Therefore, if a nonsimple project passes the net investment test (a pure investment), then the accept/reject decision rule will be the same as the simple investment case given in Section 6.4.3.

If any of the project balances calculated at the project's solving i° is positive, the project is not a pure investment. A positive project balance indicates that, at some time during the project life, the firm acts as a borrower $[PB(i^\circ)_n > 0]$ rather than an investor in the project $[PB(i^\circ)_n < 0]$. This type of investment is called a **mixed investment.**

Example 6.4   Pure versus mixed investments

Consider the four investment projects in Table 6.1 with known i° values. Determine which projects are pure investment investments.

Table 6.1 Cash Flows for Projects A, B, C, and D

n	Projects			
	A	B	C	D
0	−$1000	−$1000	−$1000	−$1000
1	−1000	1600	500	3900
2	2000	−300	−500	−5030
3	1500	−200	2000	2145
i°	33.64%	21.95%	29.95%	10%, 30%, 50%

Solution

Given: Four projects with cash flows and i^*s shown in Table 6.1

Find: Which projects are pure investments

We will first compute the project balances at the projects' respective i^*s. If there exist multiple rates of return, we may use the largest value of i^* greater than zero.[5]

- **Project A:**

$$PB(33.64\%)_0 = -\$1000$$

$$PB(33.64\%)_1 = -\$1000(1 + 0.3364) + (-\$1000) = -\$2336.40$$

$$PB(33.64\%)_2 = -\$2336.40(1 + 0.3364) + \$2000 = -\$1122.36$$

$$PB(33.64\%)_3 = -\$1122.36\,(1 + 0.3364) + \$1500 = 0$$

$(-, -, -, 0)$: Passes the net investment test (pure investment)

- **Project B:**

$$PB(21.95\%)_0 = -\$1000$$

$$PB(21.95\%)_1 = -\$1000(1 + 0.2195) + \$1600 = \$380.50$$

$$PB(21.95\%)_2 = +\$380.50(1 + 0.2195) - \$300 = \$164.02$$

$$PB(21.95\%)_3 = +\$164.02(1 + 0.2195) - \$200 = 0$$

$(-, +, +, 0)$: Fails the net investment test (mixed investment)

- **Project C:**

$$PB(29.95\%)_0 = -\$1000$$

$$PB(29.95\%)_1 = -\$1000(1 + 0.2995) + \$500 = -\$799.50$$

$$PB(29.95\%)_2 = -\$799.50(1 + 0.2995) - \$500 = -\$1,538.95$$

$$PB(29.95\%)_3 = -\$1538.95(1 + 0.2995) + \$2000 = 0$$

$(-, -, -, 0)$: Passes the net investment test (pure investment)

---

[5] In fact, it does not matter which rate we use in applying the net investment test. If one value passes the net investment test, they will all pass. If one value fails, they all fail.

- **Project D:** (There are three rates of return. We can use any of them for the net investment test.)

$$PB(50\%)_0 = -\$1000$$

$$PB(50\%)_1 = -\$1000(1 + 0.50) + \$3900 = \$2400$$

$$PB(50\%)_2 = +\$2400(1 + 0.50) - \$5030 = -\$1430$$

$$PB(50\%)_3 = -\$1430(1 + 0.50) + \$2145 = 0$$

$(-, +, -, 0)$: Fails the net investment test (mixed investment)

Comments: *As shown in Fig. 6.6, Projects A and C are the only pure investments. Project B demonstrates that the existence of a unique i° is a necessary but not sufficient condition for a pure investment.*

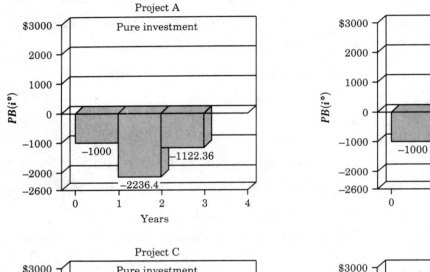

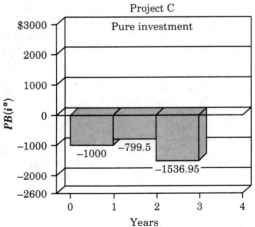

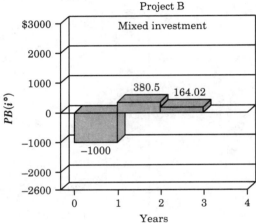

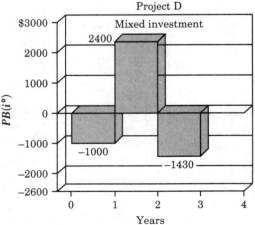

Figure 6.6 ■ Net investment test (Example 6.4)

Example 6.5 IRR for nonsimple project: pure investment

Consider Project C in Example 6.4. Assume that the project is independent and that all costs and benefits are stated explicitly. Apply the net investment test and determine the acceptability of the project using the IRR criterion.

Solution

Given: Cash flow, $i^* = 29.95\%$, MARR $= 15\%$
Find: IRR. Is the project acceptable?

The net investment test was already applied in Example 6.4, and it indicated that Project C is a pure investment. In other words, the project balances were all less than or equal to zero and the final balance was zero, as it must be if the interest rate used is an i^*. This proves that 29.95% is the true internal rate of return for the cash flow. At MARR $= 15\%$, the IRR > MARR; thus, the project is acceptable. If we compute the NPW of this project at $i = 15\%$, we obtain

$$PW(15\%) = -\$1000 + \$500(P/F, 15\%, 1) - \$500(P/F, 15\%, 2)$$
$$+ \$2000(P/F, 15\%, 3)$$
$$= \$371.74.$$

Since $PW(15\%) > 0$, the project is also acceptable under the NPW criterion. IRR and NPW criteria will always produce the same decision if the criteria are applied correctly.

Comments: *In the case of a pure investment, the firm has funds committed to the project over the life of the project and at no time withdraws money from the project. The IRR is the return earned on the funds that remain internally invested in the project.*

The Need for External Interest Rate for Mixed Investment

Thus far, when the net investment test fails, we have been implicitly accepting a simplification about interest rate that may or may not be true.[6] That is: *When we calculate the project balance at an i^* for mixed investments, we notice an important point—cash borrowed (released) from the project is assumed to earn the same interest rate through external investment as money that remains in-*

---

[6] Even for a nonsimple investment where there is only one positive rate of return, the project may fail the net investment test, as demonstrated by Project B in Example 6.4. In this case, the unique i^* still may not be a true indicator of the project's profitability.

ternally invested. In other words, solving a cash flow for an unknown interest rate assumes that money borrowed from a project can be reinvested to yield a rate of return equal to that received from the project. In fact, we have been making this assumption whether or not a cash flow produces a unique positive i°. Note that money is borrowed only when $PB(i^\circ) > 0$, and the magnitude of the borrowed amount is the project balance. When $PB(i^\circ) < 0$, no money is borrowed, even though the cash flow may be positive at that time.

In reality, it is not always possible for cash borrowed (released) from a project to be reinvested to yield a rate of return equal to that received from the project. Instead, it is very likely that the rate of return available on a capital investment in the business is much different—usually higher—from the rate of return available on other external investments. Thus, two interest rates may be necessary to compute the project balances for a project's cash flow—one on the internal investment and one on the external investment. As we will see later, by separating the interest rates, we can measure the *true* rate of return of any internal portion of investment project.

Because the net investment test is the *only* way to predict accurately project borrowing (that is, external investment) its significance now becomes very clear: In order to calculate accurately a project's true IRR, we should always test a solution by the net investment test and, when the test fails, take the further analytical step of introducing an external rate of return. Even the presence of a unique, positive i° is a necessary but not sufficient condition to predict net investment, so if we find a unique value we should still expose it to the net investment test.

Calculation of IRR for a Mixed Investment

A failed net investment test indicates a combination of internal and external investment. When this combination exists, we must calculate a rate of return on the portion of capital that remains invested internally.

How do we determine the IRR of this investment? Insofar as a project is not a net investment, there are one or more periods when the project has a net outflow of money (positive project balance) which must later be returned to the project. This money can be put into the firm's investment pool until such time as it is needed in the project. The interest rate of this investment pool is the interest rate at which the money can in fact be invested outside the project.

Recall that the NPW method assumed that the interest rate charged to any funds withdrawn from a firm's investment pool would be equal to the MARR. In this book, we will use the MARR as an established external interest rate—earned by money invested outside of the project. We can then compute IRR, as a function of MARR by finding that value of IRR that will make the terminal project balance equal to zero. (This implies that the firm wants to recover any investment made to the project fully and pays off any borrowed funds at the end of project life.) This way of computing rate of return is a correct measure of the profitability of the project represented by the cash flow. The following procedure outlines the steps for determining the IRR for a mixed investment:

Step 1: Identify the MARR (or external interest rate).

Step 2: Calculate $PB(i, \text{MARR})_n$ (or simply PB_n) according to the rule

$$PB(i, \text{MARR})_0 = A_0$$

$$PB(i, \text{MARR})_1 = \begin{cases} PB_0(1 + i) + A_1 & \text{if } PB_0 < 0 \\ PB_0(1 + \text{MARR}) + A_1 & \text{if } PB_0 > 0 \end{cases}$$

$$PB(i, \text{MARR})_N = \begin{cases} PB_{N-1}(1 + i) + A_N & \text{if } PB_{N-1} < 0 \\ PB_{N-1}(1 + \text{MARR}) + A_N & \text{if } PB_{N-1} > 0 \end{cases}$$

(As defined in the text, A_n stands for the net cash flow at the end of period n. Note also that the terminal project balance must be zero.)

Step 3: Determine the value of i by solving the equation

$$PB(i, \text{MARR})_N = 0.$$

That interest rate is the IRR for the mixed investment.

Using the MARR as an external interest rate, we may accept a single project if the IRR exceeds MARR, and should reject the project otherwise. Figure 6.7 illustrates the IRR computation for a mixed investment.

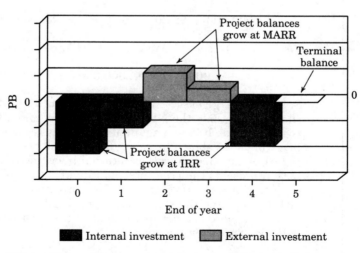

Figure 6.7 ■ Computational logic for IRR (mixed investment)

Example 6.6   IRR for nonsimple project: mixed investment

Suppose a defense contractor has received a contract worth $7,800,000 to build flight simulators for Canadian pilot training. The Canadian government will make an advance payment of $3,000,000 when the contract is signed, another $4,300,000 at the end of the first year, and the $500,000 balance at the end of the second year. The expected cash outflows required to produce these simulators are estimated to be $4,000,000 now, $2,000,000 during the first year, and $1,820,000 during the second year. The expected cash flows from this project are summarized as follows:

Year	Cash Inflow	Cash Outflow	Net Cash Flow
0	$3,000,000	$4,000,000	−$1,000,000
1	4,300,000	2,000,000	2,300,000
2	500,000	1,820,000	−1,320,000

(a) Compute the values of i°s for this project.
(b) Compute the IRR for this project, assuming MARR = 15%.
(c) Make an accept/reject decision based on the results in part (b).

Solution

Given: Cash flow shown above, MARR = 15%

Find: (a) $i°$, (b) IRR, and (c) determine whether to accept the project.

(a) Since this project has a 2-year life, we may solve the net present worth equation directly via the quadratic formula method:

$$-\$1,000,000 + \$2,300,000/(1 + i°) - \$1,320,000/(1 + i°)^2 = 0.$$

We let $X = 1/(1 + i°)$. We can rewrite the expression

$$-\$1,000,000 + \$2,300,000X - 1,320,000X^2 = 0.$$

Solving for X gives $X = 10/11$ and $10/12$, or $i° = 10\%$ and 20%. As shown in Fig. 6.8, the NPW profile intersects the horizontal axis twice, once at 10% and again at 20%. This is obviously not a net investment, as shown in Table 6.2. Because the net investment test indicates external as well as internal investment, neither 10% nor 20% represents the true internal rate of return of this government project. Since the project is a mixed investment, we need to find the IRR by applying the steps shown previously.

(b) At $n = 0$, there is a net investment to the firm so that the project balance expression becomes

$$PB(i, 15\%)_0 = -\$1,000,000.$$

Figure 6.8 ■ Multiple rates of return (Example 6.6)

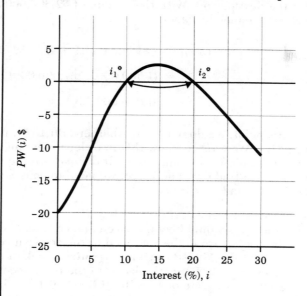

Table 6.2 Project Balance Calculations for a Mixed Investment ($000)

(a) Using $i^\circ = 10\%$

	n		
	0	*1*	*2*
Beginning balance	$ 0	−$1000	$1200
Return on investment	0	−100	120
Payment	−1000	2300	−1320
Ending balance	−1000	1200	0

(The net investment test fails.)

(b) Using $i^\circ = 20\%$

	n		
	0	*1*	*2*
Beginning balance	$ 0	−$1000	$1100
Return on investment	0	−200	220
Payment	−1000	2300	−1320
Ending balance	−1000	1100	0

(The net investment test fails.)

This net investment of $1,000,000 that remains invested grows at i for the next period. With the receipt of $2,300,000 in year 1, the project balance becomes

$$PB(i, 15\%)_1 = -\$1,000,000(1 + i) + \$2,300,000$$

$$= \$1,300,000 - \$1,000,000i$$

$$= \$1,000,000(1.3 - i).$$

At this point, we do not know whether $PB(i,15\%)_1$ is positive or negative: We want to know this in order to test for net investment and the presence of a unique $i°$. It depends on the value of i, which we want to find out. Therefore, we need to consider two situations: (1) $i < 1.3$ and (2) $i > 1.3$.

- **Case 1:** $i < 1.3 \Rightarrow PB(i, 15\%)_1 > 0$
  Since this would be a positive balance, this cash borrowed from the project would be returned to the firm's investment pool to grow at the MARR until it is required back in the project. By the end of year 2, the cash placed in the investment pool would have grown at the rate of 15% [to $1,000,000(1.3 - i)(1 + 0.15)$], and must equal the investment into the project of $1,320,000 required at that time. Then, the terminal balance must be

$$PB(i, 15\%)_2 = \$1,000,000(1.3 - i)(1 + 0.15) - \$1,320,000$$

$$= \$175,000 - \$1,150,000i$$

$$= 0.$$

Solving for i yields

$$\text{IRR} = 0.1522, \text{ or } 15.22\%.$$

The computational process is shown in graphically in Fig. 6.9.

- **Case 2:** $i > 1.3 \Rightarrow PB(i, 15\%)_1 < 0$
  The firm is still in an investment mode. Therefore, the balance at the end of year 1 that remains invested will grow at the rate of i for the next period. With the investment of $1,320,000 required in year 2 and the fact that the net investment must be zero at the end of project life, the balance at the end of year 2 should be

$$PB(i, 15\%)_2 = \$1,000,000(1.3 - i)(1 + i) - \$1,320,000$$

$$= -\$20,000 + \$300,000i - \$1,000,000i^2$$

$$= 0.$$

Figure 6.9 ■ Calculation of the IRR for a mixed investment

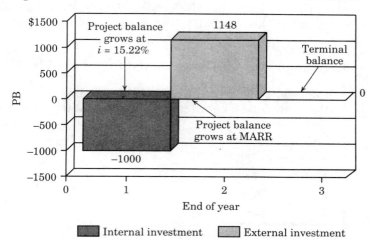

Solving for i gives

$$\text{IRR} = 0.1 \text{ or, } 0.2 < 1.3,$$

which violates the initial assumption ($i > 1.3$). Therefore, Case 1 is the correct situation.

(c) Case 1 indicates IRR > MARR, so the project would be acceptable. If we use the present worth method at MARR = 15%, we obtain

$$PW(15\%) = -\$1 + \$2.3(P/F, 15\%, 1) - \$1.32(P/F, 15\%, 2)$$

$$= \$0.0018 > 0,$$

verifying that the project is barely acceptable.

Comments: *In this example, we could have seen by inspection that Case 1 was correct. Since the project required an investment as the final cash flow, the project balance at the end of the previous period (year 1) had to be positive in order for the final balance to equal zero. Inspection does not generally work for more complex cash flows.*

A Trial-and-Error Method for Computing IRR

The trial-and-error approach for finding IRR for a mixed investment is very similar to the trial-and-error approach to finding i^*. We begin with a given MARR and a guess for IRR and solve for the project balance. (A value of IRR close to the MARR is a good starting point for most problems.) Since we desire the project balance to approach zero, we can adjust the value of IRR as needed after seeing the result of the initial guess. For example, for a given pair of interest rates (IRR, MARR), if the terminal project balance is positive, the IRR value is too low, so we raise it and recalculate. We can continue adjusting our IRR guesses in this way until we obtain a project balance equal or very close to zero.[7]

Example 6.7 IRR for mixed investment by trial and error

Consider Project D in Example 6.4, which has the following cash flow. We know from an earlier calculations that this is a mixed investment.

n	A_n
0	−$1000
1	3900
2	−5030
3	2145

Compute the IRR for this project. Assume a MARR of 6%.

Solution

Given: Cash flow as stated for mixed investment, MARR = 6%

Find: IRR

For MARR = 6%, we must compute i by trial and error. Suppose we guess $i = 8\%$:

$$PB(8\%, 6\%)_0 = -\$1000 \qquad\qquad = -\$1000$$
$$PB(8\%, 6\%)_1 = -\$1000(1 + 0.08) + \$3900 \quad = +\$2820$$
$$PB(8\%, 6\%)_2 = +\$2820(1 + 0.06) - \$5030 \quad = -\$2040.80$$
$$PB(8\%, 6\%)_3 = -\$2040.80(1 + 0.08) + \$2145 = -\$\ \ 59.06$$

The net investment is negative at the end of the project, indicating that our trial $i = 8\%$ is in error. After several trials, we conclude that for MARR = 6%, IRR is approximately 6.13%. To verify the results,

$$PB(6.13\%, 6\%)_0 = -\$1000 \qquad\qquad = -\$1000$$
$$PB(6.13\%, 6\%)_1 = -\$1000.00(1 + 0.0613) + \$3900 = \quad \$2838.66$$
$$PB(6.13\%, 6\%)_2 = +\$2820.66(1 + 0.0600) - \$5030 = -\$2021.02$$
$$PB(6.13\%, 6\%)_3 = -\$2021.02(1 + 0.0613) + \$2145 = \qquad 0$$

---

[7] For this type of problem, we may seek a computer solution that is available from the **EzCash** computer software accompanying this text.

Figure 6.10 ■ Computing the IRR for a mixed investment

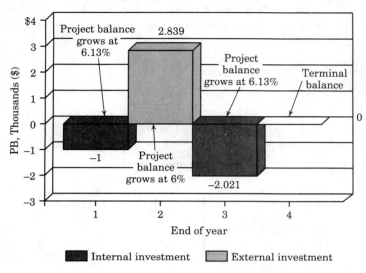

The positive balance at end of year 1 indicates borrowing from the project during year 2. However, note that the net investment becomes zero at the end of project life, confirming that 6.13% is the IRR for the cash flow. Since IRR > MARR, the investment is acceptable. Figure 6.10 is a visual representation of the occurrence of internal and external interest rates for the project.

Comments: *In the table above, at the end of year 1, the project releases +$2838.66 that must be invested outside of the project at an interest rate of 6%. The money invested externally must be returned to the project at the beginning of year 2 to meet another disbursement of $5030. At the end of year 2, there is no external investment of money and, hence, no need for an external interest rate. Instead, the amount of $2021.02 that remains invested during year 3 has to bring in a 6.13% rate of return. Finally, with the receipt of $2145, the net investment becomes zero at the end of the project's life. We conclude that i = 6.13% is a correct measure of the profitability of the project.*

Using the NPW criterion, the investment would be acceptable if the MARR were between zero and 10% or between 30% and 50%. The rejection region is 10% < i < 30% and 50% < i. Note that the project would be also accepted under the NPW analysis at MARR = i = 6%:

$$PW(6\%) = -\$1000 + \$3900(P/F, 6\%, 1)$$

$$-\$5030(P/F, 6\%, 2) + \$2145(P/F, 6\%, 3)$$

$$= \$3.55 > 0.$$

The flow chart in Fig. 6.11 summarizes how you should proceed to apply the net cash flow sign test, accumulated cash flow sign test, and net investment test to calculate an IRR, and make an accept/reject decision for a single project.

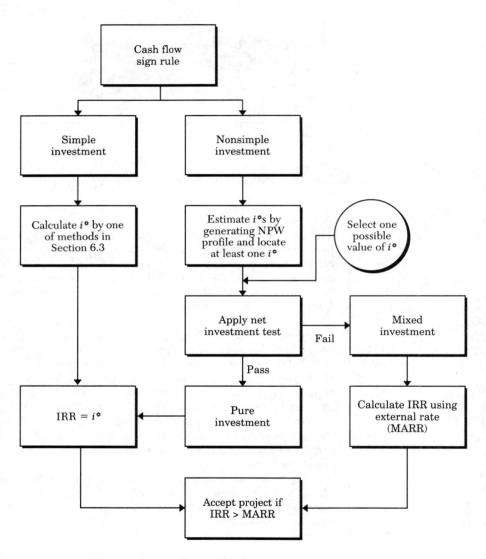

Figure 6.11 ■ Summary of IRR criterion—a flow chart that summarizes how you may proceed to apply the net cash flow sign rule and net investment test to calculate IRR for a mixed investment

■ 6.5 Comparing Mutually Exclusive Alternatives

In this section, we will present the decision procedures that should be used in comparing two or more mutually exclusive projects based on the rate of return measure. We will consider two situations: (1) alternatives that have the same economic service life and (2) alternatives that have unequal service lives.

6.5.1 Flaws in Project Ranking by IRR

Under NPW or AE analysis, the mutually exclusive project with the highest worth figure was preferred. Unfortunately, the analogy does not carry over to IRR analysis. The project with the highest IRR may *not* be the preferred alternative. To illustrate the flaws of comparing IRRs to choose from mutually exclusive projects, suppose you have two mutually exclusive alternatives, each with a one-year service life: One requires an investment of $1000 with a return of $2000 and the other requires $5000 with a return of $7000. You already obtained the IRRs and NPWs at MARR = 10% as follows:

n	A1	A2
0	−$1000	−$5000
1	2000	7000
IRR	100%	40%
PW(10%)	$818	$1364

Would you prefer the first project simply because you expect a higher rate of return?

We can see that A2 is preferred over A1 by the NPW measure. On the other hand, the IRR measure gives a *numerically* higher rating for A1. This inconsistency in ranking is due to the fact that the NPW, NFW, and AE are *absolute (dollar)* measures of investment worth while the IRR is a *relative (percentage)* measure and cannot be applied in the same way. That is, the IRR measure ignores the *scale* of the investment. Therefore, the answer is no; instead, you would prefer the second project with the lower rate of return but higher NPW. Either the NPW or the AE measure would lead to that choice, but comparison of IRRs would rank the smaller project higher. Another approach, called *incremental analysis*, is needed.

6.5.2 Incremental Analysis

In our previous ranking example, the more costly option requires an incremental investment of $4000 at an incremental return of $5000. If you decide to take the more costly option, certainly you would be interested in knowing that this additional investment can be justified at the MARR. The 10% of MARR value implies that you can always earn that rate from other investment sources— $4400 at the end of one year for $4000 investment. However, by investing the additional $4000 in the second option, you would make an additional $5000, which is equivalent to earning at the rate of 25%. Therefore, the incremental investment can be justified.

Now we can generalize the decision rule for comparing mutually exclusive projects. For a pair of mutually exclusive projects (A, B), we may rewrite B as

$$B = A + (B - A).$$

In other words, B has two cash flow components: (1) the same cash flow as A and (2) the incremental component (B − A). Therefore, the only situation in which B is preferred to A is when the rate of return on incremental component (B − A) exceeds the MARR. Therefore, for two mutually exclusive projects, rate of return analysis is done by computing the **internal rate of return on incremental investment** (IRR_Δ) between the projects. Since we want to consider increments of investment, we compute the cash flow for the difference between the projects by subtracting the cash flow for the lower investment-cost project (A) from that of the higher investment-cost project (B). Then, the decision rule is

If $IRR_{B-A} >$ MARR, select B,

If $IRR_{B-A} =$ MARR, select either one,

If $IRR_{B-A} <$ MARR, select A,

where B − A is an *investment* increment (negative cash flow). At first, it seems odd how this simple rule allows us to select the right project. Example 6.8 will illustrate the incremental investment decision rule.

Example 6.8 IRR on incremental investment: two alternatives

A pilot wants to start her own company to airlift goods to the Commonwealth of Independent States (formerly the U.S.S.R.) during their transition to a free-market economy. To economize the start-up business, she decided to purchase only one plane and fly it herself. She has two mutually exclusive options: an old aircraft (B1) or a new jet (B2), which she expects to incur higher purchase costs, but higher revenues as well because of its larger payload. In either case, she expects to fold up business in three years because of competition from larger companies.

The cash flows for the two mutually exclusive alternatives are given in thousand dollars:

n	B1	B2	B2 − B1
0	−$3,000	−$12,000	−$9,000
1	1,350	4,200	2,850
2	1,800	6,225	4,425
3	1,500	6,330	4,830

Assuming that there is no do-nothing alternative, which project would she select at MARR = 10%?

Solution

Given: Incremental cash flow between two alternatives, MARR = 10%

Find: IRR on increment, and determine which alternative is preferred

To choose the best project, we compute the incremental cash flow for B2 − B1. Then we compute the IRR on this increment of investment by solving

$$-\$9000 + \$2850(P/F, i, 1) + \$4425(P/F, i, 2) + \$4830(P/F, i, 3) = 0.$$

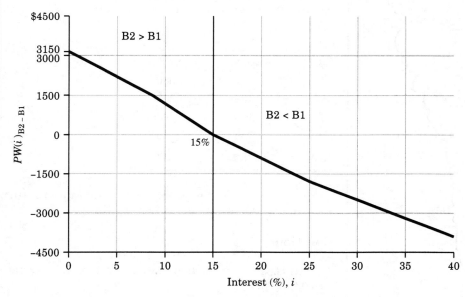

Figure 6.12 ■ NPW profile on incremental investment (B2 − B1) (Example 6.8)

We obtain $i^*_{B2-B1} = 15\%$, as plotted in Fig. 6.12. By inspection of the incremental cash flow, it is a simple investment, so $IRR_{B2-B1} = i^*_{B2-B1}$. Since $IRR_{B2-B1} > MARR$, we select B2, which is consistent with the NPW analysis.

Comments: *Why did we choose to look at the increment B2−B1 instead of B1−B2? We want the increment to have investment during at least some part of the time span so that we can calculate an IRR. Subtracting the lower initial investment project from the higher guarantees that project balance is initially less than zero, so the increment will be either a pure or a mixed investment. Ignoring the investment ranking, we might end up with an increment that involves pure borrowing and has no internal rate of return. This is the case for B1−B2. (i^*_{B2-B1} is also 15%, not − 15%.) If we erroneously compare this i^* with MARR, we might have accepted project B1 over B2. This is likely to damage our credibility with management! We will revisit this problem in Example 6.11.*

The next example will indicate that the ranking inconsistency between NPW and IRR can also occur when differences in the timing of projects' cash flows exist, even if their initial investments are the *same*.

Example 6.9   IRR on incremental investment when initial flows are equal

Consider the following two mutually exclusive investment projects that require the *same* amount of investment:

n	C1	C2
0	−$9000	−$9000
1	480	5800
2	3700	3250
3	6550	2000
4	3780	1561
IRR	18%	20%

Which project would you select based on rate of return on incremental investment assuming that MARR = 12%?

Solution

Given: Cash flows for two mutually exclusive alternatives as shown, MARR = 12%

Find: IRR on incremental investment, and determine which alternative is preferred

When initial investments are equal, we progress through the cash flows until we find the first difference, then set up the increment so that this first nonzero flow is negative, that is, an investment. Thus, we set up the incremental investment by taking (C1 − C2):

n	C1 − C2
0	0
1	−$5320
2	450
3	4550
4	2219

We next set the NPW equation equal to zero:

$$-\$5320 + \$450(P/F, i, 1) + \$4550(P/F, i, 2) + \$2219(P/F, i, 3) = 0.$$

Solving for i yields $i^\circ = 14.71\%$, which is also IRR since the increment is a pure investment. Since $\text{IRR}_{C1-C2} = 14.71\% > \text{MARR}$, we would select C1. If we used the NPW analysis, we will obtain $\text{PW}(12\%)_{C1} = \$1443$ and $\text{PW}(12\%)_{C2} = \$1185$, indicating the preference of C1 over C2.

When you have more than two mutually exclusive alternatives, you can compare them in pairs by successive examination. Example 6.10 illustrates how to compare three alternative problems. (In Chapter 13, we examine some multiple-alternative problems in the context of capital budgeting.)

Example 6.10 IRR on increment investment: three alternatives

Consider the following three mutually exclusive alternatives.

n	D1	D2	D3
0	−$2000	−$1000	−$3000
1	1500	800	1500
2	1000	500	2000
3	800	500	1000
IRR	34.37%	40.76%	24.81%

Which project would you select based on rate of return on incremental investment, assuming that MARR = 15%?

Solution

Given: Cash flows given above, MARR = 15%

Find: IRR on incremental investment, and determine which alternative is preferred

- **Step 1:** Examine the IRR for each alternative. At this point, we can eliminate any alternative that fails to meet the MARR. In this example, all three alternatives exceed the MARR.

- **Step 2:** Compare D1 and D2 in pairs.[8] Since D2 has a lower initial cost, we need to compute the rate of return on the increment (D1 − D2) that represents an increment of investment.

n	D1 − D2
0	−$1000
1	700
2	500
3	300

The incremental cash flow represents a simple investment. To find the incremental rate of return, we set up

$$-\$1000 + \$700(P/F, i, 1) + \$500(P/F, i, 2) + \$300(P/F, i, 3) = 0.$$

The solving i^*_{D1-D2} is 27.61%, which exceeds the MARR; therefore, D1 is preferred over D2.

- **Step 3:** Compare D1 and D3. Once again, D1 has a lower initial cost. We need to examine the (D3 − D1) increment.

n	D3 − D1
0	−$1000
1	0
2	1000
3	200

---

[8] If there are many alternatives, you may arrange them in order of increasing initial cost. This is not a required step, but it would make your comparison more tractable.

Again, the incremental cash flow represents a simple investment. The (D3 − D1) increment has an unsatisfactory 8.8% rate of return; therefore, D1 is preferred over D3. In summary, we conclude that D1 is the best alternative.

6.5.3 Incremental Borrowing Approach

Subtracting the less costly alternative from the more costly one is not absolutely necessary to incremental analysis. In fact, we can examine the difference between two projects A and B as either an (A − B) increment or a (B − A) increment. If the difference in flow (B − A) represents an increment of pure *investment*, then (A − B) is an increment of pure *borrowing*. When looking at increments of investment, we accepted the increment when its rate of return exceeded the MARR. When considering an increment of borrowing, however, the rate we calculate (that is, $i°$ for pure borrowing) is essentially the rate we pay to borrow money from the increment. We will call this the borrowing rate of return (BRR). Conceptually, we would prefer to get a loan from the increment rather than from our initial investment pool, if the loan rate is less than the MARR. Therefore, the decision rule is reversed:

$$\text{If BRR}_{A-B} < \text{ MARR, select A,}$$

$$\text{If BRR}_{A-B} = \text{ MARR, select either one,}$$

$$\text{If BRR}_{A-B} > \text{ MARR, select B,}$$

where A − B is a *borrowing* increment (positive cash flow).

Example 6.11 Borrowing rate of return on incremental projects

Consider Example 6.8 again, but this time compute the rate of return on the increment B1 − B2.

n	B1	B2	B1 − B2
0	−$3,000	−$12,000	+$9,000
1	1,350	4,200	−2,850
2	1,800	6,225	−4,425
3	1,500	6,330	−4,830

Note that the first incremental cash flow is positive and all others are negative, which indicates that the cash flow difference is an increment of pure *borrowing*. What is the rate of return on this increment of borrowing, and which project should we prefer?

Solution

Given: Incremental borrowing flow, MARR = 10%
Find: BRR on the increment, and determine whether it is acceptable.

Note that the signs of cash flows are simply reversed from Example 6.11. The rate of return on this increment of borrowing will be the same as that on the increment of investment, $i^*_{B1-B2} = i^*_{B1-B2} = 15\%$. However, because we are borrowing $9000, the 15% interest rate is what we are losing or paying by not investing the $9000. In effect, we are paying 15% interest on a borrowed sum. Is this an acceptable rate for a loan? Since the firm's MARR is 10%, we can assume that our maximum interest rate for borrowing should also be 10%. Certainly, borrowing under these circumstances is undesirable. Since 15% > 10%, this borrowing situation is not desirable and we should reject the lower cost alternative—B1—that produced the increment borrowing. We choose instead B2, the same result as in Example 6.8.

Since the incremental investment and incremental borrowing methods yield the same results, if you find one intuitively easier to understand, you should feel free to set up project comparisons consistently by that method. Remember that a negative increment indicates investment and a positive increment indicates borrowing. However, we recommend that alternatives be compared on the basis of incremental *investment* rather than borrowing. This strategy enables you to avoid having to remember two decision rules. It also circumvents the ranking problem of deciding which rule to apply to mixed increments. (The first rule always applies to mixed investment: If the IRR on the increment is greater than MARR, the increment is acceptable.)

6.5.4 Unequal Service Lives

In Chapters 4 and 5, we discussed the use of the NPW and AE criteria as bases for comparing projects with unequal lives. The IRR measure can also be used to compare projects with unequal lives, as long as we can establish a common analysis period. The decision procedure is then exactly the same as in the case of projects with equal life projects. It is likely, however, that we will have a multiple-root problem, which creates a substantial computational burden. As an example, suppose we apply the IRR measure to a case in which one project has a 5-year life and the other project has an 8-year life, resulting in a least common multiple of 40 years. When we determine the incremental cash flows over the analysis period, we are bound to observe many sign changes. This leads to the possibility of having many i^*s. The following two examples use IRR to compare mutually exclusive projects, one with only one sign change in the incremental cash flows, and one with several sign changes.

Example 6.12 IRR analysis for projects with different lives— where increment is a pure investment

Consider the following mutually exclusive investment projects (A, B).

n	A	B
0	−$2000	−$3000
1	1000	4000
2	1000	
3	1000	

Project A has a service life of 3 years, whereas project B has only 1 year of service life. Assume that project B can be repeated with the same investment costs and benefits over the analysis period of 3 years. Assume that the firm's MARR is 10%. Determine which project should be selected.

Solution

Given: Two alternatives with unequal lives, cash flows as shown, MARR = 10%

Find: IRR on incremental investment, and determine which is the preferred alternative

By assuming three repetitions of project B over the analysis period (as shown in Fig. 6.13) and taking the incremental cash flows (B − A), we obtain

n	A	B	(B − A)
0	−$2000	−$3000 = −$3000	−$1000
1	1000	4000 − 3000 = 1000	0
2	1000	4000 − 3000 = 1000	0
3	1000	4000 = 4000	3000

By inspection, the increment in this case is a pure investment. To compute i^*_{B-A}, we evaluate

$$-\$1000 + \frac{\$3000}{(1 + i)^3} = 0.$$

Solving for i yields

$$\text{IRR}_{B-A} = 44.22\% > 10\%.$$

Therefore, we may select project B.

Figure 6.13 ■ Comparison of unequal-lived projects when the required service period is indefinite (Example 6.12)

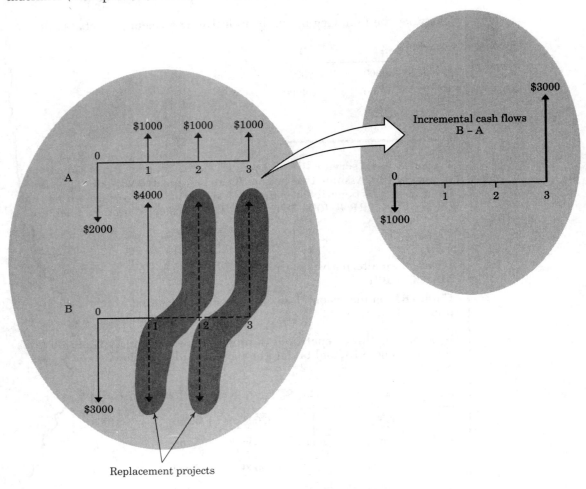

Replacement projects

Example 6.13 IRR analysis for projects with different lives where increment is a mixed investment

Consider Example 4.14 where a mail-order firm wants to install an automatic mailing system to handle product announcements and invoices. The firm had a choice between two different types of machines. Using the IRR as a decision criterion, select the best machine. Assume a MARR of 15% as before.

Solution

Given: Cash flows for two projects with unequal lives, MARR = 15%

Find: IRR on incremental investment, and determine which is the preferred alternative

Since the analysis period is equal to the least common multiple of 12 years, we may compute the incremental cash flow over this 12-year period. As shown in Fig. 6.14, we subtract cash flows of Model A from those of Model B to form the increment of investment. (We want the first cash flow difference to be a negative value.) We can then compute the IRR on this incremental cash flow.

n	Model A		Model B		Model B − Model A
0	−$12,500		−$15,000		−$2,500
1		−5,000		−4,000	1,000
2		−5,000		−4,000	1,000
3	−12,500	−3,000		−4,000	11,500
4		−5,000	−15,000	−2,500	−12,500
5		−5,000		−4,000	1,000
6	−12,500	−3,000		−4,000	11,500
7		−5,000		−4,000	1,000
8		−5,000	−15,000	−2,500	−12,500
9	−12,500	−3,000		−4,000	11,500
10		−5,000		−4,000	1,000
11		−5,000		−4,000	1,000
12		−3,000		−2,500	500

Even though there are five sign changes in the cash flow, there is only one positive i^* for this problem, which is 63.12%. Unfortunately, however, this is not a pure investment. We need to employ an external rate to compute the IRR to make a proper accept/reject decision. Assuming that the firm's MARR is 15%, we will use a trial-and-error approach: Try $i = 20\%$.

$PB(20\%, 15\%)_0 = -\$2500$

$PB(20\%, 15\%)_1 = -\$2500(1.20) + \$1000 = -\$2000$

$PB(20\%, 15\%)_2 = -\$2000(1.20) + \$1,000 = -\$1400$

$PB(20\%, 15\%)_3 = -\$1400(1.20) + \$11,500 = \$9820$

$PB(20\%, 15\%)_4 = \$9820(1.15) - \$12,500 = -\$1207$

Figure 6.14 ■ Comparison of unequal-lived projects (Example 6.13)

Model A

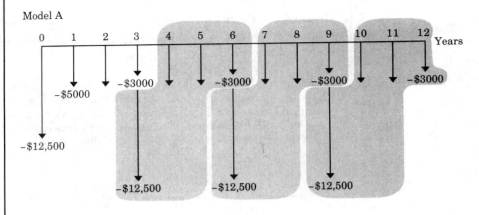

Model B

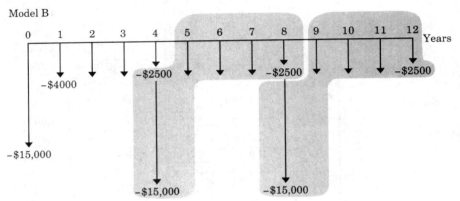

Incremental
cash flows

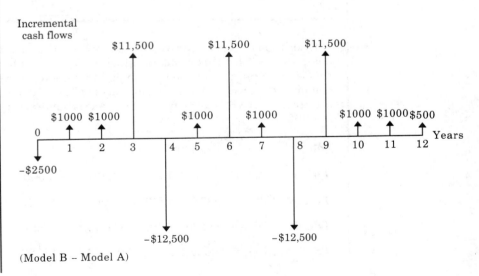

(Model B – Model A)

$$PB(20\%, 15\%)_5 = -\$1207(1.20) + \$1000 = -\$448.40$$

$$PB(20\%, 15\%)_6 = -\$448.40(1.20) + \$11,500 = \$10,961.92$$

$$PB(20\%, 15\%)_7 = \$10,961.92(1.15) + \$1000 = \$13,606.21$$

$$PB(20\%, 15\%)_8 = \$13,606.21(1.15) - \$12,500 = \$3147.14$$

$$PB(20\%, 15\%)_9 = \$3147.14(1.15) + \$11,500 = \$15,119.21$$

$$PB(20\%, 15\%)_{10} = \$15,119.21(1.15) + \$1000 = \$18,387.09$$

$$PB(20\%, 15\%)_{11} = \$18,387.09(1.15) + \$1000 = \$22,145.16$$

$$PB(20\%, 15\%)_{12} = \$22,145.16(1.15) + \$500 = \$25,966.93$$

Since $PB(20\%,15\%)_{12} > 0$, the guessed 20% is not the IRR. We may increase the value of i and repeat the calculations. After several trials, we find that the IRR is 50.68%.[9] Since $IRR_{B-A} > MARR$, Model B would be selected, which is consistent with the NPW analysis. In other words, the additional investment over the years to obtain Model B ($-\$2500$ at $n = 0$, $-\$12,500$ at $n = 4$, $-\$12,500$ at $n = 8$) yields a satisfactory rate of return: Model B, therefore, is the preferred project. (Note that Model B was picked when NPW analysis was used in Example 4.14.)

Given the complications involved in using IRR analysis to compare alternative projects, it is usually more desirable to use one of the other equivalence techniques for this purpose. As an engineering manager, you should, however, recall the intuitive appeal of the rate of return measure to many other managers and analysts within the firm. Once you have selected a project on the basis of NPW or AE analysis, you may also wish to express its worth as a rate of return for the benefit of other members of the firm.

 ## 6.6 Computer Notes

The rate of return analysis can be facilitated by using an electronic spreadsheet or **EzCash** software. We will revisit Example 6.13 to demonstrate these computational tools.

6.6.1 Incremental Analysis Using Lotus 1-2-3

For rate of return analysis, we can utilize the @IRR function of Lotus 1-2-3. The @IRR function requires two arguments: a guess concerning the resulting i^*

---

[9] It will be very tedious to solve this type of problem by a trial-and-error method on your calculator. The problem can be solved very quickly by using the **EzCash** software that accompanies this book.

and the range of cells containing the cash flows. Lotus 1-2-3 uses the guess as a first approximation of the true i°, then refines this guess until it converges into the correct value. (If you enter a guess value that deviates too much from the true value, you may see an error message. Whenever you encounter an error message using the @IRR function, you may either increase or decrease the guess value. The @IRR function will also display an error message when the cash flows contain multiple solutions.)

To solve Example 6.13, you may enter the cash flows for both models over the analysis period (12 years), as shown in Fig. 6.15. The incremental cash flows in column G are obtained by subtracting the cash flows of Model A from those of Model B. Using the @IRR function at a guessed interest rate of 50%, you may obtain the i° value of 63.31%. (Recall that the project is a mixed investment. Therefore, this solution is *not* an IRR.)

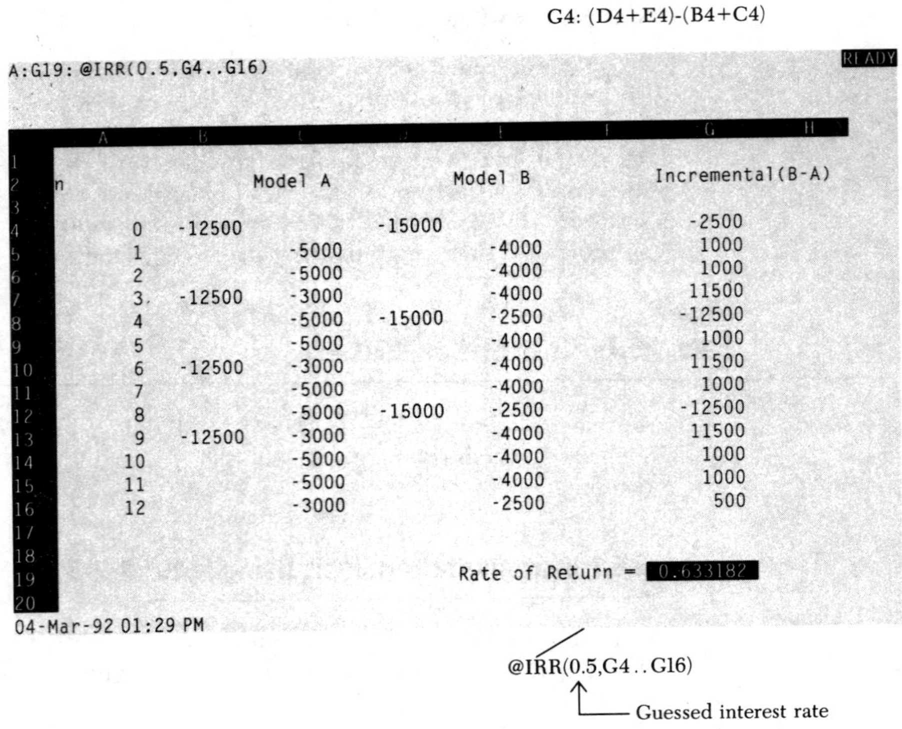

G4: (D4+E4)-(B4+C4)

A:G19: @IRR(0.5,G4..G16)   READY

	A	B		C		G	H
n		Model A		Model B		Incremental(B-A)	
0	-12500		-15000			-2500	
1		-5000		-4000		1000	
2		-5000		-4000		1000	
3	-12500	-3000		-4000		11500	
4		-5000	-15000	-2500		-12500	
5		-5000		-4000		1000	
6	-12500	-3000		-4000		11500	
7		-5000		-4000		1000	
8		-5000	-15000	-2500		-12500	
9	-12500	-3000		-4000		11500	
10		-5000		-4000		1000	
11		-5000		-4000		1000	
12		-3000		-2500		500	

Rate of Return = 0.633182

04-Mar-92 01:29 PM

@IRR(0.5,G4..G16)

└── Guessed interest rate

Figure 6.15 ■ Incremental analysis using Lotus 1-2-3 (Example 6.13)

6.6.2 Rate of Return Analysis Using EzCash

To find the ROR for the project cash flow series given in Example 6.7 (or Project D in Example 6.4) with **EzCash**, begin by entering the cash flow series using the **Cash Flow Editor**. (To access the **Cash Flow Editor**, select the **Edit** mode and the **Build** command.)

After entering the cash flow series, the **EzCash** screen will look like Fig. 6.16(a). To compute the ROR, select the **Compute** mode and the **Rate of Return** command. Specify the lower and upper bounds as 0% and 120%, respectively (as shown in Fig. 6.16b), and **EzCash** will calculate the i^* as 10%, 30%, and 50% along with the corresponding **NPW Plot** (Fig. 6.16c). The **NPW Plot** also displays three rates of return, indicating a mixed investment. As we suggested in Section 6.4.3, you may avoid mixed investment complications by selecting an alternative method such as NPW or AE analysis for determining project acceptability.

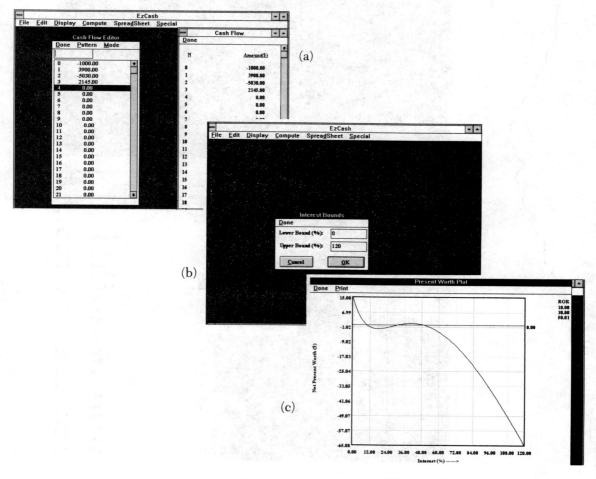

Figure 6.16 ■ NPW plot using **EzCash** (Example 6.7)

When you discover multiple i^*s, you may attempt to calculate the true IRR figure for the project. To find the IRR for Project D in Example 6.4, you follow the steps outlined in Example 6.7. Alternatively, **EzCash** has a built-in command that allows us to calculate a single, accurate rate of return (also known as the external rate of return). To use this command, return to the main menu and select **Internal Rate of Return** from the **Compute** mode

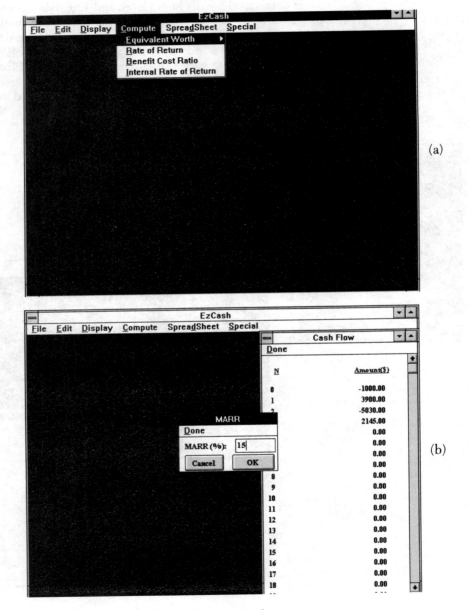

Figure 6.17 ■ Internal rate of return analysis

Figure 6.17 ■ Internal rate of return analysis (continued)

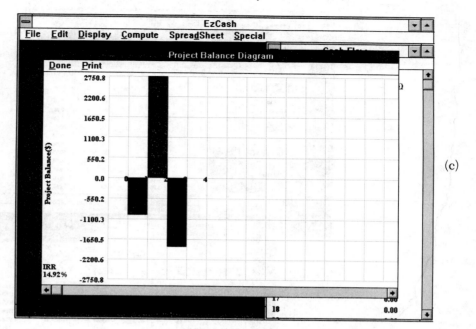

(c)

(Fig. 6.17a). You will then be prompted to input an interest rate (MARR, in our text). Using 15%, as we did in Example 6.7 (Fig. 6.17b), **EzCash** returns an IRR value of 14.92%. Figure 6.17(c) visually represents the occurrence of internal and external interest rates for the project, similar to Fig. 6.10.

When comparing two mutually exclusive alternatives based on rate of return, we need to calculate the internal rate of return on incremental investment. **EzCash** will perform an incremental analysis by subtracting a cash flow series (say, A1) for the lower investment project from that of higher investment (say, A2). The resulting cash flow series (say, A2−A1) is defined as an incremental investment series.

To solve Example 6.13 with **EzCash**, select **Special** and then **Incremental Analysis**. **EzCash** will bring up the **Incremental Analysis** and **Incremental Cash Flow** dialogue boxes (Fig. 6.18a). **Incremental Analysis** dialogue box consists of four options marked (1) **Input for Cash Flow 1**, (2) **Input for Cash Flow 2**, (3) **NPW Plot**, and (4) **Summary Report**. To continue, follow these steps:

1. Begin by selecting **Input for Cash Flow 1** to enter the cash flow series for project A. When this choice is selected the **Cash-Flow Editor** appears. Now you can enter and edit the cash flow series as explained in the main **Edit** mode in Appendix B. Your changes will appear in the **Incremental Cash Flow** dialogue box.

Figure 6.18 ■ Incremental analysis using **EzCash** (Example 6.13)

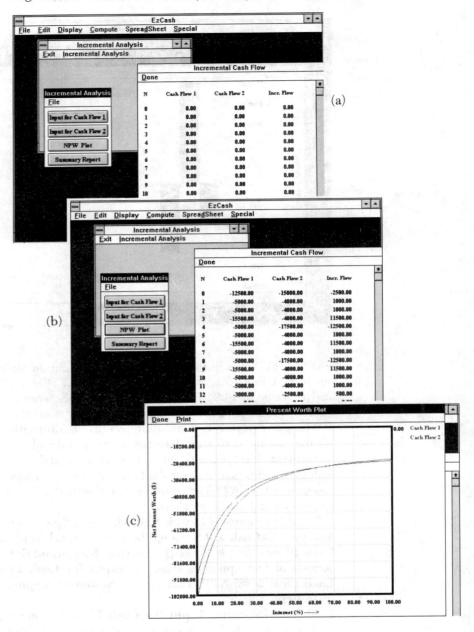

(a)

(b)

(c)

2. Repeat Step 1 to enter the cash flow series for project B by selecting **Input for Cash Flow 2**. As you complete the cash flow entries, **EzCash** will automatically create an incremental cash flow series and display it in the **Incremental Cash Flow** dialogue box (Fig. 6.18b).

Figure 6.19 ■ Report summarizing analysis of individual and incremental cash flows

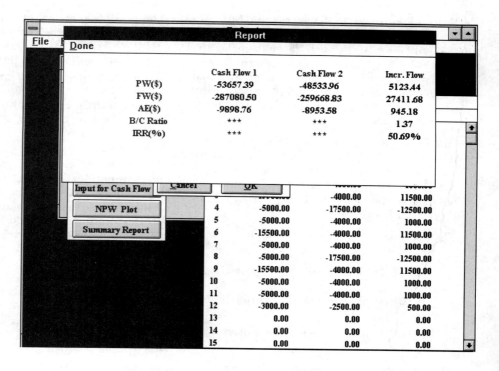

	Cash Flow 1	Cash Flow 2	Incr. Flow
PW($)	-53657.39	-48533.96	5123.44
FW($)	-287080.50	-259668.83	27411.68
AE($)	-9898.76	-8953.58	945.18
B/C Ratio	***	***	1.37
IRR(%)	***	***	50.69%

		-4000.00	11500.00
4	-5000.00	-17500.00	-12500.00
5	-5000.00	-4000.00	1000.00
6	-15500.00	-4000.00	11500.00
7	-5000.00	-4000.00	1000.00
8	-5000.00	-17500.00	-12500.00
9	-15500.00	-4000.00	11500.00
10	-5000.00	-4000.00	1000.00
11	-5000.00	-4000.00	1000.00
12	-3000.00	-2500.00	500.00
13	0.00	0.00	0.00
14	0.00	0.00	0.00
15	0.00	0.00	0.00

Buttons: Input for Cash Flow, NPW Plot, Summary Report, Cancel, OK

3. Select **NPW Plot** to view the present worth plots of the two cash flow series. You will be prompted to specify the interest range by entering its lower and upper limits. A screen similar to Fig. 6.18(c) will then be displayed. To obtain a hard copy, press the **Print** button.

4. Select **Summary Report** to obtain the comparative measures of investment worth as shown in Fig. 6.19. If there is an internal rate of return for the incremental cash flow series, **EzCash** will calculate it when you provide an external interest rate. **EzCash** will prompt you for the interest rate (MARR). If you enter 15%, as we did in Example 6.13, **EzCash** will return an IRR of 50.69%. (Note: All cash flow series are negative; there are no IRR figures for individual projects. However, the incremental cash flows which start with an investment followed by series of incremental receipts have a rate of return figure.) Since $IRR_{B-A} = 50.69\% > 15\%$, project B appears to be economically more desirable.

Summary

- **Rate of return (ROR)** is the interest rate earned on unrecovered project balances such that an investment's cash receipts make the terminal project balance equal to zero. Rate of return is an intuitively familiar and understandable measure of project profitability which many managers prefer to NPW or other equivalence measures.

- **Internal rate of return (IRR)** is another term for ROR which stresses the fact that we are concerned with the interest earned on the portion of the project that *internally* invested, not those portions that are released by (borrowed from) the project.

- The decision rule for IRR analysis is:

  If IRR > MARR, accept the project.

  If IRR = MARR, remain indifferent.

  If IRR < MARR, reject the project.

  IRR analysis yields results consistent with NPW and other equivalence methods.

- The possible presence of multiple i°s (rates of return) can be predicted by:

  - the net cash flow sign test, and
  - the accumulated cash flow sign test.

  When multiple rates of return cannot be ruled out by the two methods, it is very useful to generate a NPW profile to approximate the value of i°.

- *ALL* i° values should be exposed to the **net investment test**. Passing the net investment test indicates that the i° is an *internal* rate of return and is therefore a suitable measure of project profitability. Failure to pass the test indicates project borrowing, a situation that requires further analysis by use of an **external rate of return**.

- **External rate of return** analysis uses one rate (the firm's MARR) on externally invested balances and solves for another rate (IRR) on internally invested balances.

- When selecting between alternative projects by IRR analysis, we are concerned with **incremental investment** rather than pure project cash flows.

Table 6.3 Summary of Project Analysis Methods

Analysis Method	Equation Form(s)	Description	Comments
Payback period	—	A method for determining *when* in a project's history it breaks even (i.e., revenues repay costs).	Should be used as a screening only. Reflects liquidity, not profitability of project.
Discounted payback period	—	A variation of payback period which factors in the time value of money.	Reflects a truer measure of payback point than regular payback period.
Net present worth (NPW)	$PW(i) = \sum_{n=0}^{N} A_n (P/F, i, n)$	An *equivalence* method which translates a project's cash flows into a net present worth value.	For investments involving disbursements *and* receipts, the decision rule is: If $PW(i) > 0$, accept If $PW(i) = 0$, remain indifferent If $PW(i) < 0$, reject
			When comparing multiple alternatives, select the one with the greatest NPW value. For investments involving only disbursements (service projects), this is the project with the least negative NPW.
Net future worth (NFW)	$FW(i) = PW(i)(F/P, i, N)$	An *equivalence* method variation of NPW; a project's cash flows are translated into a net future worth value.	Most common real-world application occurs when we wish to determine a project's value at commercialization (a future date), *not* its value when we begin investing (the "present").
Capitalized equivalent (CE)	$PW(i) = \dfrac{A}{i}$	An *equivalence* method variation of NPW; calculates the NPW of a perpetual or very long-lived project that generates a constant annual net cash flow.	Most common real-world applications are civil engineering projects with lengthy service lives (i.e., $N > 50$) and *equal annual* costs (or incomes).
Annual equivalence (AE)	$AE(i) = PW(i)(A/P, i, N)$	An *equivalence* method and variation of NPW; a project's cash flows are translated into an annual equivalent sum.	AE analysis facilitates the comparison of unit costs and profits. In certain circumstances, AE may also avoid the complication of lowest common multiple of project lives which may arise in NPW analysis.

Table 6.3 Summary of Project Analysis Methods (continued)

Analysis Method	Equation Form(s)	Description	Comments
Rate of return (ROR) or internal rate of return (IRR)	$PW(i) = 0$	A relative percentage method which measures the yield as a percentage of investment over the life of a project.	Rate of return is stated in a form more intuitively understandable to most laypersons and financial administrators. Unlike equivalence methods, which are absolutes, IRR when calculated must be compared to a predetermined acceptable rate of return known as MARR. The decision rule is: If IRR > MARR, accept If IRR = MARR, remain indifferent If IRR < MARR, reject

Problems

Note: The symbol i° represents the interest rate that makes the net present value of the project equal to zero. The symbol IRR represents the *internal* rate of return of the investment. For a simple (or pure) investment, IRR $= i^{\circ}$. For a nonsimple investment, generally i° is not equal to IRR.

6.1 Assume that you are going to buy a new car worth $20,000. You will be able to make a down payment of $5000. The remaining $15,000 will be financed by the dealer. The dealer computes your monthly payment to be $609 for 48 months' financing. What is the dealer's rate of return on this loan transaction?

6.2 Mr. Smith wishes to sell a bond that has a face value of $2000. The bond bears an interest rate of 8% with bond interests payable semiannually. Four years ago, the bond was purchased at $1800. At least 9% annual return on investment is desired. What must be the minimum selling price now for the bond in order to make the desired return on investment?

6.3 Reconsider the art auction story (Vincent Van Gogh) presented in the chapter opening. John Whitney Payson, who purchased the painting for $80,000 in 1947, sold it for $53.9 million in 1988. If Mr. Payson invested his $80,000 in another investment vehicle (such as stock), how much interest would he need to earn to accumulate the same wealth from the painting investment? Assume for simplicity that the investment period is 40 years and the interest is compounded annually.

6.4 Consider two investments, A and B, with the following sequences of cash flows:

	Net Cash Flow	
n	*Project A*	*Project B*
0	−$18,000	−$20,000
1	10,000	32,000
2	20,000	32,000
3	30,000	−22,000

(a) Which project is a nonsimple investment?

(b) Compute i° for each investment.

6.5 Consider the following investment projects:

Project Cash Flows

Period	A	B	C	D	E	F
0	-$100	-$100	-$100	-$100	-$100	-$100
1	200	470	200	300	300	300
2	-300	-720	200	-300	-250	100
3	400	360	-250	50	40	-400

(a) Apply the sign rule to predict the number of possible i°s for each project.

(b) Plot the NPW profile as a function of i between 0 and 200% for each project.

(c) Compute the value(s) of i° for each project.

6.6 Consider the following infinite cash flow series with repeated cash flow patterns.

n	A_n
0	-$1000
1	800
2	800
3	500
4	500
5	800
6	800
7	500
8	500
⋮	⋮

Determine the i° for this infinite cash flow series.

6.7 An investor bought 100 shares of stock at a cost of $10 per share. He held the stock for 15 years and then sold it for a total of $4000. For the first 3 years, he received no dividends. For each of the next 7 years he received total dividends of $50 per year. For the remaining period he received total dividends of $100 per year. What rate of return did he make on the investment?

6.8 Consider the following sets of investment projects:

Project Cash Flow

n	A	B	C	D	E
0	-$100	-$100		-$200	
1	50	60	-20	120	-100
2	100	60	10	40	-50
3		30	5	40	0
4		30	-180	-20	50
5			60	40	150
6			50		
7			40		
8			30		
9			20		
10			10		

(a) Classify each project into either simple or nonsimple.

(b) Compute the i° for A using the quadratic equation.

(c) Apply the cash flow sign rules to each project and predict the number of possible positive i°s. Identify all projects having unique i°.

6.9 Consider the following sets of projects:

Net Cash Flow

n	A	B	C	D
0	-$1000	-$1000	-$1700	-$1000
1	500	800	5600	360
2	100	600	4900	4675
3	100	500	-3500	2288
4	1000	700	-7000	
5			-1400	
6			2100	
7			900	

(a) Classify each project into either simple or nonsimple.
(b) Identify all positive i°s for each project.
(c) Plot the present worth as a function of interest rate (i) for each project.

6.10 The Global Telephone Company wants to participate in the next World's Fair. To participate, the firm needs to spend $1 million at year 0 to develop a showcase. The showcase will produce a cash flow of $2.5 million at the end of year 1. Then, at the end of year 2, $1.54 million must be expended to restore the land to its original condition. Therefore, the project's expected net cash flows are as follows (in thousands of dollars):

n	Net Cash Flow
0	-$1000
1	2500
2	-1540

(a) Plot the present worth of this investment as a function of i.
(b) Compute the i°s for this investment.

6.11 Consider the following financial data for a project:

Initial investment	$10,000
Project life	8 years
Salvage value	$0
Annual revenue	$5229
Annual expenses (including income taxes)	$3000

(a) What is the i° for this project?
(b) If annual expense increases at a 7% rate over the previous year's expenses but annual income is unchanged, what is the new i°?
(c) In part (b), at what annual rate will annual income have to increase to maintain the same i° obtained in part (a)?

6.12 Consider two investments, A and B, with the following sequences of cash flows:

Net Cash Flow

n	Project A	Project B
0	-$379	-$379
1	20	100
2	60	100
3	60	100
4	100	100
5	280	100

(a) Compute the i° for each investment.
(b) Plot the present worth curve for each project on the same chart and find the interest rate that makes the two projects equivalent.

6.13 Consider an investment project with the following cash flows:

n	Cash Flow
0	−$5000
1	0
2	4840
3	1331

Compute the IRR for this investment. Is this project acceptable at MARR = 10%?

6.14 Consider the following project's cash flow and its present worth diagram in Fig. 6.20.

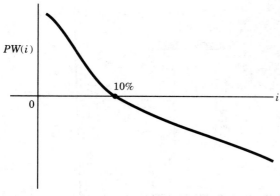

Figure 6.20 ■ Present worth function (Problem 6.14)

n	Net Cash Flow
0	−$2000
1	800
2	900
3	X

(a) Find the value of X by examining Fig. 6.20.

(b) Is this project acceptable at a MARR of 8%?

6.15 Hallmark Chemical Corporation is planning to expand one of its propylene manufacturing facilities. The land costs $300,000, the building costs $600,000, the equipment costs $250,000, and a $100,000 start-up cost is required. It is expected that the product will result in sales of $625,000 per year for 12 years, at which time the land can be sold for $300,000, the building for $200,000 (net proceeds after tax adjustment), and the equipment for $50,000 (net proceeds after tax adjustment). The annual disbursements for labor, materials, and all other expenses (including income taxes) are estimated to be $425,000. What is the IRR? If the company requires a minimum rate of return of 15% on projects, determine if this is a good investment.[10]

6.16 Recent technology has made possible a computerized vending machine that can grind coffee beans and brew fresh coffee on demand (see Fig. 6.21.) The computer also makes possible such complicated functions as changing $5 and $10 bills and tracking the age of an item, moving the oldest stock to the front of the line, thus cutting down on spoilage. With a price tag of $4500 for each unit, Easy Snack has estimated the cash flows in millions of dollars over the product's 6-year useful life, including the initial investment.

n	Net Cash Flow
0	−$20
1	8
2	17
3	19
4	18
5	10
6	3

[10] From *Engineering Economy*, 8th Ed., Problem 4.2, E. P. DeGarmo, W. G. Sullivan, and J. A. Bontadelli, Macmillan, 1988.

Figure 6.21 ■ A computerized vending machine

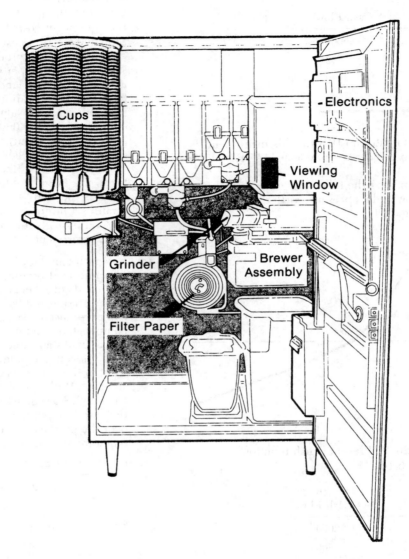

If the firm's MARR is 18%, is this product worth marketing based on the IRR criterion?

6.17 Consider two investments, A and B, with the following sequences of cash flows:

	Net Cash Flow	
n	Project A	Project B
0	−$120,000	−$100,000
1	20,000	15,000
2	20,000	15,000
3	120,000	130,000

(a) Compute the IRR for each investment.

(b) At a MARR of 15%, determine the acceptability of each project.

6.18 Consider an investment project with the following cash flows:

n	Net Cash Flow
0	−$20,000
1	94,000
2	−144,000
3	72,000

(a) Find the IRR return for this investment.

(b) Plot the present worth of the cash flow as a function of i.

(c) Using the IRR criterion, should the project be accepted at MARR = 15%?

6.19 Consider the following sets of investment projects:

	Net Cash Flow		
n	Project 1	Project 2	Project 3
0	−$1000	−$2000	−$1000
1	500	1560	1400
2	840	944	−100

Assume MARR = 12% in the following questions.

(a) Compute the $i°$ for each investment. If the problem has more than one $i°$, identify all of them.

(b) Compute the IRR for each project.

(c) Determine the acceptability of each investment.

6.20 Consider the following sets of investment projects:

	Net Cash Flow		
n	Project 1	Project 2	Project 3
0	−$1600	−$5000	−$1000
1	10,000	10,000	4,000
2	10,000	30,000	−4,000
3		−40,000	

Assume MARR = 12% in the following questions.

(a) Identify the $i°$(s) for each investment. If the problem has more than one $i°$, identify all of them.

(b) Which project(s) is (are) a mixed investment?

(c) Compute the IRR for each project.

(d) Determine the acceptability of each project.

6.21 Reconsider Problem 6.8.

(a) Compute the $i°$s for projects B through E.

(b) Identify the projects that fail the net investment test.

(c) Determine the IRR for each investment.

(d) With MARR = 12%, which projects will be acceptable using the IRR criterion?

6.22 Consider the following investment projects:

	Net Cash Flow		
n	Project A	Project B	Project C
0	−$100	−$150	−$100
1	30	50	410
2	50	50	−558
3	80	50	252
4		100	
$i°$	23.24%	21.11%	20%, 40%, 50%

Assume MARR = 12% for the following questions.

(a) Identify the simple project(s).
(b) Identify the mixed project(s).
(c) Determine the IRR for each investment.
(d) Which project would be acceptable?

6.23 Spar Aerospace would like to receive a NASA contract worth $460 million to build lifting devices for future space missions. NASA will pay $50 million when the contract is signed, another $360 million at the end of the first year, and the $50 million balance at the end of second year. The expected cash outflows required to produce these devices are estimated to be $150 million now, $100 million during the first year, and $218 million during the second year. The firm's MARR is 12%.

n	Outflow	Inflow	Net Cash Flow
0	$150	$50	-$100
1	100	360	260
2	218	50	-168

(a) Show whether or not this project is a mixed investment.
(b) Compute the IRR for this investment.
(c) Should Spar Aerospace accept the project?

6.24 Consider the following investment projects:

Net Cash Flow

n	Project A	Project B	Project C
0	-$100		-$100
1	-216	-$150	50
2	116	100	-50
3		50	200
4		40	
i°	?	15.51%	29.95%

(a) Compute the i° for Project A. If there is more than one i°, identify all of them.
(b) Identify the mixed project(s).
(c) Assuming that MARR = 10%, determine the acceptability of each project based on the IRR criterion.

6.25 Consider the following sets of investment projects:

Net Cash Flow

n	A	B	C	D	E
0	-$1,000	-$5,000	-$2,000	-$2,000	-$1,000
1	3,100	20,000	1,560	2,800	3,600
2	-2,200	-12,000	944	-200	-5,700
3		-3,000			3,600
i°	?	?	18%	32.45%	35.39%

Assume MARR = 12% in the following questions.

(a) Compute the i° for Projects A and B. If the project has more than one i°, identify all of them.
(b) Classify each project into either pure or mixed investment.
(c) Compute the IRR for each investment.
(d) Determine the acceptability of each project.

6.26 Consider an investment project whose cash flows are given as follows:

n	Net Cash Flow
0	-$5,000
1	10,000
2	30,000
3	-40,000

(a) Plot the present worth curve by varying i from 0% to 250%.

(b) Is this a mixed investment?

(c) Should the investment be accepted at MARR = 18%?

6.27 With $10,000 available, you have two investment options. The first option is to buy a certificate of deposit from a bank at an interest rate of 10% annually for 5 years. The second is to purchase a bond for $10,000 and invest the bond's interests in the bank at an interest rate of 9%. The bond pays 10% interest annually and will mature to its face value of $10,000 in 5 years. Which option is better? Assume your MARR is 9% per year.

6.28 A manufacturing firm is considering the following mutually exclusive alternatives:

Net Cash Flow

n	Project A	Project B
0	−$2000	−$3000
1	1400	3700
2	1640	320

Determine which project is a better choice at MARR = 15% based on the IRR criterion.

6.29 Consider the following two mutually exclusive alternatives:

Net Cash Flow

n	Project A1	Project A2
0	−$10,000	−$12,000
1	5,000	6,100
2	5,000	6,100
3	5,000	6,100

(a) Determine the IRR on the incremental investment in the amount of $2000.

(b) If the firm's MARR is 10%, which alternative is a better choice?

6.30 Consider the following two mutually exclusive investment alternatives.

Net Cash Flow

n	Project A1	Project A2
0	−$15,000	−$20,000
1	7,500	8,000
2	7,500	15,000
3	7,500	5,000
IRR	23.5%	20%

(a) Determine the IRR on the incremental investment in the amount of $5000. (Assume MARR = 10%.)

(b) If the firm's MARR is 10%, which alternative is a better choice?

6.31 You are considering two types of automobiles. Model A costs $18,000 and Model B costs $15,624. Although the two models are essentially the same, Model A can be sold for $9000 while Model B can be sold for $6500 after four years of use. Model A commands a better resale value because its styling is popular among young college students. Determine the rate of return on the incremental investment of $2376. For what range of values of your MARR is Model A preferred?

6.32 A plant engineer is considering two types of solar water heating system:

	Model A	Model B
Initial cost	$7,000	$10,000
Annual savings	$700	$1,000
Annual maintenance	$100	$50
Expected life	20 years	20 years
Salvage value	$400	$500

The firm's MARR is 12%. Based on the IRR criterion, which system is a better choice?

6.33 Consider the following two mutually exclusive investment projects. Assume MARR = 15%.

Net Cash Flow

n	Project A	Project B
0	−$300	−$800
1	0	1150
2	690	40
i^*	51.66%	46.31%

(a) Using the IRR criterion, which project would be selected?
(b) Sketch the $PW(i)$ function on incremental investment (B − A).

6.34 Consider Example 5.8, where Hamilton Couplings Company was considering two types of manufacturing systems to produce its shaft couplings: (1) cellular manufacturing system (CMS) and (2) a flexible manufacturing system (FMS). The relevant financial data were summarized in Fig. 5.7. Also recall that no revenues figures were given. (There is no alternative but to find the most cost-effective solution.) Based on the IRR criterion, select the most cost-effective system.

6.35 Victoria Hospital is reviewing alternative ways of cutting the stocking costs of medical supplies. Two new ways of stockless systems are considered to lower the hospital's holding and handling costs. The hospital's industrial engineer has compiled the relevant financial data for each system as follows:

	Current Practice	Just-in-Time System	Stockless Supply System
Start-up cost	0	$2.5 million	$5 million
Annual stock holding cost	$3 million	$1.4 million	$0.2 million
Annual operating cost	$2 million	$1.5 million	$1.2 million
System life	8 years	8 years	8 years

The system life of 8 years represents the contract period with medical suppliers. If the hospital's MARR is 10%, which system is more economical?

6.36 Consider the cash flows for the following projects. Assume MARR = 12%.

Project Cash Flow

n	A	B	C	D	E
0	−$1000	−$1000	−$2000	$1000	−$1200
1	900	600	900	−300	400
2	500	500	900	−300	400
3	100	500	900	−300	400
4	50	100	900	−300	400

(a) Suppose A, B, C are mutually exclusive projects. Which project would be selected based on the IRR criterion?
(b) What is the BRR (*borrowing rate of return*) for D?
(c) Would you accept D at MARR = 20%?
(d) Assume that projects C and E are mutually exclusive. Using the IRR criterion, which project would you select?

6.37 Consider the following sets of investment projects:

Net Cash Flow

n	Project 1	Project 2	Project 3
0	−$1,000	−$5,000	−$2,000
1	500	11,500	1,500
2	900	−6,600	1,000
i°	23.11%	?	17.54%

Assume MARR = 15% in the following questions.

(a) Compute the $i^\circ(s)$ for Project 2.
(b) Compute the IRR for Project 2.
(c) If the three projects are mutually exclusive investments, which project should be selected based on the IRR criterion?

6.38 Consider the following two investment alternatives:

Net Cash Flow

n	Project A	Project B
0	−$10,000	−$20,000
1	5,500	0
2	5,500	0
3	5,500	40,000
IRR	30%	?
PW(15%)	?	6,300

The firm's MARR is known to be 15%.

(a) Compute the IRR of project B.
(b) Compute the NPW of project A.
(c) Suppose that projects A and B are mutually exclusive. Using the IRR, which project would you select?

6.39 The E. F. Fedele Company is considering the acquisition of an automatic screwing machine for its assembly operation of a personal computer. Three different models with varying automatic features are under consideration. The required investments are $360,000 for Model A, $380,000 for Model B, and $405,000 for Model C, respectively. All three models are expected to have the same service life of 8 years. The following financial information is available. In the following, Model (B − A) represents the incremental cash flow determined by subtracting Model A's cash flow from Model B's.

	IRR
Model A	30%
Model B	15
Model C	25

	Incremental IRR
Model (B − A)	5%
Model (C − B)	40
Model (C − A)	15

If the firm's MARR is known to be 12%, which model should be selected?

6.40 The Flotrol Valve Company is considering three cost-reduction proposals in its batch job shop manufacturing operations. As shown on the next page, the company already calculated rates of return for the three projects along with some incremental rates of return. A0 denotes the do-nothing alternative. The required investments are $420,000 for A1, $550,000 for A2, and $720,000 for A3. If the MARR is 15%, what system should be selected?

Incremental Rate of Return, $Y - X$			
	X		
Y	A0	A1	A2
A1	18%		
A2	20	10%	
A3	25	12	13%

6.41 An electronic circuit-board manufacturer is considering six mutually exclusive cost-reduction projects for their PC-board manufacturing plant. All have lives of 10 years and zero salvage values. The required investment and the estimated after-tax reduction in annual disbursements are given for each alternative. Along with these gross rates of return, rates of return on incremental investments are also computed.

Proposal A_j	Required Investment	After-Tax Savings
A1	$ 60,000	$22,000
A2	100,000	28,200
A3	110,000	32,600
A4	120,000	33,600
A5	140,000	38,400
A6	150,000	42,200

Incremental Rate of Return, $Y - X$						
	X					
Y	A0	A1	A2	A3	A4	A5
A1	35%					
A2	25.2	9%				
A3	27		42.8%			
A4	25			0%		
A5	24				20.2%	
A6	25.1					36.3%

Which project would you select based on the rate of return on incremental

investment if it is stated that the MARR is 15%? In the table above, A0 denotes the do-nothing alternative.

6.42 Consider the following two mutually exclusive investment projects:

	Net Cash Flow	
n	*Project A*	*Project B*
0	−$100	−$200
1	60	120
2	50	150
3	50	
IRR	28.89%	21.65%

Which project would be selected under infinite planning horizon, with project repeatability likely, based on the IRR criterion? (MARR = 15%.)

6.43 Consider the following two mutually exclusive investment projects:

	Net Cash Flow	
n	*Project A1*	*Project A2*
0	−$10,000	−$15,000
1	5,000	20,000
2	5,000	
3	5,000	

(a) To use the IRR criterion, what assumption do you need to make to compare a set of mutually exclusive investments with unequal service lives?

(b) With the assumption defined above, determine the range of MARR that will indicate the selection of project A1.

6.44 Consider the following sets of investment projects:

Project Cash Flow

n	A	B	C	D	E
0	−$100	−$100	−$5	−$100	$200
1	100	30	10	30	100
2	24	30	30	30	−500
3		70	−40	30	−500
4		70		30	200
5				30	600

(a) Compute the i° for project A using the quadratic equation.
(b) Classify each project as either simple or nonsimple.
(c) Apply the cash flow sign rules to each project and determine the number of possible positive i°s. Identify all projects having unique i°.
(d) Compute the IRRs for projects B through E. (MARR = 10%.)
(e) Apply the net investment test to each project.
(f) With MARR = 10%, which projects will be acceptable using the IRR criterion?

6.45 Consider the following sets of investment projects. Assume that MARR = 15%.

Net Cash Flow

n	A	B	C	D	E	F
0	−100	−200	−4,000	−2,000	−2,000	−3,000
1	60	120	2,410	1,400	3,700	2,500
2	50	150	2,930	1,720	1,640	1,500
3	50					
i°	28.89%	21.65%	21.86%	31.10%	121.95%	23.74%

(a) Projects A and B are mutually exclusive projects. Assuming that both projects can be repeated for an indefinite period, which project would you select based on the IRR criterion?
(b) Suppose projects C and D are mutually exclusive. Using the IRR criterion, which project would be selected?
(c) Suppose projects E and F are mutually exclusive. Which project is better based on the IRR criterion?

6.46 Consider the following project's cash flows:

n	Net Cash Flow
0	−$100,000
1	310,000
2	−220,000

The project's i°s are computed as 10% and 100%, respectively. The firm's MARR is 8%.

(a) Show why this investment project fails the net investment test.
(b) Compute the IRR and determine the acceptability of this project.

6.47 Macmillan published a nonfiction mystery book after paying $350,000 to the author, who had an earlier best-seller on his résumé. The cover price was $18.95. Of that amount, $9.85 went to the publisher and the rest to the booksellers. Of the 75,000 copies printed, 32,250 were sold. The sequence of cash flow transactions was as follows:

- June 1988: A $350,000 nonrefundable advance against royalties was paid to author as a guaranteed minimum.
- March 1989: Book manuscripts were delivered and book production began.
- June 1989: 75,000 copies were printed at $179,455 (paper, printing, binding, and plant cost)

- June 1989–March 1990: 32,250 copies were sold at $317,663 (= 32,250 × $9.85).

Other expenses incurred included the market cost of $150,000 and the overhead cost (salaries, rent, etc.) of $95,337. Because of poor sales, the company sold the paperback rights to another publisher for $170,000. The remaining unsold copies had to be sold on the bargain tables as remainders or at $91,058 (= 42,750 × $2.13). The company's fiscal year ends June 30.

(a) Ignoring any tax implications and using the end-of-year convention, can you determine the rate of return for this book project?

(b) If Macmillan were to make a before-tax 20% rate of return on this book, how many copies had to be sold? (Assume that all other figures remain unchanged.)

6.48 Critics have charged that the nuclear power industry does not consider the cost of "decommissioning," or "mothballing," a nuclear power plant when doing an economic analysis and that the analysis is therefore unduly optimistic. As an example, consider a nuclear generating facility similar to those operated by Ontario Hydro: the first cost is $1.5 billion (present worth at start of operations), the estimated life is 40 years, the annual operating and maintenance costs in the first year are assumed to be 4.6% of the first cost and are expected to increase at the fixed rate of 0.05% of the first cost each year, and annual revenues have been estimated to be three times the annual operating and maintenance costs throughout plant life.

(a) The criticism of overoptimism in the economic analysis caused by omitting "mothballing" costs is

not justified since the addition of a cost to "mothball" the plant equal to 50% of the first cost only decreases the 10% rate of return to approximately 9.9%.

(b) If the estimated life of the plants is more realistically taken as 25 years instead of 40 years, then the criticism is justified. By reducing the life to 25 years, the rate of return of approximately 9% without a "mothballing" cost drops to approximately 7.7% when a cost to "mothball" the plant equal to 50% of the first cost is added to analysis.

Comment on these statements.

6.49 You have been asked by the president of the company to evaluate the proposed acquisition of a new injection molding machine for the firm's manufacturing plant. Two types of injection molding machines have been identified with the following estimated cash flows:

	Net Cash Flow	
n	Project A	Project B
0	−$30,000	−$40,000
1	20,000	43,000
2	18,200	5,000
IRR	18.1%	18.1%

You return to your office and quickly retrieve your old engineering economics text, then begin to smile: Aha—this is a classical rate of return problem! Now using a calculator, you find out that both projects have about the same rate of return, 18.1%. This rate of return figure seems to be high enough for project justification, but you recall that the ultimate justification should be done in reference to the firm's MARR. Now you call the accounting department to find out the current MARR the firm should use for project justification. "I wish I could tell you, but my boss will be back next week and he can tell you what to use," said the accounting clerk. A fellow engineer approaches you and says, "I couldn't help overhearing you talking to the clerk. I think I can help you. You see, both projects have the same IRR, and on top of that, project 1 requires less investment but returns more cash flows (−$3000 + $2000 + $1820 = $820, −$4000 + $4300 + $500 = $800), thus project 1 dominates

project 2. For this type of decision problem, you don't need to know a MARR! Comment on your fellow engineer's statement.

6.50 The B&E Cooling Technology Company, a maker of automobile air-conditioners, faces an uncertain but impending deadline to phase out the traditional chilling technique, which uses chlorofluorocarborns, or CFC's, a family of refrigerant chemicals believed to

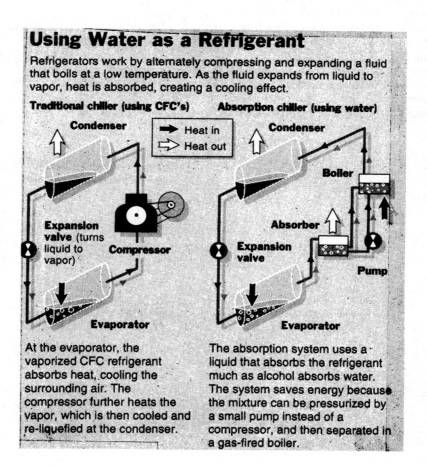

Figure 6.22 ■ Using water as a refrigerant, absorption chilling techniques

attack the earth's protective ozone layer. B&E has been pursuing other means of cooling and refrigeration. As a near-term solution, the engineers recommend a cold technology known as absorption chiller, which uses plain water as a refrigerant and semiconductors that cool down when charged with electricity (see Fig. 6.22). B&E is considering two options:

- **Option 1:** Retrofitting the plant now to adapt the absorption chiller and continuing to be a market leader in cooling technology. Because of untested technology in the large scale, it may cost more to operate the new facility while new system learning takes place.
- **Option 2:** Deferring the retrofitting until the federal deadline, which is 3 years away. With expected improvements in cooling technology and technical know-how, the retrofitting cost

will be cheaper but there will be tough market competition and the revenue would be less than in Option 1.

The financial data for the two options are as follows:

	Option 1	Option 2
Investment timing	Now	3 years from now
Initial investment	$6 million	$5 million
System life	8 years	8 years
Salvage value	$1 million	$2 million
Annual revenue	$15 million	$11 million
Annual O&M costs	$6 million	$7 million

(a) What assumptions do you need to compare these two options?
(b) If B&E's MARR is 15%, which option is a better choice based on the *IRR* criterion?

CHAPTER 7

Depreciation

S uppose that you are a design engineer for a firm that manufactures die cast auto-mobile parts. To enhance the firm's competitive position in the marketplace, management has decided to purchase a computer-aided design system which uses 3-D solid modeling and full integration with sophisticated simulation and analysis capabilities. As a member of the design team, you are excited at the prospect that the design of die cast molds, the testing of product variations, and the simulation of processing and service conditions can be made highly efficient by using this state-of-the-art system. In fact, the more you think about it, the more you wonder why this purchase wasn't made earlier.

Now ask yourself, how does the cost of this system affect the financial position of the firm? In the long run, the system promises to create greater wealth for the organization by improving design productivity, increasing product quality, and

Year						READY
0			Capital Cost Allowance (CCA):			F
1	Year					
2	1994	Rate	Welder		Welder	
3	1995		CCA		Book Value	
4	1996	0.00%				
5	1997	10.00%	0.00		46,000.00	
6	1998	20.00%	4,600.00		41,400.00	
1999	20.00%	8,280.00		33,120.00	
2000	20.00%	6,624.00		26,496.00	
Tax Rate=		20.00%	5,299.20		21,186.80	
Year End		20.00%	9.36		16,957.44	
of			49		13,565.95	

23658.63

12.00%

Income Tax	After-Cash Tax Flow
1,980.48	(10,919.52)
5,057.28	(2,286.72)

cutting down design lead time. In the short run, however, the high cost of this system has a much more negative impact on the organization's "bottom line," because it involves high initial costs that are only gradually rewarded by the benefits of the system.

Another consideration should come to mind. This state-of-the-art equipment must inevitably wear out over time, and even if its productive service extends over many years, the cost of maintaining its high level of functioning will increase as the individual pieces of hardware wear out and need to be replaced. Of even greater concern is the question of how long this system will be "state-of-the-art." When will the competitive advantage the firm has just acquired become a competitive disadvantage through obsolescence?

One of the facts of life that organizations must deal with and account for is that fixed assets lose their value—even as they continue to function and contribute to the engineering projects that use them. This loss of value, called **depreciation**, can involve deterioration and obsolescence.

The main function of **depreciation accounting** is to account for the cost of fixed assets in a pattern that matches their decline in value over time. The cost of the CAD system we have just described, for example, will be allocated over several years in the firm's financial statements so that its pattern of costs roughly matches its pattern of service. In this way, as we shall see, depreciation accounting enables the firm to stabilize the statements of financial position that it distributes to stockholders and the outside world.

On the project level, engineers must be able to assess how the practice of depreciating fixed assets influences the investment value of a given project. To do this, they need to estimate the allocation of capital costs over the life of the project, which requires understanding the conventions and techniques that accountants use to depreciate assets. This chapter will overview the conventions and techniques of asset depreciation.

We begin by discussing the nature and significance of depreciation, distinguishing its general economic definition from the related but different accounting view of depreciation. We then focus our attention almost exclusively on the rules and laws that govern asset depreciation and the methods that accountants use to allocate depreciation expenses. A knowledge of these rules will prepare you to apply them in assessing the depreciation of assets acquired in engineering projects.

Finally, we turn our attention to the subject of depletion, which utilizes similar ideas but specialized techniques to allocate the cost of the depletion of natural resource assets.

■ 7.1 Asset Depreciation

Fixed assets are the economic resources that are acquired to provide future cash flows.[1] We can define depreciation generally as the gradual decrease in utility of fixed assets with use and time. While this general definition does not adequately capture the variations in how we can define depreciation, it does provide us with a general starting point for examining the variety of underlying ideas and practices that we will discuss.

We can classify depreciation into the categories of physical and functional depreciation. **Physical depreciation** is defined as a reduction in an asset's capacity to perform its intended service due to physical impairment. Physical depreciation can occur to any fixed asset in the form of (1) deterioration from interaction with the environment, including such agents as corrosion, rotting,

---

[1] Fixed assets and other standard accounting terms are discussed in more detail in Appendix A.

and other chemical changes; and (2) wear and tear from use. Physical depreciation leads to a decline in performance and high maintenance costs.

Functional depreciation occurs as a result of changes in the organization or in technology that decrease or eliminate the need for an asset. Examples of functional depreciation include obsolescence due to advances in technology, a declining need for the services performed by an asset, or the inability to meet increased quantity and/or quality demands.

7.1.1 Economic Depreciation

This chapter is primarily concerned with accounting depreciation, which is the form of depreciation that provides us with the information used by the organization to assess its financial position. It is also useful, however, to discuss briefly the economic ideas upon which accounting depreciation is based. In doing this, we will develop a precise definition of economic depreciation which will help us distinguish between these separate conceptions of depreciation.

If you have ever owned a car, you are probably familiar with the term depreciation as it is used to describe the decreasing value of your vehicle (see example 5.7). Because its reliability and appearance usually decline with age, the vehicle is worth less with each passing year. You can calculate the economic depreciation accumulated for your car by subtracting the current market value of your car from the price you originally paid for it. We can define **economic depreciation** as follows:

$$\text{economic depreciation} = \text{purchase price} - \text{market value}.$$

Physical and functional depreciation are categories of economic depreciation.

The measurement of economic depreciation does not require that an asset be sold: The market value of an asset can be closely estimated without actually testing the value in the marketplace. The need to have a precise scheme for recording the ongoing decline in the value of an asset as a part of the accounting process leads us to explore how organizations account for depreciation.

7.1.2 Accounting Depreciation

The acquisition of fixed assets is an important activity for a business organization, whether the organization is starting up or acquiring new assets to remain competitive. Like other disbursements, the cost of these fixed assets must become expenses on a firm's balance sheet and income statement. However, unlike such costs as maintenance, material, and labor, the costs of fixed assets are not treated simply as expenses to be accounted for in the year that they are acquired. Rather, these assets are **capitalized**; that is, their costs are distributed by subtracting them as expenses from gross income—one part at a time over a number of periods. The systematic allocation of the initial cost of an asset in parts over a time known as its depreciable life is what we mean by **accounting depreciation.** Because accounting depreciation is the standard of the business world, we sometimes refer to it more generally as **asset depreciation**.

Accounting depreciation is based on the **matching concept:** A fraction of the cost of the asset is chargeable as an expense in each of the accounting periods in which the asset provides service to the firm and each charge is meant to be a percent of the whole cost which "matches" the percentage of the value utilized in the given period. The matching concept suggests that the accounting depreciation allowance generally reflects to some extent the actual economic depreciation of the asset.

We should introduce one important distinction within the general definition of accounting depreciation. Most firms calculate depreciation differently, depending on whether the calculation is for financial reports, such as the balance sheet or income statement, or for Revenue Canada for the purpose of paying taxes. In Canada, this distinction is totally legitimate under Canadian tax regulations, and many other countries permit such a distinction as well. Calculating depreciation differently for financial reports and taxes allows for the following benefits:

- It enables the firm to report depreciation to stockholders and other significant outsiders that is based on the matching concept and therefore generally reflects the actual loss in value of the assets.

- It allows the firm to benefit from the tax advantages of depreciating assets faster than would be possible using the matching concept as a guideline We will discuss these advantages in the following section.

Figure 7.1 reviews the distinct meanings of depreciation we have explored so far. We will make increasing use of the distinction between depreciation

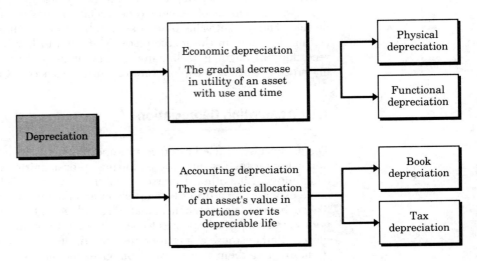

Figure 7.1 ■ Classification of depreciation

accounting for financial reporting and that used for income tax calculation as we proceed through the chapter. As we will demonstrate in Section 7.2, the manner in which depreciation is calculated has a significant influence on the financial position of a firm as measured in financial reports. (Note: Tax depreciation under Canadian tax regulations is designated as capital cost allowance, or CCA.)

■ 7.2 Net Income

Since we are interested primarily in the measurable financial aspects of depreciation, we can consider the effects of depreciation on two important measures of an organization's financial position, net income and cash flow. (We defer our discussion of methods of depreciation because it will be helpful initially to develop an organizational perspective on depreciation. Once we understand that depreciation has a significant influence on the income and cash position of a firm we will be able to appreciate fully the importance of utilizing depreciation as a means to maximize the value both of engineering projects and of the organization as a whole.)

7.2.1 Net Income = After-Tax Profit

Firms invest in a project because they expect it to increase their wealth. If the project does this, if project revenues exceed project costs, we say it has generated a **profit**, or **income**. If the project reduces the owners' wealth, if project costs exceed project revenues, we say that the project has resulted in a loss. One of the most important roles of the accounting function within an organization is to measure the amount of profit or loss a project generates each year or in any other relevant time period. Any profit generated will be taxed. The accounting measure of a project's after-tax profit during a particular time period is known as **net income.**

7.2.2 Calculation of Net Income

Accountants measure the net income of a specified operating period by subtracting expenses from revenues for that period. These terms can be defined as follows:

1. The **project revenue**[2] is the income received by a business as a result of providing products or services to outsiders. Revenue comes from sales of merchandise to customers and from fees earned by services performed for clients or others.

---

[2] Note that the cash may be received in different accounting periods.

2. The **project expenses** are the cost of doing business to generate the revenues of that period. Some common expenses are the cost of the goods sold (labor, inventory, materials, and supplies), the cost of employees' salaries, the operating cost (such as the cost of renting a building and the cost of insurance coverage), and income taxes.

The business expenses listed above are all accounted for in a straightforward fashion on the income statement and balance sheet: the amount paid by the organization for each item would translate dollar for dollar into expenses in financial reports for the period. One additional category of expenses, the purchase of new assets, is treated by depreciating the total cost gradually over time. Because capital goods are given this unique accounting treatment, depreciation is accounted for as a separate expense in financial reports. In the following section, we will discuss how this process is reflected in net income calculation.

7.2.3 Depreciation Expenses

Whether you are starting or maintaining a business, you will probably need to acquire assets (such as buildings and equipment). The cost of this property becomes part of your business expenses. The accounting treatment of these capital expenditures is different from manufacturing and operating expenses such as cost of goods sold and business operating expenses. *Capital expenditures must be capitalized,* that is, they must be systematically allocated as expenses over their depreciable lives. Therefore, when property that has a productive life extending over several years is acquired, the total costs cannot be deducted from profits in the year the asset was purchased. Instead, a depreciation allowance[3] is established over the life of the asset, and an appropriate portion of that allowance is included in the company's deductions from profit each year.

7.2.4 Depreciation and Its Relation to Income Taxes

Because it plays a role in reducing taxable income, depreciation accounting is of special concern to a company. In this section, we investigate the relationship between depreciation and income. Taxable income is defined as follows:

$$\text{taxable income} = \text{gross income (revenues)} - \text{cost of goods sold}$$

$$- \text{depreciation or CCA}$$

$$- \text{operating expenses}$$

---

[3] This allowance is based on the total cost of the property.

Once taxable income is calculated, income taxes are determined by the formula:

$$\text{income taxes} = (\text{tax rate}) \times (\text{taxable income}).$$

(We will discuss how we determine the applicable tax rate for a project in Chapter 8.) We then calculate net income with depreciable assets as follows:

$$\text{net income} = \text{taxable income} - \text{income taxes}.$$

A more common format is to present the net income in the following tabular form:

> **Gross income**
> − **Expenses**
>   cost of goods sold
>   depreciation
>   operating expenses
> _____
> **Taxable income**
> − Income taxes
> _____
> **Net income**

Our first example illustrates this relationship using numerical values.

Example 7.1 Net income within a year

A company buys a numerical control (NC) machine for $20,000 (year 0) and uses it for 5 years, after which it is scrapped. The cost of the goods produced by this NC machine should include a charge for the depreciation of the machine. Suppose the company estimates the following revenues and expenses for the first operating year.

Sales revenue	= $50,000
Cost of goods sold	= $20,000
Depreciation on NC machine	= $4,000
Operating expenses	= $6,000

If the company pays taxes at the rate of 40% on its taxable income, what is the net income during the first year from the project?

Solution

Given: Gross income and expenses as stated, income tax rate = 40%

Find: Net income

At this point, we will defer the discussion of how the depreciation amount ($4000) and tax rate (40%) are determined and treat these as givens. We consider the purchase of the machine to have been in year 0, which is also the beginning of year 1. Note that our example explicitly assumes that the only depreciation charges for year 1 are those for the NC machine, a situation that may not be typical.

Item	Amount
Gross income (revenues)	$50,000
Expenses:	
Cost of goods sold	20,000
Depreciation	4,000
Operating expenses	6,000
Taxable income	20,000
Taxes (40%)	8,000
Net income	$12,000

Comments: *In this example, the inclusion of a depreciation expense reflects the true cost of doing business. This expense is meant to match the amount of the $20,000 total cost of the machine that has been put to use or "used up" during the first year.*

This example highlights some of the reasons that income tax laws govern the depreciation of assets. If the company were allowed to claim the entire $20,000 as a year 1 expense, there would be a discrepancy between the one-time cash outlay for the machine's cost and the gradual benefits of its productive use. This discrepancy would lead to dramatic variations in the firm's net income, and net income would become a less accurate measure of the organization's performance.

On the other hand, failing to account for this cost at all would lead to increased reported profit during the accounting period. In this situation, the profit would be a "false profit" in that it would not accurately account for the real cost of the machine. Depreciating the cost over time allows the company a logical distribution of costs that matches the utilization of the machine's value.

7.2.5 Cash Flow Versus Net Income

We have just seen that the annual depreciation allowance has an important impact on both taxable and net income. However, although depreciation has a direct impact on net income, it is not a cash outlay; as such, it is important to

Figure 7.2 ■ Cash flow (capital expenditure) versus depreciation expenses

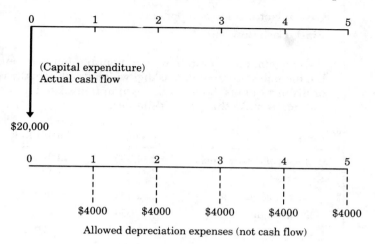

distinguish between annual income in the presence of depreciation and annual cash flow.

The situation described in Example 7.1 serves as a good vehicle to demonstrate the difference between depreciation costs as expenses and the cash flow generated by the purchase of a fixed asset. In this example, cash was expended in year 0, but the depreciation charged against the income in year 1 is not a cash outlay. Figure 7.2 summarizes the difference.

Net income (accounting profit) is important for accounting purposes, but cash flows are more important for purposes of project evaluation. However, as we will now demonstrate, net income can provide us with a starting point to estimate the cash flow of a project.

The procedure for calculating net income is identical to that for obtaining net cash flow (after-tax) from operations, with the exception of depreciation which is excluded from the net cash flow computation (it is needed only for computing income taxes). Assuming that revenues are received and expenses are paid in cash, we can obtain net cash flow by adding depreciation to net income, which cancels the operation of subtracting it from revenues:

$$\text{cash flows} = \text{net income} + \text{depreciation}$$

Example 7.2 illustrates this relationship.

Example 7.2 Cash flow versus net income

Using the situation described in Example 7.1, assume that (1) all sales are for cash, and (2) all expenses except depreciation were paid during year 1. How much cash would have been generated from operations?

Solution

Given: Net income components
Find: Cash flow

We can generate a cash flow statement by simply examining each item in the income statement and asking which items actually represent receipts or disbursements. Some of the assumptions listed in the problem statement make this process simpler.

Item	Income	Cash Flow
Gross income (revenues)	$50,000	$50,000
Expenses:		
Cost of goods sold	20,000	−20,000
Depreciation	4,000	
Operating expenses	6,000	−6,000
Taxable income	20,000	
Taxes (40%)	8,000	− 8,000
Net income	$12,000	
Net cash flow		$16,000

Here column 2 shows the income statement, and column 3 shows the statement on a cash flow basis. The sales of $50,000 are all for cash. Costs other than depreciation were $26,000 and paid in cash, leaving $24,000. Depreciation is not a cash flow—the firm did not pay out the $4,000 of

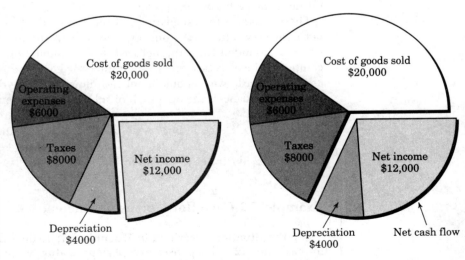

Figure 7.3 ■ Net income versus net cash flow (Example 7.2)

depreciation expenses. Taxes, however, are paid in cash, so $8,000 for taxes must be deducted from the $24,000, leaving a net cash flow from operations of $16,000. As shown in Fig.7.3, this $16,000 is exactly equal to net income plus depreciation: $12,000 + $4,000 = $16,000.

As we've just seen, depreciation has an important impact on annual cash flow in its role as an accounting expense that reduces taxable income and thus reduces taxes. (Although depreciation expenses are not actual cash flows, depreciation has a positive impact on the after-tax cash flow of the firm.) Of course, during the year in which an asset is actually acquired, the cash disbursed to purchase it creates a significant negative cash flow and, during the depreciable life of the asset, the depreciation charges will affect the taxes paid and therefore cash flows.

As shown in Example 7.2, we can see clearly that depreciation, through its influence on taxes, plays a critical role in project cash flow analysis. Because the timing of the depreciation amounts can affect the feasibility of an investment project, we will examine several alternative depreciation strategies in the following sections. Before we do this, however, we must understand the conventions and terminology that are used in depreciation accounting.

■ 7.3 Some Factors of Asset Depreciation

The process of depreciating an asset requires that we make several preliminary determinations: (1) what is the cost of the asset? (2) what is the asset's value at the end of its useful life? (3) what is the depreciable life of the asset? and, finally, (4) what method of depreciation do we choose? In this section, we will discuss each of these considerations.

7.3.1 Depreciable Property

As a starting point, it is important to recognize what constitutes a **depreciable asset,** that is, a property for which a firm may take depreciation deductions against income. For Canadian tax law purposes, any depreciable property has the following characteristics:

1. It must be used in business or held for production of income.
2. It must have a definite service life, and the life must be longer than one year.
3. It must be something that wears out, decays, gets used up, become obsolete, or loses value from natural causes.

Depreciable property includes buildings, machinery, equipment, and vehicles. Inventories are not depreciable property because they are held primarily for sale to customers in the ordinary course of business. If an asset has no definite service life, the asset cannot be depreciated. For example, *you can never depreciate land*. (This also means that you cannot depreciate the cost of clearing, grading, planting, and landscaping because these expenses are all considered part of the cost of the land.)

As a side note, we should add that while we have been focusing on depreciation within firms, individuals may also depreciate assets as long as they meet the conditions listed above. For example, an individual may depreciate a vehicle proportional to its use for business purposes.

7.3.2 Cost Base

The cost base of an asset represents the total cost that is claimed as an expense over an asset's life; that is, the sum of the asset's annual depreciation expenses.

The **cost base** generally includes the actual cost of the asset and all the other incidental expenses, such as freight, site preparation, and installation. This total cost, rather than the cost of the asset only, must be the depreciation base charged as an expense over the asset's life.

Example 7.3   Cost base

Lanier Corporation purchased machinery with an invoice price of $62,500. The vendor's invoice included a sales tax of $3263. Lanier paid the invoice and inbound transportation charges of $725 on the new machine as well as a labor cost of $2150 to install the machine in the factory. Lanier also had to prepare the site at a cost of $3500 before installation. Determine the cost base for the new machine for depreciation purposes.

Solution

Given: Invoice price = $62,500, freight = $725, installation cost = $2150, and site preparation = $3500
Find: The cost base

The cost of the machine that is applicable for depreciation is computed as follows:

Invoice price of machine, including sales tax	$62,500
Freight	725
Installation labor	2,150
Site preparation	3,500
Cost of machine (cost base)	$68,875

Comments: *Why do we include all the incidental charges relating to the acquisition of a machine in its cost? Why not treat these incidental charges as expenses of the period in which the machine is acquired? The matching of costs and revenue is the basic accounting principle. Consequently, the total costs of the machine should be viewed as an asset and allocated against the future revenue that the machine will generate. All costs incurred in acquiring the machine are costs of the services to be received from using the machine.*

Besides being used in figuring depreciation deductions, the cost base is used in calculating the gain or loss to the firm if the asset is ever sold or salvaged. (We will discuss these subjects in Chapter 8.)

7.3.3 Salvage Value and Useful Life

The salvage value is an asset's value at the end of its life; it is the amount eventually recovered through sale, trade-in, or salvage. The eventual salvage value of an asset must be estimated for an economic analysis. If this estimate subsequently differs from the undepreciated capital cost at disposal (book value), then an adjustment must be made. We will discuss these specific issues in Section 7.7.

The useful life of an asset is the period over which the asset may be expected to have value for its intended purpose. The useful life depends upon many factors such as advances in related technology, age when purchased, maintainability, obsolescence, physical deterioration, and so forth.

7.3.4 Depreciation Methods: Book and Tax Depreciation

Tax law allows firms to calculate depreciation in one way, called **book depreciation,** when reporting income to investors and another way, called **tax depreciation,** in computing taxes. (Each of these forms of depreciation can be classified as a subset of what we earlier defined as accounting depreciation.) In other words, in financial reports, such as income statements, most firms report depreciation expenses using book depreciation methods; in computing their taxes, however, they use tax depreciation methods. The distinction between tax depreciation and book depreciation methods will be explained in detail in Section 7.4. For now, the following general statements apply:

- Most firms use tax depreciation methods to calculate income tax liability because, in many cases, tax depreciation allows the firm to defer paying income taxes. This does not mean that the firm pays less taxes overall, because the total depreciation expense accounted for over time is the same in either case. However, because tax depreciation methods generally permit a higher depreciation in earlier years than book depreciation methods, the tax benefit of depreciation is enjoyed earlier, and the firm generally pays lower taxes in the initial years of an investment project. This typically leads to a better cash position in early years, and the added cash can lead to greater future wealth because of the time value of those funds.

- While we emphasize tax depreciation methods, we must also understand book depreciation methods because current tax depreciation methods are based largely on book depreciation methods.

Now that we have established the contexts for our interest in both tax and book depreciation, we can survey the methods with an accurate perspective.

■ 7.4 Book Depreciation Methods

The most widely used book depreciation methods are the straight-line method, the declining balance method, and the sum-of-years'-digits method. There are a number of reasons to study these methods of depreciation. The primary reason is that they may still be used for financial reporting to stockholders and outside parties. A company may choose to depreciate a particular type of equipment on the straight-line basis, for instance, while tax laws require it to depreciate the same asset on a declining balance basis for tax purposes. The straight line and declining balance methods of depreciation form the basis of Canadian tax law on depreciation. Our discussion of depletion in Section 7.6 is based largely on one of these book depreciation methods.

7.4.1 Straight-Line Method (SL)

The **straight-line method** of depreciation interprets a fixed asset as providing its services in a uniform fashion. The asset provides an equal amount of service in each year of its useful life. The straight-line method charges as an expense an equal fraction of the net cost of the asset each year, as expressed by the relation

$$D_n = \frac{(P - S)}{N},$$ (7.1)

where

D_n = the depreciation charge during year n

P = the cost of the asset, including installation expenses

S = the salvage value at the end of useful life

N = the useful life

$1/N$ = the straight-line depreciation rate

The book value of the asset at the end of year n is then defined as

$$\text{book value} = \text{cost base} - \text{total depreciation charges made,}$$

or

$$B_n = P - (D_1 + D_2 + D_3 + \ldots + D_n).$$

(7.2)

Example 7.4 Straight-line depreciation

Consider the following data on an automobile:

Cost of the asset, P	= $10,000
Useful life, N	= 5 years
Estimated salvage value, S	= $2,000

Compute the annual depreciation allowances and the resulting book values, using the straight-line depreciation method.

Solution

Given: $P = \$10,000$, $S = \$2,000$, $N = 5$ years
Find: D_n, B_n for $n = 1$ to 5

The straight-line depreciation rate is 1/5, or 20%. Therefore, the annual depreciation charge is

$$D_n = (0.20)(\$10,000 - \$2,000) = \$1600.$$

Then, the asset would have the following book values during its useful life:

n	B_{n-1}	D_n	B_n
1	$10,000	$1,600	$8,400
2	8,400	1,600	6,800
3	6,800	1,600	5,200
4	5,200	1,600	3,600
5	3,600	1,600	2,000

where B_{n-1} represents the book value before the depreciation charge for year n. The situation is illustrated in Fig.7.4.

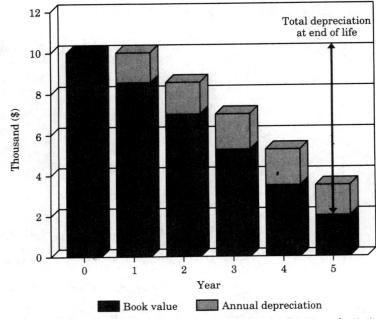

Figure 7.4 ■ Straight-line depreciation method (Example 7.4)

7.4.2 Accelerated Methods

The second depreciation concept recognizes that the stream of services provided by a fixed asset may decrease over time; in other words, it may be greatest in the first year of the asset's service life and least in the last year. This pattern may occur because the asset's mechanical efficiency tends to decline with age, because maintenance costs tend to increase with age, or because of the increasing likelihood that better equipment will become available and make it obsolete. This reasoning leads to a method that charges a larger fraction of the cost as an expense of the early years than of the later years. Any such method is called an accelerated method. The two most widely used accelerated methods are the declining balance method and the sum-of-the-years'-digits method.

Declining Balance Method (DB)

The **declining balance method** allocates for depreciation a fixed fraction of the initial book balance each year. The fraction or declining balance rate, d, is obtained as follows:

$$d = (1/N) \text{ (multiplier)} \qquad (7.3)$$

The most common multiplier is 1 (called DB). However, other multipliers are

1.5 (called 150% DB) and 2.0 (called double declining balance, or DDB). As N increases, d decreases, resulting in a situation in which depreciation is highest in the first year and decreases over the asset's depreciable life. Companies use Eq. (7.3) to determine the declining balance rate for book depreciation purposes. For tax depreciation, Revenue Canada specifies the declining balance rates companies must use.

The fractional factor can be used to determine depreciation charges for a given year, D_n, as follows:

$$D_1 = dP,$$

$$D_2 = d(P - D_1) = dP(1 - d),$$

$$D_3 = d(P - D_1 - D_2) = dP(1 - d)^2,$$

and thus for any year, n, we have a depreciation charge, D_n, of

$$D_n = dP(1 - d)^{n-1}. \qquad (7.4)$$

We can also compute the total DB depreciation (TDB) at the end of n years as follows:

$$TDB = D_1 + D_2 + \ldots + D_n$$

$$= dP + dP(1 - d) + dP(1 - d)^2 + \ldots + dP(1 - d)^{n-1}$$

$$= dP[1 + (1 - d) + (1 - d)^2 + \ldots + (1 - d)^n]. \qquad (7.5)$$

Multiplying Eq. (7.5) by $(1 - d)$, we obtain

$$TDB(1 - d) = dP[(1 - d) + (1 - d)^2 + (1 - d)^3 + \ldots + (1 - d)^n] \qquad (7.6)$$

Subtracting Eq.(7.5) from Eq. (7.6) and dividing by d gives

$$TDB = P[1 - (1 - d)^n]. \qquad (7.7)$$

The book value, B_n, at the end of n years will be the cost of the asset P minus the total depreciation at the end of n years.

$$B_n = P - TDB$$

$$= P - P[1 - (1 - d)^n]$$

$$B_n = P(1 - d)^n. \qquad (7.8)$$

Example 7.5 Declining balance depreciation

Consider the following accounting information for a computer system:

$$\text{Cost of the asset, } P = \$10,000$$

$$\text{Useful life, } N = 5 \text{ years}$$

$$\text{Estimated salvage value, } S = \$3277$$

Compute the book depreciation allowances and the resulting book values, using the 100% declining depreciation method.

Solution

Given: $P = \$10,000$, $S = \$3277$, $N = 5$ years,

Find: D_n, B_n for $n = 1$ to 5

The book value at the beginning of the first year is $10,000, and the declining balance rate is $(1/5) = 20\%$. Then, the depreciation deduction for the first year will be $2000 ($20\% \times \$10,000 = \$2000$). To figure the depreciation deduction in the second year, we must first adjust the book value for the amount of depreciation we deducted in the first year. Subtract the previous year's depreciation from the adjusted book value ($\$10,000 - \$2000 = \$8000$). Multiply this amount by the rate of depreciation ($\$8000 \times 20\% = \$1600$). By continuing the process, we obtain

n	B_{n-1}	D_n	B_n
1	$10,000	$2,000	$8,000
2	8,000	1,600	6,400
3	6,400	1,280	5,120
4	5,120	1,024	4,096
5	4,096	819	3,277

The declining balance is illustrated in terms of the book value of time in Fig. 7.5.

Figure 7.5 ■ Declining balance method (Example 7.5)

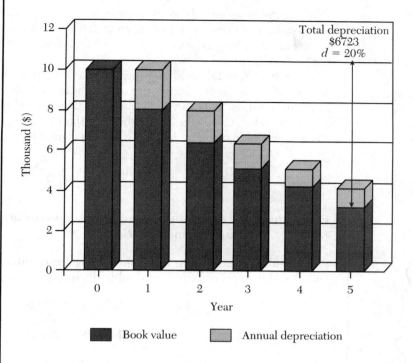

Comments: *Salvage value (S) must be estimated at the outset of depreciation analysis. In this example, the final book value (B_N) conveniently equals our estimated salvage value of $3277, an occurrence that is unusual in the real world. When B_N is not equal to S, we want to make adjustments in our depreciation methods as shown in the following section.*

When computing a depreciation allowance for tax purposes, d is specified by Revenue Canada. Whether B_N is equal to S or not, the CCA schedules cannot be changed. However, for financial reporting purposes, we want to make adjustments in our depreciation methods as shown below to make sure $B_N = S$.

When B_N is not equal to S

When $B_N > S$, we are left with a situation in which we have not depreciated the entire cost of the asset. In this case, we can switch to straight-line depreciation at an optimal time to ensure that we do depreciate all the way down to salvage value.

When $B_N < S$, we must readjust our analysis to prevent depreciation to below the salvage value. Examples of how to deal with both $B_N > S$ and $B_N < S$ follow.

Situation (1), $B_N > S$: Switch from DB to SL

If you prefer to reduce the book value of an asset to its salvage value as quickly as possible, it is done by switching from DB to SL whenever SL depreciation results in larger depreciation charges and therefore a more rapid reduction in the book value of the asset.

The switch from DB to SL depreciation can take place in any of the N years, and our objective is to identify the optimal year to switch. The switching rule is: If depreciation by DB in any year is less than it would be by SL, we should switch to and stay with the SL method for the remaining duration of the project's depreciable life. The DB and SL depreciation for any year is

$$D_n(DB) = dB_{n-1} \text{ or } D_n(SL) = \frac{B_{n-1} - S}{N - n + 1} \qquad (7.9)$$

Example 7.6 Declining balance with conversion to straight-line depreciation $(B_N > S)$

Suppose the asset given in Example 7.4 has a zero salvage instead of $2000 and is depreciated using double declining balance,

$$\text{Cost of the asset, } P = \$10,000$$

$$\text{Useful life, } N = 5 \text{ years}$$

$$\text{Salvage value, } S = \$0$$

$$d = (1/5)\,(2) = 40\%$$

Determine the optimal time to switch from DDB to SL depreciation, and the resulting depreciation schedule.

Solution

Given: $P = \$10,000$, $S = 0$, $N = 5$ years, $d = 40\%$
Find: Optimal conversion time, D_n, and B_n for $n = 1$ to 5

We will first proceed by computing the DDB depreciation for each year.

Year	D_n	B_n
1	$4000	$6000
2	2400	3600
3	1440	2160
4	864	1296
5	518	778

Figure 7.6 ■ Adjustments to the declining balance method: (a) switch from DDB to the SL after n', (b) there are no further depreciation allowances after n'' (Examples 7.6 and 7.7)

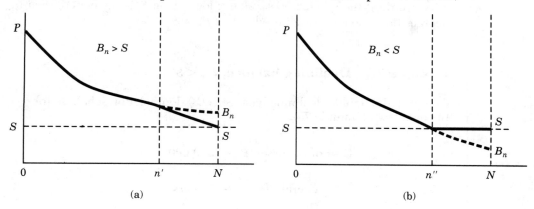

(a)   (b)

Then, we compute the SL depreciation for each year using Eq. (7.9). We compare SL to DDB depreciation for each year and use the decision rule for when to change.

If Switch to SL in Beginning Year	SL Depreciation		DDB Depreciation	Decision
2	($6000 − 0)/4 = $1500	<	$2400	Do not switch
3	($3600 − 0)/3 = $1200	<	$1440	Do not switch
4	($2160 − 0)/2 = $1080	>	$864	Switch to SL

The optimal time (year 4) in this situation corresponds to n' in Fig. 7.6(a). The resulting depreciation schedule is

Year	DDB with Switching to SL	End-of-Year Book Value
1	$4,000	$6,000
2	2,400	3,600
3	1,440	2,160
4	1,080	1,080
5	1,080	0
	$10,000	

Situation (2), $B_N < S$: Adjusting to $B_N = S$

With a relatively high salvage value, it is possible that the book value of the asset could decline below the estimated salvage value. To avoid deducting depreciation charges that would drop the book value below the salvage value, you simply stop depreciating the asset whenever you get down to $B_N = S$. In other words, if the implied book value is lower than S at any period, then the depreciation amounts are adjusted so that $B_n = S$.

Example 7.7 Declining balance, $B_N < S$

Compute the double declining balance (DB) depreciation schedule for the data from Example 7.4.

$$\text{Cost of the asset, } P \ = \ \$10,000$$

$$\text{Useful life, } N \ = \ 5 \text{ years}$$

$$\text{Salvage value, } S \ = \ \$2000$$

$$d \ = \ (1/5)(2) = 40\%$$

Solution

Given: $P = \$10,000$, $S = \$2000$, $N = 5$ years, $d = 40\%$
Find: D_n, B_n for $n = 1$ to 5

End of year	D_n		B_n
1	0.4($10,000) =	$4000	$10,000 − $4000 = $6000
2	0.4($6000) =	2400	$6000 − $2400 = $3600
3	0.4($3600) =	1440	$3600 − $1440 = $2160
4	0.4($2160) >	160	$2160 − $160 = $2000
5		0	$2000 − 0 = $2000

Note that, since B_4 is less than $S = \$2000$, we adjust D_4 to $160, making $B_4 = \$2000$. D_5 is zero and B_5 remains at $2000. The year 4 is equivalent to n'' in Figure 7.6(b).

Sum-of-Years'-Digits (SOYD) Method

Another accelerated method for allocating the cost of an asset is called sum-of-years'-digits (SOYD) depreciation. Compared with SL depreciation, SOYD results in larger depreciation charges during the early years of an asset and smaller charges as the asset reaches the end of its estimated useful life.

In the SOYD method, the numbers 1,2,3,....., N are summed, where N is the estimated years of useful life. We find this sum by the equation.[4]

$$SOYD = 1 + 2 + 3 + ... + N = \frac{N(N+1)}{2} \qquad (7.10)$$

The depreciation rate each year is a fraction in which the denominator is the SOYD and the numerator is, for the first year, N; for the second year, $N-1$; for the third year, $N-2$; and so on. Each year the depreciation charge is computed by dividing the remaining useful life by the SOYD and multiplying this ratio by the total amount to be depreciated $(P-S)$.

$$D_n = \frac{N - n + 1}{SOYD} (P - S) \qquad (7.11)$$

Example 7.8   SOYD depreciation

Compute the SOYD depreciation schedule for Example 7.4,

$$\text{Cost of the asset, } P = \$10{,}000$$

$$\text{Useful life, } N = 5 \text{ years}$$

$$\text{Salvage value, } S = \$2000$$

Solution

Given: $P = \$10{,}000$, $S = \$2000$, $N = 5$ years
Find: D_n, B_n for $n = 1$ to 5

We first compute the sum-of-years' digits:

$$SOYD = 1 + 2 + 3 + 4 + 5 = 5 (5 + 1)/2 = 15$$

---

[4] You may derive this equation by writing the SOYD expression in two ways:

$$SOYD = 1 + 2 + 3 + \cdots + N$$
$$SOYD = N + (N - 1) + \cdots + 1$$

You add these two equations and then solve for SOYD:

$$2\,SOYD = (N + 1) + (N + 1) + \cdots + (N + 1)$$
$$= N(N + 1)$$
$$SOYD = N(N + 1)/2.$$

Year	D_n	B_n
1	$(5/15)\,(\$10{,}000 - \$2000) = \$2667$	$7333
2	$(4/15)\,(\$10{,}000 - \$2000) = \$2133$	$5200
3	$(3/15)\,(\$10{,}000 - \$2000) = \$1600$	$3600
4	$(2/15)\,(\$10{,}000 - \$2000) = \$1067$	$2533
5	$(1/15)\,(\$10{,}000 - \$2000) = \quad\$533$	$2000

This situation is illustrated in Fig. 7.7.

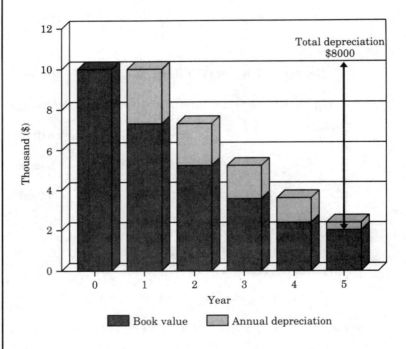

Figure 7.7 ■ Sum-of-year's-digits method (Example 7.8)

7.4.3 Units-of-Production Method

A third depreciation concept views the asset as consisting of a bundle of service units; unlike the SL and accelerated methods, however, this concept does not assume that the service units will be consumed in a time-phased pattern. The cost of each service unit is the net cost of the asset divided by the total number of such units. The depreciation charge for a period is then related to the number of service units consumed in that period. This leads to the **units-**

of-production method. By this method, the depreciation in any year is given by

$$D_n = \frac{\text{service units consumed for year } n}{\text{total service units}} (P - S) \qquad (7.12)$$

This method can be useful for depreciating equipment such as that used to exploit natural resources if the resources will be depleted before the equipment wears out. It is not, however, considered an acceptable method for general use in depreciating industrial equipment.

Example 7.9   Units-of-production depreciation

A truck for hauling coal has an estimated net cost of $65,000 and is expected to give service for 400,000 kilometres, resulting in a zero salvage value. Compute the allowed depreciation amount for the truck usage of 48,000 kilometres.

Solution

Given: $P = \$65,000$, $S = 0$, total service units = 400,000 kilometres, usage for this year = 48,000 kilometres

Find: Depreciation amount in this year

The depreciation expense in a year in which the truck traveled 48,000 kilometres would be

$$\frac{48,000 \text{ km}}{400,000 \text{ km}} (\$65,000 - 0) = (\$65,000)(0.12)$$

$$= \$7800.$$

7.4.4 Book Depreciation in Practice

Before moving on to tax depreciation, one last word about book depreciation is in order. Each category of book depreciation—straight-line, accelerated, and units-of-production—has its own conceptual basis for the pattern by which an asset provides services and is "used up." In theory, a company can match the appropriate depreciation method to the actual pattern of service provided. However, there is little evidence that companies do so. In most cases, straight-line depreciation is selected. One obvious reason for this choice is the simplicity of calculating straight-line depreciation. Another possible reason is that accelerated plans make initial reported incomes smaller, a situation that may reflect poorly on management in the eyes of stockholders.

■ 7.5 Tax Depreciation Methods

For many years, the Canadian government has used the income tax laws as a device to encourage corporations to invest in new productive assets. The two primary mechanisms that have been employed to encourage this capital investment have been depreciation allowances (capital cost allowances) and the investment tax credit.[5] Under Canadian tax law businesses must depreciate their assets using very specific guidelines, thus ensuring that they will collectively have consistent depreciation methods and values for taxation purposes. These depreciation guidelines have also been used, along with the investment tax credits, to encourage Canadian businesses to invest in certain types of assets in certain regions of the country.

7.5.1 Tax Depreciation Methods Using Capital Cost Allowances

Under Canadian tax law individuals and businesses are allowed to deduct part of the capital cost of certain depreciable property from business income that they earn during that year. As previously mentioned this tax depreciation deduction is referred to as a **capital cost allowance** (CCA). The depreciation guidelines as defined under Canadian tax law are known as the capital cost allowance system. The capital cost of a given property or asset normally comprises its complete purchase and set-up costs, including the legal expenses involved in its purchase, the engineering fees associated with its start-up, and all other costs associated with acquiring, installing, and placing the asset in operation.

Under the capital cost allowance system depreciable property or assets are grouped into specific classes. Property and assets within each class have the same method and rate of depreciation, as shown in Tables 7.1 and 7.2. Individuals or businesses may deduct any amount up to the maximum allowable in each class. The methods used to determine the maximum expense (capital cost allowance) are declining balance and straight-line depreciation.

For example, under Class 8, furniture and refrigeration equipment are both depreciated using a declining balance rate of 20%. This means that, at year-end, 20% of the undepreciated capital cost of all of the furniture and equipment in this class can be used to determine the maximum depreciation (or maximum capital cost allowance) that an individual or business may claim for the year.

7.5.2 Available for Use Rule

This rule simply indicates that an individual or business cannot claim the capital cost allowance (depreciation expense) until the property or asset becomes *available for use.*

---

[5] The investment tax credit will be discussed in Chapter 8.

Property other than a building is *available for use* at the earliest of the following dates:

- when the business first uses the asset to earn income
- when the business can use the property to either produce a saleable product or perform a saleable service
- immediately before the business disposes of the property or asset

A building or structure is available for use on the earliest of the following dates:

- when the building is used for its intended purpose
- when construction of the building is complete
- immediately before the corporation disposes of the property

Most capital cost allowance classes are calculated using a declining balance method. Table 7.1 summarizes the most common declining balance classes and their CCA rates.

Table 7.1 Capital Cost Allowance Declining Balance Classes and Rates:

Class Number	Description	CCA Rate
1	Most buildings made of brick, stone or cement acquired after 1987, including their component parts such as electrical wiring, lighting fixtures, plumbing, heating and cooling equipment, elevators and escalators.	4%
3	Most buildings made of brick, stone or cement acquired before 1988, including their component parts as listed in Class 1 above.	5%
6	Buildings made of frame, log, stucco on frame, galvanized iron or corrugated metal that are used in the business of farming or fishing, or that have no footings below ground; fences and certain greenhouses	10%
7	Canoes, boats, and most other vessels, including their furniture, fittings, or equipment	15%
8	Property that is not included in any other class such as furniture, calculators and cash registers, photocopy and fax machines, printers, telephone systems, display fixtures, refrigeration equipment, machinery, tools costing more than $200, and outdoor advertising billboards acquired after 1987.	20%

continued...

Class Number	Description	CCA Rate
9	Aircraft, including furniture, fittings, or equipment attached, and spare parts	25%
10	Automobiles (except taxis and others used for lease or rent), vans, wagons, trucks, buses, trailers, drive-in theatres, general-purpose electronic data-processing equipment (e.g., personal computers) and systems software, and timber cutting and removing equipment.	30%
10.1	Passenger vehicles costing more than $24,000 ($20,000 if acquired before September 1989) – for passenger vehicles acquired after 1990, the $24,000 cost does not include either GST or provincial sales tax.	30%
12	Chinaware, cutlery, linen, uniforms, dies, jigs, moulds or lasts, computer software (except systems software), cutting and shaping parts of a machine, certain property used for earning rental income such as apparel or costumes, and videotape cassettes; certain property costing less than $200 such as kitchen utensils, tools, and medical or dental equipment; certain property acquired after August 1989 and before 1993 for use in a business of selling or providing services such as electronic bar code scanners, and cash registers used to record multiple sales taxes.	100%
16	Automobiles for lease or rent, taxicabs, and coin-operated video games.	40%
17	Roads, sidewalks, parking-lot or storage areas, telephone, telegraph, or non-electronic data communications switching equipment.	8%
38	Most power-operated movable equipment acquired after 1987 used for moving excavating, placing, or compacting earth, rock, concrete, or asphalt.	30%
39	Machinery and equipment acquired after 1987 that is used in Canada primarily to manufacture and process goods for sale or lease.	25%
43	Manufacturing and processing machinery and equipment acquired after February 25, 1992, described in Class 39 above.	30%

Source: Revenue Canada Taxation, *1993 T2 Corporation Income Tax Guide*.

Some capital cost allowance classes are calculated using the straight-line depreciation method. Table 7.2 indicates some of these classes.

Table 7.2 Capital Cost Allowance, Straight-Line Classes

Class Number	Description	CCA Rate
13	Property that is leasehold interest (the maximum CCA rate depends on the type of leasehold and the terms of the lease)	N/A
14	Patents, franchises, concessions, and licences for a limited period - the CCA is limited to the least of: • the capital cost of the property spread out over the life of the property; and • the undepreciated capital cost of the property at the end of the taxation year.	N/A

Source: Revenue Canada Taxation, *1993 T2 Corporation Income Tax Guide.*

7.5.3 Half-Year Convention

Canadian tax regulations limit the depreciation or capital cost allowance claims of most depreciable property in the taxation year of acquisition. The rule became effective as of November 12, 1981. It prevents a significant advantage to individuals or businesses who could otherwise purchase property before their year-end and receive a full year of tax depreciation or capital cost allowance.

The depreciation deduction for new purchases is limited to one-half of the capital cost allowance normally allowed. Therefore during the first year, the capital cost allowance for all new purchases would result in a reduced deduction on a class-by-class basis.

Although most depreciable property is limited by this half-year convention, there are several depreciable properties that are exempt. They include the following:

- Some Class 12 property such as motion pictures or video tapes used for television commercial messages, non-system computer software, certified feature films, certified feature productions or certified short productions

- Class 14 patents, franchises, concessions, or licences

- Class 15 property for purpose of cutting and removing merchantable timber

- Special properties within Classes 13, 24, 27, 29 and 34 subject to 50% rules.

Capital cost allowances can be confusing even to an expert within the taxation field. For example, computer software is defined under Class 10 as a 30% declining balance for system software and under Class 12 as a 100% declining balance for all other software. As mentioned above, non-system software is also exempt from the half-year convention. An expert in taxation could easily have a problem determining which class of computer software has been purchased and whether to apply the half-year convention.

We can demonstrate the impact of the half-year rule as follows:

without the half-year rule

$$CCA_n = Pd(1 - d)^{n-1}, \text{ for } n \geq 1 \qquad (7.13)$$

$$B_n = P(1 - d)^n, \text{ for } n \geq 1, \qquad (7.14)$$

where

$$P = \text{initial cost base}$$

$$CCA_n = \text{CCA for year } n$$

$$B_n = \text{undepreciated capital cost at the end of year } n$$

$$d = \text{declining balance rate.}$$

With the half-year rule, however

$$CCA_1 = Pd/2 \qquad (7.15)$$

$$CCA_n = Pd(1 - d/2)(1 - d)^{n-2}, \text{ for } n \geq 2 \qquad (7.16)$$

$$B_n = P(1 - d/2)(1 - d)^{n-1}, \text{ for } n \geq 1 \qquad (7.17)$$

7.5.4 Calculating the Capital Cost Allowance

In order for a business to calculate its capital cost allowance it must complete a capital cost allowance (CCA) Schedule T2S(8) provided by Revenue Canada. On this schedule each declining balance class of property must be calculated separately and then totalled at the end. In the case of straight-line classes, each asset within the class involves a separate calculation. As shown in Figure 7.8 the results of the individual calculations, when summed, determine the total CCA and the undepreciated capital cost for the class.

- **Column 1: Class number** Each class of property should be identified and assigned the appropriate class number. Generally all of the depreciable property of the same class can be grouped together.

Figure 7.8 ■ Capital cost allowance

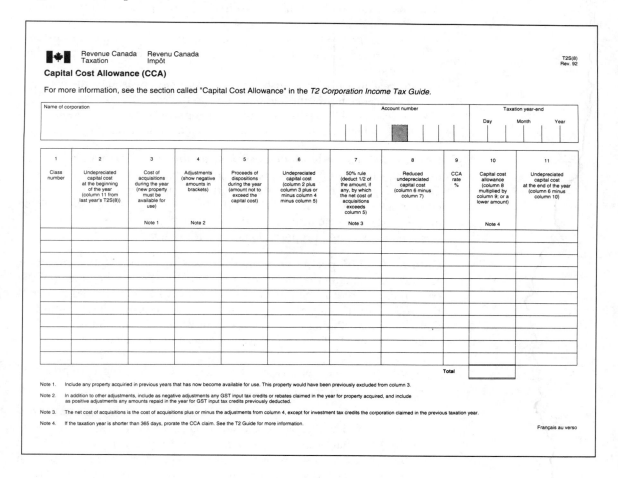

- **Column 2: Undepreciated capital cost at the beginning of the year**
  Beside each class number the total undepreciated capital cost is entered. This is the total value of all the properties within that specific class. This value equals the undepreciated capital cost at the end of the previous year.

- **Column 3: Cost of acquisitions during the year** The total cost of purchases and other acquisitions that were available for use during the year are recorded for each class. This cost includes legal fees, accounting and engineering fees, and all other fees incurred as a result of the purchase. Note that land is not a depreciable property and therefore is not eligible for capital cost allowances.

- **Column 4: Adjustments** In this column adjustments that either increase or decrease the total capital cost are recorded. The most common adjustments include
  - decreasing capital costs by any investment tax credits the business used to reduce taxes payable, or claimed as a refund in the preceding taxation year
  - decreasing the capital costs by any government assistance the corporation received or is entitled to receive in the year
  - increasing the capital cost of any property once government assistance, which previously reduced the capital cost, is repaid.

- **Column 5: Proceeds of dispositions during the year** In this column record the total property disposition proceeds that the business received during the year. If the business disposed of property for more than its capital cost, enter the capital cost. When the business disposes of a depreciable property for more than its capital cost, a capital gain results. However, losses on depreciated property do not result in capital losses.

- **Column 6: Undepreciated capital costs** The amount in this column is calculated by adding the undepreciated capital cost at the beginning of the year, plus the cost of acquisitions during the year, plus the adjustments that reduce or increase the capital cost, minus the proceeds of disposition during the year. Put another way, Column 2 plus Column 3 plus or minus Column 4 minus Column 5 equals undepreciated capital costs.

  A business cannot claim a capital cost allowance when the undepreciated capital cost is

  - positive and there is no property left in that class at the end of the taxation year. This is known as a terminal loss and is treated like an expense for that particular taxation year. A **terminal loss** results when the business sells the remaining property in a given class for less than its undepreciated value. Terminal loss is treated as a deduction from income.
  - negative at the end of the taxation year. This is known as recaptured depreciation or recaptured capital cost allowance. **Recaptured capital cost allowance** occurs when the proceeds from the disposition of property are more than the undepreciated value of the property. Recaptured depreciation must be added to the business income.

- **Column 7: Half-year rule** In most cases, property that a business acquired that is available for use during the taxation year is only eligible for 50% of the normal maximum capital cost allowance for that year. In this column, therefore, an amount equal to one-half of the net amount of additions to the class (i.e., the cost of acquisitions minus the proceeds of disposition) should be recorded. When applying this half-year rule, take into account adjustments that relate to the year's acquisitions.

- **Column 8: Reduced undepreciated capital cost** In this column, simply record the difference between the undepreciated capital cost in Column 6 prior to applying the half-year rule minus the half-year rule adjustment recorded in Column 7. The Column 8 amount is the final undepreciated capital cost value used to calculate the capital cost allowance.

- **Column 9: Capital cost allowance rate** The CCA rate for each class of capital cost allowance is specified in Canadian tax regulations. Simply enter the appropriate rate that applies to each class.

- **Column 10: Capital cost allowance** In order to claim the maximum capital cost allowance for each class, multiply the undepreciated capital cost in Column 8 by the CCA rate in Column 9, and record the result. A business may choose to claim any amount up to the maximum allowable capital cost allowance.

- **Column 11: Undepreciated capital cost at the end of the year** Calculate this value by subtracting the capital cost allowance in Column 10 from the undepreciated capital cost in Column 6. This value is the undepreciated capital cost at the beginning of the next taxation year.

Example 7.10 Calculating CCA for New Equipment Purchases

A small engineering consulting office decided to purchase a new computer to enhance their drafting skills and capabilities. The cost of the computer was $6800. The consulting firm understood this equipment was an asset in capital cost allowance declining balance Class 10, with a CCA rate of 30%. They already had $20,000 in Class 10 depreciable property at the beginning of their taxation year. What is the company's maximum capital cost allowance for Class 10 depreciable property for this year? What is their Class 10 undepreciated capital cost at the end of the year?

Solution

Given: Undepreciated capital cost at the beginning of the year = $20,000, and cost of acquisitions during the year = $6800

Find: Maximum CCA for year and undepreciated capital cost at year end

Working through the capital cost allowance schedule we determine

$$\text{Undepreciated capital cost} = \$20,000 + \$6800 = \$26,800$$

$$\text{Half-year rule adjustment} = \$6800 \times 1/2 = \$3400$$

$$\text{Reduced undepreciated capital cost} = \$26,800 - \$3400 = \$23,400,$$

and thus,

$$\text{Maximum capital cost allowance } = \$23{,}400 \times 30\% = \$7020$$

$$\begin{aligned}\text{Undepreciated capital cost at} \\ \text{the end of the year } &= \$26{,}800 - \$7020 = \$19{,}780.\end{aligned}$$

7.5.5 Capital Cost Allowance and Projects

Example 7.10 demonstrates the asset pool accounting concepts used in calculating capital cost allowance. This approach accounts for all the changes to the pool during the year and determines the tax depreciation expense for all the property included within a particular asset class.

When considering the economic attractiveness of a project, we are only concerned with the capital cost allowance that derives from the project itself. In many cases, a project includes more than one type of property, which means that the total capital cost allowance consists of a component from each asset class represented in the project. The capital cost allowance component for a given asset class does not include the effect of property additions or disposals that are unrelated to the project under consideration.

Restricting depreciation calculations to a specific project's assets presents no difficulty when calculating annual capital cost allowances up to the point of disposal of those assets. The tax implications arising from the disposal of an asset depend on whether the disposal will deplete the asset class, and the assumed timing of the disposal. Asset disposal may occur at the termination of the project when all the assets used in the project are sold, or at one or more points during the course of the project where an asset is sold and a replacement asset is purchased. If the disposal depletes an asset class, (i.e., no property remains in the class) the treatment of any terminal loss or recaptured capital cost allowance is straightforward, as explained in Section 7.5.4. However, frequently the total property within an asset class is large and not limited to property that belongs to the project under consideration. When an asset belonging to a particular project is sold without depleting the asset class, a terminal loss or recaptured CCA can also result. The tax effect of such a loss or recapture is built into the undepreciated capital cost base for the entire asset class as a result of the asset pool accounting procedure.

When the disposal of an asset does not deplete the asset class, the disposal tax effects calculation is complex and involves a specified interest rate. This complicates the application of the rate of return method because the disposal tax effects are a function of the unknown interest rate, as explained in Chapter 8, Section 8.3. Since most of the depreciable assets will have salvage values that are much smaller than the initial purchase costs and the salvage values will be realized at the end of the assets' useful lives, the disposal tax effects normally do not have a large impact on the present worth of the project under consideration. As a result, we will make the following assumptions throughout this book regarding the disposal of assets:

- Asset disposals occur just prior to the end of the year. For example, a disposal in year 5 involves a cash inflow at the end of year 5.

- The tax implications of the disposal are completely realized in the year of disposal. For example, if an asset is sold at the end of year 5 for a salvage value of S and its book value at the end of year 5 is B, then the difference $B - S$ will be subtracted from the company's taxable income in year 5. A positive $B - S$ value will result in tax savings, while a negative $B - S$ value indicates some additional tax payment.

These assumptions make the disposal tax effect independent of asset class depletion considerations. This is a reasonable approximation that simplifies the discounted cash flow analysis considerably. For a more detailed analysis of disposal tax effects and other disposal timing assumptions, including detailed formulas, see Chapter 8, Section 8.3.

Example 7.11   Calculating a Terminal Loss – No Other Property in Class

A small manufacturing company decided to sell its plant and move to a leased location. The company received $320,000 for its building. At the end of its taxation year the company had no more Class 3 depreciable properties. The undepreciated value of the property was $370,000 at the time of the sale. What is the impact on this year?

Solution

Given: Undepreciated capital cost = $370,000, and proceeds for disposition during the year = $320,000

Find: Undepreciated capital cost

$$\$370,000 - \$320,000 = \$50,000$$

Since there is an undepreciated capital cost of $50,000 and there is no property in this class remaining, this amount becomes a terminal loss. The loss will reduce income by $50,000 for tax purposes.

Example 7.12   Calculating Capital Gains and Recaptured CCA

A well-drilling company purchased some specialized drilling equipment last year for $40,000. The equipment currently has an undepreciated capital cost of $36,000. Recently, the drilling company has been offered $48,000 for this equipment. The company's owner wishes to know what the income tax considerations will be for this taxation year if he accepts the offer.

Solution

At $48,000, the sale price of the equipment will exceed the equipment's purchase price of $40,000 by $8000. Since the sale price will be greater than the purchase price, the company will have to claim an $8000 capital gain.

The undepreciated capital cost at the beginning of the year equals $36,000. Since the proceeds from disposition used in CCA calculations cannot exceed the initial acquisition price of $40,000, it is considered to be $40,000. Consequently the undepreciated capital cost at the end of the year is $-\$4000$.

Since undepreciated capital cost is a negative value, we refer to it as recaptured capital cost or recaptured depreciation expense.
Therefore this company will have a capital gain of $8000 and a recaptured capital cost allowance of $4000. Both the capital gain and the recaptured CCA will increase the company's tax payment in the year of disposal.

■ 7.6 Depletion

If you own mineral property (as distinguished from personal and real properties) such as oil, gas, or geothermal resources, you may be able to take a deduction as you deplete the resource. Any capital investment in natural resources needs to be recovered as the natural resources are removed and sold. The process of amortizing the cost of natural resources in accounting periods is called **depletion**. The objective is the same as that for depreciation, i.e., to amortize the cost in some systematic manner over the asset's useful life.

The taxpayer accomplishes this amortization using a **depletion allowance**. This allowance is applied in the same manner as the capital cost allowance just discussed. Deductions in the form of depletion allowances are contemplated for oil and gas, mining, and forestry industries, but Canadian tax regulations currently in effect apply mainly to mining and mineral processing. Prior to 1981 depletion allowance was widely available in the oil and gas industry, but since then has been mostly phased out and now applies only to certain non-conventional oil recovery projects. Currently there are no regulations providing depletion allowance to the forestry industry.

Taxpayers of all kinds, whether corporations, individuals, or trusts, are entitled to depletion allowance deductions. A trust may claim such deductions in computing its own income, but may not pass these deductions on to its beneficiaries. In the case of partnerships, however, depletion allowance deductions are taken at the partner level and not in computing partnership income.

Depletion allowance deductions must be earned by incurring specific kinds of expenditures during specific periods, as detailed by tax regulations. The expenditures are automatically deductible in computing taxable income under various other provisions of the income tax act, so that depletion allowance is a form of additional or incentive deduction. In general terms, the depletion allowance deduction that may be claimed by a taxpayer in computing income is 25% of the taxpayer's resource profits for the year to a maximum of the taxpayer's earned depletion base as at the end of the year. **Resource profits** consist of the taxpayer's income from resource-related activities. A taxpayer's **earned depletion base** is determined as one-third of the sum of those specified resource-related expenses incurred by the taxpayer.

At various times over the years, the depletion allowance deduction has been modified, usually on a temporary basis, to enhance the economic attractiveness of certain types of activities. Examples of such enhancements include the following:

- For certain expenses incurred after April 1983, regulations provide for mining exploration depletion. Under these provisions, the expenses of exploring for mineral resources in Canada earn mining exploration depletion base at the usual rate of $1.00 of base for each $3.00 of expenses incurred, but the taxpayer's deduction each year is not limited by the amount of resource profits.

- Expenses incurred between April 1, 1977 and March 31, 1980 to explore in Canada's frontier regions entitle the taxpayer to a frontier exploration allowance instead of the usual depletion allowance.

- Expenses incurred between April 10, 1978 and December 31, 1980 in connection with certain non-conventional oil projects entitle the taxpayer to a supplementary depletion allowance.

The Canadian tax regulations governing depletion are very specific and complex. For this reason it is best that you obtain professional advice when faced with a major personal or corporate investment decision concerning depletion.

■ 7.7 Additions or Improvements to Depreciable Assets

If any major repairs (e.g., engine overhauls) or improvements are made during the life of the asset, we need to determine whether these actions will extend the life of the asset or will increase the salvage value that was originally estimated. When either of these situations arises, a revised estimate of useful life should be made and the periodic depreciation expense should be updated accordingly. We will examine how the additions or improvements affect both book and tax depreciations.

7.7.1 Revision of Book Depreciation

Recall that book depreciation rates are based on estimates of the useful life of assets. These estimates of useful life are seldom precise. Therefore, after a few years of using the asset, you may find that it could last for a considerably longer or shorter period than was originally estimated. If this happens, the annual depreciation expense based on the estimated useful life may be either excessive or inadequate. (If the repairs or improvements do not extend the life or increase the salvage value of the asset, these costs may be treated as maintenance expenses during that year.) The basic procedure for correcting the book depreciation schedule is to revise the current book value and allocate this cost over the remaining years of useful life.

7.7.2 Revision of Tax Depreciation

For tax depreciation, additions or improvements you make to any property are treated as *separate* property items. The recovery period for an addition or improvement to the initial property normally begins on the date the addition or improvement is placed in service. The recovery class of the addition or improvement is the recovery class that would apply to the underlying property if it were placed in service at the same time as the addition or improvement. Example 7.13 illustrates the procedure for correcting the capital cost allowance schedule for an asset with additions or improvements made during the depreciable life.

Example 7.13

In January 1990, Kendall Manufacturing Company purchased a new numerical control machine at a cost of $60,000. It had an expected life of 10 years at the time of purchase and a zero expected salvage value at the end of the 10 years. For book depreciation purposes, it had no major overhauls planned for that period and was being depreciated using the straight-line method toward a zero salvage value, or $6000 per year. For tax purposes, this machine was a Class 8 item with a CCA of 20%. In January 1993, however, the machine was thoroughly overhauled and rebuilt at a cost of $15,000. It was estimated that the overhaul would extend the machine's useful life by 5 years.

(a) Calculate the book depreciation for 1995 on a straight-line basis.
(b) Calculate the capital cost allowance for 1995 for this machine.

Solution

Given: $P = \$60,000$, $S = \$0$, $N = 10$ years, machine overhaul = $15,000, extended life = 15 years from the original purchase

Find: D_6 for book depreciation, CCA_6 for tax depreciation

(a) Since an improvement is made at the beginning of 1993, the book value of the asset at that time consists of the original book value plus the cost added to the asset. First, we calculate the original book value at the end of 1992:

$$B_3 \text{ (before improvement)} = \$60,000 - 3(\$6000) = \$42,000.$$

After adding the improvement cost of $15,000, the revised book value is

$$B_3 \text{ (after improvement)} = \$42,000 + \$15,000 = \$57,000.$$

To calculate the book depreciation in year 1995, which is 3 years after the improvement, we need to calculate the annual straight-line depreciation amount with the extended useful life. The remaining useful life before the improvement was 7 years. Therefore, the revised remaining useful life should be 12 years. The revised annual depreciation is then $57,000/12 = $4750. With straight-line depreciation, we compute the depreciation amount for 1995:

$$D_6 \text{ (book depreciation)} = \$4750.$$

(b) For tax depreciation, the improvement is viewed as a separated property with the same recovery rate as the initial asset. Thus, we need to calculate both the tax depreciation under the original asset and that of the new asset.

The original asset would depreciate as follows:

Year	20% CCA	Book Value
0		$60,000
1990	$6000[6]	54,000
1991	10800	43,200
1992	8640	34,560
1993	6912	27,648
1994	5530	22,118
1995	4424	17,694

and the overhaul would depreciate as follows:

Year	20% CCA	Book Value
0		$15,000
1993	$1500[6]	13,500
1994	2700	10,800
1995	2160	8,640

---

[6] In the first year, only 50% of the CCA is allowed due to the half-year rule.

Therefore, the total tax depreciation for 1995 (Year 6) is $CCA_6 = \$6584$ (\$4424 plus \$2160 for a total of \$6584).

Alternatively, we may use the half-year rule CCA_n formula (Eq. 7.15, 7.16) directly. In 1995, we would be in the sixth year for the original purchase price and the third year for the improvement costs.

For the original purchase cost, therefore,

$$CCA_{1995} = CCA_6 = \$60,000(20\%) (1 - 20\%/2)(1 - 20\%)^4 = \$4424$$

and for the improvement cost,

$$CCA_{1995} = CCA_3 = \$15,000(20\%) (1 - 20\%/2)(1 - 20\%) = \$2160.$$

The total CCA for 1995 is the sum of the both of the CCA values:

$$CCA_{\text{Total}} = \$4424 + \$2160 = \$6584.$$

 ## 7.8 Computer Notes

Most depreciation calculations can be easily performed on computers. Lotus 1-2-3 has several built-in functions for depreciation calculations. **EzCash** can easily generate depreciation schedules for book as well as tax depreciation. In this section, we will briefly examine these features.

7.8.1 Generating Depreciation Schedules by Lotus 1-2-3

Table 7.3 summarizes the built-in depreciation functions for Lotus 1-2-3, Excel, and QuattroPro.

In Fig. 7.9, the worksheet columns B, C, and D illustrate how a typical book depreciation schedule can be generated by using these financial functions.

In Lotus 1-2-3, there are no financial functions that are equivalent to calculating CCA tax depreciations. Therefore, you need to provide your own programming to consider such features as the half-year convention and terminal loss or recapture. Using Example 7.7, columns E and F in Figure 7.9 show what the annual tax depreciation amounts would be for declining balance depreciation including the half-year rule and straight-line depreciation without the half-year rule.

Table 7.3 Built-in Depreciation Functions

Function Description	Lotus 1-2-3	Excel	QuattroPro
Straight-line depreciation	@SLN(cost, salvage, life)	SLN(cost, salvage, life)	@SLN(cost, salvage, life)
Double declining balance depreciation	@DDB(cost, salvage, life, period)	DDB(cost, salvage, life, period, factor)	@DDB(cost, salvage, life, period)
Sum-of-years'-digits depreciation	@SYD(cost, salvage, life, period)	SYD(cost, salvage, life, period)	@SYD(cost, salvage, life, period)
Declining balance depreciation (other than DDB; assumes no switching)		VDB(cost, salvage) like, life start, end period, factor, no switch)	

Figure 7.9 ■ Depreciation worksheet by Lotus 1-2-3

$D\$2*(\$D\$5/100) * ((1-\$D\$5/200) * (1-\$D\$5/100)^{(A11-2)}$

$\$D\$2 * \$D\$5/200$

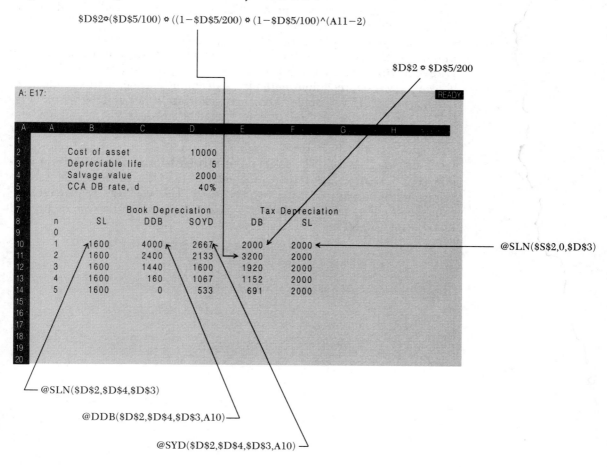

@SLN($S\$2,0,\$D\$3)

@SLN($D\$2,\$D\$4,\$D\$3)

@DDB($D\$2,\$D\$4,\$D\$3,A10)

@SYD($D\$2,\$D\$4,\$D\$3,A10)

7.8.2 Generating Depreciation Schedules with EzCash

EzCash generates both book and tax depreciation schedules from the **Special** mode. To calculate book depreciation, select the **Depreciation** command followed by **Depreciation Schedule** and **Book Depreciation**. You are prompted to enter three pieces of information concerning the asset: (1) the cost of the asset (cost base), (2) the salvage value, and (3) the depreciable life. In addition, you must select a depreciation method by scrolling the list of methods and clicking on your choice. Fig. 7.10(a) shows the window for inputting the required data. Clicking **Done** produces a depreciation schedule based on the selected depreciation method. Examples of double declining balance (200%) and straight-line schedules are shown in Fig. 7.10(b) and (c), respectively. Note that the declining balance calculation may revert to a straight-line depreciation calculation such that the book value at the end of the depreciable life equals the specified salvage value (recall Example 7.6). If you choose the **Declining balance** (**custom**) depreciation method, you must supply a fixed fraction value that represents the multiplier in Eq. (7.3). It is specified as a percentage so a value of 200 is equivalent to the double declining balance method. (Fig. 7.10(d))

To calculate tax depreciation schedules, select the **Depreciation** command under the **Special** mode followed by **Depreciation Schedule** and **Tax Depreciation**. By choosing tax depreciation, you are presented with the choice of two depreciation methods — declining balance and straight line. (Fig. 7.11(a)). If you select **Declining Balance**, a window opens (Fig. 7.11(b)) and you are required to enter the CCA cost base, the useful life, and the declining balance rate applicable to the asset. You must also choose a first year convention. Clicking **OK** produces the desired capital cost allowance schedule (Fig. 7.11(c)). The selection of **Straight-line** opens a different window (Fig. 7.11(d)) which requires the entry of the CCA cost base, the salvage value, and the useful life. After entering these data, click **OK** to generate the capital cost allowance schedule (Fig. 7.11(e)). These schedules can be printed by selecting **PRINT**.

Figure 7.10 ■ Book depreciation calculation using **EzCash**

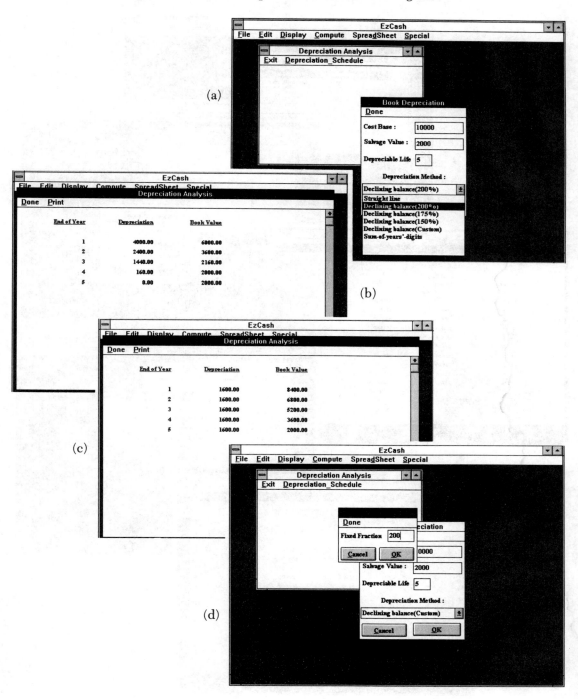

Figure 7.11 ■ Tax depreciation (CCA) calculation using **EzCash**

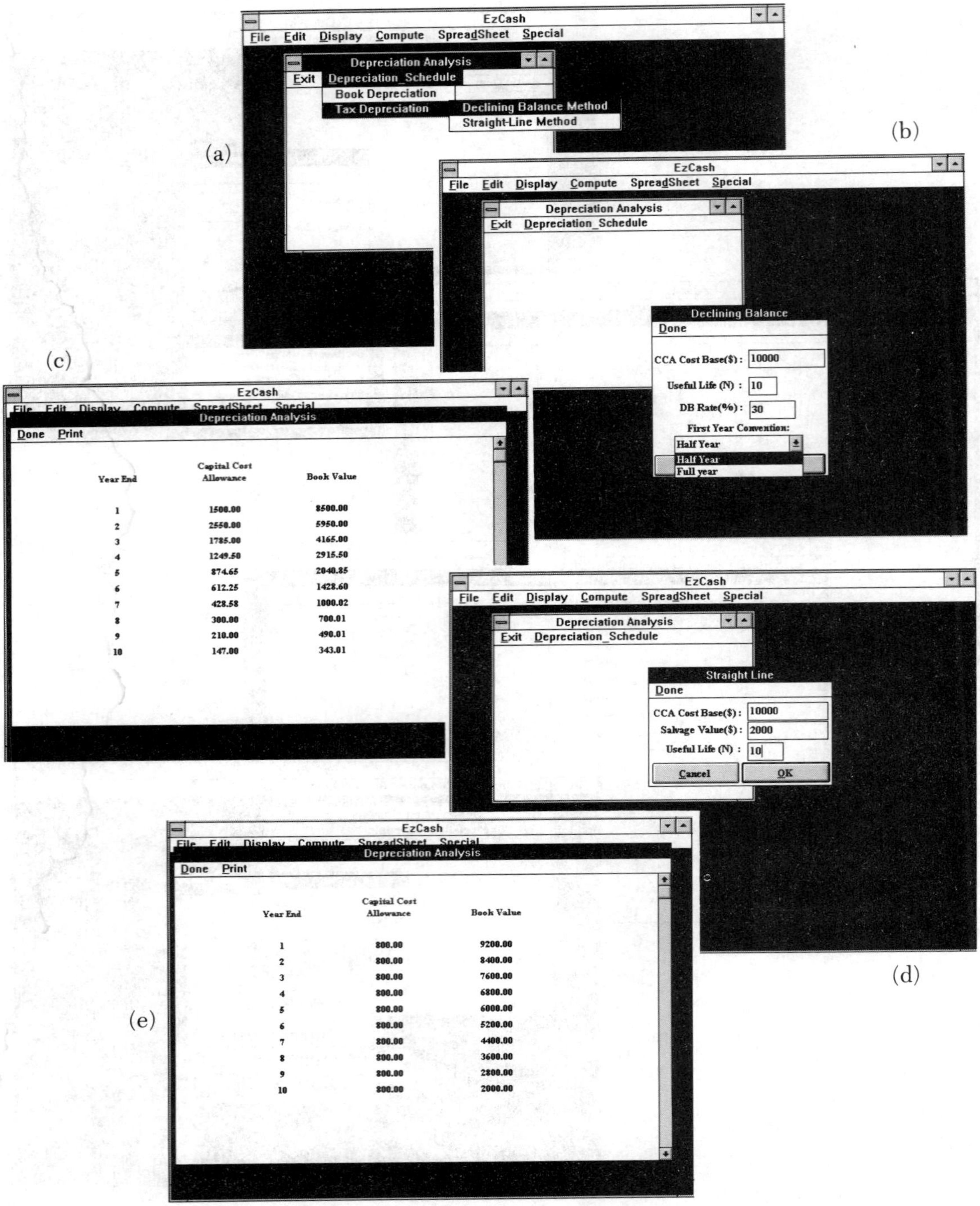

Summary

- Several different meanings and applications of depreciation have been presented in this chapter. From an engineering economics point of view, our primary concern is with **accounting depreciation:** the systematic allocation of an asset's value over its depreciable life.
- Accounting depreciation can be broken into two categories:
  1. **Book depreciation**—the method of depreciation used for financial reports;
  2. **Tax depreciation**—the method of depreciation used for calculating taxable income and income taxes; it is governed by tax legislation.
- The four components of information required to calculate depreciation are (1) the cost base of the asset, (2) the salvage value of the asset, (3) the depreciable life of the asset, and (4) the method of depreciation. Table 7.4 summarizes the differences in the way these components are treated for purposes of book and tax depreciation.
- **The capital cost allowance** gives taxpayers a break: It allows them to take earlier and faster advantage of the tax-deferring benefits of depreciation.
- Many firms select straight-line depreciation for book depreciation because of its relative ease of calculation.
- **Depletion** is a cost allocation method used particularly for natural resources. Canadian tax regulations are complex and usually require professional advice on the specific aspects of depletion allowance.
- Given the frequently changing nature of depreciation and Canadian tax law, we must use whatever percentages, depreciable lives, and salvage values are mandated *at the time an asset is acquired.*

Table 7.4 Summary of Book Versus Tax Depreciation

Component of Depreciation	Book Depreciation	Tax Depreciation (CCA)
Cost base	Based on the actual cost of the asset, plus all incidental costs such as freight, site preparation, installation, etc.	Same as for book depreciation
Salvage value	Estimated at the outset of depreciation analysis. If final book value does not equal estimated salvage value, we may need to make adjustments in our depreciation calculations	Estimated in order to calculate the terminal loss or recaptured depreciation
Depreciable life	Firms may select their own estimated useful lives	Defined by specific item under Canadian tax law
Method of depreciation	Firms may select from the following: • straight-line • accelerated methods (declining balance, double declining balance, and sum-of-years'-digits) • units-of-production	Tax legislation mandates specific depreciation rules based largely on declining balance and straight-line methods of depreciation.

Problems

(*Note:* For all of the questions involving CCA we will assume that the half-year rule applies.)

7.1 Omar Shipping Company bought a tugboat for $75,000 (year 0) and expected to use it for 5 years, after which it will be sold at $10,000. Suppose the company estimates the following revenues and expenses for the first operating year.

Operating revenue	$150,000
Operating expenses	$65,000
CCA (Tax depreciation)	$4,000

(a) If the company pays taxes at the rate of 30% on its taxable income, what is the net income during the first year?

(b) Assume for the moment (1) that all sales are for cash, (2) that all costs except capital cost allowance were paid during year 1. How much cash would have been generated from operations?

7.2 Gilbert Corporation has a gross income of $500,000 in tax year 1, and $150,000 in salaries, $150,000 in wages, $20,000 in interest, and $60,000 in capital cost allowance expenses for the asset purchased three years ago. Ajax Corporation has a gross income of $500,000 in tax year 1, and $150,000 in salaries, $110,000 in wages, and $20,000 in interest expenses. Assuming an income tax rate of 20% for both corporations, determine the net cash flows for each company.

7.3 A machine now in use which was purchased 3 years ago for $4000 has a book value of $1800. It can be sold for $2500, but could be used for 3 more years, at the end of which time it would have no salvage value. What is the amount of economic depreciation for this asset?

7.4 General Service Contractor Company paid $100,000 for a house and lot. The value of the land was appraised at $65,000 and the value of the house at $35,000. The house was then torn down at an additional cost of $5000 to build a food convenience store on the lot at a cost of $50,000. What is the total value of the property with the food store? For depreciation purposes, what is the depreciation cost base for the food store?

7.5 To automate one of their production processes, Saskatoon Corporation bought 3 flexible manufacturing cells at a price of $500,000 each. When they were delivered, Saskatoon paid freight charges of $25,000 and handling fees of $12,000. Site preparation for these cells costs $35,000. Six foremen, each earning $15 an hour, worked five 40-hour weeks setting up and testing the manufacturing cells. Special wiring and other materials applicable to the new manufacturing cells cost $1500. Determine the cost base (amount to be capitalized) for these cells.

7.6 A new drill press was purchased at $90,000 by trading in a similar machine that has a book value of $30,000. Assuming that the trade-in allowance is $25,000 and that $65,000 cash is paid for the new asset, what is the cost base for the new asset for depreciation purposes?

7.7 A lift truck priced at $35,000 is acquired by trading in a similar lift truck and paying cash for the remainder. Assuming that the trade-in allowance is $10,000 and the book value of the asset traded in is $6000, what is the cost base of the new asset for the computation of capital cost allowance for tax purposes?

7.8 Consider the following data on an asset:

Cost of the asset, P	$50,000
Useful life, N	5 years
Salvage value, S	$5,000

Compute the annual depreciation allowances and the resulting book values, using

(a) the straight-line depreciation method,
(b) double declining balance method, and
(c) sum-of-the-years'-digits method.

7.9 Consider the following data on an asset:

Cost of the asset, P	$10,000
Useful life, N	5 years
Salvage value, S	$0

Compute the annual depreciation allowances and the resulting book values using the DDB and switching to SL.

7.10 The double declining balance method is to be used for an asset with a cost of $50,000, estimated salvage value of $5000, and estimated useful life of 6 years.

(a) What is the depreciation for the first 3 fiscal years, assuming that the asset was placed in service at the beginning of the year?
(b) If switching to the straight-line method is allowed, when is the time to switch?

7.11 Compute the double declining balance (DDB) depreciation schedule for the following asset:

Cost of the asset, P	$20,000
Useful life, N	5 years
Salvage value, S	$12,000

7.12 Compute the SOYD depreciation schedule for the following asset:

Cost of the asset, P	$10,000
Useful life, N	5 years
Salvage value, S	$2,000

(a) What is the denominator of the depreciation fraction?
(b) What is the amount of depreciation for the first full year of use?
(c) What is the book value of the asset at the end of the fourth year?

7.13 Upjon Company purchased new equipment with an estimated useful life of 5 years. The cost of the equipment was $200,000, and the salvage value was estimated to be $30,000 at the end of year 5. Compute the annual depreciation expenses through the 5-year life of the equipment under each of the following methods of book depreciation:

(a) straight-line,
(b) double declining balance method (limit the depreciation expense in the fifth year to an amount that will cause the book value of the equipment at year end to equal the $30,000 estimated salvage value,)
(c) sum-of-the-years'-digits method.

7.14 If a truck for hauling coal has an estimated net cost of $80,000 and is expected to give service for 250,000 kilometres, resulting in a salvage value of $5000, depreciation would be charged at a rate of $0.30 per kilometre. Compute the allowed depreciation amount for the truck usage of 100,000 kilometres.

7.15 A bulldozer acquired at the beginning of the fiscal year at a cost of $45,000 has an estimated salvage value of $4000 and an

estimated useful life of 8 years.
Determine the following:

(a) the amount of annual depreciation by the straight-line method,
(b) the amount of depreciation for the third year computed by the double declining balance method,
(c) the amount of depreciation for the second year computed by the sum-of-the-years'-digits method.

7.16 A diesel-powered generator with a cost of $120,000 is expected to have a useful operating life of 50,000 hours. The expected salvage value of this generator would be $15,000. In its first operating year, the generator was operated 6000 hours. Determine the depreciation for the year.

7.17 Ingoll Moving Company owned four trucks dedicated primarily for its local moving business. The company's accounting record indicates the following:

Description	A	B	C	D
Truck Type				
Purchase cost ($)	50,000	25,000	18,500	35,600
Salvage value ($)	5,000	2,500	1,500	3,500
Useful life (kilometres)	200,000	120,000	100,000	200,000
Accumulated depreciation at beginning year ($)	0	1,500	8,925	24,075
Kilometres driven during the year	25,000	12,000	15,000	20,000

Determine the amount of depreciation for each truck during the year.

7.18 Zerex Furniture Company purchased a delivery truck on January 1, 1990, at a cost of $32,000. The truck has a useful life of 8 years with an estimated salvage value of $5000. The straight-line method is used for book purposes. For tax purposes, the truck would be depreciated as a CCA Class 10 asset with a declining balance rate of 30%. Determine the annual depreciation amount to be taken over the useful life of the delivery truck for both book and tax purposes.

7.19 The Harris Foundry Company purchased new casting equipment in 1990 at a cost of $180,000. Harris also paid $35,000 to have the equipment delivered and installed. The casting machine has an estimated useful life of 12 years, but it will be depreciated as a CCA Class 8 asset with a declining balance rate of 20%.

(a) What is the cost base of the casting equipment?
(b) What will be the CCA in each year in the life of the casting equipment?

7.20 A civil engineer wishes to compare the depreciation schedule of two different types of buildings. Both buildings will cost approximately $1,000,000. One building must be depreciated as a CCA Class 1 asset with a declining balance rate of 4%. The other building may be depreciated as a CCA Class 6 asset with a declining balance rate of 10%. Show the depreciation schedule for both buildings over the next 10 years. Which building has the preferred tax depreciation schedule and why?

7.21 A piece of machinery purchased at a cost of $40,000 has an estimated salvage value of $5000 and an estimated useful life of 5 years. It was placed in service on May 1 of the current fiscal year, which ends on December 31. The asset falls into CCA Class 39 with a declining balance rate of 30%. Determine the CCA amounts over the useful life.

7.22 Suppose that a taxpayer places in service a $10,000 asset that is assigned CCA Class 12 with a declining balance rate of 100%. Compute the depreciation schedule over the next 5 years.

7.23 Jim Smith purchased a farm for $350,000 and then constructed a corrugated metal building at a cost of $200,000 on the property. The building should have a useful life of about 10 years and can be depreciated since it is used as part of his farming business. Calculate the tax depreciation over the next 10 years.

7.24 Perkins Merchandise Company bought a building for $800,000 to be used as a warehouse. A number of major structural repairs completed at the beginning of the current year at a cost of $100,000 are expected to extend the life of the building 10 years beyond the original estimate. It has been depreciated by the straight-line method for 25 years. Salvage value is expected to be negligible and has been ignored. The book value of the building before the structural repairs is $400,000.

(a) What has the amount of annual book depreciation been in past years?
(b) What is the book value of the building after the repairs have been recorded?
(c) What is the amount of book depreciation for the current year, using the straight-line method? (Assume that the repairs were completed at the very beginning of the year.)

7.25 In 1994, you purchased a machine (CCA Class 9 property) for $20,000, which you placed in service in January, and a computer for $45,000 (CCA class 10 property), which you placed in service in October. You use the calendar year as your tax year. Compute the tax

depreciation allowances for each asset over the next 5 years.

7.26 In 1994, three assets were purchased and placed in service.

Asset Type	Date Placed in Service	Cost Base	CCA Property Class
Car	February 17	$15,000	Class 10
Furniture	March 25	$5,000	Class 8
Copy machine	April 3	$10,000	Class 12

Compute the CCA by year for each asset over the next 5 years.

7.27 On January 1st, 1994, three assets were purchased and placed in service.

Asset Type	Cost Base	CCA Property Class
Fishing vessel	$420,000	Class 7
Airplane	$625,000	Class 9
Licence	$50,000	Class 14 (10-year licence with no salvage value)

Compute the CCA by year for each asset over the next 5 years.

7.28 On October 1, you located your professional office in a residential home which you purchased for $120,000. The appraisal is divided into $20,000 for the land and $100,000 for the building (CCA Class 1 property).

(a) In your first year of ownership, how much CCA can you deduct for tax purposes?
(b) Suppose that the property was sold at $150,000 at the end of the 4th year of ownership. What is the book value of the property?

7.29 The Dow Ceramic Company purchased a glass molding machine in 1987 for $120,000. The company has been depreciating the machine over an estimated useful life of 10 years, assuming no salvage value, by the straight-line method of depreciation. For tax purposes, the machine has been depreciated as CCA Class 39 property with declining balance rate of 30%. At the beginning of 1990, Dow overhauled the machine at a cost of $22,000. As a result of the overhaul, Dow estimated that the useful life of the machine would extend 5 years beyond the original estimate.

    (a) Calculate the book depreciation for year 1992.
    (b) Calculate the tax depreciation for year 1992.

7.30 On January 2, 1989, Hines Food Processing Company purchased a machine priced at $75,000 that dispenses a premeasured amount of tomato juice into a can. The estimated useful life of the machine is 12 years, with a salvage value of $4500. At the time of purchase, Hines incurred the following additional expenses:

Freight-in	$ 800
Installation cost	$2500
Testing costs prior to regular operation	$1200

Book depreciation was calculated by the straight-line method, but for tax purposes the machine was classified as CCA Class 39 with a declining balance rate of 30%. In January 1991, accessories costing $5000 were added to the machine to reduce its operating costs. These accessories neither prolonged the machine's life nor provided any additional salvage value.

    (a) Calculate the book depreciation expense for 1992.
    (b) Calculate the tax depreciation expense for 1992.

7.31 On January 2, 1989, the Allen Flour Company purchased a new machine at a cost of $63,000. Installation costs for the machine were $2000. The machine was expected to have a useful life of 10 years with a salvage value of $4000. The company uses straight-line depreciation for financial reporting.

On January 3, 1991, the machine broke down and an extraordinary $6000 repair had to be made. The repair extended the machine's life to 13 years but left the salvage value unchanged. On January 2, 1992, a $3000 improvement to the machine increased its productivity. This improvement also increased the machine's salvage value to $6000, but did not affect its remaining useful life. Determine book depreciation expenses for the years ending December 31, 1989, 1990, 1991, and 1992.

7.32 On March 17, 1993, the Wildcat Oil Company began operations at its Alberta Oil Field. This field had been acquired several years earlier at a cost of $11.6 million. The field is estimated to contain 4 million barrels of oil, and to have an assessed land value of $2 million before and after its oil is pumped out.

Equipment costing $480,000 was purchased for use at the field. The equipment will have no economic usefulness once the Alberta Field is depleted; therefore, it is depreciated on the units-of-production method. Wildcat Oil also built a pipeline at a cost of $2,880,000 to serve the Alberta Field. Although this pipeline is physically capable of being used for many years, its economic usefulness is limited to the productive life of Alberta Field and thus has no salvage value. Depreciation of the pipeline, therefore, is also based on the estimated number of barrels of oil to be produced. Production at the Alberta Oil Field was 420,000 barrels in 1994 and 510,000 barrels in 1995.

(a) Compute the per-barrel book depletion rate of the oil field during years 1994 and 1995.
(b) Compute the per-barrel book depreciation rates of the equipment and the pipeline during years 1994 and 1995.

7.33 Flint Metal Shop purchased a stamping machine for $147,000 on March 1, 1992. It is expected to have a useful life of 10 years and a salvage value of $27,000, to produce 250,000 units, and to work for 30,000 hours. During 1992, Flint used the stamping machine for 2450 hours to produce 23,450 units. From the information given, compute the book depreciation expense for 1992 under each of the following methods:

(a) Straight-line
(b) Units-of-production
(c) Working hours
(d) Sum-of-the-years'-digits
(e) Double declining balance.

7.34 A company purchased a new forging machine to manufacture disks for airplane turbine engines. The new press costs $3,000,000, and it falls into a CCA Class 39 with a declining balance rate of 30%. The company has to pay property taxes for the ownership of this forging machine. The local municipality collects taxes at a rate of 1.5% on the book value at the beginning of each year.

(a) Determine the book value of the asset at the beginning of each tax year.
(b) Determine the amount of property taxes over the depreciable life.

7.35 Borland Company acquired a new lathe press at a cost of $65,000 at the beginning of the fiscal year. The equipment has an estimated life of 5 years and an estimated salvage value of $5000.

(a) Determine the annual depreciation (for financial reporting) for each of the 5 years of estimated useful life of the equipment, the accumulated depreciation at the end of each year, and the book value of the equipment at the end of each year by (1) the straight-line method, (2) the double declining balance method, and (3) the sum-of-the-years'-digits method.
(b) Determine the annual depreciation for tax purposes assuming that the equipment falls into the CCA Class 39 with declining balance rate of 30%.
(c) Assume that the equipment was depreciated under CCA Class 39 with declining balance rate of 30%. In the first month of the fourth year, the equipment was traded in for similar equipment priced at $82,000. The trade-in allowance on the old equipment was $10,000, and cash was paid for the balance. What is the cost base of the new equipment for computing the amount of depreciation for income tax purposes?

CHAPTER 8

Income Taxes

The total revenues, net income, and income taxes paid by several well-known Canadian companies during 1993 are summarized below (dollars in millions).

Company	Total Revenues	Net Income Before Taxes	Income Taxes
Canadian Imperial Bank of Commerce	$8978	$1184	$435
Imperial Oil	8903	568	289
Bell Canada	7957	1500	629
Bank of Nova Scotia	7176	1217	490
Bank of Montreal	7125	1202	487
Bombardier Inc.	4448	151	18
NOVA Corporation	3274	245	43
John Labatt	2135	194	64

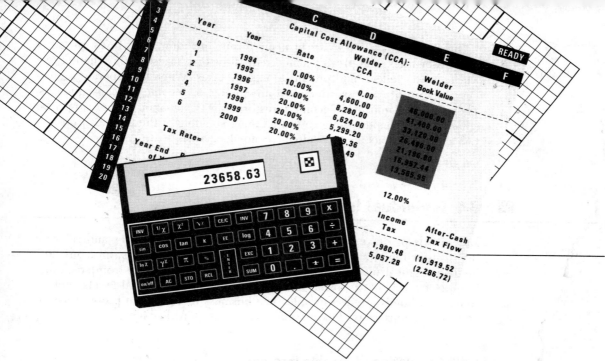

What do these companies have in common? All are among the largest of Canadian corporations, and all have a major "hidden partner" involved in every business activity. The hidden partner is Revenue Canada, which takes a share of their profits as taxes.

Although tax law is subject to frequent changes, the analytical procedures presented in this chapter provide a basis for tax analysis, and these can be adapted to reflect future changes in tax law. Thus, whereas we present many examples with 1993 tax rates, in a larger context, we present a general approach to the analysis of Canadian tax law.

Government taxation exists in many forms, including sales taxes, property taxes, user taxes, and provincial and federal income taxes. In this chapter, we will focus on federal and provincial income taxes. When you operate a business, any profits or losses incurred are subject to income tax consequences. Therefore, the impact of income taxes in project evaluation cannot simply be ignored. This chapter will give you an understanding of the Canadian tax system and of how income taxes affect economic analysis.

■ 8.1 Individual Income Tax

Individuals pay taxes on income from several sources: wages and salaries, investment income, profits from the sale of capital assets, and business income. Typical investment income may include dividends from stock and interest from bank deposits. Profits from the sale of capital assets such as stocks and bonds are also taxable. The business income from proprietorships and partnerships is taxed in the same manner as individual income.

8.1.1 Calculation of Taxable Income

In calculating taxable income, all taxpayers begin with **gross income,** which represents their earnings as determined in accordance with the provisions of the federal income tax laws. To compute income, we add

+ Wages and salary

+ Interest income

+ Dividend income

+ Rental/royalty income

+ Capital gains

+ Other income[1]

= Gross income

---

[1] Other income includes items such as consulting fees.

From gross income, Revenue Canada allows you to subtract the following amounts:

- Retirement plan contributions (registered pension plans or registered retirement savings plans)

- Annual union, professional, or like dues

- Child care expenses

- Attendant care expenses

- Business investment losses

- Moving expenses

- Alimony or maintenance paid

- Carrying charges and interest expense on money borrowed for investment purposes

- Other deductions

The result is the **net income before taxes**.

Several other deductions are also subtracted before calculating the **taxable income.** These other deductions include

- Employee home relocation loan

- Stock option and shares deduction

- Losses from previous years

- Capital gains deduction

- Northern resident's deduction

- Additional deductions

Once taxable income has been determined then federal income tax may be calculated.

8.1.2 Income Taxes on Ordinary Income

Canada has three rates for individual taxation. Rates are progressive—that is, the higher the income, the larger the percentage paid in taxes. (These rates are provided in Table 8.1.) These rates are likely to change over the years, but the progressive nature of the tax system should remain unchanged.

Table 8.1 Individual Federal Income Tax Structure for 1993

Taxable Income	Tax Rates
Less than $29,590	17%
$29,590 to $59,180	$5030 on first $29,590 plus 26% of taxable income over $29,590
More than $59,180	$12,724 on first $59,180 plus 29% of taxable income over $59,180

Once your federal income tax has been determined, you are allowed to subtract non-refundable tax credits. These credits are calculated by adding several items and then by multiplying their total by 17%. The items that are totalled include

- Basic personal amount
- Age amount
- Spousal amount
- Canada or Quebec pension plan contributions
- Unemployment insurance premiums
- Pension income amount
- Disability amount
- Disability amount transferred from dependent
- Tuition fees
- Education amount
- Tuition fees and education amount transferred from a child
- Amounts transferred from your spouse
- Medical expenses less adjustment
- Charitable donations[2]

This calculation results in the amount of your **basic federal tax**. You must then calculate your federal surtax and provincial taxes. Revenue Canada and the provincial governments frequently add individual surtaxes as a means of collecting additional revenue over the short term. However, some surtaxes last for several years and have become a longer term taxation method.

At the present time, the federal individual surtax is 3% of the **basic federal tax** payable and an additional 5% on any amounts above $12,500 of federal tax payable.

---

[2] Charitable donations over $250 earn non-refundable tax credits at 29% rather than 17%.

In addition to the federal tax payable, provincial tax on personal income must be considered. Provincial income tax is usually based upon a percentage of the **basic federal tax** payable. This basic federal tax does not include the federal individual surtax as described above. Table 8.2 provides a summary of the provincial taxes for 1993. Table 8.3 provides a summary of the individual territories and nonresident income tax rates for 1993.

Table 8.2 Individual Provincial Income Tax Rates for 1993

Province	Tax Rates
British Columbia	52.5% of basic federal tax + 20% surtax on British Columbia income tax above $5300 + 10% surtax on British Columbia income tax above $9000
Alberta	45.5% of basic federal tax + 8% surtax on Alberta income tax above $3500
Saskatchewan	50% of basic federal tax + 10% deficit reduction surtax on Saskatchewan income tax + 15% surtax on Saskatchewan income tax above $4000
Manitoba	52% of basic federal tax + 2% of net income minus various credits
Ontario	58% of basic federal tax + 17% surtax on Ontario income tax above $5500 + 8% surtax on Ontario income tax above $8000
Quebec	Quebec has a separate tax system which can be summarized as follows: 16% tax on income of less than $7000 19% tax on income above $7000 to $14,000 21% tax on income above $14,000 to $23,000 23% tax on income above $23,000 to $50,000 24% tax on income above $50,000 (Note: Quebec taxpayers receive a refundable abatement of 16.5% of the federal tax)
New Brunswick	62% of basic federal tax + 8% surtax on New Brunswick income tax above $13,500
Nova Scotia	59.5% of basic federal tax + 10% surtax on Nova Scotia income tax above $10,000
Prince Edward Island	59.5% of basic federal tax + 10% surtax on Prince Edward Island income tax above $12,500
Newfoundland	69% of basic federal tax

Table 8.3 Individual Territories and Nonresident Income Tax Rates for 1993

Category	Tax Rates
Yukon	48% of basic federal tax + 5% surtax on Yukon Territory income tax above $6000
North West Territories	45% of basic federal tax
Nonresidents and deemed residents	52% of basic federal tax (This is extra federal tax in lieu of provincial tax.)

Average Tax Rate

Despite the complexity of the rates, the tax system is basically progressive: the more you earn, the greater the percentage of income tax you pay. Average tax rate is the ratio of overall income taxes paid to taxable income. You can calculate the average tax rates from the data in Tables 8.1, 8.2, and 8.3.

Example 8.1 Average tax rate for personal income

If a person had a taxable income of $50,000 and lived in Ontario in 1993, estimate his or her personal income tax amount and average tax rate. The total amount qualified for non-refundable tax credits is $8000.

Solution

Given: Taxable income of $50,000
Find: Personal income tax amount and the average tax rate

Federal tax = $5030 + 26% × ($50,000 − $29,590)	$10,337
Minus non-refundable tax credits (17% of $8000)	$1,360
Basic federal tax payable	$8,977

and

Federal surtax (3% of $8977)	$ 269
Provincial tax in Ontario (58% × $8977)	$5,207
Total personal income taxes payable	$14,453

Therefore the average tax rate is ($14,453/$50,000)	=	28.9%

Marginal Tax Rate

Tables 8.1, 8.2, and 8.3 clearly indicate that as your personal income increases so does the tax rate being applied. The first portion of income earned is taxed at a lower rate than the higher increment of income. The rate that is applied to the last dollar earned is called the **marginal tax rate**. In engineering economics, the marginal tax rate is used to evaluate a project because most projects are incremental to base income.

Example 8.2 Marginal tax rate

Using the same data in Example 8.1, what would the person's marginal tax rate be?

Solution

Given: Basic data in Example 8.1
Find: Marginal tax rate

Marginal federal tax rate at $50,000 income	26.00%
Additional federal surtax of 3% (3% × 26%)	0.78%
Marginal Ontario provincial taxation rate (58% of basic federal rate of 26%)	15.08%
Total marginal tax rate	41.86%

This marginal tax rate indicates that for every additional dollar earned this person would pay 42 cents of income taxes.

Example 8.3 Highest marginal tax rate

What is the highest marginal tax rate in Ontario based on the data given in Tables 8.1 and 8.2?

Solution

Given: Basic data in Example 8.1
Find: The person's marginal tax rate

The highest marginal tax rate in Ontario may be calculated as follows:

Marginal federal tax rate above $59,180 income	29.00%
Additional federal surtax of 3% (3% × 29%)	0.87%
Additional federal surtax of 5% (5% × 29%)	1.45%
Marginal Ontario provincial taxation rate (58% of basic federal rate of 29%)	16.82%
Additional 17% Ontario provincial surtax	2.86%
Additional 8% Ontario provincial surtax	1.35%
Total marginal tax rate	52.35%

At this marginal tax rate, for every additional dollar earned, a person would pay 52 cents of income tax.

8.1.3 Income Taxes on Personal Capital Gains and Losses

Assets such as stocks and bonds are referred to as **financial assets,** whereas assets such as land and buildings are **fixed assets.** We commonly use the term **capital assets** to refer to everything you own and use for personal purposes, pleasure, or investment. For example, your house, furniture, car, and stocks and bonds are capital assets.

If you buy a capital asset and later sell it for more than your purchase price, the profit is defined as a **capital gain;** if you suffer a loss, it is called a **capital loss.** Thus,

$$\text{capital gain (loss)} = \text{selling price} - \text{cost base,}$$

where the selling price represents the proceeds from the sale minus any selling expense, and the cost base usually includes the purchase price plus any improvements and expenses incurred in acquiring the capital asset.

Prior to 1988, capital gains were taxable at one-half of their value. For 1988 and 1989, the taxable part of a capital gain was increased to two-thirds of its value. For 1990 and following years, the taxable part of a capital gain became three-quarters of its value. Capital gains income has been given preference under the tax laws in order to encourage capital investment. Prior to February 22, 1994, each person in Canada had a $100,000 life time capital gains exemption. This means that the first $100,000 of capital gain income is free of taxes. However this is no longer the case.

Capital gains are currently taxable at three-quarters of their actual value. Therefore the income tax payable on a capital gain is

$$\text{tax on capital gain} = \tfrac{3}{4} \times \text{capital gain} \times \text{marginal tax rate}$$

Capital losses are treated by subtracting their amounts from any capital gains during the tax year. Such losses can be used to reduce or offset capital gains from other sources to the extent that these gains exist. In a situation where capital gains CG_1, CG_2, and CG_3 are realized on three transactions and a capital loss, CL_1, is incurred on another, the total net capital gain on which tax is payable is

$$\text{net capital gain} = CG_1 + CG_2 + CG_3 - CL_1$$

(As long as the net capital gain is greater than or equal to zero.)

Therefore, a capital loss reduces the tax payable on the total capital gains and results in a savings of

$$\text{tax savings} = \tfrac{3}{4} \times \text{capital loss} \times \text{marginal tax rate}$$

which is a tax credit or cash inflow.

As a result, we can say that capital gains or losses are taxable at the capital gains tax rate, which is $\tfrac{3}{4}$ of an individual's marginal tax rate. For example, if an individual has a marginal tax rate of 41.86%, his or her capital gains tax rate is 75% × 41.86% or equal to 31.40%. The relationship between a person's marginal tax rate and his or her capital gains tax rate will change as Revenue Canada makes changes to the regulations.

Comment: *When capital losses arise, we will assume that there are always capital gains from other sources to offset the loss and result in the tax savings.*

Example 8.4 Individual capital gains income tax

Ken Smith had to calculate his expected capital gains tax payable for 1993. Unfortunately Ken had already used his $100,000 life time deduction. Ken decided to cash in his stocks in 1993 because he was concerned about rumored capital gains tax changes to be made in the February 1994 budget. Ken Smith sold 10,000 shares of Nova Scotia Power for $13 a share and sold 2000 shares of Royal Bank for $26 per share. He had purchased the Nova Scotia Power shares for $10 in 1992 and the Royal Bank shares for $28 each in 1993. Calculate the capital gain (loss) on these transactions and the tax payable if Mr. Smith was in a 50% marginal tax bracket.

Solution

Given: Shares sold during 1993 and basic data above

Find: Capital gain (loss) and income tax payable for 1993

We first determine the capital gain (loss) from the sale of stocks:

Nova Scotia Power stock purchase price (10,000 shares × $10 per share)	$100,000
Nova Scotia Power stock sold for (10,000 shares × $13 per share)	$130,000
Capital gains on Nova Scotia Power stock	$ 30,000
Royal Bank stock purchase price (2000 shares × $28 per share)	$ 56,000
Royal Bank stock sold for (2000 shares × $26 per share)	$ 52,000
Capital loss on Royal Bank stock	$ 4,000
The total capital gain for 1993	$ 26,000

Tax payable $= \$26{,}000 \times 3/4 \times 50\%$ marginal tax rate $= \$ 9{,}750$

8.1.4. Taxation of Dividends from Taxable Canadian Corporations

Dividends from taxable Canadian corporations are a source of income that are taxed in a slightly different manner. Under Canadian tax law, dividend income is given preferred taxation treatment through the following steps:

1. Dividends are first included as income at 125% of their actual dollar value.

2. After federal tax has been calculated, a federal dividend tax credit is subtracted from the federal tax payable. This tax credit is equal to 13.33% of the taxable amount of dividends reported as income.

Since this dividend tax credit reduces the basic federal tax payable, it has the effect of reducing the provincial income tax payable as well. The overall impact of this treatment is to allow dividend income to be taxed at a lower, preferred income tax rate. The federal government has provided this preferred taxation rate as a means to encourage investment in Canadian companies.

8.1.5. Timing of Personal Income Tax Payments

In order to include tax effects when evaluating personal investments, a person needs to know when taxes are payable. Depending on personal circumstances, tax payment occurs in various ways and at different times. An employer will automatically withhold a portion of an employee's salary or wage and remit that portion to Revenue Canada. However, if a person receives other income that is not taxed at the source like a salary or wage, he or she may be obliged to make monthly or quarterly payments of the estimated taxes on such income. Additionally, by April 30th of each year, an income tax return must be filed for the preceding calendar year. This return represents a formal calculation of the total taxes owed. Any difference between the taxes owed and tax payments already made is either returned as a refund or forwarded as the final payment.

The timing of the actual tax payments can be a significant complication in an economic analysis. For our purposes, we will assume that annual personal income taxes are paid on the last day of each calendar year unless indicated otherwise.

■ 8.2 Corporate Income Taxes

The corporate tax rate is applied to the taxable income of a corporation, which is defined as its gross income minus allowable deductions. As we briefly discussed in Chapter 7, the allowable deductions include the cost of goods sold, salaries and wages, rent, interest, advertising, tax depreciation, depletion, and various tax payments other than federal income tax. Taxable income is simply the difference between revenues and expenses, which includes tax depreciation or capital cost allowance (CCA).

Therefore,

$$\text{taxable income} = \text{gross income} - \text{expenses}$$

and

$$\text{corporate income tax} = \text{taxable income} \times \text{taxation rate}.$$

8.2.1 Types of Corporations

Under Canadian tax law corporations are classified as Canadian-controlled private, other private, public, corporation controlled by a public corporation, or other corporations. Examples of other corporations include credit unions, co-operatives, nonresident-owned investment corporations and Crown corporations. The type of corporation will determine whether or not the corporation is entitled to certain tax rates and deductions.

For example, Great North Inc. has been a Canadian-controlled private corporation for several years. If at some point during the taxation year, the controlling shareholder moved to the United States, the company would not be considered a Canadian-controlled private corporation. Because it is controlled by a nonresident it cannot claim the small business deduction for the entire year.

The corporation may be eligible for a **small business deduction** or **manufacturing and processing profits deduction** depending upon the nature of the business.

Small Business Deduction (SBD)

Under Canadian tax law corporations that are Canadian-controlled and private throughout the year may be able to claim the small business deduction (SBD). This deduction applies to the first $200,000 of annual taxable income earned in Canada and reduces the federal tax rate by 16%.

For a small company making less than $200,000 we can calculate the federal tax rate as follows:

Basic federal tax rate	38.00%
Minus abatement on income[3] earned in Canada	10.00%
Plus 3% additional surtax (3% of (38% − 10%))	0.84%
Minus SBD	16.00%
Total federal tax rate	= 12.84%

Manufacturing and Processing Profit Deduction (MPPD)

Corporations that derive at least 10% of their gross income from manufacturing or processing goods in Canada can claim the manufacturing and processing profits deduction (MPPD). Only income that is not eligible for the small business deduction is eligible for this deduction.

The federal taxation rate that applies to income from manufacturing and processing companies (not eligible for the SBD) can be calculated as follows:

Basic federal tax rate	38.00%
Minus abatement on income[3] earned in Canada	10.00%
Plus 3% additional surtax [3% of (38% −10%)]	0.84%
Minus MPPD	6.00%
Total federal tax rate	22.84%

---

[3] The federal tax abatement is equal to 10% of taxable income the corporation earned in the year in a Canadian province or territory. Income earned outside Canada is not eligible for the federal tax abatement.

8.2.2 Income Taxes on Operating Income

The corporate tax rate structure for 1993 is relatively simple. There are three rate brackets based on the type of business. There is a (1) general, (2) manufacturing/processing, and (3) small business taxation rate for federal corporation income tax. These rates are non-progressive, which means they do not increase with increased earnings.

Table 8.4 Federal Corporate Tax Structure for 1993

Type of Company	Tax Rate (including Surtax)
General	28.84%
Manufacturing/processing	22.84%
Small business	12.84%

Note: The manufacturing/processing deduction was changed to 7% after December 1993. Therefore, the federal rate for 1994 becomes 21.84%.

In addition to the federal corporate tax there is an additional provincial corporate tax. Table 8.5 indicates the provincial corporate taxation rates.

Table 8.5 Provincial Corporate Tax Structure for 1993

Province/ Territory	Tax Rate Standard	Tax Rate for Small Business
British Columbia	16%	10%
Alberta	15.5%	6%
Saskatchewan	15%	9.5%
Manitoba	17%	10%
Ontario	15.5%	9.5%
Quebec	16.25%	3.75%
New Brunswick	17%	9%
Nova Scotia	16%	5%
Prince Edward Island	15%	10%
Newfoundland	17%	10%
Northwest Territories	12%	5%
Yukon	10%	5%

Example 8.5 Corporate income taxes

A Canadian controlled and private mail-order computer company in Bedford, Nova Scotia sells personal computers and peripherals. The company leased showroom space and a warehouse for $20,000 a year and installed $100,000 worth of inventory checking and packaging equipment. The allowed capital cost allowance (CCA) for this capital expenditure ($100,000) within the first year will amount to $15,000. The store was completed and operations began on January 1st. The company had a gross income of $1,250,000 for the calendar year. Supplies and all operating expenses other than the lease expense were itemized as follows:

Cost of merchandise sold in the year	$500,000
Employee salaries and benefits	$250,000
Other supplies and expenses	$ 90,000
Total expenses	$840,000

Compute the taxable income for the company. How much will the company pay in federal and provincial income taxes for the year?

Solution

Given: Income, cost information, and CCA

Find: Taxable income, federal and provincial income taxes

First we compute the taxable income as follows:

Gross revenues	$1,250,000
Expenses	$ 840,000
Lease expense	$ 20,000
CCA	$ 15,000
Taxable income	$ 375,000

Now we must calculate the federal and provincial income tax.

Federal Income Tax

Taxable income	$375,000
The first $200,000 is taxable at the rate of 12.84% (Small business federal corporate tax rate)	
Federal tax $\quad\quad = \$200,000 \times 12.84\%$	$25,680
The remaining $175,000 is taxable at the rate of 28.84% (General federal corporate tax rate)	
Federal tax $\quad\quad = \$175,000 \times 28.84\%$	$50,470
Total federal tax payable	$76,150

Provincial Income Tax
(Based upon Nova Scotia tax rates)

Taxable income	$375,000
The first $200,000 is taxable at the rate of 5% (Small business provincial corporate tax rate)	
Provincial tax $\quad\quad = \$200,000 \times 5\%$	$10,000
The remaining $175,000 is taxable at the rate of 16% (Standard provincial corporate tax rate)	
Provincial tax $\quad\quad = \$175,000 \times 16\%$	$28,000
Total provincial tax payable	$38,000
Total corporate income tax payable	$114,150

8.2.3 Income Taxes on Corporate Capital Gains and Losses

Corporate capital gains are not taxed at the ordinary corporate tax rates. A company might realize capital gains from such sources as the sale of nondepreciable assets such as land or from the sale of financial assets such as stocks and bonds. Capital gains occur when an asset is sold for more than its cost base.

Corporate capital losses can be subtracted from any capital gains during the tax year in the same way as personal capital losses are netted against personal capital gains. Here, the assumption will be made that sufficient capital gains are always made to offset capital losses. For real situations where this is not the case, the net remaining losses may be carried back or over. In general, capital losses, when carried back or over, are deducted from capital gains in each year, but are not allowed to offset ordinary operating income.

Corporate capital gains are currently calculated at 75% of their actual value. They are similar to the personal capital gains previously described. Therefore, the corporate income tax payable on a capital gain is

$$\text{tax on capital gain} = \tfrac{3}{4} \times \text{capital gain} \times \text{corporate tax rate}$$

8.2.4 Consideration of Investment Tax Credit

In most recent years, the tax laws have permitted a reduction in the year's income taxes equal to a percentage of the cost of any capital assets acquired during the year. This is called the **investment tax credit.** The investment tax credit (ITC) has been used as a vehicle to encourage investments of certain types or in certain areas within Canada.

A corporation may earn investment tax credit by acquiring certain property or by making certain expenditures. For example, at the present time, Canadian-controlled private corporations can earn investment tax credits at the rate of 35% on the first $2 million of investment in scientific research and experimental development. Firms can also earn investment tax credit for investing in certain geographic regions of the country. This is done to encourage economic development within these regions. Currently corporations can earn investment tax credit on qualified property that it acquired for use in Newfoundland, Prince Edward Island, Nova Scotia, New Brunswick, Gaspe Peninsula, and prescribed offshore areas. In the past these investment tax credits have been as high as 60%.

The investment tax credit (ITC) is calculated as follows:

$$\text{investment tax credit} = \text{value of eligible property} \times \text{rate of ITC}$$

This tax savings is equivalent to reducing the net cost of the equipment by the amount of the investment tax credit. The investment tax credit is returned to the company as a tax credit at the end of the taxation year. The amount of the tax credit is subtracted from the value of the property prior to calculating the CCA.

8.2.5 Corporate Operating Losses

Ordinary corporate operating losses can be carried back to each of the preceding 3 years and forward for the following 7 years and can be used to offset taxable income in those years. The loss must be applied first to the earliest year, then to the next earliest year, and so forth. For example, an operating loss in 1994 can be used to reduce taxable income for 1991, 1992, or 1993, resulting in tax refunds or credits; any remaining losses can then be carried forward and used in 1995, 1996, and so forth, to the year 2001. If you are unable to deduct all of it as a noncapital loss within this allowed time frame, the unused part becomes a net capital loss, and you can carry it forward indefinitely to reduce taxable capital gains.

As in the disposal of capital assets, there are generally gains or losses on the sale (or exchange) of depreciable assets. To calculate a gain or loss, we first need to determine the book value of the depreciable asset at the time of disposal. When a depreciable asset used in business is sold for an amount different from its book value, this gain or loss has an effect on income taxes. The gain or loss is found as follows:

$$\text{gains (losses)} = \text{salvage value} - \text{book value}$$

where the salvage value represents the proceeds from the sale minus any selling expense or removal cost.

These gains, known as **recaptured depreciation** or **recaptured CCA**, are taxable. In the unlikely event that an asset is sold for an amount greater than its initial cost, the gains (salvage value − book value) are divided into two parts for tax purposes:

$$\begin{aligned}
\text{gains} &= \text{salvage value} - \text{book value} \\
&= (\text{salvage value} - \text{cost base}) \\
&\quad + (\text{cost base} - \text{book value}) \\
&= \text{capital gains} + \text{recaptured depreciation}
\end{aligned}$$

Recall from Section 7.3.2 that cost base means the purchase cost of an asset plus any incidental costs, such as freight and installation. As illustrated in Fig. 8.1, in this case

$$\text{capital gains} = \text{salvage value} - \text{cost base}$$

$$\text{recaptured depreciation} = \text{cost base} - \text{book value}$$

Figure 8.1 ■ Capital gains and recaptured depreciation

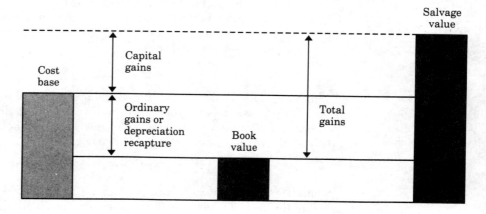

This distinction is necessary because capital gains are taxed at a capital gain tax rate and recaptured depreciation is taxed at the ordinary income tax rate. As described previously, current tax law allows a special lower rate of taxation for capital gains. When disposal of a depreciable asset results in a loss, the loss reduces taxes payable.

Section 8.3.1 describes an approximate representation of disposal tax effects for the declining balance and the straight-line depreciation methods. Section 8.3.2 provides a more rigorous analysis of disposal tax effects for the declining balance and straight-line depreciation methods. Unless otherwise stated, all examples and problems throughout this book will follow the materials covered in Section 8.3.1.

8.3.1 Calculation of Disposal Tax Effects

The manner in which gains or losses from an asset disposal affects taxes depends upon the depreciation method used, the assumed timing of the disposal, and whether or not the asset disposal depletes the asset class. As was discussed in Chapter 7, Section 7.5.5, the disposal of an asset does not normally deplete the asset class. In this case, the loss or recapture of CCA from the disposal is built into the undepreciated capital cost base for the entire asset class as a result of the asset pool accounting procedure. The tax effect attributable to the disposal will then be spread over a number of future years, which makes the calculation of the total disposal tax effect very tedious. Because of this difficulty and because disposal tax effects are normally not an important factor in determining the acceptability of a project, we assume that the tax implications of the disposal are completely realized in the year of disposal. Specifically, we assume that the disposal occurs just prior to the end of the last year of service and that the CCA in the year of asset disposal is calculated in the usual manner and without reference to the disposal. The disposal tax effect, G, then represents the tax implications over and above the tax savings realized in the year of disposal from CCA. The effects of these assumptions and variations to them will be analyzed in Section 8.3.2.

Because of the above assumptions, any gains on disposal are added to income, and any losses incurred are treated as expenses for assets depreciated by any depreciation method. This greatly simplifies the calculation of the disposal tax effect because G is now merely the extra tax payment or the tax savings realized in the year of disposal, i.e.,

$$G = t(B_{\text{Disposal}} - S),$$

where $B_{\text{Disposal}} = B_{N-1} - CCA_N$, B_{N-1} is the book value of the asset at the end of the year prior to disposal, and CCA_N is the capital cost allowance claimable in the year of disposal with the actual disposal ignored. When capital gains are involved, the total disposal tax effect G becomes

$$G = t(B_{\text{Disposal}} - P) - t_c (S - P),$$

where t_c is the capital gains tax rate and B_{Disposal} is the asset's book value at the end of the year in which the disposal occurs. The choice of $(B_{\text{Disposal}} - S)$ rather than $(S - B_{\text{Disposal}})$ is arbitrary, but more convenient because it provides the correct sign for the after-tax cash flow correction in the last year of the asset's service life.

Example 8.6   Disposal tax effect on depreciable assets

A company purchased a drill press costing $250,000. The drill press is classified as a CCA Class 43 property with a declining balance rate of 30%. If it is sold at the end of 3 years, compute the gains (losses) for the following four salvage values: (a) $150,000, (b) $104,125, (c) $90,000, and (d) $270,000. Assume that the company's combined federal and provincial tax rate is 40% and that only 3/4 of the capital gains are taxed (effectively taxed at 30%).

Solution

Given: a CCA Class 43 asset, cost base = $250,000, sold 3 years after purchase

Find: Disposal tax effects, and net salvage value from the sale if sold for $150,000, $104,125, $90,000, or $270,000

In this example, we first compute the book value of the machine at the end of year 3.

Year	Capital Cost Allowance (30% CCA rate)	Book Value
0		$250,000
1	$37,500	212,500
2	63,750	148,750
3	44,625	104,125

The CCA amount in year 1 is reduced by 50% because of the half-year rule. The book value at the end of year 3 is $104,125.

(a) Case 1: Book value < salvage value < cost base

$$\text{Disposal tax effects}\ =\ G\ =\ t(B_{\text{Disposal}} - S)$$

$$=\ 0.4 \times (\$104,125 - \$150,000)$$

$$=\ -\$18,350$$

$$\text{Net salvage value} = \text{salvage value} + \text{disposal tax effects}$$

$$= \$150,000 - \$18,350$$

$$= \$131,650$$

This situation (salvage value exceeds book value) is denoted as Case 1 in Fig. 8.2.

(b) Case 2: Salvage value = book value

In Case 2, the book value is again $104,125. Thus, if the drill press's salvage value equals $104,125, the book value, no taxes are calculated on that salvage value. Therefore, the net proceeds equals the salvage value.

(c) Case 3: Salvage value < book value

Case 3 illustrates a loss when the salvage value (say, $90,000) is less than the book value. We compute the net salvage value after tax as follows:

$$\text{Disposal tax effects} = G = t(B_{\text{Disposal}} - S)$$

$$= 0.4 \times (\$104,125 - \$90,000)$$

$$= \$5650$$

$$\text{Net salvage value} = \$90,000 + \$5650 = \$95,650$$

(d) Case 4: Salvage value > cost base

This situation is not likely for most depreciable assets (except for real property). Nevertheless, the tax treatment on this gain is as follows:

$$\text{Capital gains} = \text{salvage value (S)} - \text{cost base (P)}$$

$$= \$270,000 - \$250,000$$

$$= \$20,000$$

$$\text{Capital gains tax} = \$20,000 \times 3/4 \times 0.4$$

$$= \$6000$$

$$\text{Disposal tax effects} = \text{capital gains tax} + t(B_{\text{Disposal}} - P)$$

$$= -\$6000 + (0.4 \times (\$104{,}125 - \$250{,}000)$$

$$= -\$6000 - \$58{,}350$$

$$= -\$64{,}350$$

$$\text{Net salvage value} = \$270{,}000 - \$64{,}350$$

$$= \$205{,}650$$

Figure 8.2 ■ Calculations of gains or losses (Example 8.6)

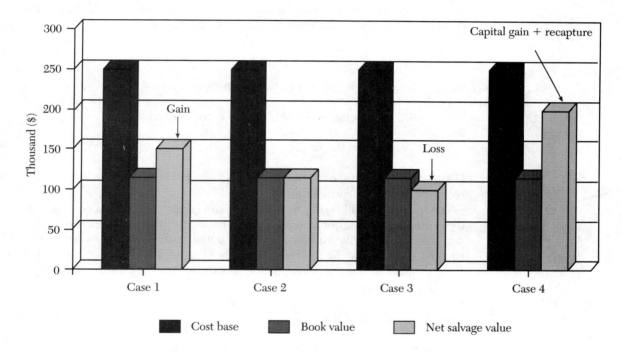

Comments: *Note that in (c) the disposal tax effects are positive, which represents a reduction in tax due to the loss and increases the net proceeds. This is realistic when the incremental tax rate (40% in this case) is positive, indicating the corporation is still paying tax, but less than if the asset had not been sold at a loss. The incremental tax rate will be discussed in Section 8.4.*

8.3.2 Advanced Topics on Disposal Tax Effects

Frequently, the disposal of an asset does not deplete the asset class. In this case, the treatment of disposal tax effects in Section 8.3.1 is only an approximation. The terminal loss or recapture of CCA due to the disposal of an asset affects future capital cost allowance claims for assets remaining in the class because of the asset pool accounting procedure. The actual tax effects over the future years after disposal are dependent on which depreciation method is used.

Removal of an asset from an asset class at any value other than its book value will change the undepreciated capital cost base. The amount of the change will be equal to the difference between the book value at disposal and the salvage value (or the cost base when capital gains are involved). We will define the change or offset in the undepreciated capital cost base as follows:

$$\Delta = (B_{\text{Disposal}} - S),$$

where, Δ represents the change in undepreciated capital cost,

B_{Disposal} represents the book value at disposal, and

S represents the salvage value.

(This choice rather than $(S - B_{\text{Disposal}})$ is convenient because it provides the correct sign for the tax correction.)

The quantity Δ will be depreciated over time within the capital cost base for the entire asset class, but actually belongs to an asset that no longer exists. The depreciation of Δ will produce a series of annual cash flows due to capital cost allowance which may be positive ($B_{\text{Disposal}} > S$) or negative ($B_{\text{Disposal}} < S$). When B_{Disposal} equals S, there is no residual effect of the asset within the capital cost base for the asset class.

You will recall from Chapter 7 that

Net income	= taxable income − income taxes
Income taxes	= tax rate × taxable income, and
Taxable income	= gross income (revenues)
	− cost of goods sold
	− operating costs
	− CCA

Therefore

$$\text{Net income} = (1 - \text{tax rate}) \times (\text{revenues} - \text{expenses})$$

$$+ \text{tax rate} \times \text{CCA}$$

$$= (1 - t)(R - E) + t \times \text{CCA}$$

and the tax rate times CCA in each year $(t \times CCA_n)$ represents a positive cash flow contribution to net income.

When an asset disposal results in a non-zero value of Δ, the CCA derived from Δ by the depreciation process will result in annual cash flows beyond the point of disposal equal to $t \times CCA_n$. These cash flows will be positive or negative depending upon the sign of Δ itself. These are the disposal tax effects which must be taken into account. The disposal tax effect which we need to quantify is simply the present value measured at the time of disposal, of all the future $t \times CCA_n$ cash flows.

In Section 8.3.1, we assumed that the disposal occurs just prior to the end of the asset's last year of service. An equally valid assumption has the disposal occurring right after the end of the asset's last year of service, i.e., at the beginning of the following year. In this section, we will analyze the disposal tax effects that occur with different disposal timing assumptions, asset addition assumptions, and depreciation methods.

8.3.2.1 Straight-Line Depreciation Method

Calculating disposal tax effects for assets depreciated on a straight-line basis is complicated by the slightly different rules applicable to the various straight-line classes. Since these types of assets arise infrequently in problems of engineering interest, we will not attempt to deal with all possible situations. Our discussion is restricted to classes that are exempt from the half-year rule and for which the depreciation process takes the asset to a zero book value over period of N_T years. N_T is also called the tax life of the asset. We will use N to indicate the service life of the asset in years. The annual CCA claimed for N years, or for N_T years, whichever is smaller, is

$$\text{CCA} = \frac{P}{N_T}, \tag{8.1}$$

where P represents the original cost of the asset. When a disposal occurs at a salvage value which differs from the book value at that point, the CCA amounts claimed in subsequent years to correct for this difference cannot exceed the annual amount claimed prior to disposal as given in Eq. (8.1). Therefore, the period of time, N^*, over which the correction is required becomes

$$N^{\ast} = \frac{\left| B_{\text{Disposal}} - S \right|}{P/N_T} = \frac{N_T \left| B_{\text{Disposal}} - S \right|}{P} \qquad (8.2)$$

This is likely to be a non-integer quantity, which can be represented as

$$N^{\ast} = N_I^{\ast} + N_{NI}^{\ast}, \; 0 \le N_{NI}^{\ast} < 1 \qquad (8.3)$$

where the subscripts I and NI refer to the integer and non-integer nature of these quantities. The maximum value of the adjustment in any year after disposal for disposal tax effects is $t \times \text{CCA}$.

If the disposal occurs immediately after the last year of service (i.e, at the beginning of year $N + 1$), the tax effects will occur over the $N_I^{\ast} + 1$ years as shown in Figure 8.3. Remember that the CCA for year N (the last year of the asset's service life) has been claimed. In this case, B_{Disposal} is the book value at the end of year N, B_N. The present value of the disposal tax effects, expressed at time of disposal or end of year N, is

$$G = \frac{t \, (B_N - S)}{|B_N - S|} \left[\sum_{n=1}^{N_I^{\ast}} \frac{P}{N_T} \Big(P/F, i, n \Big) + N_{NI} \, \frac{P}{N_T} \Big(P/F, i, N_I^{\ast} + 1 \Big) \right] \qquad (8.4)$$

Figure 8.3 ■ Straight-line disposal tax effects when disposal occurs at beginning of year $N + 1$ and $B_{\text{Disposal}} > S$.

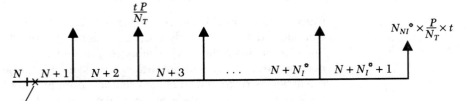

Point of disposal at beginning of year

If disposal occurs just prior to the end of the asset's last year of service (i.e., in year N), the CCA for the asset's last year of service has not been claimed. The book value at disposal, B_{Disposal}, is actually the book value of the asset at the end of year $N - 1$, B_{N-1}. The disposal tax effects will be reflected in year N and the $N_I^{\ast}$ years after the asset's service life. The series of tax effects due to the disposal in this case is shown in Figure 8.4. The present value of the disposal tax effects, expressed at time of disposal, is

$$G = \frac{t \, (B_{N-1} - S)}{|B_{N-1} - S|} \left[\sum_{n=0}^{N_I^{\ast}-1} \frac{P}{N_T} \Big(P/F, i, n \Big) + N_{NI}^{\ast} \, \frac{P}{N_T} \Big(P/F, i, N_I^{\ast} \Big) \right] \qquad (8.5)$$

Figure 8.4 ■ Straight-line disposal tax effects when disposal occurs at end of year N and $B_{\text{Disposal}} < S$.

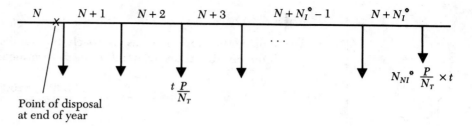

Example 8.7   Comparison of disposal tax assumptions for the straight-line depreciation method

(a) Calculate the disposal tax effects for straight-line depreciation with zero salvage value using the approximate method described in Section 8.3.1 for the following data:

$$
\begin{aligned}
P &= \$20{,}000 \\
N_T &= 8 \\
N &= 5 \\
S &= \$2000 \\
t &= 40\% \\
i &= 10\%
\end{aligned}
$$

(b) Compare these results to those obtained using the more rigorous methods developed in Section 8.3.2.

Solution

The annual CCA claim with straight-line depreciation is

$$
\text{CCA} = \frac{\$20{,}000}{8} = \$2500
$$

The book value at $N = 5$ is calculated as

$$
\begin{aligned}
B_5 = B_{\text{Disposal}} &= P - 5 \times \text{CCA} \\
&= \$7500
\end{aligned}
$$

(a) The approximate method for handling disposal tax effects uses the relationship

$$
G = t\,(B_{\text{Disposal}} - S)
$$

Then

$$G = 0.4 \, (\$7500 - \$2000) = \$2200$$

(b) When using the more rigorous approach, the number of years, N^*, over which the tax effects occur, needs to be determined:

$$N^* = \frac{|B_{\text{Disposal}} - S|}{\text{CCA}} = \frac{|\$7500 - \$2000|}{\$2500} = 2.2$$

Therefore

$$N_I^* = 2 \text{ and } N_{NI}^* = 0.2$$

If disposal occurs at the beginning of year 6, then according to Eq. (8.4):

$$G = 0.4 \times (+1) \, [\$2500 \, (P/F, 10\%, 1)$$

$$+ \$2500 \, (P/F, 10\%, 2) + 0.2 \times \$2500 \, (P/F, 10\%, 3)]$$

$$= \$1886$$

If the disposal occurs at the end of year 5, then $B_{\text{Disposal}} = B_4 = \$10,000$

and

$$N^* = \frac{|\$10,000 - \$2500|}{\$2500} = 3.2$$

or $\quad N_I^* = 3 \text{ and } N_{NI}^* = 0.2$

Using Eq. (8.5):

$$G = 0.4 \times (+1) \, [\$2500 + \$2500(P/F, 10\%, 1)$$

$$+ \$2500 \, (P/F, 10\%, 2) + 0.2 \times \$2500 \, (P/F, 10\%, 3)]$$

$$= \$2886$$

It should be noted that the G values cannot be compared directly because they include different effects. The approximate method used in (a) and the method based on the beginning of the year disposal assumption give G

values that do not include another tax effect at year 5, namely, the $t \times CCA_5$ tax saving. However, the G value calculated with the end of year assumption includes all tax effects at year 5.

The $t \times CCA_5$ tax saving is $0.4 \times \$2500$ or $\$1000$, which needs to be added to the first two G values before comparing them with the third G value.

	Total Tax Effect at Year 5
Part (a):	$\$2200 + \$1000 = \$3200$
Part (b) Beginning of year assumption:	$\$1886 + \$1000 = \$2886$
Part (b) End of year assumption:	$\$2886$

Comment: *The $314 difference between the rigorous and approximate methods is small. When discounted back to time zero at 10%, the difference is only $195 in a present value context. This demonstrates the adequacy of the simpler approach described in Section 8.3.1.*

8.3.2.2 Declining Balance Depreciation Method

As stated previously, the disposal of an asset frequently does not deplete the asset class. The loss or gain due to disposal of the asset affects the capital cost allowances of the assets remaining in the class to be claimed in future years—a result of the asset pool accounting procedure. We will still use N to indicate the service life of the asset under consideration.

When the declining balance method is used for CCA computations, we need to consider the following questions:

1. Are new assets being added to the asset class in the same year as asset disposal occurs from the class?

2. Does the disposal of the asset under consideration occur at the end of year N or the beginning of year $N + 1$?

If new assets are purchased and added to the asset class, and there is a disposal from this class in the same year, the half-year rule applies to the disposal, as well as to the new purchases. It is the difference between the total new purchases and the total disposals in a year to which the half-year rule applies (refer to Section 7.5.4). If the disposal occurs at the end of year N (the last year of the asset's service life), then in theory, the CCA for this year will not be claimed, and the first disposal tax effect will occur at the end of year N. If the disposal occurs at the beginning of year $N + 1$ (the year after the asset's service life of N years), the CCA for year N has been claimed, and the first tax effect will occur at the end of year $N + 1$. Based on the above sets of assumptions, we will discuss four cases of disposal tax effects calculation.

Case I

1. The disposal of an asset occurs at the beginning of year $N + 1$.

2. No new purchases of assets belonging to the same asset class occur in the year of disposal.

The book value to use in determining whether there is a gain or loss is the book value at the end of year N, since the CCA for year N has been claimed already. **Case I** is based on the same assumptions as those used by Edge and Irvine[4], in their derivation of capital tax factors.

When the original asset belongs to a declining balance asset class with a declining balance rate of d, the CCA values in future years resulting from the disposal are given by the usual depreciation relations:

$$CCA_n = d (1 - d)^{n-1} (B_{\text{Disposal}} - S) \tag{8.6}$$

where $n(\geq 1)$ represents the number of years from the year of disposal. The disposal tax effect (G) at the time of disposal is calculated as the present value of the $t \times CCA_n$ tax effects (see Figure 8.5):

$$G = td (B_{\text{Disposal}} - S) \sum_{n=1}^{\infty} (1 - d)^{n-1} (P/F, i, n) \tag{8.7}$$

It can be shown using the summation relationship for a geometric series that the summation term in Eq. 8.7 is equal to

$$\frac{1}{(i + d)}$$

Figure 8.5 ■ Timing of disposal tax effects for **Cases I** and **II**.

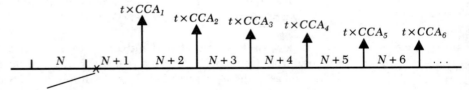

Disposal point at beginning of a year

---

[4] C. Geoffrey Edge and V. Bruce Irvine, *A Practical Approach to the Appraisal of Capital Expenditures*, Second Ed., The Society of Management Accountants of Canada, 1981.

Therefore,

$$G = \frac{td}{i+d}(B_{\text{Disposal}} - S)$$

$$= \frac{td}{i+d}(B_N - S), \tag{8.8}$$

for assets depreciated on a declining balance basis under **Case I**.

Case II

1. The disposal of an asset occurs at the beginning of year $N + 1$.

2. New purchases of assets belonging to the same asset class occur in the year of disposal.

In this case, we also have $B_{\text{Disposal}} = B_N$. As discussed in Section 7.5.4, whenever there are disposals and new purchases in the same year, the half-year rule is applied to the difference between the total purchase costs and the total salvage values from disposal. That is, the half-year rule has to be applied not only to new purchases but also to disposals. The CCA values arising from the disposal are as follows and reflect the half-year rule effect on the salvage value:

$$CCA_1 = dB_{\text{Disposal}} - \frac{d}{2}S,$$

$$CCA_n = (1-d)^{n-1}dB_{\text{Disposal}} - (1 - \frac{d}{2})(1-d)^{n-2}dS \text{ for } n \geq 2. \tag{8.9}$$

where n indicates the number of years from the year of disposal. The tax effects due to these CCAs are again represented in Figure 8.5. The present value of these tax effects, expressed at time of disposal (end of year N), is

$$G = \frac{td}{i+d}\left(B_N - S \times \frac{1+\frac{i}{2}}{1+i}\right), \tag{8.10}$$

since $B_{\text{Disposal}} = B_N$.

For $i < 25\%$, $\frac{1+\frac{i}{2}}{1+i}$ is greater than 0.9 and can be considered approximately equal to 1. Under this condition, **Case I** and **Case II** give essentially the same result. Any differences are a direct result of the half-year rule on the salvage value.

Case III

1. The disposal of an asset occurs just prior to the end of year N.

2. No new purchases of assets belonging to the same asset class occur in the year of disposal.

In this case the book value at disposal is equal to the book value at the end of year $N - 1$, i.e., $B_{\text{Disposal}} = B_{N-1}$. The CCA for year N will not be claimed specifically in this case as it forms part of the disposal tax effects realized in the year of disposal and in years that follow. The CCA values resulting from this disposal can be calculated in the same way as those for **Case I**. However, B_{Disposal} is equal to B_{N-1} rather than B_N, and these tax effects start one year earlier. These tax effects are shown in Figure 8.6. The present value of the tax effects expressed at the end of year N is

$$G = \frac{td(1 + i)}{i + d}(B_{N-1} - S) \qquad (8.11)$$

since $B_{\text{Disposal}} = B_{N-1}$.

Figure 8.6 ■ Timing of disposal tax efforts for **Cases III** and **IV**

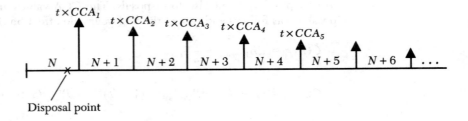

Disposal point

Case IV

1. The disposal of an asset occurs just prior to the end of year N.

2. New purchases of assets belonging to the same asset class occur in the year of disposal.

In this case, the book value at disposal is equal to the book value at the end of year $N - 1$ (i.e., $B_{\text{Disposal}} = B_{N-1}$). The CCA for year N will not be claimed in this case. The tax effects will be realized in the year of disposal and years that follow. The CCA values resulting from this disposal can be calculated in the same way as those for **Case II**. However, B_{Disposal} is equal to B_{N-1} rather than B_N in this case, and these tax effects start one year earlier as represented in Figure 8.6. The present value of the tax effects expressed at the end of year N is

$$G = \frac{td(1 + i)}{i + d}\left(B_{N-1} - S \times \frac{1 + \frac{i}{2}}{1 + i}\right), \qquad (8.12)$$

since $B_{\text{Disposal}} = B_{N-1}$.

Example 8.8   Comparison of disposal tax assumptions for the declining balance depreciation method

Repeat Example 8.6 for the four cases described in Section 8.3.2.2.

Compute the disposal tax effects for $i = 15\%$ and compare the results obtained for the various cases with those described in Example 8.6.

Solution

In this example we will require the book values at the end of year 2 and year 3 and the CCA claimed in year 3.

$$B_2 = \$148,750$$

$$CCA_3 = d\, B_2 = 0.3(\$148,750) = \$44,625$$

$$B_3 = B_2 - CCA_3 = \$104,125$$

The disposal tax effects, G, cannot be compared directly in that the G values for **Cases III** and **IV** represent the total tax effect measured at year 3, whereas G values for Example 8.6 and **Cases I** and **II** do not include the $t \times CCA_3$ tax effect which also occurs at the end of year 3.

A consistent comparison must be made on the basis of the total tax effect measured at year 3 and requires that the $t \times CCA_3$ term be added to the G values for Example 8.6 and for **Cases I** and **II**.

(a) Book value < salvage value < cost base

  Case I: $B_{\text{Disposal}} = B_3$

$$G = \frac{td}{i+d}(B_3 - S) = \frac{0.4 \times 0.3}{0.15 + 0.3} \times (\$104,125 - \$150,000)$$

$$= -\$12,233$$

  Case II: $B_{\text{Disposal}} = B_3$

$$G = \frac{td}{i+d}\left(B_3 - S \times \frac{1 + \frac{i}{2}}{1 + i}\right)$$

$$= \frac{0.4 \times 0.3}{0.15 + 0.3}\left(\$104,125 - \$150,000 \times \frac{1.075}{1.15}\right)$$

$$= -\$9,625$$

Case III: $B_{\text{Disposal}} = B_2$

$$G = \frac{td\,(1+i)}{i+d}\,(B_2 - S)$$

$$= \frac{0.4 \times 0.3 \times 1.15}{0.15 + 0.3} \times (\$148{,}750 - \$150{,}000)$$

$$= -\$383$$

Case IV: $B_{\text{Disposal}} = B_2$

$$G = \frac{td\,(1+i)}{i+d}\,\left(B_2 - S \times \frac{1 + \frac{i}{2}}{1 + i}\right)$$

$$= \frac{0.4 \times 0.3 \times 1.15}{0.15 + 0.3} \times \left(\$148{,}750 - \$150{,}000 \times \frac{1.075}{1.15}\right)$$

$$= \$2617$$

Comparing the disposal tax effects on the basis of total tax effects at year 3:

	G	+	$t \times CCA_3$	=	Total Tax Effect at Year 3
Example 8.6	$-\$18{,}350$		$\$17{,}850$		$-\$500$
Case I	$-\$12{,}233$		$\$17{,}850$		$\$5{,}617$
Case II	$-\$9{,}625$		$\$17{,}850$		$\$8{,}225$
Case III	$-\$383$		0		$-\$383$
Case IV	$\$2{,}617$		0		$\$2{,}617$

(b) Salvage value = book value at year 3

Following the same procedure as in (a), but with $S = \$104{,}125$, we find the following total tax effects at year 3.

	G	+	$t \times CCA_3$	=	Total Tax Effects at Year 3
Example 8.6	0		$17,850		$17,850
Case I	0		$17,850		$17,850
Case II	$1,810		$17,850		$19,660
Case III	$13,685		0		$13,685
Case IV	$15,768		0		$15,768

(c) Salvage value < book value

Following the same procedure as in part (a) but with $S = \$90,000$, we find the following total tax effects at year 3.

	G	+	$t \times CCA_3$	=	Total Tax Effects at Year 3
Example 8.6	$5,650		$17,850		$23,500
Case I	$3,767		$17,850		$21,617
Case II	$5,331		$17,850		$23,181
Case III	$18,016		0		$18,016
Case IV	$19,816		0		$19,816

(d) Salvage value > cost base

For most depreciable assets, this situation is not likely (except for real property). Nevertheless, the tax treatment on this gain is as follows:

$$\text{Capital gains} = \text{salvage value } (S) - \text{cost base } (P)$$

$$= \$270,000 - \$250,000$$

$$= \$20,000$$

In Example 8.6 and **Cases III** and **IV**, this capital gain is realized in year 3. For **Cases I** and **II**, the disposal at the beginning of year 4 will not result in a capital gains tax until the end of year 4. Therefore the capital gains tax measured at the end of year 3 is as follows:

$$\text{Capital gains tax} = \$20,000 \times 3/4 \times 0.4$$

$$= \$6000$$

(Example 8.6, **Cases III** and **IV**)

$$\text{Capital gains tax} = \$20,000 \times 3/4 \times 0.4 \ (P/F, 15\%, 1)$$

$$= \$5217$$

(**Cases I** and **II** measured at the end of year 3)

Substituting the cost base P for S and preceding as in part (a), we find the following total tax effects at year 3.

	Capital Gains Tax	+	G	+	$t \times CCA_3$	=	Total Tax Effects at Year 3
Example 8.6	−$6,000		−$50,350		$17,850		−$38,500
Case I	−$5,217		−$38,900		$17,850		−$26,267
Case II	−$5,217		−$34,552		$17,850		−$21,919
Case III	−$6,000		−$31,050		0		−$37,050
Case IV	−$6,000		−$26,050		0		−$32,050

Comment: *The total tax effects at the point of disposal are dependent on the specific assumptions. However, given the uncertainty in predicting a salvage value several years into the future the fact that tax effects tend to be small, and the discounting process used to move them to time zero, the additional complications demonstrated here are unwarranted in most instances. The simpler approach used in Example 8.6 is adequate for our purposes.*

■ 8.4 Income Tax Rate for Economic Analysis

As we have seen in the earlier sections, average income tax rates for corporations vary with their geographic location and the type of business. Many businesses in Canada receive the small business deduction on the first $200,000 of earned income. Many corporations also receive additional tax reductions such as the investment tax credits. Therefore the average tax paid will vary from 0% to 50% depending upon the situation. Suppose that a company now paying a tax rate of 30% on its current operating income is considering a profitable investment. What tax rate should be used in calculating the taxes on the investment's projected income?

8.4.1 Incremental Income Tax Rate

As we will explain, the choice of the rate will depend on the incremental effect on taxable income of undertaking the investment. In other words, the tax rate to use is the rate that applies to the additional taxable income projected in the economic analysis.

To illustrate, consider the ABC Corporation of Alberta, whose taxable income from operations is expected to be $200,000 for tax year 1993. The ABC management wishes to evaluate the incremental tax impact of undertaking a project during the tax year. The revenues, expenses, and taxable incomes before and after the project are estimated as follows:

	Before	After	Incremental
Gross revenue	$330,000	$400,000	$70,000
Salaries	100,000	110,000	10,000
Wages	30,000	40,000	10,000
Taxable income	$200,000	$250,000	$50,000

The base operations of ABC without the project are projected to yield a taxable income of $200,000. With the new project, the taxable income increases to $250,000. The small business corporate tax rate in Alberta is 18.84% (12.84% federal plus 6% provincial). The general business rate in Alberta is 44.34% (28.84% federal plus 15.5% provincial).

$$\text{Income tax without the project} = \$200,000 \times 18.84\%$$

$$= \$37,680$$

This is based upon the small business deduction for the first $200,000 of earned income.

$$\text{Income tax with the project} = \$200,000 \times 18.84\% + \$50,000 \times 44.34\%$$

$$= \$59,850$$

The $22,170 tax on the additional $50,000 in taxable income represents an incremental rate of 44.34%. This is the rate we should use in evaluating the project in isolation from the rest of ABC's operations.

The average tax rates before and after the new project being considered would be

	Before	**After**	**Incremental**
Taxable income	$200,000	$250,000	$50,000
Income taxes	$ 37,680	$ 59,850	$22,170
Average tax rate	18.84%	23.94%	
Incremental tax rate			44.34%

However, in conducting an economic analysis of an individual project, neither one of the company-wide average rates is appropriate—we want the incremental rate applicable to the particular project for use in generating its cash flows.

In Canada, since our corporate income tax system is not considered to be progressive a company can easily calculate its incremental taxation rate. The marginal and incremental taxation rates are identical for situations in which the project does not affect the existing tax rate calculation methodology. When a project forces a change in the method of calculating the tax rate, the marginal and incremental taxation rates will be affected.

8.4.2 Claiming Maximum Deductions

As we have seen, income taxes are calculated as a fraction of the taxable income. By definition, the fraction is the tax rate. The level of taxable income is dependent upon the gross income or revenues and the deductions that can be claimed against revenues. These deductions include all the expenses associated with delivering a good or service as well as debt interest and CCA. Revenue Canada does not permit the total deductions to exceed the revenues. In other words, the taxable income must be greater than or equal to zero.

The constraint of non-negative taxable income can limit the deductions actually used. For example, it may not be possible to claim all of the CCA in a particular year because there is insufficient income. In this case, the unused CCA is claimed in subsequent years.

Such a limitation adds considerable complexity to the calculation of taxes. However for our purposes, we will always assume that for any investment all available deductions can be claimed at their maximum level in any given year. This means that negative taxable income can arise, which then makes the taxes a positive cash flow. The underlying assumption is that there are other business operations which already pay taxes and there will be a tax savings in these other operations. These tax savings are a credit or cash inflow to the investment under consideration. This concept will be explained by way of example in Chapter 9.

8.4.3 Timing of Corporate Income Tax Payments

The taxation year for a corporation is its fiscal or business year. During its fiscal year, a corporation must make monthly payments of federal and provincial income tax based on an estimate of the taxes actually owed by the corporation. This estimate may be nothing more than a twelfth of taxes paid in the preceding fiscal year.

Although a corporation has up to 6 months after the end of its fiscal year to file federal and provincial tax returns, it must make a final tax payment within approximately 2 months of the end of its fiscal year. This payment represents the difference between total income tax payable and the amount paid in monthly installments. The actual deadline is dependent on various factors such as corporate size and type of industry, and these types of considerations can add considerable complexity to an engineering economic analysis. For our purposes, corporate investments will be assumed to occur at the beginning of a fiscal year, and income taxes will be paid as a lump sum at the end of each fiscal year unless indicated otherwise.

Summary

- Explicit consideration of taxes is a necessary aspect of any complete economic study of an investment project.
- For individuals the Canadian tax system has the following characteristics:
  1. Tax rates are progressive: The more you earn, the more you pay.
  2. Tax rates increase in stair-step fashion in three brackets
  3. Allowable exemptions and deductions may reduce the overall tax assessment.
- For corporations, the Canadian tax system has the following characteristics:
  1. Tax rates are not progressive. The rates are constant and do not increase with earnings.
  2. Tax rates depend upon the type and characteristics of the corporation.
  3. Allowable deductions and credits may reduce the overall tax assessment.
- Three distinct terms used to describe taxes were raised in this chapter: **Marginal tax rate** is the rate applied to the last dollar of income earned. **Average tax rate** is the ratio of income tax paid to net income. **Incremental tax rate** is the average rate applied to the incremental income generated by a new investment project.
- Capital gains are currently taxed at three-quarters of ordinary income. Capital losses are deducted from capital gains and net remaining losses may be carried backward and forward for consideration in other tax years.
- Disposal of depreciable assets—like disposal of capital assets—may result in gains (**recaptured depreciation**) or losses which must be considered in calculating taxes.
- Incremental tax rates should be used in evaluating investment projects for companies whose income falls below the top tax bracket or whose income varies significantly from year to year.

Problems

Assumptions for problems unless specified differently:

1. Use the 1993 tax rates provided in this chapter for corporate as well as personal taxes. Personal income tax brackets and the amount of personal exemption are updated yearly, so you need to consult the Revenue Canada tax guide for the tax rates as well as the exemptions that are applicable to your tax year.

2. Use the simplified approach when calculating disposal tax effect as described in Section 8.3.1.

3. The maximum CCA depreciation can be used in the year in which it occurs.

4. The company is profitable and can therefore take advantage of any tax credits generated from investment.

5. The company is not eligible for the manufacturing tax rate unless it specifies that it manufactures a product.

6. The half-year rule applies unless stated otherwise.

8.1 Mary Smith, a recent engineering graduate has just received her first job offer of $35,000 a year. She does not have any other income sources and feels her taxable income for 1993 will be $35,000. Mary estimated her non-refundable tax credits to be $1,360. Mary plans to work and live in Alberta. Calculate Mary's expected after tax income. What are Mary Smith's average and marginal taxation rates?

8.2 Jim Brown is a bank manager who is currently working in Nova Scotia. Jim has a taxable income of $50,000 per year. His company has promised him a big promotion if he moves to Ontario. Mr. Brown has been told he will receive an additional $15,000 and probably pay less tax. He expects that his $1,700 in non-refundable tax credit will not change if he moves to Ontario.

(a) Calculate Jim's personal income after taxes in Nova Scotia based upon his $50,000 per year taxable income. What is Jim's average and marginal tax rate?

(b) Calculate Jim's personal income after taxes if he were to move to Ontario and still make $50,000 per year taxable income. What is Jim's average and marginal tax rate?

(c) Calculate Jim's personal income after taxes if he were to move to Ontario and receive a $65,000 per year taxable income. What is Jim's average and marginal tax rate.

8.3 Gary Johnson is a civil engineer who is currently working for a company in southern Alberta. Gary is currently making a gross income of $70,000 per year. Gary felt that if he moved to Vancouver he may be able to increase his gross taxable income by $5,000. He was concerned about the cost of housing and the tax changes between properties. Gary estimates that his $1,400 in non-refundable tax credits will not change.

(a) Calculate Gary's personal income after taxes in Alberta based upon his $70,000 per year taxable income. What are Gary's average and marginal tax rates?

(b) Calculate Gary's personal income after taxes in British Columbia based upon his $75,000 per year taxable income. What are Gary's average and marginal tax rates?

(c) From a financial point of view should Gary move to British Columbia?

8.4 Roland McDonald wishes to evaluate several job proposals that he has received on an after-tax basis. He has received offers of $38,000 per year from four different companies, in four different provinces. Roland estimates his non-refundable tax credits at $1,400 per year.

Calculate Roland's after tax income (based upon his expected taxable income of $38,000) if he

(a) takes the job in Newfoundland
(b) takes the job in New Brunswick
(c) takes the job in Ontario
(d) takes the job in Alberta

8.5 Compare the tax payable between people working in Ontario and Newfoundland with the following incomes:

(a) $20,000
(b) 40,000
(c) 60,000
(d) 100,000

(For calculation purposes, assume the non-refundable tax credits are $1,500 per year in both provinces.)

8.6 Calculate the marginal tax rates of the example provided in question 8.5.

8.7 Dr.Callahan has been investing in the stock market for several years. She is currently in the highest marginal tax bracket in New Brunswick. Dr. Callahan has purchased the following stock and

wishes to calculate the tax payable if she chooses to sell her investments.

Assume that Dr. Callahan has purchased 2000 shares of each of the following stocks and there is a total of $2000 in brokerage fees for the purchase and sale of these stocks.

Stock	Purchase Price	Current Market Value
Canadian Imperial Bank of Commerce	$24.25	$34.50
Bank of Nova Scotia	28.00	32.00
GEAC	5.25	13.50
Nova Scotia Power	13.00	12.50

Calculate the tax payable if Dr.Callahan sells her stock.

8.8 Chuck Robbins wishes to evaluate his investment options. He realizes that Revenue Canada will share in any investment profits that he chooses. Chuck is currently at the top marginal rate in Saskatchewan. Calculate his after-tax rate of return if he proceeds with

(a) Savings account paying 1% per year.
(b) Guaranteed investment certificate for one year paying 5%.
(c) Industrial bond earning 8% in dividend payments with no change in market value
(d) Stock that pays no dividend and will increase in value by 8% over 1 year (assume no brokerage fees).

8.9 Bob Connors owns and operates an electrical services business. Bob is planning to expand his business over the next couple of years, but is concerned about higher taxes. Calculate his company's taxes over the next 3 years.

Assuming he is located within the province of Ontario.

	Year 1	Year 2	Year 3
Gross income	$320,000	$360,000	$420,000
Expenses:			
Salaries	$120,000	$132,000	$145,000
Utilities	5,000	6,000	6,000
Rent	15,000	16,000	16,000
Miscellaneous	5,000	6,000	6,500
Net income before tax	$175,000	$200,000	$246,500

8.10 In question 8.9 what are the corporate marginal tax rates for each year?

8.11 Tiger Construction Company has a gross income of $20,000,000 in tax year 1, and $3,000,000 in salaries, $4,000,000 in wages, $800,000 in tax depreciation expenses, loan principal payment of $200,000, and loan interest payment of $210,000. Determine the net income after tax in year 1. Tiger Construction is an Alberta-based company and qualifies for the small business deduction.

8.12 Consumers Electronics Limited (a Canadian-controlled private manufacturing company) was formed to sell a portable handset system that allows people with cellular car phones to receive calls up to 1000 feet from their vehicles. The company purchased a warehouse and converted it into a manufacturing plant for $2,000,000 (including the warehouse). It completed installation of assembly equipment worth $1,500,000 on December 31. The plant began operation on January 1. The company had a gross income of $2,500,000 for the calendar year. Manufacturing costs and all operating expenses, excluding the capital expenditures, were $1,280,000. The CCA for the capital expenditure made amounted to $128,000.
(a) Compute the taxable income for this company.

(b) How much will the company pay in federal income taxes for the year?

(c) How much will the company pay in provincial income tax for the year if the company is in New Brunswick?

(d) How much will the company pay in provincial income tax for the year if the company is in Alberta?

8.13 Consider an asset which is a CCA Class 8 property with a purchase price of $80,000. The applicable salvage values will be $40,000 in year 3, $30,000 in year 5, and $25,000 in year 7, respectively. Compute the disposal tax effect when the asset is disposed of (assume a marginal tax rate of 40%):

(a) in year 3,

(b) in year 5, and

(c) in year 7, respectively.

8.14 Consider an asset that is a CCA Class 16 property with a purchase price of $20,000. The applicable salvage values will be $15,000 in year 2, $8000 in year 4, and $1000 in year 6. Compute the disposal tax effect and net salvage value when the asset is disposed of (assume a marginal tax rate of 30%):

(a) in year 2,

(b) in year 4, and

(c) in year 6.

8.15 A bakery purchased new ovens that cost $72,000. The equipment is classified as a CCA Class 43 property with a declining balance rate of 30%. If this property is sold after 5 years, compute the disposal tax effect and net salvage value when the asset has an expected salvage of

(a) $10,000

(b) $40,000

(c) $80,000

(Assume a marginal tax rate of 30%)

8.16 Air Nova is considering the purchase of new aircraft costing $8,000,000. The equipment is classified as a CCA Class 9 property with a declining balance rate of 25%. If this property is sold after 6 years, compute the disposal tax effect and net salvage value for salvage values of

(a) $500,000

(b) $2,000,000

(c) $9,000,000

(Assume a marginal tax rate of 40%)

8.17 An electrical appliance company purchased an industrial robot costing $300,000 in year 0. The industrial robot to be used for welding operations is classified as a CCA Class 43 property. If the robot is to be sold after 5 years, compute the disposal tax effect for the following three salvage values (assume a marginal tax rate of 35%):

(a) $10,000

(b) $60,000

(c) $160,000

8.18 Laser Master, Inc., a laser-printing service company, in Bedford, Nova Scotia had a sales revenue of $1,250,000 during tax year 1993. The following represents the other financial information during the tax year.

Labor expenses	$550,000
Material costs	185,000
CCA	32,500
Interest income	6,250
Interest expenses	12,200
Rental expenses	45,000
Proceeds from the sale of old printers (already included in CCA above)	23,000

(The printers had a combined book value of $20,000 at the time of sale.) Consult the corporate tax rates for 1993 and determine

(a) The taxable income during the tax year.

(b) The disposal tax effect of selling the equipment during the tax year. (Original price was $34,000.)

(c) The amount of income taxes payable for the tax year.

8.19 Valdez Corporation (manufacturing company) will commence operations on January 1, 1994. The company projected the following financial performance during its first year of operation:

- Sales revenues for 1994 are estimated at $1,500,000.

- Labor, material, and overhead costs are projected at $600,000.

- The company will purchase a warehouse worth $500,000 in February. To finance this warehouse, the company will issue $500,000 of long-term bonds on January 1, which pay an interest at a rate of 10%. The first interest payment would occur on December 31.

- For tax depreciation purposes, the purchase cost of the warehouse is divided into $100,000 in land and $400,000 in buildings. The building will be classified as CCA Class 1 property, and will be depreciated accordingly.

- On January 5, the company purchases $200,000 of equipment, which is classified as a CCA Class 43 property.

(a) Determine the total CCA allowed in 1994.

(b) Determine Valdez's tax liability in 1994.

8.20 The Prince Edward Island Ecology Corporation expects to have a taxable income of $150,000 from its regular business in 1993. The company is considering a new venture involving the cleanup of oil spills from fishing boats on the Atlantic coast. This new venture is expected to generate an additional taxable income of $130,000.

(a) Determine the firm's marginal tax rates before and after the venture.

(b) Determine the firm's average tax rates before and after the venture.

8.21 Niagara Apple Corporation in Ontario estimates its taxable income for next year at $100,000. The company is considering expanding its product line by introducing apple-peach juice for next year. The market responses could be (1) good, (2) fair, or (3) poor. Depending on the market response, the expected additional taxable incomes are (1) $200,000 for good response, (2) $50,000 for fair response, and (3) $100,000 (loss) for poor response.

(a) Determine the marginal tax rate applicable to each situation.

(b) Determine the average tax rate that results in each situation.

8.22 A service corporation located in Alberta has the following financial information for a typical operating year:

Gross revenue	$550,000
Cost of goods sold	230,000
Operating costs	110,000

(a) Based on this financial information, determine the marginal tax rates for both federal and provincial.

(b) Determine the combined average tax rate for this corporation.

8.23 Simon Machine Tools Company of New Brunswick is considering the purchase of a new set of machine tools to process special orders. The following financial information is available.

Without the project: The company expects to have taxable incomes of $300,000 each year from its regular business over the next 3 years.

With the project: This 3-year project requires purchase of a new set of machine tools at a cost of $50,000. The equipment falls into the CCA Class 43. The tools will be sold at the end of project life for $30,000. The project will bring in an additional annual revenue of $80,000, but is expected to incur additional annual operating costs of $20,000 (not including CCA).

(a) What are the additional taxable incomes (due to undertaking the project) during years 1 through 3, respectively?

(b) What are the additional income taxes (due to undertaking the new orders) during years 1 through 3, respectively?

(c) What will be the disposal tax effect when the asset is disposed of at the end of year 3.

8.24 A machine now in use that was purchased 3 years ago at a cost of $4,000 has a book value of $1800. It can be sold for $2500, but could be used for 3 more years, at the end of which time it would have no salvage value. The annual operating and maintenance costs (O&M) amount to $10,000 for the old machine. A new machine can be purchased at an invoice price of $14,000 to replace the present equipment.

Freight-in will amount to $800 and the installation cost is $200. The new machine has an expected service life of 5 years, and will have no salvage value at the end of that time. With the new machine, the expected direct cash savings amount to $8000 the first year and $7000 in each of the next two years for O&M. Corporate income taxes are at an annual rate of 40%, and the net capital gain is taxed at 3/4 of the ordinary income tax rate. Both machines are depreciated using a CCA Class 43 property.

(*Note:* Each question should be considered independently.)

(a) If the old asset is to be sold now, what would be its economic depreciation amount?

(b) For tax depreciation purposes, what would be the first cost of the new machine (depreciation base)?

(c) If the old machine is to be sold now, what would be the disposal tax effect and net salvage value.

(d) If the old machine can be sold for $1000 now instead of $2500, then what would be the disposal tax effect and net salvage value.

8.25 Toronto Sewer Service Company expects to have taxable incomes of $300,000 from its regular residential sewer line installations over the next 2 years. The company is considering a new commercial account for a proposed shopping-mall complex during the coming year. This 2-year project requires the purchase of new digging equipment at a cost of $55,000. The equipment falls into the CCA Class 38 property with a declining balance rate of 30%. The equipment will be sold for $35,000 at the end of year 2. The project will bring in an additional annual revenue of $100,000, but it is expected to incur additional annual operating costs of $50,000.

(a) What is the marginal tax rate to use in evaluating the acquisition of the equipment during years 1 and 2, respectively?

(b) With the project, what is the average tax rate of the firm during project years 1 and 2, respectively.

(c) Calculate the disposal tax effect and net salvage value at the end of 2 years.

CHAPTER 9

Developing After-Tax Project Cash Flows

T he Canadian Telecommunications Corporation is a medium-sized manufacturer of electronic circuit boards, called "backpanels," that are sold to larger computer manufacturing firms. A critical part of their production process is the alignment of connecting leads, or pins. This alignment task is performed by two machines: an automatic pin insertion machine and a robotic secondary pinning process machine. The current production layout was designed to handle large batches of products as they moved through each of the centralized areas. Because each batch requires part-specific setups, the batches compete with one another for available work areas. After factory operation began, the orders received were not exactly as the corpora-

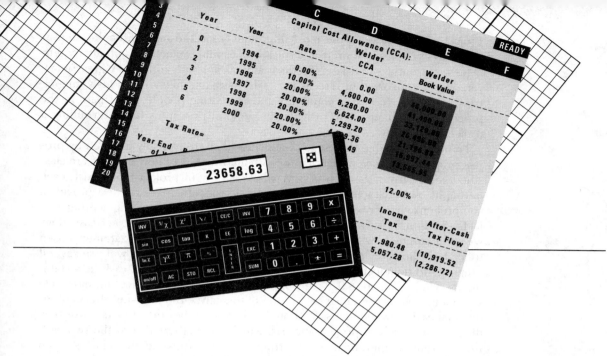

Year	Year	Rate	Capital Cost Allowance (CCA): Welder CCA	Welder Book Value
0				
1	1994	0.00%		
2	1995	10.00%	0.00	
3	1996	20.00%	4,600.00	
4	1997	20.00%	8,280.00	
5	1998	20.00%	6,624.00	
6	1999	20.00%	5,299.20	
	2000	20.00%		

Tax Rate=

Year End
of Y

```
23658.63
```

12.00%

Income
Tax

After-Cash
Tax Flow

1,980.48
5,057.28

(10,919.52
(2,286.72)

tion had anticipated. The orders were in more frequent, much smaller lot sizes. This caused scheduling difficulties, high levels of work in process, and only a limited ability to track a specific product through the process. To resolve the current problems, the corporation is considering converting from its present batch processing layout to an alternative design that consists of flexible manufacturing cells and the just-in-time manufacturing concept. The conversion from the current batch processing to flexible manufacturing would cost $12 million. However, the engineers are still uncertain whether there will be sufficient savings (cash inflows) to justify the initial investment.

Projecting cash flows is the most important—and the most difficult—step in the analysis of a capital project. Typically, a capital project will initially require investment outlays and later produce annual net cash inflows. A great many variables are involved in forecasting cash flows, and many individuals ranging from engineers to cost accountants and marketing people participate in the process. This chapter will provide the general principles for determining a project's cash flows.

To help us imagine the range of activities that are typically initiated by project proposals, we begin this chapter with an overview of how projects are classified within firms (Section 9.1). A whole variety of project types exist, each having its own characteristic set of economic concerns. We will next provide a comprehensive overview of the typical cash flow elements of engineering projects (Section 9.2). Once we have defined the elements of project cash flow, we will examine how to develop the comprehensive cash flow statements used to analyze the economic value of projects. In Section 9.3, we will use several examples to demonstrate the development of after-tax project cash flow statements. Then, in Section 9.4, we will present an alternative format for cash flow statements which can enhance our ability to analyze the after-tax effects of individual cash flow elements. Section 9.5 then describes the after-tax cash flow diagram approach to project analysis, which is an alternative to the two cash flow statement methods. By the time you have finished this chapter, you should be prepared not only to understand the format and significance of after-tax cash flow statements and after-tax cash flow diagrams, but also to develop them yourself.

■ 9.1 Project Proposals and Classifications

As we briefly mentioned in Chapter 1, in a company with capable and imaginative executives and employees and an effective incentive system, many ideas for capital investment will be advanced. Since some ideas will be good and others will not, procedures are usually established for screening projects. Firms generally classify projects as either **profit-adding** or **profit-maintaining** (see Fig. 9.1). Profit-adding projects include

- expansion projects
- product-improvement projects
- cost-improvement projects

Profit-maintaining projects include

- replacement projects
- necessity projects

In practice, projects may contain elements of more than one of these categories. Such projects are often classified according to their primary purpose.

Figure 9.1 ■ Classification of investment projects

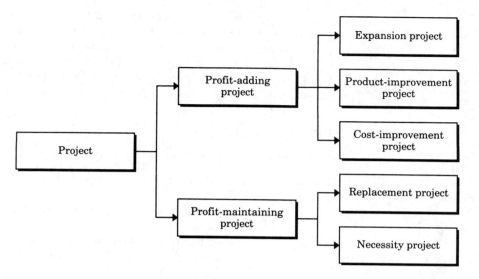

9.1.1 Profit-Adding Projects

The capital investment problems in this category all have this general form: It is proposed that a certain amount be invested now with the expectation that a return on the investment will be received in future years. The issue, then, is whether the amount of anticipated future cash flows is large enough to justify investing these funds in the project.

Expansion Projects

This project class includes expenditures intended to increase sales and profits by

1. Introducing new products. New products differ from existing products with respect to use, function, or size and are intended to increase sales by reaching new markets or customers, or by meeting end-use requirements not previously met. Sales of these products typically come in addition to existing sales. Decisions regarding new projects are ultimately based on whether the expected cash inflows from the sale of the new product are large enough to warrant the investment in equipment and working capital and the other costs required to introduce and make the product.
2. Providing facilities to meet current or forecasted sales opportunities for existing products. The issue in this case is whether to build or purchase a new facility. The expected future cash inflows are the additional incomes from the goods and services produced by the new facility.

Product-Improvement Projects

This class includes expenditures intended to improve the salability of existing products and to provide products that will displace existing products. The purpose of these expenditures is to maintain or improve the competitive position of existing products. These products differ from existing products only with respect to design, quality, color, or style and are not intended to reach new markets or customers, or to meet end-use requirements not previously met.

Cost-Improvement Projects

This class includes projects designed to

1. Reduce costs and expenses of existing operations, at current yearly production volume.
2. Avoid forecasted future cost increases that would be incurred at the current yearly production volume.
3. Avoid forecasted future cost increases that would be incurred if yearly production volume is increased.

For example, should we buy equipment to automate an operation now done manually? The expected future cash inflows on this investment would be the savings resulting from lower operating costs.

9.1.2 Profit-Maintaining Projects

Profit-maintaining projects are those whose primary purpose is not to reduce costs or increase sales but simply to maintain ongoing operations. Projects in this category are often proposed by a statement of both the reasons for the expenditure and the consequences of deferring the expenditure.

Replacement Projects

Projects in this class are those required to replace existing but obsolete or worn-out assets; failure to implement them results in a slowdown or shutdown of operations. For replacement projects, the replacement rather than the repair of existing equipment must be justified. Any incremental revenue resulting from replacement projects is considered an added benefit in the evaluation of the project. The future expected cash inflows from replacement projects are the cost savings resulting from lower operating costs, or the revenues from additional volume produced by the new equipment, or both.

Necessity Projects

Some investments are made out of necessity, rather than based on an analysis of their profitability. Such projects primarily yield intangible benefits because their economic advantages are not easily determined or are, perhaps, nonexistent. Typical examples include employee recreational facilities, child-care facilities, pollution-control equipment, and installation of safety devices. The latter two examples are projects for which capital expenditures must be made to comply with environmental control, safety, or other statutory requirements, possibly to avoid penalties. These investments use capital but provide no accountable cash inflows.

■ 9.2 The Incremental Cash Flows

When a company purchases a fixed asset such as equipment, it makes an investment. The company commits funds today in the expectation of earning a return on those funds in the future. Such an investment is similar to that made by a bank when it lends money. For the bank loan, the future cash flow consists of interest plus repayment of the principal. For the fixed asset, the future return is in the form of cash flows from the profitable use of the asset. In evaluating a capital investment, we are concerned only with those cash flows that result directly from the investment. These cash flows, called **differential** or **incremental cash flows,** represent the change in the firm's total cash flow that occurs as a direct result of the investment. In this section, we will look into some of the cash flow elements common to most investments.

9.2.1 Elements of Cash Flows

Many variables are involved in cash flow estimation, and many individuals and departments participate in the process. For example, the capital outlays associated with a new product are generally obtained from the engineering staffs, while operating costs are estimated by cost accountants and production engineers. On the cash inflow side, we may need to forecast unit sales and sales prices. These forecasts are normally made by the marketing department, giving consideration to pricing, the effects of advertising, the state of the economy, what competitors are doing, and market trends in consumer tastes.

We cannot overstate the importance and complexity of cash flow estimates. However, there are certain principles that, if observed, will help to minimize errors. In this section, we will examine some of the important cash flow elements that must be considered in project evaluation.

New Investment and Existing Assets

A typical project will usually involve a cash outflow in the form of an initial investment in equipment or other assets. The relevant investment costs are incremental costs such as the cost of the asset, any shipping and installation costs, and the cost of training in the use of the new asset.

If the purchase of a new asset results in the sale of an existing asset, the net proceeds from this sale reduce the amount of the incremental investment. In other words, the incremental investment represents the total amount of additional funds that must be committed to the investment project. When existing equipment is sold, the transaction results in either an accounting gain or loss, depending on whether the amount realized from the sale is greater or less than the equipment's book value. In any event, when existing assets are disposed of, the relevant amount by which the new investment is reduced consists of the proceeds of the sale, adjusted for tax effects.

Net Salvage Value

In many cases, the estimated salvage value of a proposed asset is so small and occurs so far in the future that it may have no significant effect on the decision. Furthermore, any salvage value that is realized may be offset by removal and dismantling costs. In situations where the estimated salvage value is significant, the net salvage value is viewed as a cash inflow at the time of disposal. The net salvage value of the existing asset is its selling price minus any costs incurred in selling, dismantling, and removing it, and an adjustment for disposal tax effects.

Investments in Working Capital

Some projects require investment in nondepreciable assets. If a project increases a firm's revenues, for example, more funds will be needed to support the higher level of operations. Investment in nondepreciable assets is often called **investment in working capital.** In accounting, working capital means the amount carried in cash, accounts receivable, and inventory that is available to meet day-to-day operating needs. For example, additional working capital may be needed to meet the greater volume of business that will be generated by a project. Part of this increase in current assets may be supplied from increased accounts payable, but the remainder must come from permanent capital. This additional working capital is as much a part of the initial investment as the equipment itself. (We will explain the amount of working capital required for a typical investment project in Section 9.3.2.)

Working Capital Release

As a project approaches termination, inventories are sold off and receivables are collected; that is, at the end of the project, these items can be liquidated at their cost. As this occurs, the company experiences an end-of-project cash flow that is about equal to the net working capital investment that was made when the project began. This recovery of working capital is not taxable income, since it merely represents a return of investment funds to the company.

Cash Revenues/Savings

A project normally will either increase revenues or reduce costs. Either way, the amount involved should be treated as a cash inflow for project analysis purposes. (A reduction in costs is equivalent to an increase in revenues.)

Manufacturing, Operating, and Maintenance Costs

The costs associated with manufacturing a new product need to be determined. Investments in fixed assets normally require periodic outlays for repairs and maintenance and for additional operating costs, all of which must be considered in the investment analysis.

Leasing Expenses

When a piece of equipment or a building is leased (instead of purchased) for business use, the leasing expenses become cash outflows. Many firms lease computers, automobiles, and industrial equipment subject to technological obsolescence. (We will discuss the leasing issue in Chapter 11.)

Interest and Repayment of Borrowed Amounts

When we borrow money to finance a project, we need to make interest payments as well as principal payments. Proceeds from both short-term borrowing (bank loans) and long-term borrowing (bonds) are treated as cash inflows, but repayments of debts are classified as cash outflows.

Income Taxes

Any income tax payments following profitable operations should be treated as cash outflows for project analysis purposes.

Tax Credit

As we learned in Chapter 8, when an investment is made in depreciable assets, Capital Cost Allowance (CCA) is an expense that offsets part of what would otherwise be additional taxable income. This is called a **tax shield,** or **tax credit,** and we must take it into account in calculating income taxes. If any investment tax credit is allowed, this tax credit will directly reduce the income taxes resulting in cash inflows.

In summary, the following types of cash flows, depicted in Fig. 9.2, are common in engineering investment projects.

- **Cash outflow**
  - Initial investment (including installation and freight costs)
  - Investment in net working capital
  - Repairs and maintenance
  - Incremental manufacturing and operating costs
  - Interest and loan repayments
  - Income taxes

Figure 9.2 ■ Types of cash flow elements in project analysis

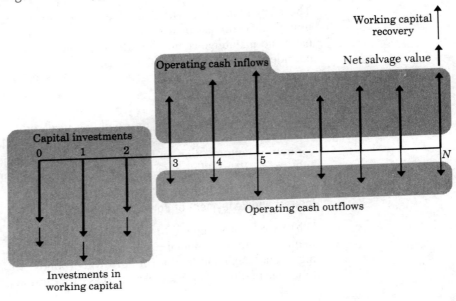

- **Cash inflow**

    - Incremental revenues

    - Reduction in costs

    - Disposal tax effect (for $B_N > S$, otherwise a cash outflow)

    - Salvage value

    - Release of net working capital

    - Proceeds from both short-term and long-term borrowing

Then, the net cash flow from the project is

$$\text{net cash flow} = \text{cash inflow} - \text{cash outflow}.$$

9.2.2 Classification of Cash Flow Elements

Once the cash flow elements are determined (both inflows or outflows), we may group them into three areas: (1) cash flow elements associated with operations, (2) cash flow elements associated with investment activities (capital expenditures), and (3) cash flow elements associated with project financing (such as borrowing). The main purpose of this grouping is to provide information about the operating, investing, and financing activities of a project.

Operating Activities

In general, cash flows from operations include current sales revenue, cost of goods sold, and operating expense. Cash flows from operations should generally reflect the cash effects of transactions entering into the determination of net income. The interest portion of a loan repayment is a deductible expense when determining net income, and it is included in the operating activities. Since we will usually look only at yearly flows, it is logical to express all cash flows on a yearly basis.

As we discussed in Chapter 7, we can determine the net cash flow from operations either (1) based on net income or (2) based on cash flow by computing income taxes directly. When we use net income as the starting point for cash flow determination, we should add any noncash expenses back to net income to compute the net cash flow. Thus,

$$\text{net operating cash flow} = \text{net income} + \text{capital cost allowance.}$$

Net Income Method	Cash Flow Method
Cash revenues (Savings)	Cash revenues (Savings)
− Cost of goods sold	− Cost of goods sold
− Capital cost allowance	
− Operating expense	− Operating expense
− Interest expense	− Interest expense
Taxable income	
− Income taxes	− Income taxes
Net income	Operating cash flow
+ Capital cost allowance	

Investing Activities

In general, four types of investment flows are associated with buying a piece of equipment: the original investment, salvage value at the end of its useful life, and working capital investment and recovery. We will assume that our outflow for both capital investment and working capital investment are as if they take place in year 0. It is quite possible that both investments will not occur instantaneously but, rather, over a few months as the project gets into gear; we could then use year 1 as an investment year. (Capital expenditures may occur over several years before a large investment project becomes fully operational. In this case, we should enter all expenditures as they occur.) For a small project, either method of timing these flows is satisfactory, because the numerical differences are likely to be insignificant.

$$- \text{ Capital expenditure}$$
$$+ \text{ Salvage value}$$
$$+ \text{ Disposal tax effect}$$
$$- \text{ Working capital investment}$$
$$\underline{+ \text{ Working capital recovery}}$$
$$\text{Net cash flow from investing activities}$$

Financing Activities

Cash flows classified as financing activities include the following:

$$+ \text{ Borrowing}$$
$$\underline{- \text{ Repayment of principal}}$$
$$\text{Net cash flow from financing activities}$$

Note that interest payments are usually classified as operating, not financing, activities. Then, net cash flow for a given year is simply the sum of the net cash flows from these three activities. Table 9.1 can be used as a checklist when you set up a cash flow statement by grouping each type of cash flow element into operating, investing, or financing activities.

Table 9.1 Classifying Cash Flow Elements into Types of Activities

Cash Flow Element	Direction	Type of Activity
Sales revenue	Inflow	Operating
Cost savings	Inflow	Operating
Manufacturing expenses	Outflow	Operating
O&M cost	Outflow	Operating
Interest payments	Outflow	Operating
Lease expenses	Outflow	Operating
Income taxes	Outflow	Operating
Capital investment	Outflow	Investing
Salvage value	Inflow	Investing
Working capital	Outflow	Investing
Working capital recovery	Inflow	Investing
Disposal tax effect	Either	Investing
Borrowed amounts	Inflow	Financing
Principal repayments	Outflow	Financing

9.3 The Income Statement Approach

In this section, we will illustrate through a series of numerical examples how we actually prepare a project's cash flow statement; a generic version is shown in Fig. 9.3. We will also consider a case in which a project generates a negative taxable income for an operating year.

9.3.1 When Projects Require Only Operating and Investing Activities

We will first illustrate the simplest case of generating after-tax cash flows for an investment project with only operating and investment activities. In the sections ahead, we will add complexities to the problem by including the working capital investment (Section 9.3.3) and borrowing activities (Section 9.3.4).

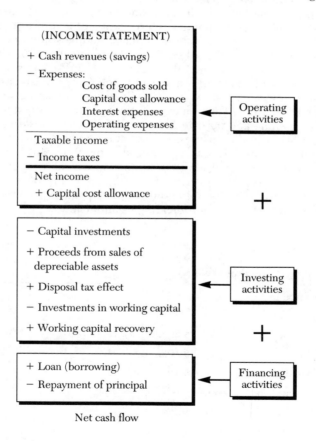

Figure 9.3 A popular format used for presenting project cash flows

Example 9.1 Cash flow statement—operating and investing activities

A computerized system has been proposed for a small tool manufacturing company. If the new system costing $125,000 is installed, it will generate annual revenues of $100,000 and require $20,000 in annual labor, $10,000 in annual material expenses, and another $10,000 in annual overhead (power and utility) expenses. The purchase will be classified as property with a declining balance CCA rate of 20%.

The company expects to phase out the facility at the end of 5 years, at which time it will be sold for $50,000. Find the year-by-year, after-tax net cash flow for the project at a 40% marginal tax rate. Find the after-tax net present worth of the project at the company's MARR of 15%.

Discussion: We can approach the problem using the format shown in Fig. 9.3 to generate an income statement and then a cash flow statement, and we will follow the form in our listing of givens and unknowns below. In year 0 (that is, at present) we have the investment cost of $125,000 for the equipment. This will be depreciated in years 1–5. The revenues and costs are uniform annual flows in years 1–5. We can see below that once we find capital cost allowances for each year, we can easily compute the results for years 1–4, which have fixed revenue and expense entries along with the variable CCA charges. In year 5 we will need to incorporate the salvage value and tax effects from the asset's disposal.

We will use the business convention that no signs (positive or negative) are used in preparing the income statement, except in the situation where we have a negative taxable income or tax savings. In this situation we will use ( ). However, in preparing the cash flow statement, we will observe explicitly the sign convention: a positive sign indicating cash inflow, a negative sign indicating cash outflow.

Income Statement (Year n)

Revenue	$100,000	
Expenses		
Labor	$ 20,000	In years 1–5
Material	$ 10,000	
Overhead	$ 10,000	
Capital cost allowance	CCA_n	
Taxable income	TI	
Income tax (40%)	T	
Net income	NI	

Cash Flow Statement (Year n)

Cash from operations	
Net income	NI
Capital cost allowance	CCA_n
Investment	$-P$ ← Equipment purchase in year 0
Salvage value	$+S$ ← Only in year 5
Disposal tax effect	$+G$ ← May come in year 5
Net cash flow	$\overline{NCF}$

Solution

Given: Cash flow information stated above

Find: After-tax cash flows

Before presenting the cash flow table, we need some preliminary calculations. The following notes explain the essential items in Table 9.2.

Table 9.2 After-Tax Cash Flow Analysis (Example 9.1)

				n		
	0	1	2	3	4	5
Income Statement						
Revenues		$100,000	$100,000	$100,000	$100,000	$100,000
Expenses						
Labor		20,000	20,000	20,000	20,000	20,000
Material		10,000	10,000	10,000	10,000	10,000
Overhead°		10,000	10,000	10,000	10,000	10,000
CCA		12,500	22,500	18,000	14,400	11,520
Taxable income		$ 47,500	$ 37,500	$ 42,000	$ 45,600	$ 48,480
Income tax (40%)		19,000	15,000	16,800	18,240	19,392
Net income		$ 28,500	$ 22,500	$ 25,200	$ 27,360	$ 29,088
Cash Flow Statement						
Cash from operation						
Net income		$ 28,500	$ 22,500	$ 25,200	$ 27,360	$ 29,088
CCA		12,500	22,500	18,000	14,400	11,520
Investment	-$125,000					50,000
Disposal tax effect						-1,568
Net cash flow	-$125,000	$ 41,000	$ 45,000	$ 43,200	$ 41,760	$ 89,040

° Overhead expenses include power, water, and other indirect costs.

- **Capital Cost Allowance Calculation:** We depreciate this property using a declining balance rate of 20%. In the first year we must apply the half-year rule. Therefore, the applicable depreciation amounts would be $12,500, $22,500, $18,000, $14,400, and $11,520. These values can easily be verified by using the following formula:

$$CCA_n = Pd(1 - d/2)(1 - d)^{n-2} \text{ for } n \geq 2$$

$$CCA_1 = Pd/2 \text{ for year } 1$$

We now have a value for our unknown CCA_n, which will enable us to complete the statement for years 1–4. The results of these simple calculations appear in Table 9.2.

- **Salvage Value and Taxes:** In year 5, we must deal with two aspects of the asset's disposal, salvage value and gains (both ordinary as well as capital). We list the estimated salvage value as a positive cash flow. Taxes are calculated as follows:

1. The total CCA in years 1–5 is $12,500 + $22,500 + $18,000 + $14,400 + $11,520 = $78,920.
2. The book value at the end of period 5 is the cost base minus the total CCA, or $125,000 − $78,920 = $46,080.
3. The loss or gain on the sale is the book value minus the salvage value, or $46,080 − $50,000 = −$3,920. (The salvage value is less than the cost base, so all of the $3,920 is recaptured CCA).
4. The tax effect on the recaptured CCA is

$$G = -\$3,920 \times (0.4)$$

$$= -\$1568.$$

This amount appears as the disposal tax effect in Table 9.2.

See Table 9.2 for the summary cash flow profile.[1]

- **Investment Analysis:** Once we obtain the project's after-tax net cash flows, we can determine their equivalent present worth at the firm's interest rate. The after-tax cash flow series from Table 9.2 is shown in Fig. 9.4. Since this series does not contain any patterns to simplify our calculations, we must find the net present worth of each payment. Using $i = 15\%$ we have

---

[1] Even though gains from equipment disposal affect income tax calculations, they are not viewed as ordinary operating income. Therefore, in preparing the income statement, the capital expenditures or related items such as disposal tax effect and salvage value are not included. Nevertheless, these items represent actual cash flows in the year they occur and must be shown in the cash flow statement.

Figure 9.4 ■ Cash flow diagram (Example 9.1)

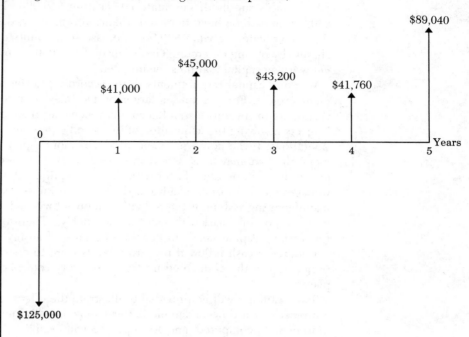

$$PW(15\%) = -\$125,000 + \$41,000(P/F, 15\%, 1) + \$45,000(P/F, 15\%, 2)$$
$$+ \$43,200(P/F, 15\%, 3) + \$41,760(P/F, 15\%, 4)$$
$$+ \$89,040(P/F, 15\%, 5)$$
$$= \$41,231.$$

This means that investing $125,000 in this automated facility would bring in enough revenue to recover the initial investment and the cost of funds with a surplus of $41,231.

9.3.2 When Projects Require Working Capital Investments

In many cases, changing a production process by replacing old equipment or adding a new product line will have an impact on cash balances, accounts receivable, inventory, and accounts payable. For example, if a company is going to market a new product, it will need inventories of the product and larger inventories of raw materials. There probably will be accounts receivable from the sales, and management might also decide to carry more cash because of the higher volume of activities. These investments in working capital are investments just like those in depreciable assets (except that they cannot be depreciated).

Consider the case of a firm planning a new product line, which will require a two-month's supply of raw materials costing $40,000. The firm could have $40,000 in cash on hand to pay for them. Alternatively, the firm could finance these raw materials via a $30,000 increase in accounts payable (60-day purchases) by buying on credit. The balance of $10,000 represents the amount of net working capital that must be invested.

Working capital requirements differ according to the nature of the investment project. For example, a larger project may require greater average investments in inventories and accounts receivable than would a smaller one. Projects involving the acquisition of improved equipment entail different considerations. If the new equipment produces more rapidly than the old equipment, the firm may be able to decrease its average inventory holdings because new orders can be filled faster with the new equipment. (One of the main advantages cited in installing advanced manufacturing systems, such as a flexible manufacturing system, is the reduction in inventory made possible by the ability to respond to market demand more quickly.) Therefore, it is also possible for working capital needs to decrease because of an investment. It would be considered a cash inflow if inventory levels were to decrease at the start of a project, since the cash freed up from inventory could be put to use in other places.

Two examples will be provided to illustrate the effects of working capital on a project's cash flows. Example 9.2 will show how the net working capital requirement is computed, and Example 9.3 will examine the effects of working capital on the computer control system project discussed in Example 9.1.

Example 9.2 Working capital requirements

Consider Example 9.1. Suppose that the tool manufacturing company's annual revenue projection of $100,000 is based on an annual volume of 10,000 units (or 833 units per month) with the following accounting information.

Price (revenue) per unit	$10
Unit variable manufacturing costs	
Labor	$ 2
Material	$ 1
Overhead	$ 1
Monthly volume	833 units
Finished goods inventory to maintain	2 months' supply
Raw materials inventory to maintain	1 month supply
Accounts payable	30 days
Accounts receivable	60 days

The accounts receivable period of 60 days means that revenues from this month's sales will be collected two months later. Similarly, accounts payable of 30 days indicates that payment for materials will be made approximately one month after the materials are received. Determine the working capital requirement for this investment.

Solution

Given: Information stated above

Find: Working capital requirement

Because there is a delay of 2 months, only 10 months of accounts receivable can be collected in year 1. That means that in year 1 the company will have cash inflows of $83,333, which is less than the projected sales of $100,000 ($8333 × 12). In years 2–5 collections will be $100,000, equal to sales, because beginning and ending accounts receivable will be $16,667 with sales of $100,000. The collections of the final accounts receivable of $16,667 would occur in the first two months of year 6, but can be added to the year 5 revenue to simplify the calculations. The important point is that cash inflow lags sales by $16,667 in the first year.

Assuming the company wishes to build up two months' inventory during the first year, it must produce 833 × 2 = 1666 more units than are sold the first year. The extra cost of these goods in the first year is 1666 units ($4 variable cost per unit) = $6664. The finished goods inventory of $6664 represents the variable cost incurred to produce 1666 more units than are sold in the first year. In years 2–4 the company will produce and sell 10,000 units per year while maintaining their 1666 units supply of finished goods. In the final year of operations, the company will produce only 8334 units (for 10 months) and will use up the finished goods inventory. As 1666 units of finished goods inventory get liquidated during the last year, there will be a working capital release in the amount of $6664. Along with the collections of the final accounts receivable of $16,667, there will be a total working capital release of $23,331 when the project terminates.

Now we can calculate the working capital requirements as follows:

Accounts receivable (833 units/month × 2 months × $10)	$16,667
Finished goods inventory (833 units/month × 2 months × $4)	6,664
Raw materials inventory (833 units/month × 1 month × $1)	833
Accounts payable (raw material purchase) (833 units/month × 1 month × $1)	(833)
Net working capital	$23,331

Figure 9.5 ■ Illustration of working capital requirement (Example 9.2)

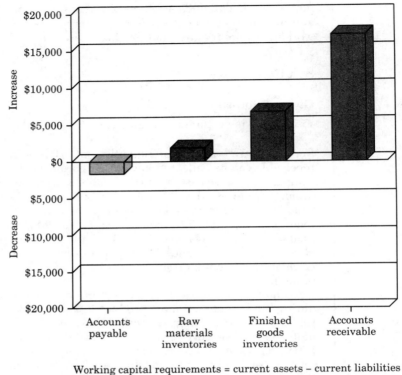

Working capital requirements = current assets – current liabilities

☐ Current liabilities   ■ Current assets

Figure 9.5 illustrates the working capital requirements in relation to finished goods inventory, raw material inventory, accounts receivable, and accounts payable.

Example 9.3   Cash flow statement—including working capital

Reconsider Example 9.1. Update the after-tax cash flows for the computer control system project by including the working capital requirement of $23,331 in year 0 and the full recovery of the working capital at the end of year 5.

Solution

Given: Flows as in Example 9.1, with the addition of the working capital requirement = $23,331

Find: Net after-tax cash flows with working capital, present worth

Using the procedure outlined above, the net after-tax cash flows for this computer control system project are grouped as shown in Table 9.3. As the table indicates, investments in working capital are cash outflows when they are expected to occur, and recoveries are treated as cash inflows at the times they are expected to materialize. In this example, we assume that the investment in working capital made at period 0 will be recovered at the end of the project's first operating cycle (say, year 1). However, the same amount of investment in working capital has to be made again at the beginning of year 2 for the second operating cycle, and the process repeats until the project terminates.

Table 9.3 After-Tax Cash Flow Analysis (Example 9.3)

				n		
	0	1	2	3	4	5
Income Statement						
Revenues		$100,000	$100,000	$100,000	$100,000	$100,000
Expenses						
Labor		20,000	20,000	20,000	20,000	20,000
Material		10,000	10,000	10,000	10,000	10,000
Overhead		10,000	10,000	10,000	10,000	10,000
CCA		12,500	22,500	18,000	14,400	11,520
Taxable income		$ 47,500	$ 37,500	$ 42,000	$ 45,600	$ 48,480
Income tax (40%)		19,000	15,000	16,800	18,240	19,392
Net income		$ 28,500	$ 22,500	$ 25,200	$ 27,360	$ 29,088
Cash Flow Statement						
Cash from operation						
Net income		$ 28,500	$ 22,500	$ 25,200	$ 27,360	$ 29,088
CCA		12,500	22,500	18,000	14,400	11,520
Investment	−$125,000					50,000
Working capital	− 23,331					23,331°
Disposal tax effect						−1,568
Net cash flow	−$148,331	$ 41,000	$ 45,000	$ 43,200	$ 41,760	$112,371

° Working capital recovery

Figure 9.6 illustrates the working capital investment cycle. Moreover, we also assumed a full recovery of the initial working capital. However, there are many situations in which the investment in working capital may not be fully recovered (for example, inventories may deteriorate in value

Figure 9.6 ■ Cash flow diagram (Example 9.3)

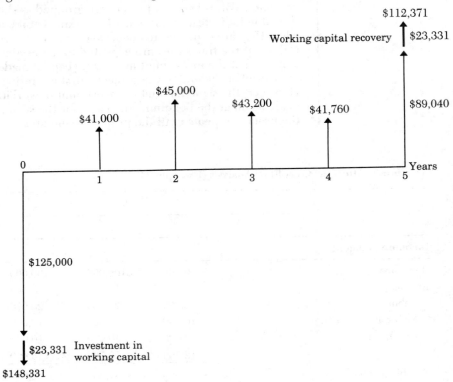

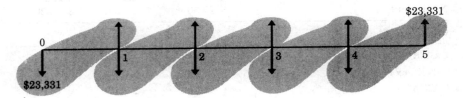

Working capital recovery cycles

or become obsolete). The equivalent net present worth of the after-tax cash flows including the effects of working capital is calculated as

$$PW(15\%) = -\$148,331 + \$41,000(P/F, 15\%, 1)$$
$$+ \$45,000(P/F, 15\%, 2) + 43,200(P/F, 15\%, 3)$$
$$+ \$41,760(P/F, 15\%, 4) + 112,371(P/F, 15\%, 5)$$
$$= \$29,500.$$

This present worth is $11,731 less than that in the situation with no working capital requirement (Example 9.1). This example demonstrates that working capital requirements must be considered when properly assessing a project's worth.

Comments: *The $11,731 reduction in present worth is just the present worth of an annual series of 15% interest payments on the working capital which is borrowed by the project at time 0 and repaid at the end of year 5.*

$$\$23,331(15\%)(P/A, 15\%, 5) = \$11,731.$$

The investment tied up in working capital results in lost earnings.

9.3.3 When Projects are Financed with Borrowed Funds

Many companies use a mixture of debt and equity to finance their physical plant and equipment. The ratio of total debt to total investment, generally called the **debt ratio,** represents the percentage of total initial investment provided by borrowed funds. For example, a debt ratio of 0.5 indicates that 50% of the initial investment is borrowed and the rest is provided from the company's earnings (equity). Since interest is a tax-deductible expense, companies in high tax brackets may incur lower after-tax financing costs by financing through debt. (Along with the effect on taxes, the method of loan repayment can also have a significant impact. We will discuss the issue of project financing in Chapter 11.)

Example 9.4 Cash flow statement—with financing (borrowing)

Rework Example 9.1, assuming that $62,500 of the $125,000 paid for the investment is obtained through debt financing (debt ratio = 0.5). The loan is to be repaid in equal annual installments at 10% interest over 5 years. The remaining $62,500 will be provided by equity (for example, from retained earnings).

Solution

Given: Same as in Example 9.1, but borrowing $62,500, equal repayment over 5 years at 10%
Find: Net after-tax cash flows in each year

We first need to compute the size of the annual loan repayment installments:

$$\$62,500(A/P,10\%,5) = \$16,487.$$

Next, we determine the amortization schedule of the loan by itemizing both the interest and the principal represented in each annual repayment. Using Eqs. (3.16) and (3.17), we obtain

Year	Beginning Balance	Interest Payment	Principal Payment	Ending Balance
1	$62,500	$6,250	$10,237	$52,263
2	52,263	5,226	11,261	41,002
3	41,002	4,100	12,387	28,615
4	28,615	2,861	13,626	14,989
5	14,989	1,499	14,988	0

Table 9.4 After-Tax Cash Flow Analysis with Borrowed Funds (Example 9.4)

				n		
	0	1	2	3	4	5
Income Statement						
Revenues		$100,000	$100,000	$100,000	$100,000	$100,000
Expenses						
Labor		20,000	20,000	20,000	20,000	20,000
Material		10,000	10,000	10,000	10,000	10,000
Overhead		10,000	10,000	10,000	10,000	10,000
CCA		12,500	22,500	18,000	14,400	11,520
Interest		6,250	5,226	4,100	2,861	1,499
Taxable income		$ 41,250	$ 32,274	$ 37,900	$ 42,739	$ 46,981
Income tax (40%)		16,500	12,910	15,160	17,096	18,792
Net income		$ 24,750	$ 19,364	$ 22,740	$ 25,643	$ 28,189
Cash Flow Statement						
Cash from operation						
Net income		$ 24,750	$ 19,364	$ 22,740	$ 25,643	$ 28,189
CCA		12,500	22,500	18,000	14,400	11,520
Investment	−$125,000					50,000
Disposal tax effect						− 1,568
Loan principal repayment[*]	$ 62,500	−10,237	−11,261	−12,387	−13,626	−14,988
Net cash flow	−$ 62,500	$ 27,013	$ 30,603	$ 28,353	$ 26,417	$ 73,153

[*] Note that the principal repayments are not tax-deductible expenses.

The resulting after-tax cash flow is detailed in Table 9.4. The present value equivalent of the after-tax cash flow series is

$$PW(15\%) = -\$62,500 + \$27,013(P/F, 15\%, 1) + \$30,603(P/F, 15\%, 2)$$

$$+ \$28,353(P/F, 15\%, 3) + \$26,417(P/F, 15\%, 4)$$

$$+ \$73,153(P/F, 15\%, 5)$$

$$= \$54,249.$$

When this amount is compared with that found in the case involving no borrowing (\$41,231), we see that debt financing actually increases present worth by \$13,018. This surprising result is largely due to the fact that the firm is able to borrow the funds at an after-tax rate of 6% (i.e., $10\%(1-t)$) which is less than its MARR (opportunity cost rate) of 15%. In general, to some extent firms can usually borrow money at lower rates than their MARR. If so, why don't they borrow all the funds they need for capital investment? We will address this question in Chapter 11.

9.3.4 When Projects Result in Negative Taxable Income

A negative taxable income does *not* mean that a firm pays no income tax. Rather, the negative figure can be used to reduce the taxable incomes from other business operations. Even if the firm does not have any other taxable income to offset in the current tax year, the operating loss can be carried back to each of the preceding 3 years and forward for the following 7 years to offset taxable income in those years. Therefore, a negative taxable income usually results in a **tax savings.** When we evaluate an investment project using an incremental tax rate, we assume that the firm has sufficient taxable income from other activities so that the changes due to the project in question will not change the incremental tax rate.

Example 9.5   Cash flow statement—negative incremental taxable income

Reconsider Example 9.1. Assume that the annual revenues are projected to be \$60,000 instead of \$100,000. Revise the after-tax cash flows with this revised revenue projection.

Solution

Given: Same figure as for Example 9.1, but revenues reduced to \$60,000 per year

Find: Taxable income, income tax, and net cash flow in each year

Table 9.5 After-Tax Cash Flow Analysis (Example 9.5)

				n		
	0	1	2	3	4	5
Income Statement						
Revenues		$60,000	$60,000	$60,000	$60,000	$60,000
Expenses						
Labor		20,000	20,000	20,000	20,000	20,000
Material		10,000	10,000	10,000	10,000	10,000
Overhead		10,000	10,000	10,000	10,000	10,000
CCA		12,500	22,500	18,000	14,400	11,520
Taxable income		$ 7,500	($ 2,500)°	$ 2,000	$ 5,600	$ 8,480
Income tax (40%)		3,000	(1,000)†	800	2,240	3,392
Net income		$ 4,500	($ 1,500)	$ 1,200	$ 3,360	$ 5,088
Cash Flow Statement						
Cash from operation						
Net income		$ 4,500	−$ 1,500	$ 1,200	$ 3,360	$ 5,088
CCA		12,500	22,500	18,000	14,400	11,520
Investment	−$125,000					50,000
Disposal tax effect						−1,568
Net cash flow	−$125,000	$ 17,000	$ 21,000	$ 19,200	$ 17,760	$ 65,040

° Negative taxable income

† Tax savings

With this reduced annual revenue, the taxable incomes in year 2 become negative. The tax consequences are illustrated in Table 9.5. Here we assume that the company has other profitable operations whose taxable incomes more than offset the negative values for this project so that the company's incremental tax rate is still 40%.

9.3.5 When Projects Require Multiple Assets

Up to this point, our examples were limited to the situations where only one asset was employed in the project. In many situations, however, a project may require the purchase of multiple assets with different property classes. For example, a typical engineering project may involve more than purchasing equipment—it may need a building for the manufacturing operation. Even the various assets may be placed in service in different points in time. What we have to do is to itemize the timing of investment requirement and the capital cost allowances according to the asset placement. Example 9.6 will illustrate the development of project cash flows with multiple assets.

Example 9.6 A project requiring multiple assets

Langley Manufacturing Company (LMC), a manufacturer of fabricated metal products, is considering the purchase of a new computer-controlled milling machine for $90,000 to produce a custom-ordered metal product. The costs for installation of the machine, site preparation, and wiring are expected to be $10,000. It also needs special jigs and dies, which cost $12,000. The milling machine is expected to last 10 years, and the jigs and dies to last 5 years. The machine will have a $10,000 salvage value at the end of its life, and the special jigs and dies are worth only $1000 as scrap metal at any time in their lives. The milling machine is classified as a CCA Class 39 property and the jigs and dies as CCA Class 12 property. LMC needs to either purchase or build a 1000 m² warehouse to store the product before shipping to its customers. LMC has decided to purchase a building near the plant at $160,000. For depreciation purposes, the warehouse cost of $160,000 is divided into $120,000 for building and structure and $40,000 for land. The building is classified as CCA Class 1 property. At the end of 10 years, the building will have a salvage value of $80,000, but the value of the land would appreciate to $110,000. The revenue from increased production is expected to be $140,000 per year. The additional annual production costs are estimated as follows: materials, $22,000; labor, $32,000; energy, $3500; and other miscellaneous costs, $2500. For the analysis, a 10-year life will be used. LMC has a marginal tax rate of 40% and a MARR of 18%.

Discussion: There are three types of assets in this problem: the milling machine, the jigs and dies, and the warehouse. The first two assets are personal properties and the last is a real property. The cost base for each asset has to be determined separately. For the milling machine, we need to add the site-preparation expense to the cost base, whereas we need to subtract the land cost from the warehouse cost to establish the correct cost base for the real property.

- Milling machine: $90,000 + $10,000 = $100,000
- Jigs and dies: $12,000
- Warehouse (building): $120,000
- Warehouse (land): $40,000

Since the jigs and dies last only 5 years, we need to make a specific assumption regarding the replacement cost at the end of 5 years. In this problem, we will assume that the replacement cost would be about the same as the cost of the initial purchase. We will also assume that the warehouse property will be placed in service in January.

Solution

Given: Cash flow elements provided above, $t = 40\%$, MARR = 18%
Find: Net after-tax cash flow, NPW

Table 9.6 and Fig. 9.7 summarize the net after-tax cash flows associated with the multiple-asset investment. In both the table and the chart below we see that the jigs and dies are fully depreciated during the project life.

Jigs and Dies Depreciation Schedule, CCA Class 12
(Declining balance rate of 100%)

Year	Capital Cost Allowance, CCA_n	Book Value
0		$12,000
1°	$6,000	6,000
2	6,000	0
5		12,000
6°	6,000	6,000
7	6,000	0

The milling machine and building are not fully depreciated. Thus, we need to calculate the book value of the milling machine and building at the end of the project life. Note that land is not a depreciable property and therefore retains its book value of $40,000 until sold.

Building Depreciation Schedule, CCA Class 1
(Declining balance rate of 4%)

Year	Capital Cost Allowance, CCA_n	Book Value
0		$120,000
1°	$2,400	117,600
2	4,704	112,896
3	4,516	108,380
4	4,335	104,045
5	4,162	99,883
6	3,995	95,888
7	3,836	92,052
8	3,682	88,370
9	3,535	84,835
10	3,393	81,442

° Half-year rule applies during the year of purchase.

Milling Machine Depreciation Schedule, CCA Class 39
(Declining balance rate of 30%)

Year	Capital Cost Allowance, CCA_n	Book Value
0		$100,000
1°	$15,000	85,000
2	25,500	59,500
3	17,850	41,650
4	12,495	29,155
5	8,747	20,408
6	6,122	14,286
7	4,286	10,000
8	3,000	7,000
9	2,100	4,900
10	1,470	3,430

° Half-year rule applies during the year of purchase.

We then can calculate the disposal tax effects associated with the disposal of each depreciable asset as follows:

Property (Asset)	Cost Base	Salvage Value	Book Value	Losses (Gains)	Disposal Tax Effect
Building	$120,000	$80,000	$81,442	$ 1,442	$ 577
Milling machine	100,000	10,000	3,430	(6,570)	−2,628
Jigs and dies	12,000	1,000	0	(1,000)	− 400

Then we can calculate the capital gains or losses associated with the sale of the land as follows:

Property (Asset)	Cost Base	Salvage Value	Book Value	Losses (Gains)	Disposal Tax Effect
Land	$40,000	$110,000	$40,000	($70,000)	−$21,000

Note:

Capital gains tax	$= \frac{3}{4} \times$ gain from sale $\times 40\%$ tax rate
Disposal tax effect at year 5	$= -\$400$
Disposal tax effect at year 10	$= -\$21,000 + \$577 - \$2628 - \$400$
	$= -\$23,451.$

Table 9.6 After-tax Cash Flow Analysis (Example 9.6)

	0	1	2	3	4	5	6	7	8	9	10
Income Statement											
Revenue		$140,000	$140,000	$140,000	$140,000	$140,000	$140,000	$140,000	$140,000	$140,000	$140,000
Expenses											
Materials		22,000	22,000	22,000	22,000	22,000	22,000	22,000	22,000	22,000	22,000
Labor		32,000	32,000	32,000	32,000	32,000	32,000	32,000	32,000	32,000	32,000
Energy		3,500	3,500	3,500	3,500	3,500	3,500	3,500	3,500	3,500	3,500
Others		2,500	2,500	2,500	2,500	2,500	2,500	2,500	2,500	2,500	2,500
CCA											
Building		2,400	4,704	4,516	4,335	4,162	3,995	3,836	3,682	3,535	3,393
Milling		15,000	25,500	17,850	12,495	8,747	6,122	4,286	3,000	2,100	1,470
Jigs and dies		6,000	6,000				6,000	6,000			
Taxable income		$56,600	$43,796	$57,634	$63,170	$67,091	$63,883	$65,878	$73,318	$74,365	$75,137
Income taxes		22,640	17,518	23,054	25,268	26,836	25,553	26,351	29,327	29,746	30,055
Net income		$33,960	$26,278	$34,580	$37,902	$40,255	$38,330	$39,527	$43,991	$44,619	$45,082
Cash Flow Statement											
Net income + CCA		$57,360	$62,482	$56,946	$54,732	$53,164	$54,447	$53,649	$50,673	$50,254	$49,945
Investment and salvage values											
Building	−$120,000										80,000
Land	− 40,000										110,000
Milling	− 100,000										10,000
Jigs and dies	− 12,000					−11,000					1,000
Disposal tax effect						− 400					−23,451
Net cash flow	−$272,000	$57,360	$62,482	$56,946	$54,732	$41,764	$54,447	$53,649	$50,673	$50,254	$227,494

Figure 9.7 ■ Cash flow diagram (Example 9.6)

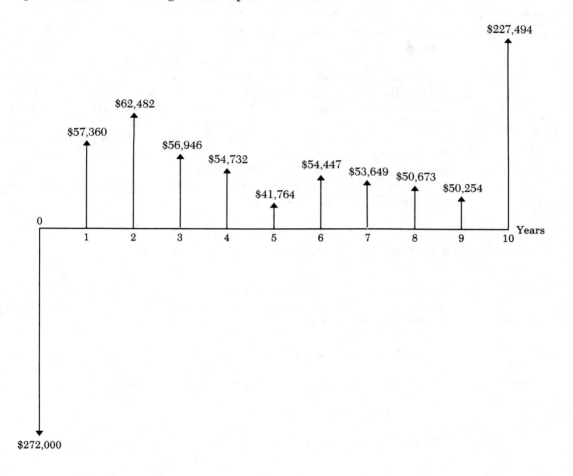

The NPW of the project is

$$PW(18\%) = -\$272,000 + \$57,360(P/F, 18\%, 1) + \$62,482(P/F, 18\%, 2)$$

$$+ ...+ \$227,494(P/F, 18\%, 10)$$

$$= \$7916 > 0.$$

This means that the project has a return on investment that is slightly greater than 18% and is therefore marginally acceptable (*IRR* = 18.7%). The return on investment may be calculated by making the above equation equal 0 and solving for *i*.

■ 9.4 Generalized Cash Flow Approach

The Canadian corporate tax system is not progressive and most companies operate with the same marginal tax rate year-after-year. If we are analyzing project cash flows for a corporation that consistently operates in the same tax bracket, we can assume that the firm's marginal tax rate will remain the same, whether the project is accepted or rejected. In this situation, we may apply the present marginal tax rate to each taxable item in the cash profile, thus obtaining the after-tax cash flows. By aggregating individual items, we have the project's net cash flows. This approach is referred to as the **generalized cash flow method.** As we shall see in later chapters, this approach affords us several analytical advantages. In particular, when we compare service projects, the generalized cash flow method is computationally efficient. (Examples will be provided in Chapters 11 and 12.)

9.4.1 Setting Up the Net Cash Flow Equation

To generate the generalized cash flow table, we first examine each cash flow element. We can do this as follows, using the scheme for classifying cash flows that we have just developed:

A_n = + Revenues at time n (R_n)

         − Expenses other than CCA (CCA_n) at time n (E_n) Operating activities

         − Interest expense at time n (I_n)

         − Income taxes at time n (T_n)

         − Investment at time n (P_n)

         + Net proceeds from sale at time n $(S_n + G_n)$ Investing activities

         − Working capital investment at time n (W_n)

         + Working capital recovery at time n (W_n')

         + Proceeds from loan (L_n) Borrowing activities

         − Repayment of principle (PP_n)

where A_n is the net after-tax cash flow at the end of period n. Capital cost allowance is *not* a cash flow and is therefore excluded from E_n (although it must be considered when calculating income taxes). Note also that $(S_n + G_n)$ represents the net salvage value after adjustments for disposal tax effect (G_n). Not all terms are relevant in calculating a cash flow in every year. The term $(S_n + G_n)$ appears only when the asset is disposed of.

In terms of symbols, we can express A_n as

$$A_n = R_n - E_n - I_n - T_n \qquad \text{Operating activities}$$

$$- P_n + (S_n + G_n) - W_n + W'_n \qquad \text{Investing activities}$$

$$+ L_n - PP_n. \qquad \text{Financing activities} \qquad (9.1)$$

If we designate T_n as the total income taxes paid at time n and t as the marginal tax rate, income taxes on this project are

$$T_n = \text{(taxable income)(marginal tax rate)}$$

$$= (R_n - E_n - I_n - CCA_n)t$$

$$= (R_n - E_n)t - (I_n + CCA_n)t. \qquad (9.2)$$

The term $(I_n + CCA_n)t$ is known as the tax shield (or tax savings) from financing and asset depreciation. Now, substitute the result of Eq. (9.2) into Eq. (9.1) to obtain

$$A_n = (R_n - E_n - I_n)(1 - t) + t \times CCA_n$$

$$- P_n + (S_n + G_n) - W_n + W'_n$$

$$+ L_n - PP_n. \qquad (9.3)$$

9.4.2 Presenting Cash Flows in a Compact Tabular Form

Table 9.7 groups the after-tax cash flow components over the life of the project by activity type. (In preparing their after-tax cash flow, however, most business firms adopt the income statement approach presented in previous sections, as they want to know the net income along with the cash flow statement.)

Table 9.7 A Compact Tabular Format for Generalized Cash Flow Approach

Cash Flow Elements	End of Period 0 1 2 ... n
Investment activities $\begin{cases} -P_n \\ +(S_n + G_n) \\ -W_n \\ +W'_n \end{cases}$	
Operating activities $\begin{cases} (1-t)(R_n) \\ -(1-t)(E_n) \\ -(1-t)(I_n) \\ +t \times CCA_n \end{cases}$	
Financing activities $\begin{cases} +L_n \\ -PP_n \end{cases}$	
Net cash flow	

Example 9.7 Generalized cash flow approach

Reconsider Example 9.1. Use the generalized cash flow approach to obtain the after-tax cash flows.

Solution

Given:

Purchase price, $P_0 = \$125,000$

Annual revenues, $R_n = \$100,000$

Annual expenses other than CCA, $E_n = \$40,000$

Capital cost allowance (CCA_n), years 1–5: $\$12,500; \$22,500; \$18,000; \$14,400; \$11,520$

Marginal tax rate, $t = 40\%$

Salvage value $= \$50,000$

Find: Annual after-tax cash flows, A_n

Equation: Equations 9.1 and 9.2, eliminating zero terms:
$A_n = (R_n - E_n)(1 - t) + t \times CCA_n - P_n + (S_n + G_n)$

- **Year 0**
  Cash flow $= -\$125,000$

- **Years 1 to 5**
  After-tax cash flow (except for net salvage in year 5)
  $= (R_n - E_n)(1 - t) + t \times CCA_n$

n	Net Operating Cash Flow
1	($100,000 − $40,000)(0.60) + $12,500(0.40) = $41,000
2	($100,000 − $40,000)(0.60) + $22,500(0.40) = $45,000
3	($100,000 − $40,000)(0.60) + $18,000(0.40) = $43,200
4	($100,000 − $40,000)(0.60) + $14,400(0.40) = $41,760
5	($100,000 − $40,000)(0.60) + $11,520(0.40) = $40,608

- **Year 5: Net Salvage Value**
  As before, in year 5 we must deal with two aspects of the asset's disposal, salvage value and taxable gains (or losses).

  1. The total capital cost allowance in years 1–5 is $12,500 + $22,500 + $18,000 + $14,400 + $11,520 = $78,920.
  2. The book value at the end of period 5 is the cost basis minus the total capital cost allowance, or $125,000 − $78,920 = $46,080.
  3. The difference between the book value and the salvage value is $46,080 − $50,000 = − $3,920.
  4. The disposal tax effect is − $3,920 × 40% = − $1568.
  5. The net proceeds from the sale of the asset are $50,000 − $1568 = $48,432.
  6. The combined net cash flow in year 5 is $40,608 + $48,432 = $89,040.

Our results and overall calculations are summarized in Table 9.8. Checking these figures versus those we obtained in Table 9.2 confirms our results.

Comments: *As a variation to Table 9.8, Table 9.9 represents another tabular format widely used in engineering economics texts. Without seeing the footnote in Table 9.9, however, it is not intuitively clear how the last column (net cash flow) is obtained. Therefore, we will use the income statement approach whenever possible throughout the text.*

Table 9.8 After-Tax Cash Flow Analysis with Generalized Cash Flow Approach (Example 9.7)

				n		
	0	1	2	3	4	5
Income Statement						
−Investment	−$125,000					
+Net proceeds from sale						$48,432
+(0.6) Revenue		$60,000	$60,000	$60,000	$60,000	60,000
−(0.6) Expenses		−24,000	−24,000	−24,000	−24,000	−24,000
+(0.4) CCA		5,000	9,000	7,200	5,760	4,608
Net cash flow	−$125,000	$41,000	$45,000	$43,200	$41,760	$89,040

Table 9.9 Net Cash Flow Table Generated by Traditional Method (Example 9.8)

1	2	3	4	5	6	7	8
Year End	Investment and Salvage	Revenue	Expense	CCA	Taxable Income	Tax	Net Cash Flow
0	−$125,000						−$125,000
1		$100,000	$40,000	$12,500	$47,500	$19,000	41,000
2		100,000	40,000	22,500	37,500	15,000	45,000
3		100,000	40,000	18,000	42,000	16,800	43,200
4		100,000	40,000	14,400	45,600	18,240	41,760
5		100,000	40,000	11,520	48,480	19,392	40,608
	50,000°				−3,920	1,568	48,432

° Salvage value
 Note that
   col. 6 = col. 3 − col. 4 − col. 5
   col. 7 = col. 6 × (0.40)
   col. 8 = col. 2 + col. 3 − col. 4 − col. 7

9.4.3 Grants and Investment Tax Credits

The federal and provincial governments will sometimes try to stimulate the economy by providing grants or investment tax credits to companies considering certain investments.

Grants

A repayable grant for the purchase of a depreciable asset results in reduced borrowing costs. For calculation purposes we treat these grants in the same manner as the financing described in Example 9.4. However, sometimes these are interest-free repayable grants which will substantially improve the company's internal rate-of-return (return on investment).

A non-repayable grant for the purchase of a depreciable asset results in reducing the cost base of the property by the amount of the grant. For example, if the company purchased a $100,000 machine and received a $40,000 grant, the cost base would be reduced to $60,000. The company would calculate the tax depreciation based on the $60,000 cost base, not the $100,000 purchase price of the machine.

Investment Tax Credits

Investment tax credits provide a reduction in the year's income tax equal to a percentage of the property acquired. This tax saving in effect reduces the net cost of the equipment to the company by the amount of the investment tax credit. For example, if a company purchased a $100,000 machine and received a 40% investment tax credit, the company would receive a $40,000 tax credit at the end of the year. The cost base of the property is also reduced by $40,000. The company would calculate the depreciation based on the $60,000 cost base, not the $100,000 purchase price of the machine.

Example 9.8 Considering grants and investment tax credits

Suppose that your company was being encouraged by the government to install a new piece of equipment. The equipment would cost $72,000 and require $4,000 in engineering and $6,500 in installation fees. The equipment is a CCA Class 43 property which is depreciated for tax purposes on a declining balance rate of 30%. Your engineers have estimated that this machine will result in net savings of $25,000 per year over the next 5 years and have a salvage value of $14,000. The company's marginal tax rate is 40%. Calculate the after-tax cash flow under the following scenarios using the generalized cash flow approach:

(a) No government assistance is provided.

(b) A $20,000 non-repayable grant is provided.

(c) A 20% investment tax credit is provided.

Solution

Given:

Equipment costs = $72,000, plus $4,000 in engineering fees and $6,500 in installation fees

Machine saves an estimated $25,000 per year

Salvage value after 5 years = $14,000

Marginal tax rate of 40%

Find: After-tax cash flows for the three scenarios

(a) No Government Assistance is Provided

$$
\begin{aligned}
\text{Cost base} &= \text{equipment cost} + \text{engineering fees} + \text{installation fees} \\
&= \$72,000 + \$4,000 + \$6,500 \\
&= \$82,500
\end{aligned}
$$

Year End	Investment and Salvage	Net Savings	CCA	Taxable Income	Tax	Net Cash Flow
0	−$82,500					−$82,500
1		$25,000	$12,375	$12,625	$5,050	19,950
2		25,000	21,038	3,962	1,585	22,415
3		25,000	14,734	10,266	4,106	20,894
4		25,000	10,303	14,697	5,879	19,121
5		25,000	7,212	17,788	7,115	17,885
5	14,000			($2,829)°	(1,132)	15,132

$$\begin{aligned} \text{° Terminal loss} &= \text{book value} - \text{salvage value} \\ &= \$16,829 - \$14,000 \\ &= \$2,829 \end{aligned}$$

Disposal tax effect $= \$2829 \times 40\% = \$1132$

(b) A $20,000 Non-repayable Grant is Provided

$$\begin{aligned} \text{Equipment cost base} &= \text{Total installed purchase price} - \text{non-repayable grant} \\ &= \$82,500 - \$20,000 \\ &= \$62,500 \end{aligned}$$

Year End	Investment and Salvage	Net Savings	CCA	Taxable Income	Tax	Net Cash Flow
0	−$62,500					−$62,500
1		$25,000	$ 9,375	$15,625	$6,250	18,750
2		25,000	15,938	9,062	3,625	21,375
3		25,000	11,156	13,844	5,538	19,462
4		25,000	7,810	17,190	6,876	18,124
5		25,000	5,467	19,533	7,813	17,187
5	14,000			1,244°	498	13,502

$$\begin{aligned} \text{° Recaptured CCA} &= \text{salvage value} - \text{book value} \\ &= \$14,000 - \$12,756 \\ &= \$1244 \end{aligned}$$

(c) A 20% Investment Tax Credit is Provided

$$\begin{aligned} \text{Equipment cost base} &= \text{installed purchase price} - 20\% \text{ investment tax credit} \\ &= \$82,500 - \$16,500 = \$66,000 \end{aligned}$$

$$\begin{aligned} \text{Tax after first year} &= \text{tax due to savings} - 20\% \text{ investment tax credit} \\ &= \$6040 - \$16,500 = (\$10,460) \end{aligned}$$

Year End	Investment and Salvage	Net Savings	CCA	Taxable Income	Tax	Net Cash Flow
0	−$82,500					−$82,500
1		$25,000	$ 9,900	$15,100	($10,460)	35,460
2		25,000	16,830	8,170	3,268	21,732
3		25,000	11,781	13,219	5,288	19,712
4		25,000	8,247	16,753	6,701	18,299
5		25,000	5,773	19,227	7,691	17,309
5	14,000			531°	212	13,788

° Recaptured CCA = salvage value − book value
= $14,000 − $13,469
= $531

9.5 The After-Tax Cash Flow Diagram Approach and Capital Tax Factors

The income statement and generalized cash flow approaches in Sections 9.3 and 9.4 require a year-by-year estimate of each cash flow element. These estimates are then combined in a manner that yields a net after-tax cash flow for each year. The resulting net cash flows are the basis for the subsequent discounted cash flow analysis.

The advantage of these approaches is that they detail the magnitude of each annual net cash flow and indicate whether its value is positive or negative. Prior to the availability of spreadsheets, problems which had a large number of cash flow elements and a long service life were very time-consuming. Therefore, an alternative approach was often employed in which the problem was represented on an after-tax cash flow diagram.

With this approach, interest factors are used to develop a relationship for the desired present equivalent, annual equivalent, or future equivalent. This alternative provides us with the required measure of investment attractiveness without making it necessary to calculate the detailed net cash flows on an annual basis.

Although most problems will be analyzed using **EzCash** or a computer spreadsheet, the after-tax cash flow diagram approach is a very useful technique which can save a considerable amount of time when a computer is not available.

9.5.1 The After-Tax Cash Flow Diagram

Net cash flows on an after-tax basis are represented by Eq. (9.3). For any specific problem, the relevant cash flow elements that compose the net cash flow can be portrayed on a cash flow diagram.

Figure 9.8 ■ A general after-tax cash flow diagram

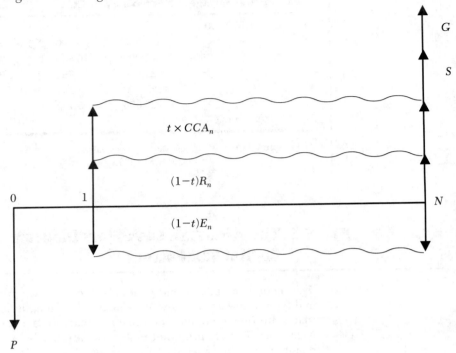

To illustrate, we will assume a typical problem, which has investment at time zero and disposal at year N. In order to simplify the problem we will also assume that there are no debt or working capital considerations, although these can be included if required.

The problem can now be represented on a cash flow diagram as shown in Figure 9.8. Here the wavy lines indicate an arbitrary pattern of cash flows and N indicates the life of the project.

The present equivalent of this cash flow line is

$$PE = -P + (1 - t) \sum_{n=1}^{N} R_n \, (P/F,i,n)$$

$$- (1 - t) \sum_{n=1}^{N} E_n \, (P/F,i,n)$$

$$+ t \sum_{n=1}^{N} CCA_n \, (P/F,i,n)$$

$$+ (S + G)(P/F,i,N). \tag{9.4}$$

If R_n and E_n follow a standard pattern (uniform series, linear gradient, geometric gradient) or a combination thereof, as they often do, we will be able to replace the summations with interest factors. However, the $t \times CCA_n$ terms are unlikely to offer any such computational efficiency except for assets depreciated on a straight-line basis. If an asset is depreciated with the declining balance method, capital tax factors can be used to avoid the difficulty.

9.5.2 Capital Tax Factors

In situations where CCA is calculated on the basis of straight-line depreciation, the annual tax savings, $t \times CCA_n$, represents a uniform series of cash flows. This fact makes calculating a present equivalent value of these cash flow elements relatively simple. However, when CCA is calculated on a declining balance basis, $t \times CCA_n$ changes each year. Using the net salvage value approach and including these tax savings explicitly on a year-by-year basis makes the present equivalent determination quite tedious. Fortunately we can employ **capital tax factors**, which handle the tax savings implicitly, to avert such problems.

Let us consider an asset of value P, which qualifies for CCA on a declining balance basis. If we ignore the half-year rule for the moment, the tax savings generated by this capital expenditure in year n is given by

$$t \times CCA_n = td(1 - d)^{n-1}P. \tag{9.5}$$

We will allow the tax savings due to this capital expenditure to last for infinite years, assuming that a correction will be exercised at the time of disposal. This situation can be represented as shown in Figure 9.9.

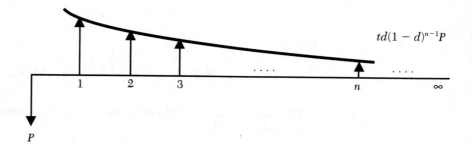

Figure 9.9 ■ Capital investment and the corresponding tax savings for the declining balance method

The present value of all the cash flows shown to time infinity is thus

$$-P + \frac{tdP}{1+i} + \frac{td(1-d)P}{(1+i)^2} + \frac{td(1-d)^2 P}{(1+i)^3} + ... + \frac{td(1-d)^{n-1}P}{(1+i)^n} + ...$$

$$= -P + \frac{tdP}{1+i}\left[1 + \frac{(1-d)}{1+i} + \frac{(1-d)^2}{(1+i)^2} + ... + \frac{(1-d)^n}{(1+i)^n}\right]$$

$$= -P + \frac{tdP}{1+i} \times \frac{1+i}{i+d}$$

$$= -P\left(1 - \frac{td}{i+d}\right)$$

$$= -P \times CTF \tag{9.6}$$

where

$$CTF = 1 - \frac{td}{i+d}$$

is the full-year capital tax factor (CTF). Note that the terms in the square brackets form an infinite geometric series with a summed value of $\frac{1+i}{i+d}$.

A similar type of analysis can be made when the half-year rule is used. In this case the tax savings in each year are

$$t \times CCA_1 = t\frac{d}{2}P \tag{9.7}$$

and

$$t \times CCA_n = td\left[1 - \frac{d}{2}\right](1-d)^{n-2}P, \text{ for } n > 1. \tag{9.8}$$

The present value of all future tax savings from year 1 to time infinity can now be shown to be

$$\frac{td}{i+d}\left(\frac{1+\frac{i}{2}}{1+i}\right)P$$

and the present value of the capital expenditure adjusted for all future tax savings is

$$-P + \frac{td}{i+d} \times \frac{1 + \frac{i}{2}}{1+i} \times P$$

$$= -P\left(1 - \frac{td}{i+d} \times \frac{1 + \frac{i}{2}}{1+i}\right)$$

$$= -P \times CTF_{\frac{1}{2}}$$

where

$$CTF_{\frac{1}{2}} = 1 - \frac{td}{i+d} \times \frac{1 + \frac{i}{2}}{1+i} \qquad (9.9)$$

is the half-year capital tax factor ($CTF_{\frac{1}{2}}$).

It is worth noting that CTF and $CTF_{\frac{1}{2}}$ are independent of any time frame considerations, as they are developed on the basis of time going to infinity. (CTF and $CTF_{\frac{1}{2}}$ tables are included in Appendix D for various values of t, d and i.)

9.5.3 Application of Capital Tax Factors

Consider the investment represented in the general after-tax cash flow diagram of Figure 9.8. The $t \times CCA_n$ cash flows and the disposal tax effects, G, have been included explicitly and appear in Eq. 9.4. The disposal tax effect, G, depends on the specific assumptions used. As discussed in Section 8.3.2, there are four possible cases where an asset disposal does not deplete the asset class.
Alternatively, the problem can be formulated with capital tax factors. Using the half-year capital tax factor, we can calculate an investment cost adjusted for all future tax savings to time infinity as $P \times CTF_{\frac{1}{2}}$. However, the investment generates tax savings only over N years, so the tax savings included for the period from N (Cases III and IV) or $N + 1$ (Cases I and II) to time infinity must be eliminated in some manner. If no new assets are added to the same class in the year of disposal, the asset class accounting procedure merely subtracts the salvage value from the undepreciated capital cost base of the asset class. The reduction of the undepreciated capital cost by S in this case will reduce the future tax savings that could be realized from this asset class. The adjustments to the future tax savings due to S are calculated with the same declining balance

scheme. However, when these adjustments are made depends on the disposal timing assumption. The situation is further complicated when new assets are added to the class in the year in which a disposal occurs. This development leads to the same four cases considered in Section 8.3.2.

Under Case I, the following assumptions are made:

1. The disposal of an asset occurs at the beginning of year $N + 1$.
2. There are no new purchases of assets belonging to the same asset class in the year of disposal.

With these assumptions, the first extra tax payment will occur in year $N + 1$ in the amount of tdS. For the nth year after N years of service of the asset, the extra tax payment will be

$$td (1 - d)^{n-1} S.$$

We don't have to consider B_{disposal} at disposal time because its effects in the subsequent years have been incorporated by applying a capital tax factor to the initial asset value P (Section 9.5.2). This is how capital tax factors allow us to simplify our numerical calculations.

The salvage value and its tax effects can thus be represented as shown in Fig. 9.10. The present values of all the cash flows in Fig. 9.10, expressed at the end of year N, is

$$S \sum_{n=1}^{\infty} \frac{td (1 - d)^{n-1} S}{(1 + i)^n}$$

$$= S\left(1 - \frac{td}{i + d}\right)$$

$$= S \times CTF_{\text{I}}$$

where

$$CTF_{\text{I}} = 1 - \frac{td}{i + d} \tag{9.10}$$

is the capital tax factor to be applied to S under Case I. It is equal to the full-year CTF developed in Section 9.5.2 and tabulated in Appendix D.

Figure 9.10 ■ Salvage value and its tax effects under Case I

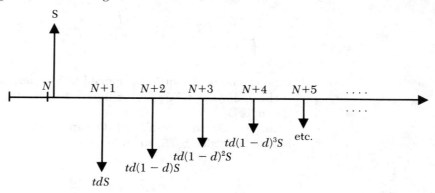

The assumptions under Case I, although not stated explicitly therein, are the basis for the capital tax factor approach recommended by a number of other authors, including Edge and Irvine[1] and Sprague and Whittaker[2].

The salvage value and corresponding tax effects under Case II are shown in Figure 9.11.

Figure 9.11 ■ Salvage value and the corresponding tax effects for Case II

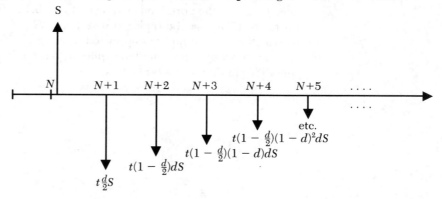

The present value of the cash flows in Fig. 9.11, expressed at the end of year n, is

$$S + \frac{t\frac{d}{2}S}{1+i} + \sum_{n=2}^{\infty} \frac{t(1-\frac{d}{2})(1-d)^{n-2}\, d\, S}{(1+i)^n}$$

$$= S\left(1 - \frac{td}{i+d} \times \frac{1+\frac{i}{2}}{1+i}\right)$$

$$= S(CTF_{II})$$

[1] C. Geoffrey Edge and V. Bruce Irvine, *A Practical Approach to the Appraisal of Capital Expenditures*, Second Edition, The Society of Management Accountants of Canada, 1981.
[2] J.C. Sprague and J.D. Whittaker, *Economic Analysis for Engineers and Managers: the Canadian Context*, Prentice Hall of Canada Ltd., 1986.

where

$$CTF_{II} = 1 - \frac{td}{i + d} \times \frac{1 + \frac{i}{2}}{1 + i} \tag{9.11}$$

is the capital tax factor to be applied to S under Case II. This factor is equal to the half-year $CTF_{\frac{1}{2}}$ factor derived in Section 9.5.2 and tabulated in Appendix D.

Now we can easily verify that the capital tax factors for Cases III and IV, as defined in Section 8.3.2, are as follows:

$$CTF_{III} = 1 - \frac{td}{i + d} (1 + i) \tag{9.12}$$

$$CTF_{IV} = 1 - \frac{td}{i + d} \left(1 + \frac{i}{2}\right) \tag{9.13}$$

Using these and the other capital tax factors derived previously, the after-tax cash flow diagram for a capital expenditure, depreciated on a declining balance basis, can be portrayed as shown in Fig. 9.12. In this diagram, the capital tax factor CTF_S may be replaced by CTF_I, CTF_{II}, CTF_{III}, or CTF_{IV}, depending on which set of assumptions apply to the disposal. The tax savings arising from CCA are no longer represented explicitly, but are implicit through their representations in the capital tax factors.

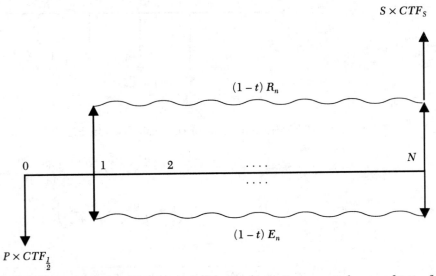

Figure 9.12 ■ The after-tax cash flow diagram for a capital expenditure for which capital tax factors can be used

When capital tax factors are used, the after-tax cash flow diagram should be developed as shown in Fig. 9.12. No $t \times CCA_n$ cash flows should appear in the diagram. If you prefer to include the $t \times CCA_n$ cash flows explicitly, then the disposal tax effects *must* be calculated as discussed in Section 8.3.2. Including the $t \times CCA_n$ cash flows in the diagram precludes the use of capital tax factors.

If a capital gain occurs when the asset is disposed, P should replace S in Fig. 9.12. The capital gains amount subtracted by the capital gains tax payment, i.e. $(S - P)(1 - t_c)$, where t_c is the capital gains tax rate, should be added to the after-tax cash flow diagram either at the end of year N (Cases III and IV) or at the end of year $N + 1$ (Cases I or II), depending on the timing of the disposal.

Coverage of capital tax factors in other texts is generally limited to Case I, with little explanation of the specific disposal timing assumption or means of treating additions to the asset class. The broader consideration given to these issues in this text yields the four cases in which the capital tax factor applied to the salvage value takes different forms.

Example 9.9 Using capital tax factors

SKS Inc. is considering the construction of a new plant to manufacture mining equipment for Alberta's oil sands industries. The following estimates apply to this investment:

Investment Category	Investment (P)	Salvage Value in 10 years (S)
Building (B)	$1,000,000	$700,000
Land (L)	400,000	600,000
Machinery (M)	1,200,000	400,000

The company plans to operate the plant for 10 years. Annual operating costs are estimated to be $800,000 and a major overhaul needed in year 6 is estimated to cost $100,000. The declining balance method with half-year rule is used for CCA purposes. The CCA rates for the building and equipment are 10% and 30%, respectively. The company's tax rate is 40% and the minimum acceptable rate of return (MARR) is 15%. What is the annual equivalent operating revenue that the plant must generate to justify the investment? Assume that disposal occurs at the beginning of year 11, and that no other purchases of building or equipment class assets take place in year 11.

Solution:

Given:

Operating cost (E) = $800,000 per year

Additional operating cost, year 6 (E_6) = $100,000

$d_B = 10\%$, $d_M = 30\%$, $t = 40\%$, MARR = 15%

Case I applies

Find: The annual equivalent revenue (*AER*) needed to generate a rate of return of 15%

The after-tax cash flow diagram of the investment is

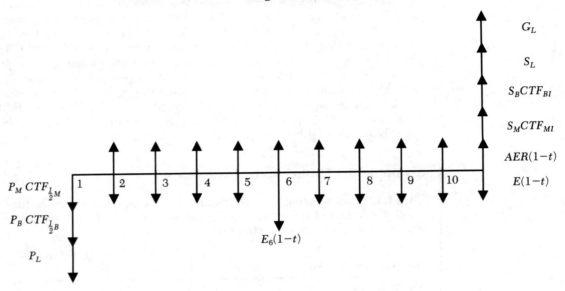

Figure 9.13 ■ After-tax cash flow diagram (Example 9.9)

where

$$CTF_{\frac{1}{2}M} = 1 - \frac{td_M}{i + d_M} \times \frac{1 + \frac{i}{2}}{1 + i} = 1 - \left[\frac{(0.4)(0.3)}{0.15 + 0.3}\right]\left[\frac{(1 + 0.075)}{1 + 0.15}\right] = 0.7507$$

$$CTF_{MI} = 1 - \frac{td_M}{i + d_M} = 0.7333$$

$$CFT_{\frac{1}{2}B} = 1 - \frac{td_B}{i + d_B} \times \frac{1 + \frac{i}{2}}{1 + i} = 1 - \left[\frac{(0.4)(0.1)}{0.15 + 0.1}\right]\left[\frac{1 + 0.075}{1 + 0.15}\right] = 0.8504$$

$$CTF_{BI} = 1 - \frac{td_B}{i + d_B} = 0.84$$

Since the disposal of land occurs at the beginning of year 11, the capital gains tax will be paid at the end of year 11. As a result, G_L expressed at the end of year 10 is calculated by moving the capital gains tax payment backward by one year with a factor of $(1 + i)^{-1}$.

Thus

$$G_L = -\frac{3}{4} t \left(\frac{S_L - P_L}{1+i} \right) \quad \text{(capital gains tax)}$$

$$= -\frac{3}{4} \times 0.4 \left(\frac{\$600,000 - \$400,000}{1.15} \right)$$

$$= -\$52,174.$$

Then

$$AE = AER(1-t) - E(1-t)$$

$$- (P_M CTF_{\frac{1}{2}M} + P_B CTF_{\frac{1}{2}B} + P_L)(A/P, 15\%, 10)$$

$$- E_6(1-t)(P/F, 15\%, 6)(A/P, 15\%, 10)$$

$$+ (S_B CTF_{BI} + S_M CTF_{MI} + S_L + G_L)(A/F, 15\%, 10)$$

$$= 0.6\,AER - 800,000 \times 0.6$$

$$- (1,200,000 \times 0.7507 + 1,000,000 \times 0.8504 + 400,000) \times 0.1993$$

$$- 100,000 \times 0.6 \times 0.4323 \times 0.1993$$

$$+ (700,000 \times 0.84 + 400,000 \times 0.7333 + 600,000 - 52,174) \times 0.0493$$

$$= 0.6\,AER - 816,859 = 0.$$

Therefore

$$AER = \frac{\$816,859}{0.6} = \$1,361,432.$$

The required annual equivalent revenue is approximately 1.361 million dollars to justify the investment.

Comment: *Although this solution provides no detailed information on the net after-tax cash flows, we are able to determine minimum annual equivalent revenue. This detail is implicit in the solution formulation.*

9.6 Computer Notes

As you have learned in the previous sections, creating an after-tax cash flow table can be a tedious and time-consuming process. An electronic spreadsheet is a perfect choice to automate much of the computation. Alternatively, **EzCash** can automate the entire analysis with built-in computational tools for depreciation and loan interest. To illustrate, we will use Example 9.4.

9.6.1 After-Tax Cash Flow Analysis with Lotus 1-2-3

One of the most popular spreadsheet applications in engineering economic analysis is found in after-tax cash flow analysis. Figure 9.14 shows how to prepare a cash flow statement for Example 9.4. In column A, you create the list of elements that are required to calculate the net income as well as the net cash flow. You enter the cell values for *Revenue, Materials, Labor, Overhead, Interest, CCA*. Then let Lotus calculate the cell values for *Taxable Income,*

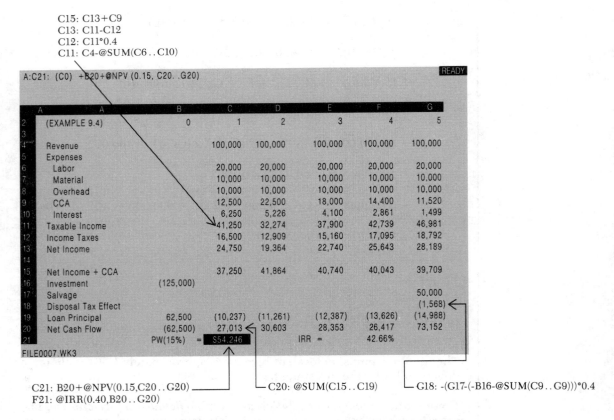

Figure 9.14 ■ After-tax cash flow analysis using Lotus 1-2-3 (Example 9.4)

Income Taxes, Net Income. On the cash flow statement side, you enter the cell values for *Investment, Salvage, Disposal Tax Effect* and *Loan Repayment*. Then Lotus will calculate the cell values for *Net Income + CCA, Net Cash Flow*. Of course, you can automate the loan interest and principal calculations using the worksheet developed in Chapter 3. Similarly, you can also automate the CCA calculations using the worksheet developed in Chapter 7.

Once we obtain the net cash flow of the investment, the next step is to measure the investment worth. Using the @NPV and @IRR functions, we can calculate the NPW as well as the IRR of the investment. These calculations are also shown in Fig. 9.14. Since this is a simple investment, the i° value determined with the @IRR function is also the IRR. This financial calculation is only the beginning of our spreadsheet utility. If we are unsatisfied with the projections for any of the income statement items, we can change the cell values. The changes will carry through to net income, net cash flow, which will yield revised NPW and IRR figures. Obviously, the potential for "what-if" analysis is almost unlimited.

9.6.2 After-Tax Cash Flow Analysis with EzCash

EzCash has a Lotus-like spreadsheet utility with many built-in financial functions to facilitate after-tax cash flow analyses. To demonstrate how we obtain a cash flow statement from **EzCash**, we will again use Example 9.4.

*   **Step 1: Working with EzCash Worksheet**

    (a) Select the **Spreadsheet** mode from the main menu of **EzCash** to initiate an after-tax cash flow analysis.

    (b) A window entitled *Project Cash Flow Analysis* will appear. The first column labels the cash flow elements and other accounting information in income statement format. To enter revenues of $100,000 from source 0 in year 1, click the mouse at that cell, type the amount, and press **Return** (see Fig. 9.15a). Since there is only one source of revenue income in Example 9.4, the rows titled *Source 1* and *Source 2* may be deleted. To do so, click the mouse at the row to be deleted and select **Edit** followed by **Delete** and **Row**.

    (c) Revenue income from *Source 0* for other years (years 2 through 5) can be input in one of two ways. For year-by-year input, move the graphic pointer to the cell, type the amount, and press the [*Return*] key. If there is a pattern (uniform, linear gradient, or geometric gradient) in the data series, you can speed up the input process with the **Copy** and **Paste** options in the **Edit** mode. With the graphic pointer at the entered revenue amount from *Source 0* in year 1, select **Edit** followed by **Copy** and **Cell** (Fig. 9.15a). The value in the cell is copied into the buffer. Then select **Edit** followed by **Paste** and **Uniform** (Fig. 9.15b). **EzCash** will prompt you to enter *To Period*. By entering 5 and clicking **Ok**, you will paste the revenue value in the buffer to cells 1 (where the pointer is) through 5. Similarly, you may enter the cash expenses (*Labor, Material,* and *Overhead*) for all 5 years

Figure 9.15 ■ Generating an after-tax cash flow by **EzCash**—Step 1: Working with **EzCash**'s worksheet

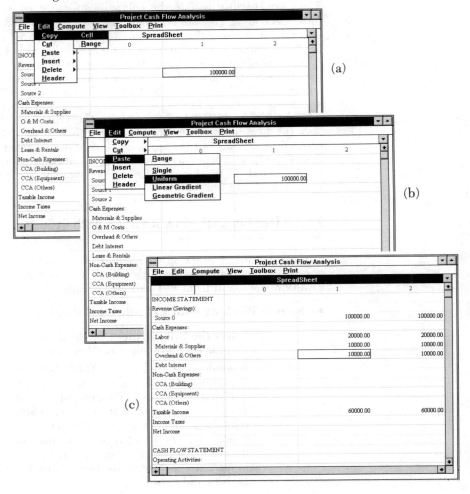

(Fig. 9.15c). The *Labor* row is inserted and the header entered from the **Edit** mode. Note that the *Taxable Income* row is automatically updated.

- **Step 2: Capital Cost Allowance Calculation**

    (a) To enter capital cost allowances, click the mouse at the row titled *CCA (Equipment)*. You may delete the other CCA rows if they do not apply. If you already know the annual CCA amounts, you may enter them year by year. **EzCash** also provides a built-in CCA generator. To access this feature, click **Toolbox** followed by **Capital Cost Allowance.** You will then be prompted with two CCA computation methods: **Declining Balance and Straight Line** (Fig. 9.16a).

Figure 9.16 ■ Generating an after-tax cash flow by **EzCash**—Step 2: Capital cost allowance calculation

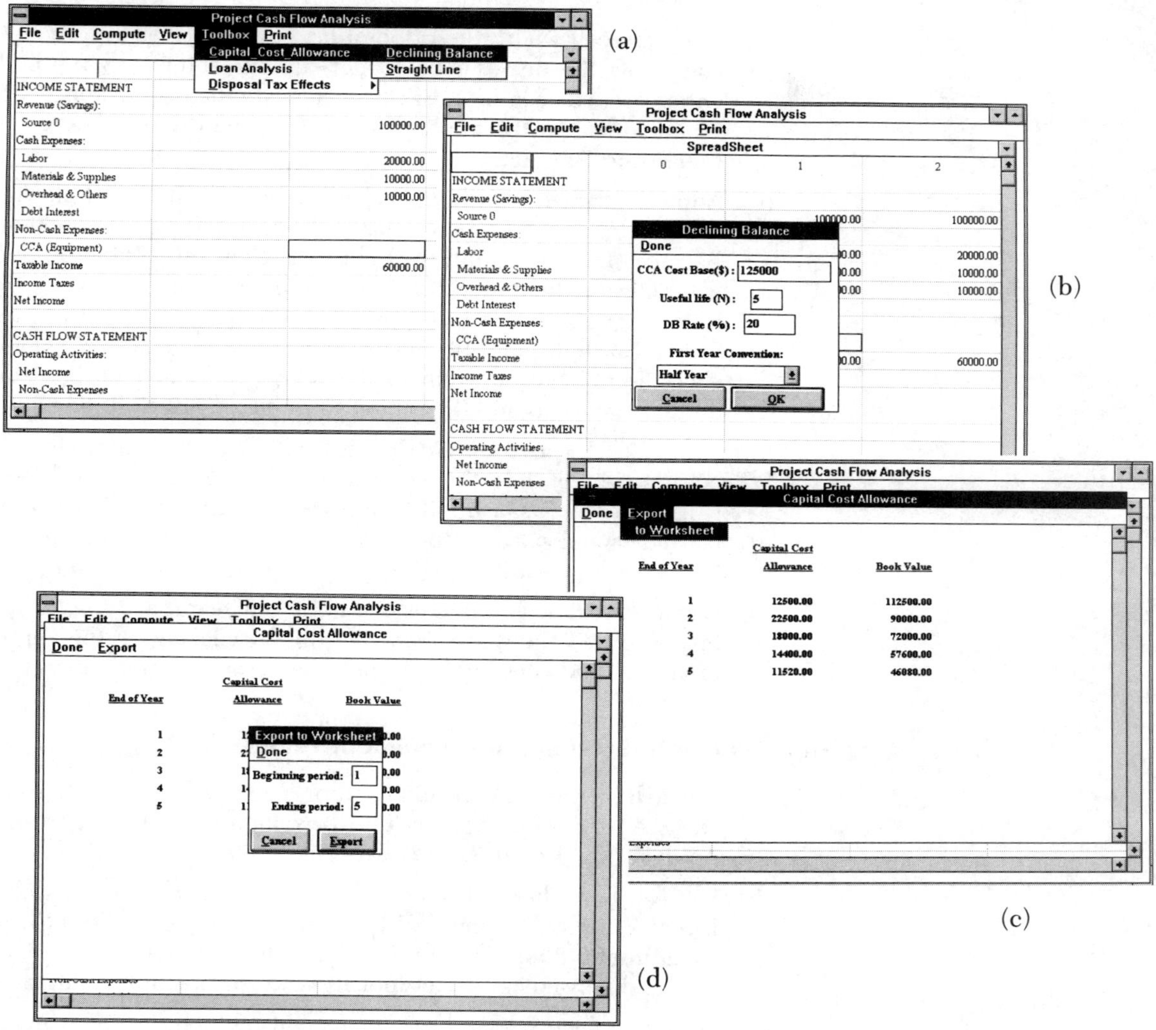

(b) After selecting **Declining Balance**, enter the following information: (1) CCA Cost Base, (2) Useful Life, (3) Declining Balance (DB) Rate, and (4) the First Year Convention (Fig. 9.16b).

(c) After entering all the information requested and clicking the **Ok** button, **EzCash** generates a CCA schedule (Fig. 9.16c). Now you may click **Export** followed by **to Worksheet** to export the calculated CCA amounts to the worksheet. Then, enter the *Beginning Period* and *Ending Period*, and click the **Export** button again to complete the export process (Fig. 9.16d). Finally, click **Done** in order to return to the *Project Cash Flow Analysis* window.

- **Step 3: Disposal Tax Effects Calculations**

    (a) Click the cell where the disposal tax effect is to be entered. If you know the disposal tax effect, simply enter it in the cell. Otherwise, you can have **EzCash** calculate the disposal tax effect by clicking **Toolbox** followed by **Disposal Tax Effect** and **Declining Balance** (Fig. 9.17a).

    (b) After selecting the declining balance method, enter the following information: (1) CCA Cost Base, (2) Useful Life, (3) Declining Balance (DB) Rate, and (4) First Year Convention (Fig. 9.17b). After clicking the **Ok** button, you will be prompted to supply the following: (1) Year to Dispose, (2) Market Value, (3) Capital Gains Tax Rate, and (4) Income Tax Rate (Fig. 9.17c). Then click the **Ok** button again, and **EzCash** will calculate both the ordinary gains tax and the capital gains tax (Fig. 9.17d).

    (c) Now click **Export to Worksheet** and **Done** to enter the disposal tax effect, into the worksheet and return to the worksheet (Fig. 9.17d). **EzCash** will automatically add the ordinary gains tax and capital gains tax together and enter the combined amount in the worksheet with the proper + or − sign (Fig. 9.17e).

    (d) Lastly, enter the equipment investment in year 0 and the salvage in year 5 into the worksheet. Do this by clicking on the appropriate cell and entering the value in the usual manner.

- **Step 4: Loan Interest and Repayment of Principal**

    (a) **EzCash** provides a built-in loan repayment generator that can be accessed by selecting the **Toolbox** mode followed by **Loan Analysis** and **Fixed Rate Loan** (Fig. 9.18a).

    (b) You can now choose from three common repayment schedules: **Equal Principal Payment**, **Equal Interest Payment**, and **Equal Installment Plan** (Fig. 9.18b). These different repayment methods are discussed in the **Computer Notes** section in Chapter 3.

    (c) Once you have selected the repayment method, in this case the **Equal Installment Plan**, you are prompted to enter the following information: (1) APR (%), (2) Loan period, (3) Loan amount, (4) Compounding Frequency, and (5) Payment Frequency (Fig. 9.18c).

    (d) The loan repayment schedule is displayed after the required information is entered and the **Ok** button is clicked (Fig. 9.18d).

    (e) Upon viewing the loan repayment schedule, you can close the window by clicking **Done**. You may also copy the interest charges and the principal repayments back to the worksheet by clicking the **Export** button and specifying the starting and ending periods (Fig. 9.18d).

Figure 9.17 ■ Generating an after-tax cash flow by **EzCash**—Step 3: Disposal tax effects calculations

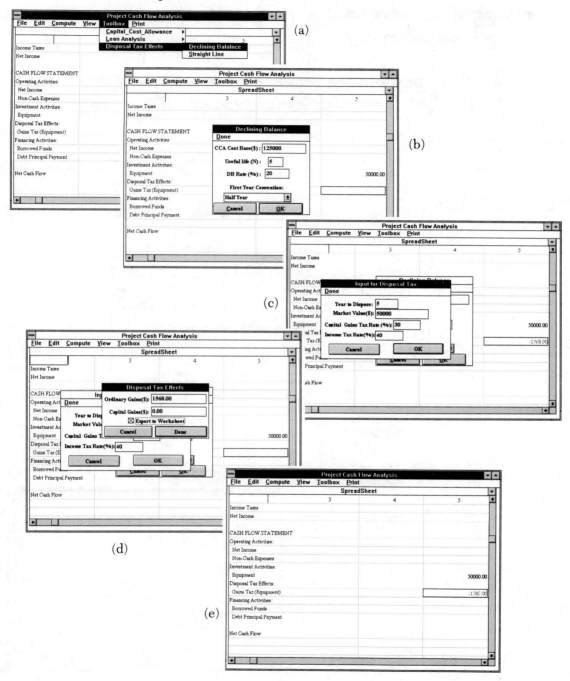

Figure 9.18 ■ Generating an after-tax cash flow by **EzCash**—Step 4: Loan interest and repayment of principal

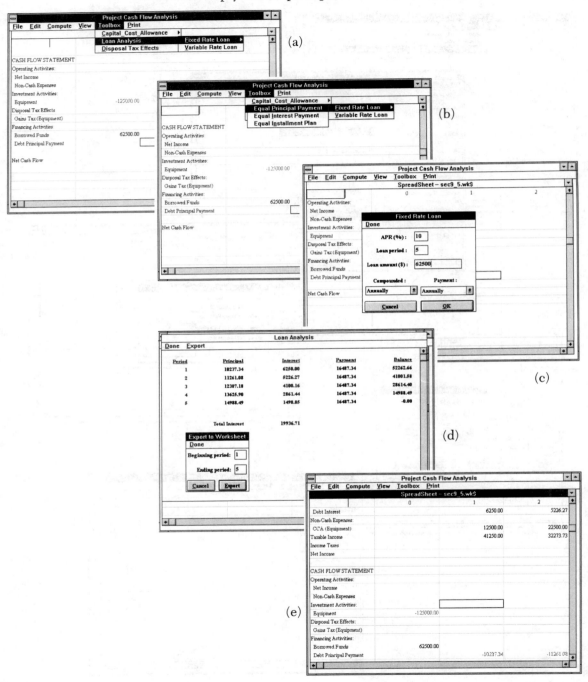

(f) Note that the Borrowed Funds amount in year 0 has to be entered separately. Once you have completed the data entry for the financing activity, the worksheet should look like Fig. 9.18e. **EzCash** automatically enters the data in both the *Debt Interest* row and the *Debt Principal Payment* row.

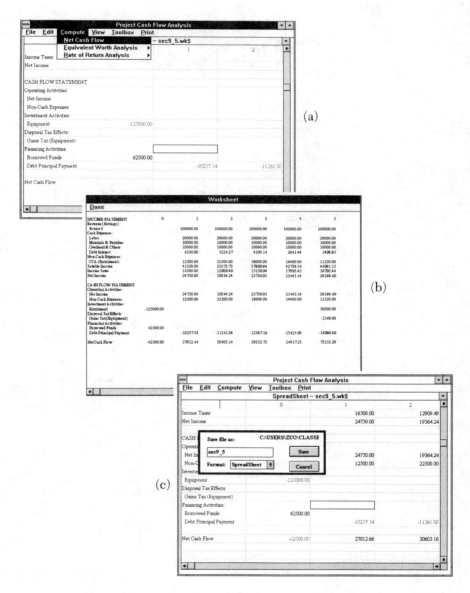

Figure 9.19 ■ Generating an after-tax cash flow by **EzCash**—Step 5: Preparing the net cash flow statement

- **Step 5: Preparing the Net Cash Flow Statement**

    (a) You are now ready to compute the taxable income, income taxes, net income, and net cash flows. To generate the net cash flow, select **Compute** followed by **Net Cash Flow** (Fig. 9.19a). **EzCash** prompts you to enter the tax rate. In our example, type 40.

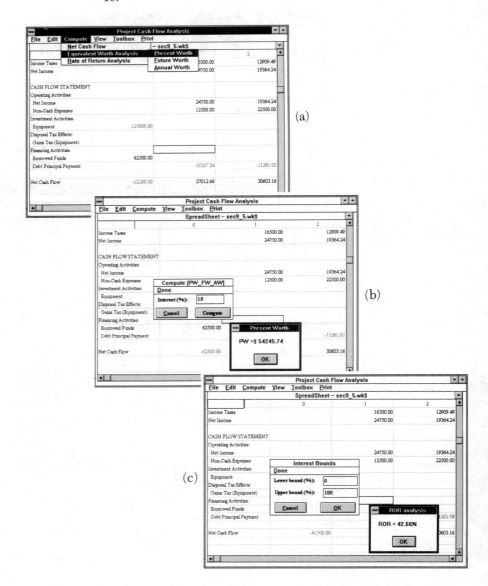

Figure 9.20 ■ Generating an after-tax cash flow by **EzCash**—Step 6: Computing investment worth and rate of return

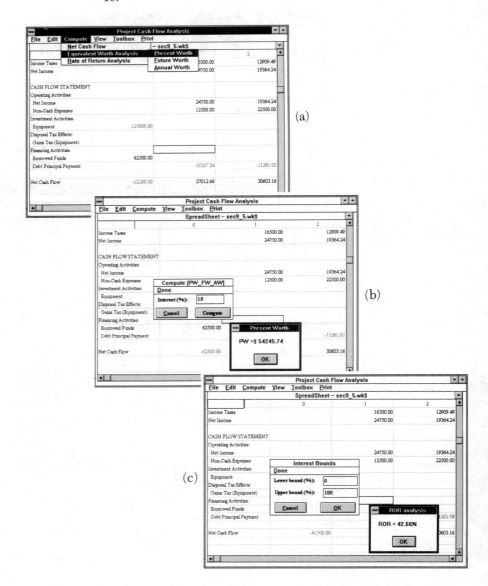

(b) After you specify the tax rate, **EzCash** automatically calculates the taxable income, income taxes, net income, and net cash flows. If you change any data elements in the worksheet, **EzCash** will automatically update the affected columns. To make sure this update is done correctly, select **Compute** and **Net Cash Flow** again to recalculate the worksheet. To view the whole worksheet, select **View** followed by **Worksheet**. The whole worksheet from Example 9.4 is shown in Fig. 9.19(b).

(c) If you are satisfied with the outcome, save the worksheet by selecting **File** followed by **Save As.** You will be prompted to enter a file name. The *Format* of the file should be *SpreadSheet*. The file name entered in Fig. 9.19(c) is sec9_5 and the full file name sec9_5.wk$ is displayed at the top of the worksheet.

(d) You may now send the worksheet to a printer by clicking the **Print** option.

- **Step 6: Computing Investment Worth and Rate of Return**

    (a) By clicking **Compute**, **Equivalent Worth Analysis,** and **Present Worth** in sequence (Fig. 9.20a), you will be prompted to specify the interest rate. After entering this rate (15% is shown) and clicking the **Compute** button, **EzCash** provides a present worth of $54,245.74 (Fig. 9.20b).

    (b) You may also calculate the annual worth and future worth of the project by selecting these options from the **Equivalent Worth Analysis** mode.

    (c) If you select **Compute**, **Rate of Return Analysis**, and **Rate of Return** in sequence, you will be prompted to enter the lower and upper bounds for the interest rate. With a lower bound of 0 and an upper bound of 100% entered, **EzCash** will provide an ROR of 42.66% for the project (Fig. 9.20c).

    (d) If you select **IRR** from the **Rate of Return Analysis** mode, you will be prompted to provide the MARR value. **EzCash** then calculates the IRR value, which is identical to the rate of return value for the simple investment considered in Example 9.4.

Summary

- Identifying and estimating the relevant project cash flows is perhaps the most challenging aspect of an engineering economic analysis. All cash flows can be organized into one of the following three categories:

  1. Investment
  2. Financing
  3. Operations

- The following cash flow types account for the most common flows a project may generate:

  1. New investment and existing assets
  2. Net salvage value
  3. Working capital
  4. Working capital release
  5. Cash revenues/savings
  6. Manufacturing, operating, and maintenance costs
  7. Interest and loan payments
  8. Taxes

- In addition, although they are not cash flows, the following elements may exist in a project analysis and they must be accounted for:

  1. Tax credits

  2. Capital cost allowance

  Table 9.1 summarizes these elements and organizes them as investment, financing, or operations elements.

- The **income statement approach** is typically used in organizing project cash flows. It groups cash flows according to whether they are operating, investing, or financing functions.

- The **generalized cash flow approach** (shown in Table 9.8) to organizing cash flows can be used when a project does not change a company's marginal tax rate. The cash flows can be generated more quickly and the formatting of the results is less elaborate than the income statement approach. There are also analytical advantages which we will discover in later chapters. However, the generalized approach is less intuitive and not commonly understood by business people.

- The **after-tax cash flow diagram approach** is a useful alternative which offers computational efficiency in assessing the economic attractiveness of a project. However, it does not provide information on annual cash flows and is sometimes considered too simplistic.

Problems

9.1 Dartmouth Environmental Company builds residential solar homes. Because of an anticipated increase in business volume, the company is considering the acquisition of environmental equipment that will cost $54,000. The acquisition cost includes delivery charges and applicable taxes. The firm estimated that if the equipment is acquired, the following additional revenues and operating expenses (excluding book and tax depreciation) should be expected:

End of Year	Additional Operating Revenue	Additional Operating Expenses	Allowed CCA
1	$34,000	$24,000	$10,800
2	42,000	24,500	17,280
3	45,000	25,000	10,368
4	56,000	25,800	6,221
5	58,000	26,000	6,221
6	50,000	25,000	3,110

The projected revenue is assumed to be in cash in the year indicated and all the additional operating expenses are expected to be paid in the year in which they incur. The estimated salvage value for the equipment at the end of the sixth year is $5000. The firm's marginal tax rate is 35%. What is the after-tax cash flow if the equipment is acquired?

9.2 An automobile manufacturing company is considering the purchase of an industrial robot for spot welding currently done by skilled labor. The initial cost of the robot is $200,000 and the annual labor savings are projected to be $100,000. If purchased, the robot will be depreciated under CCA Class 43 property. This robot will be used for 7 years, at the end of which time the firm expects to sell the robot for $30,000. The company's marginal tax rate is 40% over the project period. Determine the net after-tax cash flows for each period over the project life.

9.3 An Edmonton company is planning to market an answering device for people working alone who want the prestige that comes with having a secretary but cannot afford one. The device, called Tele-Receptionist, is similar to a voice-mail system. It uses digital recording technology to create the illusion that a person is operating the switchboard at a busy office. The company purchased a 5000 m² building and converted it to an assembly plant for $600,000 ($100,000 for land and $500,000 for the building), and completed installation of assembly equipment worth $500,000 on December 31. The plant will begin operation on January 1. The company expects to have a gross annual income of $2,500,000 over next 5 years. Annual manufacturing costs and all other operating expenses (excluding tax depreciation) are projected to be $1,280,000. For depreciation purposes, the assembly plant building will be classified as CCA Class 1 property and

the assembly equipment as CCA Class 43 property. The property value of the land and building at the end of year 5 would appreciate as much as 15% over the initial purchase cost. The residual value of the assembly equipment is estimated to be about $50,000 at the end of year 5. The firm's marginal tax rate is expected to be about 40% over the project period. Determine the project's after-tax cash flows over the period of 5 years.

9.4 Moncton Plumbing Company is considering a new commercial account for a proposed shopping-mall complex. This 2-year project requires purchase of a new set of plumbing tools worth $55,000. If purchased, the set of tools falls into CCA Class 8 property. At the end of project, the tools will be retained for future business use. The project will bring in an additional annual revenue of $150,000, but it is expected to incur additional annual operating costs of $70,000. The firm's marginal tax rate is known to be 35%. Determine the net after-tax cash flows over the project life.

9.5 A small children's clothing manufacturer is considering an investment in computerizing its management information system for material requirement planning, piece-goods coupon printing, and invoice and payroll. An outside consultant has been asked to estimate the initial hardware requirement and installation costs. He suggests the following:

PC systems (5 PCs, 2 printers)	$60,000
Local area networking system	$12,000
System installation and testing	$3,000

The expected life of these computer systems is 6 years, with an estimated salvage value of $10,000. The proposed system is classified as CCA Class 10 property. A group of computer consultants need to be hired to develop various customized software to run on these systems. These software development costs will be $20,000 and can be expensed during the first tax year. The new system will eliminate two clerks whose combined annual payroll expenses would be $45,000. Additional annual expenses to run this computerized system are expected to be $10,000. No borrowing is considered for this investment. No tax credit is available for this system. The firm's expected marginal tax rate over the next 6 years will be 40%. The firm's interest rate is 20%. Compute the after-tax cash flows over the investment's life, using the income statement approach.

9.6 The Manufacturing Division of Ontario Vending Machine Company is considering Toronto Plant's request for an automatic screw-cutting machine to be included in the division's 1994 capital budget.

- Name of project : Mazda Automatic Screw Machine
- Project cost : $48,018
- Purpose of project: To reduce the cost of some of the parts that are now being subcontracted by this plant, to cut down on inventory due to a shorter lead time, and to better control the quality of the parts. The proposed equipment includes the following cost base:

Machine cost	$35,470
Accessory cost	6,340
Tooling	2,356
Freight	980
Installation	1,200
Sales tax	1,672
Total cost	$48,018

- Anticipated savings: See Table 9.10
- Tax depreciation method: CCA Class 43

- Marginal tax rate: 40%
- MARR: 15%

   (a) Determine the net after-tax cash flows over the project life of 6 years. Assume a salvage value of $3500.

   (b) Is this project acceptable based on NPW criterion?

   (c) Determine the IRR for this investment.

9.7 A firm has been paying a print shop $14,000 annually to print the company's monthly newsletter. The agreement with this print shop has now expired but could be renewed for another 5 years. The new subcontract charges are expected to be 10% higher than what they were in the previous contract. The company is also considering the purchase of a desktop publishing system with a high-quality laser printer driven by a microcomputer. With an appropriate word/graphics software, the newsletter can be composited and printed in a near typeset quality. A special device is also required to print photos on the newsletter. The following estimates have been quoted from a computer vendor:

Microcomputer	$ 5,500
Laser printer	8,500
Photo device/scanner	10,000
Software	2,000
Total cost base	$26,000
Annual O&M costs	$10,000

At the end of 5 years each equipment piece is expected to retain only 10% of its original cost in salvage value. The

Table 9.10 Comparative Cash Flow Elements

Item	Hours		Present Subcontracting	Proposed
	Present M/C labor	Proposed M/C labor		
Setup	350	350		$ 7,700
Run	2410	800		17,600
Overhead				
Indirect labor				3,500
Fringe benefits				8,855
Maintenance				1,350
Tooling				6,320
Repair				890
Supplies				4,840
Power				795
Taxes and insurance				763
Other relevant cost				
Floor space				3,210
Subcontracting			$122,468	
Material				27,655
Other				210
		Total	$122,468	$83,688
Operating advantage				$38,780

company's marginal tax rate is 40%, and the desktop publishing system will be depreciated as CCA Class 10 property.

(a) Determine the projected net after-tax cash flows for the investment.
(b) Compute the IRR for this project.
(c) Is this project acceptable at MARR = 12%?

9.8 An asset in the CCA Class 9 property costs $100,000 and has zero estimated salvage value after 6 years of use. The

asset will generate annual revenues of $300,000 and require $100,000 annual labor and $50,000 annual material expenses. There are no other revenues and expenses. Assume a tax rate of 40%.

(a) Calculate the after-tax cash flows.
(b) Calculate the NPW with MARR = 12%.

9.9 An automaker is considering installing a three-dimensional (3-D) computerized car-styling system at a cost of $200,000 (including hardware and software). With the 3-D computer modeling, designers have the ability to view their design from many angles and to fully account for the space required for the engine and passengers. The digital information that is used to create the computer model can be revised in consultation with engineers, and the data can be used to run milling machines that make physical models quickly and precisely. The automaker expects to decrease turnaround time by 22% for designing a new auto model (from configuration to final design). The expected savings in dollars are $250,000 per year. The training and operating maintenance cost for the new system is expected to be $50,000 per year. The system has a 5-year useful life and can be depreciated as CCA Class 10 property. The system will have an estimated salvage value of $5000. The automaker's marginal tax rate is 40%. Determine the annual cash flows for this investment. What is the return on investment for this project?

9.10 Refer to the data for Dartmouth Environmental Company in Problem 9.1. If the firm expects to borrow the initial investment ($54,000) at 10% over two year periods (equal annual payments of $31,114), determine the project's net cash flows.

9.11 In Problem 9.2, to finance the industrial robot, the company will borrow the entire amount from a local bank and the loan will be paid off at the rate of $50,000 per year plus 10% on the unpaid balance. Determine the net after-tax cash flows over the project life.

9.12 Refer to the data for the children's clothing company in Problem 9.5. Suppose that the initial investment of $75,000 will be borrowed from a local bank at an interest rate of 12% over 5

years (5 equal annual payments). Then recompute the after-tax cash flow.

9.13 Saskatchewan Casting Company is considering the installation of a new process machine for their manufacturing facility. The machine costs $200,000 installed, will generate additional revenues of $50,000 per year, and will save $40,000 per year in labor and material costs. The machine will be financed by a $150,000 bank loan repayable in three equal, annual principal installments, plus 15% interest on the outstanding balance. The machine will be depreciated as CCA Class 43 property. The useful life of this process machine is 10 years, at which time it will be sold for $20,000. The marginal tax rate is 40%.

(a) Find the year-by-year after-tax cash flow for the project using the income statement approach.

(b) Compute the IRR for this investment.

(c) At MARR = 18%, is this project economically justifiable?

9.14 Consider the following financial information about a retooling project at a computer manufacturer:

1. The project costs $1 million and at the end of the fifth year, any assets held for the project will be sold. The expected salvage value will be about 10% of the initial project cost.

2. The retooling project will be considered within CCA Class 43.

3. The firm will finance 40% of the project money from an outside financial institution at an interest rate of 12%. The firm is required to repay the loan with 5 equal annual payments.

4. The firm's marginal tax rate on this investment is 40% and the MARR is 18%.

(a) Determine the after-tax cash flows.

(b) Compute the annual equivalent worth for this project.

9.15 An asset is to be purchased for $23,000, and this amount is to be borrowed with the stipulation that it be repaid by 6 equal end-of-year payments at 10% compounded annually. This asset is expected to provide a net income of $6000 per year for 6 years and is to be depreciated at the CCA Class 38 rate. The salvage value at the end of 6 years is expected to be $3000. Assume a marginal tax rate of 36% and a MARR of 15%.

(a) Determine the after-tax cash flow for this asset through 6 years.

(b) Is this project acceptable based on the IRR criterion?

9.16 A facilities engineer is considering a $50,000 investment in an energy management system (EMS). It is expected to save $15,000 annually in utility bills for 5 years. After 5 years the EMS will have a zero salvage value. Using after-tax analysis, calculate the rate of return on this investment. Assume that this equipment is CCA Class 8 property and the company has a 40% marginal tax rate.

9.17 A manufacturing company is considering the acquisition of a new injection molding machine at a cost of $100,000. Because of a rapid change in product mix, the need for this particular machine is expected to last only 8 years, after which time the machine is expected to

have a salvage value of $10,000. The annual operating cost is estimated to be $5000. The addition of this machine to the current production facility is expected to generate an annual revenue of $40,000. The firm has only $60,000 available from equity funds, so it must borrow the additional $40,000 required at an interest rate of 10% per year with repayment of principal and interest in eight equal annual amounts. The marginal income tax rate applicable for this firm is 40%. Assume that the asset is classified as CCA Class 43 property.

(a) Determine the after-tax cash flows.

(b) Determine the NPW of this project at MARR = 14%.

9.18 Reconsider Problem 6.15 in which a manufacturing company is planning to build a new facility. The land costs $300,000, the building costs $600,000, the equipment costs $250,000, and $100,000 in working capital is required. It is expected that the product will result in additional sales of $625,000 per year for 12 years, at which time the land can be sold for $300,000 after taxes, the building for $200,000 after taxes, the equipment for $50,000 after taxes, and all of the working capital recovered. The annual disbursements for labor, materials, and all other expenses are estimated to be $425,000. The incremental income taxes due to this project income, considering all expenses and CCA, are estimated to be $50,000 per year.[4]

(a) Determine the projected net after-tax cash flows from this investment.

(b) Compare the IRR of this project with that of a no working capital situation.

---

[4] From *Engineering Economy*, 8th Ed., Problem 4.2, E. P. DeGarmo, W. J. Sullivan, and J. A. Bontadelli, Macmillan, 1988.

9.19 An industrial engineer at New Brunswick Textile Mill proposed the purchase of scanning equipment for the warehouse and weave rooms. He felt that the purchase would ensure a better system of locating cartons in the warehouse by recording their location and storing this data in the computer. The estimated investment and annual operating and maintenance costs, and expected annual savings are as follows:

- Cost of equipment and installation: $44,480
- Project life: 7 years
- Expected salvage value: $0
- Investment in working capital (fully recoverable at the end of project life): $10,000
- Expected annual labor and materials savings: $62,755
- Expected annual expenses: $8035
- Depreciation method: CCA Class 43 property

The firm's marginal tax rate is 30%.

(a) Determine the net after-tax cash flows over the project life.

(b) Compute the IRR for this investment.

(c) At MARR = 18%, is this project acceptable?

9.20 Mississauga Chemical Corporation is considering investing in a new composite material. R&D engineers are investigating exotic metal-ceramic and ceramic-ceramic composites to develop materials that will withstand high temperatures, like those to be encountered in the next generation of jet fighter engines. The company expects a 3-year R&D period before applying new materials to commercial products. The following financial information is presented for management review:

1. *R&D cost:* $5 million over a 3-year period, with annual R&D expenditure of $0.5 million at the beginning of year 1, $2.5 million at the beginning of year 2, and $2 million at the beginning of year 3. These R&D expenditures will be expensed rather than amortized for tax purposes.

2. *Capital investment:* $5 million at the beginning of year 4. This investment consists of $2 million in building and $3 million in plant equipment.

3. *Depreciation method:* Building (CCA Class 1 property) and plant equipment (CCA Class 43 property).

4. *Project life:* 10 years after a 3-year R&D period

5. *Salvage value:* 10% of the initial capital investment for the equipment and 50% for the building (after 10 years of operation).

6. *Total sales:* $50 million (at the end of year 4) with an annual sales growth rate of 10% per year (compound growth) during the first 6 years (year 5 through year 11) and −10% (negative compound growth)/year for the remaining process life.

7. *Out-of-pocket expenditures:* 80% of annual sales

8. *Working capital:* 10% of annual sales (considered as investment at the beginning of each production year and recovered investments fully after 10 years of operation)

9. *Marginal tax rate:* 40%

(a) Determine the net after-tax cash flows over the project life.

(b) Determine the IRR for this investment.

(c) Determine the equivalent annual worth for this investment at MARR = 20%.

9.21 Refer to the data in Problem 9.1. Determine the project's net after-tax cash flows using the generalized cash flow approach.

9.22 Refer to the data in Problem 9.9. Determine the project's net after-tax cash flows using the generalized cash flow approach.

9.23 Refer to the data in Problem 9.13. Determine the project's after-tax cash flows using the generalized cash flow approach.

9.24 Refer to the data in Problem 9.16. Determine the project's after-tax cash flows using the generalized cash flow approach.

9.25 Refer to the data in Problem 9.17. Determine the project's after-tax cash flows using the generalized cash flow approach.

9.26 A highway contractor is considering buying a new trench excavator. This machine would be required to dig 12,000 metres of trenching per year. The excavator costs $200,000 and can dig a 1-metre wide trench at the rate of 3 metres per hour. With adequate maintenance, the production rate will remain constant for the first 1200 hours of operation, then decrease by 0.6 metres per hour for each 400 hours thereafter. Maintenance and operating costs will be $15 per hour. The contractor will depreciate the equipment using CCA Class 38. At the end of 5 years, the excavator will be sold for $40,000. Assuming that the contractor's marginal tax rate is 34% per year, determine the annual after-tax cash flow using the generalized cash flow approach.

9.27 Suppose an asset has a first cost of $6000, a life of 5 years, a salvage value of $2000 at the end of 5 years, and a net annual before-tax revenue of $1500. The firm's marginal tax rate is 35%. By using CCA Class 12, determine the cash flow after taxes using the generalized cash flow approach.

9.28 A corporation is considering the purchase of a machine that will save $130,000 per year before taxes. The cost of operating the machine, including maintenance, is $20,000 per year. The machine will be needed for 5 years after which it will have a zero salvage value. CCA Class 8 depreciation will be used. The marginal income tax rate is 40%. If the firm wants 12% IRR after taxes, how much can they afford to pay for this machine?

9.29 Your company needs an air compressor and has narrowed the choice to two alternatives, A and B. The following data have been collected.

	A	B
First cost	$15,000	$25,000
Annual O&M cost	4,600	3,000
Salvage value	0	3,000
Service life	5 years	10 years
Depreciation method	Class 43	Class 43

The marginal income tax rate is 40%. Assuming repeatability of each alternative, which alternative should be selected at MARR = 20%?

9.30 Sydney Mining Company is considering a new mining method at its Sydney mine, called longwall mining by a robot, in which coal is removed not by tunneling like a worm through an apple, leaving more of the target than is removed, but by methodically shuttling back and forth across the width of the deposit and devouring nearly everything. The method can now extract about 75%

of the available coal, compared with 50% for conventional mining, which is done largely with machines that dig tunnels. Moreover, the coal can be recovered far more inexpensively. Currently, the company mines 5 million tonnes a year with 2200 workers. By installing 2 longwall robot machines, the company can mine 5 million tonnes with only 860 workers. (A robot miner can dig more than 6 tonnes of coal every minute.) Despite the loss of employment, the United Mine Workers union generally favors longwall mines for two reasons: The union officials say, "(1) it would be far better to have highly productive operations that were able to pay our folks good wages and benefits than to have 2200 shovelers living in poverty, and (2) longwall mines are inherently safer in their design." The company projected the following financial data upon installing the longwall mining:

Robot installation (2 units)	$9.3 million
Total amount of coal deposit	50 million tonnes
Annual mining capacity	5 million tonnes
Project life	10 years
Estimated salvage value	$0.5 million
Working capital requirement	$2.5 million
Expected additional revenues	
Labor savings	$6.5 million
Accident prevention	$0.5 million
Productivity gain	$2.5 million
Expected additional expenses:	
O&M costs	$2.4 million
Marginal tax rate	40%

Estimate the firm's net after-tax cash flows if the firm chooses to depreciate the robots based on CCA Class 43 property.

9.31 Ampex Corporation produces a wide variety of tape cassettes for the commercial and government markets. Due to the increased competition in producing VHS cassettes, Ampex is concerned with pricing the product competitively. Currently, Ampex has 18 loaders that load cassette tapes in VHS cassette shells. Each loader is manned by one operator per shift. Ampex currently produces 25,000 tapes per week and operates 15 shifts per week, 50 weeks per year.

As a means of reducing the unit cost, Ampex can purchase the cassette shells for $0.15 less each than they can currently produce them. A supplier has guaranteed a price of $0.77 per cassette for the next 3 years. However, Ampex's current loaders will not properly load the proposed shells. To accommodate the vendor's shells, Ampex must purchase 8 KING-2500 VHS loaders at a cost of $40,000 each. In order for the machines to operate properly, Ampex will also have to purchase $20,827 worth of conveyor equipment, which will be included in the overall depreciation base of $340,827. The new machines are much faster and will handle more than the current demand of 25,000 cassettes per week. The new loaders will require 2 people per machine per shift, 3 shifts per day, 5 days a week. The new machines fall into the CCA Class 8 equipment and will have an approximate life of 8 years. At the end of project life, Ampex expects the market value for each loader to be $3000.

The average pay of these employees is $8.27 per hour, and adding 23% for benefits will make it $10.17 per hour. The new loaders are simple to operate, therefore, the training impact of the alternative is minimal. The operating cost, including maintenance, is expected to stay the same for the new loaders. This cost will not be considered in the analysis. The cash inflows from the project will be the material savings per cassette of $0.15 and the labor savings of 2 employees per shift. This gives an annual savings in materials and labor costs of $187,500 and $122,065, respectively. If the new loaders are purchased, the old machines will be shipped to other plants for standby use. Therefore, there will be no disposal tax effects. Ampex's combined marginal tax rate is running at 40%.

(a) Determine the after-tax cash flows over the project life.

(b) Determine the IRR for this investment.

(c) Is this investment profitable at MARR = 15%?

9.32 Halifax Developments is planning to purchase a new piece of equipment for $62,000. The equipment is a CCA Class 38 asset with a declining balance rate of 30%. This unit is expected to produce net annual revenues of $20,000 per year for the next 4 years. At the end of 4 years it will have an expected salvage value of $20,000. The company has a marginal tax rate of 40% and requires an after-tax MARR of 8%.

a) Calculate the present worth after tax by using the general cash flow approach.

b) Calculate the present worth after tax by using the after-tax cash flow diagram approach (i.e., using capital tax factors).

9.33 An Edmonton construction company is planning to purchase a new $30,000 truck (CCA Class 10 property with a 30% declining balance rate). The vehicle will have a $5,000 salvage value at the end of its 6-year life. Net revenues as a result of this purchase are expected to be $10,000 per year. The company has a marginal tax rate of 40%.

a) Calculate the IRR using the general cash flow approach.

b) Calculate the IRR using the after-tax cash flow diagram approach (i.e., using capital tax factors).

CHAPTER 10

Inflation and Economic Analysis

Y ou may have heard your parents fondly remembering the "good old days" of penny candy or 17 cent-per litre gasoline. But even a college student in his or her early twenties can relate to the phenomenon of escalating costs. Do you remember when a postage stamp cost 30 cents? When the price of a first-run movie was under $5?

The accompanying table demonstrates price differences between 1979 and 1993 for some fundamental human needs. For example, a loaf of bread cost 54 cents in 1979, whereas it cost $1.13 in 1993. The same 54 cents in 1993 would buy only a fraction of the bread it bought in 1979—specifically, about 47.8% of a package (0.54/1.13). The 54-cent sum had lost 52.2% of its purchasing power from 1979 to 1993. On a brighter note, we open this chapter with some statistics concerning engineers' pay rates.

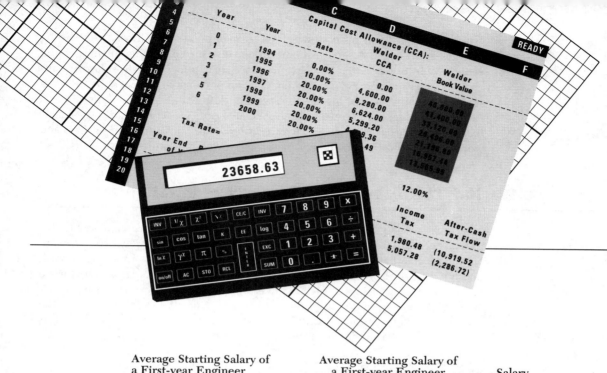

Average Starting Salary of a First-year Engineer in 1979	Average Starting Salary of a First-year Engineer in 1993	Salary Increase (%)
$18,318	$35,691	94.8%

Item	1979 Price	1993 Price	Price Increase(%)
Annual housing expense	$3429.00	$8102.00	136
Annual automobile expense	1067.00	2921.00	174
Store brand bread	0.54	1.13	109
Large eggs	0.91	1.44	58
Milk	0.56	1.85	230
Soft drink	0.51	0.87	71
Round steak	2.88	10.26	256
Postage	0.17	0.43	153
College tuition	895.00	2026.00	126
College room and board	1800.00	5050.00	181

Up to this point, we have shown how to manipulate cash flows in a variety of ways and how to compare them under constant conditions in the general economy. In other words, we have assumed that prices remain relatively unchanged over long periods. As you know from personal experience, this is not a realistic assumption. In this chapter, we will define and quantify *inflation* and then go on to apply it in several economic analyses. We will demonstrate its effect on capital cost allowance, borrowed funds, rate of return of a project, and working capital within the bigger picture of developing project cash flows.

■ 10.1 The Meaning and Measure of Inflation

Historically, the general economy has usually fluctuated in such a way that it experiences **inflation**, a loss in the purchasing power of money over time. Inflation means that the cost of an item tends to increase over time, or, to put it another way, the same dollar amount buys less of an item over time. **Deflation** is the opposite of inflation, in that prices decrease over time and hence a specified dollar amount gains in purchasing power. Inflation is far more common than deflation in the real world, so we will restrict our consideration in this chapter to accounting for inflation in economic analyses.

10.1.1 Measuring Inflation

Before we can introduce inflation into an engineering economic problem, we need a means of isolating and measuring the effect of inflation. Consumers usually have a relative if not a precise sense of how their purchasing power is declining, based on their experience shopping for food, clothing, transportation, and housing over the years. Economists have developed a measure called the **consumer price index (CPI)**, which is based on a typical **market basket** of goods and services required by the average consumer. This market basket normally consists of items from seven major groups: (1) food, (2) housing, (3) clothing, (4) transportation, (5) health and personal, (6) recreation, reading and education, and (7) tobacco products and alcoholic beverages.

The CPI compares the cost of the market basket of goods and services this month with its cost 1 month ago, or 1 year ago, or 10 years ago. The point in the past to which the current prices are compared is called the **base period.** The base period for the current index is 1986. That is, the total costs of the prescribed market basket in 1986 are given a price index of 100. *The Consumer Price Index* (Catalogue No. 62-001), a monthly publication prepared by Statistics Canada, lists various consumer price indexes. From Table 10.1, the CPI for the year 1992 is 128.1 and for 1993 is 130.4. Thus the consumer product inflation rate from 1992 to 1993 can be calculated as

$$\frac{130.4 - 128.1}{128.1} = 0.01795 = 1.795\%$$

This method of assessing inflation does not imply, however, that consumers will actually purchase the same goods and services year after year. Consumers tend to adjust their shopping practices to changes in relative prices and to substitute for items whose prices have greatly increased in relative terms. We must understand that the CPI does not take into account this sort of consumer behavior, because it is predicated on the purchase of a fixed market basket of the same goods and services, in the same proportions, month after month. For this reason, the CPI is called a **price index** rather than a **cost-of-living index**, although the general public often refers to it as a cost-of-living index.

Industry Price Index

The consumer price index is a good measure of the general price increases of consumer products. However, it is not a good measure of industry price increases. When doing engineering economy analysis, one has to select proper price indexes to estimate the price increases of raw materials, finished products, and operating costs. *Industry Price Indexes* (Catalogue No. 62-011), a monthly publication prepared by Statistics Canada, provides the Industrial Product Price Index (IPPI) and the Raw Material Price Index (RMPI) for various industrial goods. Table 10.1 lists the CPI together with several price indexes over a number of years.

Table 10.1 Selected Price Indexes (1986 = 100)

Year	CPI	IPPI	RMPI	Lumber & Timber	Primary Steel Products
1986	100.0	100.0	100.0	100.0	100.0
1987	104.4	102.8	107.3	99.0	101.5
1988	108.6	107.2	103.8	99.0	107.5
1989	114.0	109.5	107.2	99.7	110.1
1990	119.5	109.8	111.6	98.1	107.9
1991	126.2	108.7	104.7	96.8	105.6
1992	128.1	109.2	105.7	112.2	99.5
1993	130.4	112.9	111.9	153.2	102.2

From Table 10.1, one can easily calculate the inflation rate of primary steel products from 1991 to 1992:

$$\frac{99.5 - 105.6}{105.6} = -0.05777 = -5.777\%$$

Since the inflation rate calculated is negative, this means that the price of primary steel products actually decreased rather than increased over the year 1992.

Average Inflation Rate (f)

To account for the effect of varying yearly inflation rates over a period of several years, we can compute a single rate that represents an **average inflation rate.** Since each individual year's inflation rate is based on the previous year's rate, they have a compounding effect. As an example, suppose we want to calculate the average inflation rate for a 2-year period: The first year's inflation rate is 4% and the second year's is 8%, using a base price of $100.

- Step 1: We find the price at the end of the second year by the process of compounding:

$$\underbrace{\$100(1 + 0.04) \overbrace{(1 + 0.08)}^{\text{Second year}}}_{\text{First year}} = \$112.32.$$

- Step 2: To find the average inflation rate f, we establish the following equivalence equation:

$$\$100(1 + f)^2 = \$112.32 \leftarrow 100(F/P, f, 2) = \$112.32.$$

Solving for f yields

$$f = 5.98\%.$$

We can say that the price increases in the last 2 years are equivalent to an average annual percentage rate of 5.98% per year. Note that the average is a geometric not arithmetic average over a several-year period. Our computations are simplified by using a single average rate such as this rather than a different rate for each year's cash flows.

Example 10.1 Average inflation rate

Reconsider the price increases for the ten items listed in the table at the beginning of the chapter. Determine the average inflation rate for each item over the 14-year period.

Solution

Given: Prices for 10 items in 1979 and 1993
Find: Average inflation rate for each item

Let's take the first item, annual housing expense, for a sample calculation. Since we know the prices during both 1979 and 1993, we can use the appropriate equivalence formula (single-payment compound amount factor, or growth formula).

Given: $P = \$3429.00$, $F = \$8102.00$, and $N = 1993 - 1979 = 14$

Find: f

Equation: $F = P(1 + f)^N$

$$\$8102.00 = \$3429.00(1 + f)^{14}$$

$$f = (2.3628)^{\frac{1}{14}} - 1$$

$$= 0.06334 = 6.334\%$$

In a similar fashion, we obtain the average inflation rates for the remaining items. These are summarized as follows:

Item	1979 Price	1993 Price	Average Inflation Rate
Annual housing expense	$3429.00	$8102.00	6.33%
Annual automobile expense	1067.00	2921.00	7.46
Store brand bread	0.54	1.13	5.42
Large eggs	0.91	1.44	3.33
Milk	0.56	1.85	8.91
Soft drink	0.51	0.87	3.89
Round steak	2.88	10.26	9.50
Postage	0.17	0.43	6.85
College tuition	895.00	2026.00	6.01
College room and board	1800.00	5050.00	7.65

General Inflation Rate ($\bar{f}$)

When we use the CPI as a base to determine the average inflation rate, we obtain the **general inflation rate.** We need to distinguish carefully between the general inflation rate and the average inflation rate for specific goods.

- **General inflation rate ($\bar{f}$):** This average inflation rate is calculated based on the CPI. The market interest rate is expected to respond to this general inflation rate.

- **Average inflation rate (f_j):** This rate is based on some indexes other than the CPI; that is, an index specific to segment j of the economy, for example, IPPI or RMPI. Additionally, we must often estimate the future cost for an item such as labor, material, housing, or gasoline. (When we refer to the average inflation rate for just one item, we will drop the subscript j for simplicity.)

In terms of CPI, we define the general inflation rate as

$$CPI_n = CPI_0(1 + \bar{f})^n, \tag{10.1}$$

or

$$\bar{f} = \left[\frac{CPI_n}{CPI_0}\right]^{1/n} - 1, \tag{10.2}$$

where

$$\bar{f} = \text{general inflation rate,}$$

$$CPI_n = \text{the consumer price index at the end period } n,$$

$$CPI_0 = \text{the consumer price index for the base period.}$$

Knowing the CPI values for 2 consecutive years, we can calculate the annual general inflation rate:

$$\bar{f}_n = \frac{CPI_n - CPI_{n-1}}{CPI_{n-1}}, \tag{10.3}$$

where

$$\bar{f}_n = \text{the general inflation rate for period } n.$$

As an example, let us calculate the general inflation rate for the year 1992, where $CPI_{1991} = 126.2$ and $CPI_{1992} = 128.1$.

$$\frac{128.1 - 126.2}{126.2} = 0.0151 = 1.51\%$$

Example 10.2   Yearly and average inflation rates

The following table shows a utility company's cost to supply power to a new housing development; the indices are specific to the utilities industry. Assume that year 0 is the base period.

Year	Cost
0	$504,000
1	538,400
2	577,000
3	629,500

Determine the inflation rate for each period, and calculate the average inflation rate over the 3 years.

Solution

Given: Prices over a period of 3 years

Find: Inflation rates in each of the years, average inflation rate f

Inflation rate during year 1 (f_1): ($538,400 − $504,000)/$504,000 = 6.83%

Inflation rate during year 2 (f_2): ($577,000 − $538,400)/$538,400 = 7.16%

Inflation rate during year 3 (f_3): ($629,500 − $577,000)/$577,000 = 9.10%

The average inflation rate over 3 years is

$$f = \left(\frac{\$629,500}{\$504,000}\right)^{1/3} - 1 = 0.0769 = 7.69\%$$

Note that, although the average inflation rate[1] is 7.69% for the period taken as a whole, none of the years within the period had this inflation rate.

10.1.2 Actual Versus Constant Dollars

To introduce the effect of inflation into our economic analysis, we need to define several inflation-related terms.[2]

- **Actual (current) dollars** (A_n): Estimates of future cash flows for year n which take into account any anticipated changes in amount due to inflationary or deflationary effects. Usually these amounts are determined by applying an inflation rate to base-year dollar estimates.

- **Constant (real) dollars** (A'_n): Dollars of constant purchasing power independent of the passage of time. In situations where inflationary or deflationary effects have been assumed when cash flows were estimated, those estimates can be converted to constant dollars (base-year dollars) by adjustment using some readily accepted **general inflation/deflation rate**. We will assume that the base year is always time 0 unless we specify otherwise.

Conversion from Constant to Actual Dollars

Since constant dollars represent dollar amounts expressed in terms of purchasing power of the base year, we may find the equivalent dollars in year n using the general inflation rate $(\overline{f})$.

$$A_n = A'_n(1 + \overline{f})^n \leftrightarrow A'_n(F/P, \overline{f}, n), \tag{10.4}$$

---

[1] Since we obtained this average rate based on costs that are specific to the utility industry, it is not the general inflation rate.

[2] Based on the ANSI Z94 Standards Committee on Industrial Engineering Terminology, *The Engineering Economist*, Vol. 33, No.2. pp. 145–171, 1988.

where

A'_n = constant-dollar expression for the cash flow occurring at the end of year n,

A_n = actual-dollar expression for the cash flow at the end of year n.

If the future price of a specific cash flow element (j) is not expected to follow the general inflation rate, we need to use the appropriate inflation rate applicable to this cash flow element, f_j, instead of $\bar{f}$.

Example 10.3 Conversion from constant to actual dollars

Transco Company is considering making and supplying computer-controlled traffic-signal switching boxes to be used throughout Western Canada. Transco has estimated the market for its boxes by examining data on new road construction, and on deterioration and replacement of existing units. The current price per unit is $550; the before-tax manufacturing cost per unit for Transco would be $450. The start-up investment cost is $250,000. The projected sales and net before-tax cash flows in constant dollars are as follows.

Period	Unit Sales	Net Cash Flow in Constant $
0		−$250,000
1	1,000	100,000
2	1,100	110,000
3	1,200	120,000
4	1,300	130,000
5	1,200	120,000

Assume that the price per unit as well as the manufacturing cost will keep up with the general inflation rate, which is projected to be 5% annually. Convert the project's before-tax cash flows into the equivalent actual dollars.

Solution

Given: Cash flow in constant dollars, $\bar{f}$ = 5% per year

Find: Actual dollar flows

We first convert the constant dollars into actual dollars. Using Eq. (10.4), we obtain the following. (Note that the cash flow in year 0 is not affected by inflation.)

Period	Net Cash Flow in Constant $	Conversion Factor	Net Cash Flow in Actual $
0	−$250,000	$(1 + 0.05)^0$	−$250,000
1	100,000	$(1 + 0.05)^1$	105,000
2	110,000	$(1 + 0.05)^2$	121,280
3	120,000	$(1 + 0.05)^3$	138,920
4	130,000	$(1 + 0.05)^4$	158,020
5	120,000	$(1 + 0.05)^5$	153,150

Figure 10.1 illustrates this graphically.

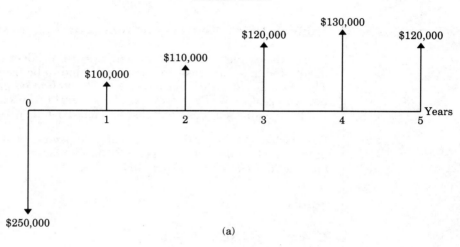

(a)

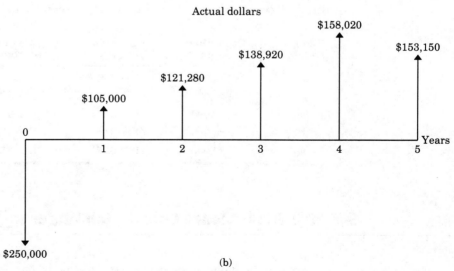

(b)

Figure 10.1 ■ Converting constant dollars (a) to equivalent actual dollars (b) (Example 10.3)

Conversion from Actual to Constant Dollars

This is the reverse process of the conversion from constant to actual dollars. Instead of using the compounding formula, we use a discounting formula (single-payment present worth factor).

$$A'_n = \frac{A_n}{(1 + \bar{f})^n} \leftrightarrow A_n(P/F, \bar{f}, n) \tag{10.5}$$

Once again, we may substitute f_j for $\bar{f}$ if the future prices are not expected to follow the general inflation rate.

Example 10.4   Conversion from actual to constant dollars

Jagura Creek Fish Company, an aquicultural production firm, has negotiated a 5-year lease on 20 acres of land to be used for fish ponds. The annual cost stated in the lease is $20,000, to be paid at the beginning of each of the 5 years. The general inflation rate is $\bar{f} = 5\%$. Find the equivalent constant cost in each period.

Discussion: Although the $20,000 annual payments are *uniform*, they are not in constant dollars. Unless an inflation clause is built into a contract, any stated amounts refer to *actual dollars*.

Solution

Given: $A_n = \$20,000, \bar{f} = 5\%$
Find: A'_n in each year. Using Eq. (10.5), we can find:

End of Period	Cash Flow in Actual $	Conversion at f	Cash Flow in Constant $
0	−$20,000	$(1 + 0.05)^0$	−$20,000
1	−20,000	$(1 + 0.05)^{-1}$	−19,048
2	−20,000	$(1 + 0.05)^{-2}$	−18,141
3	−20,000	$(1 + 0.05)^{-3}$	−17,277
4	−20,000	$(1 + 0.05)^{-4}$	−16,454

■ 10.2 Equivalence Calculation Under Inflation

In previous chapters, our equivalence analyses have taken into consideration changes in the **earning power** of money—that is, interest effects. To factor in changes in **purchasing power** as well—that is, inflation—we may use either (1) constant dollar analysis or (2) actual dollar analysis. Either method will produce the same solution; however, each uses a different interest rate and procedure. Before presenting the two procedures for integrating interest and inflation, we will give a precise definition of each interest rate.

10.2.1 Market and Inflation-Free Interest Rates

There are two types of interest rate for equivalence calculation: (1) the market interest rate and (2) the inflation-free interest rate. The interest rate that is applicable depends on the assumptions used in estimating the cash flow.

- **Market interest rate (i):** This interest rate takes into account the combined effects of the earning value of capital (earning power) and any anticipated inflation or deflation (purchasing power). Virtually all interest rates stated by financial institutions for loans and savings accounts are market interest rates. Most firms use a market interest rate (also known as **inflation-adjusted MARR**) in evaluating their investment projects.

- **Inflation-free interest rate (i'):** An estimate of the true earning power of money when inflation effects have been removed. This rate is commonly known as **real interest rate,** and it can be computed if the market interest rate and inflation rate are known. In fact, all the interest rates mentioned in the previous chapters are inflation-free interest rates. As you will see later in this chapter, the market interest rate will be the same as the inflation-free interest rate in the absence of inflation.

In calculating any cash flow equivalence, we need to identify the nature of project cash flows. There are three common cases:

- **Case 1:** All cash flow elements are estimated in constant dollars.
- **Case 2:** All cash flow elements are estimated in actual dollars.
- **Case 3:** Some of the cash flow elements are estimated in constant dollars and others are estimated in actual dollars.

For case 3, we simply convert all cash flow elements into one type—either constant or actual dollars. Then we proceed with either constant-dollar analysis for case 1 or actual-dollar analysis for case 2. Therefore, we will only discuss the first two situations in the following.

10.2.2 Constant-Dollar Analysis

Suppose that all the cash flow elements are already given in constant dollars. Now you want to compute the equivalent present worth of the constant dollars (A'_n) occurring in year n. Since there is no inflationary effect, we should use i'

to account for only the earning power of the money. To find the present worth equivalent of this constant dollar at i', we use

$$P_n = \frac{A_n'}{(1 + i')^n}.$$ (10.6)

Example 10.5 Equivalence when flows are stated in constant dollars

Consider the constant-dollar flows originally presented in Example 10.3. If Transco managers want the company to earn a 12% inflation-free rate of return (i') on any investment, what would be the present worth of this project?

Solution

Given: Constant-dollar flows, $i' = 12\%$

Find: Present worth

Since all values are in constant dollars, we can use the inflation-free interest rate. We simply discount the dollar inflows at 12% to obtain the following:

$$PW(12\%) = -\$250,000 + \$100,000(P/A, 12\%, 5)$$

$$+ \$10,000(P/G, 12\%, 4) + \$20,000(P/F, 12\%, 5)$$

$$= \$163,099 \text{ (in year 0 dollars)}.$$

Since the equivalent net receipts exceed the investment, the project can be justified.

10.2.3 Actual-Dollar Analysis

Now let us assume that all cash flow elements are estimated in actual dollars. To find the equivalent present worth of this actual dollar (A_n) in year n, we may use either the **deflation method** or the **adjusted-discount method**.

Deflation Method

The deflation method requires two steps to convert the actual dollars into equivalent present worth dollars. First we convert actual dollars into equivalent constant dollars by discounting by the general inflation rate, a step that removes the inflationary effect. Now we can find the equivalent present worth using i'.

Example 10.6 Equivalence when flows are in actual dollars: deflation method

Applied Instrumentation, a small manufacturer of custom electronics, is contemplating an investment to produce sensors and control systems that have been requested by a fruit drying company. The work would be done under a proprietary contract, which would terminate in 5 years. The project is expected to generate the following cash flows in actual dollars.

n	Net Cash Flow in Actual Dollars
0	−$75,000
1	32,000
2	35,700
3	32,800
4	29,000
5	58,000

(a) What are the equivalent year 0 dollars (constant dollars) if the general inflation rate ($\bar{f}$) is 5% per year?
(b) Compute the present worth of these cash flows at $i' = 10\%$.

Solution

Given: Actual dollar flows, $\bar{f} = 5\%$, $i' = 10\%$
Find: Present worth

The net cash flows in actual dollars will be converted to constant dollars by deflating them, again assuming a 5% yearly deflation factor. The deflated or constant-dollar cash flows will then be used to determine the NPW at i'.

(a) We convert the actual dollars into constant dollars as follows:

n	Cash Flows in Actual Dollars	Multiplied by	Cash Flows in Constant Dollars
0	−$75,000	1	−$75,000
1	32,000	$(1 + 0.05)^{-1}$	30,476
2	35,700	$(1 + 0.05)^{-2}$	32,381
3	32,800	$(1 + 0.05)^{-3}$	28,334
4	29,000	$(1 + 0.05)^{-4}$	23,858
5	58,000	$(1 + 0.05)^{-5}$	45,445

Figure 10.2(a) shows the cash flows measured in constant (year 0) dollars.

Figure 10.2 ■ Illustration of deflation method (Example 10.6)

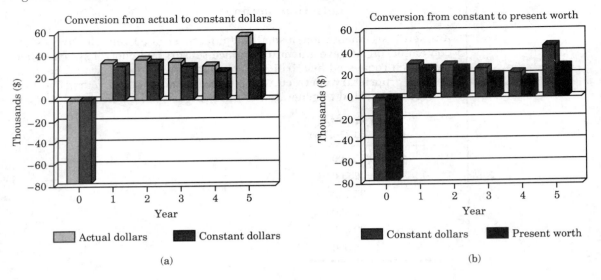

(b) We compute the equivalent present worth of constant dollars using $i' = 10\%$:

n	Cash Flows in Constant Dollars	Multiplied by	Equivalent Present Worth
0	−$75,000	1	−$75,000
1	30,476	$(1 + 0.10)^{-1}$	27,706
2	32,381	$(1 + 0.10)^{-2}$	26,761
3	28,334	$(1 + 0.10)^{-3}$	21,288
4	23,858	$(1 + 0.10)^{-4}$	16,295
5	45,445	$(1 + 0.10)^{-5}$	28,218
			$45,268

Figure 10.2(b) shows conversion from the cash flows stated in year 0 dollars to equivalent present worth.

The Adjusted-Discount Method

The two-step process shown in Example 10.6 can be greatly streamlined by the efficiency of the **adjusted-discount method**, which performs deflation and discounting in one step. Mathematically we can combine this two-step procedure into one by

$$P_n = \frac{\dfrac{A_n}{(1 + \overline{f})^n}}{(1 + i')^n}$$

$$= \frac{A_n}{(1 + \overline{f})^n (1 + i')^n}$$

$$= \frac{A_n}{[(1 + \overline{f})(1 + i')]^n}. \qquad (10.7)$$

Since the market interest rate (i) reflects both the earning power and the purchasing power, we have the following relationship:

$$P_n = \frac{A_n}{(1 + i)^n}. \qquad (10.8)$$

The equivalent present worth values in Eqs. (10.7) and (10.8) must be equal at year 0. Therefore,

$$\frac{A_n}{(1 + i)^n} = \frac{A_n}{[(1 + \overline{f})(1 + i')]^n}.$$

This leads to the following relationship among $\overline{f}$, i', and i:

$$(1 + i) = (1 + \overline{f})(1 + i')$$
$$= 1 + i' + \overline{f} + i'\overline{f}.$$

Simplifying the terms yields

$$i = i' + \overline{f} + i'\overline{f}. \qquad (10.9)$$

This implies that the market interest rate is a nonlinear function of two terms, i' and $\overline{f}$. Note that if there is no inflationary effect, the two interest rates are the same ($\overline{f} = 0 \rightarrow i = i'$). As either i' or $\overline{f}$ increases, i also increases. For example, we can easily observe that, when prices are increasing due to inflation, bond rates climb, because lenders (that is anyone who invests in a money-market fund, bond, or certificate of deposit) demand higher rates to protect themselves against the eroding value of their dollars. If inflation were at 3%, you might be satisfied with an interest rate of 7% on a bond because your return would more than beat inflation. If inflation were running at 10%, however, you would not buy a 7% bond; you might insist instead on a return of at least 14%. On the other hand, when prices are coming down, or at least are stable, lenders do not fear the loss of purchasing power with the loans they make, so they are satisfied to lend at lower interest rates.

Example 10.7 Equivalence when flows are in actual dollars: adjusted-discount method

Consider the cash flows in actual dollars in Example 10.6. Compute the equivalent present worth of these cash flows using the adjusted-discount method.

Solution

Given: Actual dollar flows, $\bar{f} = 5\%$, $i' = 10\%$
Find: Present worth

First, we need to determine the market interest rate (i). With $\bar{f} = 5\%$ and $i' = 10\%$, we obtain

$$i = i' + \bar{f} + i'\bar{f}$$
$$= 0.10 + 0.05 + 0.005$$
$$= 15.5\%.$$

n	Cash Flows in Actual Dollars	Multiplied by	Equivalent Present Worth
0	$-\$75,000$	1	$-\$75,000$
1	32,000	$(1 + 0.155)^{-1}$	27,706
2	35,700	$(1 + 0.155)^{-2}$	26,761
3	32,800	$(1 + 0.155)^{-3}$	21,288
4	29,000	$(1 + 0.155)^{-4}$	16,295
5	58,000	$(1 + 0.155)^{-5}$	28,218
			$\$45,268$

The conversion is shown in Fig. 10.3. Note that the equivalent present worth that we obtain using the adjusted-discount method ($i = 15.5\%$) is exactly the same as our result in Example 10.6.

Figure 10.3 ■ Adjusted-discount method (Example 10.7)

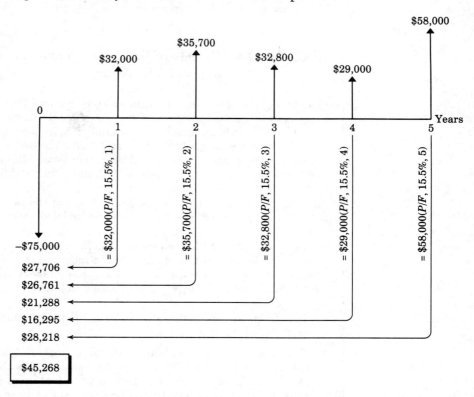

In summary, if the cash flow is estimated in terms of actual dollars, the market interest rate (i) should be used. If the cash flow is estimated in terms of constant dollars, the inflation-free interest rate (i') should be used.

■ 10.3 Effects of Inflation on Project Cash Flows

We will now introduce inflation into some fairly complex investment projects. We are especially interested in two elements of project cash flows—capital cost allowances and interest expenses—that are essentially immune to the effects of inflation. We will also consider the complication of how to proceed when multiple price indexes have been used in generating various project cash flows.

10.3.1 Capital Cost Allowances Under Inflation

Because capital cost allowances are calculated on some base-year purchase amount, they do not increase over time to keep pace with inflation. Thus, they lose some of their value to defer taxes as inflation drives up the general price level and hence taxable income. Similarly, salvage values of depreciable assets can increase with the general inflation rate and, because any gains on salvage values are taxable, they can result in increased taxes.

Example 10.8 Effects of inflation on projects with depreciable assets

Consider the investment situation given in Example 9.1. What will happen to this investment project if the general inflation rate during the next 5 years is expected to be 5% annually? Sales and operating costs are assumed to increase accordingly. Capital cost allowance will be unchanged, but taxes, profits, and thus cash flow, will be higher. The firm's inflation-free interest rate (i') is known to be 15%.

(a) Determine the NPW of the project, using the deflation method.
(b) Compare the NPW with that in the inflation-free situation.

Discussion: All cash flow elements except capital cost allowances are assumed to be in constant dollars. Since income taxes are levied on actual taxable income, we will use the actual-dollar analysis, which requires all cash flow elements be expressed in actual dollars.

• For the purposes of this illustration, all inflationary calculations are made as of year end.

• Cash flow elements such as sales, labor, material, overhead, and salvage value will be inflated at the same rate as the general inflation rate.[3] For example, whereas annual sales had been $100,000, under conditions of inflation they are 5% greater in year 1, or $105,000; 10.25% greater in year 2, and so forth.

Period	Sales in Constant $	Conversion $\bar{f}$	Sales in Actual $
1	100,000	$(1 + 0.05)^1$	105,000
2	100,000	$(1 + 0.05)^2$	110,250
3	100,000	$(1 + 0.05)^3$	115,762
4	100,000	$(1 + 0.05)^4$	121,551
5	100,000	$(1 + 0.05)^5$	127,628

The future cash flows in actual dollars for other elements can be obtained in a similar way.

• There is no change in the investment in year 0 or in capital cost allowances since these items are unaffected by expected future inflation.

---

[3] This is a simplifying assumption. In practice, these elements may have price indices other than the CPI. These differential price indices will be treated in Example 10.9.

- The salvage value is expected to increase at the general inflation rate. Therefore, the salvage in actual dollars will be

$$\$50,000(1 + 0.05)^5 = \$63,814.$$

This increase in salvage value will also increase the taxable gains as the book value remains unchanged. The calculations for both the book value and gains tax are shown in Table 10.2.

Solution

Given: Flows originally stated in constant dollars, converted to actual dollars at $\bar{f} = 5\%$, $i' = 15\%$

Find: NPW

Table 10.2 After-Tax Cash Flow Analysis Under Inflation (Example 10.8)

				n		
	0	1	2	3	4	5
Income Statement						
Revenues		105,000	110,250	115,762	121,551	127,628
Expenses						
Labor		21,000	22,050	23,152	24,310	25,526
Material		10,500	11,025	11,576	12,155	12,763
Overhead[1]		10,500	11,025	11,576	12,155	12,763
CCA		12,500	22,500	18,000	14,400	11,520
Taxable income		50,500	43,650	51,457	58,530	65,057
Income tax (40%)		20,200	17,460	20,583	23,412	26,023
Net income		30,300	26,190	30,874	35,118	39,034
Cash Flow Statement						
Cash from operation						
Net income		30,300	26,190	30,874	35,118	39,034
CCA		12,500	22,500	18,000	14,400	11,520
Investment	(125,000)					
Salvage						63,814
Disposal tax effect[2]						(7,094)
Net cash flow	(125,000)	42,800	48,690	48,874	49,518	107,275

[1] Overhead expense includes power, water, and other indirect costs.

[2] Disposal tax effect on proceeds from asset disposal:
   Total CCA claimed = $78,920
   Book value at the end of year 5 = $125,000 − $78,920 = $46,080
   Inflated salvage value = $50,000 (1 + 0.05)^5 = $63,814
   Gains on the sale = $63,814 − $46,080 = $17,734
   Disposal tax effect = −$17,734 × t = −$17.734 × 0.4 = −$7,094

Table 10.2 shows the after-tax cash flows in actual dollars. Using the deflation method, we convert the cash flows to constant dollars with the same purchasing power as those used to make the initial investment (year 0), assuming a general inflation rate of 5%. Then, we discount these constant-dollar cash flows at i' to determine the NPW.

Year	Net Cash Flow in Actual $	Conversion f	Net Cash Flow in Constant $	NPW at 15%
0	(125,000)	$(1 + 0.05)^0$	(125,000)	(125,000)
1	42,800	$(1 + 0.05)^{-1}$	40,762	35,445
2	48,690	$(1 + 0.05)^{-2}$	44,163	33,394
3	48,874	$(1 + 0.05)^{-3}$	42,220	27,760
4	49,518	$(1 + 0.05)^{-4}$	40,739	23,293
5	107,275	$(1 + 0.05)^{-5}$	84,052	41,789
				$36,680

Since NPW = $36,680 > 0, the investment is still economically attractive.

Comments: *Note that the NPW in the absence of inflation was $41,231 in Example 9.1. The $4551 decline (known as inflation loss) in the NPW under inflation, illustrated above, is due entirely to income tax considerations. Capital cost allowance is a charge against taxable income, which reduces the amount of taxes paid, and as a result, increases the cash flow attributable to an investment by the amount of those taxes saved. But capital cost allowance under existing tax laws is based on historic cost. As time goes by, capital cost allowance is charged to taxable income in dollars of declining purchasing power; as a result, the "real" cost of the asset is not totally reflected in the capital cost allowance. Depreciation costs are thereby understated, and the taxable income is overstated, resulting in higher taxes. In "real" terms, the amount of this additional income tax is $4551, which is also known as the "inflation tax."[4] In general, any investment that, for tax purposes, is to be expensed over time rather than immediately suffers the inflation tax.*

10.3.2 Multiple Inflation Rates

As we noted previously, the inflation rate (f_j) represents an average rate applicable to a specific segment j of the economy. For example, if we are estimating the future cost of a piece of machinery, we should use the inflation rate appropriate for that item. Furthermore, we may need to use several rates to accommodate the different costs and revenues in our analysis. The following example introduces the complexity of multiple inflation rates.

---

[4] This term and the explanation given here were introduced by Professor Brandt Allen in his article "Evaluating Capital Expenditures Under Inflation: A Primer," *Business Horizon*, 1976.

Example 10.9 Applying specific inflation rates

We will rework Example 9.1 using different annual changes (differential escalation rates) in the prices of cash flow components. Suppose that we expect the general rate of inflation ($\bar{f}$) to average 6% during the next 5 years. We also expect that the salvage value of the equipment will increase 3% per year, that wages (labor) and overhead will increase 5% per year, and that the cost of material will increase 4% per year. We expect sales revenue to climb at the general inflation rate. Table 10.3

Table 10.3 After-Tax Cash Flow Analysis Under Inflation (Example 10.9) (Multiple Price Indexes)

	0	1	2	3	4	5
Income Statement						
Revenues[1]		106,000	112,360	119,102	126,248	133,823
Expenses						
Labor[2]		21,000	22,050	23,152	24,310	25,526
Material[3]		10,400	10,816	11,249	11,699	12,167
Overhead[2]		10,500	11,025	11,576	12,155	12,763
CCA		12,500	22,500	18,000	14,400	11,520
Taxable income		51,600	45,969	55,124	63,684	71,848
Income tax (40%)[4]		20,640	18,388	22,050	25,474	28,739
Net income		30,960	27,581	33,075	38,210	43,109
Cash Flow Statement						
Cash from operation						
Net income		30,960	27,581	33,075	38,210	43,109
CCA		12,500	22,500	18,000	14,400	11,520
Investment	(125,000)					
Salvage[5]						57,964
Disposal tax effect[6]						(4,753)
Net cash flow	(125,000)	43,460	50,081	51,075	52,610	107,839

[1] The firm's revenues increase at an annual rate of 6%.

[2] The labor and overhead costs increase at an annual rate of 5%.

[3] The material cost increases at an annual rate of 4%.

[4] The firm's marginal tax rate remains unchanged.

[5] The salvage value is expected to increase at an annual rate of 3%: $50,000 (1 + 0.03)^5 = \$57,964$.

[6] Disposal tax effect on proceeds from asset disposal:
   Total CCA claimed = $78,920
   Book value at the end of year 5 = \$125,000 − \$78,920 = \$46,080
   Gains on the sale = \$57,964 − \$46,080 = \$11,884
   Disposal tax effect = $-\$11,884 \times t = -\$11,884 \times 0.4 = -\$4,753$

shows the relevant calculations using the income statement format. For simplicity, all cash flows and inflation effects are assumed to occur at year's end. Determine the net present worth of this investment, using the adjusted-discount method.

Solution

Given: Flows originally stated in constant dollars, converted to actual dollars at specific inflation rates f_j; $\bar{f} = 6\%$, $i' = 15\%$
Find: NPW

From Table 10.3, the after-tax cash flows in actual dollars are as follows:

n	Net Cash Flow in Actual Dollars
0	$-\$125,000$
1	43,460
2	50,081
3	51,075
4	52,610
5	107,839

To evaluate the present worth using actual dollars, we must adjust the original discount rate of 15%, which is an inflation-free interest rate i'. The appropriate interest rate to use is the market interest rate:[5]

$$i = i' + \bar{f} + i'\bar{f}$$
$$= 0.15 + 0.06 + (0.15)(0.06)$$
$$= 21.90\%.$$

The equivalent present worth is obtained as follows:

$$
\begin{aligned}
PW(21.90\%) = {} & -\$125,000 + \$43,460(P/F, 21.90\%, 1) \\
& + \$50,081(P/F, 21.90\%, 2) + \$51,075(P/F, 21.90\%, 3) \\
& + \$52,610(P/F, 21.90\%, 4) \\
& + \$107,839(P/F, 21.90\%, 5) \\
= {} & \$36,442
\end{aligned}
$$

---

[5] In practice, the market interest rate is usually given and the inflation-free interest rate can be calculated when the general inflation rate is known for years past or is estimated for time in the future. In our example, we are considering the opposite situation.

10.3.3 Effects of Borrowed Funds Under Inflation

Loan repayment is based on the historical contract amount, and the payment size does not change with inflation. Yet the value of these future payments, computed in year 0 dollars, is greatly affected by the inflation. First, we shall look at how the values of loan payments would change under inflation. Interest expenses are usually already stated in the loan contract in actual dollars and need not be adjusted. With inflation, the constant-dollar costs of both interest and debt principal repayments are reduced.

Example 10.10 Effects of inflation on payments with financing (borrowing)

Let us rework Example 9.4 with a debt-to-equity ratio of 0.50, where the debt portion of the initial investment is borrowed at 10% annual interest. Recall that the project had a net present worth of $54,249 at an interest rate of 15%. Now we assume, for simplicity, that the general inflation rate ($\bar{f}$) of 5% during the project period will affect all revenues, all expenses and salvage value except for capital cost allowance and loan payments. Determine the NPW of this investment.

Solution

Given: Flows from Example 9.4, in constant dollars except for loan payments, $\bar{f} = 5\%$, $i' = 15\%$
Find: NPW

For equal future payments, the actual-dollar cash flows for the financing activity are represented by the solid bars in Fig. 10.4. If there were inflation, the cash flow, measured in year 0 dollars, would be represented by the shaded bars in Fig. 10.4. Table 10.4 summarizes the after-tax cash flows under this situation. For simplicity, we assume that all cash flows and inflation effects occur at year's end. To evaluate the present worth using actual dollars, we must adjust the original discount rate (MARR) of 15%, which is an inflation-free interest rate i'. The appropriate interest rate to use is thus the market interest rate:

$$i = i' + \bar{f} + i'\bar{f}$$
$$= 0.15 + 0.05 + (0.15)(0.05)$$
$$= 20.75\%.$$

Then, from Table 10.4, we compute the equivalent present worth of the after-tax cash flow as follows:

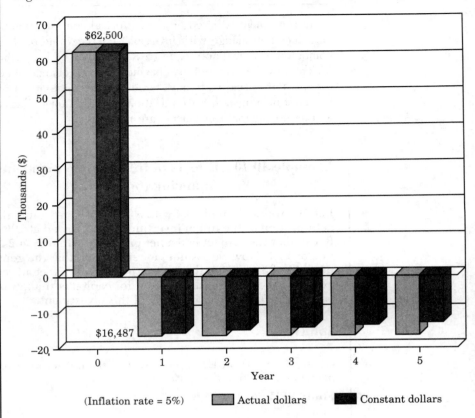

Figure 10.4 ■ Loan cash flow measured in constant dollars (Example 10.10)

(Inflation rate = 5%) ▨ Actual dollars ■ Constant dollars

n	Net Cash Flow in Actual Dollars
0	−$62,500
1	28,813
2	34,294
3	34,028
4	34,176
5	91,387

The equivalent present worth is

$$PW(20.75\%) = -\$62,500 + \$28,813(P/F, 20.75\%, 1)$$
$$+ \$34,294(P/F, 20.75\%, 2) + \dots$$
$$+ \$91,387(P/F, 20.75\%, 5)$$
$$= \$55,885.$$

Table 10.4 After-Tax Cash Flow Analysis with Borrowed Funds Under Inflation (Example 10.10)

	0	1	2	3	4	5
				n		
Income Statement						
Revenues (Savings)[1]		105,000	110,250	115,762	121,551	127,628
Expenses[1]						
Labor		21,000	22,050	23,152	24,310	25,526
Material		10,500	11,025	11,576	12,155	12,763
Overhead		10,500	11,025	11,576	12,155	12,763
Interest payment		6,250	5,226	4,100	2,862	1,499
CCA		12,500	22,500	18,000	14,400	11,520
Taxable income		44,250	38,424	47,357	55,669	63,558
Income tax (40%)[2]		17,700	15,369	18,943	22,268	25,423
Net income		26,550	23,054	28,414	33,401	38,135
Cash Flow Statement						
Cash from operation						
Net income		26,550	23,054	28,414	33,401	38,135
CCA		12,500	22,500	18,000	14,400	11,520
Investment	(125,000)					
Salvage[3]						63,814
Disposal tax effect[4]						(7,094)
Loan repayment[5]	62,500	(10,237)	(11,261)	(12,387)	(13,625)	(14,988)
Net cash flow	(62,500)	28,813	34,294	34,028	34,176	91,387

[1] The firm's revenues and expenses (excluding interest and CCA) increase at an annual rate of 5%.

[2] The firm's marginal tax rate remains unchanged.

[3] The salvage value is expected to increase at an annual rate of 5%: $50,000 (1 + 0.05)^5 = \$63,814$

[4] Disposal tax effect on proceeds from asset disposal:
    Total CCA claimed = \$78,920
    Book value at the end of year 5 = \$125,000 − \$78,920 = \$46,080
    Gains on the sale = \$63,814 − \$46,080 = \$17,734
    Disposal tax effect = $-\$17,734 \times t = -\$17,734 \times 0.4 = -\$7,094$

[5] Loan payment is not affected by inflation.

Comments: *When compared with the result of Example 9.4 (no inflation), the present worth gain due to inflation is $55,885 − $54,249 = $1636. This increase in NPW is primarily due to the debt-financing. An inflationary trend decreases the purchasing power of future dollars, which helps long-term borrowers because they can repay a loan with dollars of reduced buying power. That is, the debt-financing cost is reduced in an inflationary environment. In this case, the benefits of financing under inflation have more than offset the inflation tax effect on capital cost allowance and salvage value. The gain is not totally realistic: The interest rate for borrowing is generally higher during periods of inflation because it is a market rate.*

■ 10.4 Rate of Return Analysis Under Inflation

In addition to affecting individual aspects of a project's income statement, inflation can have a profound effect on the overall return—that is, the very acceptability—of an investment project. In this section, we will explore the effects of inflation on return on investment and show several examples.

10.4.1 Effects of Inflation on Return on Investment

The effect of inflation on the computed rate of return for an investment depends on how future revenues respond to the inflation. Under inflation, a company is usually able to compensate for increasing material and labor prices by raising its selling prices. However, even if future revenues increase to match the inflation rate, the allowable depreciation schedule, as we have seen, does not increase. The result will be increased taxable income and income-tax payments. This increase reduces the available constant-dollar after-tax benefits and, therefore, the inflation-free after-tax rate of return (IRR'). The next example will help us to understand this situation.

Example 10.11   IRR analysis with inflation

Hartsfield Company, a manufacturer of auto parts, is considering the purchase of a set of machine tools costing $30,000 that will generate increased sales of $30,000 per year and increased operating costs of $10,000 per year in each of the next 4 years. Additional profits will be taxed at a rate of 40%. The asset has a 30% capital cost allowance rate. The project has a 4-year life with zero salvage value. (All dollar figures represent constant dollars.)

(a) What is the expected internal rate of return?
(b) What is the expected IRR' if the general inflation rate is 10% during each of the next 4 years? (Here also assume that $f_j = \bar{f} = 10\%$.)
(c) If this is an independent alternative and the company has an inflation-free MARR (or MARR') of 32%, should the company invest in the equipment?

Solution

Given: Cash flows and assets with (a) no inflation, (b) $\bar{f} = 10\%$
Find: IRR

(a) **Rate of Return Analysis Without Inflation**
We find the expected rate of return by first computing the after-tax cash flow by the income statement approach, as shown in Table 10.5.

Table 10.5 Rate of Return Calculation with No Inflation (Example 10.11)

	0	1	2	3	4
			n		
Income Statement					
1. Additional sales		30,000	30,000	30,000	30,000
2. Operating costs		10,000	10,000	10,000	10,000
3. CCA		4,500	7,650	5,355	3,749
4. Taxable income		15,500	12,350	14,645	16,252
5. Income tax (40%)		6,200	4,940	5,858	6,501
6. Net income		9,300	7,410	8,787	9,751
Cash Flow Statement					
7. Cash from operation					
Net income		9,300	7,410	8,787	9,751
CCA		4,500	7,650	5,355	3,749
8. Investment	(30,000)				
9. Salvage					0
10. Disposal tax effect					3,499
11. Net cash flow	(30,000)	13,800	15,060	14,142	16,998

Lines 1 through 5 show the calculation of additional sales, operating costs, capital cost allowance, and taxes. As we emphasized in Chapter 7, capital cost allowance is not a cash expense, although it affects taxable income and thus affects cash flow indirectly. Therefore, we must add capital cost allowance (line 7) to net income to determine the net cash flow.

n	Net Cash Flow in Constant Dollars
0	−$30,000
1	13,800
2	15,060
3	14,142
4	16,998

Thus, if the investment is made, we expect to receive additional annual cash flows of $13,800, $15,060, $14,142, and $16,998.

This is a pure investment, so we can calculate the IRR for the project as follows.

$$PW(i) = -\$30,000 + \frac{13,800}{1+i} + \frac{15,060}{(1+i)^2} + \frac{14,142}{(1+i)^3}$$

$$+ \frac{16,998}{(1+i)^4} = 0$$

Since this is a simple investment, we obtain

$$\text{IRR}' = i'^* = 33.67\%.$$

The project has an iflation-free rate of return of 32.67%; that is, the company will recover its original investment ($30,000) plus interest at 33.67% each year for each dollar still invested in the project. Since the IRR' > MARR' of 32%, the company should buy the equipment.

(b) **Rate of Return Analysis Under Inflation**

With inflation, we assume that sales, operating costs, and future salvage value of the asset will increase. Capital cost allowance will be unchanged, but taxes, profits, and cash flow will be higher. We might think that higher cash flows will mean an increased rate of return. Unfortunately, this is not the case. We must recognize that cash flows for each year are stated in dollars of declining purchasing power. When the net after-tax cash flows are converted to dollars with the same purchasing power as those used to make the original investment, the resulting rate of return decreases. These calculations, assuming an inflation rate of 10% in sales and operating expenses and a 10% annual decline in the purchasing power of the dollar, are shown in Table 10.6. For example, whereas additional

Table 10.6 Rate of Return Calculation Under Inflation (Example 10.11)

			n		
	0	1	2	3	4
Income Statement					
1. Additional sales		33,000	36,300	39,930	43,923
2. Operating costs		11,000	12,100	13,310	14,641
3. CCA		4,500	7,650	5,355	3,749
4. Taxable income		17,500	16,550	21,265	25,534
5. Income tax (40%)		7,000	6,620	8,506	10,213
6. Net income		10,500	9,930	12,759	15,320
Cash Flow Statement					
7. Cash from operation					
Net income		10,500	9,930	12,759	15,320
CCA		4,500	7,650	5,355	3,749
8. Investment	(30,000)				
9. Salvage					0
10. Disposal tax effect					3,499
11. Net cash flow (actual dollars)	(30,000)	15,000	17,580	18,114	22,567
12. Net cash flow (constant dollars)	(30,000)	13,636	14,529	13,609	15,414

sales had been $30,000 yearly, under conditions of inflation they are 10% greater in year 1, or $33,000; 21% greater in year 2, and so forth. There is no change in investment or capital cost allowance since these items are unaffected by expected future inflation. We have restated the actual dollar after-tax cash flows (line 11) in dollars of a common purchasing power (constant dollars) by deflating them, again assuming an annual deflation factor of 10%. The constant-dollar cash flows (line 12) are then used to determine the rate of return.

$$PW(i) = -30,000 + \frac{13,636}{1 + i'} + \frac{14,529}{(1 + i')^2} + \frac{13,609}{(1 + i')^3}$$

$$+ \frac{15,414}{(1 + i')^4}$$

$$= 0$$

$$i^* = 31.32\%$$

The rate of return for the project's cash flows in constant dollars (year 0 dollars) is 31.32%, which is less than the 33.67% rate of return in the inflation-free case. The expected IRR' < MARR', so the investment is not acceptable anymore.

Comments: *We could also calculate the rate of return by setting the PW of the actual dollar cash flows to year 0. This would give a value of IRR = 44.46%, but this is an inflated IRR. We could then convert to an IRR' by removing inflationary effect:*

$$i' = \frac{(1 + i)}{(1 + f)} - 1,$$

which gives the same final result of IRR' = 31.32%.

10.4.2 Effects of Inflation on Working Capital

The loss of tax savings from capital cost allowance is not the only way that inflation may distort an investment's rate of return. Another source of decrease in a project's rate of return is working-capital drain. Capital projects requiring increased levels of working capital suffer from inflation because additional cash must be invested to maintain these items at the new price levels. For example, if the cost of inventory increases, additional outflows of cash are required to maintain appropriate inventory levels over time. A similar phenomenon occurs with funds committed to accounts receivable. These additional working-capital requirements can significantly reduce a project's rate of return. The next example will illustrate the effects of working-capital drain on a project's rate of return.

Example 10.12   Effect of inflation on profits with working capital

Consider Example 10.11. Suppose that a $5000 investment in working capital is expected, and that all the working capital will be recovered at the end of the project's 4-year life. Determine the rate of return on this investment.

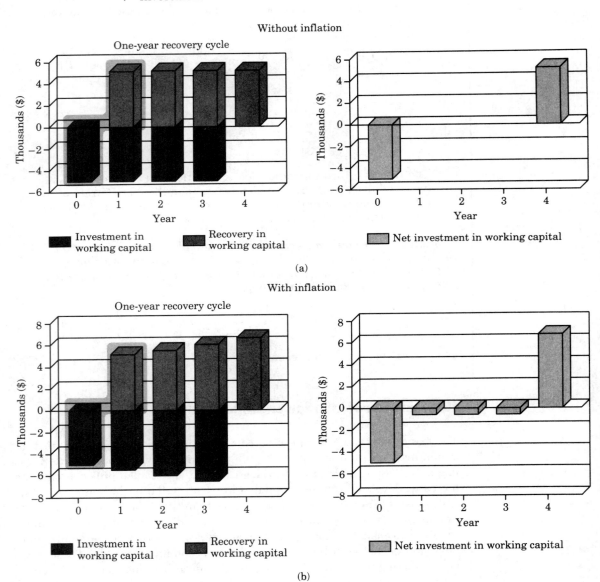

Figure 10.5 ■ Working capital requirements under inflation (Example 10.12)

Table 10.7 Effects of Inflation on Working Capital (Example 10.12)

CASE 1: Without Inflation	0	1	2	3	4
Income Statement					
1. Additional sales		30,000	30,000	30,000	30,000
2. Operating costs		10,000	10,000	10,000	10,000
3. CCA		4,500	7,650	5,355	3,749
4. Taxable income		15,500	12,350	14,645	16,252
5. Income tax (40%)		6,200	4,940	5,858	6,501
6. Net income		9,300	7,410	8,787	9,751
Cash Flow Statement					
7. Cash from operation					
Net income		9,300	7,410	8,787	9,751
CCA		4,500	7,650	5,355	3,749
8. Investment	(30,000)				
9. Salvage					0
10. Disposal tax effect					3,499
11. Working capital	(5,000)				5,000
12. Net cash flow	(35,000)	13,800	15,060	14,142	21,998

CASE 2: With Inflation	0	1	2	3	4
Income Statement					
1. Additional sales		33,000	36,300	39,930	43,923
2. Operating costs		11,000	12,100	13,310	14,641
3. CCA		4,500	7,650	5,355	3,749
4. Taxable income		17,500	16,550	21,265	25,534
5. Income tax (40%)		7,000	6,620	8,506	10,213
6. Net income		10,500	9,930	12,759	15,320
Cash Flow Statement					
7. Cash from operation					
Net income		10,500	9,930	12,759	15,320
CCA		4,500	7,650	5,355	3,749
8. Investment	(30,000)				
9. Salvage					0
10. Disposal tax effect					3,499
11. Working capital	(5,000)	(500)	(550)	(605)	6,655
12. Net cash flow (actual dollars)	(35,000)	14,500	17,030	17,509	29,222
13. Net cash flow (real dollars)	(35,000)	13,182	14,074	13,155	19,959

Solution

Given: Figures from Example 10.11, working capital = $5000 (constant dollars), (a) no inflation, (b) $\overline{f}$ = 10%

Find: IRR' and NPW at MARR' = 25%

Using the data in the upper part of Table 10.7, we can calculate the IRR' = IRR of 27.86% in the absence of inflation. The $PW(25\%)$ = $1929. The lower part of Table 10.7 includes the effect of inflation on the proposed investment. As illustrated in Fig. 10.5, working-capital levels can be maintained only by additional infusions of cash—the working-capital injected at the end of years 1, 2, and 3. For example, the $5000 investment in working capital made in year 0 will be recovered at the end of the first year. However, due to 10% inflation, the required working capital for the second year increases to $5500. In addition to reinvesting the $5000 revenues, an additional investment of $500 must be made. This $5500 will be recovered at the end of the second year. However, the project will need a 10% increase, or $6050, for the third year, and so forth.

As Table 10.7 illustrates, the effect of the working-capital drain is significant. Given an inflationary economy and investment in working capital, the project's IRR' drops to 24.19% $PW(25\%)$ = $-$536. By using either IRR analysis or NPW analysis we end up with the same result (as we must): Alternatives that are attractive when no inflation exists may not be acceptable with inflation.

10.5 Computer Notes

Many of the examples in this chapter have been presented in tabular format—for example, Table 10.3. This same tabular format lends itself very easily to electronic spreadsheet analysis. The effects of inflation can be easily taken into account by using either an electronic spreadsheet or **EzCash**. We will apply both methods to the following example:

Example 10.13   After-tax cash flow analysis including differential inflation

A construction firm is offered a fixed-price contract for a 5-year period. The firm will be paid $23,500 per year for the contract period. In order to accept the contract, the firm must purchase equipment costing $15,000 and requiring $13,000 (constant dollars) per year to operate. The equipment has a 30% capital cost allowance rate and is expected to have a salvage value of $1000 at the end of 5 years. Use a tax rate of 40% and an inflation-free interest rate of 20%. If the general inflation rate $\overline{f}$ is expected to average 5% over the next 5 years, salvage value at the annual rate of 5%, and operating expenses at 8% per year for the project duration, should the contractor accept the contract?

Solution—Lotus 1-2-3

The analysis of this situation should explicitly consider inflation, since the contractor is being offered a fixed-price contract and cannot increase the fee to compensate for increased costs. In this problem, as is common practice, we assume that all estimated costs are expressed in today's dollars, and actual costs will be increased due to inflation.

The Lotus 1-2-3 spreadsheet analysis is shown in Figure 10.6. Note that the salvage value has been increased at 5%. (When no inflation rate is given for a specific cost category, it is normally assumed that the general inflation rate holds.) Disposal tax effects are based on the difference between book value and salvage value, in this case, $3061 - $1276 = $1785, i.e., a loss of $1785. This results in a tax saving of $1785 \times 0.4 = 714.

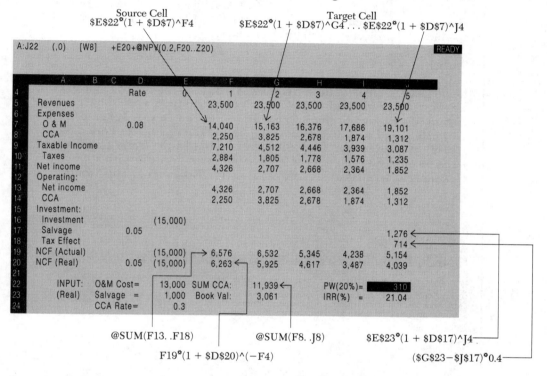

Steps to enter the growth factor into the specific cash flow element (O&M Cost):

Step 1: Enter cell formula for F7, the O&M cost in year 1 as: $E$22*(1+$D$7)^F4
Step 2: Type **/Copy** to initiate the Copy command.
Step 3: Press **ENTER** to designate cell F7 as the source.
Step 4: Move the pointer to cell G7 (year 2), the first cell in the range.
Step 5: Type the . key to anchor the range.
Step 6: Move the pointer to J7 (year 5), the last cell in the range.
Step 7: Press **ENTER** to activate the copy.

Figure 10.6 ■ Lotus 1-2-3 example of an after-tax cash flow analysis including differential inflation (Example 10.13)

In row 7, operating expenses have been increased at 8%. As explained in this chapter, actual-dollar cash flows are converted to constant-dollar flows by "deflating" at the general inflation rate. The IRR is calculated at 21.04%, based on the constant-dollar cash flows. This is compared to the MARR' of 20%, the inflation-free interest rate, indicating that this is an acceptable contract as long as the 5% and 8% rates are correct.

Since no one can predict inflation rates in any precise manner, it is always wise to investigate the effects of changes in these rates. This is very easy to do with a spreadsheet. For example, we can easily determine the value of the general inflation rate at which this project exactly earns 20%. This is shown in Fig. 10.7. The value of $\bar{f}$ was adjusted manually until the IRR was exactly 20%. From Fig. 10.7, we see that this would not be a good project if $\bar{f}$ were greater than 5.912%. (The cursor was set on cell D20 so that the exact value of $\bar{f}$ (0.05912) shows on the top of the spreadsheet.) Similarly, you can vary the O&M cost and salvage value to see how the project's profitability changes.

Solution—EzCash

Reviewing the process for developing a cash flow statement in Chapter 9, we select **Spreadsheet** from **EzCash** to obtain the screen shown in Fig. 10.8(a). We next enter the *Revenue* from *Source* 0 in year 1, i.e. $23,500. This value can be copied into the buffer and then pasted to Period 5 with a **Uniform** series from the **Edit** menu. Similarly, the *O&M Cost* for year 1 can be entered and then copied and pasted to Period 5 with a **Geometric Gradient** series (8%) from the **Edit** menu. The *Capital*

										READY
A:020 [W7] 0.05912										

	A	B C	D	E	F	G	H	I	J
			Rate	0	1	2	3	4	5
4									
5	Revenues				23,500	23,500	23,500	23,500	23,500
6	Expenses								
7	O & M		0.08		14,040	15,163	16,376	17,686	19,101
8	CCA				2,250	3,825	2,678	1,874	1,312
9	Taxable Income				7,210	4,512	4,446	3,939	3,087
10	Taxes				2,884	1,805	1,778	1,576	1,235
11	Net income				4,326	2,707	2,668	2,364	1,852
12	Operating:								
13	Net income				4,326	2,707	2,668	2,364	1,852
14	CCA				2,250	3,825	2,678	1,874	1,312
15	Investment:								
16	Investment			(15,000)					
17	Salvage		0.05						1,276
18	Tax Effect								714
19	NCF (Actual)			(15,000)	6,576	6,532	5,345	4,238	5,154
20	NCF (Real)		0.05912	(15,000)	6,209	5,823	4,499	3,368	3,868
21									
22	INPUT:	O&M Cost=	13,000	SUM CCA:		11,939		PW(20%)=	0
23	(Real)	Salvage =	1,000	Book Val:		3,061		IRR(%) =	20.00
24		CCA Rate=	0.3						

Figure 10.7 ■ Lotus 1-2-3 application to "what-if" questions (Example 10.13)

Figure 10.8 ■ **EzCash** screen output for Example 10.13

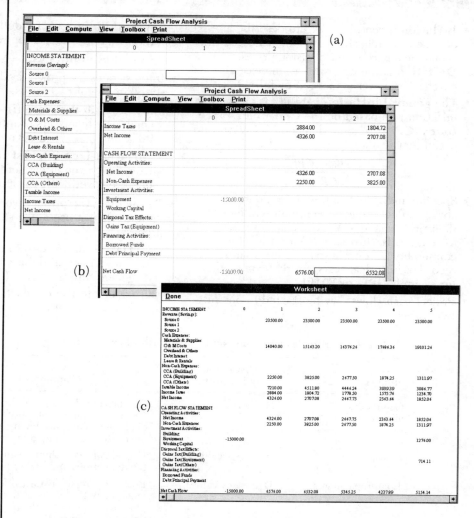

Cost Allowances and the *Disposal Tax Effects* can be calculated and exported to the spreadsheet from the **Toolbox** menu. Using the data from Example 10.13, the final screen with all the data entered and the *Net Cash Flows* calculated in actual dollars should look like Fig. 10.8(b). The screen can be scrolled left and right or up and down to view various cells. You can also view the whole worksheet by clicking **View** and then **Worksheet** in sequence (Fig. 10.8c).

Once you obtain the project's cash flows in actual dollars, you may simply select **Compute**, **Equivalent Worth Analysis**, and **Present Worth** in sequence, and then specify the market interest rate of 26% (obtained from $i = i' + f + i'f = 0.20 + 0.05 + 0.01 = 0.26$). You will find the project's NPW to be approximately $310.

Summary

- **Inflation** is the term used to describe a decline in purchasing power evidenced by an economic environment of rising prices. **Deflation** is the opposite: an increase in purchasing power evidenced by falling prices.
- The **general inflation rate** ($\bar{f}$) is an average inflation rate based on the **Consumer Price Index** (**CPI**). An annual general inflation rate $\bar{f}_n$ can be calculated from:

$$\bar{f}_n = \frac{CPI_n - CPI_{n-1}}{CPI_{n-1}}.$$

- Specific, individual commodities do not always reflect the general inflation rate in their price changes. We can claculate an **average inflation rate** (f_j) for a specific commodity (j) if we have an index (that is, a record of historical costs) for that commodity, for example, IPPI or RMPI.
- Project cash flows may be started in one of two forms.

- **Actual dollars** (A_n)—dollars that reflect the inflation or deflation rate
- **Constant dollars** (A'_n)—year 0 dollars
- Interest rates for project evaluation may be stated in one of two forms.
- **Market interest rate** (i)—a rate that combines the effects of interest and inflation; used with **actual dollar** analysis
- **Inflation-free interest rate** (i')—a rate from which the effects of inflation have been removed with constant-dollar analysis
- To calculate the present worth of actual dollars, we can use a two-step or a one-step process.
- **Deflation method—2 steps**
  1. Convert actual dollars by deflating by general inflation rate ($\bar{f}$)
  2. Calculate PW of constant dollars by discounting at i'

Table 10.8 Effects of Inflation on Project Cash Flows and Return

Item	Distortion by Inflation	Comments
Capital cost allowance	Decreases value	Capital cost allowance is charged to taxable income in dollars of declining values; taxable income is overstated, resulting in higher taxes.
Salvage values	Increases value	Higher salvage values mean lower tax deductions and higher taxes.
Loan repayments	Decreases value	Borrowers repay historical loan amounts with dollars of decreased purchasing power, reducing the debt-financing cost.
Interest expense	Decreases value	Interest expense is reduced from taxable income in dollars of declining value; taxable income is overstated, resulting in higher taxes.
Working capital requirement	Increases value	Known as *working-capital drain*, the cost of working capital increases in an inflationary economy.
Rate of return	Decreases value	Unless revenues are sufficiently increased to keep pace with inflation, tax effects and/or working-capital drain result in lower rate of return.

- **Adjusted-discount method—1 step**

$$PW(\bar{f},i') = \frac{A_n}{[(1 + \bar{f})(1 + i')]^n}$$

- A number of individual elements of project evaluations can be distorted by inflation. They are summarized in Table 10.8.

Problems

Note: In problem statements, the term "market interest rate" represents the "inflated MARR" for project evaluation or "interest rate" quoted by financial institutions for commercial loans.

10.1 The following data indicate the median price of houses in a major city during the last 5 years.

Period	Price ($)
-4	$178,200
-3	183,600
-2	185,700
-1	190,100
0	192,500

Assuming that the base period (price index = 100) is period -4, compute the average price index for the median house price.

10.2 The following indicates the price of heating oil (no. 2) in cents per litre during the last 4 years.

Period	Price
-4	62¢
-3	65
-2	77
-1	78
0	82
1	?

(a) Assuming that the base period (price index = 100) is period -4, compute the average price index for this heating oil.

(b) If the past trend is expected to continue, how would you estimate the price of heating oil at time period 1?

10.3 For prices that are increasing at an annual rate of 7% the first year and 9% the second year, determine the average inflation rate (f) over the two years.

10.4 Because of general price inflation in our economy, the purchasing power of the dollar shrinks with the passage of time. If the average general inflation rate is expected to be 8% per year for the foreseeable future, how many years will it take for the dollar's purchasing power to be one-half what it is now?

10.5 An annuity provides for 10 consecutive end-of-the-year payments of $3000. The average general inflation rate is estimated to be 4% annually, and the market interest rate is 15% annually. What is the annuity worth in terms of a single equivalent amount of today's dollars?

10.6 A company is considering buying workstation computers to support its engineering staff. It is estimated that in today's dollars the maintenance costs for the computers (paid at the end of each year) are $15,000, $20,000, $22,000, $25,000, and $30,000 for years 1–5 respectively. The general inflation rate ($\bar{f}$) is estimated to be 8% per year, and the company can receive 12% per year on its invested funds during the inflationary period. The company wants to pay for its maintenance expenses in equivalent equal payments (in actual dollars) at the end of each of the 5 years. Find the amount of the payment.

10.7 Given the following cash flows in actual dollars, convert to an equivalent cash flow in constant dollars if the base year is time 0. Keep cash flows at same point in time, that is, year 0, 4, 5, and 7. Assume that the market interest rate is 13% and the general inflation rate ($\bar{f}$) is estimated to be 4% per year.

n	Cash Flow
0	$1000
4	2000
5	3000
7	4000

10.8 The purchase of a car requires a $5000 loan repaid in monthly installments for 2 years at 12% interest compounded monthly. If the general inflation rate is 6% compounded monthly, find the actual- and constant-dollar value of the 20th payment.

10.9 A series of four annual constant-dollar payments beginning with $7000 at the end of the first year is growing at a rate of 5% per year, assuming that the base year is the current year ($n = 0$). If the market interest rate is 12% per year and the general inflation rate ($\bar{f}$) is 7% per year, find the present worth of this series of payments based on:

(a) constant-dollar analysis
(b) actual-dollar analysis

10.10 Consider Fig. 10.9, where the cash flow in actual dollars (b) is converted from the cash flow in constant dollars (a), at annual general inflation rate of $\bar{f} = 3.8\%$ and $i = 9\%$. What is the amount A in actual dollars that is equivalent to $A' = \$1000$ in constant dollars?

10.11 A 10-year $1000 bond pays a nominal rate of 12% compounded semiannually. If the market interest rate is 15% compounded annually and the general inflation rate is 6% per year, find the actual and constant dollar amount (time 0 dollars) of the 16th interest payment of the bond.

10.12 Suppose that you borrow $10,000 at 12% compounded monthly over 5 years. Knowing that the 12% represents the market interest rate, the monthly payment in actual dollars will be $222.45. If the average monthly general inflation rate is expected to be 0.5%, then determine the equivalent equal monthly payment series in constant dollars.

10.13 The annual fuel costs to operate a small solid-waste treatment plant are

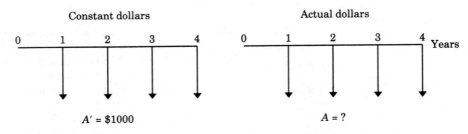

Figure 10.9 ■ Cash flow diagrams (Problem 10.10)

projected to be $1.5 million without considering any future inflation. The best estimates indicate that the annual inflation-free interest rate (i') will be 5% and the general inflation rate ($\bar{f}$) 6%. If the plant has a remaining useful life of 5 years, what is the present equivalent of its operating costs using actual-dollar analysis?

10.14 Suppose that you just purchased a used car worth $5000 in today's dollars. Assume also that you borrowed $5000 from a local bank at 9% compounded monthly over 2 years. The bank calculated your monthly payment of $228. Assume that there will be an average general inflation of 0.5% per month over the next 2 years.

(a) Determine the annual inflation-free interest rate (i') for the bank.

(b) What equal monthly payments in terms of constant dollars over the next 2 years is equivalent to the series of actual payments to be made over the life of the loan?

10.15 A man is planning to retire in 20 years. Money can be deposited at 8% compounded quarterly, and it is estimated that the future general inflation ($\bar{f}$) rate will be at 5% compounded annually. What quarterly deposit must be made each quarter until he retires so that he can make annual withdrawals of $40,000 in terms of today's dollars over 10 years after his retirement? (Assume that his first withdrawal occurs at the end of 6 months after his retirement.)

10.16 A young female engineer on her 23rd birthday decides to start saving to build up a retirement fund that pays 8% interest compounded quarterly. She feels that $500,000 worth of purchasing power in today's dollars will be adequate to see her through her sunset years after her 63rd birthday.

Assume a general inflation rate of 6% per year.

(a) If she plans to save by making 160 equal quarterly deposits, what should be the amount of the quarterly deposit?

(b) If she plans to save by making end-of-the-year deposits increasing by $1000 over each subsequent year, how much would her first deposit amount to?

10.17 A father wants to save in advance for his 8-year-old son's college expenses. The son will enter the college 10 years from now. An annual amount of $20,000 in constant dollars will be required to support the college expenses for 4 years. Assume that these college payments will be made at the beginning of the school year. The future general inflation rate is estimated to be 6% per year, and the interest rate on the savings account will average 9% compounded annually.

(a) What is the amount of the son's freshman-year expense in terms of actual dollars?

(b) What is the equivalent single-sum amount at present time for the college expenses?

(c) What is the equal amount the father must save each year until his son goes to college?

10.18 Consider the following project's after-tax cash flow and the expected annual general inflation rate during the project period.

End of Year	Cash Flow in Actual Dollars	Expected General Inflation Rate
0	−$20,000	
1	8,000	6.5%
2	8,000	7.7
3	8,000	8.1

(a) Determine the average annual general inflation rate over the project period.

(b) Convert the cash flows in actual dollars into equivalent constant dollars with the base year 0.

(c) If the annual inflation-free interest rate is 5%, what is the present worth of the cash flow? Is this project acceptable?

10.19 The Gentry Machines, Inc., has just received a special job order from one of its clients. The following financial data have been collected.

- This 2-year project requires purchase of special-purpose equipment costing $50,000. The declining balance CCA rate for the equipment is 30%.
- The machine will be sold at the end of 2 years for $30,000 (today's dollars).
- The project will bring in an additional annual revenue of $100,000 (actual dollars), but it is expected to incur additional annual operating costs of $40,000 (today's dollars).
- The project requires an investment in working capital in the amount of $10,000 (today's dollars). This working capital investment will be recovered at the end of project termination.
- The firm expects to borrow the initial investment at 10% over a 2-year period (equal annual payments of $28,810 (actual dollars)).
- The firm expects a general inflation rate of 5% per year during the project period. The firm's tax rate is 40% and its market interest rate is 18%.

(a) Compute the after-tax cash flows in actual dollars.

(b) What is the equivalent present value of this amount at time 0?

10.20 Hugh Health Product Corporation is considering the purchase of a computer to control a plant packaging for a spectrum of health products. The following data have been collected.

- First cost = $100,000 to be borrowed at 10% interest where only interest is paid each year and the principal is due in a lump sum at end of year 2
- Required service life (project life) = 6 years
- Estimated salvage value in year 6 is $15,000 (today's dollars)
- Capital cost allowance rate = 30%
- Marginal income tax rate = 40% (remains constant)
- Annual revenue = $150,000 (today's dollars)
- Annual expense (not including depreciation and interest) = $80,000 (today's dollars)
- Market interest rate = 20%

(a) With an average general inflation rate expected during the project period of 5%, affecting all revenues, expenses, and salvage value, determine the cash flows in actual dollars.

(b) Compute the net present value of the project under inflation.

(c) Compute the net present value loss (gain) due to inflation.

(d) In (c), how much is the present value loss (or gain) due to borrowing?

10.21 A textile manufacturing company is considering automating their piece-goods screen-printing system at a cost of $10,000. The firm expects to phase out this automated printing system at the end of 3 years due to changes in style. At that time, the firm will be able to scrap the system at $2000 in today's dollars. The expected net savings due to the automation are in today's dollars (constant dollars) as follows.

End of year	Cash Flows
1	$5000
2	7000
3	4000

The cost of the system may be written off with a 30% declining balance rate for tax purposes. The expected average general inflation rate over the next 3 years is around 6% per year. The firm will finance the entire project by borrowing at 10%. The scheduled repayment of the loan will be as follows.

End of Year	Principal Payment	Interest Payment
1	0	$1000
2	0	1000
3	10,000	1000

The firm's market interest rate for project evaluation during this inflation-ridden time is 20%. Assume that the net savings and salvage value will be responsive to this average inflation rate. The firm's marginal tax rate is known to be 40%.

(a) Determine the after-tax cash flows of this project in actual dollars.

(b) Determine the net present value reduction (or gains) in profitability due to inflation.

10.22 The Brown Metal Product Company is considering the purchase of a new milling machine during year 0. The machine's base price is $180,000, and it will cost another $20,000 to modify it for special use by the firm. This results in a $200,000 cost base for capital cost allowance calculation. The declining balance method with 1/2 year rule is used for capital cost allowance ($d = 30\%$). The machine will be sold after 3 years for $80,000 (actual dollars). Use of the machine will require an increase in net working capital (inventory) of $10,000 at the beginning of the project year. The machine will have no effect on revenues, but it is expected to save the firm $80,000 (today's dollars) per year in before-tax operating costs, mainly labor. The firm's marginal tax rate is 40%, and it is expected to remain unchanged over the duration of the project. However, the company expects labor costs to increase at an annual rate of 5%, but the working-capital requirement will grow at an annual rate of 8% due to inflation. The salvage value of the milling machine is not affected by the inflation. The general inflation rate is estimated to be 6% per year over the project period. The firm's market interest rate is 20%.

(a) Determine the project cash flows in actual dollars.

(b) Determine the project cash flows in constant (time 0) dollars.

(c) Is this project acceptable?

10.23 Fuller Ford Company is considering purchasing a vertical drill machine, which costs $50,000 with a 6-year service life. The salvage value of the machine at the end of 6 years is expected to be $5000. The machine will generate annual revenues of $20,000 (today's dollars) but it expects to have an annual expense (excluding depreciation) of $8000 (today's dollars). The asset has a 30% rate for capital cost allowance calculation. The project requires a working-capital investment of $5000 at year 0. The marginal income tax rate for the firm is averaging 35%. The firm's market interest rate is 18%.

(a) Determine the internal rate of return of this investment.

(b) Assume that the firm expects a general inflation rate of 5% but it also expects an 8% annual increase in revenue and working capital and a 6% annual increase in expenses due to inflation. Compute the real (inflation-free) internal rate of return. Is this project acceptable?

10.24 Sonja Jensen is considering the purchase of a fast-food franchise. Sonja will be operating on a lot that is to be converted into a parking lot in 5 years, but that may be rented in the interim for $800 per month. The franchise and necessary equipment will have a total first cost of $55,000 and a salvage value of $10,000 (in today's dollars) after 5 years. Sonja is told that the future annual general inflation rate will be 5%. The projected operating revenues and expenses in actual dollars other than rent and depreciation for the business are as follows:

End of Year	Revenue	Expenses
1	$30,000	$15,000
2	35,000	21,000
3	55,000	25,000
4	70,000	30,000
5	70,000	30,000

Assume that the initial investment will be written off with a 20% rate, and Sonja's tax rate is 30%. Sonja will be able to invest her money at a rate of at least 10% in other investment activities during this inflation-ridden period.

(a) Determine the cash flows associated with the investment over the investment life.
(b) Compute the projected after-tax rate of return (real) for this investment opportunity.

10.25 You have $10,000 cash which you want to invest. Normally, you would deposit the money in a savings account that pays an annual market interest rate of 6%. However, you are now considering the possibility of bond investment. Your alternatives are either a provincial bond paying 9% or a corporate bond paying 8.4%. Your marginal tax rate is 40% for interest income and 30% for capital gains income. You expect the general inflation rate to be 3% during the investment period. Both bonds pay interest once at the end of each year. The provincial bond matures at the end of year 5. The corporate bond is expected to have a selling price of $11,000 at the end of year 5. The equivalent before-tax cash flows for these two bonds are shown in Figure 10.10.

1. Determine the real (inflation-free) rate of return for each bond.

2. Without knowing your MARR, can you make a choice between these two bonds?

10.26 The Johnson Chemical Company has just received a special subcontracting job from one of its clients. This 2-year project requires purchase of a special-purpose painting sprayer costing $60,000. This equipment has a 30% declining balance rate. After the subcontracting work, the painting sprayer will be sold at the end of 2 years for $40,000 (actual dollars). The painting system will require an increase in net working capital (spare parts inventory such as spray nozzles) of $5000. This investment in working capital will be fully recovered at the end of project termination. The project will bring in an additional annual revenue of $120,000 (today's dollars), but it is expected to incur an additional

Figure 10.10 ■ Cash flow diagrams (Problem 10.25)

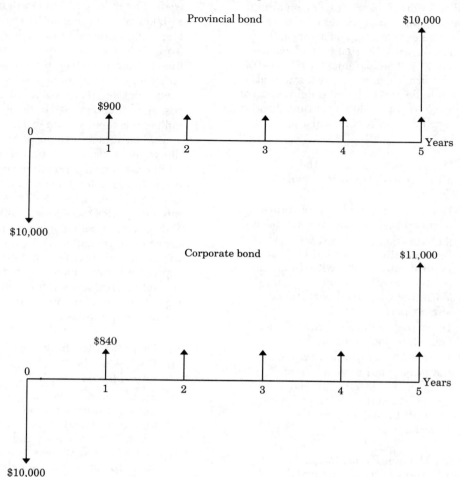

annual operating cost of $60,000 (today's dollars). It is projected that, due to inflation, there will be sales price increases at the annual rate of 5% (This implies that annual revenues will increase at an annual rate of 5%.) There will be an annual increase of 4% for expenses and working-capital requirement. The company has a tax rate of 30% and market interest rate of 15% for project evaluation during the inflationary period. If the firm expects a general inflation of 8% during the project period:

(a) Compute the after-tax cash flows in actual dollars.

(b) What is the rate of return on this investment (real earnings)?

(c) Is special order profitable?

10.27 Land Development Corporation is considering the purchase of a bulldozer which costs $100,000 and has an estimated salvage value of $30,000 at the end of 6 years. The asset will generate annual before-tax revenues of $80,000 over the next 6 years. The asset can be written off with a capital cost allowance rate of 30%. The marginal tax rate is 40%, and the firm's market interest rate is also known to be 18%. All dollar figures above represent constant dollars at time 0 and are responsive to general inflation rate $\bar{f}$.

(a) With $\bar{f} = 6\%$, compute the after-tax cash flows in actual dollars.

(b) Determine the real rate of return of this project on an after-tax basis.

(c) Suppose the initial cost will be financed through a local bank at an interest rate of 12% with an annual payment of $24,323 over 6 years. With this additional condition, answer part (a) above.

(d) In part (a), determine the present value loss due to inflation.

(e) In part (c), determine how much the project has to generate in additional before-tax annual revenue in actual dollars (equal amount) to make up for the inflation loss.

10.28 Wilson Machine Tools, Inc., a manufacturer of fabricated metal products, is considering the purchase of a high-tech computer-controlled milling machine for $95,000. The costs for installing the machine, site preparation, wiring, and rearranging other equipment are expected to be $15,000. These installation costs will be added onto the machine cost to determine the total cost base for capital cost allowance. Special jigs and tool ties for the particular product will also be required at $10,000. The milling machine is expected to last 10 years, and the jigs and dies only 5 years. Therefore, another set of jigs and dies

has to be purchased at the end of 5 years. The milling machine will have a $10,000 salvage value at the end of its life, and the special jigs and dies are worth only $300 as scrap metal at any time in their lives. The machine has a 30% capital cost allowance rate, and the special jigs and dies have a 100% capital cost allowance rate. With the new milling machine, Wilson expects an additional annual revenue of $80,000 due to increased production. The additional annual production costs are estimated as follows: materials, $9000; labor, $15,000; energy, $4500; and miscellaneous O&M costs, $3000. Wilson's marginal income tax rate is expected to remain at 35% over the project life of 10 years. All dollar figures represent today's dollars. The firm's market interest rate is 18%, and the expected general inflation rate during the project period is estimated at 6%.

(a) Determine the project cash flows.

(b) Determine the internal rate of return for the project.

(c) Suppose that Wilson expects price increases during the project period: material at 4% per year, labor at 5% per year, and energy and other O&M costs at 3% per year. To compensate for these increases in prices, Wilson is planning to increase annual revenue at the rate of 7% per year by charging a higher price to its customers. No changes in salvage value are expected for the machine as well as the jigs and dies. Determine the project cash flows in actual dollars.

(d) In (c), determine the real (inflation-free) rate of return of the project.

(e) Determine the economic loss (or gain) in present worth due to inflation.

10.29 Recent biotechnological research has made possible the development of a sensing device that uses living cells on a silicon chip (see Fig. 10.11). The chip is capable of detecting physical and chemical changes in cell processes. Proposed uses include researching the mechanisms of disease on a cellular level, developing new therapeutic drugs, and replacing the use of animals in cosmetic and drug testing. Biotech Device Corporation (BDC) has just perfected a process for mass-producing the chip. The following information has been compiled for the board of directors.

- BDC's marketing department has plans to target sales of the device to the larger chemical and drug manufacturers. They estimate that annual sales would be 2000 units if the devices were priced at $95,000 each (today's dollar).

- To support that level of sales volume, BDC would need a new manufacturing plant. Once the "go" decision is made, this plant could be built and made ready for production within one year. BDC would need a 10-hectare tract of land that would cost $1.5 million. The building would cost $5 million and has a 4% capital cost allowance rate. The first payment of $1 million would be due to the contractor on December 31, 1993, and the remaining $4 million on December 31, 1994.

- The required manufacturing equipment would be installed late in 1994 and would be paid for on December 31, 1994. BDC would have to purchase the equipment at an estimated cost of $7.5 million, including transportation, plus another $500,000 for installation. The equipment has a 30% capital cost allowance rate.

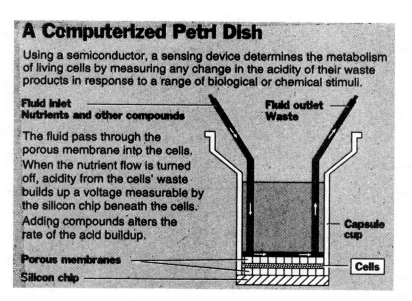

Figure 10.11 ■ Pairing chips and living cells to test drugs, Cytosensor Microphysiometer

- The project would require an initial investment of $1 million in working capital. This initial working-capital investment would be made on December 31, 1994, and on December 31 of each following year. Net working capital would be increased by an amount equal to 15% of any sales increase expected during the coming year.

- The project's estimated economic life is 6 years. At that time, the land is expected to have a market value of $2 million, the building a value of $3 million, and the equipment a value of $1.5 million. The estimated variable manufacturing costs would total 50% of the dollar sales. Fixed costs, excluding capital cost allowance, would be $5 million for the first year of operations. Since the plant would begin operations on January 1, 1995, the first operating cash flows would thus occur on December 31, 1995.

- Sales prices and fixed overhead costs, other than capital cost allowance, are projected to increase with general inflation, which is expected to average 5% per year over the 6-year life of the project.

- BDC's marginal tax rate is 40%, and its market interest rate is 20%.

    (a) Determine the after-tax cash flows of the project in actual dollars.

    (b) Determine the inflation-free (real) IRR of the investment.

    (c) Would you recommend that the firm accept the project?

CHAPTER 11

Methods of Financing Projects

Hibernia is a massive oil field located under 80 metres of water 315 kilometres east-southeast of St. John's, Newfoundland. After 25 years of exploration and pre-development engineering, the parties involved decided in 1990 to proceed with its development. Though undersea oil production is now an established technology, Hibernia's development nevertheless represents some unique technical and environmental challenges. With Hibernia's infrastructure now under construction at an estimated cost of $5.2 billion, production is scheduled to start in 1997 and continue for 18 years at an average production rate of 110,000 barrels of oil per day.

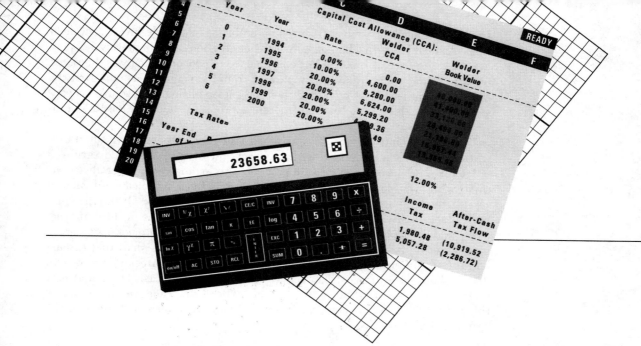

How will the Hibernia Management and Development Company Ltd., comprised of Chevron Canada Resources, Murphy Oil Corporation, Mobil Oil Canada Properties, Petro-Canada Hibernia Partnership, and the federal government, pay for this technically and financially ambitious project? Its owners are each responsible for their share of the costs and thereby earn an equivalent share of the future benefits. The seed money provided by each owner will be some combination of debt and equity funds. By sharing the ownership, each owner's financial risk becomes manageable.

In previous chapters, we have focused on problems relating to investment decisions. As mentioned in Chapter 1, investment decisions are not always independent of the source of finance. For convenience in economic analysis, however, investment decisions are usually separated from financing decisions—first the investment project is selected, and then the choice of financing sources is considered. After the source has been chosen, appropriate modifications to the investment decision are made.

Up to this point, we have also assumed that the assets employed in an investment project are obtained with the firm's own capital (retained earnings) or from a short-term borrowing. In practice, this arrangement is not always attractive or even possible. If the investment calls for a significant infusion of capital, the firm may raise the needed capital by issuing stock. Or the firm may borrow the funds by issuing bonds to finance such purchases. A third alternative—and arguably the one of most immediate concern to project engineers—is to lease the facilities and equipment. In this chapter, we shall first discuss how a typical firm may raise new capital from external sources. Then we will discuss how the external financing affects the after-tax cash flows and how the decision to borrow affects the investment decision. Then we shall examine the decision to lease, as compared to purchasing with borrowed funds or the firm's capital.

■ 11.1 Methods of Financing

The two broad choices a firm has for financing an investment project are *equity financing* and *debt financing*. (There is a hybrid financing method, known as *lease financing*, which we will discuss in the next section.) In this section, we shall look briefly at the two options for obtaining external investment funds, and also examine their effects on after-tax cash flows.

11.1.1 Equity Financing

Equity financing can take two forms: (1) use of retained earnings otherwise paid to stockholders or (2) issuance of stock. Both forms use funds invested by current or new owners of the company.

Until now, most of our economic analyses have presumed that the company had cash on hand to make capital investments—implicitly, we were dealing with cases of financing by retained earnings. If it had not reinvested these earnings, the firm might have paid them to the company's owners—the stockholders—in the form of a dividend or kept them on hand for future needs.

If a company does not have sufficient cash on hand to make an investment and does not wish to borrow in order to fund it, financing can be arranged by selling common stock to raise the required funds. (Many small biotechnology and computer firms raise capital by going public and selling common stock.)

The company has to decide how much money to raise, the type of securities to issue (common stock or preferred stock), and the basis for pricing the issue.

Once the company has decided to issue common stock, the firm must estimate **flotation costs**—the expenses it will incur in connection with the issue, such as investment bankers' fees, lawyers' fees, accountants' costs, and printing and engraving. Usually the investment banker will buy the issue from the company at a discount below the price at which the stock is to be offered to the public. (The discount usually represents the flotation costs.) If the company is already publicly owned, the offering price will commonly be based on the existing market price of the stock. If the company is going public for the first time, there will be no established price, so the investment bankers will have to estimate the expected market price at which the stock will sell after issue.

Example 11.1   Issuing common stock

Scientific Sports, Inc. (SSI), a golf club manufacturer, has developed a new metal club (Driver) made out magnesium alloy—the lightest engineering metal having good vibration-damping characteristics (see Fig. 11.1). The company expects to acquire considerable market penetration with this new product. To produce it, the company needs a new manufacturing facility, which will cost $10 million. The company decided to raise the $10 million in common stock. The firm's current

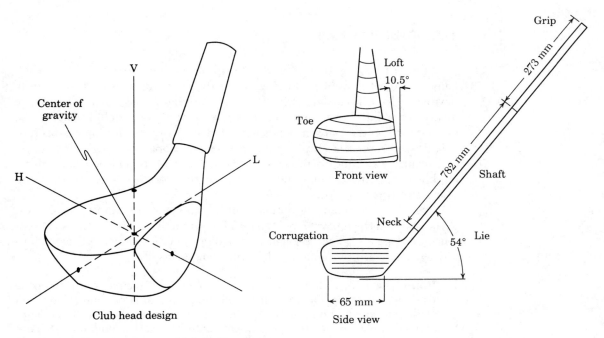

Figure 11.1 ■ SSI's new golf club (Driver)

stock price is $30 per share. The investment bankers have informed the management that it must price the new issue to the public at $28 per share because of a decreasing demand as more shares are available on the market. The flotation costs will be 6% of the issue price, so SSI will net $26.32 per share. How many shares must SSI sell to net $10 million after flotation expenses?

Solution

Let X be the number of shares to be sold. The total flotation cost will be

$$(0.06)\,(\$28)\,(X) = \$1.68X.$$

To net $10 million, we establish the following relationship:

$$\text{sales proceeds} - \text{flotation cost} = \text{net proceeds}$$
$$28X - 1.68X = 10,000,000$$
$$26.32X = 10,000,000$$
$$X = 379,940 \text{ shares.}$$

Now we can figure out the flotation costs for issuing the common stock as

$$\$1.68(379,940) = \$638,300.$$

11.1.2 Debt Financing

In addition to equity financing, the second major type of financing the company can select is **debt financing.** Debt financing includes both short-term borrowing from financial institutions and the sale of long-term bonds where money is borrowed from investors for a fixed period. With debt financing, the interest paid on the loans or bonds is treated as an expense for income-tax purposes. Since interest is a tax-deductible expense, companies in high tax brackets may incur lower after-tax financing costs with a debt. In addition to influencing borrowing interest rate and tax bracket, the loan-repayment method can also affect financing costs.

When the debt-financing option is used, we need to separate the interest payments from the repayment of the loan for our analysis. The interest-payment schedule depends on the repayment schedule established at the time of borrowing. There are two common debt-financing methods.

1. **Bond financing:** There is no partial payment of principal; only interest is paid each year (or semiannually), and the principal is paid in a lump sum at the end of the bond maturity. (See Section 3.6 for bond terminologies and valuation.) Bond financing is similar to equity financing in that there are flotation costs when issuing bonds.

2. **Term loans:** There is equal repayment, where the sum of the interest payments and the principal payments is uniform; interest payments decrease while principal payments increase over the life of the loan. Term loans are usually negotiated directly between the borrowing company and a financial institution—generally a commercial bank, an insurance company, or a pension fund.

Example 11.2 Debt financing

Referring to Example 11.1, suppose SSI has instead decided to raise the $10 million by debt financing. SSI could issue a bond or secure a term loan. Conditions for each option are as follows.

- Bond financing: The flotation cost is 1.407% of the $10 million, 5-year bond issue. The investment bankers have indicated that the bonds would sell for $995 and have a face value of $1000. The bond would require annual interest payments of 12%.

- Term loan: A $10 million bank loan can be secured at an annual interest rate of 11% for 5 years, requiring 5 equal annual installments.

(a) How many $1000 par value bonds would SSI have to sell to raise the $10 million?
(b) What are the annual payments (interest and principal) for the bond?
(c) What are the annual payments (interest and principal) for the term loan?

Solution

(a) To net $10 million, SSI would have to sell

$$\$10,000,000/(1 - 0.01407) = \$10,142,708$$

and pay $142,708 in flotation costs. Since the $1000 bond will be sold at a 0.5% discount, the total number of bonds to be sold would be

$$\$10,142,708/995 = 10,193.676.$$

(b) For the bond financing, the annual interest is equal to

$$\$10,193,676(0.12) = \$1,223,241.$$

Only the interest is paid each period, and thus the principal amount owed remains unchanged.

(c) For the term loan, the annual payments are

$$10,000,000(A/P, 11\%, 5) = \$2,705,703.$$

The principal and interest components of each annual payment are summarized in Table 11.1.

Table 11.1 Two Common Methods of Debt Financing

Plan	0	1	2	3	4	5
1. Bond financing: no principal repayments until end of life						
Beginning balance	$ 10,193,676	$ 10,193,676	$ 10,193,676	$ 10,193,676	$ 10,193,676	$ 10,193,676
Interest owed	0	1,223,241	1,223,241	1,223,241	1,223,241	1,223,241
Total payment	0	−1,223,241	−1,223,241	−1,223,241	−1,223,241	−11,416,917
Ending balance	$ 10,193,676	$ 10,193,676	$ 10,193,676	$ 10,193,676	$ 10,193,676	0
Principal payment	$ 0	$ 0	$ 0	$ 0	$ 0	$ 10,193,676
Interest payment	0	1,223,241	1,223,241	1,223,241	1,223,241	1,223,241
Total payment	$ 0	$ 1,223,241	$ 1,223,241	$ 1,223,241	$ 1,223,241	$ 11,416,917
2. Term loan: equal annual repayment ($10,000,000(A/P, 11\%, 5) = \$2,705,703$)						
Beginning balance	$10,000,000	$10,000,000	$ 8,394,297	$ 6,611,966	$ 4,633,579	$ 2,437,570
Interest owed	0	1,100,000	923,372	727,316	509,694	268,133
Total payment	0	−2,705,703	−2,705,703	−2,705,703	−2,705,703	−2,705,703
Ending balance	$10,000,000	$ 8,394,297	$ 6,611,966	$ 4,633,579	$ 2,437,570	$ 0
Principal payment	$ 0	$ 1,605,703	$ 1,782,331	$ 1,978,387	$ 2,196,009	$ 2,437,570
Interest payment	0	1,100,000	923,372	727,316	509,694	268,133
Total payment	$ 0	$ 2,705,703	$ 2,705,703	$ 2,705,703	$ 2,705,703	$ 2,705,703

11.1.3 Capital Structure

The ratio of total debt to total capital, generally called the **debt ratio,** or **capital structure,** represents the percentage of the total capital provided by borrowed funds. For example, a debt ratio of 0.5 indicates that 50% of the capital is borrowed and the remaining funds are provided from the company's equity (retained earnings or stock offerings). This type of financing is called **mixed financing.**

Borrowing affects the firm's capital structure, and the firm must determine the effects of a change in the debt ratio on its market value before making the ultimate financing decision. Even if debt financing is attractive, you should understand that companies do not simply borrow funds to finance projects. The firm usually establishes a **target capital structure,** or **target debt ratio,** after considering the effects of various financing methods. This target may change over time as business conditions vary, but the firm's management always strives to achieve this target whenever individual financing decisions are considered. If the actual debt ratio is below the target level, any new capital will probably be raised by issuing debt. On the other hand, if the debt ratio is currently above the target, expansion capital would be raised by issuing stock.

How does a typical firm set the target capital structure? This is a rather difficult question to answer, but we may list several factors that affect the capital structure policy. First, capital structure policy involves a trade-off between risk and return. As you use more debt for business expansion, the inherent business risk (uncertainty in projections of future operating income) will also increase, but investors view the business expansion as a healthy indicator for a corporation with higher expected earnings. When investors perceive higher business risk, the firm's stock price tends to be depressed. On the other hand, when they perceive higher expected earnings, the firm's stock price tends to increase. The optimal capital structure is thus the one that strikes a balance between business risk and expected future earnings. The greater the firm's business risk, the lower its optimal debt ratio.

Second, a major reason for using debt is that interest is a deductible expense for business operations, which lowers the effective cost of borrowing. On the other hand, dividends paid to common stockholders are not deductible. If a company uses debt, it must pay interest on this debt, while if it uses equity, it pays dividends to the equity investors (shareholders). A company needs $1 in before-tax (B/T) income to pay $1 of interest, but if the company is in the 34% tax bracket, it needs $1/(1 - 0.34) = 1.52 of B/T income to pay $1 of dividend.

Third, financial flexibility, or the ability to raise capital on reasonable terms from the financial market, is an important consideration. Firms need a steady supply of capital for stable operations. When money is tight in the economy, investors prefer to advance funds to companies with a healthy capital structure (lower debt ratio). These three elements (business risk, taxes, and financial flexibility) are major factors that determine the firm's optimal capital structure.

Example 11.3   Project financing based on optimal capital structure

Reconsider SSI's $10 million venture project in Example 11.1. Suppose that SSI's optimal capital structure calls for a debt ratio of 0.5. After reviewing SSI's capital structure, the investment banker convinced management that it would be better off, in view of current market conditions, to limit the stock issue to $5 million and to raise the other $5 million as debt by issuing bonds. Because the amount of capital to be raised in each category is reduced by half, the flotation cost would also change. The flotation cost for common stock will be 8.1%, whereas the flotation cost for bonds is 3.2%. The 5-year, 12% bond will have a par value of $1000 and will be sold at $985.

Assuming that the $10 million capital would be raised successfully, the engineering department has detailed the following financial information.

- The new venture will have a 5-year project life.

- The $10 million capital will be used to purchase land for $1 million, the building for $3 million, and equipment for $6 million. The plant site and building are already available, and production can begin during the first year. The building and the equipment fall into declining balance Class 1 ($d = 4\%$) and Class 43 ($d = 30\%$), respectively. The half-year rule is applicable. At the end of year 5, the salvage value for each asset is: the land for $1.5 million, the building for $2 million, and the equipment for $2.5 million.

- For common stockholders, an annual cash dividend in the amount of $2 per share is planned over the project life. This steady cash dividend payment is deemed necessary to maintain the market value of the stock.

- The unit production cost is $50.31 (material, $22.70; labor and overhead (excluding depreciation), $10.57; and tooling, $17.04).

- The unit price is $250, and SSI expects an annual demand of 20,000 units.

- The operating and maintenance cost, including advertising expense, would be $600,000 per year.

- An investment of $500,000 in working capital is required at the beginning of the project, and the amount will be fully recovered when the project terminates.

- The firm's marginal tax rate is 40%, and this rate will remain throughout the project period.

- The firm's MARR is 20%. No inflation is considered during the project life.

(a) Determine the after-tax cash flows for this investment with external financing.

(b) What is the rate of return on this investment?

Discussion: As the amount of financing and flotation costs change, we need to recalculate the number of shares (or bonds) to be sold in each category. For a $5 million common stock issue, the flotation cost increases to 8.1% (flotation costs are higher for small issues than for large ones due to existence of fixed costs—certain costs must be incurred regardless of the size of the issue, so the percentage of flotation costs increases as the size of issue gets smaller.) The number of shares to be sold to net $5 million is $5,000,000/(0.919)(28) = 194,311$ shares. For a $5 million bond issue, the flotation cost is 3.2%. Therefore, to net $5 million, SSI has to sell $5,000,000/(0.968)(985) = 5244$ units of $1000 par value. This implies that SSI is effectively borrowing $5,243,948, upon which the annual bond interest will be calculated. The annual bond interest payment is $5,243,948(0.12) = \$629,274$.

Solution

(a) **After-tax cash flows**

Table 11.2 summarizes the after-tax cash flows for the new venture. The following calculations and assumptions were used in developing the cash flow table.

- Revenue: $250 × 20,000 = $5,000,000 per year
- Costs of goods: $50.31 × 20,000 = $1,006,200 per year

Table 11.2 Effects of Project Financing on After-Tax Cash Flows

	0	1	2	3	4	5
Income Statement						
Revenue		$5,000,000	$5,000,000	$5,000,000	$5,000,000	$5,000,000
Expenses						
Cost of goods		1,006,200	1,006,200	1,006,200	1,006,200	1,006,200
O & M		600,000	600,000	600,000	600,000	600,000
Bond interest		629,274	629,274	629,274	629,274	629,274
CCA						
Building		60,000	117,600	112,896	108,380	104,045
Equipment		900,000	1,530,000	1,071,000	749,700	524,790
Taxable income		1,804,526	1,116,926	1,580,630	1,906,446	2,135,691
Income taxes		721,810	446,770	632,252	762,578	854,726
Net income		1,082,716	670,156	948,378	1,143,868	1,281,415
Cash Flow Statement						
Operating						
Net income		1,082,716	670,156	948,378	1,143,868	1,281,415
CCA		960,000	1,647,600	1,183,896	858,080	628,835
Investment						
Land	($1,000,000)*					1,500,000
Building	(3,000,000)					2,000,000
Equipment	(6,000,000)					2,500,000
Working capital	(500,000)					500,000
Disposal tax effect						(461,365)
Financing						
Common stock	5,000,000					(5,440,708)
Bond	5,000,000					(5,243,948)
Cash dividend		(388,622)	(388,622)	(388,622)	(388,622)	(388,622)
Net cash flow	($500,000)	$1,654,094	$1,929,134	$1,743,652	$1,613,326	($3,124,393)

* Figure in ( ) denotes an expenditure (cash outflow)

- Bond interest: $5,243,948 \times 0.12 = $629,274 per year
- Capital cost allowance: Assuming that the building is placed in service in January, the first year's CCA percentage is 2.0%. Therefore, the allowed CCA amount is $3,000,000 \times 0.02 = $60,000. The CCA in each subsequent year would be 4.0% of the book value at the beginning of that year. For the second year, the CCA would be

$$(\$3,000,000 - \$60,000) \times 0.04 = \$117,600.$$

CCA on equipment is calculated in a similar manner using a 30% rate (Year 1 = 15%).

- Disposal tax effect:

$$\text{Capital gain on land} = \$1,500,000 - \$1,000,000 = \$500,000$$

$$\text{Tax on capital gain} = -0.4 \times \tfrac{3}{4} \times \$500,000 = -\$150,000$$

For the building and equipment, the tax effect at the time of disposal equal (book value $-$ salvage value) $\times t$

Property	Salvage Value	Book Value	Gains (Losses)	Disposal Tax Effect, G
Building	$2,000,000	$2,497,079	$ 497,079	$ 198,831
Equipment	2,500,000	1,224,510	(1,275,490)	(510,196)

The total disposal tax effect, G, is then

$$G = -\$150,000 + \$198,831 - \$510,196 = -\$461,365$$

- Cash dividend: 194,311 shares $\times$ $2 = $388,622
- Common stock: When the project terminates and the bonds are retired, the debt ratio is no longer 0.5. If SSI wants to maintain the constant capital structure (0.5), SSI would repurchase the common stock in the amount of $5,440,708 at the prevailing market price. In developing Table 11.2, we assumed this repurchase of common stock had taken place at the end of the project year. In practice, a firm may or may not repurchase the common stock. As an alternative means of maintaining the desired capital structure, the firm may use this extra debt capacity released to borrow for other projects.
- Bond: When the bonds mature at the end year 5, the total face value in the amount of $5,243,948 must be paid to the bondholders.

(b) **Measure of project worth**
The NPW for this project is then

$$PW(20\%) = -\$500,000 + \$1,654,094(P/F, 20\%, 1) + \cdots$$
$$-\$3,124,393\ (P/F, 20\%, 5)$$
$$= \$2,749,554.$$

The investment is nonsimple and has an IRR of 330%. Even though the project requires a significant amount of cash expenditure at the end of project life, it appears to be a profitable one.

■ 11.2 Methods of Lease Financing

An alternative way of obtaining the use of facilities and equipment is to lease them. Prior to the 1950s, leasing was generally associated with real estate—land and buildings. Today, however, it is possible to lease virtually any kind of fixed asset. In 1990, about 35% of all new capital equipment acquired by businesses was financed through lease arrangements.

11.2.1 General Elements of Leases

Before explaining leasing methods, we must define the terms lessee and lessor. The **lessee** is the party leasing the property, and the **lessor** is the owner of property that is leased. Leasing takes several different forms, the most important of which are (1) **operating (or service) leases** and (2) **financial (or capital) leases.**

Operating Leases

Automobiles and trucks are frequently acquired under operating leases, and the operating-lease contract for office equipment such as computers and copying machines is becoming more common. These operating leases have three unique characteristics.

1. The lessor maintains and services the leased equipment, and the cost of this maintenance is figured into the lease payments.
2. Operating leases are not fully amortized—in other words, the payments required under the lease contract are not sufficient to recover the full cost of the equipment.
3. Operating leases frequently contain a cancellation clause, which gives the lessee the right to cancel the lease before the expiration of the basic agreement.

These operating-lease contracts are written for a period considerably shorter than the expected economic service life of the leased equipment, and the lessor expects to recover all investment costs through subsequent renewal payments, through subsequent leases to other lessees, or by sale of the leased equipment. The cancellation clause is an important consideration to the lessee, because the equipment can be returned if it is rendered obsolete by technological developments or is no longer needed because of a decline in business.

Financial Leases

In contrast to operating leases, financial leases are usually drawn up as "net" leases; that is, the lessee assumes responsibility for paying most of the operating costs of the equipment, and his/her payment to the lessor is primarily for principal and interest. The lessee's obligations under a net financial lease are, therefore, similar to the obligations incurred under a debt financial instrument.

11.2.2 Potential Benefits of Leasing

Leases are gaining popularity beyond the traditional market of high-technology areas, such as computer, aircraft, and medical equipment. Where rapid obsolescence of the asset was once a primary reason for entering into a lease, more and more companies are now requesting leases that offer economic flexibility—whether in the form of early cancellations or of upgrades—and that places the burden of ownership on the lessor. The greater propensity shown by many companies to limit their longer-term commitments probably is linked to uncertain economic conditions compounded by the potential tax liability of ownership. An example of this influence, in perhaps its most extreme form, is the case of lease financing of Boeing and Airbus jetliners. This practice is common with major airlines like Air Canada and Canadian Airlines International Ltd. In one instance, an airline negotiated a lease in which it had the option, during the first 10 years, to walk away from any or all of the aircraft on just 30 days' notice. This leasing arrangement is hardly a typical one and is unlikely to be repeated, but it does illustrate the growing trend toward shifting the financial risk of ownership from the lessee to the lessor.

Leasing is nowhere in more demand than in the transportation industries, led by trucking. Convenience of equipment disposal, as well as a reluctance to commit irrevocably to a particular product (particularly with constantly changing federal and provincial laws specifying the allowable mass, width, and length of trucks), make this option attractive.

To understand the potential source of any benefits from leasing, we shall consider the following transactions involved in the purchase scenario illustrated in Fig. 11.2. First, the funds needed to acquire the asset must be raised. Second, the asset must be purchased. Third, the asset must be disposed of at

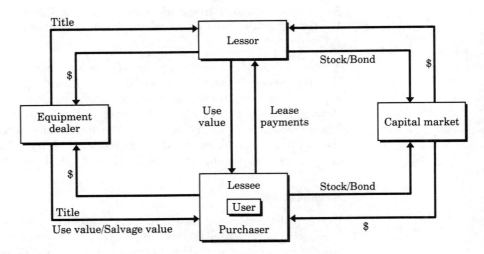

Figure 11.2 ■ A diagram describing the leasing market

the end of its useful life. These three phases of the process represent the potential sources of economic benefit from leasing. That is, if the lessor can acquire funds from the capital market on more favorable terms, or purchase the asset at a lower price, or achieve a higher salvage value for the asset than the lessee can, then some of these savings may be passed to the lessee in the form of lease payments that are less than purchase payments.

Leases are evaluated by both the lessee and the lessor. The lessee must determine whether leasing an asset is less costly than buying the asset, and the lessor must decide what the lease payments must be to produce a reasonable rate of return. Since our focus in this book is primarily on the economic utilization of fixed assets in production, we restrict our analysis to that conducted by the lessee.

11.2.3 Lease-or-Buy Decision by the Lessee

A lease-or-buy decision begins only after a company has decided that the acquisition of a piece of equipment is necessary to carry out an investment project. Having made this decision, the company may be faced with several alternative methods of financing the acquisition: cash purchase, debt purchase, or acquisition via a lease.

In a situation of debt purchase, the present worth expression is similar to that of a purchase for cash, except that it has additional items—loan repayments and a tax shield on interest payments. The only way that the lessee can evaluate the cost of a lease is to compare it against the best available estimate of what the cost would be if the lessee owned the equipment.

To lay the groundwork for a more general analysis, we shall first consider how to analyze the lease-or-buy decision for a project with a single fixed asset for which the company expects a service life of N periods. Since the net after-tax revenue is the same for both alternatives, we can only consider the incremental cost of owning the asset and the incremental cost of leasing. Using the generalized cash flow approach presented in Section 9.4, the incremental cost of owning the asset by borrowing may be expressed

$$
\begin{aligned}
PW(i)_{buy} = \ & - \text{PW of loan repayment} \\
& - \text{PW of after-tax O\&M costs} \\
& + \text{PW of tax credit on CCA and interest} \\
& + \text{PW of net proceeds from sale.} \qquad (11.1)
\end{aligned}
$$

Note that the asset acquisition (investment) cost is offset by the same amount of borrowing at time 0, so that we only need to consider the loan repayment series.

Let us assume that the firm can lease the asset at a constant amount per period. The project's incremental cost of leasing, $PW(i)_{lease}$, then becomes

$$PW(i)_{lease} = - \text{PW of after-tax lease expenses.} \tag{11.2}$$

If the lease does not provide for the maintenance of the equipment leased, then the lessee must assume this responsibility. In this situation, the maintenance term in Eq. (11.1) can be dropped from calculating the incremental cost of owning the asset.

The criterion for the decision to lease as opposed to purchase thus reduces to a comparison between $PW(i)_{buy}$ and $PW(i)_{lease}$. In our terms, purchase is preferred if the combined present value of loan repayment series and after-tax O&M expense, reduced by the present values of the depreciation tax shield and the net proceeds from disposal of the asset, is greater (less negative) than the present value of the net lease costs.

Example 11.4   Lease-or-buy decision

The Kanata Electronics Company is considering replacing an old, industrial forklift truck whose capacity is 1000 pounds (454 kilograms). The truck has been used primarily to move goods from production machines into storage. The company is working nearly at capacity and is operating on a 2-shift basis, 6 days per week. Kanata management is considering the possibility of either owning or leasing the truck. The plant engineer has compiled the following data for the management.

- The initial cost of a gas-powered truck is $20,000. The new truck would use about 20 litres of gasoline (single shift for 8 hours per day) at a cost of $0.44 per litre. If the truck is operated 16 hours per day, its expected life will be 4 years, and an engine overhaul at a cost of $1500 will be required at the end of 2 years.

- The Ottawa Industrial Truck Company was servicing the old forklift truck, and Kanata would buy the new truck through Ottawa. Ottawa offers a service agreement to users of its trucks, which provides for a monthly visit by an experienced service representative to lubricate and tune the trucks and costs $120 per month. Insurance and property taxes for the truck are $650 per year.

- The truck falls into a 30% declining balance CCA class. Kanata is in the 40% marginal tax bracket; the estimated resale value of the truck at the end of 4 years will be 15% of the original cost.

- Ottawa also has offered to lease a truck to Kanata. Ottawa will maintain the equipment and guarantee to keep the truck in serviceable condition at all times. In the event of a major breakdown, Ottawa will provide a replacement truck, at its expense. The cost of the operating lease plan is $850 per month. The contract term is 3 years' minimum, with the option to cancel on 30 days' notice thereafter.

- Based on recent experience, the company expects that funds committed to new investments should earn at least a 12% rate of return after taxes.

Compare the cost of owning versus leasing the truck.

Discussion: We may calculate the fuel costs for two-shift operations as

$$(20 \text{ litres/shift}) \, (2 \text{ shifts}) \, (\$0.44/\text{litre}) = \$17.60 \text{ per day.}$$

The truck will operate 300 days per year, so the annual fuel cost will be $5280. However, both alternatives require the company to supply its own fuel, so the fuel cost is not relevant for our decision making. Therefore, we may drop this common cost item from our calculation.

Solution

(a) **Incremental cost of owning the truck:** To compare the incremental cost of ownership with the incremental cost of leasing, we make the following additional estimates and assumptions.

- The preventive-maintenance contract, which costs $120 per month (or $1440 per year) will be adopted. With the annual insurance and taxes of $650, the equivalent present worth of the after-tax O&M is

$$P_1 = -(\$1440 + \$650) \, (1 - 0.40) \, (P/A, 12\%, 4) = -\$3809.$$

- The engine overhaul is not expected to increase either the salvage value or the service life. Therefore, the overhaul cost ($1500) will be expensed all at once rather than capitalized. The equivalent present worth of this after-tax overhaul expense is

$$P_2 = -\$1500(1 - 0.40) \, (P/F, 12\%, 2) = -\$717.$$

- If Kanata decided to purchase the truck through debt financing, the first step in determining costs would be to compute the annual installments of the debt-repayment schedule. Assuming that the entire investment of $20,000 is financed at a 10% interest rate, the annual payment would be

$$A = \$20,000(A/P, 10\%, 4) = \$6309.$$

- The equivalent present worth of this loan-payment series is

$$P_3 = -\$6309(P/A, 12\%, 4) = -\$19,163.$$

The interest payment each year (10% of the beginning balance) is calculated as follows:

Year	Beginning Balance	Interest Charged	Annual Payment	Ending Balance
1	$20,000	$2,000	−$6,309	$15,691
2	15,691	1,569	−6,309	10,951
3	10,951	1,095	−6,309	5,737
4	5,737	573	−6,309	0

With a 30% CCA declining balance and the half-year rule, the combined tax savings due to CCA and interest payments can be calculated as follows:

n	CCA_n	I_n	Combined Tax Savings
1	$3000	$2000	$5000(0.40) = $2000
2	5100	1569	6669(0.40) = 2668
3	3570	1095	4665(0.40) = 1866
4	2499	573	3072(0.40) = 1229

Therefore, the equivalent present worth of the combined tax credit is

$$P_4 = \$2000(P/F, 12\%, 1) + \$2668(P/F, 12\%, 2)$$

$$+ \$1866(P/F, 12\%, 3) + \$1229(P/F, 12\%, 4)$$

$$= \$6022.$$

- With the estimated salvage value of 15% of the initial investment ($3000) and with the CCA schedule above, we compute the net proceeds from the sale of the truck at the end of 4 years as follows:

$$\text{book value} = (\$20,000 - \$14,169) = \$5831$$

$$\text{losses (gains)} = \$5831 - \$3000 = \$2831$$

$$\text{disposal tax effects} = 0.4\ (\$2831) = \$1132.$$

The present equivalent amount of the net salvage value is

$$P_5 = \$4132\ (P/F, 12\%, 4) = \$2626.$$

The net incremental cost of owning the truck through 100% debt financing is thus

$$PW(12\%_{buy} = P_1 + P_2 + P_3 + P_4 + P_5$$

$$= -\$15,041.$$

The annual equivalent after-tax incremental cost of owning the truck for 4 years is

$$AE(12\%)_{buy} = -\$15,041(A/P, 12\%, 4)$$

$$= -\$4952.$$

Table 11.3 summarizes the after-tax cash flows associated with ownership in income statement format.

(b) Incremental cost of leasing the truck: How does the cost of acquiring a forklift truck under the lease compare with the cost of owning the truck?

- Since the lease payment is also a tax-deductible expense, however, the net cost of leasing has to be computed explicitly on an after-

Table 11.3 Incremental Cost of Owning the Truck: Buy Option (Example 11.4)

	0	1	2	3	4
			n		
Income Statement					
Operating cost					
Debt interest		$2,000	$1,569	$1,095	$ 573
CCA		3,000*	5,100	3,570	2,499
Maintenance		1,440	1,440	1,440	1,440
Insurance/taxes		650	650	650	650
Engine overhaul			1,500		
Taxable income		(7,090)	(10,259)	(6,755)	(5,162)
Income taxes		(2,836)†	(4,104)	(2,702)	(2,065)
Net income		−$4,254	−$6,155	−$4,053	−$3,097
Cash Flow Statement					
Net income		−$4,254	−$6,155	−$4,053	−$3,097
CCA		3,000	5,100	3,570	2,499
Investment	−$20,000				3,000
Disposal tax effect					1,132
Loan repayment	20,000	−4,309	−4,740	−5,214	−5,736
Net cash flow	0	−$5,563	−$5,795	−$5,697	−$2,202

* Half-year rule
† Tax savings

$$PW(12\%) = -\$5,563(P/F, 12\%, 1) - \$5,795(P/F, 12\%, 2) - \$5,697(P/F, 12\%, 3)$$
$$- \$2,202(P/F, 12\%, 4)$$
$$= -\$15,041$$

tax basis. The calculation of the annual incremental leasing costs is as follows.

Annual lease payments (12 months)	$10,200
Less 40% taxes	4,080
Annual net costs after taxes	$ 6,120

- Table 11.4 summarizes the after-tax cash flows associated with the lease option in income-statement format. The total net present equivalent incremental cost of leasing is

$$PW(12\%)_{lease} = -\$6120(P/A, 12\%, 4)$$

$$= -\$18,589.$$

Purchasing the truck with debt financing would save Kanata $3548 in NPW, or $1168 per year.

- Here, we have assumed that the lease payments occur at the *end* of the period; however, many leasing contracts require the payments to be made at the *beginning* of each period. In the latter situation, we can easily modify the present worth expression of the lease expense to reflect this cash-flow timing:

$$PW(12\%)_{lease} = -\$6120 - \$6120(P/A, 12\%, 3)$$

$$= -\$20,819.$$

Table 11.4 Incremental Cost of Leasing the Truck: Operating Lease Option (Example 11.4)

			n		
	0	1	2	3	4
Income and Cash Flow Statement					
Expenses					
Lease payment		$10,200	$10,200	$10,200	$10,200
Taxable income		(10,200)	(10,200)	(10,200)	(10,200)
Income taxes		(4,080)	(4,080)	(4,080)	(4,080)
Net income	0	− 6,120	− 6,120	− 6,120	− 6,120
Net cash flow	0	− $6,120	− $6,120	− $6,120	− $6,120

Note: For leasing there are no noncash expenses such as CCA; therefore, the net cash flow after taxes will be the same as the net income after taxes.

Comments: *In our example, leasing the truck appears to be more expensive than purchasing the truck with debt financing. You should not conclude, however, that leasing is always more expensive than owning. In many cases, analysis favors a lease option. The interest rate, salvage value, lease-payment schedule, and debt financing all have an important effect on decision making.*

11.2.4 The Lessor's Point of View

Why are operating leases more expensive in the preceding example? To answer this question, we need to look at the transaction from the lessor's point of view. The lessor is assured of receiving only a finite period revenue. At the expiration of the lease, the lessee may return the equipment, and the lessor will receive no further revenue from it unless the lessor is able to rent it to someone else or sell it as used equipment. The risk to the lessor is that a particular lessee will not have a continued need for the truck and that the lessor will not be able to dispose of the returned equipment profitably. We described this function as absorbing the risk of obsolescence.

Another reason an operating lease option may be expensive is that the lessor may accept responsibility for maintaining the equipment in good operating condition. The lessor must charge a price sufficient to provide a cushion to absorb any unusual maintenance costs that might be incurred. If Kanata intends to use the truck for 4 years, it would be wiser to seek a financial lease arrangement instead of the operating lease.

Summary

- Methods of financing fall into two broad categories.
  1. **Equity financing** uses earnings or funds raised from the issuance of stock to finance a capital investment.
  2. **Debt financing** uses money raised through loans or by the issuance of bonds to finance a capital investment.
- Companies do not simply borrow funds to finance projects. Well-managed firms usually establish a **target capital structure** and strive to maintain the **debt ratio** when individual projects are financed.
- **Leases** are financial instruments by which a firm obtains and uses capital equipment without purchasing and owning it outright. The specifics of lease arrangements may be negotiated to offer mutual benefits to the lessee and the lessor, but the typical elements of a lease contract include:
  1. **Rental payments**—Payments from lessee to lessor in exchange for use of the equipment.
  2. **Expiration dates**—The lease is negotiated for a finite period of time after which it may be renewed or renegotiated.
  3. **Return of equipment to lessor**—Typically, at the expiration of the lease, the leased item returns to the lessor who may re-lease it or dispose of it.

Optional aspects of leases include:
  1. **Cancellation clauses**—The lessee may have the right to cancel the lease prior to its expiration date.

2. **Maintenance clauses**—Maintenance costs may be borne by the lessee or the lessor, depending on how the contract is negotiated.

3. **Ownership privileges**—Some leases transfer ownership of the equip-ment to the lessee at the expiration date.

4. **Obsolescence clauses**—The lease may include the flexibility of upgrading equipment to keep up with technological advances during the lease period.

Problems

Note: Unless otherwise stated, the half-year rule applies.

11.1 Optical World Corporation, a manufacturer of peripheral vision storage systems, needs $10 million to market its new robotic-based vision systems. The firm is considering both financing options: common stock and bonds. If the firm decides to raise the capital through issuing common stock, the flotation costs will be 6% and the share price will be $25. If the firm decides to use debt financing, it can sell a 10-year, 12% bond with a par value of $1000. The bond flotation costs will be 1.9%.

(a) For equity financing, determine the flotation costs and the number of shares to be sold to net $10 million.

(b) For debt financing, determine the flotation costs and the number of $1000 par value bonds to be sold to net $10 million. What is the required annual interest payment?

11.2 Consider a project with an initial investment of $300,000 that must be financed at an interest rate of 12% per year. Assuming that the required repayment period is 6 years, determine the repayment schedule by identifying the principal as well as the interest payments for each of the following methods:

(a) equal repayment of the principal

(b) equal repayment of the interest

(c) equal annual installment

11.3 A construction company is considering the proposed acquisition of a new earthmover. The purchase price is $100,000, and an additional $25,000 is required to modify the equipment for special use by the company. The equipment falls into a declining balance CCA class for which $d = 30\%$. It will be sold after 5 years (project life) for $50,000. Purchase of the earthmover will have no effect on revenues, but it is expected to save the firm $60,000 per year in before-tax operating costs, mainly labor. The firm's marginal tax rate is 40%. Assume that the initial investment is to be financed by a bank loan at an interest rate of 10%, payable annually. Determine the after-tax cash flows and the worth of investment for this project.

11.4 A chemical plant is considering the purchase of a computerized control system. The initial cost is $200,000 and will produce net savings of $100,000 per year. If purchased, the system will be written off at a declining balance CCA rate of 30%. This system will be used for 4 years, at the end of which time the firm expects to sell the system for $30,000. The firm's marginal tax rate on this investment is 35%. The firm is considering the purchase of the computer-control system either through its retained earnings or by borrowing from a local bank. Two commercial banks are willing to lend the $200,000 at an interest rate of 10%, but they have different repayment plans. Bank A requires 4 equal annual principal payments with interest calculated based on the unpaid balance.

Bank A's Repayment Plan

End of Year	Principal	Interest
1	$50,000	$20,000
2	50,000	15,000
3	50,000	10,000
4	50,000	5,000

Bank B offers a payment plan that extends over 5 years with 5 equal annual payments.

Bank B's Repayment Plan

End of Year	Principal	Interest	Total
1	$32,759	$20,000	$52,759
2	36,035	16,724	52,759
3	39,638	13,121	52,759
4	43,602	9,157	52,759
5	47,962	4,796	52,759

(a) Determine the cash flows if the computer-control system is bought through the company's retained earnings (equity financing).

(b) Determine the cash flows if the asset is financed through either bank A or bank B.

(c) Recommend the best course of financing action. (Assume that the firm's MARR is also known to be 10%.)

11.5 Air Alberta, a regional airline carrying passengers within Alberta, is considering the possibility of adding a new long-range aircraft to its fleet. The aircraft being considered for purchase is the McDonnell Douglas DC-9-532 "Funjet," which is quoted at $60 million per unit. McDonnell Douglas requires 10% down payment at the time of delivery and the balance to be paid over a 10-year period at an interest rate of 12% compounded annually. The actual payment schedule calls for only interest payments over the 10-year period and the original principal amount to be paid off at the end of 10th year. Air Alberta expects to generate $35 million per year by adding this aircraft to its current fleet but also estimates an operating and maintenance cost of $20 million per year. The aircraft is expected to have a 15-year service life with the salvage value worth 15% of the original purchase price. If the aircraft is bought, it will be written off at a declining balance CCA rate of 25%. The firm's combined federal and provincial marginal tax rate is 38% and its required minimum attractive rate of return is 18%.

(a) Determine the cash flow associated with the debt financing.

(b) Is this project acceptable?

11.6 Consolidated Power Company currently owns and operates a coal-fired combustion turbine plant, which was installed 20 years ago. Because of degradation of the system, there were 65 forced outages during the last year alone and two boiler explosions during the last 7 years. Consolidated is planning to scrap the current plant and install a new, improved gas turbine which produces more energy per unit of fuel than typical coal-fired boilers (see Fig. 11.3). The 50-megawatt gas-turbine plant, which runs on gasified coal, wood, or agricultural wastes, will cost Consolidated $65 million. Consolidated wants to raise the capital from three financing

Figure 11.3 ■ A new coal-processing treatment that makes turbines efficient and versatile
(Problem 11.6)

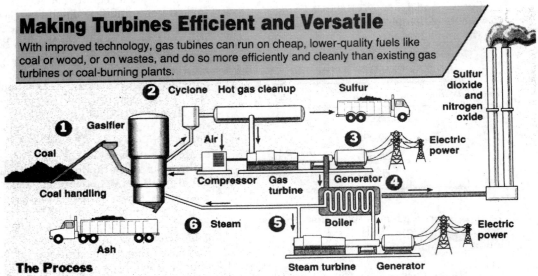

Making Turbines Efficient and Versatile

With improved technology, gas tubines can run on cheap, lower-quality fuels like coal or wood, or on wastes, and do so more efficiently and cleanly than existing gas turbines or coal-burning plants.

The Process

1. Coal or other hydrocarbon material is treated with steam under pressure, releasing combustible gases.

2. The gas is cleaned in a cyclone and with other processes, removing the sulfur.

3. Carbon monoxide and hydrogen are burned in a gas turbine, which spins a generator to make electricity and also powers the compressor.

4. Hot exhaust from the turbine goes to a boiler, which makes steam, and then goes up the stack.

5. The steam drives another turbine, which also turns a generator to make electricity.

6. The remaining steam and air from the compressor are sent to the gasifier to process more fuel into combustible gases.

sources: 45% common stock, 10% preferred stock, and 45% borrowed funds. Consolidated's investment banks quote Consolidated the following flotation costs:

Financing Source	Flotation Costs	Selling Price	Par Value
Common stock	4.6%	$32 per share	$ 10
Preferred stock	8.1	55 per share	15
Bond	1.4	$980	1000

(a) What are the total flotation costs to raise $65 million?

(b) How many shares (both common and preferred) or bonds must be sold to raise $65 million?

(c) If Consolidated will make annual cash dividends of $2 per common share, and annual bond interest payment at the rate of 12%, how much cash should Consolidated have available to meet both the equity and debt obligation? The preferred stockholders receive annual dividends at 6% of par value. Consolidated's marginal tax rate is 40%.

11.7 The Jacob Company needs to acquire a new lift truck for transporting its final product to the warehouse. One alternative is to purchase the lift truck for $40,000, which will be financed by the bank at an interest rate of 12%. The loan must be repaid in 4 equal installments, payable at the end of each year. Under the borrow-to-purchase arrangement, Jacob would have to maintain the truck at a cost of $1200 payable at year-end. Alternatively, Jacob could lease the truck on a 4-year contract for a lease payment of $11,000 per year. Each annual lease payment must be made at the *beginning* of each year. The truck would be maintained by the lessor. The truck falls into the declining balance CCA class for which $d = 30\%$. It has a salvage value of $10,000, which is the expected market value after 4 years, at which time Jacob plans to replace the truck irrespective of whether it leases or buys. Jacob has a marginal tax rate of 40% and a MARR of 15%.

(a) What is Jacob's cost of leasing in present worth?

(b) What is Jacob's cost of owning in present worth?

(c) Should the truck be leased or purchased?

11.8 Janet Wigandt, an electrical engineer for Instrument Control, Inc. (ICI), has been asked to perform a lease-buy analysis on a new pin-inserting machine for its pc-board manufacturing.

- **Buy option:** The equipment costs $120,000. To purchase it, ICI could obtain a term loan for the full amount at 10% interest with 4 equal annual installments (end-of-year payment). The machine falls into the 30% declining balance CCA class. Annual revenues of $200,000 and operating costs of $40,000 are anticipated. It also requires an annual maintenance cost of $10,000. Because technology is changing rapidly in pin-inserting machines, the salvage value of the machine is expected to be only $20,000.

- **Lease option:** Business Leasing, Inc. (BLI) is willing to write a 4-year operating lease on the equipment for payments of $44,000 at the *beginning* of each year. Under this operating-lease arrangement, BLI will maintain the asset so that the maintenance cost of $10,000 will be saved annually. ICI's marginal tax rate is 40%, and its MARR is 15% during the analysis period.

(a) What is ICI's present value (incremental) cost of owning the equipment?

(b) What is ICI's present value (incremental) cost of leasing the equipment?

(c) Should ICI buy or lease the equipment?

11.9 Consider the following lease versus borrow-and-purchase problem.

- **Borrow-and-purchase option**
  1. Jensen Manufacturing Company plans to acquire sets of special industrial tools with a 4-year life and a cost of $200,000, delivered and installed.
  2. The firm's MARR is known to be 15%.
  3. Jensen can borrow the required $200,000 on a 10% loan over 4 years. Four equal annual payments (end-of-year) would be made in the amount of

  $$\$63,094 = \$200,000(A/P, 10\%, 4).$$

  The annual interest and principal payment schedule along with the equivalent present worth of these payments is as follows:

End of Year	Interest	Principal
1	$20,000	$43,094
2	15,961	47,403
3	10,950	52,144
4	5,736	57,358

4. The estimated salvage value for the tool sets at the end of 4 years is $20,000.

5. If Jensen borrows and buys, it will have to bear the cost of maintenance, which will be performed by the tool manufacturer at a fixed contract rate of $10,000 per year.

6. The tools will be written off with a declining balance CCA rate of 20%, and Jensen's marginal tax rate is 40%.

- **Lease option**
  1. Jensen can lease the tools for 4 years at an annual rental charge of $70,000, payable at the end of each year.
  2. The lease contract specifies that the lessor will maintain the tools at no additional charge to Jensen.

  (a) What is Jensen's PW of after-tax cash flow of leasing?
  (b) What is Jensen's PW of after-tax cash flow of owning?

11.10 The Quebec Division of Precision Machinery, Inc. manufactures drill bits. One of the production processes of a drill bit requires tipping, where carbide tips are inserted into the bit to make it stronger and more durable. This tipping process usually requires 4 or 5 operators, depending on the weekly work load. The same operators were assigned to a stamping operation in which the size of the drill bit and the company's "logo" are imprinted into the bit. Precision is considering acquiring three automatic tipping machines to replace the manual tipping and stamping operations. If the tipping process is automated, Precision engineers will have to redesign the shapes of the carbide tips to be used in the machine. The new design requires less carbide, resulting in material savings. The following financial data have been compiled.

- Project life: 6 years
- Expected annual savings: reduced labor, $56,000; reduced material, $75,000; other benefits (reduced carpal tunnel syndrome and related problems), $28,000; reduced overhead, $15,000
- Expected annual O&M costs: $22,000
- Tipping machines and site preparation: equipment (3 machines) costs including delivery, $180,000; site preparation, $20,000
- Salvage value: $30,000 (3 machines) at the end of 6 years
- CCA: 30% declining balance
- Investment in working capital: $25,000 at the beginning of the project year, and that same amount will be fully recovered at the end of project year
- Other accounting data: marginal tax rate of 39%, MARR of 18%

To raise $200,000, Precision is considering the following financing options.

- Option 1: Finance the tipping machines using their retained earnings.
- Option 2: Secure a 12% term loan over 6 years (6 equal annual installments).
- Option 3: Lease the tipping machines. Precision can obtain a 6-year financial lease on the equipment (no maintenance) for payments of $55,000 at the *beginning* of each year.

(a) Determine the net after-tax cash flows for each financing option.
(b) What is Precision's present value cost of owning the equipment by borrowing?
(c) What is Precision's present value cost of leasing the equipment?
(d) Recommend the best course of action for Precision.

11.11 Tom Hagstrom has decided to acquire a new car for his business. One alternative is to purchase the car outright for $16,170, financing with a bank loan for the net purchase price. The bank loan calls for 36 equal monthly payments of $541.72 at an interest rate of 12.6% compounded monthly. Payments must be made at the end of each month.

Buy or Lease Your New '92

Buy	Lease
$16,170	$425 per month
	36-month open-end lease
	An annual metrage allowed —24,000 kilometres

If Tom takes the lease option, he is required to pay $500 for a security deposit refundable at the end of the lease, and $425 a month at the beginning of each month for 36 months. If the car is purchased, it will be written off at a declining balance CCA rate of 30%. It has a salvage value of $5800, which is the expected market value after 3 years, at which time Tom plans to replace the car irrespective of whether he leases or buys. Tom's marginal tax rate is 35%. His MARR is known to be 13% per year.

(a) Determine the annual cash flows for each option.

(b) Which option is the better choice?

11.12 The Boggs Machine Tool Company has decided to acquire a pressing machine. One alternative is to lease the machine on a 3-year contract for a lease payment of $15,000 per year, with payments to be made at the beginning of each year. The lease would include maintenance. The second alternative is to purchase the machine outright for $100,000, financing with a bank loan for the net purchase price and amortizing the loan over a 3-year period at an interest rate of 12% per year (annual payment = $41,635). The

yearly interest and principal payments would be

End of Year	Interest	Principal
1	$12,000	$29,635
2	8,444	33,191
3	4,461	37,174

Under the borrow-to-purchase arrangement, the company would have to maintain the machine at an annual cost of $5000 payable at year-end. The machine falls into a 30% declining balance CCA class. It has a salvage value of $50,000, which is the expected market value at the end of year 3. At that time the company plans to replace the machine irrespective of whether it leases or buys. Boggs has a tax rate of 40% and a MARR of 15%.

(a) What is Boggs' PW cost of leasing?
(b) What is Boggs' PW cost of owning?
(c) From the financing analysis in (a) and (b) above, what are the advantages and disadvantages of leasing and owning?

11.13 An asset is to be purchased for $25,000. The asset is expected to provide revenue of $10,000 a year, and have operating costs of $2500 a year. The asset falls into the 30% declining balance CCA class. The company is planning to sell the asset at the end of year 5 for $5000. Given that the company's marginal tax rate is 30% and that it has a MARR of 10% for any project undertaken, answer the following questions.

(a) Find the net cash flow for each year, given that the asset is purchased with borrowed funds at an interest rate of 12% with repayment in 5 equal end-of-year payments.

(b) Find the net cash flow for each year, given that the asset is leased at a rate of $3500 a year (financial lease).

(c) Which method (if any) should be used to obtain the new asset?

11.14 Enterprise Capital Leasing Company is in the business of leasing tractors for construction companies. The firm wants to set a 3-year lease-payment schedule for a tractor purchased at $53,000 from the equipment manufacturer. The asset falls into a 30% declining balance CCA class. The tractor is expected to have a salvage value of $22,000 at the end of 3 years' rental return. Enterprise will require a lessee to make a security deposit in the amount of $1500 that is refundable at the end of the lease term. Enterprise's marginal tax rate is 35%. If Enterprise wants an after-tax return of 10%, what lease-payment schedule should be set?

11.15 The headquarters building owned by a rapidly growing company is not large enough for current needs. A search for additional space revealed the following alternatives that would provide sufficient room, enough parking, and the desired appearance and location.

- Option 1: Lease extra space for $144,000 per year.
- Option 2: Purchase extra space for $800,000 including a $150,000 cost for land.
- Option 3: Remodel the current headquarters building.

It is believed that land values will not decrease over the ownership period, but the value of all structures will decline to 10% of the initial cost in 30 years. Annual property-tax payments are expected to be 5% of the initial cost. The present headquarters can be remodeled at a cost of $300,000 to make it comparable to other alternatives. However, the remodeling will occupy part of the existing parking lot. An adjacent, privately owned parking lot can be leased for 30 years under an agreement that the first year's rental of $9000 will increase by $500 each year. The study period for the comparison is 30 years, and the desired rate of return on investment is 12%. Assume that the firm's marginal tax rate is 40% and the new building and remodeled structure will be written off at a declining balance CCA rate of 4%. If the annual upkeep costs are the same for all three alternatives, which one is preferable?

11.16 The National Parts, Inc., an auto-parts manufacturer, is considering purchasing a rapid prototyping system to reduce prototyping time for form, fit, and function applications in auto-parts manufacturing. An outside consultant has been called in to estimate the initial hardware requirement and installation costs. He suggests the following:

- Prototyping equipment: $187,000
- Postcuring apparatus : $10,000
- Software: $15,000
- Maintenance: $36,000 per year by the equipment manufacturer
- Resin: Annual liquid polymer consumption, 2000 litres at $70 per litre.
- Site preparation: Some facility changes are required when installing the rapid prototyping system. (Certain liquid resins contain a toxic substance, so the work area must be well vented.)

The expected life of the system is 6 years with an estimated salvage value of $30,000. The proposed system falls into the 30% declining balance CCA class.

A group of computer consultants must be hired to develop various customized software to run on these systems. These software development costs will be $20,000 and can be expensed during the first tax year. The new system will reduce prototype development time by 75% and the material waste (resin) by 25%. This reduction in development time and material waste will save the firm $114,000 and $35,000 annually. The firm's expected marginal tax rate over the next 6 years will be 40%. The firm's interest rate is 20%.

(a) Assuming that the entire initial investment will be financed from the firm's retained earnings (equity financing), determine the after-tax cash flows over the investment life. Then compute the NPW of this investment.

(b) Assuming the entire initial investment will be financed through a local bank at an interest rate of 13% compounded annually, determine the net after-tax cash flows for the project. Then compute the NPW of this investment.

(c) Suppose that a financial lease is available for the prototype system at $62,560 per year, payable at the beginning of each year. Compute the NPW of this investment with lease financing.

(d) Select the best financing option based on the rate of return on incremental investment.

11.17 Anglo Chemical Corporation (ACC) is a multinational manufacturer of industrial chemical products. ACC has made great progress in energy-cost reduction by implementing several cogeneration projects in the United States and Canada, including the completion of a 35 megawatt (MW) unit in Chicago and a 29 MW unit at Baton Rouge. The division of ACC that is being considered for one of their more recent cogeneration projects is a chemical plant located in Sarnia. The plant has a power usage of 80 million kilowatt hours (kWh) annually. However, on the average, it uses 85% of its 10 MW capacity, which would bring the average power usage to 68 million kWh annually. Ontario Hydro presently charges $0.09 per kWh of electric consumption for the ACC plant, a rate that is considered high throughout the industry. Because ACC's power consumption is so large, the purchase of a cogeneration unit is considered to be desirable. This installation of the cogeneration unit would allow ACC to generate their own power and to avoid the annual $6,120,000 expense to Ontario Hydro. The total initial investment cost would be $10,500,000: $10,000,000 for the purchase of the power unit itself, a gas-fired 10 MW Allison 571, and engineering, design, and site preparation, and $500,000 for the purchase of interconnection equipment such as poles and distribution lines used to interface the cogenerator with the existing utility facilities. ACC is considering two financing options.

- ACC could finance $2,000,000 through the manufacturer at 10% for 10 years, and will finance the remaining $8,500,000 by issuing common stock. The flotation cost for common stock offering is 8.1% and the stock will be priced at $45 per share.
- The investment bankers have indicated that 10-year 9% bonds could be sold at a price of $900 for each $1000 bond. The flotation costs would be 1.9% to raise $10.5 million.

(a) Determine the debt-repayment schedule for the term loan from the equipment manufacturer.

(b) Determine the flotation costs and the number of common stocks to sell to raise the $8,500,000.

(c) Determine the flotation costs and the number of $1000 par value bonds to be sold to raise $10.5 million.

11.18 (Continuation of Problem 11.17) As ACC management has decided to raise the $10.5 million by selling bonds, ACC engineers have estimated the operating costs of the cogeneration project.

- The annual cash flow is comprised of many factors: maintenance, standby power, overhaul costs, and other miscellaneous expenses. Maintenance costs are projected to be approximately $500,000 per year. The unit must be overhauled every 3 years at a cost of $1.5 million. Miscellaneous expenses, such as additional personnel and insurance, are expected to total $1 million. Another annual expense is that for standby power, which is the service provided by the utility in the event of a cogeneration unit trip or scheduled maintenance outage. Unscheduled outages are expected to occur 4 times annually, each outage averaging 2 hours in duration at an annual expense of $6400. Overhauling the unit takes approximately 100 hours and occurs every 3 years, requiring another triennial power cost of $100,000. Fuel (spot gas) will be consumed at a rate of 8.44 megajoules per kWh, including the heat recovery cycle. At $1.896 per gigajoule, the annual fuel cost will reach $1,280,000.
- Due to obsolescence, the expected life of the cogeneration project will be 12 years, after which Allison will pay ACC $1 million for salvage of all equipment.
- Revenue will be realized from the sale of excess electricity to the utility company at a negotiated rate. Since the chemical plant will consume on average 85% of the unit's 10 MW output, 15% of the output will be sold

at $0.04 per kWh, bringing in an annual revenue of $480,000.
- ACC's marginal tax rate (combined federal and provincial) is 36% and their minimum required rate of return for any cogeneration project is 27%. The anticipated costs and revenues are summarized as follows.

Initial investment	
Cogeneration unit, engineering, design and site preparation (10% declining balance CCA)	$10,000,000
Interconnection equipment (30% declining balance CCA)	500,000
Salvage after 12-year's use	
Cogeneration unit	975,000
Interconnection equipment	25,000
Annual expenses	
Maintenance	500,000
Misc. (additional personnel and insurance)	1,000,000
Standby power	6,400
Fuel	1,280,000
Other operating expenses	
Overhaul every 3 years	1,500,000
Standby power during overhaul	100,000
Revenues	
Sale of excess power to Ontario Hydro	480,000

(a) If the cogeneration unit and other connecting equipment could be financed by issuing corporate bonds at an interest rate of 9% compounded annually with the flotation expenses as indicated in Problem 11.17, determine the net cash flow from the cogeneration project.

(b) If the cogeneration unit can be leased, what would be the maximum annual lease amount that ACC is willing to pay?

11.19 Northern Office Automation, Inc. (NOAI) is a leading developer of imaging systems, controllers, and related accessories. The company's product line consists of systems for desktop publishing, automatic identification, advanced imaging, and office information markets. The firm's manufacturing plant in Winnipeg, Manitoba, consists of 8 different functions: cable assembly, board assembly, mechanical assembly, controller integration, printer integration, production repair, customer repair, and shipping. The process to be considered is the transportation of pallets loaded with 8 packaged desktop printers from printer integration to the shipping department. Several alternatives have been examined to minimize operating and maintenance costs. The two most feasible alternatives are:

- Use of gas-powered lift trucks to transport pallets of packaged printers from printer integration to shipping. The truck also transports printers that must be reworked on the return trip. The trucks can be leased at a cost of $5465 per year. With a maintenance contract costing $6317 per year, the dealer will maintain the trucks. A fuel cost of $1660 per year is also expected. The truck requires a driver for each of the 3 shifts at a total cost of $58,653 per year for labor. It is also estimated that transportation by truck would cause damages to material and equipment totaling $10,000 per year.

- Installing an Automatic Guided Vehicle System (AGVS) to transport pallets of packaged printers from printer integration to shipping. Additionally, this AGVS will transport products requiring rework on its trip back to printer integration. The AGVS, using an electrical powered cart and embedded wire-guidance system, would be able to do the same job that the truck currently does, but without drivers. The total investment costs including installation are itemized as follows:

Vehicle and system installation	$ 97,255
Staging conveyer	24,000
Power supply lines	5,000
Transformers	2,500
Floor surface repair	6,000
Batteries and charger	10,775
Shipping	6,500
Sales tax	6,970
Total AGVS system cost	$159,000

NOAI could obtain a term loan for the full investment amount ($159,000) at a 10% interest. The loan would be amortized over 5 years, with payments made at the end of each year. The AGVS falls into a 30% declining balance CCA class and it has an estimated service life of 10 years and no salvage value. If the AGVS is installed, a maintenance contract would be obtained at a cost of $20,000, payable at the beginning of each year.

The firm's marginal tax rate is 35% and its MARR is 15%.

(a) Determine the net cash flows for each alternative over 10 years.

(b) Compute the incremental cash flows (AGVS option—Gas truck option) and determine the rate of return on this incremental investment.

(c) Determine the best course of action based on the rate of return criterion.

(d) Repeat (c) based on the present worth criterion.

CHAPTER 12

Replacement Decisions

Three printing technologies largely dominate the news industry: letterpress, offset, and flexography. Letterpress is the oldest of the three. Once the predominant method, in recent decades, letterpress has been replaced at many newspapers by offset, which provides better reproduction quality. Even more recently, flexography seemed to herald the next technological advance. It promised the economic advantage of lower operating costs and the noneconomic advantage of a water-based ink that would not rub onto readers' hands. However, some users reported that the reproduction quality of flexography was inferior to that of offset.

In 1989, a new process called *keyless offset* was being tested, and it promised to combine the best qualities of flexography and traditional offset. At that time, the *Washington Post* was still being printed by the letterpress process.

If a company decides to evaluate its current printing process, it should consider the following questions:

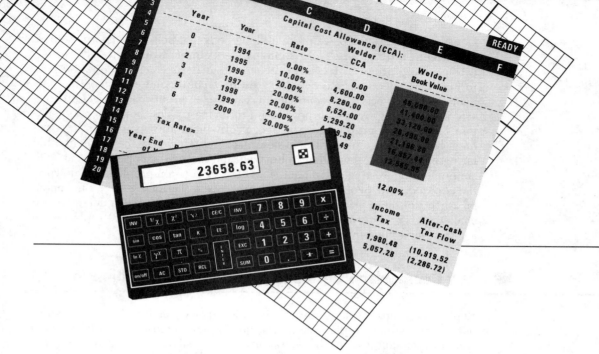

The spreadsheet shows:

Year	Year	Year	Rate	Capital Cost Allowance (CCA): Welder CCA	Welder Book Value
0		1994	0.00%	0.00	46,000.00
1		1995	10.00%	4,600.00	41,400.00
2		1996	20.00%	8,280.00	33,120.00
3		1997	20.00%	6,624.00	26,496.00
4		1998	20.00%	5,299.20	21,196.80
5		1999	20.00%	9.36	16,957.44
6		2000	20.00%	49	13,565.95

Tax Rate=

Year End of Y...

Calculator display: 23658.63

12.00%

Income Tax	After-Cash Tax Flow
1,980.48	(10,919.52)
5,057.28	(2,286.72)

READY

1. Should the letterpress printing press be replaced by one of the more advanced technologies?

2. If so, when should the press be replaced—now or sometime in the future?

The answer to the first question is almost certainly *yes*. Even if the press is in perfect working order, at some point, it will fail to meet the company's needs due to its physical or functional depreciation. When the need arises to replace the press, the company will probably want to take advantage of the technological advances of a newer printing method.

The answer to the second question is less clear. Presumably, the current press can be maintained in serviceable condition for some time—what will be the economically optimal time to make the replacement? Furthermore, when is the technologically optimal time to switch? By waiting for the promised keyless process to evolve and gain commercial acceptance, the company may be able to obtain even better results than either traditional offset or flexography can offer.

In Chapters 4 through 6, we presented methods that help one choose the best investment alternative. The problems we examined primarily concerned profit-adding projects. However, economic analysis is also frequently performed on projects carried out with existing facilities and equipment (profit-maintaining projects). As mentioned in Chapter 9, profit-maintaining projects are those whose primary purpose is not to increase sales, but rather to maintain ongoing operations. In practice, profit-maintaining projects less frequently involve the comparison of new machines. However, the problem often facing management is whether to buy new and more efficient equipment or to continue using existing equipment. This class of decision problems is known as the **replacement problem.** In this chapter, we will examine three fundamental aspects of the replacement problem: (1) comparison between defender and challenger; (2) determination of economic service life; and (3) replacement analysis when the required service period is long.

■ 12.1 Comparison Between Defender and Challenger

Replacement projects are those required to replace existing obsolete or worn-out assets. If these replacements are not implemented, the result will be a slowdown or shutdown of the operations. The question raised is whether existing equipment should be replaced with more efficient equipment. This situation has given rise to the use of two terms commonly used in the boxing world, **defender** and **challenger**. In each boxing class is a current defending champion. The current champion is constantly faced with new challengers who want to depose him from his position. In replacement analysis, the defender is the existing machine (or system), and the challenger is the best available replacement equipment.

12.1.1 Current Market Value and Sunk Costs

An existing piece of equipment will be removed at some future time, either when the task it performs is no longer necessary or when the task can be performed better by different equipment. Therefore, the question is not whether the existing equipment will be removed, but *when* it will be removed. Another variation of this same question is why we should replace existing equipment *at this time* rather than postponing replacement by repairing or overhauling the existing equipment. Another aspect of the defender-challenger comparison concerns determination of exactly which equipment *is* the challenger. If the defender is to be replaced by the challenger, we would want to install the best alternative.

Current Market Value

The most common problem in considering the replacement of existing equipment is determination of what financial information is actually relevant to the analysis. Often, there is a tendency to include irrelevant information in the analysis. To illustrate the decision problem, let us consider Example 12.1.

Example 12.1 Relevant information for replacement analysis

Macintosh Printing, Inc. purchased a $20,000 printing machine two years ago. The company expected this machine to have a 5-year life and a salvage value of $5000. The $20,000 capital expenditure was set up to be depreciated using a capital cost allowance (CCA) declining balance rate of 20% with the half-year rule. (Annual CCA: $2000, $3600, $2880, $2304, and $1843). The company has just spent $5000 on repairs, and current operating costs are now running at $4000 per year. Furthermore, the anticipated salvage value is reduced to $2000 at the end of the 5-year period. In addition, the company finds that the current machine has a market value of $10,000 today, and the manufacturer will allow the company that full amount in trade for the new machine. What value(s) for the defender should be relevant in our analysis?

Solution

In this example, four different dollar amounts relating to the defender have been presented:

1. *Original cost*: The printing machine was purchased for $20,000 two years ago.
2. *Book value*: The original cost minus the accumulated CCA with half-year convention (if sold now) is

$$\$20,000 - (\$2000 + \$3600) = \$14,400.$$

3. *Market value* of the old machine: The company estimates it to be $10,000.
4. *Trade-in allowance:* In this example, it is the same as the market value. (This value could be different from the market value.)

In this example, and in all defender analyses, the relevant cost is the current market value for the equipment. The original cost, book value, repair cost, and trade-in value are irrelevant. There is a common misunderstanding that the trade-in value is the **current market value** of the equipment and thus could be used as a suitable current value for the equipment. This is not always true, however. For example, a car dealer typically offers a trade-in value on a customer's old car that will reduce the price of the new car. Would the dealer offer the same value on the old car without also selling the new one? This is not generally the case. In many instances, the trade-in allowance is inflated to make the deal look good, but the price of the new car is also inflated to compensate for the dealer's trade-in cost. In this type of situation, the trade-in value does not represent the true value of the item, so we should not use it in economic analysis.[1]

Sunk Cost

A **sunk cost** is any past cost that will be unaffected by a future investment decision. In Example 12.1, the current book value is $14,400, and the market value is $10,000. Therefore, a loss of $4400 will occur if the machine was taken out of service. With the $5000 repair cost added, the total unrecovered investment if the trade were made would be $9,400. In economic analysis, it is tempting to add this loss on disposal to the cost of the challenger. This is *incorrect* (Fig. 12.1). (Note, however, that the loss will be recovered to the extent allowed by the tax laws.) Only future costs should be considered, and past or sunk costs should be ignored. Thus, the value of the defender that should be used in a replacement analysis is its current market value adjusted for disposal tax effects, not the cost when it was originally purchased and not the cost of any repairs that may have been made.

12.1.2 Gains (Losses) on Disposal of Defender

Replacement studies require a knowledge of the tax depreciation schedule and of taxable gains or losses. Note that the depreciation schedule is in effect determined at the time of acquisition, whereas the disposal tax effects are determined by the relevant tax law at the time of disposal.

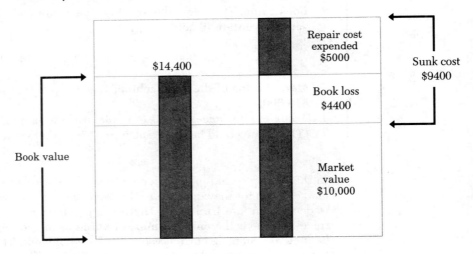

Figure 12.1 ■ Sunk cost associated with an asset's disposal (Example 12.1)

---

[1] If we do make the trade, however, the actual net cash flow at this time, properly used, is certainly relevant.

Example 12.2 Net salvage value from disposal of old machine

In Example 12.1, determine the taxable gains (or losses) and the net salvage value from the disposal of the old machine if the firm's marginal income tax rate is 40%. Assume that the MARR = 12%.

Solution

We compute the following:

Current book value	= $14,400
Current market value	= $10,000
Losses	= $4400
Disposal tax effects	= $(B_{\text{Disposal}} - \text{salvage value}) \times t$
	= $4400 \times 40\%$
	= $1760
Net salvage value	= $10,000 + $1760
	= $11,760

This calculation is illustrated in Fig. 12.2.

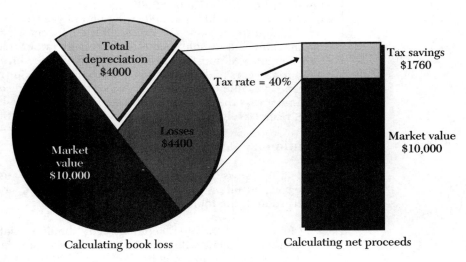

Depreciation base = $20,000

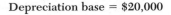

Figure 12.2 ■ Net salvage value from sale of defender (Example 12.2)

12.1.3 Basic Replacement Decision

Although replacement projects are a subcategory of the mutually exclusive project decisions we studied in Chapters 4, 5, and 6, they do possess unique characteristics that allow us to use specialized concepts and analysis techniques in their evaluation. We will consider two basic approaches to analyzing replacement problems: the **cash flow approach** and the **opportunity cost approach.**

Cash Flow Approach

We will start with a replacement-decision problem where both the defender and the challenger have the *same useful life,* which begins now. Presumably, the defender has been around for a while, and the required service period for the defender is relatively short. In this situation, we can use the cash flow approach, as long as *the analysis period is the same* for all alternatives.

Example 12.3 Replacement analysis using the cash flow approach

Consider Example 12.1. The company has been offered a chance to purchase another machine for $15,000. Over its 3-year useful life, the machine will reduce labor and use of raw materials sufficiently to cut operating costs from $8000 to $6000. This reduction in costs will allow before-tax profit to rise by $2000 per year. It is estimated that the new machine can be sold for $6000 at the end of year 3. The new machine would fall into the same CCA class. If the new machine is purchased, the old machine will be sold to another company rather than exchanged for the new machine. Suppose that the company will need either machine (old or new) for 3 years and that it does not expect a superior new machine to become available on the market during this required service period. Assuming that the firm's interest rate and tax rate are known to be 12% and 40%, respectively, decide whether replacement is justified *now.*

Solution

Table 12.1 shows the worksheet format the company uses to analyze a typical replacement project. Each line is numbered, and a line-by-line description of the table follows.

- *Lines 1 - 4:* If the old machine is kept, the depreciation schedule would be ($2880, $2304, and $1843). This results in total CCA of $12,267 and a remaining book value of $20,000 − $12,267 = $7373.

- *Lines 5 - 6:* The repair cost in the amount of $5000 was already incurred before the replacement decision, and thus this is a sunk cost. If the old machine is retained for the next 3 years, the before-tax annual O&M costs are as shown in Line 6.

Table 12.1 Replacement Analysis Worksheet (Example 12.3)

Option 1: Keep the Defender (old machine)

n:	−2	−1	0	1	2	3
			Investment Data/Cost Information			
(1) CCA		$2,000	$3,600	$2,880	$2,304	$1,843
(2) Book value	$20,000	$18,000	$14,400	$11,520	$9,216	$7,373
(3) Salvage value						$2,000
(4) Loss from sale						$5,373
(5) Repair cost			$5,000			
(6) O&M costs				$8,000	$8,000	$8,000
			Cash Flow Statement			
(7) + (0.4)(CCA)				$1,152	$ 922	$ 737
(8) − (0.6)(O&M)				−4,800	−4,800	−4,800
(9) Net salvage value (salvage value + disposal tax effects)						4,149
(10) Net cash flow			0	−$3,648	−$3,878	$86

Option 2: Replace the Defender (new machine)

			Investment Data/Cost Information			
(11) Net proceeds from sale of old machine			$11,760			
(12) Cost of new equipment			$15,000			
(13) CCA				$1,500	$2,700	$2,160
(14) Book value				$13,500	$10,800	$8,640
(15) Salvage value						$6,000
(16) Losses from sale of new machine						$2,640
(17) O&M costs				$6,000	$6,000	$6,000
			Cash Flow Statement			
(18) Net investment			−$3,240			
(19) + (0.4)(CCA)				$ 600	$1,080	$ 864
(20) − (0.6)(O&M)				−$3,600	−$3,600	−$3,600
(21) Net salvage value from sale of New machine (salvage value + disposal tax effect[*])						$7,056
(22) Net cash flow			−$3,240	−$3,000	−$2,520	$4,320

[*] Disposal tax effects = loss × *t* = $2640 × 40% = $1056

- *Lines 7 - 8:* The CCA results in a tax reduction that is equal to the CCA amount multiplied by the tax rate. Since there are no revenues, the operating cash flows can be easily computed using the generalized cash flow approach discussed in Chapter 9.

$$+ \text{(tax rate)(CCA)}$$

$$\underline{- (1 - \text{tax rate})(\text{O\&M})}$$

Net operating cash flows

- *Line 9:* Since the old equipment would be sold at less than book value, the sale would create a loss that could reduce the firm's taxable income and hence its tax payment. The tax savings should be equal to

$$\text{(book value } - \text{ market value) (tax rate)}$$

$$= (\$5373)(40\%) = \$2149$$

This loss is an operating loss because it reflects the fact that inadequate depreciation was taken on the old machine. An operating loss results in tax savings.

Net salvage value from the sale of old equipment at the end of year 3

$$= \text{market value} + \text{tax savings due to loss}$$

$$= \$2000 + \$2149 = \$4149$$

- *Line 10:* Since no new investment is required to keep the old machine, the net cash flows would consist of the operating cash flows and the net proceeds from the sale of the old machine at the end of year 3.

- *Line 11:* Here we show the price received from the sale of the old equipment at $n=0$. Example 12.2 illustrated the calculation steps.

- *Lines 12 - 15:* Here we show the purchase price of the new machine, including installation and freight charges. Lines 13 - 14 show the depreciation schedule along with the book values for the new machine (20% CCA rate), but the depreciation amount of \$1,500 in year 1 reflects the half-year convention.

- *Line 16:* Since we expect a loss on the sale of the new machine at the end of year 3 (that is, the book value exceeded the sale price), the taxable loss would be \$2640.

- *Line 18:* Here we show the total net cash outflow at the time the replacement is made. The company writes a cheque for \$15,000 to pay for the machine. However, this outlay is partially offset by the proceeds from the sale of the old equipment and tax savings in the amount of \$11,760.

- *Lines 19 - 20:* Here we show the net operating cash flows over the project's 3-year life. These cash flows represent the tax savings from the CCA claims and the after-tax O&M cost.

- *Line 21:* This line shows the cash flows associated with the termination of the new machine. To begin, Line 15 shows the estimated salvage value of the new machine at the end of its 3-year life, $6000. Since the book value of the new machine at the end of year 3 is $8640, the company will have tax savings as calculated in the disposal tax effects plus salvage value, leaving the net proceeds of $7056.
- *Line 22:* This line represents the net cash flows associated with replacement of the old machine.

The actual cash flow diagrams are shown in Fig. 12.3. Since both the defender and challenger have the same service life, we can use either PW or AE analysis.

$$PW(12\%)_{old} = -\$3648(P/F, 12\%, 1) - \$3878(P/F, 12\%, 2) + \$86(P/F, 12\%, 3)$$

$$= -\$6288$$

$$AE(12\%)_{old} = -\$6288(A/P, 12\%, 3) = -\$2618$$

$$PW(12\%)_{new} = -\$3240 - \$3000(P/F, 12\%, 1) - \$2520(P/F, 12\%, 2)$$

$$+ \$4320(P/F, 12\%, 3)$$

$$= -\$4853$$

$$AE(12\%)_{new} = -\$4853(A/P, 12\%, 3) = -\$2020$$

There is an annual difference of $598 in favour of the challenger; therefore, the replacement should be made now. Figure 12.4 shows the incremental net cash flows that are expected if the replacement is made.

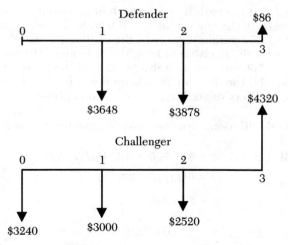

Figure 12.3 ■ Comparison of defender and challenger based on cash flow approach (Example 12.3)

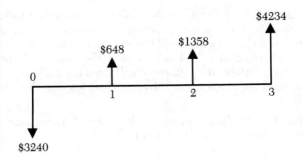

Figure 12.4 ■ Incremental cash flow (challenger-defender) based on cash flow approach (Example 12.3)

Opportunity Cost Approach

In the previous example, $11,760 in net salvage value was forgone by not selling the defender. Therefore, another way to analyze such a problem is to charge the $11,760 as an **opportunity cost** of keeping the asset. That is, instead of deducting the net salvage value from the purchase cost of the challenger, it is considered as a cash outflow for the defender.

Example 12.4 Replacement analysis using an opportunity cost approach

Rework Example 12.3 by using an opportunity cost approach.

Solution

Recall that the cash flow approach in Example 12.3 credited the net proceeds in the amount of $11,760 from the sale of the defender toward the $15,000 purchase price of the challenger, and there was no initial outlay for the decision to keep the defender. The opportunity cost treats the $11,760 current net salvage value of the defender as a cost that is incurred if the decision is to keep the defender. Figure 12.5 illustrates the cash flows related to these decision options.

Since the lifetimes are the same, we can use either PW or AE analysis.

$$PW(12\%)_{old} = -\$11,760 - \$3648(P/F, 12\%, 1) - \$3878(P/F, 12\%, 2)$$
$$+ \$86(P/F, 12\%, 3)$$
$$= -\$18,048$$

$$AE(12\%)_{old} = -\$18,048(A/P, 12\%, 3) = -\$7514$$

$$PW(12\%)_{new} = -\$15,000 - \$3000(P/F, 12\%, 1)$$

$$-\$2520(P/F, 12\%, 2) + \$4320(P/F, 12\%, 3)$$
$$= -\$16,613$$

$$AE(12\%)_{new} = -\$16,613(A/P, 12\%, 3)$$
$$= -\$6916$$

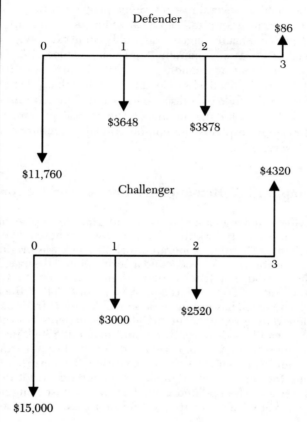

Figure 12.5 ■ Comparison of defender and challenger based on opportunity cost approach (Example 12.4)

The result is the same as Example 12.3. Since both the challenger and defender cash flows were adjusted by the same amount—$11,760—at time 0, this should not come as a surprise.

Comments: *Recall that in Examples 12.3 and 12.4, we assumed the same service life for both defender and the challenger. In general, however, old equipment has a relatively short remaining life compared with new equipment, so this assumption is overly simplistic. In the following sections, we will tackle some of the complexities of defender/challenger service lives.*

12.2 Economic Service Life

You have probably seen a 50-year-old automobile still in service. Provided it receives the proper repair and maintenance, almost anything can be kept operating for an extended period of time. If one can keep any car operating for an almost indefinite period, why aren't more old cars spotted on the street? There could be several reasons. Some people get tired of driving the same old car. Others may want to keep a car as long as it will last, but they realize that the repair and maintenance costs will become excessive.

Because an asset's operating and maintenance (O&M) costs increase with age, in engineering economic analysis, we are more interested in knowing what an asset's practical service life is, rather than what is its remaining physical life. In the following discussions, we define **economic service life** as the period of useful life that minimizes the equivalent annual cost of an asset. The next example explains the computational procedure for obtaining the economic service life.

Example 12.5 Remaining economic service for the defender

Oakville Trucking Company is considering the replacement of a 500-kilogram-capacity forklift truck that was purchased 3 years ago at a cost of $15,000. The diesel-operated forklift truck was originally expected to have a useful life of 8 years and a zero estimated salvage value at the end of that period. The truck is being depreciated using a CCA declining balance rate of 20%. It has a book value of $8640 if the asset is disposed now. The truck has not been dependable and is frequently out of service while awaiting repairs. The maintenance expenses of the truck have been rising steadily and currently amount to about $3000 per year. The truck could be sold for $6000. If retained, it would require an immediate $1500 overhaul to keep it in operable condition. This overhaul will neither extend the service life originally estimated nor will it increase the value of the truck. The updated annual operating costs, engine overhaul cost, and market values over the next 5 years are estimated as follows:

n	O&M	CCA	Engine Overhaul	Market Value
−3				
−2		$1500		
−1		2700		
0		2160	$1500	$6000
1	$3000	1728		4000
2	3500	1382		3000
3	3800	1106		1500
4	4500	885		1000
5	4800	708	$5000	0

The shaded area represents irrelevant information (sunk costs, except the history of depreciation schedule). A drastic increase in costs during the fifth year is expected due to another overhaul that will be required to keep the truck in operating condition. The firm's marginal tax rate is 40%, and its MARR is 15%. Compute the economic service life of this lift truck.

Solution

To determine the economic service life, we will first list the gains or losses to be realized if the truck is disposed of at the end of each year. In doing so, we need to compute the book values at the end of each year, assuming that the asset will be disposed of at that time. The book value calculation is illustrated in Table 12.2. We may use two approaches in finding the economic life of an asset: (1) generalized cash flow approach and (2) tabular approach. We will demonstrate both approaches.

(a) **Generalized Cash-Flow Approach**

We need to estimate the net salvage values and the operating costs over the operating life of the asset. These values provide a basis for identifying the relevant after-tax cash flows at the end of an assumed operating period. The overhaul (repair) cost of $1500 can be treated as a deductible operating expense for tax purposes, as long as it does not add value to the property. (Any repair or improvement expenses that increase the value of the property must be capitalized by following proper CCA methods.)

If the company retains the forklift truck, it is in fact deciding to overhaul the truck and invest the machine's current market value (after taxes) in that alternative. Although there is no physical cash flow transaction, the firm is withholding from investment the market value of the truck (opportunity cost).

If the lift truck was retained for one more year, the relevant after-tax cash flows would be determined as shown in Table 12.3(a). Since there are few cash-flow elements (O&M, depreciation, and salvage value), a more efficient way to obtain the after-tax cash flow is to use the generalized cash-flow approach discussed in Chapter 9. For the situation where we retain the defender 1 year, Table 12.3(b) summarizes the cash flows using the generalized cash-flow approach. The annual equivalent value for this 1-year operating period is

$$AE(15\%) = [-\$7956 + \$4056(P/F, 15\%, 1)](A/P, 15\%, 1)$$

$$= -\$5093.$$

Table 12.2 Forecasted Capital and Operating Costs—Defender (Example 12.5)

Holding Period N	O&M	Permitted Annual CCA Amounts over the Holding Period								Total CCA Value	Book Value	Expected Market Value	Loss	Disposal Tax Effects	Net A/T Salvage Value
		-2	-1	0	1	2	3	4	5						
0	$1500[1]	$1500	$2700	$2160						$6360	$8640	$6000	$2640	$1056	$7056
1	3000	1500	2700	2160	$1728					8088	6912	4000	2912	1165	5165
2	3500	1500	2700	2160	1728	$1382				9470	5530	3000	2530	1012	4012
3	3800	1500	2700	2160	1728	1382	$1106			10576	4424	1500	2924	1170	2670
4	4500	1500	2700	2160	1728	1382	1106	$885		11461	3539	1000	2539	1016	2016
5	9800[2]	1500	2700	2160	1728	1382	1106	885	$708	12169	2831	0	2831	1132	1132

[1] In year 0, engine overhaul ($1500) if the asset is to be kept.

[2] In year 5, normal operating expense ($4800) + overhaul ($5000)

If the firm retained the truck for 2 more years, the annual equivalent value would be

$$AE(15\%) = [-\$7956 - \$1109(P/F, 15\%, 1)$$
$$+ \$2465(P/F, 15\%, 2)](A/P, 15\%, 2)$$
$$= (-\$7057)(0.6151)$$
$$= -\$4340$$

Table 12.3 After-Tax Cash Flow Calculation for Retaining the Defender for One More Year (Example 12.5)

(a) Income Statement Approach

Year	0	1	
Operating expense	$1500	$3000	
CCA	_____	1728	Net income calculations
Taxable income	(1500)	(4728)	
Tax savings (40%)	+600	+1891	
Net income	−900	−2837	
CCA added		+1728	
Investment/salvage	−7056*	4000	Adjustment to obtain cash flow
Disposal tax effect	_____	1165†	
After-tax cash flow	−$7956	$4056	

(b) Generalized Cash Flow Approach

	0	1
Investment	−$7056	
+(0.40) (CCA)		+$691
−(0.60) (operating expense)	−900	−1800
Net proceeds from sale	_____	+5165
Net cash flow	−$7956	$4056

* The current after-tax market value is treated as an investment required to retain the defender.

† Disposal tax effect $= (B_{\text{Disposal}} - \text{salvage value}) \times t$
$$= (\$6912 - \$4000)(40\%)$$
$$= \$1165$$

For using the truck more than 2 years, the relevant cash flows are shown in Fig. 12.6, and the annual equivalent calculations are as follows:

- $n = 3$:

$$
\begin{aligned}
AE(15\%) &= [-\$7956 - \$1109(P/F,15\%,1) \\
&\quad -\$1547(P/F,15\%,2) \\
&\quad +\$832(P/F,15\%,3)](A/P,15\%,3) \\
&= (-\$9543)(0.4380) \\
&= -\$4180
\end{aligned}
$$

- $n = 4$:

$$
\begin{aligned}
AE(15\%) &= [-\$7956 - \$1109(P/F, 15\%, 1) \\
&\quad -\$1547(P/F, 15\%, 2) - \$1838(P/F, 15\%, 3) \\
&\quad -\$330(P/F, 15\%, 4)](A/P, 15\%, 4) \\
&= (-\$11,487)(0.3503) \\
&= -\$4024
\end{aligned}
$$

Figure 12.6 ■ Defender's net cash flows when it is retained for 1, 2, 3, 4, or 5 years (Example 12.5)

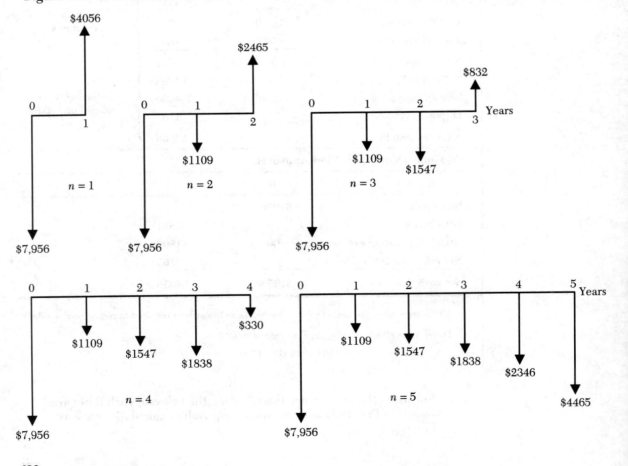

- $n=5$:

$$
\begin{aligned}
AE(15\%) &= [-\$7956 - \$1109(P/F, 15\%, 1) \\
&\quad -\$1547(P/F, 15\%, 2) - \$1838(P/F, 15\%, 3) \\
&\quad -\$2346(P/F, 15\%, 4) \\
&\quad -\$4465(P/F, 15\%, 5)](A/P, 15\%, 5) \\
&= (-\$14,859)(0.2983) \\
&= -\$4432
\end{aligned}
$$

If the truck were sold after 4 years, it would have a minimum annual cost of \$4024 per year, and that is the life that is most favourable for comparison purposes. Therefore, the economic life for this truck is 4 years (see Fig. 12.7).

(b) Tabular Approach

The tabular approach is to separate the annual cost elements into two parts, one associated with the capital recovery of the asset and the other associated with operating the asset. In computing the capital recovery cost, we need to determine the after-tax salvage values at the end of each holding period as shown in Table 12.2. Then we compute the total annual equivalent value of the asset for any given year's operation:

> Total annual equivalent
>
> value = − capital recovery cost
>
> − equivalent annual operating cost

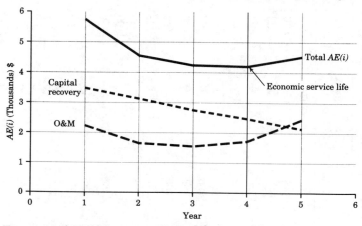

Figure 12.7 ■ Economic service life (Example 12.5)

Table 12.4 illustrates the tabular method for determining the economic life of the asset. We see that, as the asset ages, O&M costs increase. At the same time, the capital recovery cost decreases with prolonged use of the asset. The combination of decreasing capital recovery costs and increasing annual O&M costs results in the total annual equivalent cost taking on a form similar to that depicted in Fig. 12.7, which indicates that the economic life of the asset is 4 years.

Table 12.4 Tabular Calculation of Economic Life (Example 12.5)

(1) Holding Period N		(2) Before-Tax Operating Expenses		(3) After-Tax Cash Flow if the Asset Is Kept for N More Years				(4) Net Cash Flow A/T	(5) Annual Equivalent Value If the Asset Is Kept for N More Years		
N	n	O&M[1]	CCA	Investment & Net Salvage	A/T O&M[2]	CCA Tax Savings[3]	Net O&M Flow		−CR(i)[4]	+Net O&M[5]	Total[6] AE(i)
1	0	−$1500		−$7056	−$900		−$900	−$7956	−$2949	−$2144	−$5093
	1	−3000	$1728	5165	−1800	+691	−1109	+4056			
2	0	−1500		−7056	−900		−900	−7956	−2474	−1866	−4340
	1	−3000	1728	0	−1800	+691	−1109	−1109			
	2	−3500	1382	4012	−2100	+553	−1547	+2465			
3	0	−1500		−7056	−900		−900	−7956	−2322	−1858	−4180
	1	−3000	1728	0	−1800	+691	−1109	−1109			
	2	−3500	1382	0	−2100	+553	−1547	−1547			
	3	−3800	1106	2670	−2280	+442	−1838	+832			
4	0	−1500		−7056	−900		−900	−7956	−2068	−1956	−4024 Economic life
	1	−3000	1728	0	−1800	+691	−1109	−1109			
	2	−3500	1382	0	−2100	+553	−1547	−1547			
	3	−3800	1106	0	−2280	+442	−1838	−1838			
	4	−4500	885	2016	−2700	+354	−2346	−330			
5	0	−1500		−7056	−900		−900	−7956	−1936	−2496	−4432
	1	−3000	1728	0	−1800	+691	−1109	−1109			
	2	−3500	1382	0	−2100	+553	−1547	−1547			
	3	−3800	1106	0	−2280	+442	−1838	−1838			
	4	−4500	885	0	−2700	+354	−2346	−2346			
	5	−9800	708	1132	−5880	+283	−5597	−4465			

[1] O&M + engine overhaul

[2] After-tax operating cost in year n = O&M$_n(1-t)$, where t is marginal tax rate

[3] CCA tax savings in year n = $CCA_n(t)$

[4] Capital recovery cost = [(investment − salvage) $(A/P, i, N)$ + salvage (i)]

[5] Annual equivalent O&M = (PW net O&M flow over N) $(A/P, i, N)$

[6] Total annual equivalent cost = $-CR(i) + AE(i)$ of O&M

As in the case of the defender, we need to consider explicitly how long the challenger should be held once it is placed in service. For instance, in the case of a truck-rental firm that frequently purchases a fleet of identical trucks, the firm might wish to arrive at a policy decision on how long to keep each vehicle before replacement. With that life span computed, the firm could then stagger the schedule of purchase and replacement of trucks to smooth out the annual capital expenditures for truck purchases. For the challenger whose revenues are either unknown or irrelevant, we can compute its economic life based on the operating costs for the challenger and its year-by-year salvage values. This is demonstrated in Example 12.6.

Example 12.6   Economic service life of the challenger

As a challenger to the equipment described in Example 12.5, consider a new electric-lift truck that costs $18,000, which has operating costs of $1000 in the first year and a salvage value of $10,000 at the end of the first year. For the remaining years, operating costs increase each year by 15% of the previous year's operating costs. Similarly, the salvage value declines each year by 25% of the previous year's salvage value. The lift truck has an operating life of 7 years, with an overhaul costing $3000 required during the fifth year of service and another overhaul costing $4500 during the seventh year of service. The asset is depreciated using a CCA declining balance rate of 20%. The firm's marginal income tax rate is 40% and its MARR is 15%. Find the economic service life of this new machine.

Solution

We may project the expected operating costs and the salvage value over the next 8-year period as shown in Table 12.5. With these figures, we are now ready to generate the after-tax entries. For the first two operating years, we compute as follows:

- $n = 1$:

$$AE(15\%) = \{-\$18{,}000 + [-(0.6)(\$1000)$$
$$+ (0.40)(\$1800)$$
$$+ \$12{,}480](P/F,\ 15\%,\ 1)\}(A/P,\ 15\%,\ 1)$$
$$= -\$8099$$

- $n = 2$:

$$AE(15\%) = \{-\$18{,}000 + [-0.6(\$1000)$$
$$+ 0.4(\$1800)](P/F,\ 15\%,\ 1) + [0.6(\$1150)$$
$$+ 0.4(\$3240) + \$9684](P/F,\ 15\%,\ 2)\}(A/P,\ 15\%,\ 2)$$
$$= -\$6222$$

Table 12.5 Forecasted Capital and Operating Costs—Challenger (Example 12.6)

Holding Period N	O&M	Permitted Annual CCA Amounts over the Holding Period							Total CCA	Book Value	Expected Market Value	Losses	Disposal Tax Effects	Net A/T Salvage Value
		1	2	3	4	5	6	7+						
1	1,000	$1800							$1800	$16,200	$10,000	$6200	$2480	$12,480
2	1,150	1800	$3240						5040	12,960	7,500	5460	2184	9684
3	1,323	1800	3240	$2592					7632	10,368	5,625	4743	1897	7522
4	1,521	1800	3240	2592	$2074				9706	8294	4,219	4075	1630	5849
5	4,749[1]	1800	3240	2592	2074	$1659			11,365	6635	3,164	3471	1388	4552
6	2,011	1800	3240	2592	2074	1659	$1327		12,692	5308	2,373	2935	1174	3547
7	6,813[2]	1800	3240	2592	2074	1659	1327	$1062	13,754	4246	1,780	2466	986	2766

[1] In year 5, normal operating expense ($1749) + overhaul ($3000)

[2] In year 7, normal operating expense ($2313) + overhaul ($4500)

Table 12.6 Tabular Calculation of Economic Life—Challenger (Example 12.6)

Holding Period N	n	O&M[1]	CCA	Investment & Net Salvage	A/T O&M[2]	CCA Credit	Net O&M Flow	Net Cash Flow A/T	Total AE(i)
	0			−$18,000				−$18,000	
1	1	−$1,000	$1,800	+12,480	−$600	+$720	$120	+12,600	−$8099
	0			−18,000				−18,000	
2	1	−1,000	1800	0	−600	+720	120	120	−6222
	2	−1,150	3240	+9684	−690	+1296	606	+10,290	
	0			−18,000				−18,000	
	1	−1,000	1800	0	−600	+720	120	120	−5402
3	2	−1,150	3240	0	−690	+1296	606	606	
	3	−1,323	2592	+7522	−794	+1037	243	7765	
	0			−18,000				−18,000	
	1	−1,000	1800	0	−600	+720	120	120	−4898
4	2	−1,150	3240	0	−690	+1296	606	606	
	3	−1,323	2592	0	−794	+1037	243	243	
	4	−1,521	2074	+5849	−913	+830	−83	5766	
	0			−18,000				−18,000	
	1	−1,000	1800	0	−600	+720	120	120	
5	2	−1,150	3240	0	−690	+1296	606	606	−4617
	3	−1,323	2592	0	−794	+1037	243	243	
	4	−1,521	2074	0	−913	+830	−83	−83	
	5	−4,749[1]	1659	+4552	−2,849	+664	−2185	2367	
	0			−18,000				−18,000	
	1	−1,000	1800	0	−600	+720	120	120	−4536
	2	−1,150	3240	0	−690	+1296	606	606	Economic
6	3	−1,323	2592	0	−794	+1037	243	243	service
	4	−1,521	2074	0	−913	+830	−83	−83	life
	5	−4,749[1]	1659	0	−2,849	+664	−2185	−2185	
	6	−2,011	1327	+3547	−1,207	+539	−676	2871	
	0			−18,000				−18,000	
	1	−1,000	1800	0	−600	+720	120	120	
	2	−1,150	3240	0	−690	+1296	606	606	
	3	−1,323	2592	0	−794	+1037	243	243	
7	4	−1,521	2074	0	−913	+830	−83	−83	−4577
	5	−4,749[1]	1659	0	−2,849	+664	−2185	−2185	
	6	−2,011	1327	0	−1,207	+531	−676	−676	
	7	−6,813[1]	1062	+2766	−4,088	+425	−3663	−897	

[1] O&M + engine overhaul ($3000 in year 5 and $4500 in year 7, respectively)

[2] After-tax operating cost in year $n = -O\&M_n(1-t)$, where t is marginal tax rate

Similarly, we can compute the annual equivalent value for the subsequent years as shown in Table 12.6. The economic service life of the new machine appears to be 6 years with an $AE(15\%) = -\$4536$ indicating that, even though an expensive overhaul is required during the fifth year of service, it is more economical to keep the equipment over a 6-year life.

The rationale for finding the economic life of a new asset is that, by replacing perpetually according to the economic life, we obtain the minimum infinite AE cost stream. Figure 12.8 illustrates this concept. Of course, we should be envisioning a long period of required service of the asset. This life is no doubt heavily influenced by the tax allowances for depreciation.

One-year replacement cycle

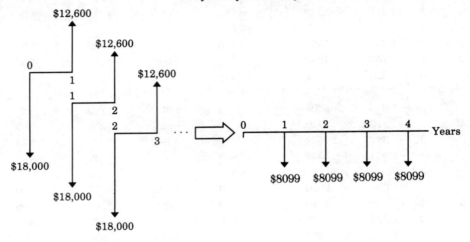

Two-year replacement cycle

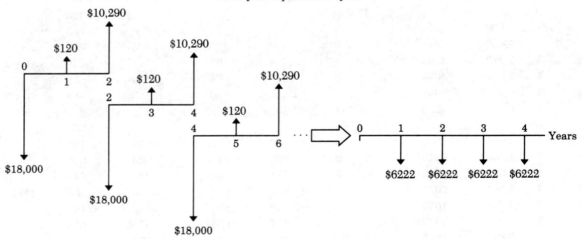

Figure 12.8 ■ Conversion of infinite number of replacement cycles to infinite AE cost streams (Example 12.6)

■ 12.3 Replacement Analysis When Required Service Is Long

Now that we have determined the economic service lives for both the defender and the challenger, the next question is how to use these pieces of information in deciding whether *now* is the time to replace the defender. Before presenting an analytical approach to answer that question, we will consider several important assumptions.

12.3.1 Required Assumptions and Decision Framework

In deciding whether now is the time to replace the defender, we need to consider the following three factors:

- Planning horizon (study period)
- Technology
- Relevant cash flow information

Planning Horizon (Study Period)

By the planning horizon, we simply mean the service period required by the defender and a sequence of future challengers. The infinite planning horizon is used when we are simply unable to predict when the activity under consideration will be terminated. In other situations, it may be clear that the project will have a definite and predictable duration. In these cases, replacement policy should be formulated more realistically based on a finite planning horizon.

Technology

Predictions of technological patterns over the planning horizon refer to the development of types of challengers that may replace those under study. An infinite variety of predictions might be made of purchase cost, salvage value, and operating cost dictated by the efficiency of the machine over the life of an asset. If we assume that all future machines will be the same as those now in service, we are saying that there will be no technological progress in the area. In other cases, we may explicitly recognize that machines that may become available in the future will be significantly more efficient, reliable, or productive than those now on the market. (Personal computers are a good example.) This situation leads to a recognition of a technological change or obsolescence. Clearly, if the best available machine gets better and better all the time, our best decision may be to delay replacement in contrast to the situation where no technological change is likely.

Revenue and Cost Patterns over Asset Life

There are many varieties of predictions possible in estimating patterns of revenue, cost, and salvage value over the life of the asset. Sometimes revenue is increasing, but costs and salvage value are not increasing over the life of the machine. In other situations, a decline in revenue over equipment life can be expected. This will determine whether the replacement analysis is directed toward cost minimization (with constant revenue) or profit maximization (with varying revenue). We will formulate a replacement policy for the asset for which salvage values do not increase with age.

Decision Framework

To illustrate the decision framework, we will indicate a sequence of assets by the notation (j_0,n_0), (j_1,n_1), (j_2,n_2),.... Each pair of numbers (j,n) indicates an asset type and the lifetime for which that asset will be retained. The defender, asset 0, is listed first; if the defender is replaced now, $n_0 = 0$. A sequence of pairs may cover a finite period or an infinite period. For example, the sequence of $(j_0,2)$, $(j_1,5)$, $(j_2,3)$ indicates retaining the defender for 2 years, replacing the defender with an asset of type j_1, using it for 5 years, replacing j_1 with an asset of type j_2, and using it for 3 years. In this situation, the total planning horizon covers 10 years (2+5+3). The special case of keeping the defender for n_0 periods, followed by infinitely repeated purchases and the use of an asset of type j for n_j years, is represented by (j_0,n_0), $(j,n_j)_\infty$. This sequence covers an infinite period, and the relationship is illustrated in Fig. 12.9.

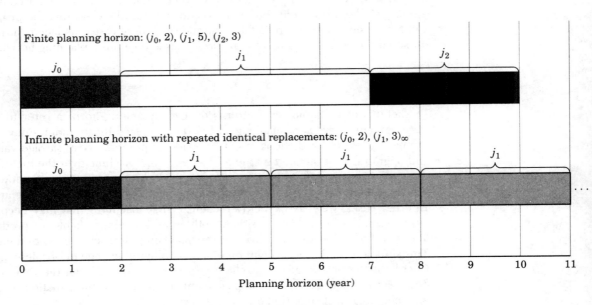

Figure 12.9 ■ Types of typical replacement decision framework

12.3.2 Replacement Decisions Using the AE Method

The annual equivalent approach is frequently used in replacement analysis, but it is important to know that the use of AE generally stems from tradition rather than any compelling logic. In other words, the present worth approach is equally applicable in replacement analysis. In this section, however, we will examine the situation when the use of the annual cash flow analysis can be computationally advantageous.

Different Economic Service Lives

In replacement analysis, it is likely that the defender and the challenger have different economic service lives. The use of the AE method in replacement analysis, however, is not because we have to deal with this unequal service life problem, but because it provides us a marginal analysis for determining when to replace the defender on a year-to-year basis.

In Chapter 4, we discussed the general principle for comparing alternatives with unequal service lives. In particular, we pointed out that the use of the AE method relies on the concept of *repeatability* of projects and one of two assumptions, an infinite planning horizon or a common service period. For defender-challenger situations, however, repeatability of the defender cannot be assumed. In fact, by virtue of our problem definition, we are not *repeating* the defender, but *replacing* it with the challenger, an asset that in some way constitutes an improvement over the current equipment. Thus, the assumptions we made for using an annual cash flow analysis on unequal service life alternatives are frequently not valid in the usual defender-challenger situation.

The complication—the unequal life problem—can be resolved, however, if we recall that the replacement problem at hand is to decide between two alternatives, replacing the defender or retaining the defender for the present time. That is, the question is not whether to replace the defender, but when to do so. When the defender is replaced, it will always be by the challenger—the best available equipment. In this situation, the use of the annual cash flow analysis in replacement decisions provides us with a convenient way to perform marginal analysis to see whether the cost of extending the use of the defender for an additional year exceeds the savings resulting from delaying the purchase of the challenger.

If the defender-challenger problem assumes a continuing requirement for the equipment, we can show that the AE method also provides us with a convenient way to determine when is the best time to replace the defender.

Decision Procedure Under Infinite Planning Horizon

We will now consider a situation where the future challenger remains unchanged in terms of costs and efficiency over its infinite planning horizon. Here we are considering the special case of keeping the defender n_0 years, followed by infinitely repeated purchases once every n_j periods and use of the challenger $[(j_0, n_0), (j, n_j)_\infty]$.

Example 12.7 Enumerating replacement-decision alternatives

Suppose the defender in Example 12.5 has a 4-year economic life (AE defender = −$4024), whereas the challenger in Example 12.6 has a 6-year economic life (AE challenger = $4536). In terms of decision alternatives, how many options do we have?

Solution

There are five decision options:

- Replace now
- Replace at the end of year 1 by the challenger
- Replace at the end of year 2 by the challenger
- Replace at the end of year 3 by the challenger
- Replace at the end of year 4 by the challenger

If the costs and efficiency of the current challenger remain unchanged in future years, Fig. 12.10 shows the possible replacement patterns associated with each alternative. From the figure, we observe two things: First, on an annual basis, the cash flows after the remaining economic life of the defender are the same. Second, even though there are five

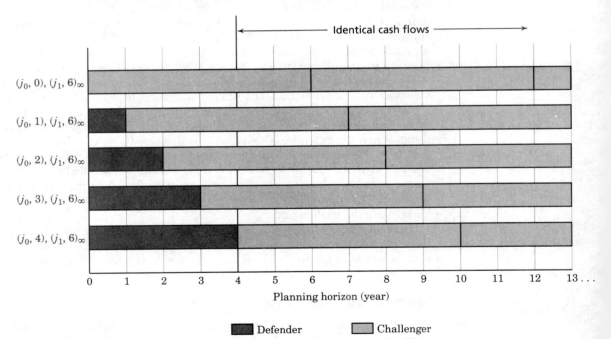

Figure 12.10 ■ Replacement sequence with identical future challengers under infinite planning horizon (Example 12.7)

decision alternatives, it is not necessary to look at them all when we decide to determine if now is the time to replace the defender. With a remaining economic life set at 4 years and with the same cost after this economic life, we need to examine the annual cash flow only during the first 4 years. If the magnitude of AE for the defender over its remaining economic life is less than the magnitude absolute value of AE for the challenger, we know that retaining the defender for less than 4 years (then replacing it by the challenger) will always be a more costly option. Therefore, we only need to compare the AEs of the defender and challenger over their respective economic service lives. Thus, in our example, the old truck should be kept for now.

The analysis procedure for an infinite planning period is summarized in Fig. 12.11.

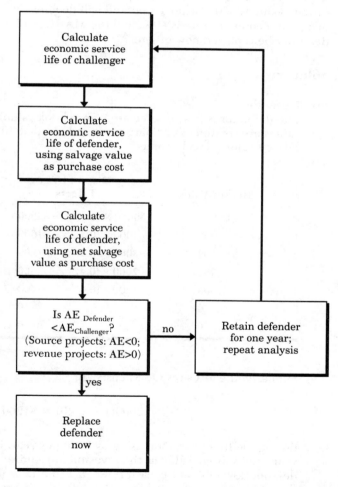

Figure 12.11 ■ Analysis procedure for infinite planning period

Example 12.8   Replacement analysis under infinite planning horizon (AE approach)

Advanced Electrical Insulator Company is considering replacing one of its inspection machines that have been used to test the mechanical strength of electrical insulators with a newer and more efficient one. The old machine is fully depreciated, and it has a remaining useful life of 5 years. The firm does not expect to realize any salvage value from scrapping the old machine in 5 years, but it can sell it now to another firm in the industry for $5000. If the machine is kept, its O&M costs are $2000 this year and are expected to increase by $1500 per year from now on. Future market values are expected to decline by $1000 per year.

   The new machine costs $10,000 and has O&M costs of $2000 the first year, increasing by $1000 per year thereafter. The new machine is depreciated using a CCA declining balance rate of 30%. The expected salvage value is $6000 after 1 year and will decline 15% each year. The marginal income tax rate is 40%, and the MARR is 15%. Should the defender be replaced now or later?

Solution

(a)   **Defender**
The defender is fully depreciated (zero book value), so all salvage values are treated as ordinary gains and taxed at 40%. The after-tax salvage values thus become

n	Market Value	Disposal Tax Effects	Net Salvage Value
0	$5000	$-\$5000 \times (0.4) = -\$2000$	$3000
1	4000	$-4000 \times (0.4) = -1600$	2400
2	3000	$-3000 \times (0.4) = -1200$	1800
3	2000	$-2000 \times (0.4) = -800$	1200
4	1000	$-1000 \times (0.4) = -400$	600
5	0	0	
6	0	0	

Similarly, the after-tax O&M costs are

$$\$2000(1 - 0.40) = \$1200$$

during the first year, increasing by $900 per year. Using the current year's net salvage value as the investment required to retain the defender, we obtain the cash flows in Table 12.7, which indicate that the remaining economic life of the defender is 1 year.

Table 12.7 Economic Service Life: Defender (Example 12.8)

n	Net Salvage	A/T O&M	Capital Recovery	AE of O&M	Total AE for n years
0	$3000				
1	2400	−$1200	−$1050	−$1200	−$2250*
2	1800	−2100	−1080	−1619	−2627
3	1200	−3000	−968	−2016	−2984
4	600	−3900	−931	−2394	−3325
5	0	−4800	−895	−2751	−3646

Note: The symbol * indicates total AE at the end of the economic service life.

(b) Challenger

Because the challenger will be depreciated over its life, we must determine the book value of the asset at the end of each period to compute the after-tax salvage value. This is shown in Table 12.8(a). With the after-tax salvage values computed in Table 12.8(a), we are now ready to find the economic service life of the challenger by generating AE values. These calculations are summarized in Table 12.8(b). The economic life of the challenger is 4 years with an annual equivalent cost of $4063. Since the annual equivalent cost for the defender's remaining economic life (1 year) is $2250, which is less than $4063, the decision will be to keep the defender at least one more year.

When to Replace the Defender

While the economic life of the defender is defined as the year of service that minimizes the annual equivalent cost (or maximizes the annual equivalent revenue), this is not necessarily the *optimal time* to replace the defender. The correct replacement time depends on data for the challenger as well as on data for the defender. We will explain this point using the defender-challenger situation in Example 12.8.

For the defender, we could perform another economic life calculation at the end of the first year. However, it is simpler to do a **marginal analysis**—calculate the incremental cost of operating the defender for just *one more year*. We will use an opposite sign convention for convenience, with outflows being positive and inflows negative. The costs of owning and operating the defender from $n = 1$ to $n = 2$, expressed at time $n = 2$, are

$$\overbrace{\$2400(1.15) - \$1800} + \underbrace{\$2100} = \$3060.$$

The amount indicated by an overbrace represents the ownership cost, and the amount indicated by an underbrace represents the O&M cost. The marginal costs of owning and operating ($3060) are still lower than $4063, so we would

Table 12.8 Economic Service Life Calculation for Challenger (Example 12.8)

(a) After-Tax Salvage Value Calculation

Holding Period N	O&M	Permitted Annual CCA Amounts over the Holding Period							Total CCA	Book Value	Expected Market Value	Losses	Disposal Tax Effects	Net A/T Salvage Value
		1	2	3	4	5	6	7+						
1	$2000	$1500							$1500	$8500	$6000	$2500	$1000	$7000
2	3000	1500	$2550						4050	5950	5100	850	340	5440
3	4000	1500	2550	$1785					5835	4165	4335	−170	−68	4267
4	5000	1500	2550	1785	$1250				7085	2915	3685	−770	−308	3377
5	6000	1500	2550	1785	1250	$875			7960	2040	3132	−1092	−437	2695
6	7000	1500	2550	1785	1250	875	$612		8572	1428	2662	−1234	−494	2168
7	8000	1500	2550	1785	1250	875	612	$428	9000	1000	2263	−1263	−505	1768

Disposal tax effects = $(B_{\text{Disposal}} - \text{salvage value}) \times t$

(b) Annual Equivalent Value Calculation

n	A/T O&M Expenses	CCA Tax Savings	Net O&M	AE of Net O&M	AE of Capital Recovery	Total AE for n years
1	−$1200	$600	−600	−600	−$4834	−$5434
2	−1800	1020	−720	−656	−3674	−4330
3	−2400	714	−1686	−953	−3145	−4098
4	−3000	500	−2500	−1263	−2800	−4063*
5	−3600	350	−3250	−1557	−2562	−4119
6	−4200	245	−3955	−1831	−2376	−4207
7	−4800	171	−4629	−2085	−2230	−4315

* Economic service life

keep the defender at least one more year. At the end of year 2, the marginal costs of owning and operating the defender one more year are

$$\$1800(1.15) - \$1200 + \$3000 = \$3870$$

For the period from $n = 3$ to $n = 4$, the costs are

$$\$1200(1.15) - \$600 + \$3900 = \$4680.$$

The marginal costs are greater than \$4063, so we should plan to replace the defender at the end of year 3 and acquire the challenger. Then, the challenger would be replaced at 4-year intervals in our planning period. This strategy is based upon salvage value and O&M cost estimates for the defender and challenger over several years. If these estimates are correct, then the strategy is valid. At the end of each year of operation, these estimates need to be revisited in relation to the most recent operating costs for the defender and the new challenger. Figure 12.12 illustrates this concept.

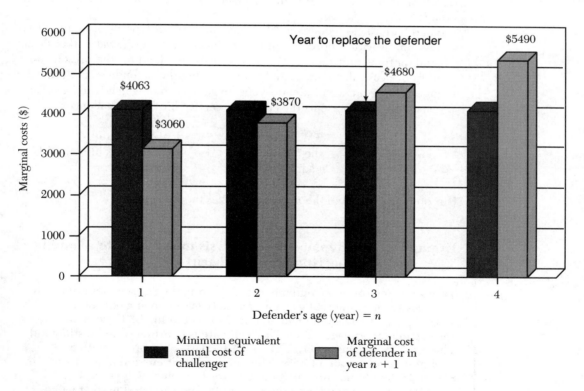

Figure 12.12 ■ Marginal costs of operating the defender for one additional year (Example 12.8)

12.3.3 Replacement Decisions Using the PW Method

The PW method is applicable to a more general class of replacement decisions than the AE method discussed in the previous section. The PW method provides a more direct solution to a replacement problem with either an infinite or a finite planning horizon or a technological change in the future challenger. We will develop the replacement decision procedure for both situations: (1) the infinite planning horizon and (2) the finite planning horizon. We begin by analyzing an infinite planning horizon with no technological change. While such a simplified situation is not likely to occur in real life, the analysis of this replacement situation introduces methods useful for analyzing infinite horizon replacement problems with technological change.

Infinite Planning Horizon

We will consider the situation where a firm has a machine in use in a process that is expected to continue for an indefinite period. There is presently on the market a new machine that is in some ways more effective for the application being considered than the defender. The problem is when, if at all, the defender should be replaced with the challenger. Here we are considering the situation $[(j_0, n_0), (j, n_j)_\infty]$.

When there is no technological change, an asset of type j can be purchased at any time n with a purchase price, revenues, and salvage value that do not depend on time, only on age. We now let $PW(i,n)$ stand for the present worth of all future cash flows associated with retaining the defender for n years and an indefinite sequence of identical challengers, each of which is replaced after n_j years. Here n_j represents the economic service life of the challenger.

$$PW(i,n) = \begin{array}{c} \text{PW of retaining} \\ \text{the defender for} \\ n \text{ additional years} \end{array} + \begin{array}{c} \text{PW of repeated} \\ \text{replacements of} \\ \text{the challenger} \end{array} \qquad (12.1)$$

Our objective is to find the n that maximizes the $PW(i,n)$.

Example 12.9 Replacement analysis under infinite planning horizon (PW approach)

Trenton Construction Company is considering the replacement of an air-compressor machine that was purchased 6 years ago at a cost of $10,000. At that time, the machine had an initial expected life of 11 years and a zero estimated salvage value. That estimate is considered to be still good. The machine has been classified as a CCA Class 8 property with a declining balance rate of 20%. The old machine's current market value is $2000. If the old machine is retained over the remaining years, the revenue, operating and maintenance costs, and expected market value over the years are as follows:

n	Revenue	O&M	Market Value
0			$2,000
1	$10,800	$7,000	1,500
2	10,800	8,000	1,000
3	10,800	8,000	1,000
4	10,800	9,000	1,000
5	10,800	9,000	0

A new air-compressor machine can be bought and installed for $12,000 which, over its 5-year economic life, will expand sales somewhat from $10,800 to $11,500 a year. However, they will drop to $10,000 after that and continue at that rate as the machine gets older. Furthermore, the new machine will reduce labor and power usage sufficiently to cut the first year's operating costs to $4000, followed by a modest annual increase of $500 for 4 years and a $1000 increase thereafter.

The new machine is classified as a CCA Class 8 property with a declining balance rate of 20%. It has an estimated salvage value of $8000 after one year, $6000 after two years, $4000 after three years, $3000 after four years, $2000 after five years, $1000 after six years and zero at the end of the seven-year life. The firm's marginal income tax rate is 40%. No special tax credit is expected. Should the firm buy the new machine at MARR = 15%?

Solution

We will first determine the after-tax cash flows associated with the new machine over its economic service life. These are shown in Table 12.9. (Note that, in this problem, we have already computed the economic service life at 5 years using a procedure similar to the one described in Example 12.6.) The PW of this cash flow series is $4891, and its equivalent annual worth is $1459. Similar calculations can be made for other years of use for the challenger and the defender. These are summarized as follows:

Annual Equivalent Worth

n	Defender	Challenger
1	$1623	$301
2	1382	920
3	1409	1128
4	1307	1348
5	1160	1459
6		1379
7		1281

Table 12.9 After-Tax Cash Flows over the Asset's Economic Life (Example 12.9)

				n		
	0	1	2	3	4	5
Income Statement						
Revenue		$11,500	$11,500	$11,500	$11,500	$11,500
Expenses						
O&M		4000	4500	5000	5500	6000
CCA		1200	2160	1728	1382	1106
Taxable income		6300	4840	4772	4618	4394
Income tax		2520	1936	1909	1847	1758
Net income		$3780	$2904	$2863	$2771	$2636
Cash Flow Statement						
Net income		$3780	$2904	$2863	$2771	$2636
CCA		1200	2160	1728	1382	1106
Investment	−$12,000					
Salvage						2000
Disposal tax effects						970
Net cash flow	−$12,000	$4980	$5064	$4591	$4153	$6712

$$PW(15\%) = -\$12{,}000 + \$4980(P/F, 15\%, 1) + ...\$6712(P/F, 15\%, 5)$$

$$= \$4891$$

$$AE(15\%) = \$4891(A/P, 15\%, 5)$$

$$= \$1459.$$

The defender has a remaining useful life of 1 year ($n_0 = 1$), whereas the challenger has a 5-year economic service life ($n_j = 5$). (Note that these annual equivalent values represent the equivalent annual after-tax cash inflows for a revenue project.)

For an immediate replacement of the defender by the challenger, under an infinite planning horizon we obtain

- $n = 0$:

$$PW(15\%,0) = \$1459 \, (P/A, 15\%, \infty)$$

$$= \$1459 \, (1/0.15) = \$9727$$

Suppose we retain the old machine one more year and then replace it with the new one. Now we will compute $PW(i,n)$ using Eq. (12.1).

- $n = 1$:

$$PW(15\%, 1) = \$1623(P/F, 15\%, 1) + (P/F, 15\%, 1)(\$9727)$$
$$= \$9870$$

- $n = 2$:

$$PW(15\%, 2) = \$1382(P/A, 15\%, 2) + (P/F, 15\%, 2)(\$9727)$$
$$= \$9602$$

- $n = 3$:

$$PW(15\%, 3) = \$1409(P/A, 15\%, 3) + (P/F, 15\%, 3)(\$9727)$$
$$= \$9613$$

- $n = 4$:

$$PW(15\%, 4) = \$1307(P/A, 15\%, 4) + (P/F, 15\%, 4)(\$9727)$$
$$= \$9293$$

- $n = 5$:

$$PW(15\%, 5) = \$1160(P/A, 15\%, 5) + (P/F, 15\%, 5)(\$9727)$$
$$= \$8725$$

As shown in Fig. 12.13, this leads to the conclusion that the defender should be kept for only one more year. The present worth of $9870 represents the net benefit associated with replacing the defender in 1 year, replacing it with the challenger, and then replacing the challenger every 5 years for an indefinite period.

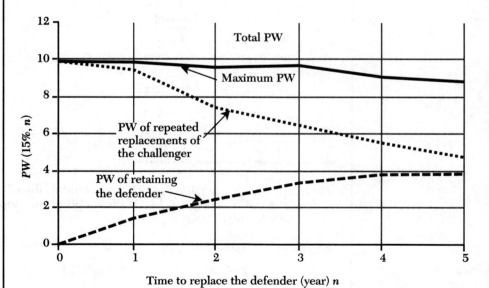

Figure 12.13 ■ PW analysis under infinite planning horizon (Example 12.9)

Finite Planning Horizon

If the planning period is finite (for example, 10 years), a comparison based on the AE method over its economic service life does not generally apply. The procedure for solving such a problem with a finite planning horizon is to establish *all* "reasonable" replacement patterns and then use the PW for the planning period to select the most economical pattern. To illustrate the procedure, let us consider Example 12.10.

Example 12.10 Replacement analysis under finite planning horizon (PW approach)

Reconsider the defender and the challenger in Example 12.8. Suppose that the firm has a contract to perform a given service on the current defender for the next 7 years. After the contract work, neither the defender nor the challenger will be retained. What is the best replacement strategy?

Solution

Before we examine the possible replacement-decision options, recall the annual equivalent values for the defender and challenger under the assumed service lives (a figure in the box denotes the value at the economic service life, $n_0 = 1$ and $n_j = 4$).

Annual Equivalent Worth

n	Defender	Challenger
1	$\boxed{-\$2250}$	$-\$5434$
2	-2627	-4330
3	-2984	-4098
4	-3325	$\boxed{-4063}$
5	-3646	-4119
6		-4207
7		-4315

There are many ownership options that would fulfill a 7-year planning horizon, as shown in Fig. 12.14. Of these options, six appear to be the most likely by inspection.

- **Option 1:** $(j_0,0)$, $(j,4)$, $(j,3)$

$$PW(15\%)_1 = 0 - \$4063(P/A, 15\%, 4)$$

$$- \$4098(P/A, 15\%, 3)(P/F, 15\%, 4)$$

$$= -\$16,950$$

Figure 12.14 ■ Some possible replacement patterns under finite planning horizon (Example 12.10)

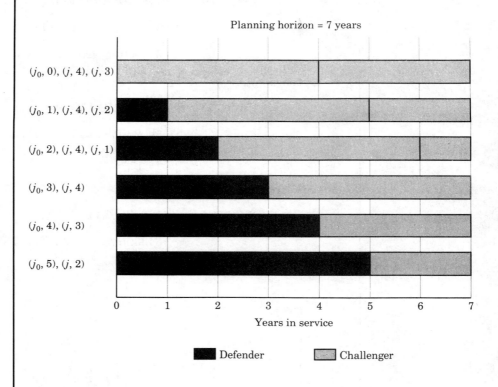

Planning horizon = 7 years

$(j_0, 0), (j, 4), (j, 3)$

$(j_0, 1), (j, 4), (j, 2)$

$(j_0, 2), (j, 4), (j, 1)$

$(j_0, 3), (j, 4)$

$(j_0, 4), (j, 3)$

$(j_0, 5), (j, 2)$

Years in service

■ Defender   ☐ Challenger

- **Option 2:** $(j_0, 1), (j, 4), (j, 2)$

$$PW(15\%)_2 = -\$2250(P/F, 15\%, 1)$$
$$-\$4063(P/A, 15\%, 4)(P/F, 15\%, 1)$$
$$-\$4330(P/A, 15\%, 2)(P/F, 15\%, 5)$$
$$= -\$15,544$$

- **Option 3:** $(j_0, 2), (j, 4) (j, 1)$

$$PW(15\%)_3 = -\$2627(P/A, 15\%, 2)$$
$$-\$4063(P/A, 15\%, 4)(P/F, 15\%, 2)$$
$$-\$5434 (P/F, 15\%, 7)$$
$$= -\$15,085$$

- **Option 4:** $(j_0, 3), (j, 4)$

$$PW(15\%)_4 = -\$2984(P/A, 15\%, 3)$$
$$-\$4063(P/A, 15\%, 4)(P/F, 15\%, 3)$$
$$= -\$14{,}440 \leftarrow \text{minimum cost}$$

- **Option 5:** $(j_0, 4), (j, 3)$

$$PW(15\%)_5 = -\$3325(P/A, 15\%, 4)$$
$$-\$4098(P/A, 15\%, 3)(P/F, 15\%, 4)$$
$$= -\$14{,}843$$

- **Option 6:** $(j_0, 5), (j, 2)$

$$PW(15\%)_6 = -\$3646(P/A, 15\%, 5)$$
$$-\$4330(P/A, 15\%, 2)(P/F, 15\%, 5)$$
$$= -\$15{,}722$$

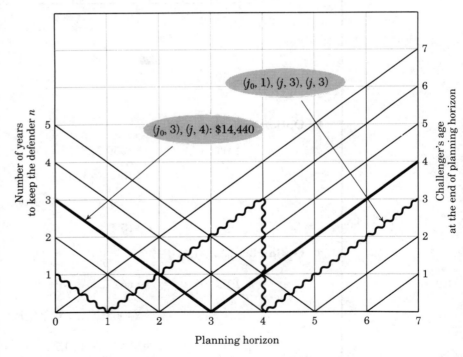

Figure 12.15 ■ Graphical representation of replacement strategies under finite planning horizon (Example 12.10)

An examination of the present equivalent cost of a planning horizon of 7 years indicates that the least-cost solution appears to be option 4: Retain the defender for 3 years, purchase the challenger, and keep it for 4 years.

Comments: *In this example, we examined only six possible decision options that would be likely to lead to the best solution, but it is important to note that there are several other possibilities that we have not examined. To explain, consider Fig. 12.15, which shows a graphical representation of various replacement strategies under finite planning horizon.*

For example, the replacement strategy (shown as a wiggly line in Fig. 12.15), $[(j_0,1), (j,3), (j,3)]$, is certainly feasible, but we did not include it in the previous computation. Naturally, as we extend the planning horizon, the number of possible decision options can easily multiply. To make sure that we indeed find the optimal solution for such a problem, an optimization technique such as dynamic programming[2] can be used.

12.3.4 Consideration of Technological Change

Thus far, we have defined the challenger simply as the best available replacement for the defender. It is more realistic to recognize that often the replacement decision involves an asset now in use versus a candidate for replacement that is in some way an improvement on the present asset. This, of course, reflects the technological progress that is continually going on. The future models of a machine are likely to be more effective than the current model. In most areas, this technological change appears as a combination of gradual advances in effectiveness, with an occasional technological breakthrough that revolutionizes the character of the machine.

The prospect of improved future challengers makes the current challenger a less desirable alternative. By retaining the defender, we may have a later opportunity to acquire an improved challenger. If this is so, the prospect of improved future challengers may affect the present decision between the defender and the challenger. It is difficult to forecast future technological trends in any precise fashion. However, in developing a long-term replacement policy, we need to take technological change into consideration.

12.4 Computer Notes

Perhaps the most tedious aspect of replacement analysis is calculating economic service lives. **EzCash** has a built-in function for automating this calculation—once economic service lives have been determined—either **EzCash** or a spreadsheet can accommodate the rest of the replacement analysis. By now you should have a good idea of how to tailor one of these two computer methods to a specific problem, so in this section, we will focus on a screen format that you can use to determine economic service lives, using the data from Example 12.6.

---

2 See F.S. Hillier and G. J. Lieberman, *Introduction to Operations Research*, Fifth Edition, McGraw Hill, 1990 (Chapter 11).

12.4.1 Economic Service Life Calculations on Lotus 1-2-3

Even programming the procedure on the spreadsheet can take a lot of time. Figure 12.16 shows a typical Lotus application to determine the economic service life for Example 12.6. On the top portion of the screen, you will enter a series of financial data, such as first cost, tax rate, current book value, and MARR. In the middle of the screen, you enter the market values, depreciation expenses, and O&M costs in columns B, C, and D. These input data represent before-tax figures. Then, the corresponding after-tax figures are displayed in columns E, F, and G, respectively. In the lower portion of the screen, you will find the NPW figures on operating costs by subtracting the depreciation tax credits from the *A/T* O&M cost. The cumulative operating costs in column C

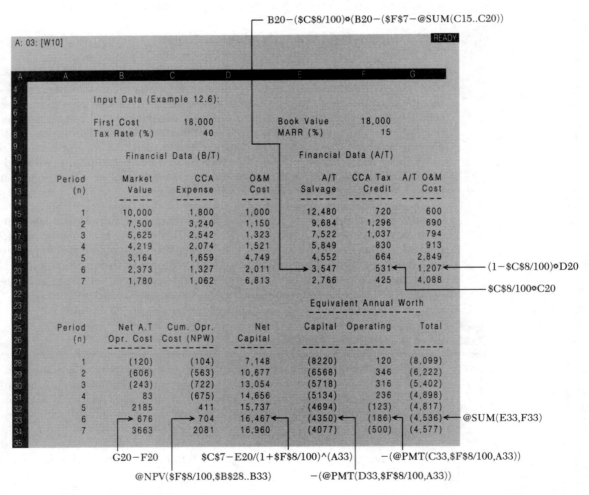

Figure 12.16 ■ Annotated cell formulas for Lotus 1-2-3 screen output (Example 12.6)

represent the NPW figure for the total operating cost up to the holding period. The net capitals in column D represent the first cost less the NPW of the salvage value at the end of each holding period. Finally, columns E and F represent the annualized capital recovery and operating cash flows for each holding period. From the total column in G, the minimum cost occurs at the end of year 6, which is the economic service life.

12.4.2 Economic Service Life Calculation with EzCash

From the opening menu, select the **Special** mode followed by the **Economic Service Life** command. Click on **Economic Service Life** and when the message window (Fig. 12.17(a)) appears, enter all worksheet values as positive quantities. When you click on **OK** in this window, the **Economic Service Life** worksheet appears with menu choices that correspond to the worksheet columns: **Market Value, O & M Cost**, and **Revenue** (Figure 12.17(b)). Selection of any of these choices will open up a **Cash Flow Editor** window through which data can be entered. These data are then automatically transferred to the **Economic Service Life** worksheet (Fig. 12.17(c)). Enter the challenger data for each worksheet column and then select the last menu choice, **Compute Economic Service Life**, which opens up a new window. The user is then prompted to enter four pieces of information: (1) the tax rate, (2) the interest rate, (3) the current book value of the asset, and (4) the remaining physical life of the asset. In addition, the user must specify whether the half-year convention is applicable and select either to do a declining balance or straight-line CCA calculation (Fig. 12.17(d)). For the challenger in Example 12.6, choose the **declining balance** calculation. Since the challenger represents a new asset, the half-year convention is also selected.

After entering all the required information, click **OK.** The user is asked to specify the declining balance rate (Fig. 12.17(e)). Clicking on **OK** in this window produces a table showing the economic service life calculations. The table includes the after-tax annual equivalent values for revenue, O & M cost, capital cost, and the CCA tax credit in addition to the total annual equivalent cost for holding time periods of length N (Fig. 12.17(f)).

The procedure for determining the economic service life of the defender is identical to the one followed for the challenger. The defender data from Example 12.5 are entered into the **Economic Service Life Worksheet** (Fig. 12.18(a)) and the **Input for Economic Service Life Calculation** window (Fig. 12.18(b)). Since the defender is an existing asset which is more than one year old, the half-year convention does not apply. Clicking **OK** prompts the user to enter the declining balance rate (Figure 12.18(c)). The economic service life calculations for the defender are shown in Figure 12.18(d).

If an asset qualifies for the straight-line method of CCA, the CCA calculations will require different information concerning the asset. These data are entered through a separate window (Fig. 12.19). This window is accessed from the **Input for Economic Service Life Calculation** window (Fig. 12.17(d)) by scrolling to, and selecting, **Straight Line** as the CCA method.

Figure 12.17 ■ Economic service life calculation for the challenger using **EzCash** (Example 12.6)

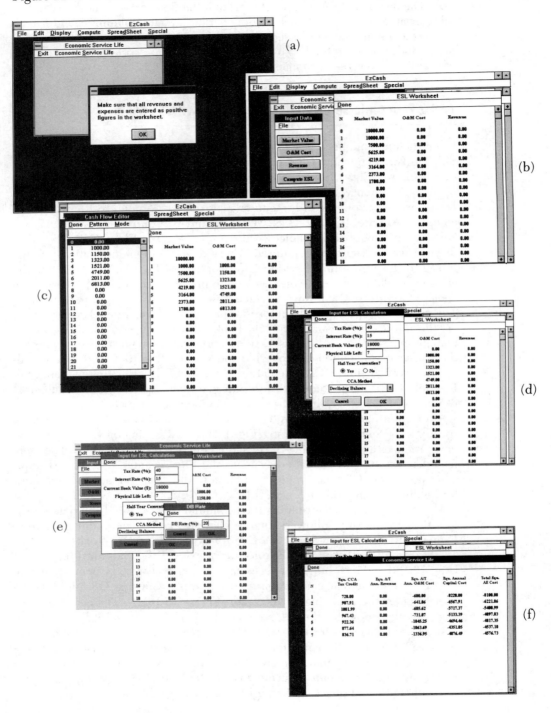

Figure 12.18 ■ Economic service life calculations for the defender by using **EzCash** (Example 12.5)

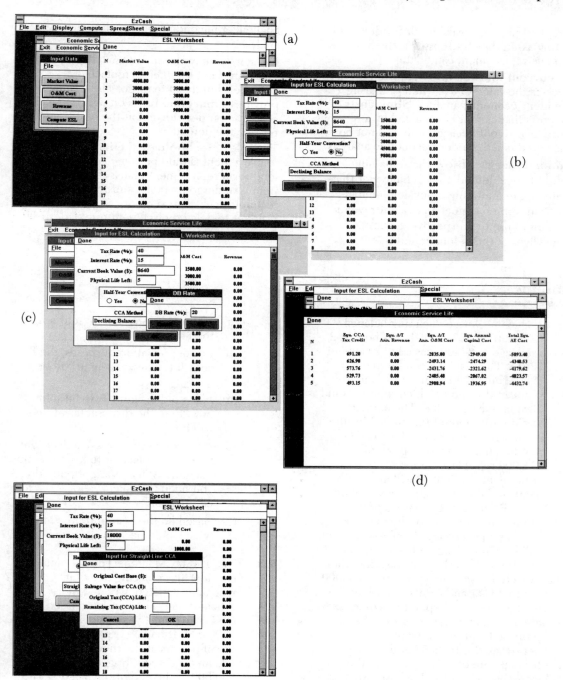

Figure 12.19 ■ Straight-line method input for economic service life calculation using **EzCash**

Summary

- In replacement analysis, the **defender** is an existing asset; the **challenger** is the best available replacement candidate.
- The **current market value** adjusted for tax effects is the value to use when preparing a defender's economic analysis. **Sunk costs**—past costs that cannot be changed by any future investment decision—should not be considered in a defender's economic analysis.
- **Economic service life** is the remaining useful life of a defender *or* a challenger that results in the minimum equivalent annual cost or maximum annual equivalent revenue. We want to use the respective economic service lives of the defender and the

challenger in conducting a replacement analysis.
- Ultimately, in replacement analysis, the question is not *whether* to replace the defender, but *when* to do so. The AE method provides a **marginal basis** on which to make a year-by-year decision about the best time to replace the defender.
- The role of **technological change** in asset improvement should be weighed in making long-term replacement plans: If a particular item is undergoing rapid, substantial technological improvements, it may be prudent to delay replacement until a desired future model is available.

Problems

12.1 A machine with a declining balance CCA rate of 30%, which was bought 2 years ago at $4400, has a book value of $2618. The machine can be sold for $2500, but could be used for 3 more years, at the end of which time it would have no salvage value. The annual O&M costs amount to $10,000 for the old machine. A new machine can be purchased at an invoice price of $14,000 to replace the present equipment. Freight-in will amount to $800, and the installation cost will be $200. Because of the nature of the product manufactured, the machine also has an expected life of 3 years, and will have no salvage value at the end of that time. With the new machine, the expected cash savings amount to $8000 the first year and $7000 in each of the next two years. Corporate income taxes are at an annual rate of 40%. The proposed equipment would be depreciated using a CCA declining balance rate of 30%. The MARR is assumed to be 12%. Each question should be considered independently.

(a) If the old asset is to be sold now, then its sunk cost is _____ .

(b) For depreciation purposes, the first cost of the new machine will be _____ .

(c) If the old machine is to be sold now, then the disposal tax effects will be _____ .

(d) Now you need to make a decision to either replace or to keep the old machine. What assumption(s) do you need to make a valid comparison?

(e) Based on the assumption(s) in (d), what would be your decision?

12.2 The Halifax Machine Tool Company is considering replacing one of its CNC machines with a newer and more efficient one. The firm purchased the CNC machine 4 years ago at a cost of $200,000. It had an expected economic life of 8 years at the time of purchase and an expected salvage value of $20,000 at the end of the 8 years. The original salvage estimate is still good, and the machine still has a remaining economic life of 4 years. It is being depreciated using a CCA declining balance rate of 20%. The firm can sell this old machine now to another firm in the industry for $80,000. The new machine can be purchased for $150,000, including installation costs. It may be depreciated using a CCA declining balance rate of 20%, and cash operating expenses would be reduced by $25,000 per year over its 6-year economic life. At the end of its economic life, the machine is estimated to be worthless. The company has a marginal tax rate of 40%, and it has a MARR of 12%.

(a) If the old asset is sold now, compute the following: Current book value, loss (gain) on disposal, and disposal tax effects.

(b) If you decide to retain the old machine, what is the opportunity (investment) cost of retaining the old asset?

(c) Compute the cash flows associated with retaining the old machine in years 1 to 4.

(d) Compute the cash flows associated with purchasing the new machine in years 1 to 6 (using the opportunity cost concept).

(e) If the firm will need the service of these machines for an indefinite period and no technology improvement in future machines is expected, what will be your decision?

12.3 A commuter airline company is considering replacing one of its baggage loading/unloading machines with a newer and more efficient one. The current book value of the old machine is $50,000, and it has a

remaining economic life of 5 years. The expected salvage from scrapping the old machine at the end of 5 years is zero, but it can be sold now to another firm in the industry for $10,000. The old machine is being depreciated using a CCA declining balance rate of 20%. The new baggage-handling machine has a purchase price of $120,000 and an estimated economic life of 7 years. It will be depreciated using a CCA declining balance rate of 20%, and it has an estimated salvage value of $30,000. It is expected to economize on electric power usage, labor, and repair costs, and also to reduce the amount of damaged luggage. In total, an annual savings of $50,000 will be realized if it is installed. The company's marginal tax rate is 40%, and the firm uses a 15% of MARR.

(a) What is the initial cash outlay required for the new machine?

(b) Calculate the annual CCA amounts for both machines and compute the disposal tax effects.

(c) What are the cash flows for the defender in years 0 to 5?

(d) Should the airline purchase the new machine?

12.4 Toronto Fashion Company purchased a computer-aided design machine 5 years ago at a cost of $50,000. It had an expected life of 10 years at the time of purchase and an expected salvage value of $5000 at the end of the 10 years. It has been depreciated using a CCA declining balance rate of 30%. A new machine can be purchased for $75,000, including installation costs. Over its 5-year life, it will reduce cash operating expenses by $30,000 per year. Sales are not expected to change. At the end of its economic life, the machine is estimated to be worthless. It will be depreciated using a CCA declining balance rate of 30%. The old machine can be sold today for $25,000. The firm's marginal tax rate is 35%. The firm's interest rate for project justification is known to be 15%. The firm does not expect any better machine (than the current challenger) to be available for the next 5 years.

(a) Determine the cash flows associated with each option (keeping the defender versus purchasing the challenger).

(b) Should Toronto Fashion replace the defender now?

12.5 The Northwest Manufacturing Company is currently manufacturing its product on a machine that is fully depreciated for tax purposes. The cost of the product is $10 per unit, and in the past year, 3000 units were produced and sold for $18 per unit. It is expected that both the future demand for the product and the unit price will stay steady at 3000 units per year and $18 per unit. The old machine has a remaining economic life of 3 years. The old machine could be sold on the open market now for $5000. Three years from now it is expected to have a salvage value of $1000. The new machine would cost $35,000, and the unit manufacturing cost on the new machine is projected to be $9 per unit. The new machine has an expected economic life of 5 years and an expected salvage value of $5000. For tax purposes, the entire cost can be depreciated according to a CCA declining balance rate of 20%. The current corporate income tax rate is 40%, and the appropriate MARR is 10%. The firm does not expect a significant improvement in technology for future challengers, and it needs the service of either machine for an indefinite period of time.

(a) Compute the after-tax cash flows over the remaining economic life if the firm decides to retain the old machine.

(b) Compute the after-tax cash flows over the economic service life if the firm decides to purchase the machine.

(c) Should the equipment be acquired now?

12.6 A firm is considering replacing a machine that has been used for making a certain kind of packaging material. The new improved machine will cost $31,000 installed and will have an estimated economic life of 10 years with a salvage value of $2500. Operating costs are expected to be $1000 per year throughout the service life. The new equipment will be classified as a CCA Class 43 asset with a declining balance rate of 30%. The old machine in use had an original cost of $25,000 four years ago, and at the time it was purchased, its service life (physical life) was estimated to be 7 years with a salvage value of $5000. The old machine has a current realizable value of $7500. The old machine was classified as a CCA Class 39 property with a declining balance rate of 25%. If the firm retains the old machine further, its updated market values and operating costs for the next 4 years are as follows:

Year End	Market Value	Book Value	Operating Costs
0	$7,500	$9,228	
1	4,000	6,921	$3,000
2	3,000	5,191	3,500
3	1,000	3,893	4,600
4	0	2,920	5,800

The firm's minimum attractive rate of return is 12% and the marginal tax rate is 35%.

(a) With the updated estimates of the market values and operating costs over the next 4 years, determine the economic life of the old machine.

(b) Determine whether it is economical to make the replacement now.

(c) If the firm's decision is not to replace the old machine now, then when should the firm do it?

12.7 A certain type of machine has a first cost of $10,000. End-of-year book values (given a declining balance CCA rate of 20%), salvage values, and annual O&M costs are provided over its useful life as follows:

Year End	Book Value	Salvage Value	Operating Costs
1	$9000	$5300	$1500
2	7200	2900	2000
3	5760	2100	2500
4	4608	1600	3000
5	3686	1200	3500
6	2949	900	4000

(a) Determine the economic life if the MARR is 20% and the marginal tax rate is 40%.

(b) Determine the economic life if the MARR is 10% and the marginal tax rate remains at 40%.

12.8 Given the following data:

$$P = \$20,000$$

$$S_n = \$12,000 - \$2000n, \ n \le 6$$

$$d = 30\%$$

$$O\&M_n = \$3000 + \$1000(n - 1)$$

$$t = 40\%$$

where,

$$P = \text{asset purchase price}$$

$$S_n = \text{market value at the end of year } n$$

$$d_n = \text{declining balance rate}$$

$$O\&M_n = \text{O\&M cost during year } n$$

$$t = \text{marginal tax rate}$$

(a) Determine the economic service life if $i = 10\%$.

(b) Determine the economic service life if $i = 25\%$.

12.9 A special-purpose machine is to be purchased at a cost of $15,000. The following table shows the expected annual operating and maintenance cost and the salvage values for each year of service. The asset will be depreciated using the CCA declining balance rate of 20%.

Year of Service	O&M Costs	Market Value
1	$2,000	$10,000
2	3,000	8,000
3	5,000	5,000
4	6,000	3,000
5	8,000	0

(a) If the interest rate is 10%, what is the economic life for this machine assuming a marginal income tax rate of 40%?

(b) Repeat (a) above using $i = 15\%$ and the marginal tax rate of 30%.

12.10 Quintana Electronic Company is considering the purchase of new equipment to perform operations currently being performed on different, less efficient equipment. The purchase price of the new machine is $150,000, delivered and installed. A Quintana industrial engineer estimates that the new equipment will produce savings of $30,000 in labor and other direct costs annually, as compared with the present equipment. The engineer estimates the proposed equipment's economic life at 10 years, with zero salvage value. The present equipment is in good working order and will last, physically, for at least 20 more years. Quintana Company expects to pay income taxes of 40%. Quintana uses a 10% discount rate for analysis performed on an after-tax basis. Depreciation of the new equipment for tax purposes is computed on the basis of a CCA declining balance rate of 25%.

(a) Assuming that the present equipment has zero book value and zero salvage value, should the company buy the proposed equipment?

(b) Assuming that the present equipment is being depreciated at a declining balance rate of 25%, has a book value of $78,750 (cost $120,000; accumulated depreciation $41,250), and zero salvage value today, should the company buy the proposed equipment?

(c) Assuming the present equipment, as is, is being depreciated at a declining balance rate of 25%, that it has a book value of $78,750 (cost $120,000; accumulated depreciation $41,250) and a salvage value today of $85,000, and that, if retained for 10 more years, its salvage value will be zero, should the company buy the proposed equipment?

(d) Assume that the new equipment will save only $15,000 a year, but that its economic life is expected to be 20 years. If other conditions are as described in (a) above, should the company buy the proposed equipment?

12.11 Quintana Company decided to purchase the equipment described in Problem 12.10 (hereafter called "model A" equipment). Two years later, even better equipment (called "model B") comes on the market. This new model makes model A obsolete, with no resale value. The model B equipment costs $300,000 delivered and installed, but it is expected to result in annual savings of $75,000 over the cost of operating the model A equipment. The economic life of model B is estimated to be 10 years with a zero salvage value. (The model B also has a CCA declining balance rate of 25%.)

(a) What action should the company take?

(b) If the company decides to purchase the model B equipment, a mistake has been made somewhere, because good equipment, bought only two years previously, is being scrapped. How did this mistake occur?

12.12 A special-purpose machine was purchased 4 years ago for $20,000. At the time, it was estimated that this machine would have a life of 10 years and a salvage value of $1000, with the cost of removal being $1500. These estimates are still good. This machine has annual operating costs of $2000, and its present book value is $3456. The asset has been depreciated using a CCA declining balance rate of 40%. A new machine, which is more efficient, will reduce the operating costs to

$1000, but will require an investment of $20,000, plus $1000 for installation. The life of the new machine is estimated to be 11 years with a salvage of $1000 and a cost of removal of $1500. An offer of $6000 has been made for the old machine, and the purchaser would pay for its removal. The new asset has a CCA declining balance rate of 40%. The marginal tax rate is 30%. Find the economic advantage of replacement or of continuing with the present machine. State any assumptions that you make. (Assume MARR = 8%).

12.13 The 5-year-old defender has a current market value of $4000, with O&M costs of $3000 this year, further increasing by $1500 per year. Future market values are expected to decline by $1000 per year. The machine is fully depreciated (with no remaining assets left in its CCA class), but it can be used another 3 years. The challenger costs $6000 and has O&M costs of $2000 per year, increasing by $1000 per year. The challenger will be depreciated using a CCA declining balance rate of 50%. The machine will be needed for only 3 years, and the salvage value at the end of 3 years is expected to be $2000. The marginal income tax rate is 40%, and the MARR is 15%.

(a) Determine the annual after-tax cash flows for retaining the old machine for 3 years.

(b) Determine if now is the time to replace the old machine. In doing so, show the annual after-tax cash flows for the challenger.

12.14 A Bedford-based metal-cutting firm is considering the purchase of a new machine tool to replace an obsolete one. The machine being used for the operation has a book value of $4000, but a market value of zero; however, it is in good working order, and it will last physically for at least an additional 5 years. The machine will continue to be depreciated using a declining balance rate of 20%, if the firm decides to keep the old machine for the additional 5 years. The new machine will perform the operation so much more efficiently that the engineers estimate that labor, material, and other direct costs will be reduced $3000 a year if it is installed. The new machine costs $10,000 delivered and installed, and its economic life is estimated to be 5 years with zero salvage value. The firm's MARR is 10%, and its marginal tax rate is 40%. Assume the firm will depreciate the asset using a CCA declining balance rate of 20%.

(a) What is the investment required to keep the old machine?

(b) Compute the after-tax cash flow to use in the analysis for each option.

(c) If the firm uses the internal rate of return method in analyzing investments, should the firm buy the new machine on that basis?

12.15 Eastern Lighting Company is considering the replacement of an old, relatively inefficient vertical drill machine that was purchased 5 years ago at a cost of $10,000. The machine had an original expected life of 10 years and a zero estimated salvage value at the end of that period. It has been depreciated on a declining balance rate of 30% and now has a book value of $5000. The division manager reports that a new machine can be bought and installed for $12,000 which, over its 5-year life, will expand sales from $10,000 to $11,500 a year and, furthermore, which will reduce labor and raw materials usage sufficiently to cut annual operating costs from $7000 to $5000. The new machine, which will be depreciated using a CCA declining balance rate of 30%, has an estimated salvage value of $2000 at the end of its 5-year life. The old machine's current market value is $1000; the firm's marginal tax rate is 40%, and its MARR is 15%.

(a) Should the new machine be purchased now?

(b) What current market value of the old machine would make the two options equal?

12.16 The annual equivalent after-tax value of retaining the defender machine over 4 years (physical life) or operating its challenger over 6 years (physical life) are as follows:

n	Defender	Challenger
1	−$3000	−$5000
2	−2500	−4000
3	−2800	−3000
4	−3500	−4500
5		−5000
6		−5500

If you need the service of either machine only for the next 10 years, what is the best replacement strategy? Assume that your MARR is 12% and there will be no technology improvement in the future challengers.

12.17 The after-tax annual equivalent worth of retaining the defender over 4 years (physical life) or operating its challenger over 6 years (physical life) are as follows:

n	Defender	Challenger
1	$3000	$2000
2	3500	3000
3	3800	3600
4	3200	3400
5		3000
6		2500

If you need the service of either machine only for the next 8 years, what is the best replacement strategy? Assume that your MARR is 12% and there will be no technology improvement in the future challengers.

12.18 An existing asset that cost $16,000 two years ago has a market value of $12,000 today, an expected salvage value of $2000 at the end of its economic life of 6 more years, and annual operating costs of $4000. The old asset has been depreciated using a CCA declining balance rate of 20%. A new asset under consideration as a replacement has an initial cost of $10,000, an expected salvage value of $4000 at the end of its economic life of 5 years, and annual operating costs of $2000. The new asset will be depreciated using a CCA declining balance rate of 20%. It is assumed that this new asset could be replaced by another one identical in every respect after 3 years at a salvage value of $5000, if desired. The firm's marginal tax rate is 30%.

(a) By using a MARR of 11%, a 6-year study period, and present-worth calculations, decide whether the existing asset should be replaced by the new one.

(b) Repeat (a) above based on the annual equivalent criterion.

12.19 Advanced Robotics Company is faced with the prospect of having to replace its old call-switching systems, which have been used in its headquarters office for 10 years. This particular system was installed at a cost of $100,000, and it was assumed that it would have a 15-year life with no appreciable salvage value. The current annual operating costs are $20,000 for this old system, and these costs would be the same for the rest of its life. A sales representative from Funny Bell is trying to sell a computerized switching system to this firm. The new system would require an investment of $200,000 for installation. The economic life of this computerized system is estimated to be 10 years with a salvage value of $18,000, and the system will reduce the annual operating costs to $5000. No detailed agreement has been made with the sales representative on the disposal of the old system. The old switching system has been fully depreciated and the new system can be depreciated by using a CCA declining balance rate of 10%. Determine the ranges of resale value associated with the old system that can justify the installment of the new system at a MARR of 14%. Assume that the firm's marginal tax rate is 40%.

12.20 Five years ago, a conveyor system was installed in a manufacturing plant at a cost of $35,000. It was estimated that the system, which is still in operating condition, would have a useful life of 8 years. The equipment was depreciated using a CCA declining balance rate of 20%. If the firm continues to operate the system, its estimated market values and operating costs for the next 3 years are as follows:

Year End	Market Value	Book Value	Operating Costs
0	$10,500	$12,902	
1	5,000	10,322	$4,200
2	2,500	8,258	4,900
3	1,000	6,606	6,500

A new system can be installed for $43,500; it would have an estimated economic life of 10 years with a salvage value of $3500. The operating costs are expected to be $1500 per year throughout the service life. This firm's MARR is 18%. The system would be depreciated using a CCA declining balance rate of 30%. The firm's marginal tax rate is 35%.

(a) Decide whether to replace the existing system now.

(b) If the decision is not to replace now, when should replacement occur?

12.21 A company is currently producing chemical compounds by a process installed 10 years ago at a cost of $100,000. It was assumed that the process would have a 20-year life with a zero salvage value, and it is being depreciated using a CCA declining balance rate of 25%. The current market value of this equipment, however, is $60,000, and the initial estimate about its economic life is still good. The annual operating costs associated with this process are $18,000. A sales representative from Canadian Instrument Company is trying to sell a new chemical compound-making process to the company. This new process will cost $200,000, have an economic life of 10 years with a salvage value of $20,000, and reduce annual operating costs to $4000. The new process will be depreciated using a CCA declining balance rate of 30%. Assuming the company desires a return of 12% on all investments, should it invest in the new process? The firm's marginal income tax rate is 40%.

12.22 Eight years ago a lathe was purchased for $45,000, and the lathe is fully depreciated. Its operating expenses were $8700 for a year. A sales representative offers a new machine for $53,500, and its operating costs are $5700 per year. An allowance of $8500 would be made for the old machine on purchase of the new one. The old machine is expected to be scrapped at the end of 5 years. The new machine's economic service life is 5 years with a salvage value of $12,000. The new machine will be depreciated using a CCA declining balance rate of 30%. The new machine's O&M cost is estimated to be $4200 for the first year, increasing at an annual rate of $500 thereafter. The firm's marginal income tax rate is known to be about 35%, and its MARR is 12%. What option would you recommend?

12.23 Four years ago a machine was bought for $23,000. It has been depreciated using a CCA declining balance rate of 30%. If sold now, it will bring $2000. If sold at the end of the year, it will bring $1500. Annual operating costs are $3800. A new machine will cost $50,000 with a 12-year life and have a $3000 salvage value, depreciated using a CCA declining balance rate of 30%. The operating cost will be $3000 as of the end of each year with $6000 per year savings due to better quality control. If the firm's MARR is 10% and its marginal tax rate is 40%, should the machine be purchased now?

12.24 London Ceramic Company has an automatic glaze sprayer that has been used for the past 10 years. The sprayer can be used for another 10 years and will have a zero salvage value at that time. The annual operating and maintenance costs for the sprayer amount to $15,000 per year. Due to an increase in business, a new sprayer must be purchased.

- **Option 1:** If the old sprayer is retained, a new sprayer of smaller capacity will be purchased at a cost of $48,000. The new sprayer will have a $6000 salvage value in 10 years. This new sprayer will also have annual operating and maintenance costs of $12,000. The old sprayer has been fully depreciated, and it has a current market value of $6000. The new sprayer may be depreciated using a CCA declining balance rate of 20%.

- **Option 2:** If the old sprayer is sold, a new sprayer of larger capacity will be purchased for $84,000. This sprayer will have a $9000 salvage value in 10 years and will have annual operating and maintenance costs of $24,000. This large-capacity sprayer may be depreciated using a CCA declining balance rate of 20%.
The marginal tax rate is known to be about 40%. Which option should be selected at MARR = 12%?

12.25 National Woodwork Company, a manufacturer of window frames, is considering replacing the conventional manufacturing system by a flexible manufacturing system (FMS) shown in Fig. 12.19. The company cannot produce rapidly enough to meet the demand. Some manufacturing problems identified are listed below:

- The present system is expected to be useful for another 5 years, but will require an estimated $105,000 per year in maintenance, which will increase $10,000 each year as parts become more scarce. The current market value of the existing system is $140,000 and the machine has been fully depreciated.

- The proposed system will reduce or entirely eliminate the set-up times, and each window can be made as it is ordered by the customer. Customers phone their orders into the head office, where details are fed into the company's main computer. These manufacturing details will then be dispatched to computers on the manufacturing floor, which are in turn connected to the computer that controls the proposed FMS. This will eliminate the warehouse space and material handling time that are needed when using the conventional system.

- Before installing the FMS, the old equipment will be removed from the job-shop floor at an estimated cost of $100,000. This cost includes doing the electrical work needed for the new system. The proposed FMS will cost $1,200,000. The economic life of the machine is expected to be 10 years, and the salvage value is expected to be

$120,000. The change in the style of windows has been minimal in the past few decades and should continue to remain stable in the future. The proposed equipment falls in Class 43 with a declining balance rate of 30%. The total annual savings will be $664,243: $12,000 from the reduction of defective windows, $511,043 from the lay off of 13 workers, $100,200 from the increase in productivity, and $41,000 from the near elimination of warehouse space and material handling. The O&M costs will be only $15,000, increasing by $2000 per year. The National Woodwork's MARR is about 15%, and the expected marginal tax rate over the project years is 40%.

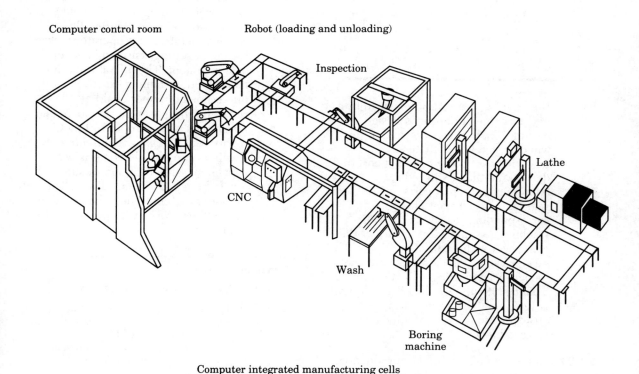

Computer control room   Robot (loading and unloading)

Inspection

Lathe

CNC

Wash

Boring machine

Computer integrated manufacturing cells

Figure 12.19 ■ A conceptual flexible manufacturing system (FMS) adopted by National Woodwork Company (Problem 12.25)

(a) What assumptions are required to compare the conventional system with the FMS?

(b) With the assumptions defined in (a) should the FMS be installed now?

12.26 A 6-year-old machine that originally cost $8000 has a current market value of $1500. It has been depreciated using a CCA declining balance rate of 30%. If the machine is continued in service for the next 5 years, its O&M costs and salvage value are estimated as follows:

O&M costs

End of Year	Operation and Repairs	Delays Due to Breakdowns	Salvage Value
1	$1300	$600	$1200
2	1500	800	1000
3	1700	1000	500
4	1900	1200	0
5	2000	1400	0

It is suggested that it be replaced by a new machine of improved design costing $6000. It is believed that this will completely eliminate any breakdowns and the resulting cost of delays, and that it will reduce operation and repair costs to $200 a year less at each age than the corresponding costs with the old machine. Assume a 5-year life for the challenger with a $1000 terminal salvage value. The new machine may be depreciated using a CCA declining balance rate of 30%. The firm's MARR is 12%, and its marginal tax rate is 30%. Should the old machine be replaced now?

12.27 In 2 × 4 and 2 × 6 lumber production, significant amounts of wood are present in the sideboards produced after the initial log cutting. Instead of processing the sideboards into wood chips for the paper mill, an "edger" is used to reclaim additional lumber, thus resulting in savings for the company. An edger is capable of reclaiming lumber by any of the following three methods: (1) removing rough edges, (2) splitting large sideboards, and (3) salvaging 2 × 4 lumber from low-quality 4 × 4 boards. Quebec Forest Product's engineers discovered that a significant reduction in production costs could be achieved simply by replacing the original "edger" machine with a newer laser-controlled model.

- **Old edger:** The old edger was placed in service 12 years ago and fully depreciated. Any machine scrap value would offset the removal cost of the equipment. No user market exists for this obsolete equipment. The old edger needs two operators. During the cutting operation, the operator makes the edger settings based on his/her own judgment. The operator has no means of determining exactly what dimension of lumber could be recovered from a given sideboard and must guess at the proper setting to recover the highest grade of lumber. Furthermore, the old edger is not capable of salvaging good-quality 2 × 4s from poor-quality 4 × 4s. The defender can continue to be in service for another 5 years with proper maintenance.

Current market value	$0
Current book value	0
Annual maintenance cost	$2500 in year 1 increasing at the rate of 15% each year over the previous year's cost
Annual operating costs (labor and power)	$65,000

- **New laser-controlled edger:** The new edger has numerous advantages over its defender, including laser beams that indicate where cuts should be made to obtain the maximum yield by the edger. The new edger requires a single operator, and this labor savings is reflected in the lower operating and maintenance costs of $35,000 a year.

Estimated Cost

Equipment	$35,700
Equipment installation	21,500
Building	47,200
Conveyor modification	14,500
Electrical (wiring)	16,500
Subtotal	$135,400
Engineering	7,000
Construction management	20,000
Contingency	16,200
Total	$178,600

Useful life of new edger	10 years
Salvage value	
Building (tear down)	$0
Equipment	10% of the original cost
Annual O&M costs	$35,000

Depreciation Methods

Building	4% declining balance
Equipment and installation	30% declining balance

Twenty-five percent of total mill volume passed through the edger. A 12% yield improvement is expected to be realized on this production, resulting in an improvement of total mill volume of $(0.25)(0.12) = 3\%$.

Annual Wood Savings

	Gross Wood	Chips	Retained
1994 budget	2.35	1.25	1.10
With edger improvement	2.28	1.18	1.10
CDS/MBF improvement	(0.07)	(0.07)	0
Wood costs/CD	$72.60	$53.00	
Wood costs/MBF where MBF = 1,000 board feet	$(5.08)	(3.71)	$(1.37)

Annual wood
savings = (1994 budgeted volume)

$\times$ (net woods cost reduction)

= 42,258 MBF $\times$ $1.37/MBF

= $57,895

(a) Should the defender be replaced now if the mill's MARR and marginal tax rate are 16% and 40%, respectively?

(b) Since the defender will be replaced eventually by the current challenger, when is the optimal time to replace?

12.28 Tiger Construction Company purchased the current bulldozer (Caterpillar D8H) and placed it in service 5 years ago. Since the purchase of the Caterpillar, new technology has produced changes in machines, which resulted in an increase in productivity of approximately 20%. The Caterpillar worked in a system with a fixed (required) production level to maintain overall system productivity. As the Caterpillar aged and had more downtime, it had to be scheduled more hours to maintain the required production. Tiger is considering the purchase of a new bulldozer (Komatsu

K80A) to replace the Caterpillar. The following data have been collected by Tiger's civil engineer.

	Defender	Challenger
	(Caterpillar D8H)	*(Komatsu K80A)*
Useful life	Not known	Not known
Purchase price		$400,000
Salvage value (if kept until):		
0 year	$75,000	$400,000
1 year	60,000	300,000
2 year	50,000	240,000
3 year	30,000	190,000
4 year	30,000	150,000
5 year	10,000	115,000
Fuel use (litres/hr)	66.50	90
Operating hours (hr/yr):		
1	1,800	2,500
2	1,800	2,400
3	1,700	2,300
4	1,700	2,100
5	1,600	2,000
Productivity index	1.00	1.20

Other Relevant Information

Fuel cost ($/litre)	$.50
Operator's wages ($/hr)	$23.40
Required rate of return (MARR)	15%
Marginal tax rate	40%
Depreciation methods:	
Defender (D8H)	Fully depreciated
Challenger (K80A)	Declining balance rate of 20%

(a) The civil engineer notices that both machines have different working hours and hourly production capacity. To compare these different units of capacity, the engineer needs to devise a combined index to reflect the machine's productivity as well as actual operating hours. Develop such a combined productivity index for each period.

(b) Adjust the operating and maintenance costs by this index.

(c) Compare the two alternatives. Should the defender be replaced now?

12.29 Reconsider Problem 12.28 with the following price index to be forecasted during the next 5 project years.

Forecasted Price Index

Year	Inflation	Fuel	Wage	General Maintenance
0	100	100	100	100
1	108	110	115	108
2	116	120	125	116
3	126	130	130	124
4	136	140	135	126
5	147	150	140	128

With these inflationary trends predicted, should the defender be replaced now?

12.30 Rivera Industries, a manufacturer of home heating appliances, is considering the purchase of an Amada turret punch press, a more advanced piece of machinery, to replace the current system that involves four old presses. Presently, the four smaller presses are used (in varying sequences, depending on the product) to produce one component of a product until a scheduled time when all machines must retool to set up for a different component. Because the setup cost is high, production runs of individual components are long. To prevent extended backlogging while other products are being manufactured, large inventory buildups of one component commonly occur.

- The four presses currently in use were purchased 3 years ago at a price of $100,000. The manufacturing engineer expects that these machines can be used for 8 more years, but will have no market value after that. These presses have been depreciated using a CCA declining balance rate of 30%. The current book value is $41,650, and the present market value is estimated to be $40,000. The average setup cost, which is determined by the number of required labor hours times the labor rate for the old presses is $80/hour, and the number of setups per year expected by the Production Control Department is 200. This yields a yearly setup cost of $16,000. The expected operating and maintenance cost for each year in the remaining life of this system are estimated as follows:

Year	Setup Costs	O&M Costs
1	$16,000	$15,986
2	16,000	16,785
3	16,000	17,663
4	16,000	18,630
5	16,000	19,692
6	16,000	20,861
7	16,000	22,147
8	16,000	23,562

These costs, estimated by the manufacturing engineer with the aid of data provided by the vendor, represent a reduction in efficiency and an increase in needed service and repair over time.

- The price of the 2-year-old Amada turret punch press is $135,000. The new press would be paid for with cash from the company's capital fund. Additionally, the company would incur installation costs which would total $1200. An expenditure of $12,000 would be required to recondition the press at its original condition. This recondition would extend the Amada's economic service life to 8 years with no salvage value at that time. The Amada would be depreciated using a CCA declining balance rate of 30%. The cash savings of the Amada over the present system are due to the reduced setup time. The average setup cost of the Amada is $15, and the Amada would incur 1000 setups per year, yielding a yearly setup cost of $15,000. The savings due to the reduced setup time are experienced because of the reduction in carrying costs associated with that inventory level when the production run and ordering quantity are reduced. The Inventory Control Department has estimated that at least $26,000 and probably $36,000 per year could be saved by the shortening of the production runs. The operating and maintenance costs of the Amada as estimated by the manufacturing engineer are similar but somewhat less than the O&M costs for the present system.

Year	Setup Costs	O&M Costs
1	$15,000	$11,500
2	15,000	11,950
3	15,000	12,445
4	15,000	12,990
5	15,000	13,590
6	15,000	14,245
7	15,000	14,950
8	15,000	15,745

The reduction in the O&M costs is due to the age difference of the machines and the reduced power requirements of the Amada.

- If Rivera Industries delays the replacement of the current four presses for another year, the second-hand Amada machine will no longer be available, and the company will have to buy a brand-new machine at an installed price of $200,450. The expected setup costs would be the same as those for the second-hand machine, but the annual operating and maintenance costs would be about 10% lower than the estimated O&M costs for the second-hand machine. The expected economic service life of the brand-new press would be 8 years with no salvage value. The brand-new press falls into CCA Class 43 property with a declining balance rate of 30%.

Rivera's MARR is 12% after taxes, and the marginal income tax rate is expected to be 40% over the life of the project.

(a) Assuming that the company would need the service of either press for an indefinite period, what do you recommend?

(b) Assuming that the company would need the press for only 5 more years, what do you recommend?

CHAPTER 13

Capital Budgeting Decisions

Can-Alberta, an Alberta-based bottling company, stores the gasoline used by its fleet of trucks in underground storage tanks. Current Alberta Fire Code regulations require all such tanks to be upgraded to meet 1999 standards or to be replaced by 1999. Tanks installed prior to 1969 cannot be upgraded and must be replaced by 1996. Tanks installed between 1970 and 1979 may be upgraded by 1997 and then must be replaced within a 10-year period. Tanks installed during 1980 and thereafter may be completely upgraded to 1999 standards and then have an expected life of 30 years.

Several tanks may occupy one underground pit. All tanks in a pit are either upgraded or replaced simultaneously. Prior to the upgrade or replacement, yearly inspections of each tank are undertaken at a cost of approximately $800 per tank. Some locations may have more than one pit. The pits on a given site may be upgraded or replaced individually or as a group. There are certain economies of scale associated with upgrading or replacing all tanks on a site at the same time.

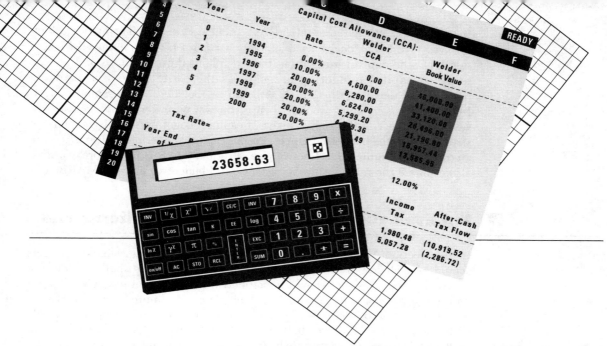

The company has over 17 pits in 11 different locations throughout the province. Given the upgrade/replacement option for each pit as well as the option of working on pits individually or in combination with others at the same site, there are over 340 alternatives to be considered. The company wants to meet all Fire Code requirements over a 25-year time horizon.

If there were no budget limitations, the replacement problem would be easy; simply select for each pit or combination of pits the least-cost alternative. However, to replace or upgrade all tanks over the 25-year time horizon will cost in excess of $750,000, and the company anticipates having an annual tank-replacement budget of $200,000 or less. Given these budget restrictions, it would like to determine the least-cost replacement/upgrade strategy for the 25-year period. It is also interested in minimizing the maximum budget allocations required in any one year.

In this chapter, we shall present the basic framework of capital budgeting, which involves investment decisions related to fixed assets. As mentioned in Section 1.1.2, the **capital budget** outlines the planned expenditures on fixed assets, and **capital budgeting** encompasses the entire process of analyzing projects and deciding whether they should be included in the capital budget. In the previous chapters, we have focused on how we evaluate and compare investment projects—the analysis part of capital budgeting. In this chapter, we shall focus on the budgeting aspect. Proper capital budgeting decisions require an estimate of minimum attractive rate of return (MARR). Therefore, we shall also examine the relationship between capital budgeting and the MARR.

■ 13.1 Evaluation of Multiple Investment Alternatives

In this section, we shall examine the process of deciding whether projects should be included in the capital budget. In particular, we shall consider the decision procedures that should be applied when we have to evaluate a set of multiple investment alternatives for which we have a limited capital budget.

In Chapters 4, 5, and 6, we learned how to compare two or more mutually exclusive projects. Now, we shall extend the comparison techniques to a set of multiple decision alternatives which are not necessarily mutually exclusive. Here, we distinguish a **project** from an **investment alternative,** which is a decision option. For a single project, we have two investment alternatives: to accept or reject the project. For two independent projects, we can have four investment alternatives: (1) to accept both projects, (2) to reject both projects, (3) to accept only the first project, and (4) to accept only the second project. As we add interrelated projects, the number of investment alternatives to consider grows exponentially.

13.1.1 Independent Versus Dependent Projects

For a proper capital budgeting analysis, a firm must group all projects being considered into decision alternatives. This grouping requires the firm to distinguish between projects that are independent of one another and those that are dependent in order to formulate alternatives correctly.

Independent Projects

An **independent project** is one that may be accepted or rejected without influencing the accept-reject decision of another independent project. For example, the purchase of a milling machine, office furniture, and a forklift truck constitutes three independent projects. Only projects that are economically independent of one another can be evaluated separately. (Budget constraints may prevent us from selecting one or more of several independent projects; this external constraint does not alter the fact that the projects are independent.)

Dependent Projects

In many decision problems, several investment projects are related to one another such that the acceptance or rejection of one project influences the probability of acceptance of others. There are two such types of dependencies: mutually exclusive projects and contingent projects. We say that two or more projects are **contingent** if the acceptance of one requires the acceptance of another. For example, the purchase of a computer printer is dependent upon the purchase of a computer, but the computer may be purchased without considering the purchase of the printer.

13.1.2 Formulating Mutually Exclusive Alternatives

We can view the selection of investment projects as a problem of selecting a single decision alternative from a set of alternatives. Note that each investment project is an investment alternative, but a single investment alternative may entail a whole group of investment projects. The common method of handling various project relationships is to arrange the investment projects so that the selection decision involves only mutually exclusive alternatives. To obtain this set of mutually exclusive alternatives, we need to enumerate all of the feasible combinations of the projects under consideration.

In enumerating the number of investment alternatives, it is often useful to visualize the project relationships in graphic form. For this purpose, we will introduce a set of symbols to represent interdependence as shown in Fig. 13.1. If projects are mutually exclusive, they are connected by a solid line; if one project is contingent on another project, they are connected by a line with an arrow pointing to the project that the other is contingent upon; if projects are independent, no lines are drawn between them. We will use this symbolism in developing the mutually exclusive alternatives for three different investment situations: independent projects, mutually exclusive projects, and contingent projects.

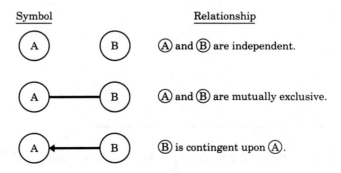

Figure 13.1 ■ A graphical representation of project relationships

Independent Projects

With a given number of independent investment projects, we can easily enumerate mutually exclusive alternatives. For example, in considering two projects, A and B, we have four decision alternatives, including a "do nothing" alternative.

Alternative	Description	X_A	X_B
1	Reject A, reject B	0	0
2	Accept A, reject B	1	0
3	Reject A, accept B	0	1
4	Accept A, accept B	1	1

Ⓐ Ⓑ

In our notation, X_j is a decision variable associated with investment project j. If $X_j = 1$, project j is accepted; if $X_j = 0$, project j is rejected. Since the acceptance of one of these alternatives will exclude any other, the alternatives are considered to be mutually exclusive.

Mutually Exclusive Projects

Suppose we are considering two independent sets of mutually exclusive projects. Each pair of projects (A1, A2), and (B1, B2), is mutually exclusive. The selection of either A1 or A2, however, is also independent of the selection of any project from the set (B1, B2). For this set of investment projects, the mutually exclusive alternatives are

Alternative	(X_{A1}, X_{A2})	(X_{B1}, X_{B2})
1	(0, 0)	(0, 0)
2	(1, 0)	(0, 0)
3	(0, 1)	(0, 0)
4	(0, 0)	(1, 0)
5	(0, 0)	(0, 1)
6	(1, 0)	(1, 0)
7	(0, 1)	(1, 0)
8	(1, 0)	(0, 1)
9	(0, 1)	(0, 1)

Ⓐ1——Ⓐ2

Ⓑ1——Ⓑ2

Note that, with two independent sets of mutually exclusive projects, we can have nine different alternatives.

Contingent Projects

Suppose C is contingent on the acceptance of both A and B, and acceptance of B is contingent on acceptance of A. The possible number of decision alternatives can be formulated as follows:

Alternative	X_A	X_B	X_C
1	0	0	0
2	1	0	0
3	1	1	0
4	1	1	1

Thus, we can easily formulate a set of mutually exclusive investment alternatives with a limited number of projects that are independent, mutually exclusive, or contingent merely by arranging the projects in a logical sequence.

One difficulty with the enumeration approach is that, as the number of projects increases, the number of mutually exclusive alternatives increases exponentially. For example, for 10 independent projects, the number of mutually exclusive alternatives is 2^{10}, or 1024. For 20 independent projects, there are 2^{20}, or 1,048,576, alternatives. As the number of decision alternatives increases, we may have to resort to mathematical programming to find the solution. Fortunately, in real-world business, the number of engineering projects to consider at any one time is usually manageable, so the enumeration approach is a practical one.

Example 13.1 Formulating mutually exclusive alternatives

Consider the following five investment projects with estimated cash flows. Suppose projects A and B are mutually exclusive. Project C is independent, but project D is contingent on project C. Project E is contingent on project B.

Project	0	1	2	3
A	−100	50	50	50
B	−200	50	100	200
C	−100	30	80	120
D	−300	100	100	100
E	−200	100	100	150

Formulate the total number of investment alternatives and tabulate their cash flows.

Solution

From the graphic relationship in our example, we can deduce two sets of independent project groups, (A, B, BE) and (C, CD). (No projects in the two groups are connected by arrows.) In the first group, there are four mutually exclusive alternatives, counting the "do-nothing" alternative; in the second group, there are another three.

- **First group (A, B, BE):**

Alternative	X_A	X_B	X_{BE}
1	0	0	0
2	1	0	0
3	0	1	0
4	0	0	1

- **Second group (C, CD):**

Alternative	X_C	X_{CD}
1	0	0
2	1	0
3	0	1

Table 13.1 Total Number of Mutually Exclusive Alternatives (Example 13.1)

Alternative	Projects	Cash Flow 0	1	2	3
1	–	$ 0	$ 0	$ 0	$ 0
2	A	−100	50	50	50
3	C	−100	30	80	120
4	B	−200	50	100	200
5	A, C	−200	80	130	170
6	B, C	−300	80	180	320
7	B, E	−400	150	200	350
8	C, D	−400	130	180	220
9	A, C, D	−500	180	230	270
10	B, C, E	−500	180	280	470
11	B, C, D	−600	180	280	420
12	B, C, D, E	−800	280	380	570

By combining the two groups, the total number of mutually exclusive alternatives that can be obtained from the group of projects in this example is 12, as shown in Table 13.1.

After formulating all possible combinations, we treat them as mutually exclusive alternatives. The cash flow for any alternative is simply the sum of the cash flows of the included projects.

Now that we have the list of alternatives to consider, the next step is to apply a specific decision criterion to measure the investment worth associated with each alternative.

13.1.3 Multiple-Alternative Comparison by NPW Method

The two-alternative comparison problem in Example 4.10 showed us some aspects of total investment analysis. The first step for the proper use of any selection criterion is to formulate mutually exclusive decision alternatives. Before applying any investment criterion to select projects, we must follow this procedure:

- Step 1: Remove any individual project that fails to meet the present worth criterion (or an equivalent criterion).

- Step 2: Develop a list of all possible project combinations. The resulting list contains all the mutually exclusive alternatives from which we must choose.

- Step 3: Order the alternatives by increasing order of investment. The ordering is normally done by the investment required at time 0. (This step is not critical, but it helps us organize the decision problem. If there is a tie, it is broken arbitrarily.)

- Step 4: If budget constraints exist, remove any project that exceeds the budget.

- Step 5: Remove any alternatives that are dominated by another alternative. Project domination means that one alternative requires less investment yet generates more revenue for every period. (This step may be omitted, but it makes our analysis more efficient by reducing the number of investment alternatives to be considered.)

We now shall work through a multiple-alternative problem in Example 13.2.

Example 13.2   Applying NPW to more than two alternatives

Consider Example 13.1 again. Use the PW measure to select the best alternative at MARR = 10%.

Solution

There are 12 alternatives, including the "do-nothing" option. By convention we organize and analyze them in order of increasing initial investment. Since the mutually exclusive alternatives in Table 13.1 are already listed in increasing order of initial investment, we will simply apply the procedure to this list. Computing the NPW for each alternative at MARR = 10% yields Table 13.2. With no budget constraint imposed, alternative A10 (projects B, C, and E) best maximizes the present worth at $PW(10\%) = \$248.15$.

Table 13.2 Net Present Worth Calculations of All Decision Alternatives (Example 13.2)

Alternative	Projects	Cash Flow				PW(10%)
		0	1	2	3	
1	–	$ 0	$ 0	$ 0	$ 0	$ 0
2	A	−100	50	50	50	24.34
3	C	−100	30	80	120	83.54
4	B	−200	50	100	200	78.36
5	A, C	−200	80	130	170	107.88
6	B, C	−300	80	180	320	161.90
7	B, E	−400	150	200	350	164.61
8	C, D	−400	130	180	220	32.23
9	A, C, D	−500	180	230	270	56.57
10	B, C, E	−500	180	280	470	248.15
11	B, C, D	−600	180	280	420	110.59
12	B, C, D, E	−800	280	380	570	196.84

13.1.4 Multiple-Alternative Comparison by IRR on Incremental Investment

The two-alternative problems in Example 6.11 showed some aspects of incremental rate of return analysis. We now extend the incremental analysis to multiple alternatives.

As mentioned in Example 6.11, to expedite our analysis, we should first order the alternatives so that we are always examining increments of investment (instead of increments of borrowing). Then we perform repeated two-alternative comparisons. This procedure can be summarized as follows.

- Step 1: Order the investment alternatives by increasing order of initial investment. If ties occur or if more than one period of investment flows occurs, order the contenders such that, when the lower-ordered option is subtracted from the higher-ordered one, an increment of investment is produced (that is, the first nonzero incremental cash flow is negative).

- Step 2: Compute the cash flow difference between the next alternative in order (y) and the one most recently selected (z). Make a two-alternative analysis based on this pair. Compute the IRR_{y-z} on the **increment of investment.** If the incremental cash flows indicate a nonsimple investment, you need to find the IRR using the procedures in Section 6.4.4:

$$\text{if } \text{IRR}_{y-z} \geq \text{MARR, retain alternative } y.$$

$$\text{if } \text{IRR}_{y-z} < \text{MARR, retain alternative } z.$$

- Step 3: Repeat step 2 until all alternatives have been compared. Accept the last alternative for which the cash flow difference was evaluated favorably.

We now have a conceptual framework and a set of decision rules for selecting the best investment alternatives via incremental rate of return analysis. As an example of this method, we consider Example 13.3.

Example 13.3 Applying IRR to more than two alternatives

Consider the 12 investment alternatives given in Table 13.1. Select the best alternative based on the IRR on incremental investment using MARR = 10%.

Solution

Implementing the incremental analysis successfully differs from the NPW approach in that the ordering of investment alternatives is a necessary step. All the investment alternatives have an investment required only at time 0. Ordering by increasing of initial investment yields the list in Table 13.2. If there is no budget limit, at this stage we need to examine whether there is any project dominance (a useful step in reducing the alternatives). Note that A7 dominates A8 because it requires less investment yet generates more revenue for every period. A10 also dominates both A9 and A11. After we have eliminated these three alternatives, the investment alternatives are as shown in Table 13.3.

Now the initial investment values for A4 and A5 are the same, so we may look at two possible differential cash flows and select for analysis the one that represents an increment of investment.

	A5–A4	A4–A5
n	*Increment of Borrowing*	*Increment of Investment*
0	0	0
1	30	−30
2	30	−30
3	−30	+30

Therefore, we place A5 before A4.

Table 13.3 Nondominated Alternatives Ordered by Increasing Order of Initial Investment (Example 13.3)

Alternative	Projects	Cash Flow 0	1	2	3	
1	–	$ 0	$ 0	$ 0	$ 0	
2	A	−100	50	50	50	
3	C	−100	30	80	120	Reordered
5	A, C	−200	80	130	170	to create
4	B	−200	50	100	200	incremental
6	B, C	−300	80	180	320	investment.
7	B, E	−400	150	200	350	
10	B, C, E	−500	180	280	470	
12	B, C, D, E	−800	280	380	570	

We are now in a position to apply incremental analysis. Since we know how to choose between two alternatives, we can choose among multiple alternatives by successive examinations. Figure 13.2 illustrates the following procedure.

- A2 vs. A1 (do nothing):
  $IRR_{2-1} = 23.38\% > 10\%$, so A2 is selected.

- A3 vs A2:
  Incremental cash flow is 0, −20, 30, 70.
  $IRR_{3-2} = 176.56\% > 10\%$, so A3 is selected.

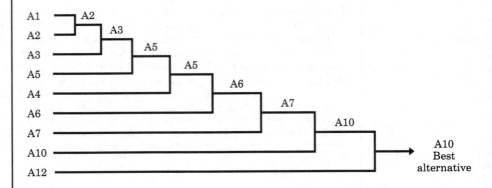

Figure 13.2 ■ Solution of a multiple-alternative problem by successive two-alternative analyses. Here, alternatives 8, 9, and 11 were dominated. (Example 13.3)

- **A5 vs. A3:**
  Incremental cash flow is $-100, 50, 50, 50$.
  $IRR_{5-3} = 23.38\% > 10\%$, so A5 is selected.

- **A4 vs. A5:**
  Incremental cash flow is $0, -30, -30, 30$.
  $IRR_{4-5} = -38.20\% < 10\%$, so A5 is selected.

- **A6 vs. A5:**
  Incremental cash flow is $-100, 0, 50, 150$.
  $IRR_{6-5} = 28.96\% > 10\%$, so A6 is selected.

- **A7 vs. A6:**
  Incremental cash flow is $-100, 70, 20, 30$.
  $IRR_{7-6} = 11.86\% > 10\%$, so A7 is selected.

- **A10 vs. A7:**
  Incremental cash flow is $-100, 30, 80, 120$.
  $IRR_{10-7} = 43.74\% > 10\%$, so A10 is selected.

- **A12 vs A10:**
  Incremental cash flow is $-300, 100, 100, 100$.
  $IRR_{12-10} = 0\% < 10\%$, so A10 is selected.

Since there are no alternatives left to compare, A10 is the ultimate choice. Compare this result with the result of the PW analysis: The PW criterion also picked A10 at MARR = 10%.

Comments: *As seen above, the use of IRR criterion gives a solution consistent with that of NPW, but creates a considerable computational burden. Therefore, if you are only interested in finding the best alternative, the NPW approach is recommended over IRR analysis.*

■ 13.2 Capital Rationing

Capital rationing refers to situations where the funds available for capital investment are not sufficient to cover all the projects having either positive NPWs or IRRs (or return on invested capital) greater than the MARR. In this situation, we enumerate all investment alternatives as before, but eliminate from consideration any mutually exclusive alternatives exceeding the budget. The most efficient way to proceed in a capital rationing situation is to select the group of projects that maximizes the total NPW of future cash flows over required investment outlays.

Example 13.4 Three projects under budget constraints

Yukon Mining Company is considering the following independent projects:

Year	A	B	C
0	−$1000	−$4000	−$6000
1	800	2000	3000
2	800	2000	3000
3	800	2000	3000
PW(15%)	$827	$566	$850

Suppose the firm has only $7000 to spend initially. Formulate all mutually exclusive investment alternatives under this budget constraint; indicate the feasible and the optimal investment(s).

Solution

The possible decisions are

j	(X_A, X_B, X_C)	Funds Required	Budget Remaining	Total NPW
1	0 0 0	$ 0	$7,000	$ 0
2	1 0 0	1,000	6,000	827
3	0 1 0	4,000	3,000	566
4	1 1 0	5,000	2,000	1,393
5	0 0 1	6,000	1,000	850
6	1 0 1	7,000	0	1,677
7	0 1 1	10,000	−3,000	1,416
8	1 1 1	11,000	−4,000	2,243

Of the eight possible choices, only the first six would be feasible because of the budget constraint. Further, A6 would be the best combination of projects because of its maximum NPW.

Example 13.5 Four projects under budget constraint

The facilities department at Manitoba Telephone had under consideration four energy-efficiency projects.

- **Project 1 (electrical):** This project requires replacing the existing standard efficiency motors in the air conditioners and exhaust blowers of a particular building with high-efficiency motors.

- **Project 2 (building envelope):** This project involves coating the inside surface of existing fenestration in a building with low-emissivity solar film.

- **Project 3 (air conditioning):** This project requires the installation of heat exchangers between a building's existing ventilation and relief air ducts.

- **Project 4 (lighting):** This project requires the installation of specular reflectors and delamping of a building's existing ceiling grid-lighting troffers.

These projects require capital outlays in the $25,000 to $50,000 range and have useful lives of about 8 years. The facilities department's first task was to estimate the annual savings that can be realized by these energy-efficiency projects. Currently, the company pays 7.68 cents per kilowatt-hour (kWh) for electricity and $150 per thousand cubic metres (m^3). Assuming that the current energy prices would continue for the next 8 years, the company has estimated the cash flows and the NPW (at 15%) for each project as follows:

Project	Investment	Annual O&M Cost	Annual Savings (Energy)	Annual Savings (Dollars)	NPW
1	$15,600	$400	51,120 kWh	$ 3,926	$ 222
2	34,950	350	173,700 kWh	13,340	23,340
3	45,160	450	72,207 m^3	10,831	1,423
4	31,410	314	130,666 kWh	10,035	12,211

As each project could be adopted by itself in isolation, there were at least as many alternatives as there were projects. With $100,000 approved for energy improvement funds during the current fiscal year, the department did not have sufficient capital on hand to undertake all four projects without any additional allocation from headquarters. Enumerate the total number of decision alternatives and select the best decision alternative.

Solution

Since the NPW for each project is greater than zero, all four projects pass the initial project screening. With four projects, we will have 16 decision alternatives. However, the department cannot select alternatives 15 or 16 because both alternatives exceed the budget. Alternatives 13 and 14 require more investment than does alternative 12, but their NPWs are less than that of alternative 12. Therefore, alternative 12 is the best decision alternative.

j	(X_1, X_2, X_3, X_4)	Funds Required	Total NPW
1	0 0 0 0	$ 0	$ 0
2	1 0 0 0	15,600	222
3	0 0 0 1	31,410	12,211
4	0 1 0 0	34,950	23,340
5	0 0 1 0	45,160	1,423
6	1 0 0 1	47,010	12,433
7	1 1 0 0	50,550	23,562
8	1 0 1 0	60,760	1,645
9	0 1 0 1	66,360	35,551
10	0 0 1 1	76,570	13,634
11	0 1 1 0	80,110	24,763
12	1 1 0 1	81,960	35,773
13	1 0 1 1	92,170	13,856
14	1 1 1 0	95,710	24,985
15	0 1 1 1	111,520	36,974
16	1 1 1 1	127,120	37,196

■ 13.3 Choice of Minimum Attractive Rate of Return

Thus far, we have said little about what interest rate, or MARR, is suitable for use in a particular investment situation. Choosing the MARR is indeed a difficult problem; there is no single rate that is always appropriate. In this section, we shall discuss briefly how to select a MARR for project evaluation. Our first topic will be the logic of the cost of capital. Next, we shall consider the costs of the major types of capital. Then, we shall see how the costs of the individual components of the capital structure are brought together to form a weighted average cost of capital. Finally, we shall examine the relationship between capital budgeting and the cost of capital.

13.3.1 Logic of Tax-Adjusted Weighted-Average Cost of Capital

In most of the capital budgeting examples in the earlier chapters, we assumed that the firms under consideration were financed entirely with equity funds. In those cases, the cost of capital may represent the firm's required return on equity. However, as discussed in Chapter 11, most firms finance a substantial portion of their capital budget with long-term debt (bonds), and many also use preferred stock as a source of capital. In these cases, a firm's cost of capital must reflect the average cost of the various sources of long-term funds that the firm uses, and not just the cost of equity. In this section, we shall discuss the ways in which the cost of each individual type of financing (retained earnings, common stock, preferred stock, and debt) can be estimated, given the firm's target capital structure.

Cost of Equity (i_e)

Whereas debt and preferred stocks are contractual obligations that have easily determined costs, it is not easy to measure the cost of equity. In principle, the cost of equity capital involves the **opportunity cost.** In fact, the firm's after-tax cash flows belong to the stockholders. Management may either pay out the earnings in the form of dividends, or retain earnings and reinvest them in the business. If management decides to retain earnings, there is an opportunity cost involved—stockholders could have received the earnings as dividends and invested this money in other financial assets. Therefore, the firm should earn on its retained earnings at least as much as the stockholders themselves could earn in alternative but comparable investments.

What rate of return can stockholders expect to earn on these retained earnings? This question is difficult to answer, but the value sought is often regarded as the rate of return stockholders require on the firm's common stock. If the firm cannot invest retained earnings so as to earn at least the rate of return on equity, it should pay these funds to these stockholders and let them invest directly in other assets that do provide this return.

When investors are contemplating buying a firm's stock, they have two things in mind; (1) cash dividends and (2) gains (share appreciation) at the time of sale. From a conceptual standpoint, they determine market values of stocks by discounting expected future dividends at a rate that takes into account any future growth. Since investors seek growth companies, a desired growth factor for future dividends is usually included in the calculation.

To illustrate, let's take a simple numerical example. Suppose investors in the common stock of ABC Corporation expect to receive a dividend of $5 by the end of the first year. The future annual dividends will grow at an annual rate of 10%. They will hold the stock for two more years, and expect the market price of the stock will rise to $120 by the end of the third year. Given these hypothetical expectations, ABC expects that investors would be willing to pay $100 for this stock in today's market. What is the required rate of return on ABC's common stock (k_r)? We may answer this question by solving the following equation for k_r.

$$\$100 = \frac{\$5}{(1 + k_r)} + \frac{5(1 + 0.1)}{(1 + k_r)^2} + \frac{5(1 + 0.1)^2 + 120}{(1 + k_r)^3}$$

In this case, $k_r = 11.44\%$. This implies that, if ABC finances a project by retaining its earnings or by issuing additional common stock at the going market price of $100 per share, it must realize at least 11.44% on new investment just to provide the minimum rate of return required by the investors. Therefore, 11.44% is the specific cost of equity that should be used when calculating the weighted average cost of capital. If there are flotation costs in issuing a new stock, the cost of equity will increase. If investors view ABC's stock as risky and, therefore, are willing to buy the stock at a lower price than $100 (but with the same expectations), the cost of equity will also increase. Now we can generalize the result as follows.

Cost of Retained Earnings (k_r): Let's assume the same hypothetical situation for ABC. Recall that ABC's retained earnings belong to holders of its common stock. If ABC's current stock is traded for a market price of P_0, with the first-year dividend of $D°$ but growing at the annual rate of g thereafter, the specific cost of retained earnings for an infinite period of holding (stocks will change hands over the years but it does not matter who holds the stock) can be calculated as

$$P_0 = \frac{D°}{(1 + k_r)} + \frac{D°(1 + g)}{(1 + k_r)^2} + \frac{D°(1 + g)^2}{(1 + k_r)^3} + \cdots$$

$$= \frac{D°}{1 + k_r} \sum_{n=0}^{\infty} \left[\frac{(1 + g)}{(1 + k_r)} \right]^n$$

$$= \frac{D°}{1 + k_r} \left[\frac{1}{1 - \dfrac{1 + g}{1 + k_r}} \right].$$

Solving for k_r, we obtain

$$k_r = \frac{D°}{P_0} + g. \tag{13.1}$$

If we use k_r as the discount rate for evaluating the new project, it will have a positive NPW only if the project's IRR exceeds k_r. Therefore, any project with a positive NPW, calculated at k_r, induces a rise in the market price of the stock. Hence, by definition, k_r is the rate of return required by shareholders and should be used as the cost of the equity component when calculating the weighted average cost of capital.

Issuing New Common Stock (k_e): If there are flotation costs in issuing a new stock, we can modify the cost of retained earnings (k_r) by

$$k_e = \frac{D°}{P_0(1 - f_c)} + g, \tag{13.2}$$

where k_e = the cost of common equity, and f_c = the flotation cost as a percentage of stock price.

Either calculation is deceptively simple because in fact, there are several ways to determine the cost of equity. In reality, the market price fluctuates constantly, as do the firm's future earnings. Thus, future dividends may not grow at a constant rate as the model indicates. For a stable corporation with moderate growth, however, the cost of equity as calculated by evaluating either Eq. (13.1) or Eq. (13.2) can serve as a good approximation.

Cost of Preferred Stock (k_p): A preferred stock is a hybrid security in the sense that it has some of the properties of bonds and others that are similar to common stock. Like bondholders, holders of preferred stock receive a fixed annual dividend, in fact, many firms view the payment of the preferred dividend as an obligation just like interest payments to bondholders. It is therefore relatively easy to determine the cost of preferred stock. For the purposes of calculating the weighted average cost of capital, the specific cost of a preferred stock will be defined as

$$k_p = \frac{D^\circ}{P^\circ},$$ (13.3)

where D° = the fixed annual dividend, and P° = the net issuing price after deducting the flotation cost.

Cost of Equity (i_e): Once we have determined the specific cost of each equity component, we can determine the weighted average cost of equity (i_e) for a new project.

$$i_e = ak_r + bk_e + (1 - a - b)k_p,$$ (13.4)

where a = a fraction of total equity financed from retained earnings, b = a fraction of total equity financed from issuing new stock. Then $(1 - a - b)$ represents the fraction of equity financed from issuing preferred stock.

Example 13.6   Determining cost of equity

Alpha Corporation needs to raise $10 million for plant modernization. Alpha's target capital structure calls for a debt ratio of 0.4, indicating that $6 million has to be financed from equity.

• Alpha is planning to raise $6 million from the following equity sources:

Source	Amount	Fraction of Total Equity
Retained earnings	$1 million	0.167
New common stock	4 million	0.667
Preferred stock	1 million	0.167

- Alpha's current common stock price is $40—the market price that reflects the firm's future plant modernization. Alpha is planning to pay an annual cash dividend of $5 at the end of first year, and the annual cash dividend will grow at an annual rate of 8% thereafter.

- Additional common stock can be sold at the same price of $40, but there will be 12.4% flotation costs.

- Alpha can issue $100 par preferred stock with a 9% dividend. (This means that Alpha will calculate the dividend based on the par value, which is $9 per share.) The stock can be sold on the market for $95, and Alpha must pay flotation costs of 6% of the market price.

Determine the cost of equity to finance the plant modernization.

Solution

We will itemize the cost of each equity component.

- **Cost of retained earnings:** With $D° = \$5$, $g = 8\%$, and $P_0 = \$40$,

$$k_r = \frac{\$5}{\$40} + 0.08 = 20.5\%.$$

- **Cost of new common stock:** With $D° = \$5$, $g = 8\%$, $f_c = 12.4\%$,

$$k_e = \frac{\$5}{\$40(1 - 0.124)} + 0.08 = 22.27\%.$$

- **Cost of preferred stock:** With $D° = \$9$, $P° = \$95(1 - 0.06)$,

$$k_p = \frac{\$9}{\$95(1 - 0.06)} = 10.08\%.$$

- **Cost of equity:** With $a = 0.167$, $b = 0.667$, and $(1 - a - b) = 0.167$,

$$i_e = (0.167)(0.205) + (0.667)(0.2227) + (0.167)(0.1008)$$

$$= 19.96\%.$$

Cost of Debt

Now let us consider the calculation of the specific cost that is to be assigned to the debt component of the weighted average cost of capital; it is relatively straightforward and simple. As we discussed in Chapter 11, there are two types of debt financing: term loans and bonds. Because the interest payments on both are tax deductible, the effective cost of debt will be reduced. To determine the after-tax cost of debt (i_d), we can evaluate the following expression:

$$i_d = ck_s(1 - t_m) + (1 - c)k_b(1 - t_m), \tag{13.5}$$

where c = the fraction of the term loan over the total debt, k_s = the before-tax interest rate on the term loan, t = the firm's marginal tax rate, and k_b = the before-tax interest on the bond.

As for bonds, a new issue of long-term bonds will incur flotation costs. These costs reduce the proceeds to the firm, thereby raising the specific cost of the capital raised. For example, when a firm issues a $1000 par bond but nets only $940, the flotation cost will be 6%. Therefore, the effective after-tax cost of the bond component will be higher than the nominal interest rate specified on the bond. We will examine this problem with a numerical example.

Example 13.7 Determining cost of debt

In Example 13.6, suppose that Alpha has decided to finance the remaining $4 million by securing a term loan and issuing 20-year $1000 par bonds for the following condition.

Source	Amount	Fraction	Interest Rate	Flotation Cost
Term loan	$1 million	0.25	12% per year	
Bonds	$3 million	0.75	10% per year	6%

If the bond can be sold to net $940 (after deducting the 6% flotation cost), determine the cost of debt to raise $4 million for the plant modernization. Alpha's marginal tax rate is 38% and is expected to be remain constant in the future.

Solution

First, we need to find the effective after-tax cost of issuing the bond with a flotation cost of 6%. The before-tax specific cost is found by solving the following equivalence formula (see Section 3.6.2).

$$\$940 = \frac{\$100}{(1 + k_d)} + \frac{\$100}{(1 + k_d)^2} + \cdots + \frac{\$100 + \$1000}{(1 + k_d)^{20}}$$

Solving for k_d, we obtain k_d = 10.74%. Note that the cost of the bond component increases from 10% to 10.74% after considering the 6% flotation cost.

The after-tax cost of debt is the interest rate on debt, multiplied by $(1 - t_m)$. In effect, the government pays part of the cost of debt because interest is tax deductible. Now we are ready to compute the after-tax cost of debt as follows.

$$i_d = (0.25)(0.12)(1 - 0.38) + (0.75)(0.1074)(1 - 0.38)$$

$$= 6.85\%$$

Tax-Adjusted Weighted-Average Cost of Capital

Now we have determined the specific cost of each financing component. Assuming that a firm raises capital based on the target capital structure and that the target capital structure remains unchanged in the future, we can determine a **tax-adjusted weighted-average cost of capital** (or, simply stated, the **cost of capital**). This cost of capital represents a composite index reflecting the cost of raising funds from different sources. The cost of capital is defined as

$$k = \frac{i_d D}{V} + \frac{i_e E}{V}, \qquad (13.6)$$

where

D = total debt capital (such as bonds) in dollars,

E = total equity capital in dollars,

$V = D + E$,

i_e = average equity interest rate per period considering all equity sources,

i_d = after-tax average borrowing interest rate period considering all debt sources, and

k = tax-adjusted weighted-average cost of capital.

Note that the cost of equity is already expressed in terms of after-tax cost, because any return to holders of either common stock or preferred stock is made after payment of income taxes. The following example works through the computations for finding the cost of capital (k).

Example 13.8 Calculating cost of capital

Reconsider Examples 13.6 and 13.7. The marginal income tax rate (t) for Alpha is expected to remain at 38% in the future. Assuming that Alpha's capital structure (debt ratio) remains also unchanged in the future, determine the cost of capital (k).

Solution

With D = $4 million, E = $6 million, V = $10 million, i_d = 6.92%, i_e = 19.96%, and Eq. (13.6), we calculate

$$k = \frac{(0.0685)(4)}{10} + \frac{(0.1996)(6)}{10}$$

$$= 14.71\%$$

Comments: *Note that the cost of debt is the interest rate on new debt, not that on outstanding debt; in other words, we are interested in the marginal cost of debt. Our primary concern with the cost of capital is to use it in evaluating a new investment project. The rate at which the firm has borrowed in the past is less important for this purpose. This 14.74% would be the cost of capital that a company with this financial structure would expect to pay to its investors.*

13.3.2 Choice of MARR When Project Financing Is Known

In Chapters 9 and 11, we focused on calculating after-tax cash flows, including situations involving debt financing. When the cash flow computations reflect interest, taxes, and debt repayment, what is left is called **net equity flow.** If the goal of the firm is to maximize the wealth of the stockholders, why not focus only on the after-tax cash flow to equity, instead of on the flow to all suppliers of capital? Focusing on only the equity flows will permit us to use the cost of equity as the appropriate discount rate. In fact, we have implicitly assumed that all after-tax cash flow problems in this book represent net equity flows, so the MARR used represents the **cost of equity** (i_e).

Example 13.9   Project evaluation by net equity flow

Suppose the Alpha Corporation, which has the capital structure described in Example 13.6, wishes to install a new set of machine tools, which is expected to increase revenues over the next 5 years. The tools require an investment of $150,000, to be financed with 60% equity and 40% debt. The equity interest rate (i_e) that combines both sources of common and preferred stocks is 19.96%. Alpha will use a 12% short-term loan to finance the debt portion of the capital ($60,000), with the loan to be repaid in equal annual installments over 5 years. CCA is claimable at a declining balance rate of 30% and the half-year rule applies. Salvage value is expected to be zero. Additional revenues and operating costs are expected to be

n	Revenues	Operating Cost
1	$68,000	$20,500
2	73,000	20,000
3	79,000	20,500
4	84,000	20,000
5	90,000	20,500

The marginal tax rate (combined federal and provincial rate) is 38%. Evaluate this venture by using net equity flows.

Table 13.4 After-Tax Cash Flow Analysis When Project Financing Is Known: Net Equity Cash Flow Method (Example 13.9)

	0	1	2	3	4	5
				n		
Income Statement						
Revenues		$68,000	$73,000	$79,000	$84,000	$90,000
Expenses						
Operating cost		20,500	20,000	20,500	20,000	20,500
Interest payment		7,200	6,067	4,797	3,376	1,783
CCA		22,500	38,250	26,775	18,743	13,120
Taxable income		17,800	8,683	26,928	41,881	54,597
Income tax (38%)		6,764	3,300	10,233	15,915	20,747
Net income		$11,036	$5,383	$16,695	$25,966	$33,850
Cash Flow Statement						
Cash from operation						
Net income		11,036	$5,383	$16,695	$25,966	$33,850
CCA		22,500	38,250	26,775	18,743	13,120
Investment and salvage	−$150,000					0
Disposal tax effects						11,633
Loan principal repayment	+60,000	−9,445	−10,578	−11,847	−13,269	−14,861
Net cash flow	−$90,000	$24,091	$33,055	$31,623	$31,440	$43,742

$i_e = 19.96\%$:

$$PW(19.96\%) = -\$90,000 + \$24,091(P/F, 19.96\%, 1) + \$33,055(P/F, 19.96\%, 2)$$
$$+ \$31,623(P/F, 19.96\%, 3) + \$31,440(P/F, 19.96\%, 4)$$
$$+ \$43,742(P/F, 19.96\%, 5)$$
$$= \$4,163$$

$$IRR = 21.88\% > 19.96\%$$

$$\text{Book value at disposal} = \$150,000(0.85)(0.7)^4$$
$$= \$30,613$$

$$G = (\$30,613 - 0)0.38$$
$$= \$11,633$$

Solution

The calculations are shown in Table 13.4, following the principles given in Chapter 11. The internal rate of return for this cash flow is 21.88%, which exceeds $i_e = 19.96\%$. Thus, the project would be profitable.

13.3.3 Choice of MARR When Project Financing Is Unknown

If we are to use exclusively net equity flows using i_e, what is the use of the k? The answer is that we may evaluate investments without explicitly treating the debt flows (both interest and principal) by using the value of k. In this case, we make the tax adjustment to the discount rate by employing the effective after-tax cost of debt. This approach recognizes that the net interest cost is effectively transferred from the tax collector to the creditor in the sense that there is a dollar-for-dollar reduction in taxes up to this amount of interest payments. Therefore, the debt financing is treated implicitly. This method would be appropriate when debt financing is not identified with individual investments, but rather enables the company to engage in a set of investments. (All previous examples in this book have implicitly assumed net equity cash flows with known financing sources. Therefore, the MARRs represented the cost of equity (i_e), not the cost of capital (k).)

Example 13.10 Project evaluation by cost of capital

In Example 13.9, suppose that Alpha Corporation has not decided how the $150,000 will be financed. However, Alpha believes that the project will be financed according to its target capital structure. Evaluate Example 13.9 using k.

Solution

By not accounting for the cash flows related to debt financing, we calculate the after-tax cash flows as shown in Table 13.5. Notice that, when we use this procedure, interest and the resulting tax shield are ignored in deriving the net incremental after-tax cash flow. In other words, there is no cash flow related to financing activity. Thus, taxable income is overstated, and then income taxes are also overstated. To compensate for this overstatement, the discount rate is reduced accordingly. The implicit assumption is that the tax overpayment is exactly equal to the reduction in interest implied by i_d.

The time 0 flow is simply the total investment—$150,000 in this example. Recall that Alpha's k was calculated to be 14.71%. The internal rate of return for the after-tax flow in the last line of Table 13.5 is 16.54%, which exceeds the value of k. Thus, the investment would be profitable. Here, we evaluated the after-tax flow by using the value of k, and we reached the same conclusion about the desirability of the investment.

Comments: *The net equity flow and cost of capital methods usually will come to the same accept/reject decision for independent projects (assuming the amortization schedule for debt repayment, such as term loans), and usually will rank projects identically for mutually exclusive alternatives. Any differences that arise are likely to be insignificant. For example, special financing arrangements may increase (or even decrease) the attractiveness of a project by manipulating tax shields and the timing of financing inflows and payments.*

Table 13.5 After-Tax Cash Flow Analysis When Project Financing Is Unknown (Example 13.10)

	0	1	2	3	4	5
Income Statement						
Revenues		$68,000	$73,000	$79,000	$84,000	$90,000
Expenses						
Operating cost		20,500	20,000	20,500	20,000	20,500
CCA		22,500	38,250	26,775	18,743	13,120
Taxable income		25,000	14,750	31,725	45,257	56,380
Income tax (38%)		9,500	5,605	12,055	17,198	21,424
Net income		$15,500	$9,145	$19,670	$28,059	$34,956
Cash Flow Statement						
Cash from operation						
Net income		$15,500	$9,145	$19,670	$28,059	$34,956
CCA		22,500	38,250	26,775	18,743	13,120
Investment and salvage	−$150,000					0
Disposal tax effects						11,663
Net cash flow	−$150,000	$38,000	$47,395	$46,445	$46,802	$59,709

$k = 14.71\%$:

$$PW(14.71\%) = -\$150,000 + \$38,000(P/F, 14.71\%, 1) + \$47,395(P/F, 14.71\%, 2)$$
$$+ \$46,445(P/F, 14.71\%, 3) + \$46,802(P/F, 14.71\%, 4)$$
$$+ \$59,709(P/F, 14.71\%, 5)$$
$$= \$7,010$$
$$IRR = 16.54\% > 14.71\%$$

In summary, in cases where the exact debt-financing and repayment schedules are known, we recommend the use of the net equity flow method. The proper MARR would be the cost of equity, i_e. If no specific assumption is made about the exact instruments that will be used to finance a particular project (but we do assume that the given capital structure proportions will be maintained), we may determine the after-tax cash flows without incorporating any debt cash flows. Then, we use the cost of capital (k) as the proper MARR.

13.3.4 Effects of Capital Rationing on the Choice of MARR

It is important to distinguish between the cost of capital (k), as calculated in Section 13.3.1 and the MARR (i) used in project evaluation under capital rationing. When investment opportunities exceed the available money supply, we must decide which opportunities are preferable. Obviously, we want to ensure that all the selected projects are more profitable than the best rejected project. The best rejected project is the best opportunity foregone, and its value is called the **opportunity cost.** When there is a limit on capital, the MARR is assumed to be equal to this opportunity cost, which is greater than the cost of capital. In other words, the value of i represents the corporation's time-value trade-offs and reflects partially the investment opportunities available. Thus, there is nothing illogical about borrowing money at k and evaluating cash flows using the different rate, i. Presumably, the money will be invested to earn a rate i or greater. In the following example, we shall provide guidelines for selecting a MARR for project evaluation under capital rationing.

A company may borrow funds to invest in profitable projects, or it may return (invest) to its **investment pool** any unused funds until they are needed for other investment activities. (Here, we may view the borrowing rate as a cost of capital (k), as defined in Eq. 13.6.) Suppose that all available funds can be placed in investments yielding a return equal to l, the **lending rate.** We view these funds as an investment pool. The firm may withdraw funds from this investment pool for other investment purposes, but if left in the pool, the funds will earn at the rate of l (which is thus the opportunity cost). The MARR is thus related to either the borrowing interest rate or the lending interest rate. To illustrate the relationship among the borrowing interest rate, lending interest rate, and MARR, let us define the following:

$$k = \text{the borrowing rate (or the cost of capital)}$$

$$l = \text{the lending rate (or opportunity cost)}$$

$$i = \text{the MARR.}$$

Generally (but not always), we might expect k to be greater than or equal to l. We must pay more for the use of someone else's funds than we can receive for "renting out" our own funds (unless we are running a lending institution).

Example 13.11   Selecting MARR

Consider a firm faced with the following investment opportunities. For simplicity, all projects last only 1 year:

	Cash Flow		
Project	A_0	A_1	IRR
1	−$10,000	$12,000	20%
2	−10,000	11,500	15
3	−10,000	11,000	10
4	−10,000	10,800	8
5	−10,000	10,600	6
6	−10,000	10,400	4

Suppose $k = 10\%$ and $l = 6\%$. Assuming that the firm has available (a) $40,000, (b) $60,000, and (c) $0 for investments, what is the reasonable choice for the MARR in each case?

Solution

(a) If the firm has $40,000 available for investing, it should invest in projects 1, 2, 3, and 4. Clearly, it should not borrow at 10% to invest in either project 5 or 6. In these cases, $l < \text{MARR} < k$.

(b) If the firm has $60,000 available, it should invest in projects 1, 2, 3, 4, and 5 (although it would probably see no difference between project 5 and lending). It would lend the remaining $10,000 rather than invest it in project 6. For this new situation, we have $\text{MARR} = l$.

(c) If the firm has no funds available, it would probably borrow to invest in projects 1 and 2. It might also borrow to invest in project 3, but it would be indifferent to that alternative, unless some other consideration was involved. In this case, $\text{MARR} = k$; therefore, we can say that $l \leq \text{MARR} \leq k$ when we have a complete certainty about future investment opportunities. Figure 13.3 illustrates the concept of selecting a MARR under capital rationing.

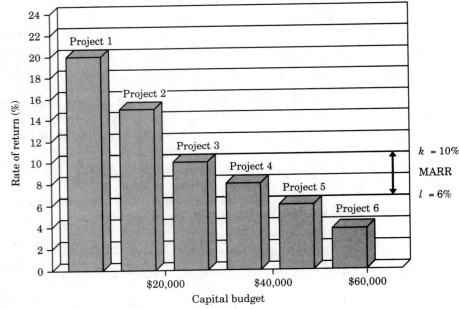

Figure 13.3 ■ A choice of MARR under capital rationing (Example 13.11)

Now we can generalize what we have learned. If a firm finances investments through borrowed funds, it should use the MARR = k; if the firm is a lender, it should use MARR $= l$. A firm may be a lender in one period and a borrower in another; consequently, the appropriate rate to use may vary from period to period. In fact, whether the firm is a borrower or a lender may well depend on its investment decisions.

In practice, most firms establish a MARR for all investment projects. Note the assumption that we made in Example 13.11: *Complete certainty* about investment opportunities. Generally, the MARR will be much greater than k, the firm's cost of capital. For example, if $k = 10\%$, a MARR of 15% would not be considered excessive. Few firms are willing to invest in projects earning only slightly more than their cost of capital due to elements of *risk* in the project.

If the firm has a large number of current and future opportunities that will yield the desired return, we can then view the MARR as the minimum rate at which the firm is willing to invest, and we can also assume that proceeds from current investments can be reinvested to earn at the MARR. Furthermore, *if we choose the "do-nothing" alternative, all available funds are invested at the MARR.* In engineering economics, we also normally separate the risk issue from our concept of MARR. As we shall show in Chapter 15, we do treat the effects of risk explicitly when we must. Therefore, any future reference to the MARR in this book will refer strictly to the risk-free interest rate.

Summary

- When dealing with multiple investment alternatives with various interdependencies, we need to organize them into a set of mutually exclusive projects that cover all feasible investment combinations.

- **Independent projects** are those for which acceptance or rejection of one project does not affect acceptance or rejection of the other(s). If only independent projects are being considered, the maximum number of investment alternatives is 2^x where $x =$ the number of independent projects.

- **Dependent projects** are those for which acceptance or rejection of one project affects acceptance or rejection of the other(s). Dependent projects can be:

  1. **Mutually exclusive**—accepting one project requires rejecting the other(s); or

  2. **Contingent**—accepting one project is contingent upon having first accepted another.

- Under conditions of **capital rationing**, insufficient funds are available to accept all the projects that pass the NPW or incremental IRR tests. In this situation, we should select the project (or combination of projects) that maximizes NPW while remaining within budget constraints.

- The selection of an appropriate MARR depends generally upon the **cost of capital**— the rate the firm must pay to various sources for the use of capital.

  1. The **cost of equity** (i_e) is used when we know the debt-financing methods and repayment schedules.

  2. The **tax-adjusted weighted-average cost of capital** (k) is used when exact financing methods are unknown but the firm keeps its capital structure on target. In this situation, the project's after-tax cash flows contain no debt cash flows.

- The tax-adjusted weighted-average cost of the capital formula is a composite index reflecting the cost of funds raised from different sources. The formula is:

$$k = \frac{i_d D}{V} + \frac{i_e E}{V}.$$

- Under conditions of capital rationing, the selection of MARR is more difficult, but generally the following possibilities exist:

Conditions	MARR
The firm borrows some capital from lending institutions at the borrowing rate, k, and some from its investment pool at the lending rate, l	$l < \text{MARR} < k$
The firm borrows all capital from lending institutions at the borrowing rate, k	$\text{MARR} = k$
The firm borrows all capital from its investment pool, at the lending rate, l	$\text{MARR} = l$

Problems

13.1 Consider the 5 investment projects with project interdependencies as outlined below:

- (A1 and A2) are mutually exclusive.
- (B1 and B2) are mutually exclusive and either project is contingent on the acceptance of A2.
- C is contingent on the acceptance of B1.

Which of the following statements is (are) correct?

(a) There are total of 5 feasible decision alternatives.
(b) (A1, B2) is feasible.
(c) (A2, B1) and (A2, B2) are both feasible.
(d) (B1, C) is feasible.
(e) (B1, B2, C) is feasible.
(f) (A1, B1, C) is feasible.
(g) All of the above.
(h) None of the above.

13.2 Four investment proposals have been submitted by the Corporate Planning Committee. The cash flow profiles for the 4 investment proposals are summarized as follows. Each project has a 5-year service life. The firm's MARR is 10%. (All cash flows are given in after-tax figures.)

	Project Profiles			
	A	B	C	D
Initial investment	$60,000	$40,000	$80,000	100,000
Annual net operating cash flow	20,000	10,000	15,000	25,000
Salvage value	0	20,000	20,000	30,000

(a) Suppose that the 4 projects are mutually independent and there is no limit on the budget. Formulate all possible decision alternatives along with the composite cash flows.

(b) Suppose that projects A and B are mutually exclusive. Formulate all possible decision alternatives.

13.3 Consider the following set of investment projects.

Project	Initial Investment	PW(10%)
A	$100	55
B	200	63
C	400	95
D	300	70
E	150	60
F	250	75

- Projects A and E are mutually exclusive.
- Project B is contingent upon project A.
- Project D is contingent upon project E.
- Project C is contingent upon project A.
- Project F is an independent project.
- Projects B and D are mutually exclusive.

(a) Formulate all mutually exclusive alternatives.
(b) With no restriction on budget, which alternative is the best?

13.4 Consider the following set of investment projects.

Project	Initial Investment	PW(10%)
A	$400	65
B	550	70
C	620	95
D	580	75
E	380	60
F	600	80

- Projects A and B are mutually exclusive.
- Project C is contingent upon project A.
- Project D is contingent upon project C.
- Project F is contingent upon project B.
- Project E is contingent upon project F.

(a) Draw a diagram that represents the interrelationships among the projects.
(b) With no capital limit, list all combinations of decision alternatives and then find the best decision alternative based on the present worth criterion.

13.5 Consider the following 5 mutually exclusive investment projects each having project life of 15 years.

After-Tax Cash Flows

Project	n = 0	n = 1 to 15	IRR
A	−$335	$ 50	12.40%
B	−500	110	20.69
C	−725	149	19.05
D	−885	170	17.50
E	−940	184	17.92

(a) Using the net present worth criterion, which project would be selected at $i = 15\%$?
(b) Using the rate of return on incremental investment, which project would be selected?

13.6 A construction firm is considering the following mutually exclusive alternatives.

Project Cash Flows

n	A1	A2	A3	A4
0	−$2000	−$1000	−$3000	−$4000
1	1000	500	1000	1000
2	1000	500	2000	2000
3	1000	1000	1500	3000

(a) Assuming no capital limit, select the best project based on the NPW criterion using MARR = 10%.

(b) Select the best project based on the rate of return on incremental investment at an interest rate of 10%.

13.7 Consider the following set of investment projects.

Project	Initial Investment	PW(10%)
A	$500	570
B	700	630
C	900	980
D	650	770
E	600	650
F	750	820

- Projects A and E are mutually exclusive.
- Project B is contingent upon project F.
- Project D is contingent upon project E.
- Project C is contingent upon project B.
- Projects B and D are mutually exclusive.

(a) With no capital limit, list all combinations of decision alternatives.

(b) Find the best decision alternative based on the present worth criterion.

13.8 Consider the following 6 investment projects.

Project Cash Flows

n	A	B	C	D	E	F
0	-$100	-$100	-$100	-$100	-$300	-$350
1	-50	0	30	40	-200	300
2	50	0	50	80	250	350
3	50	200	80	30	500	100

(a) Plot the PW(i) of each project on the same chart as a function of interest rate between 0 and 40%.

(b) If all projects are mutually exclusive, which project would be selected using the present worth criterion? Assume $i = 15\%$.

(c) In (b), using the rate of return criterion, which project would be selected? Assume $i = 15\%$.

13.9 Consider the following 6 mutually exclusive investment projects. Assume that MARR = 15%.

Project Cash Flows

n	A	B	C	D	E	F
0	-$100	-$200	-$4000	-$2000	-$2000	-$3000
1	60	120	2410	1310	-1400	3700
2	50	150	2930	1720	3640	1000
3	50					
IRR (%)	28.89	21.65	20.86	31.10	☐	☐

(a) Compute the rate of return for projects E and F.

(b) Under infinite planning horizon with project repeatability likely, which project would you select based on the rate of return principle?

13.10 Consider the following 4 investment projects. Assume that $i = 15\%$.

Project Cash Flows

n	A	B	C	D
0	-$100	-$200	-$300	-$ 800
1	60	120	0	1150
2	30	100	690	30
3	50			
IRR	19.84%	6.81%	51.66%	46.31%

(a) If projects A, B, and C are mutually exclusive, which project would be selected under the infinite planning horizon with project repeatability likely?

(b) Suppose project D is contingent on project B. Using the rate of return criterion, which project(s) would be selected?

13.11 Six cost-reduction independent proposals have been forwarded by the industrial engineering department. A 12% MARR is expected. Using the facts below, which plan, if any, should you accept to get the largest net present worth? Salvage values and disposal tax effects are assumed to be zero after a ten-year service life.

Proposal	Investment	Annual Savings	Life
I	$14,000	$1,800	10 yrs.
II	17,000	2,100	12
III	12,000	1,400	10
IV	10,000	1,300	10
V	20,000	2,500	12
VI	16,000	1,900	15

13.12 Six mutually exclusive projects have been proposed for cost reduction in a chemical processing plant. All have lives of ten years, zero salvage value, and no disposal tax effects. The required investment and the estimated after-tax reduction in annual disbursements are given for each alternative.

A_j	Required Investment	Annual Savings	IRR
A1	$ 60,000	$22,000	35%
A2	100,000	28,200	25.2
A3	110,000	32,600	27
A4	120,000	33,600	25
A5	140,000	38,400	24
A6	150,000	42,200	25.1

(a) Which project would you select based on rate of return on incremental investment if it is stated that the MARR is 15% after taxes? Assume that there is no budget limitation. Show any additional calculations needed to support your answer.

(b) Which project would be selected using the present worth on total investment approach?

13.13 DNA Corporation, a biotech engineering firm, has identified 7 R&D projects for funding. Each project will last 3 to 5 years in R&D and the benefit figure represents the royalty income in present worth from selling the R&D results to pharmaceutical companies.

Project Type	Investment Required	Net Benefits (in Present Worth)
1. Vaccines	$15 million	$22 million
2. Carbohydrate chemistry	25 million	40 million
3. Antisense	35 million	80 million
4. Chemical synthesis	5 million	15 million
5. Antibodies	60 million	90 million
6. Peptide chemistry	23 million	30 million
7. Cell transplant/ gene therapy	19 million	32 million

DNA Corporation can only raise $100 million. Because of lack of researchers in chemistry areas, projects 2 and 6 cannot be funded simultaneously. Which R&D projects should be funded?

13.14 Eight investment proposals have been identified for cost reduction in a certain metal fabricating plant. All have lives of ten years, zero salvage values, and no disposal tax effects. The required investment and the estimated after-tax reduction in annual disbursements are given for each proposal. Gross rates of return are also shown for each proposal: Proposals A1, A2, A3, and A4 are mutually exclusive because they are alternative ways of changing operation A. Similarly B1, B2, and B3 are mutually exclusive.

Proposal	Required Investment	Annual Savings	Rate of Return
A1	$20,000	$ 5,600	25%
A2	30,000	6,900	19
A3	40,000	8,180	16
A4	50,000	10,070	15
B1	10,000	1,360	6
B2	20,000	3,840	14
B3	30,000	5,200	12
C	20,000	5,440	24

(a) Which proposals would you select if it is stated that the MARR is 10% after taxes and if there is no limitation on available funds, using the rate of return criterion?

(b) Which proposals would you select at a MARR of 10% using the present worth on total investment approach?

13.15 Ten potential investments in energy conservation have been identified for possible funding. All have lives of 10 years and are assumed to have 100% resale value after 10 years. All investments are made in nondepreciable assets. The required investment, the estimated after-tax savings in annual expense, and the rates of return on total investment are also shown for each proposal. The proposals related to a particular activity are identified by the same letter and they are mutually exclusive.

Proposal	Required Investment	Annual Savings	Rate of Return
A1	$20,000	$5,000	25%
A2	30,000	5,700	19
A3	40,000	8,400	16
A4	50,000	9,300	18.6
B1	10,000	600	6
B2	20,000	2,800	14
B3	30,000	4,800	16
C1	20,000	4,800	24
C2	60,000	10,200	17
D	5,000	700	14

(a) List all possible decision alternatives.

(b) At a MARR of 12%, which alternative would you select?

13.16 Given the following 4 independent sets of proposals (proposals with the same letter within each group are mutually exclusive), a budget of $50,000, and a MARR of 8%, choose the best combination of alternatives, including the do-nothing alternatives.

Alternative	First Cost	PW(8%)
A1	$100,000	$40,000
A2	20,000	−5,000
B1	15,000	8,000
B2	20,000	9,000
B3	25,000	−10,000
C1	40,000	20,000
C2	50,000	15,000
D1	10,000	2,000
D2	20,000	3,000
D3	15,000	4,000
D4	5,000	5,000

13.17 Given the following set of alternatives for 4 independent proposals (the alternatives related to a particular proposal are identified by the same letter and are mutually exclusive), a budget of $60,000, and a MARR of 10%:

Alternative	First Cost	PW(10%)
A1	$50,000	$18,000
A2	20,000	10,000
B1	16,000	12,000
B2	20,000	11,000
C1	40,000	15,000
C2	30,000	10,000
D1	10,000	2,000
D2	20,000	4,000
D3	15,000	3,000

(a) Disregard as many of the activities as possible. Explain why, briefly.
(b) What would be the maximum number of combinations of alternatives to be considered after discarding the ones from part (a)?
(c) Determine the best project combination within the budget constraint.

13.18 Consider the following set of investment projects, each having a service life of 10 years.

Projects	First Cost	Net Annual Revenue	Net Salvage Value
A1	$10,000	$2,000	$ 1,000
A2	12,000	2,100	2,000
B1	20,000	3,100	5,000
B2	30,000	5,000	8,000
C	35,000	4,500	10,000

- A1 and A2 are mutually exclusive.
- B1 and B2 are mutually exclusive.

With a budget limit of $50,000, which investment projects should be selected? Assume that the firm's MARR is 8%.

13.19 XYZ Company is considering the following 4 investment projects.

Project Cash Flows

n	A	B	C	D
0	−$1000	−$1500	−$3000	−$2000
1	500	1000	1300	1000
2	500	200	1500	2000
3	500	1000	1300	5000

Projects A and B are mutually exclusive and project D is contingent on project A. Project C is independent.

(a) With a capital budget of $3000 at $n = 0$ and MARR = 8%, which project(s) should be included in the budget?
(b) How much money would you expect to accumulate at $n = 3$ if you had a budget of $5000 at $n = 0$?

13.20 Gene Fowler owns a house that contains 20.2 square metres of windows and 4.0 square metres of doors. Electricity usage totals 46,502 kilowatt hours: 7960 kWh for lighting and appliances, 5500 kWh for water heating, 30,181 kWh for space heating to 20°C, and 2861 kWh for space cooling to 25°C. The energy-savings alternatives shown in Table 13.6 have been suggested by the local power company for Fowler's 162.0-square-metre home.

(a) If Fowler will stay in the house for next 10 years, which alternatives would be selected with no budget constraint? Assume that his interest rate is 8%. Assume also that all installations would last 10 years. Fowler will be conservative in

Table 13.6 Energy-Saving Alternatives

No.	Structural	Annual Savings ($)	Estimated Costs ($)	Payback Period
1	Add storm windows	128–156	455–556	3.5 yrs
2	Insulate ceilings to R–30	149–182	408–499	2.7
3	Insulate floors to R–11	158–193	327–399	2.1
4	Caulk windows and doors	25–31	100–122	4.0
5	Weatherstrip windows and doors	31–38	224–274	7.2
6	Insulate ducts	184–225	1677–2049	9.1
7	Insulate space heating water pipes	41–61	152–228	3.7
8	Install heat retardants on E, SE, SW, W windows	37–56	304–456	8.2
9	Install heat-reflecting film on E, SE, SW, W windows	21–31	204–306	9.9
10	Install heat-absorbing film on E, SE, SW, W windows	5–8	204–306	39.5
11	Upgrade 6.5 EER A/C to 9.5 EER unit	21–32	772–1158	36.6
12	Install heat pump water-heating system	115–172	680–1020	5.9
13	Install water heater jacket	26–39	32–48	1.2
14	Install clock thermostat to reduce heat from 20°C to 16°C for 8 hours each night	96–144	88–132	1.1

Note: EER (Energy Efficiency Ratio). R-value indicates the degree of resistance to heat. The higher the number, the greater the insulating quality.

calculating the net present worth of each alternative (using the minimum annual savings at the maximum cost). Ignore any tax credits available to energy-saving installations.

(b) If he wants to limit his energy savings investments to $1800, which alternatives should he select?

13.21 Calculate the after-tax cost of debt under each of the following conditions.

(a) Interest rate, 12%; tax rate, 25%.
(b) Interest rate, 14%; tax rate, 34%.
(c) Interest rate, 15%; tax rate, 40%.

13.22 Sweeney Paper Company is planning to sell $10 million worth of long-term bonds with a 10% interest rate. The company believes that it can sell the $1000 par value bonds at a price that will provide a yield to maturity of 13%. The flotation costs will be 1.9%. If Sweeney's marginal tax rate is 38%, what is its after-tax cost of debt?

13.23 Mobil Appliance Company's earnings, dividends, and stock price are expected to grow at annual rate of 9%. Mobil's common stock is currently traded at $18 per share. Mobil's last cash dividend was $1.00, and its expected cash dividend for the end of this year is $1.09. Determine the cost of retained earnings (k_r).

13.24 Refer to Problem 13.23. Mobil wants to raise capital to finance a new project by issuing a new common stock. With the new project, the cash dividend is expected to be $1.10 at the end of the current year, and its growth rate is 10%. The stock now sells for $18, but new common stock can be sold to net Mobil $15 per share.

(a) What is Mobil's flotation cost in percentage?
(b) What is Mobil's cost of new common stock (k_e)?

13.25 The Callaway Company's cost of equity is 18%. Its before-tax cost of debt is 12%, and its marginal tax rate is 40%. The firm's capital structure calls for a debt-to-equity ratio of 50%. Calculate Callaway's after-tax weighted-average cost of capital.

13.26 Delta Chemical Corporation is expected to have the following capital structure for the foreseeable future.

Source of Financing	Percent of Total Funds	Before-Tax Cost	After-Tax Cost
Debt	30%		
Short-term	10	14%	
Long-term	20	12	
Equity	70		
Common stock	55		30%
Preferred stock	15		12

The flotation costs are already included in each cost component. The marginal income tax rate (t) for Delta is expected to remain at 40% in the future. Determine the cost of capital (k).

13.27 Maritime Textile Company is considering the acquisition of a new knitting machine at a cost of $200,000. Because of a rapid change in fashion styles, the need for this particular machine is expected to last only 5 years, after which it is expected to have a salvage value of $50,000. The annual operating cost is estimated at $10,000. The addition of this machine to the current production facility is expected to generate an additional revenue of $90,000 annually. The declining balance CCA rate for this asset is 30%. The income tax rate applicable for Maritime is 36%. The initial investment will be financed with 60% equity and 40% debt. The before-tax debt interest rate that combines both short-term and long-term financing is 12%, with the loan to be repaid in equal annual installments. The equity interest rate (i_e) that combines both sources of common and preferred stocks is 18%.

(a) Evaluate this investment project by using net equity flows.

(b) Evaluate this investment project by using k.

13.28 The Huron Development Company is considering buying an overhead pulley system. The new system has a purchase price of $100,000, an estimated useful life of 5 years, and an estimated salvage value of $30,000. It is expected to economize on electric power usage, labor, and repair costs, as well as to reduce the number of defective products. A total annual savings of $45,000 will be realized if the new pulley system is installed. The company is in the 30% marginal tax bracket, and the system falls into a 30% declining balance CCA class. The initial investment will be financed with 40% equity and 60% debt. The before-tax debt interest rate that combines both short-term and long-term financing is 15%, with the loan to be repaid in equal annual installments over the project life. The equity interest rate (i_e) that combines both sources of common and preferred stocks is 20%.

(a) Evaluate this investment project by using net equity flows.

(b) Evaluate this investment project by using k.

13.29 Consider the following 2 mutually exclusive machines.

	Machine A	Machine B
Initial investment	$40,000	$60,000
Service life	6 years	6 years
Salvage value	$4000	$8000
Annual O&M cost	$8000	$10,000
Annual revenues	$20,000	$28,000
Declining balance CCA rate (half-year rule applicable)	30%	30%

The initial investment will be financed with 70% equity and 30% debt. The before-tax debt interest rate that combines both short-term and long-term financing is 10%, with the loan to be repaid in equal annual installments over the project life. The equity interest rate (i_e) that combines both sources of common and preferred stock is 15%. The firm's marginal income tax rate is 35%.

(a) Compare the alternatives using $i_e = 15\%$. Which alternative should be selected?

(b) Compare the alternatives using k. Which alternative should be selected?

(c) Compare the results obtained in (a) and (b).

13.30 National Food Processing Company is considering investments in plant modernization and plant expansion. All of these proposed projects would be completed in 2 years, with varying requirements of money and plant engineering. Although there is some uncertainty in the data, the management is willing to use the data in Table 3.7 in selecting the best set of proposals.

The resource limitations are:

- First-year expenditures: $450,000
- Second-year expenditures: $420,000
- Engineering hours: 11,000 hours

The situation requires that a new or modernized production line be provided (projects 1 or 2). The numerical control (project 3) is applicable only to the new line. The company obviously does not want to both buy (project 6) and build (project 5) raw material processing facilities; it can, if desirable, rely on the present supplier as an independent firm. Neither the maintenance shop project (project 4) nor the delivery-truck purchase (project 7) is mandatory.

Table 13.7 Plant Modernization and Expansion Alternatives

No.	Project	Investment		Engineering	
		First Year	Second Year	NPW	Hours
1	Modernize production line	$300,000	$ 0	$100,000	4000
2	Build new production line	100,000	300,000	150,000	7000
3	Numerical control for new production line	0	200,000	35,000	2000
4	Modernize maintenance shops	50,000	100,000	75,000	6000
5	Build raw material processing plant	50,000	300,000	125,000	3000
6	Buy present subcontractor's facilities for raw-material processing	200,000	0	60,000	600
7	Buy new fleet of delivery trucks	70,000	10,000	30,000	0

(a) Enumerate all possible mutually exclusive alternatives without considering the budget and engineering-hour constraints.

(b) Identify all feasible mutually exclusive alternatives.

(c) Using the NPW maximization principle, select the best decision alternative.

13.31 Consider the following investment projects:

Project Cash Flows

n	A	B	C	D
0	-$2000	-$3000	-$1000	
1	1000	4000	1400	-$1000
2	1000		-100	1090
3	1000			
i°	23.38%	33.33%	32.45%	9%

Suppose that you have only $3500 available at period 0. Neither additional budgets nor borrowing are allowed in any future budget period. However, you can lend out any left funds (or available funds) at 10% interest per period.

(a) If you want to maximize the future worth at period 3, which projects would you select? What is the future worth (the total amount available for lending at the end of period 3)? No partial projects are allowed.

(b) Suppose in (a) that, at period 0, you are allowed to borrow $500 at an interest rate of 13%. The loan has to be repaid at the end of year 1. Which project would you select to maximize your future worth at period 3?

(c) Considering the lending rate of 10% and the borrowing rate of 13%, what would be the reasonable MARR for project evaluation?

CHAPTER 14

Economic Analyses in the Public Sector

C anadians currently spend millions of hours per year in traffic jams. Most traffic experts believe that adding and enlarging highway systems will not alleviate the problem. As a result, current research on traffic management is focusing on three areas: (1) development of computerized dashboard navigational systems, (2) development of roadside sensors and signals that monitor and help manage the flow of traffic, and (3) development of automated steering and speed controls that might allow cars to drive themselves on certain stretches of highway.

In Toronto, perhaps the most traffic-congested city in Canada, traffic delays cost motorists millions of dollars. But Toronto has already implemented a system of

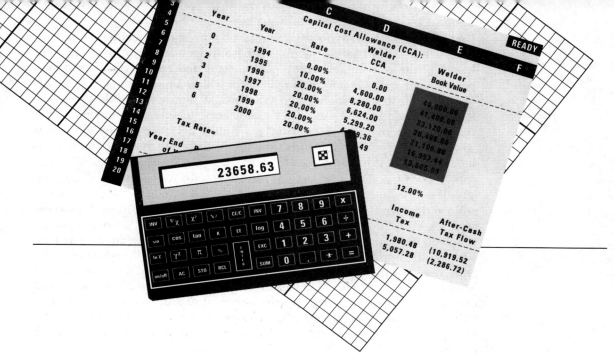

computerized traffic-signal controls that has reduced travel time, fuel consumption, and pollution. In the United States, between Santa Monica and downtown Los Angeles, testing of an electronic traffic and navigational system—including highway sensors and cars with computerized dashboard maps—is being sponsored by federal, state, and local governments and General Motors Corporation. This test program costs $40 million; to install it throughout Los Angeles could cost $2 billion.

In Canada, the estimates for implementing "smart" roads and vehicles nationwide is even more staggering. Advocates say the rewards far outweigh the costs. Do they?

Up to this point, we have focused attention on investment decisions in the private sector; the primary objective of these investments was to increase the wealth of the corporation. In the public sector, federal, provincial, and local governments spend hundreds of millions of dollars annually on a wide variety of public activities. In addition, governments, at all levels, regulate the behaviour of individuals and businesses by influencing the use of enormous quantities of productive resources. How can public decision makers determine whether their decisions that affect the use of these productive resources are, in fact, in the best public interest?

Benefit-cost analysis is a decision-making tool for systematically developing useful information about the desirable and undesirable effects of public projects. In a sense, we may view the public sector's benefit-cost analysis as the private sector's profitability analysis. In other words, benefit-cost analysis attempts to determine whether the social benefits of a proposed public activity outweigh the social costs. Usually, these public investment decisions involve a great deal of expenditure and their benefits are expected to occur over an extended period of time. Examples of benefit-cost analyses include studies of the public transportation system, environmental regulations on noise and pollution, public safety programs, education and training programs, public health programs, flood control, water resource development, and national defense programs.

There are three types of benefit-cost analysis problems: (1) to maximize the benefits for any given set of costs (or budgets), (2) to maximize the net benefits when both benefits and costs vary, and (3) to minimize costs to achieve any given level of benefits (often called "cost-effectiveness" analysis). We will consider these types of decision problems in this chapter.

■ 14.1 The Framework of Benefit-Cost Analysis

To evaluate public projects that are designed to accomplish widely differing tasks, we need to measure the benefits or costs in the same units in all projects so that we have a common perspective by which to judge the different projects. In practice this means expressing both the benefits and the costs in monetary units, a process that often must be performed without accurate data. In performing benefit-cost analysis, we normally define "users" as the public and "sponsors" as the government.

The general framework for benefit-cost analysis can be summarized as follows.

1. Identify the users' benefits that are expected from the project.
2. Quantify, as much as possible, these benefits in dollar terms so that different benefits may be compared against one another and against the costs of attaining.
3. Identify the sponsor's costs.

4. Quantify, as much as possible, these costs in dollar terms to allow comparisons.
5. Determine the equivalent benefits and costs at the base period using the interest rate appropriate for the project.
6. Accept the project if the equivalent users' benefits exceed the equivalent sponsor's costs.

We can use benefit-cost analysis to choose among such alternatives as allocating funds for construction of a mass-transit system, a dam with irrigation, highways, or an air-traffic control system. If the projects are on the same scale with respect to cost, it is merely a question of choosing the project for which the benefits exceed the costs by the greater amount.

■ 14.2 Valuation of Benefits and Costs

In the abstract, the framework we just developed for benefit-cost analysis is no different from the one we have used throughout this text to evaluate private investment projects. The complications, as we shall discover in practice, arise in trying to identify and assign values to all benefits and costs of a public project.

14.2.1 Users' Benefits

To begin a benefit-cost analysis, we identify all project **benefits** (favorable outcomes) and **disbenefits** (unfavorable outcomes) to the user. We should also consider the indirect consequences resulting from the project—the so-called **secondary effect.** For example, construction of a new highway will create new businesses such as gas stations, restaurants, and motels (benefits), but it will divert some traffic from the old road, and as a consequence some businesses would be lost (disbenefits). Once the benefits and disbenefits are quantified, we define the users' benefits as follows:

$$\text{users' benefits (B)} = \text{benefits} - \text{disbenefits}.$$

In identifying user's benefits, we should classify each one as a **primary benefit**—one directly attributable to the project—or as a **secondary benefit**—one indirectly attributable to the project. As an example, at one time, the Canadian government was considering building a superconductor research laboratory in Ottawa. If the laboratory ever materializes, it could bring many scientists and engineers, along with other supporting population, to the region. Primary national benefits may include the long-term benefits that could accrue as a result of various applications of the research to Canadian businesses. Primary regional benefits may include economic benefits created by the research laboratory activities, which will generate many new supporting businesses. The secondary benefits might include the creation of new economic wealth as a consequence of a possible increase in international trade and any increase in the incomes of various regional producers attributable to a growing population.

The reason for making this distinction is that it may make our analysis more efficient. If primary benefits alone are sufficient to justify project costs, we can save time and effort by not quantifying the secondary benefits.

14.2.2 Sponsor's Costs

We can determine the cost to the sponsor by identifying and classifying the expenditures required and any savings (or revenues) to be realized. The sponsor's costs should include both capital investment and the annual operating cost. Any sales of products or services that take place on completion of the project will generate some revenues—for example, toll revenues on highways. These revenues reduce the sponsor's costs. Therefore, we calculate the sponsor's costs by combining these cost elements:

sponsor's costs = capital costs + operating and maintenance costs − revenues.

14.2.3 Selecting an Interest Rate

As we learned in Chapter 13, the selection of an appropriate MARR for evaluating an investment project is a critical issue in the private sector. In public project analyses, we also need to select an interest rate, called the **social discount rate,** to determine the equivalent benefits as well as the equivalent costs. Selecting the social discount rate for a public project evaluation is as critical as selecting a MARR in the private sector.

Since present value calculations were initiated in the evaluation of public water resources and related land-use projects in the 1930s, there has been a tendency to use relatively low rates of discount as compared with those existing in markets for private assets. During the 1950s and into the 1960s, the rate for water resource projects was 2.63%, which for much of this period was even below the yield on long-term government securities. The persistent use of a lower interest rate for water resource projects is also a political issue. This point is best explained by the following newspaper article.

> With a $3.5 billion price tag, the Central Arizona Project (CAP) to divert Colorado River water to central and southern Arizona is a massive undertaking by anyone's standards. But back in 1981, auditors at the General Accounting Office took a look at the interest rate that the Interior Department plans to charge on federal funds advanced for the project and discovered a far more startling statistic. By interpreting the law to require only a 3.343% interest payment rather than the market rate on the reimbursable portions of the project, the Interior Department is creating a staggering interest subsidy of $175 billion... over the 50-year payback period.

Senator Howard Metzenbaum looks at those figures and cringes..."What possible reason can anyone give for providing 3.34% interest when the government itself is paying 9 and 10 and 11% for its money?"

The Interior Department is justifying its low interest charge for the CAP on a legal interpretation that classifies the entire Central Arizona Project as one "unit" of the Colorado River Basin Project— meaning all features of the project, even though they will be constructed in phases over a long period of time, should enjoy the same 3.343% interest rate that was charged when the first construction began in the early 1970s.[1]

In recent years, many public economists have argued that the appropriate rate of discount for government investments should be such as to maximize efficiency in the use of the nation's total economic resources. But with the growing interest in performance budgeting and systems analysis in the 1960s, there has been a tendency on the part of government agencies to examine the appropriateness of the discount rate in the public sector in relation to the efficient allocation of resources in the economic system as a whole.[2]

There are two prevailing views on the basis for determining the social discount rate:

1. **Projects without private counterparts:** *The social discount rate should reflect only the prevailing government borrowing rate.* Projects such as dams designed purely for flood control, access roads for noncommercial uses, and reservoirs for community water supply may have no corresponding private counterparts. In those areas of government activity where benefit-cost analysis has been employed in evaluation, the rate of discount traditionally used has been the cost of government borrowing. In fact, water resource project evaluations follow this view exclusively.

2. **Projects with private counterparts:** *The social discount rate should represent the rate that could have been earned had the funds not been removed from the private sector.* If all public projects were financed by borrowing at the expense of private investment, we may focus on the opportunity cost of capital in alternative investments in the private sector in determining the social discount rate. In the case of public capital projects similar to some in the private sector that produce a commodity or a service (such as electric power) that is sold on the market, the rate of discount employed would be the average cost of capital discussed in Chapter 13. The reasons for using the private rate of return as the opportunity cost of capital in projects similar to those in the private sector are (1) to prevent the public sector from transferring capital from higher-yielding to lower-yielding investments, and (2) to force public project evaluators to employ market standards in justifying projects.

---

[1] *Atlanta Journal and Constitution*, Sunday, July 7, 1985. Reprinted with permission from the Atlanta Journal and The Atlanta Constitution.

[2] R. F. Mikesell, *The Rate of Discount for Evaluating Public Projects*, American Enterprise Institute for Public Policy Research, 1977.

14.2.4 Motor Vehicle Inspection Program

Now that we have defined the general framework for benefit-cost analyses and have discussed the appropriate discount rate, we will illustrate the process of quantifying the benefits and costs associated with a public project.

Some provincial governments employ inspection systems for motor vehicles. Critics have often charged that these programs lack efficiency and have a poor benefit-to-cost ratio in terms of reducing fatalities, injuries, accidents, and pollution.

Elements of Benefits and Costs

The primary and secondary benefits of the motor vehicle inspection program are identified as follows:

- **Users' Benefits**

  Primary benefits: Deaths and injuries related to motor-vehicle accidents impose specific financial costs on individuals and society. Preventing such costs through the inspection program has the following primary benefits.

  1. Retention of contributions to society that might be lost due to individuals' deaths.
  2. Retention of productivity that might be lost while individuals recuperate from accidents.
  3. Savings of medical, legal, and insurance services.
  4. Savings on property replacement or repair costs.

  Secondary benefits: Some of the secondary benefits are not measurable (for example, avoidance of pain and suffering); others can be quantified. We should consider both types.

  1. Savings of income of families and friends of accidents victims who might otherwise be tending to accident victims.
  2. Avoidance of air and noise pollution; savings on fuel costs.
  3. Savings on enforcement and administrative costs related to the investigation of accidents.
  4. Pain and suffering.

- **Users' Disbenefits**
  1. Cost of spending time having a vehicle inspected (including travel time), as opposed to devoting that time to an alternative endeavor (opportunity cost).
  2. Cost of inspection fee.
  3. Cost of repairs that would not have been made if the inspection had not been performed.
  4. Value of time expended in repairing the vehicle (including travel time).
  5. Cost in time and direct payment for reinspection.

- **Sponsor's Cost**
  1. Capital investments in inspection facilities.
  2. Operating and maintenance costs associated with inspection facilities. These include all direct and indirect labor, personnel, and administrative costs.

- **Sponsor's Savings**
  1. Inspection fee.

Valuation of Benefits and Costs

The aim of benefit-cost analysis is to maximize the equivalent value of all benefits less that of all costs (expressed either in present values or annual values). This objective is in line with promoting the economic welfare of citizens. In general, the benefits of public projects are difficult to measure, whereas the costs are more easily determined. For simplicity, we will only attempt to quantify the primary users' benefits and sponsor's costs on an annual basis.

(a) Calculation of Primary Users' Benefits
  1. *Benefits due to reduction of deaths:* The equivalent value of the average income stream lost by victims of fatal accidents[3] was estimated at $820,000 per victim in 1994 dollars. Ontario estimated that the inspection program would reduce the number of annual fatal accidents by 250, resulting a potential savings of

$$(250)(\$820,000) = \$205,000,000$$

  2. *Benefits due to reduction of damage to property:* The average cost of damage to property per accident was estimated at $3800. This figure includes the cost of repairs for damages to the vehicle, the cost of insurance, the cost of legal and court administration, the cost of police accident investigation, and the cost of traffic delay due to accidents. Ontario estimated that accidents would be reduced by 26,400 per year, and that about 63% of all accidents would result in damage to

---

[3] These estimates were based on the total average income that these victims could have generated if they had lived. This average value on human life was calculated by considering several factors, such as age, sex, and income group.

property only. Therefore, the estimated present value of benefits due to reduction of property damage was estimated at

$$(\$3800)(26,400)(0.63) = \$63,201,600.$$

The overall annual primary benefits are estimated as sum of

Value of reduction in fatalities	$205,000,000
Value of reduction in property damage	$ 63,201,600
Total	$268,201,600

(b) Calculation of Primary Users' Disbenefits

1. *Opportunity cost associated with time spent bringing vehicles for inspection:* This cost is estimated as

$$C_1 = (\text{REG})(\text{TR})(\text{WRATE}),$$

where

$$\text{REG} = \text{number of cars inspected in 1994,}$$

$$\text{TR} = \text{average duration involved in travel time,}$$

$$\text{WRATE} = \text{average wage rate.}$$

With an estimated TR of 1.02 hours, WRATE of $16.50, and REG of 5,136,224, we obtain

$$C_2 = 5,136,224(1.02)(\$16.50)$$

$$= \$68,442,650.$$

2. *Cost of inspection fee:* This cost may be calculated as

$$C_3 = (\text{Fee})(\text{REG}),$$

where *Fee* = the amount of inspection fee to be paid by each owner. Assuming an inspection fee of $10 for each car, the cost of inspection fee is estimated as

$$C_3 = (\$10)(5,136,224)$$

$$= \$51,362,240$$

3. *Opportunity cost associated with time spent waiting during the inspection process:* This cost may be calculated by the formula

$$C_4 = (\text{WT})(\text{WRATE})(\text{REG}),$$

where WT = average waiting time = 9 minutes = 0.15 hours. Thus,

$$C_4 = 0.15(\$16.50)(5,136,224) = \$12,712,154.$$

4. *Vehicle usage costs for the inspection process:* These costs are estimated as

$$C_5 = (REG)(VC)(MI),$$

where

$$VC = \text{cost per kilometre of operating an automobile}$$

$$= \$0.28 \text{ per kilometre}$$

$$MI = \text{average round trip to inspection station}$$

$$= 30 \text{ kilometres}$$

Thus,

$$C_5 = 5,136,224(\$0.28)(30) = \$43,144,282.$$

The overall primary annual disbenefits are estimated as

Item	Amount
C_1	$ 68,442,650
C_2	51,362,240
C_3	12,712,154
C_4	43,144,282
Total annual disbenefits	$175,661,326

(c) Calculation of Primary Sponsor's Costs

Ontario estimates an expenditure of $82,000,000 for inspection facilities (this value represents the annualized capital expenditure) and another annual operating expenditure of $12,000,000 for inspection, adding up to $94,000,000.

(d) Calculation of Primary Sponsor's Savings

The sponsor's costs are offset to a large degree by annual inspection revenue, which must be subtracted to avoid double counting. Annual fee revenues are the same as the direct cost of inspection incurred by the users (C_3), which was calculated as $51,362,240.

Reaching a Final Decision

Finally, a discount rate of 6% was deemed appropriate because Ontario currently finances most projects by issuing a 6% long-term bond. In fact, the time streams of costs and benefits were already discounted so as to obtain their present and annual equivalent values.

From the estimates, the primary benefits of inspection are valued at $268,201,600 as compared to the primary disbenefits of inspection, which total $175,661,326. Therefore, the user's net benefits are:

$$\text{user's net benefits} = \$268,201,600 - \$175,661,326$$

$$= \$92,540,274$$

Now the sponsor's net costs are:

$$\text{sponsor's net costs} = \$94,000,000 - \$51,362,240$$

$$= \$42,637,760$$

Since all benefits and costs are expressed in annual equivalents, we can use these values directly in computing the degree of benefits that exceeds the sponsor's costs:

$$\$92,540,274 - \$42,637,760 = \$49,902,514 \text{ per year.}$$

This positive AE amount indicates that the provincial inspection system is economically justifiable. We can assume the AE amount would have been even greater if we had also factored in secondary benefits. (For simplicity, we have not explicitly considered the vehicle growth in the province. For a complete analysis, this growth factor must be considered to account for all related benefits and costs in equivalence calculations.)

14.2.5 Difficulties of Public Project Analysis

As we have observed in the motor-vehicle inspection program in the previous section, public benefits are generally very difficult to quantify in a convincing manner. For example, consider the valuation of a saved human life in any category. Conceptually, the total benefit associated with saving a human life may include the avoidance of the costs of insurance administration, legal and court costs, as well as the average potential income lost, considering age and sex, because of premature death. Obviously, any attempt to put precise numbers on human life will turn out to be an insurmountable task.

As an example, a few years ago, a 50-year-old business executive was killed in a plane accident. The investigation indicated that the plane was not properly maintained according to the federal guidelines. The executive's family sued the airline, and the court eventually ordered the airline to pay $5,250,000 to the victim's family. The judge calculated the value of the lost human life based on the assumption that, if the executive had lived and worked in the same capacity until his retirement, his remaining lifetime earnings would be equivalent to $5,250,000 at the time of award. This is an example of how an individual human life was assigned a dollar value, but clearly any attempt to establish an average amount that represents the general population would be potentially controversial. We might even take exception to this individual case: Does the executive's salary adequately represent his worth to his family? Should we also assign a dollar value to their emotional attachment to him, and if so, how much?

Consider the situation in which a local government is planning to widen a typical municipal highway to relieve chronic traffic congestion. Knowing that the project will be financed by local and provincial taxes and that many out-of-province travelers are expected to benefit, should the project be justified solely on the benefits to the local residents? Which point of view should we take in measuring the benefits—the municipal level, the provincial level, or both? It is important that any benefit measure be done from the proper *point of view*.

In addition to valuation and point-of-view issues, there are also many possibilities for tampering with the results of benefit-cost analyses. Unlike in the private sector, many public projects are undertaken based on political pressure rather than on their economic benefits alone. In particular, whenever the benefit-cost ratio becomes marginal or less than one, there is tendency to inflate the benefit figures to make the project look good.

■ 14.3 Benefit-Cost Ratios

An alternative way of expressing the worthiness of a public project is to compare the user's benefits (B) to the sponsor's cost (C) by taking the ratio B/C. In this section, we shall define the benefit-cost (B/C) ratio, and explain the relationship between the conventional NPW criterion and the B/C ratio.

14.3.1 Definition of Benefit-Cost Ratio

For a given benefit-cost profile, let B and C be the present values of benefits and costs defined by

$$B = \sum_{n=0}^{N} b_n (1 + i)^{-n} \tag{14.1}$$

$$C = \sum_{n=0}^{N} c_n (1 + i)^{-n}, \tag{14.2}$$

where

$$b_n = \text{benefit at the end of period } n, \; b_n \geq 0$$

$$c_n = \text{expense at the end of period } n, \; c_n \geq 0$$

$$A_n = b_n - c_n$$

$$N = \text{project life}$$

$$i = \text{the sponsor's interest rate (discount rate)}$$

Note that the sponsor's costs (C) consist of the equivalent capital expenditure (I) and the equivalent annual operating costs (C') accrued in each successive period. (Note the sign convention in calculating a benefit-cost ratio. Since we are using a ratio, all benefits and cost flows are expressed in positive units. Recall that in previous equivalent worth calculation our sign convention was to explicitly assign "+" for cash inflows and "−" for cash outflows.) Let's assume that a series of initial investments is required during the first K periods, while annual operating and maintenance costs accrue in each period following. Then, the equivalent present value for each component is

$$I = \sum_{n=0}^{K} c_n (1 + i)^{-n} \tag{14.3}$$

$$C' = \sum_{n=K+1}^{N} c_n (1 + i)^{-n} \tag{14.4}$$

and $C = I + C'$.

The B/C ratio is defined as

$$BC(i) = \frac{B}{C} = \frac{B}{I + C'}, \qquad I + C' > 0. \tag{14.5}$$

If we are to accept a project, the $BC(i)$ must be greater than 1.

An alternative measure, called the **net B/C ratio, $B'C(i)$,** considers only the initial capital expenditure as a cash outlay, and uses annual net benefits:

$$B'C(i) = \frac{B - C'}{I} = \frac{B'}{I}, \qquad I > 0. \tag{14.6}$$

The decision rule has not changed—the ratio must be greater than 1. However, some analysts prefer this measure because it indicates the **net benefit (B')** expected per dollar invested. Note that we must express the values of B, C', and I in present worth equivalents. Or, we can compute these values in terms of annual equivalents, and use them in calculating the B/C ratio. The resulting B/C ratio is not affected.

Example 14.1 Benefit-cost ratio

A public project being considered by a local government has the following estimated benefit-cost profile (see Fig. 14.1).

n	b_n	c_n	A_n
0		$10	−$10
1		10	−10
2	$20	5	15
3	30	5	25
4	30	8	22
5	20	8	12

Assume that $i = 10\%$, $N = 5$, and $K = 1$. Compute B, C, I, C', $BC(10\%)$ and $B'C(10\%)$.

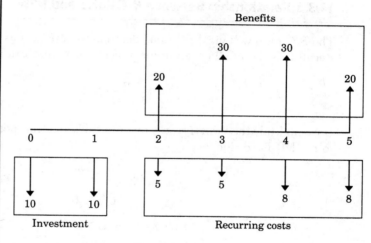

Figure 14.1 ■ Classification of project's cash flow elements

Solution

$$B = \$20(P/F, 10\%, 2) + \$30(P/F, 10\%, 3)$$
$$+ \$30(P/F, 10\%, 4) + \$20(P/F, 10\%, 5)$$
$$= \$71.98$$

$$C = \$10 + \$10(P/F, 10\%, 1) + \$5(P/F, 10\%, 2) + \$5(P/F, 10\%, 3)$$
$$+ \$8(P/F, 10\%, 4) + \$8(P/F, 10\%, 5)$$
$$= \$37.41$$

$$I = \$10 + \$10(P/F, 10\%, 1)$$

$$= \$19.09$$

$$C' = C - I$$

$$= \$18.32$$

Using Eqs. (14.5) and (14.6), we can compute the B/C ratios as

$$BC(10\%) = \frac{\$71.98}{\$19.09 + \$18.32} = 1.92 > 1$$

$$B'C(10\%) = \frac{\$71.98 - \$18.32}{\$19.09} = 2.81 > 1$$

The B/C ratios exceed 1, so the users' benefit exceeds the sponsor's cost.

14.3.2 Relationship Between B/C Ratio and NPW

The B/C ratio will yield the same decision for a project as the NPW criterion. Recall that the $BC(i)$ criterion for project acceptance can be stated as

$$\frac{B}{I + C'} > 1. \tag{14.7}$$

If we multiply the term $(I + C')$ on both sides and transpose the term $(I + C')$ to the left-hand side, we have

$$B > (I + C')$$

$$B - (I + C') > 0 \tag{14.8}$$

$$PW(i) = B - C > 0, \tag{14.9}$$

which is the same decision rule as that of accepting a project by the NPW criterion. We can easily verify the similar relationship between the netted B/C ratio and the NPW criterion. This implies that we could use the benefit-cost ratio in evaluating private projects instead of using the NPW criterion, or we could use the NPW criterion in evaluating the public projects. Either approach will signal consistent project selection. Recall that, in Example 14.1, $PW(10\%) = B - C = \$34.57 > 0$, and the project would be acceptable under the NPW criterion.

14.3.3 Incremental Analysis

Let us now consider how we choose among public projects that are mutually exclusive. As we explained in Chapter 6, we must use the incremental investment approach in comparing alternatives based on any relative measure such as IRR or B/C.

Incremental Analysis Based on $BC(i)$

To apply the incremental analysis, we compute the incremental differences for each term $(B, I, \text{and } C')$, and take the B/C ratio based on these differences. To use the $BC(i)$ on incremental investment, we may proceed as follows.

1. Remove any alternatives with a B/C ratio less than 1.
2. Arrange the remaining alternatives in the increasing order of the denominator $(I + C')$. Thus, the alternative with the smallest denominator should be first (j), the alternative with the next smallest second (k), and so forth.
3. Compute the incremental differences for each term $(B, I, \text{and } C')$ for the paired alternatives (j, k) in the list.

$$\Delta B = B_k - B_j$$

$$\Delta I = I_k - I_j$$

$$\Delta C' = C'_k - C_j$$

4. Compute the $BC(i)$ on incremental investment by evaluating

$$BC(i)_{k-j} = \frac{\Delta B}{\Delta I + \Delta C'}.$$

If $BC(i)_{k-j} > 1$, select the k alternative. Otherwise select the j alternative.
5. Compare the selected alternative with the next one on the list by computing the incremental benefit-cost ratio. Continue the process until you reach the bottom of the list. The alternative selected during the last pairing is the best.

We may modify the decision procedures when we encounter the following situations:

- If $\Delta I + \Delta C' = 0$, we cannot use the benefit-cost ratio. This implies that both alternatives require the same initial investment and operating expenditure. If this happens, we simply select the alterative with the largest B value.

- In situations where we have to compare public projects with unequal service lives but they can be repeatable, we may compute all component values $(B, C', \text{and } I)$ on an annual basis and use them in incremental analysis.

- If we use the netted B/C ratio $(B'C(i))$ as a basis, we need to order the alternatives in increasing order of I and compute the netted B/C ratio on incremental investment.

Example 14.2 Incremental benefit-cost ratios

Consider three investment projects, A1, A2, and A3. Each project has the same service life, and the present worth of each component value (B, I, and C') is computed at 10% as follows.

PW	A1	A2	A3
		Projects	
I	$ 5,000	$20,000	$14,000
B	12,000	35,000	21,000
C'	4,000	8,000	1,000
PW(i)	$ 3,000	$ 7,000	$ 6,000

(a) If all three projects are independent, which projects would be selected based on $BC(i)$ and $B'C(i)$ criteria, respectively?

(b) If the three projects are mutually exclusive, which project would be the best alternative? Show the sequence of calculations required to produce the correct results, using (1) the aggregate B/C ratio, and (2) the netted B/C ratio.

Solution

(a) Since $PW(i)_1$, $PW(i)_2$, and $PW(i)_3$ are positive, all projects would be acceptable if they were independent. Also, $BC(i)$ and $B'C(i)$ values for each project are greater than 1, so the use of either ratio will lead to the same accept-reject conclusion under the NPW criterion.

	A1	A2	A3
$BC(i)$	1.33	1.25	1.40
$B'C(i)$	1.60	1.35	1.43

(b) If these projects are mutually exclusive, we must use the principle of incremental analysis. If we attempt to rank the projects according to the size of their B/C ratio, obviously we will observe a different project preference. For example, using the $BC(i)$ ratio on total investment, we see that A3 appears to be the most desirable and A2 the least desirable. Certainly, with $PW(i)_2 > PW(i)_3 > PW(i)_1$, project A2 would be selected under the NPW criterion. By computing the incremental B/C ratios, we can eliminate this inconsistency problem and conclude with project selection consistent with the NPW criterion.

We will first arrange the projects by increasing order of their denominator $(I + C')$ for the $BC(i)$ criterion, and I for the $B'C(i)$ criterion:

Ranking Base	A1	A3	A2
$I + C'$	$9,000	$15,000	$28,000
I	5,000	14,000	20,000

- Using the aggregate B/C ratio: With the do-nothing alternative, we first drop from consideration any project that has a B/C ratio smaller than 1. In our example, the B/C ratios of all three projects exceed 1, so the first incremental comparison is between A1 and A3:

$$BC(i)_{3-1} = \frac{\$21,000 - \$12,000}{(\$14,000 - \$5000) + (\$1000 - \$4000)}$$

$$= 1.5 > 1.$$

Since the ratio is greater than 1, we prefer A3 over A1. Therefore, A3 becomes the "current best" alternative.

Next, we must determine whether the incremental benefits to be realized from A2 would justify the additional expenditure. Therefore, we need to compare A2 and A3 as follows:

$$BC(i)_{2-3} = \frac{\$35,000 - \$21,000}{(\$20,000 - \$14,000) + (\$8000 - \$1000)}$$

$$= 1.08 > 1.$$

The incremental B/C ratio again exceeds 1, and therefore we prefer A2 over A3. With no further projects to consider, A2 becomes the ultimate choice.

- Using the net B/C ratios: If we had to use the net B/C ratio on this incremental investment decision, we would obtain the same conclusion. Since all $B'C(i)$ ratios exceed 1, there will be no do-nothing alternative. By comparing the first pair of projects on this list, we obtain

$$B'C(i)_{3-1} = \frac{(\$21,000 - \$12,000) - (\$1000 - \$4000)}{(\$14,000 - \$5000)}$$

$$= 1.33 > 1.$$

Project A3 becomes the "current best." Next, comparing A2 and A3 yields

$$B'C(i)_{2-3} = \frac{(\$35,000 - \$21,000) - (\$8000 - \$1000)}{(\$20,000 - \$14,000)}$$

$$= 1.17 > 1.$$

Therefore, A2 becomes the best choice by the net B/C criterion.

14.4 Analysis of Public Projects Based on Cost-Effectiveness

In evaluating public investment projects, we may encounter situations where the competing alternatives have the same goals but the effectiveness with which those goals can be met may or may not be measurable in dollars. In these situations, we can compare decision alternatives directly based on their **cost-effectiveness.** Here we judge the effectiveness of an alternative in dollars or some nonmonetary measure by the extent to which that alternative, if implemented, will attain the desired objective. The preferred alternative is then either the one that produces the maximum effectiveness for a given level of cost, or the minimum cost for a fixed level of effectiveness.

14.4.1 The General Procedure for Cost-Effectiveness Studies

A typical cost-effectiveness analysis procedure involves the following steps.

- Step 1: Establish the goals to be achieved by the analysis.
- Step 2: Identify the imposed restrictions on achieving the goals, such as budget or weight.
- Step 3: Identify all the feasible alternatives to achieve the goals.
- Step 4: Identify the social interest rate to use in the analysis.
- Step 5: Determine the equivalent life-cycle cost of each alternative, including research and development, testing, capital investment, annual operating and maintenance costs, and salvage value.
- Step 6: Determine the basis for developing the cost-effectiveness index. There are two approaches: (1) the fixed-cost approach and (2) the fixed-effectiveness approach. In the fixed-cost approach, determine the amount of effectiveness obtained at a given cost. In the fixed-effectiveness approach, determine the cost to obtain the predetermined level of effectiveness.
- Step 7: Compute the cost-effectiveness index for each alternative based on the selected criterion in Step 6.
- Step 8: Select the alternative with the maximum cost-effective index.

14.4.2 A Cost-Effectiveness Case Example

To illustrate the procedures involved in cost-effectiveness analysis, we shall present an example of selecting the most cost-effective program for developing an adverse-weather precision-guided weapon system.[4]

---

[4] The case example is provided by Frederick A. Davis. All numbers herein do not represent actual values.

Problem Statement

During the Persian Gulf War, precision-guided weapons demonstrated remarkable success and accuracy against a wide array of fixed and mobile targets. Such weapons rely upon laser designation of the target by an aircraft. The aircraft illuminates the target by laser; and the weapon, with its on-board laser sensor, locks onto the target and flies to it. During the war, aircraft were required to fly at moderate altitude (about 10,000 feet) to escape vulnerability to antiaircraft artillery batteries. At these altitudes, the aircraft were above the cloud/smoke levels. Unfortunately, laser beams cannot penetrate cloud cover, smoke, or fog. As a result, on those days when clouds, smoke, and fog were present, the aircraft were unable to deliver the precision-guided weapons (see Fig. 14.2). This led to development of a weapon system that would correct these deficiencies.

Defining the Goals

Selection of the best system will be based on cost/kill decision criteria. Cost/kill is defined as the unit cost of the weapon divided by the probability of the weapon achieving its target. Mission-effectiveness studies determine this probability.

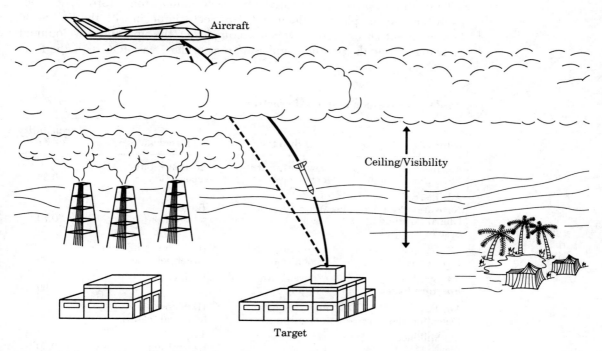

Figure 14.2 ■ A conceptual use of an adverse-weather guidance weapon by an aircraft

The purposes of this study are to evaluate these alternatives from a cost viewpoint and to determine the best option, based on a cost/kill decision criterion. Also, anticipating parliamentary scrutiny of high weapon-system costs, any option costing more than the present value of life-cycle cost of $120K per unit cannot be considered for selection. To respond rapidly to this critical military requirement, the initial operational capability (IOC) date is assumed to be 7 years. IOC is defined as that point in the project life when the first block of weapons has been delivered to the field and is ready for operational use.

Description of Precision-Weapon Alternatives

Considering numerous weapon alternatives, preferred concepts were developed and are listed in Table 14.1. Also presented in the table are the qualitative characteristics of each alternative. These six weapon system alternatives are also considered to be mutually exclusive. Only one of the alternatives will be selected to fill the mission capability void that currently exists. Each of the alternatives is at some level of technological maturity. Some are nearly off-the-shelf, while others are just emerging from laboratory development. Because of this, some of the alternatives will require considerably more up-front development funding prior to production than others. Table 14.1 also summarizes the results (probability of kill) by the mission studies for the six guidance systems before entering into laboratory research and development.

To consider the costs associated with each option, from up-front development through complete production, the project will begin with the Full Scale Development (FSD) phase, which accomplishes the up-front development prior to entering the production phase. To meet the 7-year IOC date, the FSD

Table 14.1 Weapon System Alternatives

Alternative A_j	Advantage	Disadvantage	Probability of Kill
A1: Inertial Navigation System	Low cost, mature technology	Accuracy, target recognition	0.33
A2: Inertial Navigation System: Global Positioning System	Moderate cost Mature technology	Target recognition	0.70
A3: Imaging Infrared (I^2R)	Accurate, target recognition	High cost, bunkered target detection	0.90
A4: Synthetic Aperture Radar	Accurate, target recognition	High cost	0.99
A5: Laser Detection/Ranging	Accurate, target recognition	High cost, technical maturity	0.99
A6: Millimetre Wave (MMW)	Moderate cost, accurate	Target recognition	0.80

phase must be completed in 4 years. A 10,000-unit buy over a 5-year production life is required to meet military needs. Because of the differing technological maturity of the six alternatives, widely varying FSD investments will be required.

Life-Cycle Cost for Each Alternative

The costs associated with FSD and production for each alternative will vary significantly. For systems incorporating the most current of emerging technologies, the FSD completion costs will be higher than those considered mature. Production costs will vary, depending upon the complexity of the system components. The objective of the FSD program is to rigorously test demonstrated system capability and correct any design flaws that are uncovered. The cost estimates used in both the FSD and the production phases were generated considering labor hours, material costs, equipment/tooling costs, subcontractor costs, travel, flight-test costs, documentation, and costs for program reviews. Table 14.2 summarizes the equivalent life-cycle cost for each program, estimated in constant dollars (1990), as is the standard military practice in project cost estimation. The Department of National Defence will follow government guidelines in selecting the interest rate for equivalent life-cycle cost calculation, which is 10%.

Cost-Effectiveness Index

The equivalent life-cycle costs of the system need to be divided by the 10,000 units to be purchased in order to arrive at a cost/unit figure. Then, the

Table 14.2 Life-Cycle Costs for Weapon Development Alternatives

Phase	Year	A1	A2	A3	A4	A5	A6
				Expenditures in Million Dollars			
FSD	0	$ 15	$ 19	$ 50	$ 40	$ 75	$ 28
	1	18	23	65	45	75	32
	2	19	22	65	45	75	33
	3	15	17	50	40	75	27
IOC	4	90	140	200	200	300	150
	5	95	150	270	250	360	180
	6	95	160	280	275	370	200
	7	90	150	250	275	340	200
	8	80	140	200	200	330	170
PW(10%)		$315.92	$492.22	$884.27	$829.64	$1227.23	$613.70

Sample calculation: Equivalent life-cycle cost for A1

$$PW(10\%) = \$15 + \$18(P/F, 10\%, 1) + \cdots + \$80(P/F, 10\%, 8)$$
$$= \$315.92$$

cost/kill for each alterative is computed. The following shows the resultant cost/kill and kill/cost figures:

		Probability		
Type	Cost/Unit	Kill	Cost/Kill	Kill/Cost
A1	$ 31,592	0.33	$ 95,733	0.0000104
A2	49,220	0.70	70,314	0.0000142
A3	88,427	0.90	98,252	0.0000102
A4	82,964	0.99	83,802	0.0000119
A5	122,723	0.99	123,963	0.0000081
A6	61,370	0.80	76,713	0.0000130

Figure 14.3 graphically presents the cost/kill values of the six guidance systems. Obviously, the A5 system is not a feasible solution, as its unit cost exceeds the constraint of $120K. Of the feasible alternatives, the lowest cost/kill value is that of the A2 weapon system. It is that system that will meet the requirement for an adverse-weather weapon system at the best cost/target kill value. (Note: adoption of the kill/cost criterion would have resulted in the

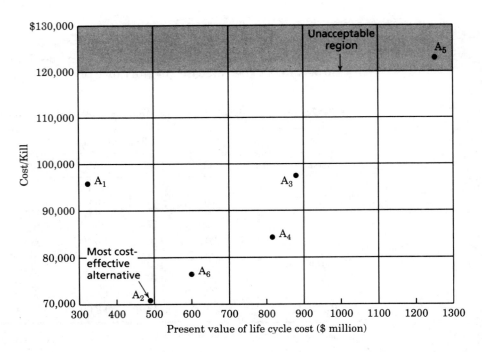

Figure 14.3 ■ Cost-effectiveness for six adverse weather guidance weapon systems

selection of the same alternative. Even though the cost estimates and their probability of kill were based on the best engineering judgment associated with the complexity of the system and their technological risk, project overrun costs are common, and, therefore, uncertainty in the estimated costs for the six weapon alternatives can be expected. The issue of handling this uncertainty is the topic for Chapter 15.)

Summary

- **Benefit-cost analysis** is commonly used to evaluate public projects; several facets unique to public project analysis are neatly addressed by benefit-cost.
  1. Benefits of a nonmonetary nature can be quantified and factored into our analysis.
  2. A broad range of project users that are distinct from the sponsor are considered— benefits and disbenefits to *all* these users can (and should) be taken into account.
- Difficulties involved in public project analysis include:
  1. Identifying all the users of the project.
  2. Identifying all the benefits and disbenefits of the project.
  3. Quantifying all the benefits and disbenefits in dollars or some other unit of measure.
  4. Selecting an appropriate interest rate at which to discount benefits and costs to a present value.

- The B/C ratio is defined as:

$$BC(i) = \frac{B}{C} = \frac{B}{I + C'}, \qquad I + C' > 0.$$

The decision rule is if $BC(i) \geq 1$, the project is acceptable.
- The net B/C ratio is defined as:

$$B'C(i) = \frac{B - C'}{I} = \frac{B'}{I}, \qquad I > 0.$$

The net ratio expresses the net benefit expected per dollar invested. The same decision rule applies as for the B/C ratio.
- The **cost-effectiveness method** allows us to compare projects on the basis of cost and nonmonetary effectiveness measures. We may either maximize effectiveness for a given cost criterion or minimize cost for a given effectiveness criterion.

Problems

14.1 The province of Nova Scotia was considering a bill that would ban the use of road salt on highways and bridges during icy conditions. Road salt is known to be toxic, costly, corrosive and caustic. XYZ Chemical Company produces a calcium magnesium acetate de-icer (CMA) and sells it for $600 a tonne as Ice-B-Gon. Road salts, on the other hand, sold for an average of $14 a tonne in 1994. Nova Scotia needs about 600,000 tonnes of road salt each year. (Nova Scotia spent $9.2 million on road salt in 1994.) XYZ Company estimates that each tonne of salt on the road cost $650 in highway corrosion, $525 in rust on vehicles, $150 in corrosion to utility lines, and $100 in damages to water supplies, for a total of $1425. There is also unknown salt damage to vegetation and soil surrounding areas of highways. The province of Nova Scotia would ban road salt (at least on expensive steel bridges or near sensitive lakes) if provincial studies support XYZ's cost claims.

(a) What would be the users' benefits and costs if there were a complete ban on road salt in Nova Scotia?

(b) How would you go about determining the salt damages (in dollars) to vegetation and soil?

14.2 A public school system in Alberta is considering the adoption of a 4-day school week as opposed to the current 5-day school week in high schools. The community is hesitant about the plan, but the superintendent of the school system envisions many benefits associated with the 4-day system, with Wednesday being the "day off." The following pros and cons have been cited:

- Experiments with the 4-day system indicate that having a day off in the middle of the week will cut down both on teacher and pupil absences.
- The longer hours on school days will require increased attention spans, which is not an appropriate expectation for younger children.
- The reduction in costs to the provincial government should be substantial as the number of subsidized lunches served would be cut by approximately 20 percent.
- The province bases its grants to the local systems largely on the average number of pupils attending school in the system. Since the number of absences will decrease, the grants to the local system should increase.
- Older students might want to work on Wednesdays. Unemployment is a problem in this region, however, and any influx of new job-seekers could aggravate an existing problem. Community centers, libraries, and other public areas may also experience increased usage on Wednesdays.

- Parents who provide transportation for their children will see a saving in fuel costs. This would primarily involve those parents whose children live less than 3 kilometres from the school. Children living more than 3 kilometres from school are eligible for free transportation provided by the local school board.
- Decreases in both public and private transportation should result in fuel conservation, decreased pollution, and less wear on the roads. Traffic congestion should ease on Wednesdays in areas where congestion caused by school traffic is a problem.
- Working parents will be forced to make child-care arrangements (and possibly payments) for one weekday per week.
- The students will benefit from wasting less time driving to and from school; Wednesdays will be available for study, thus taking the heavy demand off most nights. Bussed students will spend far less time per week waiting for buses.
- The local school board should see some ease in funding problems. The two areas most greatly impacted are the transportation and nutritional programs.

(a) For this type of public study, what do you identify as the users' benefits and costs?

(b) What items would be considered as the sponsor's costs?

(c) Discuss any other benefits or costs associated with the 4-day school week.

14.3 The Electrical Department of the City of Winnipeg, Manitoba operates generating and transmission facilities serving approximately 140,000 people in the city and surrounding areas. The city has proposed construction of a $300-million 235-MW circulating fluidized bed combustor (CFBC) to power a turbine generator currently receiving steam from an existing boiler fueled by gas or oil. Among the advantages associated with the use of CFBC systems are the following:

- A variety of fuels can be burned, including inexpensive low-grade fuels with high ash and high sulfur content.
- The relatively low combustion temperatures inhibit the formation of nitrogen oxides. Acid-gas emissions associated with CFBC units would be expected to be significantly lower than emissions from conventional coal-fueled units.
- The sulfur-removal method, low combustion temperatures, and high-combustion efficiency characteristic of CFBC units result in solid wastes that are physically and chemically more amenable to land disposal than the solid wastes resulting from conventional coal-burning boilers with flue-gas desulfurization equipment.

Based on Ministry of Energy (MOE) projections of growth in energy demand and expected market penetration, demonstration of a 235-MW unit could lead to as much as 41,000 MW of CFBC generation being constructed by the year 2010. The proposed project would reduce the city's dependency on oil and gas fuels by converting its largest generating unit to coal-fuel capability. Consequently, substantial reductions of local acid-gas emissions could be realized in comparison to the permitted emissions associated with oil fuel. The city has requested a $50 million cost share from the MOE. Cost sharing is considered attractive because the MOE's cost share would largely offset the risk of using such a new technology. To qualify for the cost-sharing money, the city has to address the following questions for the MOE.

(a) What is the significance of the project at local and national levels?

(b) What items would constitute the users' benefits and costs associated with the project?

(c) What items would constitute the sponsor's costs?

By putting yourself in the city engineer's position, respond to such questions.

14.4 The following information appeared in the *New York Times,* April 12, 1990.

Predicting a doubling of traffic in the next three decades, federal highway officials are actively promoting a major program of computerization and automation that would fundamentally alter the designs of vehicles and highways. In the latest move to assist researchers, Transportation Secretary Samuel K. Skinner announced an $8 million project today to equip 100 cars in Orlando, Fla., with computerized displays that will receive instantaneous traffic updates and detour instructions from a traffic management center. The displays will suggest the best route to take to destinations selected by the drivers and will update the information quickly in response to accidents or traffic congestion. It is expected to go into operation in January 1992, and will be evaluated for a year. The project is being sponsored by the automobile association along with the General Motors Corporation, state and local governments, and the Department of Transportation. The federal share of the project's cost is $2.5 million.[7]

Suppose that the experimental project proves to be successful and the city of Orlando will consider a full implementation of the computerized communication system.

(a) What would you identify as the primary elements of the users' costs and benefits associated with the project?

---

[7] " 'Smart' Cars and Highways to Help Unsnarl Gridlock," *New York Times,* April 12, 1990, p. A16. Copyright © 1990 by The New York Times Company. Reprinted by permission.

(b) What would you identify as the secondary elements of the users' costs and benefits?

(c) What would you identify as the elements of the sponsor's costs?

14.5 A city government is considering two types of town-dump sanitary systems. Design A requires an initial outlay of $400,000, with annual operating and maintenance costs of $50,000 for the next 15 years; design B calls for an investment of $300,000, with annual operating and maintenance costs of $80,000 per year for the next 15 years. Fee collections from the residents would be $85,000 per year. The interest rate is 8%, and there is no salvage value associated with either system.

(a) Using the benefit-cost ratio $(B'C(i))$, which system should be selected?

(b) If a new design (design C), which requires an initial outlay of $350,000 and annual operating and maintenance costs of $65,000, is proposed, would your answer in (a) change?

14.6 The Canadian government is considering building apartments for government employees working in a foreign country and living in locally owned housing. A comparison of two possible buildings indicates the following:

	Building X	Building Y
Original investment by Government agencies	$8,000,000	$12,000,000
Estimated annual maintenance costs	240,000	180,000
Savings in annual rent now being paid to house employees	1,960,000	1,320,000

Assume salvage or sale value of apartments to be 60% of the first investment. Use 10% and a 20-year study period to compute the B/C ratio on incremental investment and make a recommendation. (There is no do-nothing alternative.)

14.7 Three public investment alternatives are available, A1, A2, and A3. Their respective total benefits, costs, and first costs are given in present worth. These alternatives have the same service life.

	Proposals		
Present worth	A1	A2	A3
I	100	300	200
B	400	700	500
C'	100	200	150

Assuming no do-nothing alternative, which project would you select based on the benefit-cost ratio $(B'C(i))$ on incremental investment?

14.8 A city which operates automobile parking facilities is evaluating a proposal that it erect and operate a structure for parking in a city's downtown area. Three designs for a facility to be built on available sites have been identified. (All dollar figures are in thousands.)

	Design A	Design B	Design C
Cost of site	$ 240	$180	$ 200
Cost of building	2200	700	1400
Annual fee collection	830	750	600
Annual maintenance cost	410	360	310
Service life	30 years	30 years	30 years

At the end of the estimated service life, whichever facility had been constructed would be torn down and the land would be sold. It is estimated that the proceeds from the resale of the land will be equal to the cost of clearing the site. If the city's interest rate is known to be 10%, which design alternative would be selected based on the benefit-cost criterion?

14.9 The federal government is planning a hydroelectric project for a river basin. In addition to the production of electric power, this project will provide flood control, irrigation, and recreation benefits. The estimated benefits and costs that are expected to be derived from the three alternatives under consideration are listed below.

	Decision Alternatives		
	A	B	C
Initial cost	$8,000,000	$10,000,000	$15,000,000
Annual benefits or costs			
Power sales	1,000,000	1,200,000	1,800,000
Flood-control savings	250,000	350,000	500,000
Irrigation benefits	350,000	450,000	600,000
Recreation benefits	100,000	200,000	350,000
O&M costs	200,000	250,000	350,000

The interest rate is 10%, and the life of each of the projects is estimated to be 50 years.

(a) Find the benefit-cost ratio for each alternative.
(b) Select the best alternative based on $BC(i)$.
(c) Select the best alterative based on $B'C(i)$.

14.10 Two different routes are under consideration for a new highway.

	Length of Highway	First Cost	Annual Upkeep
The "long" route	22 kilometres	$21 million	$140,000
Transmountain shortcut	10 kilometres	$45 million	$165,000

For either route, the volume of traffic will be 400,000 cars per year, which are assumed to operate at 25 cents per kilometre. Assume a 40-year life for each road and an interest rate of 10%. Determine which route should be selected.

14.11 The government is considering undertaking the four projects listed below. These projects are mutually exclusive, and the estimated present worth of their costs and the present worth of their benefits are shown in millions of dollars. All projects have the same duration.

Projects	PW of Benefits	PW of Costs
A1	40	85
A2	150	110
A3	70	25
A4	120	73

Assuming there is no do-nothing alternative, which alternative would you select? Justify your choice by using a benefit-cost $(BC_{(i)})$ on incremental investment.

14.12 Fast growth in the population of the city of Toronto and surrounding areas, in particular, has resulted in insurmountable traffic congestion. The city has few places to turn for extra money for road improvements except to new taxes. City officials have said that the money they receive from current taxes is insufficient to widen overcrowded roads, improve roads that don't meet modern standards, and pave dirt roads. Provincial residents now pay 20 cents in taxes on every litre of gasoline. Twelve cents of that goes to the federal government and 8 cents to the provincial government. The city commissioner has suggested that the city get the money by tacking an extra penny-a-litre tax onto gasoline, bringing the total federal and provincial gas tax to 21 cents a litre. This would add about $2.6 million a year to the road-construction budget. The extra money would have a significant impact. With the additional revenue, the city could sell a $24 million bond issue. The city would then have the option of spreading that amount among many smaller projects or concentrating on a major project. Assuming that voters would approve a higher gas tax, the city engineers were asked to prepare a priority list outlining which roads would be improved with the extra money. As shown in Table 14.3, the road engineers also computed the possible public benefits associated with each road-construction project, accounting for possible reduction in travel time, a reduction in the accident rate, land appreciation, and savings in operating costs of vehicles. Assume a 20-year planning horizon and an interest rate of 10%. Which projects would be considered for funding in (a) and (b)?

(a) Due to political pressure, each area will have the same amount of funding, say, $6 million.
(b) The funding will be based on tourist traffic volumes. Areas I and II combined will get $15 million and Areas III and IV combined will get $9 million. It is desirable to have at least one four-lane project from each area.

14.13 Moncton's Public Works Department is responsible for the collection and disposal of all solid waste within the city limits. The city must collect and dispose of an average of 300 tonnes of garbage each day. The city is considering ways to improve the current solid-waste collection and disposal system.

Table 14.3 Road Widening and Construction Projects

Area	Project	Type of Improvement	Construction Cost	Annual O&M	Annual Benefits
I	Keele Street	Four-lane	$ 980,000	$ 9,800	$ 313,600
	Lawrence Avenue	Four-lane	3,500,000	35,000	850,000
	Weston Road	Four-lane	2,800,000	28,000	672,000
	Eglinton Avenue	Four-lane	1,400,000	14,000	490,000
II	Eastern Avenue	Realign	2,380,000	47,600	523,600
	Church Street	Four-lane	5,040,000	100,800	1,310,400
	Queen Street E.	Four-lane	2,520,000	50,400	831,600
	King Street W.	Four-lane	4,900,000	98,000	1,021,000
III	O'Connor Drive	Realign	1,365,000	20,475	245,700
	Woodbine Avenue	Four-lane	2,100,000	31,500	567,000
	Leslie Street	Two-lane	1,170,000	17,550	292,000
	Sloane Avenue	Four-lane	1,120,000	16,800	358,400
IV	McNicoll Avenue	Four-lane	2,800,000	56,000	980,000
	Progress Avenue	Reconstruct	1,690,000	33,800	507,000
	Tilfield Road	Widen	975,000	15,900	273,000
	Brimley Road	Widen	1,462,500	29,250	424,200

- The present collection and disposal system uses Dempster Dumpmaster Frontend Loaders for collection, and incineration or landfill for disposal. Each collecting vehicle has a load capacity of 10 tonnes, or 24 cubic metres, and dumping is automatic. The incinerator in use was manufactured in 1942, and was designed for 150 tonnes per 24 hours. A natural-gas afterburner has been added in an effort to reduce air pollution; however, the incinerator still does not meet provincial air-pollution requirements, and it is operating under a permit from the New Brunswick provincial government. Because the capacity of the incinerator is relatively low, some garbage is not incinerated, but is taken to the city landfill. The garbage landfill is located approximately 11 kilometres and the incinerator approximately 5 kilometres from the center of the city. The metrage and costs in man-hours for delivery to the disposal sites is excessive; a high percentage of empty vehicle kilometres and man-hours is required because separate methods of disposal are used and the destination sites are remote from the collection areas. The operating cost for the present system is $905,400. This includes $624,635 to operate the incinerator, $222,928 to operate the existing landfill, and $57,837 to maintain the current incinerator.
- The proposed system locates a number of portable incinerators each with 100-tonne-per-day capacity for the collection and disposal of refuse waste collected for 3 designated areas within the city. Collection vehicles will also be staged at these incineration-disposal sites with the necessary plant and support facilities for incineration operation, collection-vehicle fueling and washing, support building for stores, and shower and locker rooms for collection and site crew personnel. The pick-up and collection procedure remains essentially the same as in the existing system. The disposal-

staging sites, however, are located strategically in the city based on the volume and location of wastes collected, thus eliminating long hauls and reducing the number of kilometres the collection vehicles must retravel from pick-up to disposal site.

Four variations of the proposed system are being considered, containing 1, 2, 3, and 4 incinerator-staging areas, respectively. The type of incinerator is a modular prepackaged unit which can be installed at several sites in the city. Such units exceed all provincial and federal standards on their exhaust emissions. The city needs 24 units, each with a rated capacity of 12.5 tonnes of garbage per 24 hours. The price per unit is $137,600, which means a capital investment of about $3,302,000. The estimated plant facilities, such as housing and foundation, were estimated to cost $200,000 per facility. This is based on a plan incorporating 4 incinerator plants strategically located around the city. Each plant would house 8 units and be capable of handling 100 tonnes of garbage per day. Additional plant features, such as landscaping, were estimated to cost $60,000 per plant.

The annual operating cost of the proposed system would vary according to the type of system configuration. It takes about 1500 to 1700 cubic metres of fuel to incinerate one tonne of garbage. The conservative 1700 cubic metres figure was used for total cost. This means that fuel cost $4.25 per tonne of garbage at a cost of $2.50 per 1000 cubic metres. Electric requirements at each plant will be 230 kW per day. If the plant is operating at full capacity, that means a $0.48 per tonne cost for electricity. Two men can easily operate one plant, but safety factors dictate three operators at a cost of $7.14 per hour. This translates to a cost of $1.72 per tonne. The maintenance cost of each plant was estimated to be $1.19 per tonne. Since 3 plants will require fewer transportation kilometres, it is necessary to consider the savings accruing from this operating advantage. Three plant locations will save 6.14 kilometres per truck per day on the average. At an estimated $0.30 per kilometre cost, this would mean an annual savings of $6750 is realized when considering minimum trips to the landfill disposer, for a total annual savings in transportation of $15,300. A labor savings is also realized because of the shorter routes, which permit more pick-ups during the day. This results in an annual savings of $103,500. The following table summarizes all costs in thousands of dollars associated with the present and proposed systems.

Item	Present System	Costs for Proposed Systems			
		Site 1	Site 2	Site 3	Site 4
Capital costs					
Incinerators		$3302	$3302	$3302	$3302
Plant facilities		600	900	1260	1920
Annex buildings		91	102	112	132
Additional features		60	80	90	100
Total		4053	4384	4764	5454
Annual O&M costs	$905.4	342	480	414	408
Annual savings					
Pick-up transportation		13.2	14.7	15.3	17.1
Labor		87.6	99.3	103.5	119.40

A bond will be issued to provide the necessary capital investment at an interest rate of 8% with a maturity date 20 years in the future. The proposed systems are expected to last 20 years with negligible salvage values. If the current system is to be retained, the annual O&M costs would be expected to increase at an annual rate

of 10%. The city will use the bond interest rate as the interest rate for any public project evaluation.

(a) Determine the operating cost of the current system in terms of dollars per tonne of solid waste.
(b) Determine the economics of each solid-waste disposal alternative in terms of dollars per tonne of solid waste.

14.14 Due to a rapid growth in population, a small town in Manitoba is considering several options in establishing a wastewater-treatment facility that can handle a wastewater flow of 8,000,000 litres per day). The town has 5 treatment options available:

• **Option 1—No action:** This option leads to continued deterioration of the environment. If growth continues and pollution results, fines imposed (as high as $10,000 per day) would soon exceed construction costs.
• **Option 2—Land treatment facility:** to provide a system for land treatment of wastewater to be generated over the next 20 years. This option will require the utilization of the most land for treatment of the wastewater. In addition to finding a suitable site, pumping of the wastewater for a considerable distance out of town will be required. The land cost in the areas is $3000 per hectare. The system will use spray irrigation to distribute wastewater over the site. No more than 2.5 centimetres of wastewater can be applied in one week per hectare.
• **Option 3—Activated sludge-treatment facility:** to provide an activated sludge-treatment facility at a site near the planning area. No pumping will be required for this alternative. Only 7 hectares of land will be needed for construction of the plant at a cost of $7000 per hectare.
• **Option 4—Trickling filter-treatment facility:** to provide a trickling filter-treatment facility at the

same site selected for the activated sludge plant of option 3. The land required will be the same as used for option 3. Both facilities will provide similar levels of treatment using different units.

- **Option 5– Lagoon treatment system:** to utilize a three-cell lagoon system for treatment. The lagoon system requires substantially more land than options 3 and 4, but less than option 2. Due to the larger land requirement, this treatment system will have to be located some distance outside of the planning area and will require pumping of the wastewater to reach the site.

The following summarizes the capital expenditures and O&M costs associated with each option:

Land Cost for each Option

Option Number	Land Required (hectares)	Land Cost	Land Value (in 20 years)
2	800	$2,400,000	$4,334,600
3	7	49,000	88,500
4	7	49,000	88,500
5	80	400,000	722,400

The price of land is assumed to be appreciating at an annual rate of 3%.

Option Number	Equipment	Capital Expenditure		Total
		Structure	Pumping	
2	$500,000	$ 700,000	$100,000	$1,300,000
3	500,000	2,100,000	0	2,600,000
4	400,000	2,463,000	0	2,863,000
5	175,000	1,750,000	100,000	2,025,000

The equipment installed will require a replacement cycle of 15 years. Its replacement cost will increase at an annual rate of 5% (over the initial cost), and its salvage value at the end of the planning horizon will be 50% of the replacement cost. The structure requires replacement after 40 years and will have a salvage value of 60% of the original cost.

Option Number	Annual Operating and Maintenance			
	Energy	Labor	Repair	Total
2	$200,000	95,000	30,000	325,000
3	125,000	65,000	20,000	210,000
4	100,000	53,000	15,000	168,000
5	50,000	37,000	5,000	92,000

The cost of energy and repair will increase at an annual rate of 5% and 2%, respectively. The labor cost will increase at an annual rate of 4%.

(a) If the interest rate (including inflation) is 10%, which option is the most cost-effective?

(b) Suppose a household discharges about 1600 litres of wastewater per day through the facility selected in (a). What should be the monthly assessed bill for this household?

CHAPTER 15

Sensitivity and Risk Analyses

The prospective profits or losses of a mining investment depend upon the vagaries of such diverse matters as metal prices, ore quality, labor negotiations, and various choices made by the mine operator. Consider the choices faced by Westmin Resources Limited when it evaluated its H-W copper-zinc deposit prior to a

H-W Mineral Deposit—Impact of Metal Price Variations

	2700	1800	1350
Production rate (tonnes per day)			
Capital cost ($ millions)	150.2	128.4	98.6
Operating cost ($/tonne)	39.02	48.33	56.69
Ore reserve* (million tonnes)	13.31	9.72	7.42
Optimistic price forecast			
Total revenue[§] ($ millions)	1,022.2	877.2	728.2
Total profits[§] ($ millions)	503.0	407.7	307.7
Gross cash flow** ($ millions)	352.8	279.3	209.1
NPV[¶] ($ millions)	56.3	49.3	34.1
IRR (percent)	19.5	20.8	19.0
Payback (years)	3.2	2.8	3.3

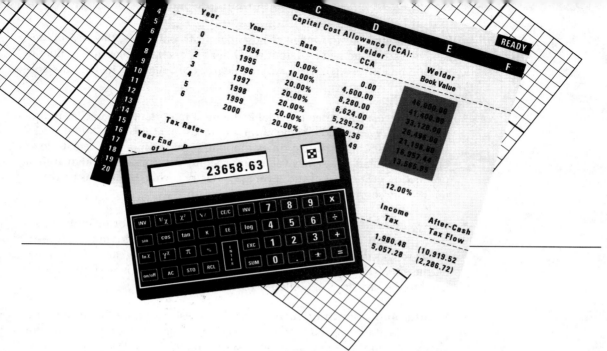

decision to proceed with development. The potential profitability of this project based on metal prices alone is illustrated at three alternate production levels below.

Pessimistic price forecast

Total revenue[§] ($ millions)	742.6	637.3	529.0
Total profits[§] ($ millions)	223.4	168.0	108.5
Gross cash flow** ($ millions)	73.2	39.6	9.9
NPV[¶] ($ millions)	(23.5)	(26.1)	(33.9)
IRR (percent)	5.6	5.0	1.6
Payback (years)	5.7	6.9	9.3

*   Represents the total ore reserve that can be utilized at the specified production level.
§   Represents the simple sums of the revenues and profits, respectively, over a 15-year project life.
** The total after-tax cash flows over the project life.
¶   Assumes an interest rate of 10% per annum.

Acknowledgement: As related by Mr. Carl C. Hunter of Dalcor Consultants Ltd., West Vancouver, B.C. (Reprinted by permission.)

While exhibiting a wide range of potential profit outcomes, the possibilities listed above are by no means definitive. Other factors that affect the profitability of a mining investment with varying degrees of reliability include the following:

- **Metal prices:** The price of metal is the most important and uncertain factor on the revenue side of metal mining investment analysis. Metal prices fluctuate from highs of five times the long term "economic price" to lows of half this value. These low values are roughly determined by the marginal production cost curve of the "average industry", while price peaks have no limit in the short term.

- **Ore grades:** It is impossible to know either the quality or quantity of ore in the ground exactly. The reliability of mining tonnage and grade estimates relates directly to the amount of drilling, sampling, and testing done on the site. The extent to which such preliminary work is performed is related to the investor's appetite for risk.

- **Capital and operating costs:** Capital cost estimates are based upon experience and the chosen level of engineering design. Typically, the investor chooses to impose a degree of accuracy on the estimate that in turn determines the level of engineering work.

  Operating costs depend on production rates and processes, the ratio of waste rock to ore, the haul distances involved, and the cost of process materials, energy, and labor. Estimating with accuracy is a function of experience, and is consequently performed by experts in various fields.

As this example illustrates, a project's economic assessment must be based on profitability calculations, which in turn are derived from estimates of variation in the significant factors that can impact on profitability. Risk analysis takes into account the range of variation as well as the likelihood of the change.

In previous chapters, the cash flows from projects were assumed to be known with complete certainty, and our analysis was concerned with measuring the economic worth of projects and selecting the best investment projects. Although these results can provide reasonable decision bases for many investment situations, we should certainly consider the more usual situation where forecasts of cash flows are subject to some degree of uncertainty. In this situation, management rarely has precise expectations regarding the future cash flows to be derived from a particular project. In fact, the best that the firm can reasonably expect to do is to estimate the range of possible future costs and benefits and the relative chances of achieving a certain return on the investment. We can use the term **risk** in describing an investment project whose cash flow is not known in advance with absolute certainty, but for which an array of alternative outcomes and their probabilities are known. We will also use the term **project risk** to refer to the variability in a project's NPW. A greater project risk means that there is a greater variability in the project's NPW. This chapter begins by exploring the origins of project risk.

■ 15.1 Origins of Project Risk

The decision to make a major capital investment such as the introduction of a new product requires cash flow information over the life of the project. The profitability estimate of the investment depends on cash flow estimations, which are generally uncertain. The factors to be estimated include the total market for the product; the market share that the firm can attain; the growth in the market; the cost of producing the product, including labor and materials; the selling price; the life of the product; the cost and life of the equipment needed; and the effective tax rates. Many of these factors are subject to substantial uncertainty. The common approach is to make single-number "best estimates" for each of the uncertain factors and then to calculate measures of profitability such as NPW or rate of return for the project. This approach has two drawbacks.

1. There is no guarantee that the "best estimates" will match actual values.
2. There is no way to measure the risk associated with the investment, or the project risk. In particular, the manager has no way of determining either the probability that the project will lose money or the probability that it will generate very large profits.

Because cash flows can be so difficult to estimate accurately, project managers frequently consider a range of possible values for cash flow elements. If there is a range of possible values for individual cash flows, it follows that there is a range of possible values for the NPW of a given project. Clearly, the analyst will want to try to gauge the probability and reliability of individual cash flows occurring and, consequently, the level of certainty about overall project worth.

Quantitative statements about risk are given as numerical probabilities, or as values for likelihoods of occurrence. Probabilities are given as decimal fractions in the interval 0.0 to 1.0. An event or outcome that is certain to occur has a probability of 1.0. As the probability of an event approaches 0, the event becomes increasingly less likely to occur. The assignment of probabilities to the various outcomes of an investment project is generally called **risk analysis;** we shall discuss this important topic in detail in the next section of this chapter. In the remaining sections, we shall assume that reasonably accurate estimates of these probabilities can be obtained, and shall focus our attention on the procedure that incorporates risk in project evaluation.

■ 15.2 Methods of Describing Project Risk

We may begin analyzing project risk by first determining the uncertainty inherent in a project's cash flows. We can do this analysis in a number of ways, ranging from making informal judgments to performing complex economic and statistical analyses. In this section, we will introduce three methods of describing project risk: (1) sensitivity analysis, (2) breakeven analysis, and (3) scenario analysis. We shall explain each method with a single example (Windsor Metal Company).

15.2.1 Sensitivity Analysis

One way to glean a sense of the possible outcomes of an investment is to perform a sensitivity analysis. This analysis determines the effect on NPW of variations in the input variables (such as revenues, operating cost, and salvage value) used to estimate after-tax cash flows. A **sensitivity analysis** reveals how much the NPW will change in response to a given change in an input variable. In a calculation of cash flows, some items have a greater influence on the final result than others. In some problems, we may easily identify the most significant item. For example, the estimate of sales volume is often a major factor in a problem in which the quantity sold varies among the alternatives. In other problems, we may want to locate the items that have an important influence on the final results so that they can be subjected to special scrutiny.

Sensitivity analysis is sometimes called "what-if" analysis because it answers questions such as: What if incremental sales are only 1000 units, rather than 2000 units? Then what will the NPW be? Sensitivity analysis begins with a base-case situation, which is developed using the most-likely values for each input. Then we change the specific variable of interest by several specific percentages above and below the most-likely value, holding other variables constant. Next, we calculate a new NPW for each of these values. A convenient and useful way to present the results of a sensitivity analysis is to plot **sensitivity graphs.** The slopes of the lines show how sensitive the NPW is to changes in each of the inputs: The steeper the slope, the more sensitive the NPW is to a change in the particular variable. It identifies the crucial variables that affect the final outcome most.

Example 15.1   Sensitivity analysis

Windsor Metal Company (WMC), a small manufacturer of fabricated metal parts, must decide whether to enter the competition to be the supplier of transmission-housings for Gulf Electric. To compete, the firm must design a new fixture for the production process and purchase a new forge. The new forge would cost $125,000, including tooling costs for the transmission-housings. If WMC gets the order, it may be able to sell as many as 2000 units per year to Gulf Electric for $50 each, and the variable production costs[1], such as direct labor and direct material costs, will be $15 per unit. The increase in fixed costs[2] other than capital cost allowance will amount to $10,000 per year. The firm expects that the proposed transmission-housings project would have about a 5-year product life. The firm also estimates that the amount ordered by Gulf Electric for the first year will be ordered in each of the subsequent 4 years. (Due to the nature of contracted production, the annual demand and unit price would remain the same over the project after the contract is signed.) The initial investment can be written off with a declining balance rate of 30%, and the firm's income tax rate is expected to remain at 40%. At the end of 5 years, the forge machine is expected to retain a market value of about 40% of the original investment. Based on this information, the engineering and marketing staffs have prepared the cash flow forecasts shown in Table 15.1. Since $PW(i)$ is positive ($43,443) at the 15% opportunity cost of capital (MARR), the project appears to be worth undertaking.

However, WMC's managers are uneasy about this project because there are too many uncertain elements that have not been considered in the analysis. Table 15.1 shows WMC's expected cash flows—but there is no guarantee that they will indeed materialize. WMC is not particularly confident in its revenue forecasts. The managers think that, if competing firms enter the market, WMC will lose a substantial portion of the projected revenues. They also are not certain about the variable and fixed cost figures. Recognizing these uncertainties, the managers want to assess the various potential future outcomes before making a final decision. Perform a sensitivity analysis to each variable and develop a sensitivity graph.

Discussion: Consider the data given in Table 15.1 and the related calculations, in which we developed projected cash flows for WMC's

---

[1] Expenses that change in direct proportion to the change in volume of sales or production.

[2] Expenses that do not vary as the volume of sales or production changes. For example, property taxes, insurance, capital cost allowance, and rent are usually fixed expenses.

Table 15.1 After-tax Cash Flow for WMC's Transmission-Housings Project (Example 15.1)

	0	1	2	3	4	5
Income Statement						
Revenues		$100,000	$100,000	$100,000	$100,000	$100,000
Variable cost		30,000	30,000	30,000	30,000	30,000
Fixed cost		10,000	10,000	10,000	10,000	10,000
CCA*		18,750	31,875	22,313	15,619	10,933
Taxable income		41,250	28,125	37,688	44,381	49,067
Income taxes		16,500	11,250	15,075	17,753	19,627
Net income		$24,750	$16,875	$22,613	$26,629	$29,440
Cash Flow Statement						
Net income		$24,750	$16,875	$22,613	$26,629	$29,440
CCA		18,750	31,875	22,313	15,619	10,933
Investment/Salvage	($125,000)					50,000
Disposal tax effect						(9,796)
Net cash flow	($125,000)	$43,500	$48,750	$44,925	$42,248	$80,578

Demand: 2000 units
Unit price: $50 per unit
Unit variable (direct labor, direct material) cost: $15 per unit
Annual fixed cost excluding CCA: $10,000 per year
* Half-year convention

Present worth = $43,443
Annual worth = $12,960
Internal ROR = 27.8%

transmission-housings project. Before undertaking the project described, the company wants to identify the key variables that determine whether the project will succeed or fail. The marketing department has estimated revenue as follows:

$$\text{annual revenue} = (\text{product demand})(\text{unit price})$$

$$= (2000)(\$50) = \$100,000$$

The engineering department has estimated variable costs such as labor and material per unit at $15. Since the projected sales volume is 2000 units per year, the total variable cost is $30,000.

Having defined the unit sales, unit price, unit variable cost, and fixed cost, we conduct a sensitivity analysis with respect to these key input variables. This is done by varying each of the estimates by a given percentage and determining what effect the variation in that item has on the final results. If the effect is large, the result is *sensitive* to that item. Our objective is to locate the most sensitive item(s).

Solution

Sensitivity analysis: We begin the sensitivity analysis with a "base-case" situation, which reflects the best estimate (expected value) for each input variable. In developing Table 15.2, we changed a given variable by 20% in 5% increments, above and below the base-case value and calculated new NPWs, holding other variables constant. The values for both sales and operating costs were the expected, or base-case, values, and the resulting $43,443 is called the base-case NPW. Now we ask a series of "what-if" questions: What if sales are 20% below the expected level? What if operating costs rise? What if the unit price drops from $50 to $45? Table 15.2 summarizes the results of varying the values of the key input variables.

Sensitivity graph: Figure 15.1 shows the transmission project's sensitivity graphs for five of the key input variables. The base-case NPW is plotted on the ordinate of the graph at the value 0% on the abscissa. Next, the value of product demand is reduced by 5% of its base-case value, and the NPW is recomputed with all other variables held at their base-case value. We repeat the process by either decreasing or increasing the relative deviation from the base case. The lines for the variable unit price, variable unit cost, fixed cost, and salvage value are obtained in the same manner. In Fig. 15.1, we see that the project's NPW is very sensitive to changes in product

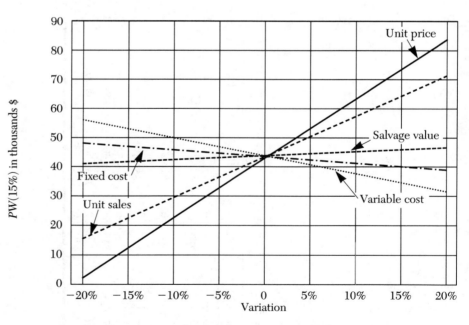

Figure 15.1 ■ Sensitivity graph—WMC's transmission-housings project (Example 15.1)

Table 15.2 Sensitivity Analysis for Five Key Input Variables (Example 15.2)

Deviation	-20%	-15%	-10%	-5%	0	5%	10%	15%	20%
Unit sales	1,600	1,700	1,800	1,900	2,000	2,100	2,200	2,300	2,400
PW(15%)	$15,285	$22,325	$29,364	$36,404	$43,443	$50,483	$57,523	$64,562	$71,602
Price ($)	40	42.5	45	47.5	50	52.5	55	57.5	60
PW(15%)	$ 3,218	$13,274	$23,331	$33,387	$43,443	$53,500	$63,556	$73,613	$83,669
Variable cost	12	12.75	13.5	14.25	15	15.75	16.5	17.25	18
PW(15%)	$55,511	$52,494	$49,477	$46,460	$43,443	$40,426	$37,410	$34,392	$31,376
Fixed cost	8,000	8,500	9,000	9,500	10,000	10,500	11,000	11,500	12,000
PW(15%)	$47,466	$46,460	$45,455	$44,449	$43,443	$42,438	$41,432	$40,427	$39,421
Salvage	40,000	42,500	45,000	47,500	50,000	52,500	55,000	57,500	60,000
PW(15%)	$40,460	$41,206	$41,952	$42,698	$43,443	$44,189	$44,935	$45,681	$46,426

demand and unit price, is fairly sensitive to changes in the variable costs, and is relatively insensitive to changes in the fixed cost and the salvage value.

Graphic displays such as that in Fig. 15.1 provide a useful means to communicate the relative sensitivities of the different variables on the corresponding NPW value. However, the sensitivity graph does not explain any variable interactions among the variables or the likelihood of realizing any specific deviation from the base case. Certainly, it is conceivable that an answer might not be very sensitive to changes in either of two items, but very sensitive to combined changes in them.

15.2.2 Breakeven Analysis

When we perform a sensitivity analysis of a project, we are asking how serious the effect of lower revenues or higher costs will be. Managers sometimes prefer to ask how much sales can decrease below forecasts before the project begins to lose money. This type of analysis is known as **breakeven analysis.** In other words, breakeven analysis is a technique for studying the effect of variations in output on the firm's NPW. We will present an approach to breakeven analysis based on the project's cash flows.

To illustrate the procedure of breakeven analysis based on NPW, we use the generalized cash flow approach discussed in Chapter 9. We compute the PW of cash inflows as a function of an unknown variable (say X)—this variable could be annual sales, for example.

$$\text{PW of cash inflows} = f(x)_1$$

Next, we compute the PW of cash outflows as a function of X.

$$\text{PW of cash outflows} = f(x)_2$$

NPW is, of course, the difference between these two numbers. Then, we look for the breakeven value of x that makes

$$f(x)_1 = f(x)_2.$$

Note that this breakeven value calculation is similar to that of the internal rate of return where we want to find the interest rate that makes the NPW equal zero and many other similar "cutoff values" where a choice changes.

Example 15.2   Breakeven analysis

Through the sensitivity analysis in Example 15.1, the WMC managers are convinced that the NPW is most sensitive to changes in annual sales volume. Determine the breakeven NPW value as a function of that variable.

Table 15.3 Breakeven Analysis with Unknown Annual Sales X (Example 15.2)

Items	0	1	2	3	4	5
				n		
Cash inflow						
Net salvage:						$40,204
Revenue:						
$X(1 - 0.4)50$		30X	30X	30X	30X	30X
CCA credit:						
$+0.4$ (CCA)		$7,500	$12,750	$8,925	$6,248	$4,373
Cash outflow						
Investment:	−$125,000					
Variable cost:						
$-X(1 - 0.4)15$		−9X	−9X	−9X	−9X	−9X
Fixed cost:						
$-0.6(10,000)$		−$6,000	−$6,000	−$6,000	−$6,000	−$6,000
Net cash flow	−$125,000	21X + $1,500	21X + $6,750	21X + $2,925	21X + $248	21X + $38,577

Solution

The analysis is shown in Table 15.3, where we set out the revenues and costs of the WMC transmission-housings project in terms of an unknown amount of annual sales X. (Notice that, if the project suffers a loss, this loss can be used to reduce the tax due on the rest of the company's business. In this case, the project produces a tax savings—the tax outflow is negative.)

$$PW \text{ of cash inflows} = (PW \text{ of after-tax net revenue})$$
$$+ (PW \text{ of net salvage value})$$
$$+ (PW \text{ of tax savings from CCA})$$
$$PW \text{ of cash outflows} = (PW \text{ of capital expenditure})$$
$$+ (PW \text{ of after-tax expenses})$$

We calculate both the PWs of cash inflow and outflows as:

- PW of cash inflows:

$$PW(15\%)_{inflow} = 30X(P/A, 15\%, 5) + \$40,204(P/F, 15\%, 5)$$
$$+ \$7500(P/F, 15\%, 1) + \$12,750(P/F, 15\%, 2)$$
$$+ \$8925(P/F, 15\%, 3) + \$6248(P/F, 15\%, 4)$$
$$+ \$4373(P/F, 15\%, 5)$$
$$= 30X(P/A, 15\%, 5) + \$47,766$$
$$= 100.566X + \$47,766$$

- PW of cash outflows:

$$PW(15\%)_{outflow} = \$125,000 + (9X + \$6000)(P/A, 15\%, 5)$$
$$= 30.1694X + \$145,113$$

The NPW of all cash flows for the WMC is thus

$$PW(15\%) = (30X - 9X - \$6000)(P/A, 15\%, 5)$$
$$+ (\$47,766 - \$125,000)$$
$$= (21X - \$6000)(3.3522) - \$77,234$$
$$= 70.3962X - \$97,347$$

In Table 15.4, we compute the PW of these revenues and costs to give us the PW of the inflows and the PW of the outflows.

Roughly estimating X, we can see that the NPW will be negative if the company produces and sells fewer than 1300 transmission-housings. The

Table 15.4 Determination of Breakeven Volume Based on Project's NPW
(Example 15.2)

Units (X)	PW of Inflow (100.566X + $47,766)	PW of Outflow (30.1694X + $145,113)	NPW (70.3962X − $97,347)
0	$ 47,766	$145,113	($97,347)
500	98,049	160,198	(62,149)
1000	148,332	175,282	(26,951)
1500	198,615	190,367	8,247
2000	248,898	205,452	43,443
2500	299,181	220,537	78,643

NPW will be just slightly positive if the company sells 1400 units.
Precisely calculated, the zero-NPW point (breakeven volume) is 1383
units:

$$PW(15\%) = 70.3962X - \$97,347$$

$$= 0$$

$$X_b = 1383 \text{ units}$$

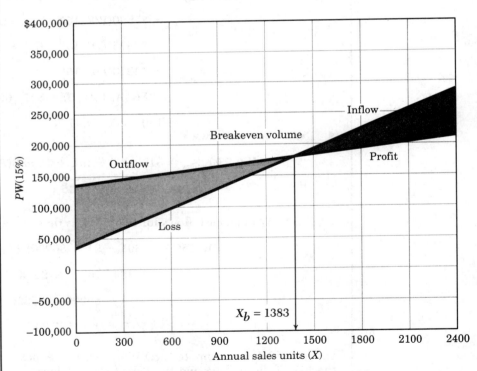

Figure 15.2 ■ Breakeven analysis based on net cash flow (Example 15.2)

In Fig. 15.2, we have plotted the PWs of the inflows and outflows under various assumptions about annual sales. The two lines cross when sales are 1383 units, the point at which the project has a zero NPW. Again we see that, as long as sales are greater than 1383, the project has a positive NPW.

15.2.3 Scenario Analysis

Although both sensitivity and breakeven analyses are useful, they do have limitations. It is often difficult to specify precisely the relationship between a particular variable and the NPW. The relationship is further complicated by interdependencies among the variables. Holding operating costs constant while varying unit sales may ease the analysis, but in reality, operating costs do not behave in that manner. Yet it may complicate the analysis too much to permit movement in more than one variable at a time. A scenario analysis is a technique that does consider the sensitivity of NPW both to changes in key variables and to the range of likely variable values. For example, the decision-maker may consider two extreme cases, a "worst-case" scenario (low unit sales, low unit price, high variable cost per unit, high fixed cost, and so on) and a "best-case" scenario. The NPWs under the worst and best conditions are then calculated and compared to the expected, or base-case, NPW.

Example 15.3   Scenario analysis

Consider again WMC's transmission-housings project in Example 15.1. Assume that the company's managers are fairly confident of their estimates of all the project's cash flow variables except that for unit sales. Further, assume that they regard a decline in unit sales to below 1600 or a rise above 2400 as extremely unlikely. Thus, decremental annual sales of 400 units defines the lower bound, or the worst-case scenario, whereas incremental annual sales of 400 units defines the upper bound, or the best-case scenario. (Remember that the most-likely value was 2000 in

Table 15.5 Scenario Analysis for WMC (Example 15.3)

Variable Considered	Worst-case Scenario	Most-likely-case Scenario	Best-case Scenario
Unit demand	1600	2000	2200
Unit price ($)	48	50	53
Variable cost ($)	17	15	12
Fixed cost ($)	11,000	10,000	8,000
Salvage value ($)	30,000	50,000	60,000
PW(15%)	($5,564)	$43,443	$91,077

annual unit sales.) Discuss the worst- and best-case scenarios, assuming that the unit sales for all 5 years would be equal.

Discussion: To carry out the scenario analysis, we ask the marketing and engineering staffs to give optimistic (best-case) and pessimistic (worst-case) estimates for the key variables. Then we use the worst-case variable values to obtain the worst-case NPW and the best-case variable values to obtain the best-case NPW. Table 15.5 summarizes the results of our analysis. We see that the base-case produces a positive NPW, the worst-case produces a negative NPW, and the best-case produces a large positive NPW.

By just looking at the results in Table 15.5, it is not easy to interpret scenario analysis or to make a decision based on it. For example, we can say that there is a chance of losing money on the project, but we do not yet have a specific probability for this possibility. Clearly, we need estimates of the probabilities of occurrence of the worst-case, the best-case, the base-case (most-likely), and all the other possibilities. This need leads us directly to the next step, developing a probability distribution. If we can predict the effects on the NPW of variations in the parameters, why should we not assign a probability distribution to the possible outcomes of each parameter and combine these distributions in some way to produce a probability distribution for the possible outcomes of the NPW? We shall consider this issue in the next sections.

■ 15.3 Probability Concepts for Investment Decisions

In this section, we shall assume that the analyst has available the probabilities of future events from either previous experience in a similar project or a market survey. The use of probability information can provide management with a range of possible outcomes and the likelihood of achieving different goals under each investment alternative.

15.3.1 Assessment of Probabilities

We will first define terms related to probability concepts, such as random variable, probability distribution, and cumulative probability distribution.

Random Variables

A **random variable** is a parameter or variable that can have more than one possible value. The value of a random variable at any one time is unknown until the event occurs, but the probability that the random variable will have a specific value is known in advance. In other words, with each possible value of the random variable, there is associated a likelihood, or probability, of occurrence. For example, when your college team plays a football game, there are only three possible *events* regarding the game outcome: win, lose, or tie. Here the game outcome is a random variable, and is largely dictated by the strength of your opponent.

To indicate random variables, we will adopt the convention of using a capital italic letter (for example, X). To denote the situation where the random variable takes a specific value, we will use a lower-case italic letter (for example, x). Random variables are classified as either *discrete* or *continuous*. Any random variables that take on only isolated values are **discrete random variables. Continuous random variables** may have any value in a certain interval. For example, the game outcome described above should be a discrete random variable. But suppose you are interested in the amount of beverage sold on a given game day. The quantity of beverage sold will depend on the weather conditions, the number of people attending the game, and other factors. In this case, the quantity is a random variable which takes a continuous value.

Probability Distributions

For a discrete random variable, you need to assess the probability figure for each random event. For a continuous random variable, you need to assess the probability function as the event takes place over a continuous domain. In either case, there is a range of possibilities for each feasible outcome which together make up the **probability distribution.**

Probability assessments may be based on past observations or historical data if the same trends or characteristics of the past are expected to prevail in the future. Forecasting weather or predicting a game outcome in many professional sports is done based on the compiled statistical data. Any probability assessments based on objective data are called **objective probabilities.** We are not restricted to objective probabilities. In many real investment situations, there are no objective data to look for. In these situations, we assign the **subjective probabilities** that we think appropriate to the possible states of nature. As long as we act consistently with our beliefs about the possible events, we may reasonably account for the economic consequences of those events in our profitability analysis.

For a continuous random variable, we usually try to establish the range of values; that is, we try to determine a minimum value (L) and a maximum value (H). Next, we determine whether any value within these limits might be more likely to occur than are the other values. That is, does the distribution have a mode (M_o), or a most likely value?

If there is a mode, we can represent the variable by a triangular distribution, such as that shown in Fig. 15.3. If we have no reason to assume that one value is any more likely to occur than any other value, perhaps the best we can do is to describe the variable as a uniform distribution, as shown in Fig. 15.4. These two distributions are frequently used to represent the variability of a random variable when the only information we have is its minimum, its maximum, and whether or not the distribution has a mode. For example, suppose it was the best judgment of the analyst that the sales revenue could vary anywhere from $2000 to $5000 per day, and any value within the range is equally likely. This

Figure 15.3 ■ A triangular probability distribution

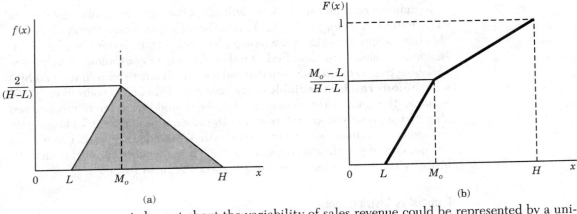

(a)   (b)

judgment about the variability of sales revenue could be represented by a uniform distribution.

For the WMC's transmission-housings project, we can think of the discrete random variables (X and Y) as those variables whose values cannot be predicted with certainty at the time of decision making. Let us assume the probability distributions in Table 15.6: We see that the product demand with the highest probability is 2000 units, whereas the unit sales price with the highest probability is $50. These, therefore, are the most-likely values. We also see that there is a substantial probability that a unit demand other than 2000 units will be realized. When we use only the most-likely values in an economic analysis, we ignore these other outcomes.

Cumulative Distribution

As we have observed in the previous section, the probability distribution provides information regarding the probability that a random variable will be some value x. We can use this information, in turn, to define the cumulative

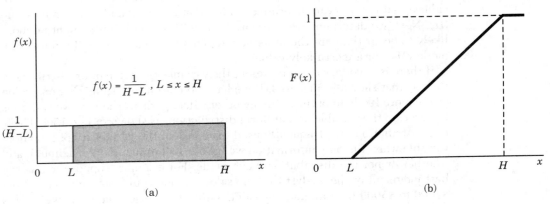

(a)   (b)

Figure 15.4 ■ A uniform probability distribution

Table 15.6 Probability Distributions for Unit Demand (X) and Unit Price (Y) for the WMC's Transmission-Housings Project (X and Y are independent random variables.)

Product Demand (X)		Unit Sale Price (Y)	
Units (x)	P(X = x)	Unit price (y)	P(Y = y)
1600	0.20	48	0.30
2000	0.60	50	0.50
2400	0.20	53	0.20

distribution function. The cumulative distribution function shows the probability that the random variable will attain a value smaller than or equal to some value x. A common notation for the cumulative distribution is

$$F(x) = P(X \leq x) = \begin{cases} \displaystyle\sum_{j=1}^{J} p_j & \text{for discrete random variable} \\ \displaystyle\int_{-\infty}^{x} f(x)\, dx & \text{for continuous random variable,} \end{cases}$$

where $x_J \leq x < x_J + 1$ p_j is the probability of occurrence of value $x_j (x_j \leq x)$ of the discrete random variable, and $f(x)$ is a probability function for a continuous variable X.

In Example 15.4, we shall explain the method of incorporating probabilistic information into our analysis by using WMC's transmission-housings project. With respect to a continuous random variable, the cumulative distribution rises continuously in a smooth (rather than stepwise) fashion. In the next section, we shall show how to compute some composite statistics using all the data.

Example 15.4 Cumulative probability distributions

Suppose that the only parameters subject to risk are the number of unit sales (X) to Gulf Electric each year and the unit sales price (Y). From experience in the market, WMC assesses the probabilities of outcomes for each variable as shown in Table 15.6. Determine the cumulative probability distributions for these random variables.

Solution

Consider the demand probability distribution (X) given previously in Table 15.6 for WMC's transmission-housings project:

Unit Demand (X)	Probability, P(X = x)
1600	0.2
2000	0.6
2400	0.2

If we want to know the probability that the demand will be less than or equal to any particular value, we can use the following cumulative probability function:

$$F(x) = P(X \le x) = \begin{cases} 0, x \le 1600 \\ 0.2, 1600 < x < 2000 \\ 0.8, 2000 \le x < 2400 \\ 1.0, x \ge 2400 \end{cases}$$

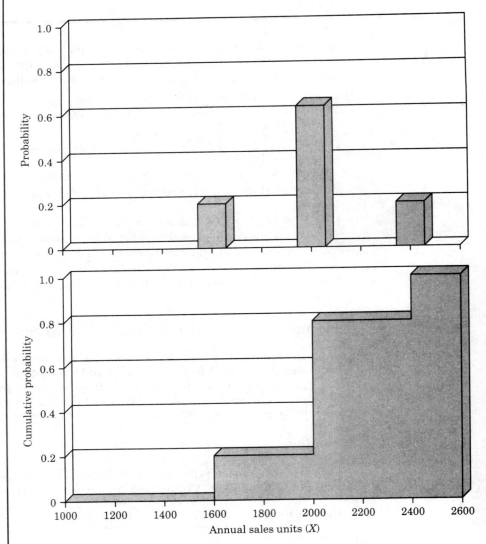

Figure 15.5 ■ Probability and cumulative probability distributions for random variable X (annual sales)

For example, if we want to know the probability that the demand will be less than or equal to 2000, we can examine the appropriate value $x = 2000$ and we shall find the probability of 80%.

We can find the cumulative distribution for Y in a similar fashion. Graphic representations of the probability distributions and the cumulative probability distributions for X and Y are given in Figs. 15.5 and 15.6, respectively.

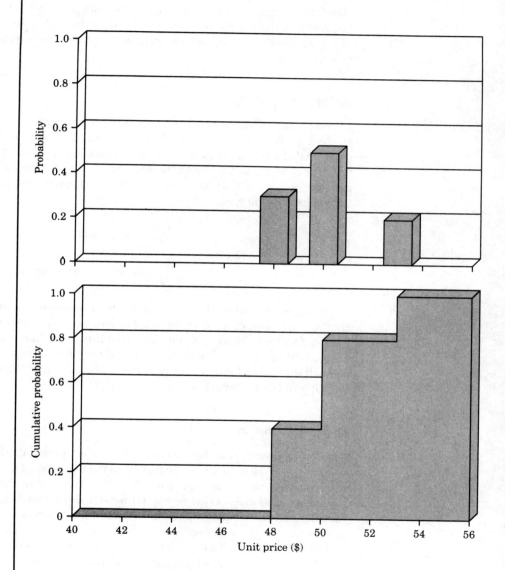

Figure 15.6 ■ Probability and cumulative probability distributions for random variable Y (unit price)

15.3.2 Summarization of Probabilistic Information

Although knowledge of the probability distribution of a random variable allows us to make a specific probability statement, a single value that may characterize the random variable and its probability distribution is often desirable. Such a quantity is the expected value of the random variable. We also want to know something about how the values of the random variable are dispersed about the mean. In investment analysis, this dispersion information is interpreted as the degree of project risk. Mean indicates the weighted average of the random variable, and the variance captures the variability of the random variable.

Measure of Expectation

The expected value (also called the **mean**) is a weighted average value of the random variable where the weighting factors are the probabilities of occurrence. All distributions (discrete and continuous) have an expected value. We will use $E[X]$ (or μ) to denote the expected value of random variable X. For a random variable X that has either discrete or continuous values, we compute an expected value with

$$E[X] = \mu = \begin{cases} \sum p_j\, x_j, & \text{discrete case} \\ \\ \int xf(x)\, dx, & \text{continuous case} \end{cases} \tag{15.1}$$

The expected value of a distribution tells us important information about the "average" or expected value of a random variable such as the NPW, but it does not tell us anything about the variability on either side of the expected value. Will the range of possible values of the random variable be very small, with all the values located at or near the expected value? We shall examine this question in the following section.

Measure of Variation

Another measure that we need in analyzing probabilistic situations is a measure of the risk due to the variability of the outcomes. There are several measures of the variation of a set of numbers that are used in statistical analysis—the **range** and the **variance** (or **standard deviation**), among others. The variance and the standard deviation are used most commonly in the analysis of risk situations. We will use $Var[X]$ or σ_x^2 to denote the variance, and σ_x to denote the standard deviation of random variable X. (If there is only one random variable in an analysis, we normally omit the subscript.)

The variance tells us the degree of spread, or dispersion, of the distribution on either side of the mean value. As the variance increases, the spread of the distribution increases; the smaller the variance, the narrower the spread about the expected value. For the unit demand (X) in Fig. 15.5 and the unit price (Y) in Fig. 15.6, the variance of the narrow distribution of Y is less than the variance of the wider distribution of X.

To determine the variance, we first calculate the deviation of each possible outcome x_j from the expected value ($x_j - \mu$), then raise the result to the second power and multiply it by the probability of x_j occurring (that is, p_j). The summation of all these products serves as a measure of the distribution's variability. For a random variable that has only discrete values, the equation to compute variance[3] is

$$Var[X] = \sigma_x^2 = \sum (x_j - \mu)^2 p_j, \qquad (15.2)$$

where p_j is the probability of occurrence of value x_j of the random variable, and μ is as defined by Eq. (15.1). To be most useful, any measure of risk should have a definite value. One such measure is the standard deviation. To calculate the standard deviation, we take the positive square root of $Var[X]$, which is measured in the same units as is X:

$$\sigma_x = \sqrt{Var[X]}. \qquad (15.3)$$

The standard deviation is a probability-weighted deviation (more precisely, the square root of the sum of squared deviations) from the expected value. Thus, it gives us an idea of how far above or below the expected value the actual value is likely to be. For most probability distributions, the actual value will be observed within the $\pm 3\sigma$ range.

In practice, the actual calculation of the variance is somewhat easier if we use the following formula.

$$Var[X] = \sum p_j x_j^2 - \left(\sum p_j x_j \right)^2$$
$$= E[X^2] - (E[X])^2 \qquad (15.4)$$

The term $E[X^2]$ in Eq. (15.4) is interpreted as the mean value of the squares of the random variable (that is, the actual values squared). The second term is simply the mean value squared. The following example will illustrate computation of the measures of variation.

---

[3] For a continuous random variable, we compute the variance as follows:

$$Var[X] = \int (x - \mu)^2 f(x)\, dx.$$

Example 15.5

Consider the WMC's transmission-housings project. Using unit sales (X) and unit price (Y) as estimated in Table 15.6, compute the means, variances, and standard deviations for the random variables X and Y.

Solution

- For the product demand variable (X):

x_j	p_j	$x_j p_j$	$(x_j - E[X])^2$	$(x_j - E[X])^2 p_j$
1600	0.20	320	$(-400)^2$	32,000
2000	0.60	1200	0	0
2400	0.20	480	$(400)^2$	32,000
		$E[X] = 2000$		$Var[X] = 64,000$
				$\sigma_x = 252.98$

- For the variable unit price (Y):

y_j	p_j	$y_j p_j$	$(y_j - E[Y])^2$	$(y_j - E[Y])^2 p_j$
\$48	0.30	14.40	$(-2)^2$	1.20
\$50	0.50	25.00	$(0)^2$	0
\$53	0.20	10.60	$(3)^2$	1.80
		$E[Y] = 50.00$		$Var[Y] = 3.00$
				$\sigma_y = 1.73$

15.3.3 Joint and Conditional Probabilities

Thus far, we have not treated the values of variables influencing the values of others. It is, however, entirely possible—indeed, it is likely—that the values of some parameters will be dependent on the values of others. We commonly express these dependencies in terms of conditional probabilities. An example is that product demand probably will be influenced by unit price.

We define a joint probability as

$$P(x,y) = P(X = x \mid Y = y)P(y) \tag{15.5}$$

where $P(X = x \mid Y = y)$ is the **conditional probability** of observing x given $Y = y$, and $P(y)$ is the **marginal probability** of observing all events Y. If X and Y are **independent**, then the joint probability is simply

$$P(x,y) = P(x)P(y). \tag{15.6}$$

The concepts of joint, marginal, and conditional distributions are best illustrated by numerical examples.

Table 15.7 Assessments of Conditional and Joint Probabilities

Unit price Y	Probability	Unit Sales X	Conditional Probability	Joint Probability
$48	0.30	1600	0.10	0.030
		2000	0.64	0.192
		2400	0.26	0.078
50	0.50	1600	0.17	0.085
		2000	0.66	0.330
		2400	0.17	0.085
53	0.20	1600	0.50	0.100
		2000	0.40	0.080
		2400	0.10	0.020

Suppose that the WMC marketing staff estimates that, for a given unit price of $48, the conditional probability that the company can sell 1600 units is 0.10. Then, the probability of this joint event (unit sales = 1600 and unit sales price = $48) is

$$P(x,y) = P(x = 1600, y = \$48)$$
$$= P(x = 1600 \mid y = \$48)P(y = \$48)$$
$$= 0.10 \times 0.30$$
$$= 0.03.$$

We can obtain the probabilities for other joint events in a similar fashion; they are presented in Table 15.7.

From Table 15.7, we can see that the unit demand (X) ranges from 1600 to 2400 units, the unit price (Y) ranges from $48 to $53, and that there are nine possible joint events. The sum of these joint probabilities must equal 1.

Joint Event (X,Y)	P(X,Y)
(1600, $48)	0.030
(2000, $48)	0.192
(2400, $48)	0.078
(1600, $50)	0.085
(2000, $50)	0.330
(2400, $50)	0.085
(1600, $53)	0.100
(2000, $53)	0.080
(2400, $53)	0.020
	Sum = 1.000

The marginal distribution for x can be developed from the joint by fixing x and summing over y:

x	$P(x_j) = \sum_y P(x, y)$
1600	$P(1600, \$48) + P(1600, \$50) + P(1600, \$53) = 0.215$
2000	$P(2000, \$48) + P(2000, \$50) + P(2000, \$53) = 0.602$
2400	$P(2400, \$48) + P(2400, \$50) + P(2400, \$53) = 0.183$

This distribution tells us that 60.2% of the time we can expect to have the demand of 2000 units, and that 21.5 and 18.3% of the time we can expect to have the demand of 1600 and 2400 units, respectively.

Suppose now that we know that the unit demand was 1600 and we want to develop a conditional distribution for the unit price (y) that accounts for this fact. The conditional distribution is

y	$P(y \mid x = 1600) = \dfrac{P(x, y)}{P(x)}$
\$48	$\dfrac{P(1600, \$48)}{P(1600)} = \dfrac{0.03}{0.215} = 0.140$
\$50	$\dfrac{P(1600, \$50)}{P(1600)} = \dfrac{0.085}{0.215} = 0.395$
\$53	$\dfrac{P(1600, \$53)}{P(1600)} = \dfrac{0.10}{0.215} = 0.465$

Note that this differs from the original unconditional distribution for Y shown in Table 15.6. For example, here, for a given demand of 1600 units, there is 0.140 probability that the unit price will be \$48.

■ 15.4 Probability Distribution of NPW

After we have identified the random variables in the project and assessed the probabilities of the possible events, the next step is to develop the NPW distribution of the project.

15.4.1 Procedure for Developing a NPW Distribution

We will consider the situation where all random variables used in calculating NPW are independent. To develop the NPW distribution, we may follow these steps.

- Express the NPW as a function of unknown random variables.
- Determine the probability distribution for each random variable.

- Determine the joint events and their probabilities.
- Evaluate the NPW equation at these joint events.
- Order the NPW values in increasing order of NPW.

These steps can best be illustrated by a numerical example.

Example 15.6

Consider WMC's transmission-housings project. With the unit demand (X) and price (Y) and their distributions given in Table 15.6, develop the NPW distribution for the WMC's transmission-housings project. Then, calculate the mean and variance of the NPW distribution.

Solution

Table 15.8 summarizes the after-tax cash flow for the WMC's transmission-housings project as functions of random variables X and Y. From this table, we can compute the PW of cash inflows as follows:

$$PW(15\%) = 0.6XY(P/A, 15\%, 5) + \$47,766$$

$$= 2.0113XY + \$47,766.$$

The PW of cash outflows is

$$PW(15\%) = \$125,000 + (9X + \$6000)(P/A, 15\%, 5)$$

$$= 30.1694X + \$145,113.$$

Thus, the NPW is

$$PW(15\%) = 2.0113X(Y - 15) - \$97,347.$$

If the product demand X and the unit price Y are random variables, the $PW(15\%)$ will also be a random variable. To determine the NPW distribution, we need to consider all the combinations of possible outcomes. The first possibility is the event where $x = 1600$ and $y = \$48$. Since X and Y are considered to be independent random variables, the probability of this joint event is

$$P(x = 1600, y = \$48) = P(x = 1600)P(y = \$48)$$

$$= (0.20)(0.30)$$

$$= 0.06.$$

Table 15.8 After-Tax Cash Flow as a Function of Unknown Unit Demand (X) and Unit Price (Y) (Example 15.6)

Items	0	1	2	3	4	5
				n		
Cash inflow						
Net salvage value						$40,204
Revenue:						
$X(1-0.4)Y$		0.6XY	0.6XY	0.6XY	0.6XY	0.6XY
CCA credit:						
0.4 (CCA)		$7,500	$12,750	$8,925	$6,248	$4,373
Cash outflow						
Investment	($125,000)					
Variable cost:						
$-X(1-0.4)15$		−9X	−9X	−9X	−9X	−9X
Fixed cost:						
$-0.6(10,000)$		($6,000)	($6,000)	($6,000)	($6,000)	($6,000)
Net cash flow	($125,000)	0.6X(Y − 15) + $1,500	0.6X(Y − 15) + $6,750	0.6X(Y − 15) + $2,925	0.6X(Y − 15) + $248	0.6X(Y − 15) + $38,577

Present worth of cash inflows:

$$PW(15\%) = 0.6XY(P/A, 15\%, 5) + \$7,500(P/F, 15\%, 1) + \$12,750(P/F, 15\%, 2) + \$8,925(P/F, 15\%, 3)$$
$$+ \$6,248(P/F, 15\%, 4) + \$4,373(P/F, 15\%, 5) + \$40,204(P/F, 15\%, 5)$$
$$= 2.0113XY + \$47,766$$

Present worth of cash outflows:

$$PW(15\%) = \$125,000 + (9X + \$6,000)(P/A, 15\%, 5)$$

NPW:

$$PW(15\%) = 2.0113XY + \$47,766 - \$125,000 - (9X + \$6,000)(P/A, 15\%, 5)$$
$$= 2.0113X(Y - 15) - \$97,347$$

With these values as input, we compute the possible NPW outcome as follows:

$$PW(15\%) = 2.0113X(Y - 15) - \$97,347$$

$$= 2.0113(1600)(48 - 15) - \$97,347$$

$$= \$8850.$$

There are eight other possible outcomes: They are summarized with their joint probabilities in Table 15.9, and depicted in Fig. 15.7.

The NPW probability distribution in Table 15.9 indicates that the project's NPW varies between \$8850 and \$86,084, but that there is no loss under any of the circumstances examined. From the cumulative distribution, we further observe that there is a 0.38 probability that the project would realize an NPW less than that forecast for the base-case situation (\$43,443). On the other hand, there is a 0.32 probability that the NPW will be greater than this value. Certainly, the probability distribution provides much more information on the likelihood of each possible event than does the scenario analysis presented in Section 15.2.3.

We have developed a probability distribution for the NPW by considering the random cash flows. As we have observed, the probability distribution helps us to see what the data imply in terms of the risk of the project. Now, we shall learn how to summarize the probabilistic information—the mean and the variance. For the WMC's transmission-housings project, we compute the expected value of the NPW distribution as shown in Table 15.10. Note that this expected value is the same as the most-likely value of the NPW distribution. This equality was expected because X and Y have a symmetrical probability distribution, respectively.

Table 15.9 The NPW Probability Distribution with Independent Random Variables (Example 15.6)

Event No.	Outcome x	Outcome y	Marginal Probability X $p(x)$	Marginal Probability Y $p(y)$	Joint Probability $P(x, y)$	Cumulative Joint Probability	NPW
1	1600	$ 48	0.200	0.300	0.060	0.060	$ 8,850
2	1600	50	0.200	0.500	0.100	0.160	15,286
3	1600	53	0.200	0.200	0.040	0.200	24,940
4	2000	48	0.600	0.300	0.180	0.380	35,399
5	2000	50	0.600	0.500	0.300	0.680	43,443*
6	2000	53	0.600	0.200	0.120	0.800	55,512
7	2400	48	0.200	0.300	0.060	0.860	61,948
8	2400	50	0.200	0.500	0.100	0.960	71,602
9	2400	53	0.200	0.200	0.040	1.000	86,084

* base-case estimate

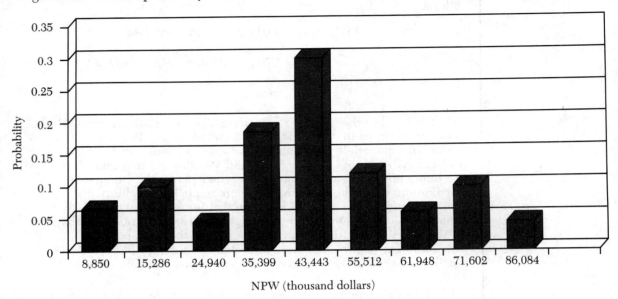

We obtain the variance of the NPW distribution, assuming independence between X and Y and using Eq. (15.2), as shown in Table 15.11. We would obtain the same result more easily by using Eq. (15.4).

Table 15.10 Calculation of the Mean of NPW Distribution (Example 15.6)

Event No.	Outcome x	Outcome y	Marginal Probability X $p(x)$	Marginal Probability Y $p(y)$	Joint Probability $P(x, y)$	NPW	Weighted NPW
1	1600	$ 48	0.200	0.300	0.060	$ 8,850	$ 531
2	1600	50	0.200	0.500	0.100	15,286	1,529
3	1600	53	0.200	0.200	0.040	24,940	998
4	2000	48	0.600	0.300	0.180	35,399	6,372
5	2000	50	0.600	0.500	0.300	43,443	13,033
6	2000	53	0.600	0.200	0.120	55,512	6,661
7	2400	48	0.200	0.300	0.060	61,948	3,717
8	2400	50	0.200	0.500	0.100	71,602	7,160
9	2400	53	0.200	0.200	0.040	86,084	3,443
						$E[PW(15\%)] =$	$43,443

Table 15.11 Calculation of the Variance of NPW Distribution (Example 15.6)

Event No.	Outcome x	Outcome y	Marginal Probability X p(x)	Marginal Probability Y p(y)	Joint Probability P(x, y)	NPW	(NPW − μ) μ = $43,443	Weighted (NPW − μ)²
1	1600	$ 48	0.200	0.300	0.060	$ 8,850	$(34,594)	$ 71,803,694
2	1600	50	0.200	0.500	0.100	15,286	(28,158)	79,285,044
3	1600	53	0.200	0.200	0.040	24,940	(18,503)	13,694,973
4	2000	48	0.600	0.300	0.180	35,399	(8,045)	11,648,806
5	2000	50	0.600	0.500	0.300	43,443	0	0
6	2000	53	0.600	0.200	0.120	55,512	12,068	17,477,553
7	2400	48	0.200	0.300	0.060	61,948	18,505	20,545,124
8	2400	50	0.200	0.500	0.100	71,602	28,159	79,291,802
9	2400	53	0.200	0.200	0.040	86,084	42,640	72,727,330

$$\text{Var}[PW(15\%)] = \$366,474,326$$
$$\sigma = \$19,144$$

15.4.2 Decision Rule

Once the expected value has been located from the NPW distribution, it can be used to make an accept-reject decision, in much the same way that a single NPW is used when a single possible outcome is considered for an investment project. The decision rule is called the **expected value criterion** and using it, we may accept a single project if its expected NPW value is positive. In the case of mutually exclusive alternatives, we select the one with the highest expected NPW. The use of expected NPW has an advantage over the use of a point estimate, such as the likely value, because it includes all the possible cash flow events and their probabilities.

The justification for the use of the expected value criterion is based on the **law of large numbers,** which states that if many repetitions of an experiment are performed, the average outcome will tend toward the expected value. This justification may seem to negate the usefulness of the expected value criterion in economic analysis, since most often in project evaluation we are concerned with a single, nonrepeatable "experiment"—that is, investment alternative. However, if a firm adopts the expected value criterion as a standard decision rule for *all* its investment alternatives, over the long term the law of large numbers predicts that accepted projects will tend to meet their expected values. Individual projects may succeed or fail, but the average project result will tend to meet the firm's standard for economic success.

The expected value criterion is simple and straightforward to use, but it fails to reflect the variability of investment outcome. We can enrich our decision by incorporating the variability information along with the expected value. Since the variance represents the dispersion of the distribution, it is desirable to minimize it. In other words, the smaller the variance, the less the variability associated with the NPW. Therefore, when we compare mutually exclusive projects, we may select the alternative with the smaller variance if its expected value is the same as or larger than those of other alternatives. In cases where there are no clear-cut preferences, the ultimate choice will depend on the decision maker's trade-offs—that is, how willing he or she is to accept variability in order to achieve a higher expected value.

■ 15.5 Risk Simulation

Thus far, we have examined analytical methods of determining the NPW distributions and computing their means and variances. As seen in Section 15.4.1, the NPW distribution offers numerous options for graphically presenting to the decision maker probabilistic information, such as the range and likelihood of occurrence of possible levels of NPW. Where we can adequately evaluate the risky investment problem by analytical methods, it is generally preferable to do so. However, there are many investment situations that we cannot solve easily by analytical methods. In these situations, we may develop the NPW distribution through computer simulation.

15.5.1 Computer Simulation

Before we examine the details of risk simulation, let us consider a situation where we wish to train a new astronaut for a future space mission. Several approaches exist for training this astronaut. One possibility is to place the trainee in an actual space shuttle, and to launch her into space. This approach would be expensive; it also would be extremely risky because any human error the trainee made might have tragic consequences. As an alternative, we can place the trainee in a flight simulator that is designed to mimic the behavior of the actual space shuttle in space, but that does so in a computer-controlled laboratory. The advantage of this approach is that the astronaut trainee learns all the essential functions of space operation in a simulated space environment. The flight simulator generates test conditions approximating operational conditions, and any human errors during training cause no harm to the astronaut or to the equipment being used.

The use of computer simulation is not restricted to simulating a physical phenomenon such as space flight. In recent years, techniques for simulating the results of some investment decisions before they are actually executed have been developed. Like the flight simulator, an investment simulator is a model for testing the results of certain decisions before they are enacted. In fact, we can analyze the WMC's transmission-housings project by building a simulation model. The general approach is to assign a subjective (or objective) probability distribution to each unknown factor and to combine these into a probability distribution for the project profitability as a whole. If we can simulate the actual state of nature for unknown investment variables on a computer, we may be able to obtain the resulting NPW distribution.

The unit demand (X) in our WMC's transmission-housings project was one of the random variables in the problem. We can know the exact value for this random variable only after the project is implemented. Can we predict the actual value before making a decision?

The following logical steps are often suggested for a computer program that simulates investment scenarios:

- Step 1: Identify all the variables that affect the measure of investment worth (for example, NPW after taxes).

- Step 2: Identify the relationships among all the variables. The relationships of interest here are expressed by the equations or series of numerical computations by which we compute the NPW of an investment project. These equations make up the model we are trying to analyze.

- Step 3: Classify the variables into two groups: the variables whose values are known with certainty, and the random variables for which exact values cannot be specified at the time of decision making.

- Step 4: Define distributions for all the random variables.

- Step 5: Perform Monte Carlo sampling (demonstrated in Section 15.5.3), and describe the resulting distribution of value (NPW).

- Step 6: Compute the distribution parameters and prepare graphic displays of simulation results.

Figure 15.8 illustrates the logical steps involved in simulating a risky investment project. The risk simulation process we have described has two impor-

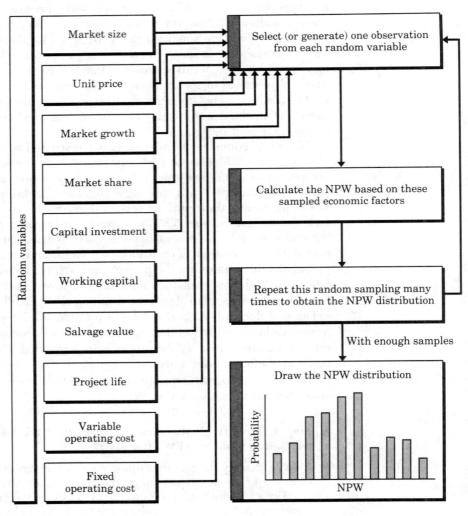

Figure 15.8 ■ Logical steps involved in simulating a risky investment

tant advantages when compared with the analytical approach discussed in the previous section.

1. There is practically no limit to the number of variables that can be considered, and the distributions used to define the possible values for each random variable can be of any type and any shape. The distributions can be based on statistical data if they are available, or, more commonly, on subjective judgment.
2. The method lends itself to sensitivity analyses. By defining some factors that have the most significant effect on the resulting NPW values and using different distributions (in terms of either shape or range) for each variable, we can observe the extent to which the NPW distribution is changed.

15.5.2 Model Building

In this section, we shall present some of the procedural details related to the first three steps (model building) outlined in Section 15.5.1. To illustrate the typical procedure involved, we shall work with an investment setting for the WMC's transmission-housings project, described in Example 15.1.

The initial step is to define the measure of investment worth and the factors that affect that measure. For our presentation, we choose the measure of investment worth as an after-tax NPW computed at a given interest rate i. In fact, we are free to choose any measure of worth, such as annual worth or future worth. In the second step, we must divide into two groups all the variables that we listed in step 1 as affecting NPW. One group consists of all the variables or parameters for which values are known. The second group includes all remaining parameters or variables, for which we do not know exact values at the time of analysis. The third step is to define the relationships that tie together all the variables. These relationships may take the form of a single equation or several equations.

Example 15.7

Reconsider the WMC's transmission-housings project in Example 15.1. Identify the input factors that are related to the project and develop the simulation model for the NPW distribution.

Discussion: For the WMC project, the variables that affect NPW value are investment required, unit price, demand, variable production cost, fixed production cost, tax rate, CCA, and the firm's interest rate. Some of the parameters that might be included in the known group are investment cost and interest rate (MARR).

If we have already purchased the equipment or have received a price quote, then we also know the CCA amount. Assuming that we are operating in a stable economy, we would probably know the tax rates for computing income taxes due.

The group of parameters with unknown values would usually include all the variables relating to costs and future operating expenses, and to future demand and sales prices. These are the random variables for which we must assess the probability distributions.

For simplicity, we classify the input parameters or variables for the WMC's transmission-housings project as follows:

Assumed to Be Known Parameters	Assumed to Be Unknown Parameters
MARR	Unit price
Tax rate	Demand
CCA amount	Salvage value
Investment amount	
Project life	
Fixed production cost	
Variable production cost	

Note that, unlike the situation in Section 15.4, here we treat the salvage value as a random variable. With these assumptions, we are now ready to build the NPW equation for the WMC project.

Solution

Recall that the basic investment parameters assumed for the WMC's project in Example 15.1 were as follows:

- Investment = $125,000
- Marginal tax rate = 0.40
- Annual fixed cost = $10,000
- Variable unit production cost = $15/unit
- MARR = i = 15%
- Annual CCA amounts:

n	CCA_n
1	18,750
2	31,875
3	22,313
4	15,619
5	10,933

The after-tax annual revenue is expressed in terms of functions of product demand (X) and unit price (Y):

$$R_n = XY(1 - t) = 0.6XY.$$

The after-tax annual expenses excluding CCA are also expressed as a function of product demand (X):

$$E_n = (\text{Fixed cost} + \text{Variable cost})(1 - t)$$

$$= (\$10,000 + 15X)(0.60)$$

$$= \$6000 + 9X.$$

Then, the net after-tax cash revenue after operating expenses are deducted is

$$V_n = R_n - E_n$$

$$= 0.6XY - 9X - \$6000.$$

The present worth of the net after-tax cash inflow from revenues is

$$\sum_{n=1}^{5} V_n(P/F, 15\% \ n) = [0.6X(Y - 15) - \$6000] \ (P/A, 15\%, 5)$$

$$= 0.6X(Y - 15) \ (3.3522) - \$20,113.$$

We will now compute the present worth of the total CCA credits:

$$\sum_{n=1}^{5} CCA_n t(P/F, i, n) = 0.40 \ [\$18,750(P/F, 15\%, 1) + \$31,875(P/F, 15\%, 2)$$

$$+ \$22,313(P/F, 15\%, 3) + \$15,619(P/F, 15\%, 4)$$

$$+ \$10,933(P/F, 15\%, 5)]$$

$$= \$27,777.$$

Since the total CCA amount is \$99,489, the book value at the end of year 5 is \$25,511 (\$125,000 − \$99,489). Any salvage value greater than this book value is treated as a taxable gain, and this gain is taxed at t. In our example, the salvage value is considered to be a random variable. Thus, the amount of gains (losses) also becomes a random variable. Therefore, the net salvage value after tax adjustment is

$$S - (S - \$25,511)t = S(1 - t) + \$25,511t$$

$$= 0.6S + \$10,204.$$

Then, the present worth equivalent of this amount is

$$(0.6S + \$10,204)(P/F, 15\%, 5).$$

Now, the NPW equation can be summarized as

$$PW(15\%) = -\$125,000 + 0.6X(Y-15)(3.3522) - \$20,113 + \$27,777$$
$$+ (0.6S + \$10,204)(0.4972)$$
$$= -\$112,263 + 2.0113X(Y-15) + 0.2983S.$$

Note that the NPW function is now expressed in terms of three random variables X, Y, and S.

15.5.3 Monte Carlo Sampling

For some variables, we may base the probability distribution on objective evidence of the past if the decision maker feels the same trend will continue to operate in the future. If not, we may use subjective probabilities as discussed in Section 15.3.1. Once we specify a distribution for a random variable, we need to determine ways to generate samples from this distribution. **Monte Carlo sampling** is a specific type of simulation method in which a random sample of outcomes is generated for a specified probability distribution. In this section, we shall discuss the Monte Carlo sampling procedure for an *independent* random variable.

Random Numbers

The sampling process is the key part of the analysis. It must be done such that the sequence of values sampled will be distributed in the same way as the original distribution. To accomplish this objective, we need a source of independent, identically distributed uniform random numbers between 0 and 1. We can use a table of random numbers, but most digital computers have programs available to generate "equally likely (uniform)" random decimals between 0 and 1. We will use $U(0, 1)$ to denote such a statistically reliable uniform random number generator, and we will use $U_1, U_2, U_3, \ldots$ to represent uniform random numbers generated by this routine.

Sampling Procedure

For any given random numbers, the question is, how are they used to sample a distribution in a simulation analysis? The first task we must do is to convert the distribution into its corresponding cumulative frequency distribution. Then, the random number generated is set equal to its numerically equivalent percentile and is used as the entry point on the $F(x)$ axis of the cumulative frequency graph. The sampled value of the random variable is the x-value corresponding to this cumulative percentile entry point.

This method of generating random values works because choosing a random decimal between 0 and 1 is equivalent to choosing a random percentile of the distribution. Then, the random value is used to convert the random percentile to a particular value. The method is general and can be used for any cumulative probability distribution, either continuous or discrete.

Example 15.8

In Example 15.7, we developed a NPW equation for the WMC's transmission-housings project as a function of three random variables—demand (X), unit price (Y), and salvage value (S):

$$PW(15\%) = -\$112,263 + 2.0113X(Y - 15) + 0.2983S.$$

- For random variable X, we will assume the same discrete distribution as defined in Table 15.6.

- For random variable Y, we will assume a triangular distribution with $L = \$48$, $H = \$53$, and $M_o = \$50$.

- For random variable S, we will assume a uniform distribution with $L = \$30,000$ and $H = \$50,000$.

With the random variables $(X, Y,$ and $S)$ distributed as above, and assuming that these random variables are *mutually independent* of each other, we need three uniform random numbers to sample one realization from each random variable. Determine the NPW distribution based on 200 iterations.

Discussion: As outlined previously, a simulation analysis consists of a series of repetitive computations of NPW. To perform the sequence of repeated simulation trials, we generate a sample observation for each random variable in the model and substitute these values into the NPW equation. Each trial requires that we use a different random number in the sequence to sample each distribution. Thus, if there are three random variables affecting the NPW, we need three random numbers for each trial. After each trial, the computed NPW is stored in the computer. As shown in Fig. 15.9, each value of NPW computed in this manner represents one state of nature. The trials are continued until a sufficient number of NPW values is available to define the NPW distribution.

Solution

Suppose the following three uniform random numbers are generated for the first iteration: $U_1 = 0.12135$ for X, $U_2 = 0.82592$ for Y, and $U_3 = 0.86886$ for S.

- **Demand (X):** The cumulative distribution for X is already given in Example 15.4. To generate one sample (observation) from this discrete distribution, we first find the cumulative probability function, as depicted in Fig. 15.10(a).

Figure 15.9 ■ A logical sequence of Monte Carlo simulation to obtain the NPW distribution for the WMC's transmission-housings project (Example 15.8)

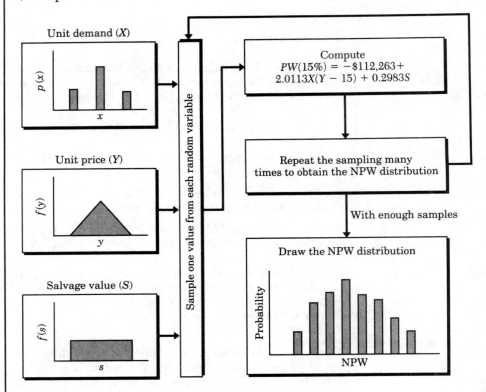

On a given trial, suppose the random number obtained from the computer is 0.12135. We then enter the vertical axis at the 12.135 percentile (the percentile numerically equivalent to the random number), read across to the cumulative curve, then read down to the x-axis to find the corresponding value of the random variable X; this value is 1600. This is the value of x that we use in the NPW equation. On the next trial, we sample another value of x by obtaining another random number, entering the ordinate at the numerically equivalent percentile, and reading the corresponding value of x from the x-axis.

- **Price (Y):** Assuming that the unit price random variable can be estimated by the three parameters, its probability distribution is shown in Fig. 15.10(b). Note that Y takes a continuous value (unlike the discrete assumption in Table 15.6). The sampling procedure is again similar to the discrete situation. Using a random number of $U = 0.82592$, we can approximate $y = \$51.38$ using a linear interpolation.

Figure 15.10 ■ Illustration of a sampling scheme for discrete and continuous random variables (Example 15.8)

x: Unit demand (discrete distribution)

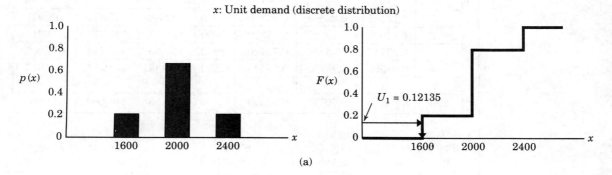

(a)

y: Unit price (triangular distribution)

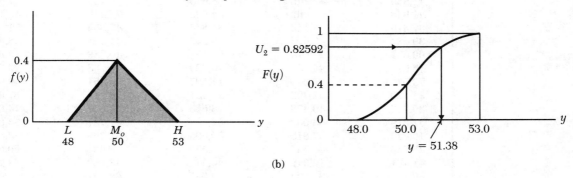

(b)

S: Salvage value (uniform distribution)

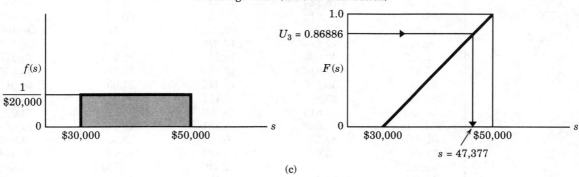

(c)

- **Salvage (S):** With the salvage value (S) distributed uniformly between $30,000 and $50,000, and a random number of $U = 0.86886$, the sample value is $s = \$47,377$, or $s = \$30,000 + (\$50,000 - \$30,000)0.86886$. The sampling scheme is shown in Fig. 15.10(c).

Table 15.12 Observed NPW Values ($) for WMC's Simulation Project

No.	U_1	U_2	U_3	x	y	s	NPW
1	0.12135	0.82592	0.86886	1600	$51.38	$47,377	$18,957
2	0.72976	0.79885	0.41879	2000	$51.26	$38,376	$45,044
3	0.23145	0.58484	0.78720	2000	$50.50	$45,744	$44,202
4	0.67520	0.17786	0.71426	2000	$49.33	$44,285	$39,057
:	:	:	:	:	:	:	:
200	0.95953	0.70178	0.84848	2400	$50.88	$46,969	$74,969

$18,957	$45,044	$44,202	$39,057	$71,736
36,369	17,840	67,478	40,753	42,239
42,459	35,708	39,624	12,535	9,968
35,024	65,755	42,994	68,907	44,485
60,623	15,053	69,225	36,067	61,402
11,589	32,746	68,015	42,129	47,656
41,946	67,002	34,236	77,236	43,142
37,875	38,143	34,848	47,353	71,315
39,484	42,678	38,540	44,435	8,890
60,681	71,514	65,181	46,359	12,030
18,071	44,182	9,329	41,763	20,554
73,988	47,975	11,597	65,551	40,034
36,129	42,403	74,451	76,318	67,673
39,728	40,378	8,086	14,677	14,616
51,819	63,063	52,028	47,737	34,661
10,843	37,815	39,093	41,768	43,886
77,240	40,536	43,540	43,634	60,644
41,481	12,312	40,272	43,568	68,213
43,863	39,637	9,035	63,691	37,972
68,937	18,021	12,831	43,560	46,653
41,042	16,140	45,007	49,947	42,839
62,297	50,912	40,781	65,079	41,965
13,007	34,089	39,758	15,572	34,167
69,448	12,223	43,275	17,208	18,434
12,493	42,280	76,754	30,754	75,072
42,833	78,019	72,864	32,782	45,630
12,217	36,548	75,412	37,354	46,015
39,434	45,845	8,127	44,344	38,522
72,550	45,618	39,300	41,816	44,292
45,735	70,083	43,591	66,513	11,656
40,944	42,755	39,983	35,732	14,628
32,043	10,031	46,821	47,448	36,499
43,683	73,704	41,446	40,931	11,638
38,667	65,243	36,958	69,238	38,658
41,642	42,189	38,494	72,450	73,042
48,075	43,216	34,298	14,778	34,031
36,012	67,325	74,614	12,795	17,090
45,000	68,793	39,600	67,268	42,771
18,049	41,051	36,344	46,269	42,956
37,718	71,594	10,198	51,866	34,321
11,207	39,140	36,741	16,672	74,969

Now we can compute the NPW equation with these sample values, yielding

$$PW(15\%) = -\$112{,}263 + 2.0113(1600)(51.3841 - 15)$$

$$+ \, 0.2983(\$47{,}377.20)$$

$$= \$18{,}957.$$

This result completes the first iteration of NPW_1 computation.

For the second iteration, we need to generate another set of three uniform random numbers (assume they are 0.72976, 0.79885, and 0.41879) to generate the respective sample from each distribution and to compute $NPW_2 = \$45{,}044$. If we repeat this process for 200 iterations, we obtain the NPW values listed in Table 15.12. By ordering the observed data by increasing NPW value and tabulating the ordered NPW values, we obtain a frequency distribution shown in Table 15.13. Such a tabulation results from dividing the entire range of computed NPWs into a series of subranges (20, in this case), and then counting the number of computed values that fall in each of the 20 intervals. Note that the sum of all the frequencies of column 3 is the total number of trials that were made.

Column 4 simply expresses the frequencies of column 3 as a fraction of the total number of trials. At this point, all we have done is to arrange the 200 numerical values of NPW into a table of relative frequencies.

15.5.4 Simulation Output Analysis

After a sufficient number of repetitive simulation trials has been run, the analysis is essentially completed. The only remaining tasks are to tabulate the computed NPW values to determine the expected value and to make various graphic displays useful to management.

Interpretation of Simulation Results

Once we obtain a NPW frequency distribution (such as that shown in Table 15.13), we need to make the assumption that the actual relative frequencies of column 4 in Table 15.13 are representative of the probability of having a NPW in each range. That is, we assume that the relative frequencies we observed in the sampling are representative of the proportions we would have obtained if we had examined all the possible combinations.

This sampling is analogous to polling the opinions of voters about a candidate for public office. We could speak to every registered voter if we had the time and resources, but a simpler procedure would be to interview a

Table 15.13 Simulated NPW Frequency Distribution for the WMC's Transmission-Housings Project (Example 15.8)

Cell No.	Cell Interval	Observed Frequency	Relative Frequency	Cumulative Frequency
1	\$ 8,086 ≤ NPW ≤ \$11,583	10	0.05	0.05
2	\$11,583 < NPW ≤ \$15,079	18	0.09	0.14
3	\$15,079 < NPW ≤ \$18,576	8	0.04	0.18
4	\$18,576 < NPW ≤ \$22,073	2	0.01	0.19
5	\$22,073 < NPW ≤ \$25,569	0	0.00	0.19
6	\$25,569 < NPW ≤ \$29,066	0	0.00	0.19
7	\$29,066 < NPW ≤ \$32,563	2	0.01	0.20
8	\$32,563 < NPW ≤ \$36,060	14	0.07	0.27
9	\$36,060 < NPW ≤ \$39,556	23	0.12	0.39
10	\$39,556 < NPW ≤ \$43,053	37	0.19	0.57
11	\$43,053 < NPW ≤ \$46,550	27	0.14	0.71
12	\$46,550 < NPW ≤ \$50,046	9	0.05	0.75
13	\$50,046 < NPW ≤ \$53,543	4	0.02	0.77
14	\$53,543 < NPW ≤ \$57,040	0	0.00	0.77
15	\$57,040 < NPW ≤ \$60,536	0	0.00	0.77
16	\$60,536 < NPW ≤ \$64,033	7	0.04	0.81
17	\$64,033 < NPW ≤ \$67,530	10	0.05	0.86
18	\$67,530 < NPW ≤ \$71,027	10	0.05	0.91
19	\$71,027 < NPW ≤ \$74,523	10	0.05	0.96
20	\$74,523 < NPW ≤ \$78,019	9	0.05	1.00

Cell width = \$ 3,497

Mean = \$42,461

Standard deviation = \$18,727

Minimum NPW value = \$ 8,086

Maximum NPW value = \$78,019

smaller group of persons selected with an unbiased sampling procedure. If 60% of this scientifically selected sample supports the candidate, it would probably be safe to assume that 60% of all registered voters support the candidate. Conceptually, we do the same thing with simulation. As long as we ensure that a sufficient number of trials have been made, we can rely on the simulation results.

Once we have obtained the probability distribution of the NPW, we face the crucial question: How do we use this distribution in decision making. Recall that the probability distribution provides information regarding the probability that a random variable will attain some value x. We can use this information, in turn, to define the cumulative distribution, which expresses the probability that the random variable will attain a value smaller than or equal to some x—that is, $F(x) = P(X \leq x)$. Thus, if the NPW distribution is known, we can also compute the probability that the NPW of a project will be negative. We use this probabilistic information in judging the profitability of the project.

With the assurance that 200 trials was a sufficient number for the WMC project, we may interpret the relative frequencies in column 4 of Table 15.13 as probabilities. The NPW values range between $8086 and $78,019, indicating no loss for any situation. The NPW distribution has an expected value of $42,461 and a standard deviation of $18,727.

Creation of Graphic Displays

Using the output data of Table 15.13, we can create the distribution in Fig. 15.11(a). A picture such as this can give the decision maker a feel for the ranges of possible NPWs, for the relative likelihoods of loss versus gain, for the range of NPWs that is most probable, and so on.

Another useful display is the conversion of the NPW distribution to the equivalent cumulative frequency, as shown in Fig. 15.11(b). Usually, the decision maker is concerned with the likelihood of attaining at least a given level of NPW. Therefore, we construct the cumulative distribution by accumulating the areas under the distribution as the *NPW* decreases. The decision maker can use Fig. 15.11(b) to answer many questions, for example: What is the likelihood of making at least a 15% return on investment, that is, the likelihood that the NPW will be at least 0? In our example, this probability is 100%.

15.5.5 Dependent Random Variables

All our simulation examples have considered independent random variables. We must recognize that some of the random variables affecting the NPW may be related to one another. If they are, we need to sample from distributions of the random variables in a manner that accounts for any dependency. For example, in the WMC project, both the demand and the unit price are not known with certainty. Both of these parameters would be on our list of variables for which we need to describe distributions, but they could be related inversely. When we describe distributions for these two parameters, we have to account for the dependency. This issue can be critical, as the results obtained from a simulation analysis can be misleading if the analysis does not account for the dependent relationships. The sampling techniques for these dependent random variables are beyond the scope of this text but can be found in many simulation textbooks.

Figure 15.11 ■ The simulation result for WMC's transmission project based on 200 iterations

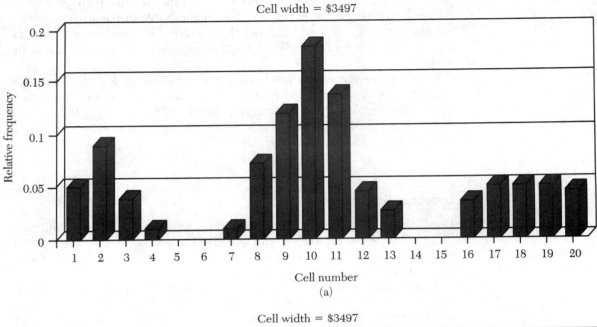

Cell width = $3497

(a)

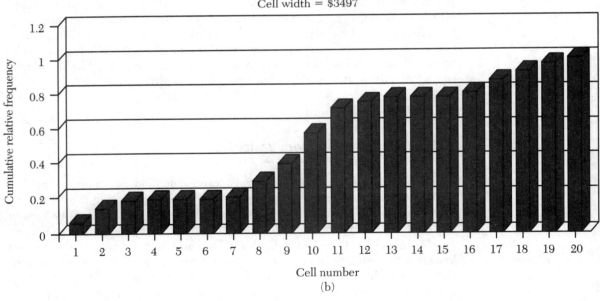

Cell width = $3497

(b)

15.6 Computer Notes

It is in the area of sensitivity analysis and breakeven analysis that the electronic spreadsheet truly reveals its power. Conceptually, a NPW distribution can be simulated using either manual labor (pencil, paper, and calculator).

However, computers have the advantage of being able to handle large amounts of data rapidly, so risk simulation is rarely done manually. We will demonstrate how an electronic spreadsheet can facilitate both sensitivity and breakeven analyses.

The spreadsheet in Fig. 15.12 has been structured in the format of Table 15.1, the after-tax cash flow for WMC's transmission-housings project. However, the sales, price, variable cost, fixed cost, and salvage have been keyed to a percent change, as shown in columns E and G of Fig. 15.12. For example, the formula for sales in location D14 is $C$8*(1 + $G$5/100)*$C$9*(1 + $G$6/100). Note that the NPW at 15% is shown in G10, and for zero change for all variables (the base-case values), the NPW is $43,443, as calculated in Example 15.1.

Since all the variables whose value we wish to investigate are keyed to the percent change, all we have to do to calculate a new NPW is to change the percent change for the variable of interest. As we change the percent deviation for price, say 20%, we see instantly the new NPW of $83,669 appears in location G10. This is the number shown in Table 15.2. The complete Table 15.2 can be computed in a matter of minutes from this spreadsheet.

D14 = $C$8*(1+$G$5/100)*$C$9*(1+$G$6/100) READY

A	A B	C	D	E	F	G	H
1							
2	Input data:			Sensitivity analysis:			
3	Revenue	10,000					
4	Var. cost (unit)	15		Category		% Change	
5	Fixed cost	10,000		Sales (unit)		0	
6	Salvage	50,000		Price (unit)		0	
7	Tax rate	0.4		Var. cost (unit)		0	
8	Sales(unit)	2,000		Fixed cost		0	
9	Price (unit)	50		Salvage		0	
10	MARR(%)	15		PW(MARR)		43,443	
11	CCA rate	0.3					
12	Income statement						
13		0	1	2	3	4	5
14	Revenues		100,000	100,000	100,000	100,000	100,000
15	Var. cost		30,000	30,000	30,000	30,000	30,000
16	Fixed cost		10,000	10,000	10,000	10,000	10,000
17	CCA		18,750	31,875	22,313	15,619	10,933
18	Taxable income		41,250	28,125	37,688	44,381	49,067
19	Income taxes		16,500	11,250	15,075	17,753	19,627
20	Net income		24,750	16,875	22,613	26,629	29,440
21							
22	Cash flow statement						
23							
24	Cash from operation:						
25	Net income		24,750	16,875	22,613	26,629	29,440
26	CCA		18,750	31,875	22,313	15,619	10,933
27	Investment activity						
28	Investment	(125,000)					
29	Salvage						50,000
30	Disposal tax effect						(9,796)
31	Net cash flow	(125,000)	43,500	48,750	44,925	42,248	80,578
32							

Sample cell formulas:

Revenues: D14 = $C$8*(1+$G$5/100)*$C$9*(1+$G$6/100)

Variable cost: D15 = $C$4*(1+$G$7/100)*$C$8*(1+$G$5/100)

Disposal tax effect: H30 = −C7*[H29+C28+@SUM(D17..H17)]

Figure 15.12 ■ Sensitivity and breakeven analyses by Lotus 1-2-3 (Example 15.2)

Summary

- Often, cash flow amounts and other aspects of investment project analysis are uncertain. Whenever such uncertainty exists, we are faced with the difficulty of **project risk**—the possibility that an investment project will not meet our minimum requirements for acceptability and success.

- Three of the most basic tools for assessing project risk are:

  1. **Sensitivity analysis**—identifying the project variables which, when varied, have the greatest effect on project acceptability.

  2. **Breakeven analysis**—identifying the value of a particular project variable that causes the project to exactly break even.

  3. **Scenario analysis**—comparing a "base-case" or expected project measurement (such as NPW) to one or more additional scenarios, such as best and worst case, to identify the extreme and most likely project outcomes.

- Sensitivity, breakeven, and scenario analyses are reasonably simple to apply, but also somewhat simplistic and imprecise in cases where

we are dealing with multifaceted project uncertainty. **Probability concepts** allow us to further refine the analysis of project risk by assigning numerical values to the likelihood that project variables will have certain values.

- The end goal of a probabalistic analysis of project variables is to produce an NPW distribution. From the distribution we can extract such useful information as the **expected value,** the extent to which other NPW values vary from or are clustered around the expected value (**variance**), and the best- and worst-case NPW's.

- **Risk simulation,** in general, is the process of modeling reality to observe and weigh the likelihood of the possible outcomes of a risky undertaking. **Monte Carlo sampling** is a specific type of randomized sampling method in which a random sample of outcomes is generated for a specified probability distribution. Because Monte Carlo sampling and other simulation techniques often rely on generating a significant number of outcomes, they can be more conveniently performed on the computer than by hand.

Problems

15.1 Ford Construction Company is considering the acquisition of a new earth-mover. The mover's basic price is $70,000, and it will cost another $15,000 to modify it for special use by the company. This earthmover falls into asset class 38 (d = 30%). It will be sold after 4 years for $30,000. The earth-mover purchase will have no effect on revenues, but it is expected to save the firm $32,000 per year in before-tax operating costs, mainly labor. The firm's tax rate (federal plus provincial) is 40%, and its MARR is 15%.

(a) Is this project acceptable based on the most-likely estimates given in the problem?

(b) Suppose that the project will require an increase in net working

capital (spare-parts inventory) of $2000. With consideration of this new requirement, would be the project still acceptable?

(c) If the firm's MARR is increased to 20%, what would be the required savings in labor so that the project is still profitable?

15.2 Alberta Metal Forming Company has just invested $500,000 of fixed capital in a manufacturing process which is estimated to generate an after-tax annual cash flow of $200,000 in each of the next 5 years. At the end of year 5, there will be no further market for the product. The net salvage value for the manufacturing process is zero. If a marketing problem delays plant start-up for 1 year (leaving only 4 years of process life), what additional after-tax cash flow will be needed to maintain the same internal rate of return as would be experienced if no delay occurred?

15.3 A real estate developer seeks to determine the most economical height for a new office building. The building will be sold after 5 years. The relevant net annual revenues and salvage values are as follows.

	Height			
	2 Floors	*3 Floors*	*4 Floors*	*5 Floors*
First cost	$500,000	$750,000	$1,250,000	$2,000,000
Net after-tax Lease revenue	199,100	169,200	149,200	378,150
Net resale value (after-tax)	600,000	900,000	2,000,000	3,000,000

(a) The developer is uncertain about the interest rate (i) to use but is certain that it is in the range 5% to 30%. Find the range of values of i for which that building height is the most economical.

(b) Suppose that the developer's interest rate is known to be 15%. What would be the cost, in terms of net present value, of a 10% overestimation in resale value?

15.4 A special-purpose machine was purchased 4 years ago for $20,000. It was estimated at that time that this machine would have a life of 10 years and a salvage value of $1000 with a cost of removal of $1500. These estimates are still good. This machine has annual operating costs of $2000 and its current book value is $13,000. If the machine is retained over the life of the asset, the remaining annual CCA would be $2000 for years 5 through 10. A new machine, which is more efficient, will reduce the operating costs to $1000 but it will require an investment of $12,000. The life of the new machine is estimated to be 6 years with a salvage value of $2000. The new machine's CCA can be

calculated with a declining balance rate of 30%. There has been an offer of $6000 for the old machine, and the purchaser would pay for removal of the machine. The firm's tax rate is 40%, and its required minimum rate of return is 10%.

(a) What incremental cash flows will occur at the end of years 0 through 6 as a result of replacing the old machine? Should the old machine be replaced now?

(b) Suppose that the annual operating costs for the old machine would increase at an annual rate of 5% over the remaining service life. With this change in future operating costs for the old machine, would the answer in (a) change?

(c) What is the minimum trade-in value for the old machine so that both alternatives are economically equivalent?

15.5 A local telephone company is installing a phone line for a new row of apartment complexes. Two types of cables are considered, conventional copper wire and fiber optics. Transmission by copper wire cables, although cumbersome, involves much less complicated and expensive support hardware than fiber optics. The local company may use five different types of copper-wire cables: 100 pairs, 200 pairs, 300 pairs, 600 pairs, and 900 pairs per cable. In calculating the cost per metre of cable the following equation is used:

$$\text{first cost} = [\text{cost per metre} + \text{cost per pair} (\text{number of pairs})](\text{length})$$

$$\begin{aligned} \text{22 gauge copper wire cost per} &= \$5.55 \text{ per metre} \\ \text{pair} &= \$0.043 \text{ per pair} \end{aligned}$$

The annual cost of the cable as a percent of the first cost is 18.4%. The life of the system is 30 years.

In considering fiber optics, a ribbon is referred to rather than a cable. One ribbon contains 12 fibers. The fibers are grouped in fours; therefore, one ribbon contains three groups of four fibers. Each group of four fibers can produce 672 lines (equivalent to 672 pairs of wires), and since each ribbon contains three groups the total capacity of the ribbon is 2016 lines. To transmit signals using fiber optics, many modulators, wave guides, and terminators are needed to convert the signals from electric currents to modulated light waves. The fiber-optics ribbon costs $9321 per kilometre. At each end of the ribbon three terminators are needed, one for each group of four fibers at a cost of $30,000 per terminator. Twenty-one modulating systems are needed at each end of the ribbon at a cost of $12,092 for a unit in the central office, and $21,217 for a unit in the field. Every 6706 metres a repeater is required to keep the modulated light waves in the ribbon at an intelligible intensity for detection. The unit cost of this repeater is $15,000. The annual cost including income taxes for the 21 modulating systems is 12.5% of the first cost for the units. The annual cost of the ribbon itself is 17.8% of its first cost. The life of the whole system is 30 years. (All figures represent after-tax costs.)

(a) Suppose that the apartments are located just 8 kilometres from the phone company's central switching system and will require about 2000 telephones. This would require either 2000 pairs of copper wire or one fiber-optics ribbon and related hardware. If the telephone company's interest rate is 15%, which option is more economical?

(b) In (a), suppose that the apartments are located 16 kilometres from the phone company's central switching system. Which option is more economically attractive? What if the apartments are located 40 kilometres from the switching system?

15.6 A small manufacturing firm is considering the purchase of a new machine to modernize one of its production lines. Two types of machine are available on the market. The lives of machine A and machine B are 8 years and 10 years, respectively. The machines have the following receipts and disbursements. Using a MARR (after tax) of 10% and tax rate of 30%:

Item	Machine A	Machine B
First cost	$6000	$8500
Service life	8 years	10 years
Salvage value	500	1000
Annual O&M		
Costs	700	520
CCA	DB (30%)	DB (30%)

(a) Which machine would be most economical to purchase under the infinite planning horizon? Explain any assumption that you need to make about future alternatives.

(b) Determine the breakeven annual O&M costs for machine A so that the present worth of machines A and B is the same.

(c) Suppose that the required service life of the machine is only 5 years. The estimated salvages at the end of the required service period are estimated to be $3000 for machine A and $3500 for machine B, respectively. Which machine is more economical?

15.7 The management of Langdale Mill must replace a number of old looms in the mill's weaveroom. The looms to be replaced are two 220-cm Prime Minister looms, sixteen 135-cm Prime Minister looms, and twenty-two 185-cm Draper X-P2 looms. The company may either replace the old looms with new ones of the same kind or buy 21 new shutterless Pignone looms. The first alternative requires the purchase of 40 new Prime Minister and Draper looms and the scrapping of the old looms. The second alternative involves scrapping the 40 old looms, relocating 12 Picanol looms, and constructing a concrete floor plus purchasing the 21 Pignone looms and various related equipment.

Description	Alternative 1	Alternative 2
Machinery and related equipment	$2,119,170	$1,071,240
Removal cost of old looms and site preparation	$26,866	$49,002
Salvage value of removed looms	$62,000	$62,000
Annual sales	$7,915,748	$7,455,084
Annual labor cost	$261,040	$422,080
Annual O&M cost	$1,092,000	$1,560,000
CCA	DB (30%)	DB (30%)
Project life	8 years	8 years
Salvage value	$169,000	$54,000

The book value of all old looms are negligible. The corporate executives feel that various investment opportunities available for the mills will guarantee a rate of return on investment of at least 18%. The mill's tax rate is 40%.

(a) Perform a sensitivity analysis on the project's data, varying net operating revenue, labor cost, annual O&M cost, and the MARR. Assume that each of these variables can deviate from its base-case expected value by ±10%, ±20%, and ±30%.

(b) From the results of part(a), prepare sensitivity diagrams and interpret the results.

15.8 Susan Campbell is thinking about going into the motel business. The cost to build the motel is $2,200,000. The lot costs $600,000. Furniture and furnishings cost $400,000 and have a 20% CCA rate, while the motel building has a 4% CCA rate. The land will appreciate at an annual rate of 5% over the project period, but the building and furnishings will have a zero salvage value after 25 years. When the motel is full (100% capacity), it takes in (receipts) $4000 per day for 365 days per year. The motel has fixed operating expenses, exclusive of CCA, of $230,000 per year. The variable operating expenses are $170,000 at 100% capacity and vary directly with percent capacity down to zero at 0% capacity. If interest is 10% compounded annually, at what percent capacity must this motel operate to break even? (Assume Susan's business tax rate is 31%.)

15.9 A plant engineer wishes to know which of two types of light bulbs should be used to light a warehouse. The bulbs currently used cost $45.90 per bulb and last 14,600 hours before burning out. The new bulb ($60 per bulb) provides the same amount of light and consumes the same amount of energy but lasts twice as long. The labor cost to change a bulb is $16.00. The lights are on 19 hours a day, 365 days a year. If the firm's MARR is 15%, what is the maximum price (per bulb) the engineer should be willing to pay to switch to the new bulb? (Assume the firm's tax rate is 40%.)

15.10 Robert Cooper is considering purchasing rental property, containing stores and offices, at a cost of $250,000. Cooper estimates that annual receipts from rentals will be $35,000 and that annual disbursements other than income taxes will be about $12,000. The property is expected to appreciate at an annual rate of 5%. Cooper expects to retain the property for 20 years once it is acquired. The CCA rate for the building is 4%. Cooper's business tax rate is 30% and his MARR is 15%. What would be the minimum annual rental receipts that make the investment break even?

15.11 The City of Guelph was having a problem locating land for a new sanitary landfill site when the solution of burning the solid waste to generate steam was proposed. At the same time, Uniroyal Tire Company seemed to be having the same problem, disposing of solid waste in the form of rubber tires. It was determined that there would be about 200 tonnes per day of waste to be burned; this included municipal and industrial waste. The city is considering building a waste-fired steam plant, which would cost $6,688,800. To finance the construction cost, the city will issue resource-recovery revenue bonds in the amount of $7,000,000 at an interest rate of 11.5%. Interest on the bonds is payable annually. The differential amount between the actual construction costs and the amount of bond financing ($7,000,000 − $6,688,800 = $311,200) will be used to settle the bond discount and expenses associated with the bond financing. The expected life of the steam plant is 20 years. The expected salvage value is estimated to be about $300,000. The expected labor costs would be $335,000 per year. The annual operating and maintenance costs (including fuel, electricity, maintenance, and water) are expected to be $175,000. The city expects 20% down-time per year for the waste-fired steam plant. This down-time would result in 4245 kilograms of waste which, along with 3265 kilograms of waste after incineration, would have to be disposed of as land fill. At the present rate of $42.88 per kilogram, this will cost the city a total of $322,000 per year. The revenues for the steam plant will come from two sources: (1) steam sales, and (2) disposal tipping fees. With an input of 200 tonnes per day and an output of 6.64 kilograms of steam per kilogram refuse, a maximum of 1,327,453 kilograms of steam can be produced per day. However, with 20% down-time, the actual output would be 1,061,962 kilograms of steam per day. The initial steam charge will be approximately $4.00 per tonne. This would bring in $1,550,520 in steam revenue the first year. The tipping fee is used in conjunction with the sale of steam to offset total plant cost. It is the goal of the Guelph steam plant to phase out the tipping fee as soon as possible. The tipping fee will be $20.85 per tonne in the first year of plant operation, and will be phased out at the end of the eighth year. The scheduled tipping fee assessment would be:

Year	Tipping Fee
1	$976,114
2	895,723
3	800,275
4	687,153
5	553,301
6	395,161
7	208,585

(a) At an interest rate of 10%, would the steam plant generate sufficient revenue to recover the initial investment?

(b) At an interest rate of 10%, what would be the minimum charge (per tonne) for steam sales to make the project break even?

(c) Perform sensitivity analysis for the input variable of the plant's down time.

15.12 Two different methods of solving a production problem are under consideration. Both methods are expected to be obsolete in 6 years. Method A would cost $80,000 initially and have annual operating costs of $22,000 a year. Method B would cost $52,000 initially and cost $17,000 a year to operate. The salvage value realized with method A would be $20,000 and with method B would be $15,000. Method A would generate $16,000 revenue income a year more than method B. Investments in both methods have a 30% CCA rate. The firm's income tax rate is 40%. The firm's MARR is 20%.
What would be the required additional annual revenue for method A such that both methods would be indifferent?

15.13 A company is currently paying a sales representative $0.25 per kilometre to drive her car for company business. The company is considering supplying the representative with a car, which would involve the following: A car costs $12,000 and has a service life of 5 years and a market value of $3500 at the end of that time. Monthly storage costs for the car are $80, and the cost of fuel, tires, and maintenance is 15 cents per kilometre. CCA for the car can be computed with a DB rate of 30%. The firm's tax rate is 40%. How many kilometres must a sales representative travel by car annually for the cost of the two methods of providing transportation to be equal if the interest rate is 15%?

15.14 Rocky Mountain Publishing Company is considering introducing a new morning newspaper in Edmonton. Its direct competitor charges $0.50 at retail, with $0.05 going to the retailer. For the level of news coverage the company desires, it determines the fixed cost of editors, reporters, rent, pressroom expenses, and wire service charges to be $300,000 per month. The variable cost of ink and paper is $0.10 per copy, but advertising revenues of $0.05 per paper will be generated. To print the morning paper, the publisher has to purchase a new printing press, which will cost $600,000. The press machine will have a 30% CCA rate. The press machine will be used for 10 years, at which time its salvage value would be about $100,000. Assume 25 weekdays in a month, 40% of tax rate, and 13% of MARR. How many copies per day must be sold to break even at a selling price of $0.50 per paper at retail?

15.15 A corporation is trying to decide whether to buy the patent for a product designed by another company. The decision to buy will mean an investment of $8 million, and the demand for the product is not known. If demand is light, the company expects a return of $1.3 million each year for 3 years. If the demand is moderate, the return will be $2.5 million each year for 4 years, and a high demand means a return of $4 million each year for 4 years. It is estimated the probability of a high demand is 0.4 and the probability of a light demand is 0.2. The firm's interest rate (risk-free) is 12%. Calculate the expected present worth. On this basis should the company make the investment? (All figures represent after-tax values.)

15.16 Juan Carlos is considering the two investment projects whose present values are described as follows.

- **Project 1:** $PW(10\%) = 2X(X - Y)$, where X and Y are statistically independent discrete random variables with the following distributions.

X		Y	
Event	Probability	Event	Probability
$20	0.6	$10	0.4
40	0.4	20	0.6

- **Project 2:**

PW(10%)	Probability
$ 0	0.24
400	0.20
1600	0.36
2400	0.20

Note: Cash flows between the two projects are also assumed to be statistically independent.

(a) Develop the NPW distribution for project 1.
(b) Compute the mean and variance of the NPW for project 1.
(c) Compute the mean and variance of the NPW for project 2.
(d) Suppose that projects 1 and 2 are mutually exclusive. Which project would you select?

15.17 A financial investor has an investment portfolio worth $350,000. A bond in his investment portfolio will mature next month and provide him with $25,000 to reinvest. The choices have been narrowed down to the following two options.

- **Option 1:** Reinvest in a foreign bond that will mature in one year. This will entail a brokerage fee of $150. For simplicity, assume that the bond will provide interest over the one-year period of $2450, $2000, or $1675, and that the probabilities of these occurrences are assessed to be 0.25, 0.45, and 0.30, respectively.
- **Option 2:** Reinvest in a $25,000 certificate with a savings and loan association. Assume this certificate has an expected effective variable annual rate of 7.5%.

(a) Which form of reinvestment should the investor choose in order to maximize his expected financial gain?
(b) If the investor solicits professional advice from an investment firm, what would be the maximum amount the investor should pay for this service?

15.18 Kellog Company is considering the following investment project and has estimated all costs and revenues in constant dollars. The project requires a purchase of a $9000 asset, which will be used for only 2 years (project life).

- The salvage value of this asset at the end of 2 years is expected to be $4000.
- The project requires an investment of $2000 in working capital, and this amount will be fully recovered at the end of project year.
- The annual revenue is a discrete random variable (X), but it can be described by the following probability distribution.

Annual Revenue (X)	Probability
$10,000	0.30
20,000	0.40
30,000	0.30

- The investment has a declining balance rate of 30% for CCA purposes.

- The firm expects a general inflation rate ($\bar{f}$) of 5% during the project period. It is assumed that the revenues, salvage value, and working capital are responsive to this general inflation rate.
- The income tax rate for the firm is 40%. The firm's inflation-free interest rate (i') is 10%.

(a) Determine the NPW as a function of X.

(b) In (a), compute the expected NPW of this investment.

(c) In (a), compute the variance of the NPW of this investment.

15.19 A manufacturing firm is considering two mutually exclusive projects. Both projects have an economic service life of one year with no salvage value. The first cost and the net year-end revenue for each project are given as follows:

First Cost	Project 1 $1000		Project 2 $800	
	Probability	Revenue	Probability	Revenue
Net revenue given in PW at time 0	0.2	$2000	0.3	$1000
	0.6	3000	0.4	2500
	0.2	3500	0.3	4500

We assume that both projects are statistically independent from each other.

(a) If you are an expected value maximizer, which project would you select?

(b) If you also consider the variance of the project, which project would you select?

15.20 A business executive is trying to decide whether to undertake one of two contracts or neither one. She has simplified the situation somewhat and feels that it is sufficient to imagine that the contracts provide the alternatives shown as follows.

Contract A		Contract B	
NPW	Probability	NPW	Probability
$100,000	0.2	$40,000	0.3
50,000	0.4	10,000	0.4
0	0.4	−10,000	0.3

(a) Should the executive undertake either one of the contracts? If so, which one? What would she do if she made decisions by maximizing her expected NPW?

(b) What would be the probability that contract A would result in a larger profit than contract B?

15.21 Two alternative machines are being considered for a cost-reduction project.

- Machine A has a first cost of $60,000 and a net salvage value of $22,000 at the end of 6 years' service life. Probabilities of annual after-tax operating costs of this machine are estimated as follows:

Annual O&M costs	Probability
$ 5,000	0.20
8,000	0.30
10,000	0.30
12,000	0.20

- Machine B has a first cost of $35,000 and its estimated net salvage value at the end of 4 years' service is estimated to be negligible. The annual after-tax operating costs are estimated to be as follows:

Annual O&M costs	Probability
$ 8,000	0.10
10,000	0.30
12,000	0.40
14,000	0.20

The MARR on this project is 10%. The required service period of these machines is estimated to be 12 years and no technological advances in either machine are expected.

(a) Assuming independence, calculate the mean and variance for the equivalent annual cost of operating each machine.

(b) From the results of part(a), calculate the probability that the annual cost of operating machine A will exceed the cost of operating machine B.

15.22 Mount Manufacturing Company produces industrial and public safety shirts. As is done in most apparel manufacturing, the cloth must be cut into shirt parts by marking sheets of paper in the way that the particular cloth is to be cut. At present, these sheet markings are done manually. Mount has the option of purchasing one of two automated marking systems, the Lectra System 305 or the Tex Corporation Marking System. The comparative characteristics of the two systems are as follows.

Most-Likely Estimates

	Manual Marking	Lectra System	Tex System
Annual labor cost	$103,218	$ 51,609	$ 51,609
Annual material savings		$230,000	$274,000
Investment cost		$136,150	$195,500
Estimated life		6 years	6 years
Salvage value		$ 20,000	$ 15,000
CCA rate		30%	30%

The firm's tax rate is 40%, and the interest rate used for project evaluation is 12% after taxes.

(a) Based on the most-likely estimates, which alternative is the best?

(b) Suppose that the company estimates the material savings during the first year for each system on the basis of following probability distribution.

Lectra System

Material Savings	Probability
$150,000	0.25
230,000	0.40
270,000	0.35

Tex Corporation

Material Savings	Probability
$200,000	0.30
274,000	0.50
312,000	0.20

Further assume that the annual material savings for both Lectra and Tex are statistically independent. Compute the mean and variance for the equivalent annual value of operating each system.

(c) In part (b), calculate the probability that the annual benefit of operating Lectra will exceed the benefit of operating Tex.

15.23 The following represents the net cash flows and their respective probabilities for an investment project over its service life.

Time 0	Cash flow (A_1)	−$1000		
	Probability	1.0		
Time 1	Cash flow (A_2)	$300	$500	$1000
	Probability	0.3	0.4	0.3
Time 2	Cash flow (A_3)	$300	$1000	$2000
	Probability	0.2	0.7	0.1

The firm's MARR (i) is known to be 10%.

(a) Express the $PW(i)$ as function of $A_1, A_2,$ and A_3.

(b) Using the following sequence of random numbers, obtain two samples of $PW(i)$.

uniform random deviates $= 0.024, 0.01, 0.13,$
$$0.45, 0.56, 0.21, \ldots$$

15.24 Consider the following investment project with three estimates for annual cash flows: low, most-likely, and high estimates. Assume that these three estimates correspond to the parameters of a triangular distribution.

End of Year	Low	Most-Likely	High
0	−1,500	−1,000	−500
1	1,000	2,000	2,500

You want to use risk simulation to estimate the mean and the standard deviation of the present value of this project at an interest rate of 10%. Suppose that first two random numbers generated from a computer are 0.20 and 0.85. Compute the cash flows generated and the net present value using these two random numbers.

15.25 In Problem 15.23, using the random number generator on your computer, obtain 100 $PW(i)$ samples. Then construct a histogram based on these samples, and find the mean and variance of $PW(i)$.

15.26 Consider the following investment project whose periodic cash flows are projected to follow a triangular distribution.

Project Cash Flow Estimates

n	Lowest	Most-Likely	Highest
0	−$40	−$35	−$30
1	15	18	25
2	10	20	25
3	25	30	40

The firm's interest rate for project evaluation is 15%. Using a spreadsheet package, obtain a NPW distribution of the project based on 100 repeated samplings. Also estimate the mean and variance of the distribution and determine the probability that the NPW is less than or equal to $30.

15.27 Consider the following project, whose net cash flows over the next 5 years are given by the three estimates. A quantity in parentheses represents a negative cash flow.

Three Estimates

End of Year	Low	Most-Likely	High
0	($1,891,610)	($1,161,215)	($562,810)
1	(1,254,410)	(499,266)	106,170
2	(326,620)	150,595	707,430
3	227,350	607,372	918,290
4	753,750	803,229	918,680
5	657,300	720,800	980,400

These three estimates are equivalent to the pessimistic, most-likely, and optimistic estimates in a triangular distribution. Using a spreadsheet package:

(a) Express the NPW as a function of random cash flows at a given interest rate of i.

(b) At $i = 12\%$, obtain 100 NPW samples on your computer and develop a NPW probability distribution.

(c) From the results of part (b), find the expected value and the variance of the NPW of this project.

(d) From the results of part (b), find the probability the NPW will exceed zero, and interpret the results.

15.28 (Embellishment of Example 1.7) A major trucking company is considering the installation of a two-way mobile satellite messaging service on their 2000 trucks. Based on tests last year on 120 trucks, the company found that satellite messaging cuts 60% from its $5 million bill for long-distance communications with truck drivers. More important, the number of "deadhead" kilometres—those driven without paying loads —was reduced by 0.5%. Applying that improvement to all 230 million kilometres covered by the fleet each year would produce an extra $1.25 million savings.

Equipping all 2000 trucks with the satellite hook-up will require an investment of $8 million and the construction of a message-relaying system costing $2 million. The equipment and on-board devices will have a service life of 8 years and negligible salvage value; they will be depreciated with a 30% CCA rate. The company's tax rate is about 38% and its required minimum attractive rate of return is 18%.

(a) Determine the annual net cash flows from the project.
(b) Perform a sensitivity analysis on the project's data, varying savings in telephone bills and savings in deadhead kilometres. Assume that each of these variables can deviate from its base-case expected value by ±10%, ±20%, and ±30%.
(c) Prepare sensitivity diagrams and interpret the results.

15.29 The following is the comparison of the cost structure of a conventional manufacturing technology (CMT) with a flexible manufacturing system (FMS) at one Canadian firm.

Most-Likely Estimates	CMT	FMS
Number of part types	3,000	3,000
Number of pieces produced/year	544,000	544,000
Variable labor cost/part	$2.15	$1.30
Variable material cost/part	$1.53	$1.10
Total variable cost/part	$3.68	$2.40
Annual overhead costs	$3.15 million	$1.95 million
Annual tooling costs	$470,000	$300,000
Annual inventory costs	$141,000	$31,500
Total annual fixed operating costs	$3.76 million	$2.28 million
Investment	$3.5 million	$10 million
Salvage value	$0.5 million	$1 million
Service life	10 years	10 years
CCA rate	30%	30%

(a) The firm's tax rate and MARR are 40% and 15%, respectively. Determine the incremental cash flow (FMS − CMT) based on the most likely estimates.

(b) Management feels confident about all of the input estimates in CMT. However, the firm does not have any previous experience in operating a FMS. Therefore, many input estimates except the investment and salvage value are subject to variations. Perform a sensitivity analysis on the project's data, varying the elements of operating costs. Assume that each of these variables can deviate from its base-case expected value by ±10%, ±20%, and ±30%.

(c) Prepare sensitivity diagrams and interpret the results.

(d) Suppose that probabilities of the variable material cost and the annual inventory cost for the FMS are estimated as follows:

Material Cost

Cost per Part	Probability
$1.00	0.25
1.10	0.30
1.20	0.20
1.30	0.20
1.40	0.05

Inventory Cost

Annual Inventory Cost	Probability
$ 25,000	0.10
31,000	0.30
50,000	0.20
80,000	0.20
100,000	0.20

What are the best and the worst cases of incremental NPW?

(e) In part (d), assuming that the random variables of the cost per part and the annual inventory cost are statistically independent, find the mean and variance of the NPW for the incremental cash flows.

(f) In parts (d) and (e), what is the probability that the FMS would be a more expensive investment option?

15.30 A conventional muffler dampens the sound produced when the engine ignites fuel, using the baffles to redirect the sound of these explosions into an enclosed chamber where much of the sound dissipates. But this conventional muffler system tends to create a build-up of exhaust that causes back pressure on the engine, reducing power and efficiency. The electronic muffler shown in Fig. 15.13 would have the same purpose as its conventional counterpart—reducing noise—but would not create the back pressure. As a result, the designers say the electronic muffler would raise power output by as much as 7 percent and increase fuel efficiency.

A company is planning to build and market an electronic muffler on a trial basis for one or two car models. Based on initial responses from automakers, the firm has estimated the project's financial performance as follows.

Most-Likely Estimates

Installed first cost	$9 million
Working capital	$1.5 million
Estimated life	5 years
Salvage value	$1 million
CCA rate	30%
Annual gross revenue	$8 million
Annual O&M costs	$3 million

It is difficult to predict the actual response from the automakers when the product is out on the market; the company has simplified the uncertain situation somewhat and feels that it is sufficient to imagine three possible outcomes. If product acceptance is poor, revenue will be only $3 million a year, but a strong response will produce gross revenue of $16 million a year. The firm estimates that there is a

Figure 15.13 ■ Making a quieter muffler (Problem 15.30)

Making a Quieter Muffler

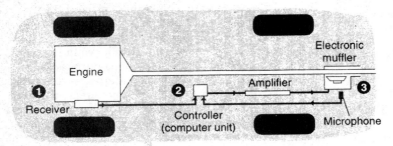

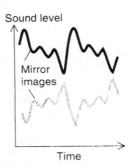

How Sound Waves Are Canceled

— Engine noise
⋯ Opposing sound
  Output sound

Sound level

Mirror images

Time

1 The receiver monitors the sounds generated by the engine, sending them to the controller.

2 The controller generates a pattern of sound waves that represent the mirror image of the engine sound waves.

3 At the muffler unit, the two patterns are combined, in effect, canceling each other. A microphone monitors the greatly reduced output sound, sending the results back to the controller so that adjustments can be made.

25% chance of poor acceptance, a 25% chance of excellent acceptance, and a 50% chance of average acceptance (the base-case). The annual O&M costs as well as the working-capital investment will vary accordingly.

Three States of Acceptance

	Poor	Most-Likely	Strong
Gross revenue	$3	$8	$16
Annual O&M cost	$2	$3	$ 5
Working capital	$1	$1.5	$ 3

The working-capital investment will be recovered at the end of project life. The firm's tax rate is 35%, and the MARR after-tax is known to be 12%. (Assume that the first-year market response will prevail for the remaining project years.)

(a) Develop a present worth expression for this investment.

(b) Assuming that the state of product acceptance is a continuous triangular random variable with parameters assessed above, develop a NPW probability distribution based on 10 Monte Carlo samplings.

(c) With the results of part (b), find the mean and variance of the NPW distribution, and interpret the results.

15.31 MBQ Corporation is considering building a production line for very large scale integrated circuits (VLSIC). The small geometries involved demand that a "dry etch" technique be used rather than wet acid baths. The corporation has some experience with plasma etching, but the desired dimensions push the limits of conventional plasma etch technology and newer reactive ion etch (RIE) equipment promises to provide higher process yields than the plasma etch equipment, but at a higher initial investment cost. Three dry etch processes are planned for the new semiconductor plant. While it is feasible to run multiple processes on one machine, separate machines will be purchased for each process to avoid possible cross contamination during normal operations. A 3-year planning horizon is believed to be reasonable, since it is anticipated that the next generation of integrated circuits will be introduced at that time. These future products may require new production technologies.

Two major product groups are planned for the new plant. One is a 64-bit microprocessor chip set (MCS), and the second is a four-megabyte (4-MB) dynamic random access memory chip (DRAM). The 64-bit chip set includes a central processing unit, a floating point coprocessor, a memory management unit, and other special-pupose support chips. These chips will be sold as matched sets which will operate at different clock speeds. Those that can operate at more than 66 megahertz (MHz) will be sold as 66 MHz chip sets, those that cannot operate at 66 MHz but can operate at more than 33 MHz will be sold as 33 MHz chip sets, and those that can operate at more than 16 MHz but less than 33 MHz will be sold as 16 MHz chip sets. Those that cannot operate at 16 MHz or above are considered scrap and are not packaged.

Based upon market studies, anticipated process, and packaging yields, the total production for the first year is planned to be 6000 wafer starts, with 12,000 the second year and 15,000 the third year. Enough dynamic RAM wafer starts will be made to fill out the planned production capacity. The annual revenue projections (in millions of dollars) are estimated to be *uniformly* distributed between the two bounds as follows:

Revenue Projection Range (RIE Option)

Year	Microprocessor Wafer Starts	4 Megabyte RAM Wafer Starts
1	$10M–$18M	$8M–$12M
2	35–45	20–26
3	30–40	25–32

Revenue Projection Range (Plasma Option)

Year	Microprocessor Wafer Starts	4 Megabyte RAM Wafer Starts
1	$9M–$12M	$7M–$10M
2	34–42	20–24
3	33–40	22–30

The investment required and other financial information for each option can be summarized as follows.

- **RIE Option:** At $400,000 each for the RIE systems, the RIE option has a total investment of $1,560,000 for three machines, again including 30% of the system cost for installation, spares, and sales tax, subject to a 30% CCA rate. The estimated salvage value at the end of year 3 is $500,000 for the three systems. The estimated expenses for repair and maintenance are $20,000 for the first year and $40,000 for each of following two years. Operator labor is estimated to be $16,000 for each of the three years in the planning period. The RIE systems are expected to require $80,000 of process engineering support the first year, $60,000 the second year, and $50,000 the third year. Process gas consumption is expected to be $1000 the first year, $2000 the second year and $2500 the third year.

- **Plasma Etch Option:** The plasma etch systems under consideration cost $250,000 each. The total initial cost of the plasma etch option is $858,000 including three machines plus 30% of the system cost for installation, spares, and sales tax. This investment also has a 30% CCA rate. At the end of year 3, it is believed that the three plasma etchers can be sold for $200,000. It is estimated that repair and maintenance expenses will total $20,000 the first year and $30,000 each year for the second and third years. The estimated operator labor cost is $20,000 per year. Process engineering support is expected to cost $80,000 the first year, $65,000 the second year, and $60,000 the third year. Process gases will cost $2000 the first year, $4000 the second year, and $5000 the third year.

(a) Assuming that the firm's combined federal and provincial tax rate is 40% and its MARR is 20%, find the NPW expression for each option.

(b) Assume that all probabilities are independent. Using the Monte Carlo method for 200 trials, estimate the expected NPW and the NPW distribution for each option.

(c) With the results of part (b), which option would you recommend?

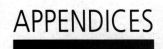
APPENDICES

Basic Financial Reports

Of the various reports corporations issue to their stockholders, the annual report is by far the most important. The annual report contains basic financial statements as well as management's opinion of the past year's operations and the firm's future prospects. There are three basic financial reports provided by any business firm: (1) the **balance sheet,** a picture of the financial status at a given point in time, (2) the **income statement,** which indicates whether the company is making or losing money during a stated period of time, and the **cash flow statement,** which details where the cash was obtained and where the cash was spent. The fiscal year can be any 12-month term, but is usually January 1 through December 31 of a calendar year.

■ A.1 The Balance Sheet

The balance sheet contains a summary listing of the assets (items owned by the firm) and the liabilities (debts owed by the firm) on a particular date, usually December 31 of each year. One of the important features of any balance sheet statement is that the assets of the firm equal the liabilities of the firm plus the owner's equity in the business.

A.1.1 Assets

Asset items are listed in the order of current assets, fixed assets, and other assets. This ordering sequence represents the length of time it typically takes to convert them into cash. **Current assets** can be converted to cash or its equivalent in less than one year. These generally include three major accounts. The first is cash. A firm typically has a cash account at a bank to provide the funds needed to conduct day-to-day business. The second account is accounts receivable, which is money that is owed to the firm but has not yet been paid. For example, when a company receives an order from a retail store, it sends an invoice along with the shipment to the retailer. The unpaid invoice falls into a category called "accounts receivable." As this bill is paid, it will be deducted from the accounts receivable account and placed into the cash account. A typical firm will have a 30- 45-day accounts receivable, depending on the frequency of its bills and the payment terms for its customers. The third account is inventories, which constitute materials and supplies on hand.

Fixed assets are relatively permanent and take time to convert into cash. The most common fixed assets include the physical investment in the business, such as land, buildings, factory machinery, office equipment, and automobiles. With the exception of land, most fixed assets have a limited useful life. For example, buildings and equipment are used up over a period of years. Each year, a portion of the usefulness of these assets expires, and a portion of their total cost should be recognized as a depreciation expense. The term **depreciation** means the accounting process for this gradual conversion of fixed assets into expenses.

Finally the balance sheet shows **other assets.** Typical assets in this category include investments made in other companies and intangible assets, such as goodwill, copyrights, and franchises. Goodwill appears on the balance sheet only when a going business is purchased in its entirety. This indicates any additional amount paid for the business above the fair market value of the business. (Here, the fair market value is defined as the price that a buyer is willing to pay when the business is offered for sale.)

A.1.2 Forms of Capital

Any business needs assets to operate. To acquire the assets, the firm must raise capital. When the firm finances its long-term needs externally, it may obtain funds from the capital markets. Capital comes in two basic forms, **debt** and **equity**. Debt capital refers to the borrowed capital from financial institutions. Equity capital refers to the capital obtained from the owners of the company.

The basic methods of **debt financing** include obtaining a bank loan and issuing bonds. For example, a firm needs $10,000 to purchase a computer. In this situation, the firm might borrow money from a bank and repay the loan with specified interest in a few years, a procedure known as **short-term debt financing.** Suppose that the firm needs $100 million for a construction project. It would normally be very expensive (or require a substantial amount of mortgage) to borrow this sum directly from a bank. In this situation, the firm might go to the public to borrow money on a long-term basis. The document that records the nature of such an arrangement between a company and its investors is called a **bond**. Raising capital through issuing bonds is called a **long-term debt financing.**

Similarly, there are different types of **equity capital.** The equity of a proprietorship represents the money provided by the owner. For a corporation, equity capital comes in two forms: **preferred** and **common stock.** Investors provide capital to a corporation, and the company agrees to provide the investor with fractional ownership in the corporation. Preferred stock pays a stated **dividend**, much like the interest payment on bonds. Preferred stock has preference over common as to the receipt of dividends and to the distribution of assets in the event of liquidation. When a company makes profits, it has to decide what to do about the profits. It can retain some of the profits for future investment and pay out the remaining profits to its common stockholders.

A.1.3 Liabilities

The liabilities of a company indicate where the funds were obtained to acquire the assets and to operate the business. Here, **liability** means an obligation to an outside party (or creditor). We usually divide liabilities into **current** and **other liabilities.** Current liabilities are those debts that must be paid in the near future (normally payable within one year). The major current liabilities include **accounts** and **notes payable** within a year. They also include **accrued expenses** (wages, salaries, interest, rent, and taxes, owed but not yet due for payment) and advance payments and deposits from customers. Other liabilities include long-term liabilities, such as bonds, mortgages, and long-term notes, which are due and payable more than one year in the future.

A.1.4 Working Capital

Current assets and current liabilities are collectively known as **working capital.** The term **net working capital** refers to the difference between current assets and current liabilities. This figure indicates the extent to which current assets can be converted to cash to meet current obligations. Therefore, we view a firm's net working capital as a measure of its liquidity position. Here liquidity means that the capital can be converted rapidly to cash. Consequently, the more net working capital a firm has, the greater its ability to satisfy creditors' demands.

A.1.5 Stockholders' Equity (Owners' Net Worth)

Stockholders' equity (or **owner's equity** for a proprietorship) indicates the portion of the assets of a company that are provided by the investors (owners). Therefore, stockholders' equity is the liability of the company to its owners. This represents the amount that is available to the owners after all other debts have been paid. It generally consists of preferred and common stock, treasury stock, capital surplus, and retained earnings. **Paid-in capital (capital surplus)** is the amount of money received from the sale of stock in excess of the par (stated) value of the stock. **Outstanding stock** is the number of shares issued that actually are held by the public. The corporation can buy back part of its issued stock and hold it as **treasury stock.** The amount of **retained earnings** represents the cumulative net income of the firm since its beginning, less the total dividends that have been paid to stockholders. In other words, retained earnings indicate the amount of assets that have been financed by plowing profits back into the business. These retained earnings belong to the stockholders.

■ A.2 The Income Statement

The **income statement** is the second key financial report; it summarizes the firm's net income (revenue minus expenses) for a specified period of time. Every business prepares an annual income statement. Most businesses prepare quarterly and monthly income statements as well. The company's **accounting period** refers to the period of time covered by an income statement.

A.2.1 Reporting Format

At the top of the income statement are revenues from the business operation. **Revenue** is the price of goods sold and services rendered during a given accounting period. **Net sales** represents the gross sales less any sales returns and allowances. Shown on the next several lines are the expenses and costs of doing business as deductions from the revenues. The largest expense for a typical manufacturing firm is its production expense, including such items as labor, materials, and overhead, known as **cost of goods sold.** Net sales less the cost of goods sold and operating expenses indicates the **net operating profit.** Operating expenses (such items as leases and administrative expenses) are subtracted from the net operating profit; the result is **taxable income** (or income before income taxes). Finally, we determine **net income** (or **net profit**) by subtracting **income taxes** from the taxable income. Net income is also known as **accounting income.**

A.2.2 Retained Earnings

As a part of the income statement, many corporations also report their retained earnings during the accounting period. When a corporation earns profits, it must decide what to do with them. The corporation may decide to pay out the profits as **dividends** to its stockholders. Or it may retain the profits in the business to finance expansion or support other business activities. When the company decides to do a combination of dividends and retained earnings, the category "available for common stockholders" reflects the net earnings of the corporation less the preferred stock dividends. When preferred and commons stock dividends are subtracted from net income, the remainder is retained earnings (profits) for the year.

A.2.3 Earnings Per Share

Additional important financial information is provided in the income statement, namely, the **earnings per share (EPS).** In simple situations, we compute this by dividing the "available earnings to common stockholders" by the number of shares of common stock outstanding. Stockholders and potential investors want to know what their share of profits is, not just the total dollar amount. Presentation of profits on a per-share basis allows the stockholders to relate earnings to what they paid for a share of stock. Naturally, companies want to report a higher EPS to their investors as a means of summarizing how well they managed their businesses for the benefits of owners.

■ A.3 The Cash Flow Statement

The income statement explained in the previous section only indicates whether the company was making or losing money during the reporting period. Its emphasis is on determining the net income (profits) of the firm, for **operating activities.** However, the income statement ignores two other important business activities for the period—**financing** and **investing activities.** The third financial statement, the cash flow statement, details how the company generated the cash and how the cash was used during the reporting period.

A.3.1 Sources and Uses of Cash

The difference between the **sources** (inflows) and **uses** (outflows) of cash represents the **net cash flow** during the reporting period. This is a very important piece of information because investors judge the value of an asset (or a whole firm) by the cash flows it generates. Certainly, a firm's net income is important, but cash flows are even more important. This is particularly true because we need cash to pay dividends and to purchase the assets required to continue operations. As we mentioned in Chapter 1, the goal of the firm should be to maximize the price of its stock. Since the value of any asset depends on the cash flows produced by the asset, managers want to maximize cash flows available to investors over the long run. Therefore, investment decisions should be based on cash flows rather than profits. For such investment decisions, it is necessary to convert profits (as determined in the income statement) to cash flows.

A.3.2 Reporting Format

In preparing the cash flow statement, companies identify the sources and uses of cash according to types of business activities, specifically, operating activities, investing activities, and financing activities. We start with the net change in operating cash flows from the income statement. Here operating cash flows represent those cash flows related to production and sales of goods or services. All **noncash expenses** are simply added back to net income (or after-tax profits). For example, an expense such as depreciation is only an **accounting expense** (bookkeeping entry). While they are charged against current income as an expense, accounting expenses do not involve an actual cash outflow. The actual cash flow may have occurred when the asset was purchased.

Once we determine the operating cash flows, we consider any cash flow transactions related to the investment activities. Investment activities include such items as purchasing new fixed assets (cash outflow) or reselling old equipment (cash inflow). Finally, we detail any cash transactions related to financing any capital used in business. For example, the company could borrow or sell more stock, resulting in cash inflows. Paying off existing debt will result in cash outflows. By summarizing cash inflows and outflows from three activities for a given accounting period, we obtain the net changes in cash flow position of the company.

■ A.4 The Gillette Company's Financial Statements

Exhibits A.1 and A.2 show a comparative balance sheet and the income statement for The Gillette Company as of December 31, 1990. From the income statement (Exhibit A.2) we see that net sales were $4.345 billion in 1990, compared with $3.819 billion in 1989, a gain of 13.78%. Profits from operations rose 16.35% to $773 million, and net income was up 29.22% to $367.9 million. Earnings per common share climbed at a faster pace to $3.20, an increase of 18.52%.

The Gillette Company and Subsidiary Companies
Consolidated Balance Sheet[1]
(Millions of dollars)

December 31, 1990 and 1989	1990	1989
Assets		
Current Assets		
Cash and cash equivalents	$ 80.5	$ 109.5
Short-term investments, at cost which approximates market value	.7	27.3
Receivables, less allowances: 1990, $46.1; 1989, $28.3.	1,015.7	828.5
Inventories	758.4	688.2
Prepaid taxes and expenses	238.2	201.0
Total Current Assets	2,093.5	1,854.5
Property, Plant, and Equipment, at cost less accumulated depreciation	861.6	744.8
Intangible Assets, less accumulated amortization	240.2	260.4
Other Assets	476.0	254.3
	$ 3,671.3	$ 3,114.0
Liabilities and Stockholders' Equity		
Current Liabilities		
Loans payable	$ 267.3	$ 325.5
Current portion of long-term debt.	103.1	15.2
Accounts payable and accrued liabilities	839.1	674.9
Income taxes.	98.4	45.7
Total Current Liabilities	1,307.9	1,061.3
Long-Term Debt.	1,045.7	1,041.0
Deferred Income Taxes	117.4	82.5
Other Long-Term Liabilities, principally accrued pensions.	326.6	245.4
Minority Interest	8.3	13.8
Redeemable Preferred Stock		
8.75% Cumulative Series B Convertible Preferred, without par value,		
600,000 shares issued and redeemable at $1,000 per share.	600.0	600.0
Stockholders' Equity		
8.0% Cumulative Series C ESOP Convertible Preferred, without par value,		
165,872 shares issued	100.0	—
Unearned ESOP compensation	(92.4)	—
Common stock, par value $1 per share		
Authorized 290,000,000 shares		
Issued: 1990, 138,083,494 shares; 1989, 137,714,292 shares.	138.1	137.7
Additional paid-in capital.	176.8	169.4
Earnings reinvested in the business	1,635.6	1,430.0
Cumulative foreign currency translation adjustments	(209.5)	(183.9)
Treasury stock, at cost: 1990: 40,865,085 shares; 1989, 40,864,775 shares.	(1,483.2)	(1,483.2)
Total Stockholders' Equity	265.4	70.0
	$ 3,671.3	$ 3,114.0

[1] Courtesy of The Gillette Company

The Gillette Company and Subsidiary Companies
Consolidated Statement of Income and Earnings Reinvested in the Business[1]
(Millions of dollars, except per share amounts)

Years Ended December 31, 1990, 1989 and 1988	1990	1989	1988
Net Sales	$4,344.6	$3,818.5	$3,581.2
Cost of Sales	1,824.6	1,581.9	1,487.4
Operating Expenses	1,747.3	1,572.5	1,479.8
	3,571.9	3,154.4	2,967.2
Profit from Operations	772.7	664.1	614.0
Nonoperating Charges (Income)			
Interest income	(49.4)	(31.1)	(37.2)
Interest expense	169.0	145.8	138.3
Other charges—net	59.9	75.8	64.3
	179.5	190.5	165.4
Income before Income Taxes	593.2	473.6	448.6
Income Taxes	225.3	188.9	180.1
Net Income	367.9	284.7	268.5
Preferred Stock dividends, net of tax benefit	57.3	23.5	—
Net Income Available to Common Stockholders	310.6	261.2	268.5
Earnings Reinvested in the Business at beginning of year	1,430.0	1,261.6	1,083.8
	1,740.6	1,522.8	1,352.3
Common Stock dividends declared	105.0	92.8	90.7
Earnings Reinvested in the Business at end of year	$1,635.6	$1,430.0	$1,261.6
Net Income per Common Share	$ 3.20	$ 2.70	$ 2.45
Dividends declared per common share	$ 1.08	$ .96	$ .86
Average number of common shares outstanding (thousands)	97,057	96,722	109,559

[1] Courtesy of The Gillette Company

A.4.1 The Balance Sheet

The $3671.3 million of assets shown in the balance sheet (Exhibit A.1) were necessary to support sales. Gillette obtained the bulk of the funds used to buy assets by: (1) buying on credit from their suppliers (accounts payable), (2) borrowing from financial institutions (notes payable and long-term bonds), (3) selling preferred and common stock to investors, and (4) plowing earnings into business as reflected in the retained earnings account. We interpret the 1990 data from the balance sheet to mean that the company still has invested $861.6 million in plant and equipment as of the reporting date. Depreciation expenses for 1990 were $138.4 million.[1] In the beginning of 1990, Gillette had total fixed assets of $744.8 million. This implies that Gillette added new fixed assets costing $255.2 million during the year.

---

[1] Source: The Gillette Company's 1990 Annual Report. Courtesy of The Gillette Company.

Gillette had a total long-term debt of $1045.7 million, which consists of the several bonds issued during previous years. The interest payments associated with these long-term debts were about $169 million, as shown in Exhibit A.2. The preferred stockholders stand next in line to receive a portion of the firm's income. Gillette has 600,000 shares of preferred stock. Gillette paid a cash dividend of $57.3 million (Exhibit A.2) to the preferred stockholders during 1990. After all creditors and preferred stockholders have been paid, the remaining income of $310.6 million (Exhibit A.2) belongs to the common stockholders. This income may be fully retained for reinvestment in the firm, or it may be paid out as dividends to the common stockholders. Gillette retained $205.60 million to support future growth and paid the rest ($105 million) out as dividends. Gillette had 138.08 million shares of common stock outstanding (Exhibit A.1). Investors actually have provided the company with a total capital of $314.9 million ($138.1 million of common stock and $176.8 million of paid-in capital). That is, each common stockholder provides about $2.28 per share ($314.9/138.08 = $2.28 per share). However, Gillette has retained the current ($205.60 million) as well as previous earnings of $1430 million, or $11.84 per share since it was incorporated. These earnings belong to The Gillette Company's common stockholders. Therefore, stockholders on the average have a total investment of $2.28 + $11.84 = $14.12 per share in the company. This figure, $14.12, is known as the **stock's book value.** In the fall of 1990, the stock traded in the general range of $52 to $65 per share.[2] Note that this market price is different from the stock's book value. Many factors affect the market price; the most important is how investors expect the company to do in the future. Certainly, the company's Sensor project and the actual performance of the project have had a major influence on the market value of its stock.

A.4.2 Income Statement

The company's income statement for 1990 is shown in Exhibit A.2. We can see that Gillette had earnings available to common stockholders of $310.6 million, so the company earned $3.20 per share of average stock outstanding.[3] Of this amount, it paid out $105 million, or $1.08 per share, in dividends, and it retained $205.60 million. As we noted previously, an income statement provides information on how the firm's operations affect retained earnings; it does not say, however, how and where these cash flows were invested. In addition, it ignores the inflows and outflows from nonoperating activities. The cash flow statement is a tool for overcoming these deficiencies in the basic financial statements.

A.4.3 Cash Flow Statement

Exhibit A.3 contains The Gillette Company's cash flow statement for 1990. As shown in the exhibit, cash flow from operations amounted to $447.9 million. Note that this is significantly less than that found under the shorthand rule for calculating net cash flow

---

[2] Source: The Gillette Company's 1990 Annual Report. Courtesy of The Gillette Company.

[3] Note that Gillette originally issued 138.08 million shares of common stock as shown in Exhibit A.1. However, Gillette repurchased its own stock of 40.87 million shares during 1990 and put them under treasury stock. Therefore, the average outstanding common stock held by the public amounts to 97.05 million shares, as shown in Exhibit A.2.

The Gillette Company and Subsidiary Companies
Consolidated Statement of Cash Flows[1]
(Millions of dollars)

Years Ended December 31, 1990, 1989 and 1988	1990	1989	1988
Operating Activities			
Net income.	$ 367.9	$ 284.7	$ 268.5
Adjustments to reconcile net income to net cash provided by operating activities:			
Depreciation and amortization	177.0	149.1	141.2
Other	21.1	53.1	34.8
Changes in assets and liabilities, net of effects from acquisition of businesses:			
Accounts receivable.	(170.8)	(186.2)	(100.6)
Inventories.	(47.3)	(54.3)	(63.3)
Accounts payable and accrued liabilities	82.3	84.1	24.6
Other working capital items	41.9	18.6	(43.7)
Other noncurrent assets and liabilities	(24.2)	(7.6)	(22.5)
Net cash provided by operating activities	447.9	341.5	239.0
Investing Activities			
Additions to property, plant and equipment	(255.2)	(222.6)	(189.0)
Disposals of property, plant and equipment.	25.5	8.1	30.8
Acquisition of businesses, less cash acquired	(113.5)	(72.3)	(29.6)
Sale of businesses	122.1	—	20.8
Other	27.3	(24.1)	(13.6)
Net cash used in investing activities.	(193.9)	(310.9)	(180.6)
Financing Activities			
Issuance of preferred stock	100.0	600.0	—
Purchase of treasury stock	—	—	(855.2)
Proceeds from exercise of stock option and purchase plans	7.8	5.1	10.2
Proceeds from long-term debt	31.0	27.6	949.8
Reduction of long-term debt	(196.8)	(649.7)	(129.2)
Increase (decrease) in loans payable	(59.4)	71.2	107.8
Dividends paid.	(159.2)	(113.9)	(94.7)
Net cash used in financing activities.	(276.6)	(59.7)	(11.3)
Effect of Exchange Rate Changes on Cash	(6.4)	(17.8)	(9.8)
Increase (Decrease) in Cash and Cash Equivalents	(29.0)	(46.9)	37.3
Cash and Cash Equivalents at Beginning of Year.	109.5	156.4	119.1
Cash and Cash Equivalents at End of Year	$ 80.5	$ 109.5	$ 156.4

[1] Courtesy of The Gillette Company

[that is, adding back depreciation and amortization expenses ($177 million) to net income, or $367.9 + $177 = $544.9 million]. The main reason for the difference is that the changes in working-capital items have been taken into account in the cash flow statement. For example, an increase in accounts receivable ($170.8 million) represents the amount of total sales on credit. Since this figure is included in the total sales in determining the net income, we need to subtract this figure to determine the true cash position. Similar adjustments are also necessary for inventories, accounts payable, and other working capital items as summarized in Exhibit A.3. For the investment activities, there was an investment outflow of $255.2 million for new plant and equipment. On the other hand, the financing decision produced a net outflow of $276.6 million due to $196.8 million reduction in a long-term debt, $59.4 million reduction in a short-term debt, and $159.2 million in cash dividends. However, they only generated $138.8 million in cash inflows through issuance of new preferred stocks and bonds. Finally, there was the effect of exchange-rate changes on cash at foreign subsidiaries. This amounts to a net decrease of $6.4 million. Together, the three types of cash flow generated a negative cash flow of $29 million. This net decrease of $29 million denotes the change in The Gillette Company's cash position as shown in the cash accounts in the balance sheet. Although this was already known from the balance sheet, one cannot explain this change in cash without the aid of a cash flow statement.

EzCash Software

This section describes the contents of the **EzCash** software package and various commands provided in the software.

■ B.1 What Is EzCash?

EzCash is a Windows-based integrated computer-aided economic analysis package designed to solve the most frequently used engineering economic decision problems. It is menu driven for convenience and flexibility, and it provides the following features.

- A flexible and easy-to-use graphic editor for data input and modifications.
- A logical sequence of windows within the economic computing environment which enhances the student's understanding.
- A Lotus-like data entry for after-tax cash flow analysis.
- An extensive array of computational modules and user selected graphics outputs.

EzCash makes engineering economic analyses relatively simple to perform on the computer. The user-friendliness, efficient computational algorithms, and the graphics presentation are areas in which **EzCash** offers significant advantages relative to spreadsheet approaches.

B.1.1 Systems Requirements

EzCash may be used with the IBM PC as well as strictly IBM-compatible computers with the following additional features.

- Microsoft MS-DOS version 3.1 or later.
- Microsoft Windows Operating System (Version 3.1 or later).
- A display adapter that is supported by Windows.

- The minimum amount of net memory of 400KB.
- A mouse that is supported by Windows.
- A printer that is supported by Windows, if you want to print with **EzCash**.

B.1.2 Typographical Conventions

EzCash commands and instructions are issued by pressing particular keys or sequences of keys. These keys are designated either by symbols (for arrow keys, function keys, and return or enter key). This manual uses the following convention:

- *Click* or *Press*: This tells you to press a designated key or mouse button. To perform an operation, use the left mouse button, the arrow keys, and the **Return** or **Enter** key. All **EzCash** commands will be highlighted in boldface.

- *Select*: This tells you to position the cursor pointer to a particular location in the **EzCash** screen, usually with the mouse, and click the left mouse button (or hit the **Return** key) to activate the command.

- *Enter*: This tells you to type numerical values as an input.

■ B.2 Installing EzCash

It is important to make a copy of **EzCash** from your original distribution diskette. This backup copy should be used to run **EzCash**, and the original diskette should be stored in a safe place.

B.2.1 Installing EzCash to a Hard Disk

You will need the backup copy of your **EzCash** diskette to accomplish this task. You should install Microsoft Windows on your computer before installing **EzCash**. **EzCash** requires MS Windows Version 3.1 or later and 800KB of hard disk storage space. If there is not enough space available in your hard disk, you may run **EzCash** directly from Drive A. To install **EzCash** on your computer, insert the **EzCash** diskette in Drive A and type

```
C> A:
A> install
```

Running the installation program will create a subdirectory called **EzCash** in your hard disk (or C Drive) and copy all the necessary files into this subdirectory.

If you have any difficulty installing **EzCash**, you may create a subdirectory on your own and then copy the files in the distribution diskette into the subdirectory. Assuming that the distribution diskette is in Drive A, do the following:

```
C>MD\EzCash
C>CD\EzCash
C> XCOPY A:*.* C:
```

EzCash and all its font files will be installed in your subdirectory.

B.2.2 Starting EzCash

After installation, you can start Windows from the MS-DOS prompt. To start Windows from the MS-DOS prompt, type

```
C>win
```

and press **Return**. Windows should appear on your screen. Now you are ready to run **EzCash**. Move the mouse cursor and select the **File/Run** command menu from the Program Manager. Type C:\ezcash\ezcash and press **Return**. This will start the program. (You can start **EzCash** from drive A if you have not installed, or do not wish to install, **EzCash** on your hard drive. To do this, insert the disk provided into drive A. Select the **File/Run** command sequence, type A:*ezcash.*)

B.2.3 To Create a Program Icon

You can also create a program icon for **EzCash**. By having a unique program icon available for **EzCash**, you can run **EzCash** directly by double-clicking the icon from Windows. To create a program icon,

1. Choose the **New** command under the **File** menu in the *Program Manager*.

2. From the pop-up window labelled *New Program Object*, click the **Program Group** and then **Ok**.

3. A pop-up window labelled *Program Group Properties* will appear. In the *Description* box, type ezcash and press **Return**. You will see a new window labelled **EzCash**.

4. Return to the *Program Manager* and select **File/New** to open a pop-up window labelled *New Program Object*.

5. Select the *Program Item* and then **Ok**.

6. A pop-up window labelled *Program Item Properties* will appear. Type **EzCash** in the *Description* box. Move to the *Command Line* box, then type C:\ezcash\ezcash.

7. In the *Working Directory* box, type C:\ezcash, the complete path of the directory in which the executable files are placed. Now you will see an icon placed in the program group **EzCash**.

8. Double click the **EzCash** icon to start the program.

Figure B.1 ■ Opening screen for **EzCash**

■ B.3. An Overview of EzCash Command Modes

As you run **EzCash**, your first opening screen will appear as shown in Figure B.1. This opening screen contains two main operating command modes: **Cash** and **Quit**.

- **Cash** : This command performs various engineering economic calculations.
- **Quit** : This command exits the **EzCash** program.

To select either mode, point to the menu item and click the left mouse button.

If you click **Cash**, you will see the **EzCash** window which displays six command modes (Fig. B.2): **File**, **Edit**, **Display**, **Compute**, **SpreadSheet**, and **Special**. A menu selection is done by positioning the mouse cursor over the menu to be selected and clicking the left mouse button.

B.3.1 File Mode

In **File** mode, you can open a net cash flow file created previously by the **EzCash** program using the **Open** command. Any EzCash net flow file created by the **EzCash** **Edit** mode will have a file extension **.nc$**. However, you do not enter this file extension when opening the file. The **Save** command will save the net cash flow created by the **Edit** and **Build** commands. If no path is specified when opening or saving a cash flow file, **EzCash** will search for the file from or save it in the current directory.

Figure B.2 ■ Main window of EzCash

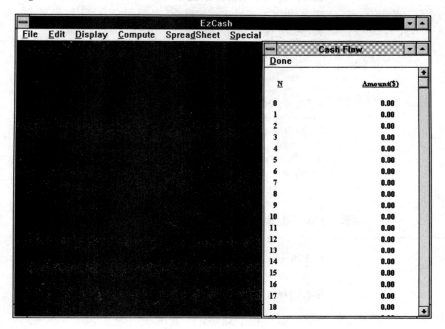

B.3.2 Edit Mode

With the **Edit** mode, you can create a new cash flow or edit the cash flow that was previously created. There are five subcommands that you can use in creating or editing the cash flows. These include **Build**, **Copy**, **Cut**, **Paste**, and **Scale**.

B.3.3 Display Mode

Once you have created a cash flow file, you can display various economic properties of the cash flow series in tabular and graphical form. Specifically, you can obtain (1) net cash flow table, (2) a cash flow diagram, (3) a present worth table as a function of interest rate, (4) a net present worth plot, (5) a project balance table, and (6) a project balance diagram.

B.3.4 Compute Mode

With the **Compute** mode, you can obtain the various measures of investment worth for the project cash flows created by the **Edit** mode. Typical measures include net present worth, net future worth, equivalent annual worth, rate of return, internal rate of return, and benefit-cost ratio.

B.3.5 SpreadSheet Mode

With the **SpreadSheet** mode, you can prepare an after-tax cash flow for an investment project. There are several built-in commands that allow you to generate depreciation schedules, loan repayment schedules, investment tax credits, and gains taxes. You can also view or print the entire worksheet.

B.3.6 Special Mode

Under the **Special** mode, five computational modules are provided: (1) loan analysis, (2) depreciation, (3) economic service life, (4) equivalence, and (5) incremental analysis. These commands are independent of the *Cash Flow Editor* and can be accessed during any computational mode.

■ B.4 Using File Mode

In **File** mode, there are five subcommands: **New**, **Open**, **Save**, **Save As**, and **Exit**.

B.4.1 New

The **New** command will clear the cash flow data from the screen as well as from memory. It also resets all the system variables to their initial states. When a new cash flow is created, it is a good idea to invoke the **New** command to clear the memory as well as screen.

B.4.2 Open

The **Open** command allows you to open a net cash flow file (with the file extension **.nc$**) created previously by the *Cash Flow Editor*. You can switch between drives and directories to select the file that needs to be opened. Directories are enclosed in square brackets. Trying to open any non–**.nc$** files will result in an error message.

B.4.3 Save

The **Save** command will save the net cash flow that is already open under the current file name. If you do not have an open file but attempt to save the existing cash flow by selecting this command, **EzCash** will bring up the *Save As* dialogue box. If no path is specified, **EzCash** will save the current file in the current directory.

B.4.4 Save As

The **Save As** command allows you to save the data in the current cash editor buffer under a different name. Before **EzCash** allows you to save any cash flow, it checks to see if another file with the same name exits on the disk. If one does not exist, the cash flow file is saved automatically. If one does exist, **EzCash** prompts you to select either **Cancel** or **Replace**. All filenames will be saved with the file extension **.nc$**.

B.4.5 Exit

The **Exit** command will terminate the program.

■ B.5 Using Edit Mode

With the **Edit** mode, it is possible to create a new cash flow or edit the cash flow that was previously created. There are five subcommands that can be used in creating or editing the cash flows. These include **Build**, **Copy**, **Cut**, **Paste**, and **Scale**.

B.5.1 Build Command

The **Build** command allows you to enter or edit the cash flow by selecting a cash flow pattern. As you select the **Build** command, the *Cash Flow Editor* window will appear on the screen as shown in Figure B.3. To initiate the data entry, follow these steps.

1. Use the mouse to move the cursor to the period where the cash flow amount needs to be entered and click to confirm.
2. Click the prompt area to see a blinking cursor.
3. Enter the amount. For a negative amount (outflow), enter a "−" sign followed by the dollar amount. For a positive amount, no "+" sign is required. Then press **Return**. This completes the entry of a single amount.
4. After entering the first input, the cursor advances one period automatically. You can repeat step 3 to enter more cash flows for sequential time periods. If you wish to skip one or more periods before entering the next cash flow, you repeat steps 1,2 and 3.
5. If you press **Tab**, you will be able to scroll the *Cash Flow Editor* vertically using the arrow keys. To return to input mode from scroll, press **Tab** again.

Pattern Command

Within the *Cash Flow Editor*, the **Pattern** command is provided to facilitate entering a cash flow series that is repeated or has a special pattern. There are four patterns to choose from: *single, uniform, gradient,* and *geometric*.

- *Single:* If there is no pattern in the cash flow series, you may choose the *single* pattern. To enter the single period of cash flow, the following steps should be taken.
  - Step 1: Enter the period of the cash flow at the *Period* box.
  - Step 2: Press **Tab** to move to *Amount* box and enter the amount. (For an outflow, enter a "−"sign.)
  - Step 3: Click the **Ok** button if these entries are correct.

  Clicking the **Done** button will clear the dialogue box and return the control to the *Pattern* dialogue box.

Figure B.3 ■ Cash Flow Editor

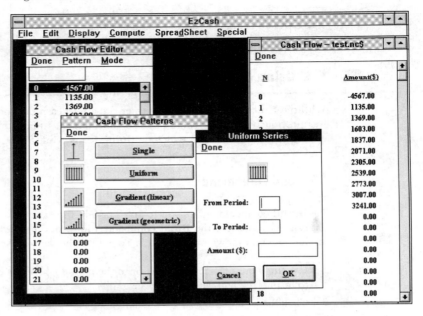

- *Uniform:* To enter a uniform series, select the **Uniform** button to bring up the *Uniform Cash Flow* dialogue box.
  - Step 1: Enter the period where the uniform series begins at the *From Period* box.
  - Step 2: Enter the period where the uniform series ends at the *To Period* box.
  - Step 3: Specify the constant amount at the *Amount* box.
  - Step 4: Click the **Ok** button if these entries are correct.

  To enter another uniform series while still in the uniform pattern, press the **Cancel** button to clear the previous entries and repeat steps 1 through 4.

- *Linear Gradient:* To enter a linear gradient series, select **Gradient (Linear)** which will bring up the *Linear Gradient Cash Flow* dialogue box.
  - Step 1: Enter the period where the gradient series begins at the *From Period* box.
  - Step 2: Enter the period where the gradient series terminates at the *To Period* box.
  - Step 3: Enter the amount of the first cash flow in the gradient series at the *Initial Flow* box.
  - Step 4: Move to the *Gradient* box, enter the gradient amount. To enter a decreasing gradient series, enter a negative gradient amount.
  - Step 5: Click the **Ok** button if these entries are correct.

  To enter another gradient series while still in the gradient pattern, press the **Cancel** button to clear the entries and repeat steps 1 through 5.

- *Geometric:* To enter a cash flow series with geometric gradient, take the following steps.

  - Step 1: Select the **Gradient (geometric)** from the **Pattern** menu.

  - Step 2: Enter the period where the geometric gradient series begins at the *From Period* box.

  - Step 3: Enter the period where the geometric gradient series terminates at the *To Period* box.

  - Step 4: Enter the amount of the first cash flow in the geometric series at the *Initial Flow*.

  - Step 5: Enter the gradient amount in percentage at the *Gradient* box. To enter a decreasing geometric gradient series, enter a negative gradient amount in percentage.

  - Step 6: Click the **Ok** button if these entries are correct.

  To enter another geometric gradient series while still in the geometric pattern, press the **Cancel** button to clear the previous entries and repeat steps 2 through 6.

Mode Command

As a default, any cash flow entered previously is overwritten. If a cash flow needs to be added on the existing cash flow instead of being overwritten, you may select the **Accumulate** mode. To return to the default mode (**Overwrite**), select the **Overwrite** mode.

B.5.2 Copy or Cut Command

A fast way to enter the same data into several periods or to delete a block of data is to use the **Copy** or **Cut** command. It is possible to copy or cut cash flows by specifying the range to be copied or cut. Once a block of data has been copied or cut, it will remain in the resident memory until another copy or cut action is taken. To copy or cut the block of cash flow, you may issue the following sequence of commands.

- Step 1: Select the **Copy** or **Cut** command from the **Edit** mode.

- Step 2: Enter the beginning period of the data range to be copied or cut at the *From Period* box.

- Step 3: Move to *To Period* box and enter the ending period of the data range to be copied or cut.

- Step 4: Clicking the **Ok** button will copy or cut the selected cash flows into the memory buffer and clear the contents of the edit boxes.

The cash flow block that is copied or cut resides in the temporary buffer. If another copy/cut operation is performed, the old cash flow block in the temporary buffer will be replaced with the new cash flow block that was copied or cut.

Figure B.4 ■ Dialogue box to paste cash flow range

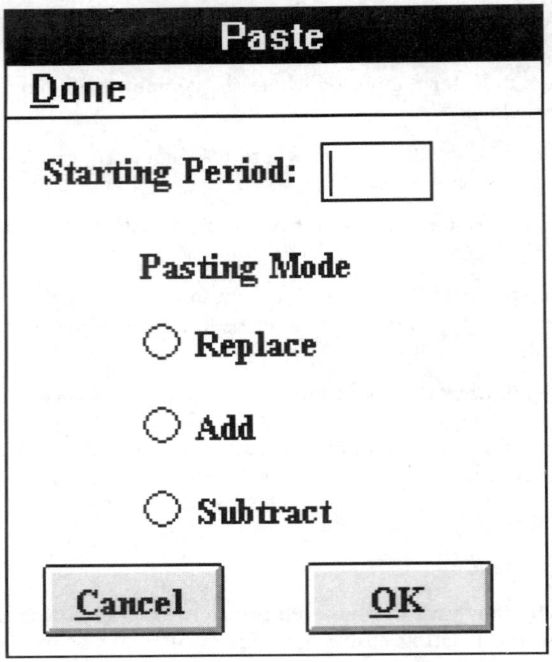

```
┌────────────────────────────────────┐
│              Paste                  │
├────────────────────────────────────┤
│ Done                                │
├────────────────────────────────────┤
│                                     │
│  Starting Period:  ┌──────────┐     │
│                    │          │     │
│                    └──────────┘     │
│                                     │
│         Pasting Mode                │
│                                     │
│         ◯ Replace                   │
│                                     │
│         ◯ Add                       │
│                                     │
│         ◯ Subtract                  │
│                                     │
│   ┌──────────┐    ┌──────────────┐  │
│   │ Cancel   │    │      OK       │  │
│   └──────────┘    └──────────────┘  │
└────────────────────────────────────┘
```

B.5.3 Paste Command

The **Paste** command allows you to paste the data copied (or cut) with the **Copy** or **Cut** command. You can paste over, add to, or subtract from the current cash flows by specifying the starting period of the pasting operation. The *Paste* dialogue box appears as shown in Figure B.4.

At the *From Period* box, enter the beginning period of the cash flow where it is to be pasted. Notice that there are three radio buttons marked **Replace**, **Add**, and **Subtract** under the label *Pasting Mode*. Selecting the **Replace** radio button and clicking **Ok** will replace the existing cash flow with the cash flow range copied. Selecting the **Add** radio button and pressing **Ok** will replace the existing cash flow with the sum of the cash flow range copied and the existing cash flow. Selecting the **Subtract** radio button and pressing **Ok** will replace the existing cash flow with the difference of the existing cash flow and the cash flow range copied.

B.5.4 Scale Command

The **Scale** command allows you to alter the magnitude of the current cash flows with four mathematical operations. The dialogue box appears as shown in Figure B.5. The scaling operations consist of four buttons marked **Addition**, **Subtraction**, **Multiplication**, and **Division**. As you select any one of these buttons, you need to specify the range of cash flows that need to be scaled and the amount or scaling factor.

Figure B.5 ■ Dialogue box to scale cash flow

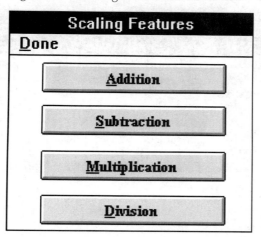

■ B.6 Using Display Mode

Once you have created a cash flow series using the **Edit** mode, you can obtain various economic characteristics of the cash flow series in tabular as well as graphical form. There are several plotting functions provided by **EzCash**. Specifically, you can obtain the net cash flow plot, the net present worth plot, and the project balance diagram for the net cash flow series in the *Cash Flow Editor*. You can also obtain either the present worth table or the project balance table at specified interest rates.

B.6.1 Cash Flow

This **Cash Flow** command allows you to scan the actual cash flow amount through the *Cash Flow* window. You can always visually check the cash flow pattern as well as the actual amount by scanning the *Cash Flow* window.

B.6.2 Cash Flow Plot

This **Cash Flow Plot** command will bring up a plot of the existing cash flow. The plot consists of the time or period plotted against the x-axis and the amount in dollars plotted against the y-axis. All inflows are shown as bars above the x-axis and all outflows are displayed as bars below the x-axis. (See Figure B.6.)

B.6.3 Present Worth Table

With the **Present Worth Table** command, you can obtain the NPW at varying interest rates. You need to specify the lower and upper bounds of the interest rate and the increment size to evaluate the NPW values. If you enter a range of interest between 0% and 100% with an increment of 1%, **EzCash** will calculate and display the NPW of the cash flows at interest rates of 0%, 1%, 2%, ... 99%, and 100%.

Figure B.6 ■ A typical cash flow plot

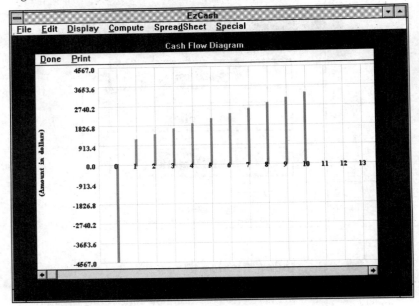

B.6.4 Present Worth Plot

With the **Present Worth Plot** command, you can obtain the NPW plot within the specified interest bounds. The acceptable value for the interest rate is between −99% and 900%.

B.6.5 Project Balance Table

The **Project Balance Table** command allows you to calculate the project balance at a specified interest rate and display the results in tabular form. You can also use this command to obtain the loan balances for a commercial loan that requires an unusual repayment plan.

B.6.6 Project Balance Plot

The **Project Balance Plot** command allows you to calculate the project balance at a specified interest rate and displays the results in graphical form. (See Figure B.7.)

■ B.7 Using Compute Mode

Once you have created a cash flow series using the **Edit** mode, you can obtain various measures of investment worth. There are six computational commands, as shown in Figure B.8.

Figure B.7 ■ A typical project balance plot

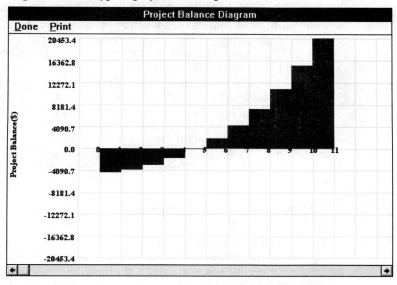

Figure B.8 ■ Computational commands available in **EzCash**

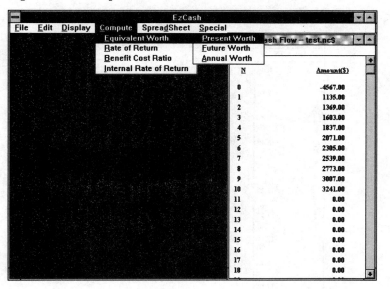

B.7.1 Equivalent Worth Command

This command allows you to calculate the net present worth, net future worth at the specified base period, and equivalent annual worth over the project life. Here the base period represents the period that is used for establishing the equivalent worth. For present worth calculations, the base period is 0.

B.7.2 Rate of Return Command

With the **Rate of Return** command you can calculate the project's rate of return (i°), including any multiple rates, within the lower and upper bounds of the interest rate specified. You can also obtain the NPW plot within the specified interest range. The acceptable value for the interest range is between -99% and 900%. It is always a good idea to plot the NPW function to see if it crosses the horizontal axis. If it does not, change the range and plot the NPW again. (Note that, if i° happens to fall on either the lower bound or upper bound of the interest range, **EzCash** may not identify all the rates of return. For example, a cash flow of -100, 225, -126 has roots 5% and 20%. With the interest range of 0% and 20% entered, **EzCash** may fail to detect the second rate of return (20%). To avoid this problem, it is best to keep the range wide (say 0 to 100%) or switch to the NPW plot to see if any roots are there. (See Figure B.9.)

B.7.3 Benefit-Cost Ratio Command

The **Benefit-Cost Ratio** command allows you to calculate the benefit-cost ratio when you specify the interest rate. First, the present worth of all cash inflows (B) is calculated at a given interest rate. Then, the present worth of all cash outflows (C) is calculated. Finally, a ratio of B over C is calculated and displayed.

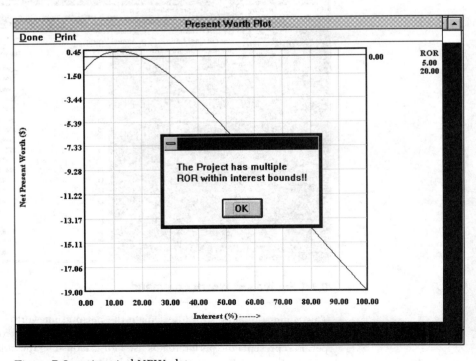

Figure B.9 ■ A typical NPW plot

B.7.4 Internal Rate of Return Command

The **Internal Rate of Return** command allows you to calculate the internal rate of return for pure investment and the portion of the rate of return *internal* to the mixed investment. Recall the difference between ROR and IRR. ROR represents the interest rate, i°, that makes the net present worth zero. Recall also that, if the project is a pure investment, the rate of return i° is also the internal rate of return to the project. For a mixed investment, the portion of rate of return that is *internal* to the project is calculated with the aid of MARR. (Here MARR is viewed as an interest earned on external investment for the funds withdrawn from the project.) **EzCash** will automatically plot the project balance diagram as functions of dual interest rates (IRR and MARR) respectively.

For a borrowing project such as ($+1000$, -400, -400, -400), it is not likely to find an IRR. In fact, it has a borrowing rate of return of 9.7%. To find such a borrowing rate of return, you must use the **Rate of Return** command.

■ B.8 Using SpreadSheet Mode

With the **SpreadSheet** mode, you can obtain an after-tax project cash flow. As you select the **SpreadSheet** mode, you will see a Lotus-like worksheet screen as shown in Fig. B.10. This worksheet functions just like an electronic spreadsheet, and you can edit, compute, scroll, and update the information as needed. Its six major commands are **File**, **Edit**, **Compute**, **View**, **Toolbox**, and **Print**.

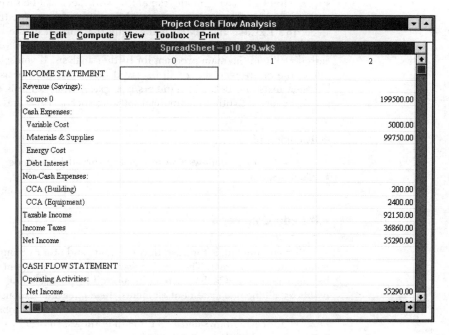

Figure B.10 ■ A Lotus-like worksheet used in project cash flow analysis

B.8.1 File Command

The **File** command has five subcommands: **New**, **Open**, **Save**, **Save As**, and **Quit**.

New Command

The **New** command will clear the worksheet from the screen as well as from memory and reset all the variables to their initial states.

Open Command

The **Open** command allows you to open a worksheet file that was previously created by **EzCash**. You can only open a worksheet file with the file extension **.wk$**. You can switch between drives and directories to select the file that needs to be opened. Directories are enclosed in square brackets.

Save Command

The **Save** command will save the current worksheet under the current file name. If you do not have an open file but attempt to save the existing spreadsheet by selecting this menu item, the spreadsheet will bring up the *Save As* dialogue box.

Save As Command

There are three file saving modes: **Spreadsheet**, **EzCash**, and **Text**. If you select the **spreadsheet** file saving mode, the entire worksheet will be saved under the file extension **.wk$**. Whenever you save any worksheet file, do not type the file extension. If you select the **EzCash** file saving mode, only the portion of the net cash flow from the worksheet will be saved under the file extension **.nc$**. Then you can retrieve this net cash flow file in the main program for further analysis. If you select the **Text** file saving mode, the entire worksheet will be saved under the file extension **.txt**. The delimiter between texts and between numbers is a space. Once a file is saved in **Text** mode, it can be exported to other commercial software such as *Excel* and *WordPerfect*.

Quit Command

The **Quit** command allows you to exit the spreadsheet mode and return to the main **EzCash** window.

B.8.2 Edit Command

The first column labels the cash flow elements and other accounting information in income statement format. The row labels in blue indicate the entries where the data will be calculated by **EzCash** or cannot be altered by the user. Following common business practices, all income statement elements (Revenues and Expenses) are entered in positive numbers. However, all cash flow data elements should observe sign conventions, i.e.: "+" for cash inflows and "−" for cash outflows. In order to facilitate data entry operation, several editing subcommands are provided.

Mouse Buttons

As in Lotus 1-2-3, you can move around the worksheet in the **Edit** mode by using the mouse. A cell pointer indicates the current cell location within the worksheet. You can scroll the worksheet as needed by using the mouse. In the **SpreadSheet** mode, the mouse buttons have the following functional assignment:

- **Left Mouse** button: Click this button to select commands or position the cell pointer to a specific worksheet cell.

- **Right Mouse** button: Click this button to enter the numerical data into the cell indicated by the cell pointer. (You can also press **Return** to enter the data into the cell.)

Copy or Cut

With the **Copy** or **Cut** command, you can copy or cut the content in a single cell (or range of cells) indicated by the cell pointer. Either operation must be performed before undertaking a subsequent Paste operation. The net effect of **Copy** or **Cut** and **Paste** is to copy or cut the contents of the copy area to the paste area.

- **Cell:** This command will copy or cut the value in the cell highlighted by the cell pointer.

- **Range:** This command will copy or cut a range of cells when the starting cell of the range is highlighted by the cell pointer. Upon selecting **Range**, the *Range* dialogue box will appear. In the *To* box, enter the ending period of the range you wish to copy along with the row. (You cannot copy or cut a series of data in a column.) Selecting the **Ok** button will copy the range selected and clear the *Range* dialogue box.

Paste

To use the **Paste** command, you need to specify the target cell or the beginning cell of the range you wish to paste. It is essential to keep in mind that pasting takes place along the row.

- **Range:** This command is highlighted only if a range has been copied/cut. If a cell has been copied, the **Range** command is gray, indicating that it is no longer active. Before selecting the **Range** command, you need to move the pointer to the start of the destination range. After the command is selected, the range copied/cut is pasted along the row.

- **Single:** This command is highlighted only if a cell has been copied/cut. If a range is copied, the **Single** command is gray, indicating that it is no longer active. Upon selecting the **Single** command, the cell copied/cut is pasted into the cell that is highlighted by the cell pointer.

- **Uniform:** To paste the same value of the single cell copied/cut over a range of cells along the row, select this command. Suppose you want to enter an O&M cost element as a uniform series in the amount of $1,000 from $n = 1$ to $n = 10$. Take the following steps to input this O&M cost element:

  - Step 1: Position the cell pointer in the starting cell (O&M Costs, 1).

  - Step 2: Enter the amount of 1000.

- Step 3: Select **Edit/ Copy/ Cell** to copy the amount in the cell (O&M Costs, 1).
- Step 4: Select **Edit/ Paste/ Uniform** to bring up the *Range* dialogue box. Enter 10 in the *To* box, as the end period of the range to which you wish to paste the value of the cell copied.
- Step 5: Click **Ok** to confirm the operation.

- **Linear Gradient:** This command allows you to paste the cell copied/cut as a starting value for a linear gradient series. Take the following steps to input a linear gradient series:
  - Step 1: Position the cell pointer in the starting period of the linear gradient series.
  - Step 2: Enter the amount for the first period of the linear gradient series.
  - Step 3: Select **Edit/ Copy/ Cell** to copy the cell content in Step 2.
  - Step 4: Select **Edit/ Paste/ Linear Gradient** to bring up the *Linear Gradient* range dialogue box. Specify the ending period of the linear gradient series and the gradient amount.
  - Step 5: Click **Ok** to confirm the operation.

- **Geometric Gradient:** This command will paste the cell copied/cut as a starting value for a geometric gradient series. The input procedure is very similar to the situation for the linear gradient series. Upon selecting this command, you will see the *Geometric Gradient* dialogue box. In the *To Period* box, enter the ending period of the range in which you wish to copy the values. In the *Gradient* box, enter the geometric gradient percentage. Clicking **Ok** will replace the existing values with the values computed over the range selected and clear the *Geometric Gradient* dialogue box.

Insert Row

The **Insert** command from the **Edit** menu adds blank rows into the worksheet. To insert a row, follow these steps:

- Position the cell pointer in the row where you want the new blank row to appear.
- Select the **Insert** command and press the **Return** key. A blank row will be inserted at the cell pointer.

When you insert rows, existing rows move down to make room for the new rows, and **EzCash** adjusts any cell formulas so they continue to refer to the same data.

Delete Row

The **Delete** command removes an entire row from the worksheet. To delete a row, follow these steps:

- Position the cell pointer in the row you want to delete.
- Select the **Delete** command and press the **Return** key.

All rows below a deleted row will shift up.

Header

The **Header** command from the Edit menu will allow you to edit the row label in the first column. To edit the text field in a selected row, follow these steps:

- Position the cell pointer in the row you want to edit the label (the text field in the first column).
- Select **Edit/ Header** to bring up a header dialogue box.
- In the *Header* box, type a header text that you wish to change or add. The maximum text field size that you can enter is 16 character positions.
- Click **Ok** to replace the existing header with the new header.

B.8.3 Compute Command

Once you complete all the data entry for the worksheet, you can select the **Net Cash Flow** command to calculate the taxable income, income taxes, net income and net cash flow. With the **Compute** command, you can also calculate the equivalent worth as well as the rate of return figures for the project cash flows.

Net Cash Flow

To compute the net cash flow, follow these steps:

- Select **Compute/Net Cash Flow** to bring up the *Tax Rate* dialogue box. In the *Tax Rate* box, enter the income tax rate (or incremental tax rate) in percentage.
- Click **Ok** to calculate the taxable income, income taxes, and net cash flows. From this point on, when you change any data in the cells, **EzCash** will automatically update the column sums and net cash flows.

Equivalent Worth Analysis

Once you have obtained a net cash flow, you may wish to calculate the various measures of investment worth. The **Equivalent Worth Analysis** command from **Compute** allows you to obtain the net present worth, future worth, or annual worth.

Rate of Return Analysis

Along with the equivalent worth measures, the second primary measure of investment worth is rate of return. To initiate the rate of return search, specify the lower as well as upper bounds of the interest rate in percentage. If the project has multiple rates of return, a message box will pop up to inform you so. To obtain the internal rate of return for a project cash flow, you need to specify the external interest rate, MARR.

B.8.4 View

If you create a worksheet that is larger than the entire screen, it is still possible to see all (or most) of the worksheet on the screen at one time.

Cash Flow

Before calculating an economic worth of the project in the worksheet, you may wish to preview the entire project cash flow series. The **Cash Flow** subcommand allows you to view the entire project cash flow series by opening a cash flow summary window. To calculate the net present worth of the cash flow series, you may select **Compute/ Equivalent Worth/ Present Worth** commands.

Worksheet

On the other hand, it is often desirable to scan the entire worksheet to see if there are missing entries. If you select the **Worksheet** subcommand, you will be able to view all (or most) of the entire worksheet as shown in Figure B.11.

B.8.5 Toolbox

The **Toolbox** command has several built-in functions that allow you to calculate the capital cost allowances, loan-payment schedule, and disposal tax effects (gains taxes). Once you preview the computational results, you can copy back the results to the main worksheet.

Worksheet							
Done							
INCOME STATEMENT	0	1	2	3	4	5	
Revenue (Savings):							
Source 0			199500.00	209475.00	219948.75	230946.19	2424
Cash Expenses:							
Variable Cost			5000.00	5250.00	5512.50	5788.12	6(
Materials & Supplies			99750.00	104737.50	109974.38	115473.09	1212
Energy Cost							
Debt Interest							
Non-Cash Expenses:							
CCA (Building)			200.00	192.00	184.32	176.95	1
CCA (Equipment)			2400.00	1680.00	1176.00	823.20	5
Taxable Income			92150.00	97615.50	103101.55	108684.83	1144
Income Taxes			36860.00	39046.20	41240.62	43473.93	457
Net Income			55290.00	58569.30	61860.93	65210.89	686
CASH FLOW STATEMENT							
Operating Activities:							
Net Income			55290.00	58569.30	61860.93	65210.89	686
Non-Cash Expenses			2600.00	1872.00	1360.32	1000.15	7
Investment Activities:							
Building	-1000.00	-4000.00					
Equipment		-8000.00					
Land	-1500.00						
Working Capital		-1000.00	-1496.00	-1571.00	-1650.00	-1732.00	-18
Disposal Tax Effects:							
Gains Tax(Building)							
Gains Tax(Equipment)							
Gains Tax(Land)							
Financing Activities:							
Borrowed Funds							
Debt Principal Payment							
Net Cash Flow	-2500.00	-13000.00	56394.00	58870.30	61571.25	64479.04	675

Figure B.11 ■ A zoom out feature to view the entire worksheet

Capital Cost Allowances

In preparing a typical income statement, you will normally enter the revenues series and costs of goods (labor, material and overhead). Then you may need to enter the capital cost allowances (depreciation data). If you already know the annual depreciation amounts, you may enter them manually. However, **EzCash** provides a built-in capital cost allowance generator and you can copy the results back to the cells in the worksheet. For example, to enter the series of capital cost allowances for a building, you may follow these steps:

- Position the cell pointer in the row labeled *CCA Building* where you wish to enter the depreciation data.

- Select the type of depreciation method. This should be either *Declining Balance* or *Straight-Line*.

- Enter the required data, such as cost base, useful life, DB rate, and whether or not the half-year convention should apply.

- Click **Ok** to bring up the *Depreciation Analysis* dialogue box. The *Depreciation Analysis* dialogue box has two commands. Selecting the **Done** button will clear the *Depreciation Analysis* dialogue box. Selecting the **Export** button will bring up the *Export* dialogue box.

- Enter the beginning and ending periods of the CCA values that you wish to export.

- Click **Ok** to export the range specified. The CCA values will be posted in the row where the cell pointer is located.

Loan Analysis

If you finance a project by borrowing, there will be interest as well as principal payments that must be considered in your project cash flows. To obtain the interest and principal payments associated with financing activity, **EzCash** also provides a built-in loan repayment generator that can be accessed by opening the **Toolbox** command. To initiate the loan data entry, follow these steps:

- Position the cell pointer in the row labeled *Interest* where you wish to enter the interest payments.

- Select the type of loan. This can be either *Fixed-Rate* or *Variable-Rate* loan.

- If you select the *Fixed Rate Loan*, you need to choose one of the three common repayment schedules (which were discussed in Chapter 3). Once you select the repayment method, enter the following information: (1) loan interest rate in APR, (2) loan period, (3) amount of loan, and (4) interest (compounding) and payment periods.

- Click **Ok** to view the loan repayment schedule.

- To copy the loan transactions back to the main worksheet, specify the starting and ending periods of the cell range. Click **Ok** to post the interest payments in the row where the cell pointer is located. The principal payments will automatically be posted in the row labeled *Debt Principal Repayment*. (Note: The *loan amount* must be input separately. Move the cell pointer to the row labeled *Borrowed Funds* and position it to the cell where the actual borrowing will take place. Enter the *loan amount*.)

Disposal Tax Effect

To determine the disposal tax effect for the asset placed in service, you may follow these steps:

- Move the cursor to point the cell where you wish to post the gains tax value.
- Select the **Disposal Tax Effect** command from the **Toolbox** menu.
- Specify the type of CCA method and enter all required CCA data for the asset.
- Open the *Input for Disposal Tax* dialogue box and enter all required data (such as year to dispose, income tax rate, capital gains tax rate and market value) to generate the depreciation schedule for the asset. Then click **Ok** to bring up the *Disposal Tax Effect* dialogue box. In the *Ordinary Gains Tax* box, the ordinary gains tax value will be listed. In the *Capital Gains Tax* box, the capital gains tax will be shown.
- To export these values to the main worksheet, select *Export* to Worksheet and click **Ok**.

B.8.6 Print

You can obtain a hard copy of the worksheet by selecting the **Print** command. The **Print** command will follow the printing format and procedures required under MicroSoft Window 3.1. Therefore, you must configure your printer according to the requirements set by Windows.

■ B.9 Using Special Mode

With the **Special** mode, you can perform five types of common engineering economic calculations: loan-repayment plan, depreciation calculations (book depreciation as well as tax depreciation), economic service life (before- or after-tax basis), economic equivalence, and incremental analysis.

B.9.1 Loan Analysis

Using the **Loan Analysis** command, you can enter up to 400 loan periods. This number will be sufficient, as a typical home mortgage loan of 30 years (or 360 months) is common. The loan repayment schedule includes the principal payment, interest payment, cumulative interest payment, and loan balance for each period over the life of the loan. You can scroll the data pages to view the complete repayment schedule. There are two loan types: (1) fixed-rate loan and (2) variable-rate loan.

Fixed Rate Loan

Under a fixed rate loan arrangement, the periodic interest rate is fixed over the life of the loan in calculating the loan payment. In **EzCash**, three common types of loan repayment method are provided.

- **Equal principal payment** uses equal principal payments, decreasing interest payments and thus, decreasing periodic payments over the life of the loan.

- **Equal interest payment** uses constant interest payments over the life of the loan, with principal repaid as a lump sum at the final period.

- **Equal installment payment** uses increasing principal payments, decreasing interest payments, and equal periodic payments over the life of the loan.

Selecting any of the above three menu items will bring up the *Fixed Rate Loan* dialogue box. Enter the annual percentage rate (APR) of the loan, the period of loan, and the amount of loan. From the pull down menus, select the compounding period as well as the payment period. (See Figure B.12.) Click **Ok** to view the loan repayment schedule or press **Print** to obtain a hard copy of the repayment schedule. (Note: If you enter a compounding period that is less frequent than the payment period, e.g, monthly payment with annual compounding, an effective interest rate will be calculated based on the payment period (monthly in this case). Note also that the loan periods should agree with the total number of payment periods.)

Variable Rate Loan

Selecting this menu item will bring up the *Variable Rate Loan* dialogue box along with the *Variable Rate Interest* window. Enter the time interval in which the first loan rate (APR) is applicable and the corresponding annual percentage rate of the loan. Select the **Ok** button to accept these values. All changes will reflect in the *Variable Interest Rate* window. Continue this process of input until you have finished entering the various interest rates for the entire loan period. Select the **Compute** command to bring up the *Payment Schedule* dialogue box. From the pull down menus, specify the compounding period as well as the payment period. Then enter the amount of the loan. Click **Ok** to see the loan payment schedule.

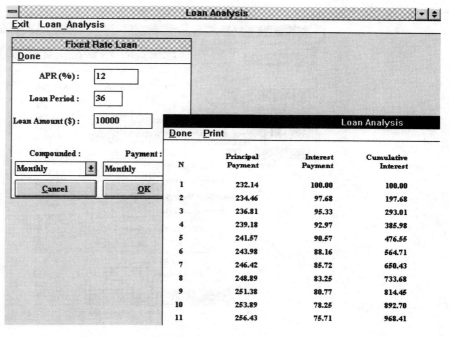

Figure B.12 ■ An input window for loan analysis

B.9.2 Depreciation Command

With the **Depreciation** command, you can obtain a depreciation schedule for conventional book depreciation methods (straight-line, declining balance, and sum-of-the-years' digit), as well as the tax depreciation methods. For a book depreciation method, you will be asked to enter the depreciation base, useful life, salvage value, and declining balance rate. Then **EzCash** will generate three types of conventional depreciation schedules: straight-line method, declining balance method, and sum-of-the-years' digit method. If you choose the tax depreciation option, you will be asked to choose between the declining balance method and straight-line method. If you select the declining balance option, you need to enter the depreciation (CCA) base, useful life, declining balance rate, and whether or not to apply the half-year convention. Upon entering these data, **EzCash** will generate a tax depreciation (or capital cost allowance) schedule. (With the flexibility built into the depreciation generator, **EzCash** will be able to adapt to any future changes in depreciation tax law.)

B.9.3 Economic Service Life Command

With the **Economic Service Life** command, you can calculate the economic service life on an after-tax basis for a defender as well as a challenger, assuming an infinite sequence of replacements. If you prefer a before-tax analysis, you simply enter a tax rate of 0%. The economic service life calculation is done using the *Cash Flow Editor* as an input tool, so you simply enter the required information, such as salvage value, O&M, and revenues (if any) over the life of the project. Then, specify the tax rate, the interest

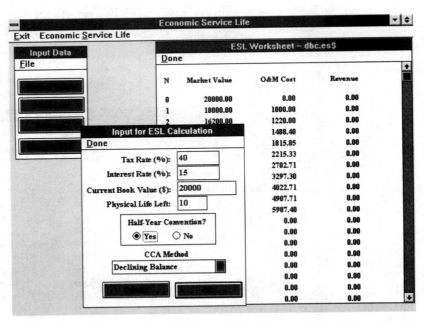

Figure B.13 ■ An opening window to determine economic service life calculation

rate, the current book value, and the depreciation (CCA) method. You can edit the data as needed to answer many "what-if" questions. The input worksheet will be saved with the file extension **.es$**. (Do not type the file extension when you specify the filename. **EzCash** will assign the file extension whenever you save an economic service life worksheet, or will only retrieve a file with that extension.)

B.9.4 Equivalence Command

With the **Equivalence** command, you will be able to calculate various economic equivalence values upon specifying proper input parameters. For example, with an equal-payment series with the interest rate, constant amount and present worth specified, you can solve for an unknown period. Four major cash flow patterns are provided: (1) single-payment transaction, (2) equal-payment series, (3) linear gradient series, and (4) geometric gradient series.

The screen output in Figure B.14 illustrates the situation where you wish to find the gradient(%) or the growth factor for a geometric gradient payment series. You will find that the dialogue box has two sections, *Given* and *Find*. Unlike other edit boxes, these edit boxes display both input and output results. In this example, you deposit $174,506 at an interest rate of 8.5% and a series of annual withdrawals will be made over 12 years. The first withdrawal size is given to be $45,000. Now you wish to determine the unknown growth factor (g) to establish such economic equivalence.

In the *Given* section, move along the edit boxes using the **Tab** key and input the four known parameters (P = $174,506, i = 8.5%, N = 12 years, A_1 = $45,000). Leave the edit box of the unknown parameter that needs to be calculated (g = ?). From the *Find* section, click the **Gradient** radio button and press **Ok** to calculate the gradient factor, which is −16.11%.

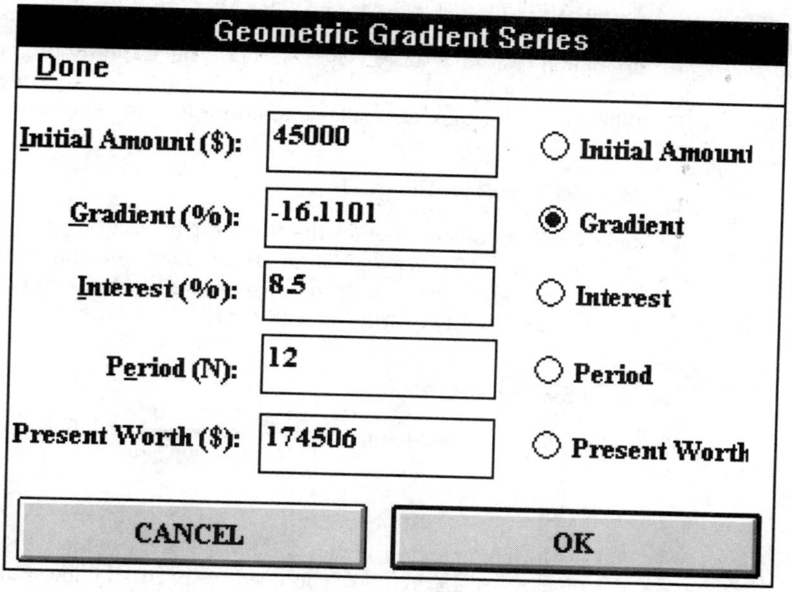

Geometric Gradient Series
Done

Initial Amount ($):	45000	○ Initial Amount
Gradient (%):	-16.1101	● Gradient
Interest (%):	8.5	○ Interest
Period (N):	12	○ Period
Present Worth ($):	174506	○ Present Worth

CANCEL	OK

Figure B.14 ■ Equivalence dialogue box for geometric gradient series

Figure B.15 ■ An incremental analysis window

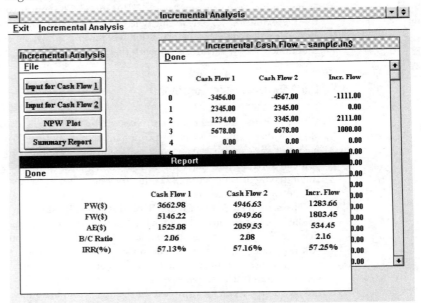

B.9.5 Incremental Analysis Command

When comparing mutually exclusive investment alternatives, we often want to calculate the internal rate of return on incremental investment. When this command is invoked, **EzCash** will perform an incremental analysis by subtracting a cash flow series (say A1) for the lower investment project from that of higher investment (say A2) to form an incremental cash flow series (say A2 − A1). Then, the various economic measures on individual investments as well as the incremental investment are calculated. The following commands are available to perform an incremental analysis (Figure B.15).

Input for Cash Flow 1

Select this command to enter the first project cash flow series. Clicking this command will bring up the *Cash Flow Editor*. If there are some patterns in your cash flow series, you can facilitate the input process by using the **Pattern** command. The input worksheet can be saved or retrieved with the file extension **.in$**.

Input for Cash Flow 2

Select this command to enter the second project cash flow series.

NPW Plot

Select this command to obtain the present worth plots of the two cash flow series on the same chart. In doing so, you need to specify the lower and upper bound of the interest rate.

Report

Select this command to see the summary of various economic measures (present worth, future worth, annual worth, benefit-cost ratio, and internal rate of return) for the two projects as well as the incremental investment. You need to specify the interest rate to obtain this summary report.

■ B.10 EzCash Applications

Demonstrations of the various **EzCash** commands introduced above are found in the following applications.

1. Equivalence calculations (Chapter 2)
2. Loan analysis (Chapter 3)
3. NPW, NFW, and AW analyses (Chapter 4)
4. Rate of return analysis (Chapter 6)
5. Project balance analysis (Chapter 6)
6. Incremental analysis (Chapter 6)
7. Depreciation analysis (Chapter 7)
8. After-tax **EzCash** flow analysis (Chapter 9)
9. Inflation analysis (Chapter 10)
10. Economic service life (Chapter 12)

Interest Factors for Discrete Compounding

Table C.1 Interest Rate Factors (0.25%)

	Single Payment		Equal Payment Series				Gradient Series		
N	Compound Amount Factor (F/P,i,N)	Present Worth Factor (P/F,i,N)	Compound Amount Factor (F/A,i,N)	Sinking Fund Factor (A/F,i,N)	Present Worth Factor (P/A,i,N)	Capital Recovery Factor (A/P,i,N)	Gradient Uniform Series (A/G,i,N)	Gradient Present Worth (P/G.i.N)	N
1	1.0025	0.9975	1.0000	1.0000	0.9975	1.0025	0.0000	0.0000	1
2	1.0050	0.9950	2.0025	0.4994	1.9925	0.5019	0.4994	0.9950	2
3	1.0075	0.9925	3.0075	0.3325	2.9851	0.3350	0.9983	2.9801	3
4	1.0100	0.9901	4.0150	0.2491	3.9751	0.2516	1.4969	5.9503	4
5	1.0126	0.9876	5.0251	0.1990	4.9627	0.2015	1.9950	9.9007	5
6	1.0151	0.9851	6.0376	0.1656	5.9478	0.1681	2.4927	14.8263	6
7	1.0176	0.9827	7.0527	0.1418	6.9305	0.1443	2.9900	20.7223	7
8	1.0202	0.9802	8.0704	0.1239	7.9107	0.1264	3.4869	27.5839	8
9	1.0227	0.9778	9.0905	0.1100	8.8885	0.1125	3.9834	35.4061	9
10	1.0253	0.9753	10.1133	0.0989	9.8639	0.1014	4.4794	44.1842	10
11	1.0278	0.9729	11.1385	0.0898	10.8368	0.0923	4.9750	53.9133	11
12	1.0304	0.9705	12.1664	0.0822	11.8073	0.0847	5.4702	64.5886	12
13	1.0330	0.9681	13.1968	0.0758	12.7753	0.0783	5.9650	76.2053	13
14	1.0356	0.9656	14.2298	0.0703	13.7410	0.0728	6.4594	88.7587	14
15	1.0382	0.9632	15.2654	0.0655	14.7042	0.0680	6.9534	102.2441	15
16	1.0408	0.9608	16.3035	0.0613	15.6650	0.0638	7.4469	116.6567	16
17	1.0434	0.9584	17.3443	0.0577	16.6235	0.0602	7.9401	131.9917	17
18	1.0460	0.9561	18.3876	0.0544	17.5795	0.0569	8.4328	148.2446	18
19	1.0486	0.9537	19.4336	0.0515	18.5332	0.0540	8.9251	165.4106	19
20	1.0512	0.9513	20.4822	0.0488	19.4845	0.0513	9.4170	183.4851	20
21	1.0538	0.9489	21.5334	0.0464	20.4334	0.0489	9.9085	202.4634	21
22	1.0565	0.9466	22.5872	0.0443	21.3800	0.0468	10.3995	222.3410	22
23	1.0591	0.9442	23.6437	0.0423	22.3241	0.0448	10.8901	243.1131	23
24	1.0618	0.9418	24.7028	0.0405	23.2660	0.0430	11.3804	264.7753	24
25	1.0644	0.9395	25.7646	0.0388	24.2055	0.0413	11.8702	287.3230	25
26	1.0671	0.9371	26.8290	0.0373	25.1426	0.0398	12.3596	310.7516	26
27	1.0697	0.9348	27.8961	0.0358	26.0774	0.0383	12.8485	335.0566	27
28	1.0724	0.9325	28.9658	0.0345	27.0099	0.0370	13.3371	360.2334	28
29	1.0751	0.9301	30.0382	0.0333	27.9400	0.0358	13.8252	386.2776	29
30	1.0778	0.9278	31.1133	0.0321	28.8679	0.0346	14.3130	413.1847	30
31	1.0805	0.9255	32.1911	0.0311	29.7934	0.0336	14.8003	440.9502	31
32	1.0832	0.9232	33.2716	0.0301	30.7166	0.0326	15.2872	469.5696	32
33	1.0859	0.9209	34.3547	0.0291	31.6375	0.0316	15.7736	499.0386	33
34	1.0886	0.9186	35.4406	0.0282	32.5561	0.0307	16.2597	529.3528	34
35	1.0913	0.9163	36.5292	0.0274	33.4724	0.0299	16.7454	560.5076	35
36	1.0941	0.9140	37.6206	0.0266	34.3865	0.0291	17.2306	592.4988	36
40	1.1050	0.9050	42.0132	0.0238	38.0199	0.0263	19.1673	728.7399	40
48	1.1273	0.8871	50.9312	0.0196	45.1787	0.0221	23.0209	1040.0552	48
50	1.1330	0.8826	53.1887	0.0188	46.9462	0.0213	23.9802	1125.7767	50
60	1.1616	0.8609	64.6467	0.0155	55.6524	0.0180	28.7514	1600.0845	60
72	1.1969	0.8355	78.7794	0.0127	65.8169	0.0152	34.4221	2265.5569	72
80	1.2211	0.8189	88.4392	0.0113	72.4260	0.0138	38.1694	2764.4568	80
84	1.2334	0.8108	93.3419	0.0107	75.6813	0.0132	40.0331	3029.7592	84
90	1.2520	0.7987	100.7885	0.0099	80.5038	0.0124	42.8162	3446.8700	90
96	1.2709	0.7869	108.3474	0.0092	85.2546	0.0117	45.5844	3886.2832	96
100	1.2836	0.7790	113.4500	0.0088	88.3825	0.0113	47.4216	4191.2417	100
108	1.3095	0.7636	123.8093	0.0081	94.5453	0.0106	51.0762	4829.0125	108
120	1.3494	0.7411	139.7414	0.0072	103.5618	0.0097	56.5084	5852.1116	120
240	1.8208	0.5492	328.3020	0.0030	180.3109	0.0055	107.5863	19398.9852	240
360	2.4568	0.4070	582.7369	0.0017	237.1894	0.0042	152.8902	36263.9299	360

Table C.2 Interest Rate Factors (0.50%)

	Single Payment		Equal Payment Series				Gradient Series		
N	Compound Amount Factor (F/P,i,N)	Present Worth Factor (P/F,i,N)	Compound Amount Factor (F/A,i,N)	Sinking Fund Factor (A/F,i,N)	Present Worth Factor (P/A,i,N)	Capital Recovery Factor (A/P,i,N)	Gradient Uniform Series (A/G,i,N)	Gradient Present Worth (P/G.i.N)	N
1	1.0050	0.9950	1.0000	1.0000	0.9950	1.0050	0.0000	0.0000	1
2	1.0100	0.9901	2.0050	0.4988	1.9851	0.5038	0.4988	0.9901	2
3	1.0151	0.9851	3.0150	0.3317	2.9702	0.3367	0.9967	2.9604	3
4	1.0202	0.9802	4.0301	0.2481	3.9505	0.2531	1.4938	5.9011	4
5	1.0253	0.9754	5.0503	0.1980	4.9259	0.2030	1.9900	9.8026	5
6	1.0304	0.9705	6.0755	0.1646	5.8964	0.1696	2.4855	14.6552	6
7	1.0355	0.9657	7.1059	0.1407	6.8621	0.1457	2.9801	20.4493	7
8	1.0407	0.9609	8.1414	0.1228	7.8230	0.1278	3.4738	27.1755	8
9	1.0459	0.9561	9.1821	0.1089	8.7791	0.1139	3.9668	34.8244	9
10	1.0511	0.9513	10.2280	0.0978	9.7304	0.1028	4.4589	43.3865	10
11	1.0564	0.9466	11.2792	0.0887	10.6770	0.0937	4.9501	52.8526	11
12	1.0617	0.9419	12.3356	0.0811	11.6189	0.0861	5.4406	63.2136	12
13	1.0670	0.9372	13.3972	0.0746	12.5562	0.0796	5.9302	74.4602	13
14	1.0723	0.9326	14.4642	0.0691	13.4887	0.0741	6.4190	86.5835	14
15	1.0777	0.9279	15.5365	0.0644	14.4166	0.0694	6.9069	99.5743	15
16	1.0831	0.9233	16.6142	0.0602	15.3399	0.0652	7.3940	113.4238	16
17	1.0885	0.9187	17.6973	0.0565	16.2586	0.0615	7.8803	128.1231	17
18	1.0939	0.9141	18.7858	0.0532	17.1728	0.0582	8.3658	143.6634	18
19	1.0994	0.9096	19.8797	0.0503	18.0824	0.0553	8.8504	160.0360	19
20	1.1049	0.9051	20.9791	0.0477	18.9874	0.0527	9.3342	177.2322	20
21	1.1104	0.9006	22.0840	0.0453	19.8880	0.0503	9.8172	195.2434	21
22	1.1160	0.8961	23.1944	0.0431	20.7841	0.0481	10.2993	214.0611	22
23	1.1216	0.8916	24.3104	0.0411	21.6757	0.0461	10.7806	233.6768	23
24	1.1272	0.8872	25.4320	0.0393	22.5629	0.0443	11.2611	254.0820	24
25	1.1328	0.8828	26.5591	0.0377	23.4456	0.0427	11.7407	275.2686	25
26	1.1385	0.8784	27.6919	0.0361	24.3240	0.0411	12.2195	297.2281	26
27	1.1442	0.8740	28.8304	0.0347	25.1980	0.0397	12.6975	319.9523	27
28	1.1499	0.8697	29.9745	0.0334	26.0677	0.0384	13.1747	343.4332	28
29	1.1556	0.8653	31.1244	0.0321	26.9330	0.0371	13.6510	367.6625	29
30	1.1614	0.8610	32.2800	0.0310	27.7941	0.0360	14.1265	392.6324	30
31	1.1672	0.8567	33.4414	0.0299	28.6508	0.0349	14.6012	418.3348	31
32	1.1730	0.8525	34.6086	0.0289	29.5033	0.0339	15.0750	444.7618	32
33	1.1789	0.8482	35.7817	0.0279	30.3515	0.0329	15.5480	471.9055	33
34	1.1848	0.8440	36.9606	0.0271	31.1955	0.0321	16.0202	499.7583	34
35	1.1907	0.8398	38.1454	0.0262	32.0354	0.0312	16.4915	528.3123	35
36	1.1967	0.8356	39.3361	0.0254	32.8710	0.0304	16.9621	557.5598	36
40	1.2208	0.8191	44.1588	0.0226	36.1722	0.0276	18.8359	681.3347	40
48	1.2705	0.7871	54.0978	0.0185	42.5803	0.0235	22.5437	959.9188	48
50	1.2832	0.7793	56.6452	0.0177	44.1428	0.0227	23.4624	1035.6966	50
60	1.3489	0.7414	69.7700	0.0143	51.7256	0.0193	28.0064	1448.6458	60
72	1.4320	0.6983	86.4089	0.0116	60.3395	0.0166	33.3504	2012.3478	72
80	1.4903	0.6710	98.0677	0.0102	65.8023	0.0152	36.8474	2424.6455	80
84	1.5204	0.6577	104.0739	0.0096	68.4530	0.0146	38.5763	2640.6641	84
90	1.5666	0.6383	113.3109	0.0088	72.3313	0.0138	41.1451	2976.0769	90
96	1.6141	0.6195	122.8285	0.0081	76.0952	0.0131	43.6845	3324.1846	96
100	1.6467	0.6073	129.3337	0.0077	78.5426	0.0127	45.3613	3562.7934	100
108	1.7137	0.5835	142.7399	0.0070	83.2934	0.0120	48.6758	4054.3747	108
120	1.8194	0.5496	163.8793	0.0061	90.0735	0.0111	53.5508	4823.5051	120
240	3.3102	0.3021	462.0409	0.0022	139.5808	0.0072	96.1131	13415.5395	240
360	6.0226	0.1660	1004.5150	0.0010	166.7916	0.0060	128.3236	21403.3041	360

Table C.3 Interest Rate Factors (0.75%)

	Single Payment		Equal Payment Series				Gradient Series		
N	Compound Amount Factor (F/P,i,N)	Present Worth Factor (P/F,i,N)	Compound Amount Factor (F/A,i,N)	Sinking Fund Factor (A/F,i,N)	Present Worth Factor (P/A,i,N)	Capital Recovery Factor (A/P,i,N)	Gradient Uniform Series (A/G,i,N)	Gradient Present Worth (P/G.i.N)	N
1	1.0075	0.9926	1.0000	1.0000	0.9926	1.0075	0.0000	0.0000	1
2	1.0151	0.9852	2.0075	0.4981	1.9777	0.5056	0.4981	0.9852	2
3	1.0227	0.9778	3.0226	0.3308	2.9556	0.3383	0.9950	2.9408	3
4	1.0303	0.9706	4.0452	0.2472	3.9261	0.2547	1.4907	5.8525	4
5	1.0381	0.9633	5.0756	0.1970	4.8894	0.2045	1.9851	9.7058	5
6	1.0459	0.9562	6.1136	0.1636	5.8456	0.1711	2.4782	14.4866	6
7	1.0537	0.9490	7.1595	0.1397	6.7946	0.1472	2.9701	20.1808	7
8	1.0616	0.9420	8.2132	0.1218	7.7366	0.1293	3.4608	26.7747	8
9	1.0696	0.9350	9.2748	0.1078	8.6716	0.1153	3.9502	34.2544	9
10	1.0776	0.9280	10.3443	0.0967	9.5996	0.1042	4.4384	42.6064	10
11	1.0857	0.9211	11.4219	0.0876	10.5207	0.0951	4.9253	51.8174	11
12	1.0938	0.9142	12.5076	0.0800	11.4349	0.0875	5.4110	61.8740	12
13	1.1020	0.9074	13.6014	0.0735	12.3423	0.0810	5.8954	72.7632	13
14	1.1103	0.9007	14.7034	0.0680	13.2430	0.0755	6.3786	84.4720	14
15	1.1186	0.8940	15.8137	0.0632	14.1370	0.0707	6.8606	96.9876	15
16	1.1270	0.8873	16.9323	0.0591	15.0243	0.0666	7.3413	110.2973	16
17	1.1354	0.8807	18.0593	0.0554	15.9050	0.0629	7.8207	124.3887	17
18	1.1440	0.8742	19.1947	0.0521	16.7792	0.0596	8.2989	139.2494	18
19	1.1525	0.8676	20.3387	0.0492	17.6468	0.0567	8.7759	154.8671	19
20	1.1612	0.8612	21.4912	0.0465	18.5080	0.0540	9.2516	171.2297	20
21	1.1699	0.8548	22.6524	0.0441	19.3628	0.0516	9.7261	188.3253	21
22	1.1787	0.8484	23.8223	0.0420	20.2112	0.0495	10.1994	206.1420	22
23	1.1875	0.8421	25.0010	0.0400	21.0533	0.0475	10.6714	224.6682	23
24	1.1964	0.8358	26.1885	0.0382	21.8891	0.0457	11.1422	243.8923	24
25	1.2054	0.8296	27.3849	0.0365	22.7188	0.0440	11.6117	263.8029	25
26	1.2144	0.8234	28.5903	0.0350	23.5422	0.0425	12.0800	284.3888	26
27	1.2235	0.8173	29.8047	0.0336	24.3595	0.0411	12.5470	305.6387	27
28	1.2327	0.8112	31.0282	0.0322	25.1707	0.0397	13.0128	327.5416	28
29	1.2420	0.8052	32.2609	0.0310	25.9759	0.0385	13.4774	350.0867	29
30	1.2513	0.7992	33.5029	0.0298	26.7751	0.0373	13.9407	373.2631	30
31	1.2607	0.7932	34.7542	0.0288	27.5683	0.0363	14.4028	397.0602	31
32	1.2701	0.7873	36.0148	0.0278	28.3557	0.0353	14.8636	421.4675	32
33	1.2796	0.7815	37.2849	0.0268	29.1371	0.0343	15.3232	446.4746	33
34	1.2892	0.7757	38.5646	0.0259	29.9128	0.0334	15.7816	472.0712	34
35	1.2989	0.7699	39.8538	0.0251	30.6827	0.0326	16.2387	498.2471	35
36	1.3086	0.7641	41.1527	0.0243	31.4468	0.0318	16.6946	524.9924	36
40	1.3483	0.7416	46.4465	0.0215	34.4469	0.0290	18.5058	637.4693	40
48	1.4314	0.6986	57.5207	0.0174	40.1848	0.0249	22.0691	886.8404	48
50	1.4530	0.6883	60.3943	0.0166	41.5664	0.0241	22.9476	953.8486	50
60	1.5657	0.6387	75.4241	0.0133	48.1734	0.0208	27.2665	1313.5189	60
72	1.7126	0.5839	95.0070	0.0105	55.4768	0.0180	32.2882	1791.2463	72
80	1.8180	0.5500	109.0725	0.0092	59.9944	0.0167	35.5391	2132.1472	80
84	1.8732	0.5338	116.4269	0.0086	62.1540	0.0161	37.1357	2308.1283	84
90	1.9591	0.5104	127.8790	0.0078	65.2746	0.0153	39.4946	2577.9961	90
96	2.0489	0.4881	139.8562	0.0072	68.2584	0.0147	41.8107	2853.9352	96
100	2.1111	0.4737	148.1445	0.0068	70.1746	0.0143	43.3311	3040.7453	100
108	2.2411	0.4462	165.4832	0.0060	73.8394	0.0135	46.3154	3419.9041	108
120	2.4514	0.4079	193.5143	0.0052	78.9417	0.0127	50.6521	3998.5621	120
240	6.0092	0.1664	667.8869	0.0015	111.1450	0.0090	85.4210	9494.1162	240
360	14.7306	0.0679	1830.7435	0.0005	124.2819	0.0080	107.1145	13312.3871	360

Table C.4 Interest Rate Factors (1.0%)

	Single Payment		Equal Payment Series				Gradient Series		
N	Compound Amount Factor (F/P,i,N)	Present Worth Factor (P/F,i,N)	Compound Amount Factor (F/A,i,N)	Sinking Fund Factor (A/F,i,N)	Present Worth Factor (P/A,i,N)	Capital Recovery Factor (A/P,i,N)	Gradient Uniform Series (A/G,i,N)	Gradient Present Worth (P/G.i.N)	N
1	1.0100	0.9901	1.0000	1.0000	0.9901	1.0100	0.0000	0.0000	1
2	1.0201	0.9803	2.0100	0.4975	1.9704	0.5075	0.4975	0.9803	2
3	1.0303	0.9706	3.0301	0.3300	2.9410	0.3400	0.9934	2.9215	3
4	1.0406	0.9610	4.0604	0.2463	3.9020	0.2563	1.4876	5.8044	4
5	1.0510	0.9515	5.1010	0.1960	4.8534	0.2060	1.9801	9.6103	5
6	1.0615	0.9420	6.1520	0.1625	5.7955	0.1725	2.4710	14.3205	6
7	1.0721	0.9327	7.2135	0.1386	6.7282	0.1486	2.9602	19.9168	7
8	1.0829	0.9235	8.2857	0.1207	7.6517	0.1307	3.4478	26.3812	8
9	1.0937	0.9143	9.3685	0.1067	8.5660	0.1167	3.9337	33.6959	9
10	1.1046	0.9053	10.4622	0.0956	9.4713	0.1056	4.4179	41.8435	10
11	1.1157	0.8963	11.5668	0.0865	10.3676	0.0965	4.9005	50.8067	11
12	1.1268	0.8874	12.6825	0.0788	11.2551	0.0888	5.3815	60.5687	12
13	1.1381	0.8787	13.8093	0.0724	12.1337	0.0824	5.8607	71.1126	13
14	1.1495	0.8700	14.9474	0.0669	13.0037	0.0769	6.3384	82.4221	14
15	1.1610	0.8613	16.0969	0.0621	13.8651	0.0721	6.8143	94.4810	15
16	1.1726	0.8528	17.2579	0.0579	14.7179	0.0679	7.2886	107.2734	16
17	1.1843	0.8444	18.4304	0.0543	15.5623	0.0643	7.7613	120.7834	17
18	1.1961	0.8360	19.6147	0.0510	16.3983	0.0610	8.2323	134.9957	18
19	1.2081	0.8277	20.8109	0.0481	17.2260	0.0581	8.7017	149.8950	19
20	1.2202	0.8195	22.0190	0.0454	18.0456	0.0554	9.1694	165.4664	20
21	1.2324	0.8114	23.2392	0.0430	18.8570	0.0530	9.6354	181.6950	21
22	1.2447	0.8034	24.4716	0.0409	19.6604	0.0509	10.0998	198.5663	22
23	1.2572	0.7954	25.7163	0.0389	20.4558	0.0489	10.5626	216.0660	23
24	1.2697	0.7876	26.9735	0.0371	21.2434	0.0471	11.0237	234.1800	24
25	1.2824	0.7798	28.2432	0.0354	22.0232	0.0454	11.4831	252.8945	25
26	1.2953	0.7720	29.5256	0.0339	22.7952	0.0439	11.9409	272.1957	26
27	1.3082	0.7644	30.8209	0.0324	23.5596	0.0424	12.3971	292.0702	27
28	1.3213	0.7568	32.1291	0.0311	24.3164	0.0411	12.8516	312.5047	28
29	1.3345	0.7493	33.4504	0.0299	25.0658	0.0399	13.3044	333.4863	29
30	1.3478	0.7419	34.7849	0.0287	25.8077	0.0387	13.7557	355.0021	30
31	1.3613	0.7346	36.1327	0.0277	26.5423	0.0377	14.2052	377.0394	31
32	1.3749	0.7273	37.4941	0.0267	27.2696	0.0367	14.6532	399.5858	32
33	1.3887	0.7201	38.8690	0.0257	27.9897	0.0357	15.0995	422.6291	33
34	1.4026	0.7130	40.2577	0.0248	28.7027	0.0348	15.5441	446.1572	34
35	1.4166	0.7059	41.6603	0.0240	29.4086	0.0340	15.9871	470.1583	35
36	1.4308	0.6989	43.0769	0.0232	30.1075	0.0332	16.4285	494.6207	36
40	1.4889	0.6717	48.8864	0.0205	32.8347	0.0305	18.1776	596.8561	40
48	1.6122	0.6203	61.2226	0.0163	37.9740	0.0263	21.5976	820.1460	48
50	1.6446	0.6080	64.4632	0.0155	39.1961	0.0255	22.4363	879.4176	50
60	1.8167	0.5504	81.6697	0.0122	44.9550	0.0222	26.5333	1192.8061	60
72	2.0471	0.4885	104.7099	0.0096	51.1504	0.0196	31.2386	1597.8673	72
80	2.2167	0.4511	121.6715	0.0082	54.8882	0.0182	34.2492	1879.8771	80
84	2.3067	0.4335	130.6723	0.0077	56.6485	0.0177	35.7170	2023.3153	84
90	2.4486	0.4084	144.8633	0.0069	59.1609	0.0169	37.8724	2240.5675	90
96	2.5993	0.3847	159.9273	0.0063	61.5277	0.0163	39.9727	2459.4298	96
100	2.7048	0.3697	170.4814	0.0059	63.0289	0.0159	41.3426	2605.7758	100
108	2.9289	0.3414	192.8926	0.0052	65.8578	0.0152	44.0103	2898.4203	108
120	3.3004	0.3030	230.0387	0.0043	69.7005	0.0143	47.8349	3334.1148	120
240	10.8926	0.0918	989.2554	0.0010	90.8194	0.0110	75.7393	6878.6016	240
360	35.9496	0.0278	3494.9641	0.0003	97.2183	0.0103	89.6995	8720.4323	360

Table C.5 Interest Rate Factors (1.25%)

	Single Payment		Equal Payment Series				Gradient Series		
N	Compound Amount Factor (F/P,i,N)	Present Worth Factor (P/F,i,N)	Compound Amount Factor (F/A,i,N)	Sinking Fund Factor (A/F,i,N)	Present Worth Factor (P/A,i,N)	Capital Recovery Factor (A/P,i,N)	Gradient Uniform Series (A/G,i,N)	Gradient Present Worth (P/G.i.N)	N
1	1.0125	0.9877	1.0000	1.0000	0.9877	1.0125	0.0000	0.0000	1
2	1.0252	0.9755	2.0125	0.4969	1.9631	0.5094	0.4969	0.9755	2
3	1.0380	0.9634	3.0377	0.3292	2.9265	0.3417	0.9917	2.9023	3
4	1.0509	0.9515	4.0756	0.2454	3.8781	0.2579	1.4845	5.7569	4
5	1.0641	0.9398	5.1266	0.1951	4.8178	0.2076	1.9752	9.5160	5
6	1.0774	0.9282	6.1907	0.1615	5.7460	0.1740	2.4638	14.1569	6
7	1.0909	0.9167	7.2680	0.1376	6.6627	0.1501	2.9503	19.6571	7
8	1.1045	0.9054	8.3589	0.1196	7.5681	0.1321	3.4348	25.9949	8
9	1.1183	0.8942	9.4634	0.1057	8.4623	0.1182	3.9172	33.1487	9
10	1.1323	0.8832	10.5817	0.0945	9.3455	0.1070	4.3975	41.0973	10
11	1.1464	0.8723	11.7139	0.0854	10.2178	0.0979	4.8758	49.8201	11
12	1.1608	0.8615	12.8604	0.0778	11.0793	0.0903	5.3520	59.2967	12
13	1.1753	0.8509	14.0211	0.0713	11.9302	0.0838	5.8262	69.5072	13
14	1.1900	0.8404	15.1964	0.0658	12.7706	0.0783	6.2982	80.4320	14
15	1.2048	0.8300	16.3863	0.0610	13.6005	0.0735	6.7682	92.0519	15
16	1.2199	0.8197	17.5912	0.0568	14.4203	0.0693	7.2362	104.3481	16
17	1.2351	0.8096	18.8111	0.0532	15.2299	0.0657	7.7021	117.3021	17
18	1.2506	0.7996	20.0462	0.0499	16.0295	0.0624	8.1659	130.8958	18
19	1.2662	0.7898	21.2968	0.0470	16.8193	0.0595	8.6277	145.1115	19
20	1.2820	0.7800	22.5630	0.0443	17.5993	0.0568	9.0874	159.9316	20
21	1.2981	0.7704	23.8450	0.0419	18.3697	0.0544	9.5450	175.3392	21
22	1.3143	0.7609	25.1431	0.0398	19.1306	0.0523	10.0006	191.3174	22
23	1.3307	0.7515	26.4574	0.0378	19.8820	0.0503	10.4542	207.8499	23
24	1.3474	0.7422	27.7881	0.0360	20.6242	0.0485	10.9056	224.9204	24
25	1.3642	0.7330	29.1354	0.0343	21.3573	0.0468	11.3551	242.5132	25
26	1.3812	0.7240	30.4996	0.0328	22.0813	0.0453	11.8024	260.6128	26
27	1.3985	0.7150	31.8809	0.0314	22.7963	0.0439	12.2478	279.2040	27
28	1.4160	0.7062	33.2794	0.0300	23.5025	0.0425	12.6911	298.2719	28
29	1.4337	0.6975	34.6954	0.0288	24.2000	0.0413	13.1323	317.8019	29
30	1.4516	0.6889	36.1291	0.0277	24.8889	0.0402	13.5715	337.7797	30
31	1.4698	0.6804	37.5807	0.0266	25.5693	0.0391	14.0086	358.1912	31
32	1.4881	0.6720	39.0504	0.0256	26.2413	0.0381	14.4438	379.0227	32
33	1.5067	0.6637	40.5386	0.0247	26.9050	0.0372	14.8768	400.2607	33
34	1.5256	0.6555	42.0453	0.0238	27.5605	0.0363	15.3079	421.8920	34
35	1.5446	0.6474	43.5709	0.0230	28.2079	0.0355	15.7369	443.9037	35
36	1.5639	0.6394	45.1155	0.0222	28.8473	0.0347	16.1639	466.2830	36
40	1.6436	0.6084	51.4896	0.0194	31.3269	0.0319	17.8515	559.2320	40
48	1.8154	0.5509	65.2284	0.0153	35.9315	0.0278	21.1299	759.2296	48
50	1.8610	0.5373	68.8818	0.0145	37.0129	0.0270	21.9295	811.6738	50
60	2.1072	0.4746	88.5745	0.0113	42.0346	0.0238	25.8083	1084.8429	60
72	2.4459	0.4088	115.6736	0.0086	47.2925	0.0211	30.2047	1428.4561	72
80	2.7015	0.3702	136.1188	0.0073	50.3867	0.0198	32.9822	1661.8651	80
84	2.8391	0.3522	147.1290	0.0068	51.8222	0.0193	34.3258	1778.8384	84
90	3.0588	0.3269	164.7050	0.0061	53.8461	0.0186	36.2855	1953.8303	90
96	3.2955	0.3034	183.6411	0.0054	55.7246	0.0179	38.1793	2127.5244	96
100	3.4634	0.2887	197.0723	0.0051	56.9013	0.0176	39.4058	2242.2411	100
108	3.8253	0.2614	226.0226	0.0044	59.0865	0.0169	41.7737	2468.2636	108
120	4.4402	0.2252	275.2171	0.0036	61.9828	0.0161	45.1184	2796.5694	120
240	19.7155	0.0507	1497.2395	0.0007	75.9423	0.0132	67.1764	5101.5288	240
360	87.5410	0.0114	6923.2796	0.0001	79.0861	0.0126	75.8401	5997.9027	360

Table C.6 Interest Rate Factors (1.50%)

	Single Payment		Equal Payment Series				Gradient Series		
N	Compound Amount Factor (F/P,i,N)	Present Worth Factor (P/F,i,N)	Compound Amount Factor (F/A,i,N)	Sinking Fund Factor (A/F,i,N)	Present Worth Factor (P/A,i,N)	Capital Recovery Factor (A/P,i,N)	Gradient Uniform Series (A/G,i,N)	Gradient Present Worth (P/G.i,N)	N
1	1.0150	0.9852	1.0000	1.0000	0.9852	1.0150	0.0000	0.0000	1
2	1.0302	0.9707	2.0150	0.4963	1.9559	0.5113	0.4963	0.9707	2
3	1.0457	0.9563	3.0452	0.3284	2.9122	0.3434	0.9901	2.8833	3
4	1.0614	0.9422	4.0909	0.2444	3.8544	0.2594	1.4814	5.7098	4
5	1.0773	0.9283	5.1523	0.1941	4.7826	0.2091	1.9702	9.4229	5
6	1.0934	0.9145	6.2296	0.1605	5.6972	0.1755	2.4566	13.9956	6
7	1.1098	0.9010	7.3230	0.1366	6.5982	0.1516	2.9405	19.4018	7
8	1.1265	0.8877	8.4328	0.1186	7.4859	0.1336	3.4219	25.6157	8
9	1.1434	0.8746	9.5593	0.1046	8.3605	0.1196	3.9008	32.6125	9
10	1.1605	0.8617	10.7027	0.0934	9.2222	0.1084	4.3772	40.3675	10
11	1.1779	0.8489	11.8633	0.0843	10.0711	0.0993	4.8512	48.8568	11
12	1.1956	0.8364	13.0412	0.0767	10.9075	0.0917	5.3227	58.0571	12
13	1.2136	0.8240	14.2368	0.0702	11.7315	0.0852	5.7917	67.9454	13
14	1.2318	0.8118	15.4504	0.0647	12.5434	0.0797	6.2582	78.4994	14
15	1.2502	0.7999	16.6821	0.0599	13.3432	0.0749	6.7223	89.6974	15
16	1.2690	0.7880	17.9324	0.0558	14.1313	0.0708	7.1839	101.5178	16
17	1.2880	0.7764	19.2014	0.0521	14.9076	0.0671	7.6431	113.9400	17
18	1.3073	0.7649	20.4894	0.0488	15.6726	0.0638	8.0997	126.9435	18
19	1.3270	0.7536	21.7967	0.0459	16.4262	0.0609	8.5539	140.5084	19
20	1.3469	0.7425	23.1237	0.0432	17.1686	0.0582	9.0057	154.6154	20
21	1.3671	0.7315	24.4705	0.0409	17.9001	0.0559	9.4550	169.2453	21
22	1.3876	0.7207	25.8376	0.0387	18.6208	0.0537	9.9018	184.3798	22
23	1.4084	0.7100	27.2251	0.0367	19.3309	0.0517	10.3462	200.0006	23
24	1.4295	0.6995	28.6335	0.0349	20.0304	0.0499	10.7881	216.0901	24
25	1.4509	0.6892	30.0630	0.0333	20.7196	0.0483	11.2276	232.6310	25
26	1.4727	0.6790	31.5140	0.0317	21.3986	0.0467	11.6646	249.6065	26
27	1.4948	0.6690	32.9867	0.0303	22.0676	0.0453	12.0992	267.0002	27
28	1.5172	0.6591	34.4815	0.0290	22.7267	0.0440	12.5313	284.7958	28
29	1.5400	0.6494	35.9987	0.0278	23.3761	0.0428	12.9610	302.9779	29
30	1.5631	0.6398	37.5387	0.0266	24.0158	0.0416	13.3883	321.5310	30
31	1.5865	0.6303	39.1018	0.0256	24.6461	0.0406	13.8131	340.4402	31
32	1.6103	0.6210	40.6883	0.0246	25.2671	0.0396	14.2355	359.6910	32
33	1.6345	0.6118	42.2986	0.0236	25.8790	0.0386	14.6555	379.2691	33
34	1.6590	0.6028	43.9331	0.0228	26.4817	0.0378	15.0731	399.1607	34
35	1.6839	0.5939	45.5921	0.0219	27.0756	0.0369	15.4882	419.3521	35
36	1.7091	0.5851	47.2760	0.0212	27.6607	0.0362	15.9009	439.8303	36
40	1.8140	0.5513	54.2679	0.0184	29.9158	0.0334	17.5277	524.3568	40
48	2.0435	0.4894	69.5652	0.0144	34.0426	0.0294	20.6667	703.5462	48
50	2.1052	0.4750	73.6828	0.0136	34.9997	0.0286	21.4277	749.9636	50
60	2.4432	0.4093	96.2147	0.0104	39.3803	0.0254	25.0930	988.1674	60
72	2.9212	0.3423	128.0772	0.0078	43.8447	0.0228	29.1893	1279.7938	72
80	3.2907	0.3039	152.7109	0.0065	46.4073	0.0215	31.7423	1473.0741	80
84	3.4926	0.2863	166.1726	0.0060	47.5786	0.0210	32.9668	1568.5140	84
90	3.8189	0.2619	187.9299	0.0053	49.2099	0.0203	34.7399	1709.5439	90
96	4.1758	0.2395	211.7202	0.0047	50.7017	0.0197	36.4381	1847.4725	96
100	4.4320	0.2256	228.8030	0.0044	51.6247	0.0194	37.5295	1937.4506	100
108	4.9927	0.2003	266.1778	0.0038	53.3137	0.0188	39.6171	2112.1348	108
120	5.9693	0.1675	331.2882	0.0030	55.4985	0.0180	42.5185	2359.7114	120
240	35.6328	0.0281	2308.8544	0.0004	64.7957	0.0154	59.7368	3870.6912	240
360	212.7038	0.0047	14113.5854	0.0001	66.3532	0.0151	64.9662	4310.7165	360

Table C.7 Interest Rate Factors (1.75%)

	Single Payment		Equal Payment Series				Gradient Series		
	Compound Amount Factor	Present Worth Factor	Compound Amount Factor	Sinking Fund Factor	Present Worth Factor	Capital Recovery Factor	Gradient Uniform Series	Gradient Present Worth	
N	(F/P,i,N)	(P/F,i,N)	(F/A,i,N)	(A/F,i,N)	(P/A,i,N)	(A/P,i,N)	(A/G,i,N)	(P/G,i,N)	N
1	1.0175	0.9828	1.0000	1.0000	0.9828	1.0175	0.0000	0.0000	1
2	1.0353	0.9659	2.0175	0.4957	1.9487	0.5132	0.4957	0.9659	2
3	1.0534	0.9493	3.0528	0.3276	2.8980	0.3451	0.9884	2.8645	3
4	1.0719	0.9330	4.1062	0.2435	3.8309	0.2610	1.4783	5.6633	4
5	1.0906	0.9169	5.1781	0.1931	4.7479	0.2106	1.9653	9.3310	5
6	1.1097	0.9011	6.2687	0.1595	5.6490	0.1770	2.4494	13.8367	6
7	1.1291	0.8856	7.3784	0.1355	6.5346	0.1530	2.9306	19.1506	7
8	1.1489	0.8704	8.5075	0.1175	7.4051	0.1350	3.4089	25.2435	8
9	1.1690	0.8554	9.6564	0.1036	8.2605	0.1211	3.8844	32.0870	9
10	1.1894	0.8407	10.8254	0.0924	9.1012	0.1099	4.3569	39.6535	10
11	1.2103	0.8263	12.0148	0.0832	9.9275	0.1007	4.8266	47.9162	11
12	1.2314	0.8121	13.2251	0.0756	10.7395	0.0931	5.2934	56.8489	12
13	1.2530	0.7981	14.4565	0.0692	11.5376	0.0867	5.7573	66.4260	13
14	1.2749	0.7844	15.7095	0.0637	12.3220	0.0812	6.2184	76.6227	14
15	1.2972	0.7709	16.9844	0.0589	13.0929	0.0764	6.6765	87.4149	15
16	1.3199	0.7576	18.2817	0.0547	13.8505	0.0722	7.1318	98.7792	16
17	1.3430	0.7446	19.6016	0.0510	14.5951	0.0685	7.5842	110.6926	17
18	1.3665	0.7318	20.9446	0.0477	15.3269	0.0652	8.0338	123.1328	18
19	1.3904	0.7192	22.3112	0.0448	16.0461	0.0623	8.4805	136.0783	19
20	1.4148	0.7068	23.7016	0.0422	16.7529	0.0597	8.9243	149.5080	20
21	1.4395	0.6947	25.1164	0.0398	17.4475	0.0573	9.3653	163.4013	21
22	1.4647	0.6827	26.5559	0.0377	18.1303	0.0552	9.8034	177.7385	22
23	1.4904	0.6710	28.0207	0.0357	18.8012	0.0532	10.2387	192.5000	23
24	1.5164	0.6594	29.5110	0.0339	19.4607	0.0514	10.6711	207.6671	24
25	1.5430	0.6481	31.0275	0.0322	20.1088	0.0497	11.1007	223.2214	25
26	1.5700	0.6369	32.5704	0.0307	20.7457	0.0482	11.5274	239.1451	26
27	1.5975	0.6260	34.1404	0.0293	21.3717	0.0468	11.9513	255.4210	27
28	1.6254	0.6152	35.7379	0.0280	21.9870	0.0455	12.3724	272.0321	28
29	1.6539	0.6046	37.3633	0.0268	22.5916	0.0443	12.7907	288.9623	29
30	1.6828	0.5942	39.0172	0.0256	23.1858	0.0431	13.2061	306.1954	30
31	1.7122	0.5840	40.7000	0.0246	23.7699	0.0421	13.6188	323.7163	31
32	1.7422	0.5740	42.4122	0.0236	24.3439	0.0411	14.0286	341.5097	32
33	1.7727	0.5641	44.1544	0.0226	24.9080	0.0401	14.4356	359.5613	33
34	1.8037	0.5544	45.9271	0.0218	25.4624	0.0393	14.8398	377.8567	34
35	1.8353	0.5449	47.7308	0.0210	26.0073	0.0385	15.2412	396.3824	35
36	1.8674	0.5355	49.5661	0.0202	26.5428	0.0377	15.6399	415.1250	36
40	2.0016	0.4996	57.2341	0.0175	28.5942	0.0350	17.2066	492.0109	40
48	2.2996	0.4349	74.2628	0.0135	32.2938	0.0310	20.2084	652.6054	48
50	2.3808	0.4200	78.9022	0.0127	33.1412	0.0302	20.9317	693.7010	50
60	2.8318	0.3531	104.6752	0.0096	36.9640	0.0271	24.3885	901.4954	60
72	3.4872	0.2868	142.1263	0.0070	40.7564	0.0245	28.1948	1149.1181	72
80	4.0064	0.2496	171.7938	0.0058	42.8799	0.0233	30.5329	1309.2482	80
84	4.2943	0.2329	188.2450	0.0053	43.8361	0.0228	31.6442	1387.1584	84
90	4.7654	0.2098	215.1646	0.0046	45.1516	0.0221	33.2409	1500.8798	90
96	5.2882	0.1891	245.0374	0.0041	46.3370	0.0216	34.7556	1610.4716	96
100	5.6682	0.1764	266.7518	0.0037	47.0615	0.0212	35.7211	1681.0886	100
108	6.5120	0.1536	314.9738	0.0032	48.3679	0.0207	37.5494	1816.1852	108
120	8.0192	0.1247	401.0962	0.0025	50.0171	0.0200	40.0469	2003.0269	120
240	64.3073	0.0156	3617.5602	0.0003	56.2543	0.0178	53.3518	3001.2678	240
360	515.6921	0.0019	29410.9747	0.0000	57.0320	0.0175	56.4434	3219.0833	360

Table C.8 Interest Rate Factors (2.0%)

	Single Payment		Equal Payment Series				Gradient Series		
N	Compound Amount Factor (F/P,i,N)	Present Worth Factor (P/F,i,N)	Compound Amount Factor (F/A,i,N)	Sinking Fund Factor (A/F,i,N)	Present Worth Factor (P/A,i,N)	Capital Recovery Factor (A/P,i,N)	Gradient Uniform Series (A/G,i,N)	Gradient Present Worth (P/G.i.N)	N
1	1.0200	0.9804	1.0000	1.0000	0.9804	1.0200	0.0000	0.0000	1
2	1.0404	0.9612	2.0200	0.4950	1.9416	0.5150	0.4950	0.9612	2
3	1.0612	0.9423	3.0604	0.3268	2.8839	0.3468	0.9868	2.8458	3
4	1.0824	0.9238	4.1216	0.2426	3.8077	0.2626	1.4752	5.6173	4
5	1.1041	0.9057	5.2040	0.1922	4.7135	0.2122	1.9604	9.2403	5
6	1.1262	0.8880	6.3081	0.1585	5.6014	0.1785	2.4423	13.6801	6
7	1.1487	0.8706	7.4343	0.1345	6.4720	0.1545	2.9208	18.9035	7
8	1.1717	0.8535	8.5830	0.1165	7.3255	0.1365	3.3961	24.8779	8
9	1.1951	0.8368	9.7546	0.1025	8.1622	0.1225	3.8681	31.5720	9
10	1.2190	0.8203	10.9497	0.0913	8.9826	0.1113	4.3367	38.9551	10
11	1.2434	0.8043	12.1687	0.0822	9.7868	0.1022	4.8021	46.9977	11
12	1.2682	0.7885	13.4121	0.0746	10.5753	0.0946	5.2642	55.6712	12
13	1.2936	0.7730	14.6803	0.0681	11.3484	0.0881	5.7231	64.9475	13
14	1.3195	0.7579	15.9739	0.0626	12.1062	0.0826	6.1786	74.7999	14
15	1.3459	0.7430	17.2934	0.0578	12.8493	0.0778	6.6309	85.2021	15
16	1.3728	0.7284	18.6393	0.0537	13.5777	0.0737	7.0799	96.1288	16
17	1.4002	0.7142	20.0121	0.0500	14.2919	0.0700	7.5256	107.5554	17
18	1.4282	0.7002	21.4123	0.0467	14.9920	0.0667	7.9681	119.4581	18
19	1.4568	0.6864	22.8406	0.0438	15.6785	0.0638	8.4073	131.8139	19
20	1.4859	0.6730	24.2974	0.0412	16.3514	0.0612	8.8433	144.6003	20
21	1.5157	0.6598	25.7833	0.0388	17.0112	0.0588	9.2760	157.7959	21
22	1.5460	0.6468	27.2990	0.0366	17.6580	0.0566	9.7055	171.3795	22
23	1.5769	0.6342	28.8450	0.0347	18.2922	0.0547	10.1317	185.3309	23
24	1.6084	0.6217	30.4219	0.0329	18.9139	0.0529	10.5547	199.6305	24
25	1.6406	0.6095	32.0303	0.0312	19.5235	0.0512	10.9745	214.2592	25
26	1.6734	0.5976	33.6709	0.0297	20.1210	0.0497	11.3910	229.1987	26
27	1.7069	0.5859	35.3443	0.0283	20.7069	0.0483	11.8043	244.4311	27
28	1.7410	0.5744	37.0512	0.0270	21.2813	0.0470	12.2145	259.9392	28
29	1.7758	0.5631	38.7922	0.0258	21.8444	0.0458	12.6214	275.7064	29
30	1.8114	0.5521	40.5681	0.0246	22.3965	0.0446	13.0251	291.7164	30
31	1.8476	0.5412	42.3794	0.0236	22.9377	0.0436	13.4257	307.9538	31
32	1.8845	0.5306	44.2270	0.0226	23.4683	0.0426	13.8230	324.4035	32
33	1.9222	0.5202	46.1116	0.0217	23.9886	0.0417	14.2172	341.0508	33
34	1.9607	0.5100	48.0338	0.0208	24.4986	0.0408	14.6083	357.8817	34
35	1.9999	0.5000	49.9945	0.0200	24.9986	0.0400	14.9961	374.8826	35
36	2.0399	0.4902	51.9944	0.0192	25.4888	0.0392	15.3809	392.0405	36
40	2.2080	0.4529	60.4020	0.0166	27.3555	0.0366	16.8885	461.9931	40
48	2.5871	0.3865	79.3535	0.0126	30.6731	0.0326	19.7556	605.9657	48
50	2.6916	0.3715	84.5794	0.0118	31.4236	0.0318	20.4420	642.3606	50
60	3.2810	0.3048	114.0515	0.0088	34.7609	0.0288	23.6961	823.6975	60
72	4.1611	0.2403	158.0570	0.0063	37.9841	0.0263	27.2234	1034.0557	72
80	4.8754	0.2051	193.7720	0.0052	39.7445	0.0252	29.3572	1166.7868	80
84	5.2773	0.1895	213.8666	0.0047	40.5255	0.0247	30.3616	1230.4191	84
90	5.9431	0.1683	247.1567	0.0040	41.5869	0.0240	31.7929	1322.1701	90
96	6.6929	0.1494	284.6467	0.0035	42.5294	0.0235	33.1370	1409.2973	96
100	7.2446	0.1380	312.2323	0.0032	43.0984	0.0232	33.9863	1464.7527	100
108	8.4883	0.1178	374.4129	0.0027	44.1095	0.0227	35.5774	1569.3025	108
120	10.7652	0.0929	488.2582	0.0020	45.3554	0.0220	37.7114	1710.4160	120
240	115.8887	0.0086	5744.4368	0.0002	49.5686	0.0202	47.9110	2374.8800	240
360	1247.5611	0.0008	62328.0564	0.0000	49.9599	0.0200	49.7112	2483.5679	360

Table C.9 Interest Rate Factors (3.0%)

N	Single Payment Compound Amount Factor (F/P,i,N)	Present Worth Factor (P/F,i,N)	Equal Payment Series Compound Amount Factor (F/A,i,N)	Sinking Fund Factor (A/F,i,N)	Present Worth Factor (P/A,i,N)	Capital Recovery Factor (A/P,i,N)	Gradient Series Gradient Uniform Series (A/G,i,N)	Gradient Present Worth (P/G.i.N)	N
1	1.0300	0.9709	1.0000	1.0000	0.9709	1.0300	0.0000	0.0000	1
2	1.0609	0.9426	2.0300	0.4926	1.9135	0.5226	0.4926	0.9426	2
3	1.0927	0.9151	3.0909	0.3235	2.8286	0.3535	0.9803	2.7729	3
4	1.1255	0.8885	4.1836	0.2390	3.7171	0.2690	1.4631	5.4383	4
5	1.1593	0.8626	5.3091	0.1884	4.5797	0.2184	1.9409	8.8888	5
6	1.1941	0.8375	6.4684	0.1546	5.4172	0.1846	2.4138	13.0762	6
7	1.2299	0.8131	7.6625	0.1305	6.2303	0.1605	2.8819	17.9547	7
8	1.2668	0.7894	8.8923	0.1125	7.0197	0.1425	3.3450	23.4806	8
9	1.3048	0.7664	10.1591	0.0984	7.7861	0.1284	3.8032	29.6119	9
10	1.3439	0.7441	11.4639	0.0872	8.5302	0.1172	4.2565	36.3088	10
11	1.3842	0.7224	12.8078	0.0781	9.2526	0.1081	4.7049	43.5330	11
12	1.4258	0.7014	14.1920	0.0705	9.9540	0.1005	5.1485	51.2482	12
13	1.4685	0.6810	15.6178	0.0640	10.6350	0.0940	5.5872	59.4196	13
14	1.5126	0.6611	17.0863	0.0585	11.2961	0.0885	6.0210	68.0141	14
15	1.5580	0.6419	18.5989	0.0538	11.9379	0.0838	6.4500	77.0002	15
16	1.6047	0.6232	20.1569	0.0496	12.5611	0.0796	6.8742	86.3477	16
17	1.6528	0.6050	21.7616	0.0460	13.1661	0.0760	7.2936	96.0280	17
18	1.7024	0.5874	23.4144	0.0427	13.7535	0.0727	7.7081	106.0137	18
19	1.7535	0.5703	25.1169	0.0398	14.3238	0.0698	8.1179	116.2788	19
20	1.8061	0.5537	26.8704	0.0372	14.8775	0.0672	8.5229	126.7987	20
21	1.8603	0.5375	28.6765	0.0349	15.4150	0.0649	8.9231	137.5496	21
22	1.9161	0.5219	30.5368	0.0327	15.9369	0.0627	9.3186	148.5094	22
23	1.9736	0.5067	32.4529	0.0308	16.4436	0.0608	9.7093	159.6566	23
24	2.0328	0.4919	34.4265	0.0290	16.9355	0.0590	10.0954	170.9711	24
25	2.0938	0.4776	36.4593	0.0274	17.4131	0.0574	10.4768	182.4336	25
26	2.1566	0.4637	38.5530	0.0259	17.8768	0.0559	10.8535	194.0260	26
27	2.2213	0.4502	40.7096	0.0246	18.3270	0.0546	11.2255	205.7309	27
28	2.2879	0.4371	42.9309	0.0233	18.7641	0.0533	11.5930	217.5320	28
29	2.3566	0.4243	45.2189	0.0221	19.1885	0.0521	11.9558	229.4137	29
30	2.4273	0.4120	47.5754	0.0210	19.6004	0.0510	12.3141	241.3613	30
31	2.5001	0.4000	50.0027	0.0200	20.0004	0.0500	12.6678	253.3609	31
32	2.5751	0.3883	52.5028	0.0190	20.3888	0.0490	13.0169	265.3993	32
33	2.6523	0.3770	55.0778	0.0182	20.7658	0.0482	13.3616	277.4642	33
34	2.7319	0.3660	57.7302	0.0173	21.1318	0.0473	13.7018	289.5437	34
35	2.8139	0.3554	60.4621	0.0165	21.4872	0.0465	14.0375	301.6267	35
40	3.2620	0.3066	75.4013	0.0133	23.1148	0.0433	15.6502	361.7499	40
45	3.7816	0.2644	92.7199	0.0108	24.5187	0.0408	17.1556	420.6325	45
50	4.3839	0.2281	112.7969	0.0089	25.7298	0.0389	18.5575	477.4803	50
55	5.0821	0.1968	136.0716	0.0073	26.7744	0.0373	19.8600	531.7411	55
60	5.8916	0.1697	163.0534	0.0061	27.6756	0.0361	21.0674	583.0526	60
65	6.8300	0.1464	194.3328	0.0051	28.4529	0.0351	22.1841	631.2010	65
70	7.9178	0.1263	230.5941	0.0043	29.1234	0.0343	23.2145	676.0869	70
75	9.1789	0.1089	272.6309	0.0037	29.7018	0.0337	24.1634	717.6978	75
80	10.6409	0.0940	321.3630	0.0031	30.2008	0.0331	25.0353	756.0865	80
85	12.3357	0.0811	377.8570	0.0026	30.6312	0.0326	25.8349	791.3529	85
90	14.3005	0.0699	443.3489	0.0023	31.0024	0.0323	26.5667	823.6302	90
95	16.5782	0.0603	519.2720	0.0019	31.3227	0.0319	27.2351	853.0742	95
100	19.2186	0.0520	607.2877	0.0016	31.5989	0.0316	27.8444	879.8540	100

Table C.10 Interest Rate Factors (4.0%)

N	Single Payment Compound Amount Factor (F/P,i,N)	Single Payment Present Worth Factor (P/F,i,N)	Equal Payment Series Compound Amount Factor (F/A,i,N)	Equal Payment Series Sinking Fund Factor (A/F,i,N)	Equal Payment Series Present Worth Factor (P/A,i,N)	Equal Payment Series Capital Recovery Factor (A/P,i,N)	Gradient Series Gradient Uniform Series (A/G,i,N)	Gradient Series Gradient Present Worth (P/G,i,N)	N
1	1.0400	0.9615	1.0000	1.0000	0.9615	1.0400	0.0000	0.0000	1
2	1.0816	0.9246	2.0400	0.4902	1.8861	0.5302	0.4902	0.9246	2
3	1.1249	0.8890	3.1216	0.3203	2.7751	0.3603	0.9739	2.7025	3
4	1.1699	0.8548	4.2465	0.2355	3.6299	0.2755	1.4510	5.2670	4
5	1.2167	0.8219	5.4163	0.1846	4.4518	0.2246	1.9216	8.5547	5
6	1.2653	0.7903	6.6330	0.1508	5.2421	0.1908	2.3857	12.5062	6
7	1.3159	0.7599	7.8983	0.1266	6.0021	0.1666	2.8433	17.0657	7
8	1.3686	0.7307	9.2142	0.1085	6.7327	0.1485	3.2944	22.1806	8
9	1.4233	0.7026	10.5828	0.0945	7.4353	0.1345	3.7391	27.8013	9
10	1.4802	0.6756	12.0061	0.0833	8.1109	0.1233	4.1773	33.8814	10
11	1.5395	0.6496	13.4864	0.0741	8.7605	0.1141	4.6090	40.3772	11
12	1.6010	0.6246	15.0258	0.0666	9.3851	0.1066	5.0343	47.2477	12
13	1.6651	0.6006	16.6268	0.0601	9.9856	0.1001	5.4533	54.4546	13
14	1.7317	0.5775	18.2919	0.0547	10.5631	0.0947	5.8659	61.9618	14
15	1.8009	0.5553	20.0236	0.0499	11.1184	0.0899	6.2721	69.7355	15
16	1.8730	0.5339	21.8245	0.0458	11.6523	0.0858	6.6720	77.7441	16
17	1.9479	0.5134	23.6975	0.0422	12.1657	0.0822	7.0656	85.9581	17
18	2.0258	0.4936	25.6454	0.0390	12.6593	0.0790	7.4530	94.3498	18
19	2.1068	0.4746	27.6712	0.0361	13.1339	0.0761	7.8342	102.8933	19
20	2.1911	0.4564	29.7781	0.0336	13.5903	0.0736	8.2091	111.5647	20
21	2.2788	0.4388	31.9692	0.0313	14.0292	0.0713	8.5779	120.3414	21
22	2.3699	0.4220	34.2480	0.0292	14.4511	0.0692	8.9407	129.2024	22
23	2.4647	0.4057	36.6179	0.0273	14.8568	0.0673	9.2973	138.1284	23
24	2.5633	0.3901	39.0826	0.0256	15.2470	0.0656	9.6479	147.1012	24
25	2.6658	0.3751	41.6459	0.0240	15.6221	0.0640	9.9925	156.1040	25
26	2.7725	0.3607	44.3117	0.0226	15.9828	0.0626	10.3312	165.1212	26
27	2.8834	0.3468	47.0842	0.0212	16.3296	0.0612	10.6640	174.1385	27
28	2.9987	0.3335	49.9676	0.0200	16.6631	0.0600	10.9909	183.1424	28
29	3.1187	0.3207	52.9663	0.0189	16.9837	0.0589	11.3120	192.1206	29
30	3.2434	0.3083	56.0849	0.0178	17.2920	0.0578	11.6274	201.0618	30
31	3.3731	0.2965	59.3283	0.0169	17.5885	0.0569	11.9371	209.9556	31
32	3.5081	0.2851	62.7015	0.0159	17.8736	0.0559	12.2411	218.7924	32
33	3.6484	0.2741	66.2095	0.0151	18.1476	0.0551	12.5396	227.5634	33
34	3.7943	0.2636	69.8579	0.0143	18.4112	0.0543	12.8324	236.2607	34
35	3.9461	0.2534	73.6522	0.0136	18.6646	0.0536	13.1198	244.8768	35
40	4.8010	0.2083	95.0255	0.0105	19.7928	0.0505	14.4765	286.5303	40
45	5.8412	0.1712	121.0294	0.0083	20.7200	0.0483	15.7047	325.4028	45
50	7.1067	0.1407	152.6671	0.0066	21.4822	0.0466	16.8122	361.1638	50
55	8.6464	0.1157	191.1592	0.0052	22.1086	0.0452	17.8070	393.6890	55
60	10.5196	0.0951	237.9907	0.0042	22.6235	0.0442	18.6972	422.9966	60
65	12.7987	0.0781	294.9684	0.0034	23.0467	0.0434	19.4909	449.2014	65
70	15.5716	0.0642	364.2905	0.0027	23.3945	0.0427	20.1961	472.4789	70
75	18.9453	0.0528	448.6314	0.0022	23.6804	0.0422	20.8206	493.0408	75
80	23.0498	0.0434	551.2450	0.0018	23.9154	0.0418	21.3718	511.1161	80
85	28.0436	0.0357	676.0901	0.0015	24.1085	0.0415	21.8569	526.9384	85
90	34.1193	0.0293	827.9833	0.0012	24.2673	0.0412	22.2826	540.7369	90
95	41.5114	0.0241	1012.7846	0.0010	24.3978	0.0410	22.6550	552.7307	95
100	50.5049	0.0198	1237.6237	0.0008	24.5050	0.0408	22.9800	563.1249	100

Table C.11 Interest Rate Factors (5.0%)

	Single Payment		Equal Payment Series				Gradient Series		
N	Compound Amount Factor (F/P,i,N)	Present Worth Factor (P/F,i,N)	Compound Amount Factor (F/A,i,N)	Sinking Fund Factor (A/F,i,N)	Present Worth Factor (P/A,i,N)	Capital Recovery Factor (A/P,i,N)	Gradient Uniform Series (A/G,i,N)	Gradient Present Worth (P/G.i.N)	N
1	1.0500	0.9524	1.0000	1.0000	0.9524	1.0500	0.0000	0.0000	1
2	1.1025	0.9070	2.0500	0.4878	1.8594	0.5378	0.4878	0.9070	2
3	1.1576	0.8638	3.1525	0.3172	2.7232	0.3672	0.9675	2.6347	3
4	1.2155	0.8227	4.3101	0.2320	3.5460	0.2820	1.4391	5.1028	4
5	1.2763	0.7835	5.5256	0.1810	4.3295	0.2310	1.9025	8.2369	5
6	1.3401	0.7462	6.8019	0.1470	5.0757	0.1970	2.3579	11.9680	6
7	1.4071	0.7107	8.1420	0.1228	5.7864	0.1728	2.8052	16.2321	7
8	1.4775	0.6768	9.5491	0.1047	6.4632	0.1547	3.2445	20.9700	8
9	1.5513	0.6446	11.0266	0.0907	7.1078	0.1407	3.6758	26.1268	9
10	1.6289	0.6139	12.5779	0.0795	7.7217	0.1295	4.0991	31.6520	10
11	1.7103	0.5847	14.2068	0.0704	8.3064	0.1204	4.5144	37.4988	11
12	1.7959	0.5568	15.9171	0.0628	8.8633	0.1128	4.9219	43.6241	12
13	1.8856	0.5303	17.7130	0.0565	9.3936	0.1065	5.3215	49.9879	13
14	1.9799	0.5051	19.5986	0.0510	9.8986	0.1010	5.7133	56.5538	14
15	2.0789	0.4810	21.5786	0.0463	10.3797	0.0963	6.0973	63.2880	15
16	2.1829	0.4581	23.6575	0.0423	10.8378	0.0923	6.4736	70.1597	16
17	2.2920	0.4363	25.8404	0.0387	11.2741	0.0887	6.8423	77.1405	17
18	2.4066	0.4155	28.1324	0.0355	11.6896	0.0855	7.2034	84.2043	18
19	2.5270	0.3957	30.5390	0.0327	12.0853	0.0827	7.5569	91.3275	19
20	2.6533	0.3769	33.0660	0.0302	12.4622	0.0802	7.9030	98.4884	20
21	2.7860	0.3589	35.7193	0.0280	12.8212	0.0780	8.2416	105.6673	21
22	2.9253	0.3418	38.5052	0.0260	13.1630	0.0760	8.5730	112.8461	22
23	3.0715	0.3256	41.4305	0.0241	13.4886	0.0741	8.8971	120.0087	23
24	3.2251	0.3101	44.5020	0.0225	13.7986	0.0725	9.2140	127.1402	24
25	3.3864	0.2953	47.7271	0.0210	14.0939	0.0710	9.5238	134.2275	25
26	3.5557	0.2812	51.1135	0.0196	14.3752	0.0696	9.8266	141.2585	26
27	3.7335	0.2678	54.6691	0.0183	14.6430	0.0683	10.1224	148.2226	27
28	3.9201	0.2551	58.4026	0.0171	14.8981	0.0671	10.4114	155.1101	28
29	4.1161	0.2429	62.3227	0.0160	15.1411	0.0660	10.6936	161.9126	29
30	4.3219	0.2314	66.4388	0.0151	15.3725	0.0651	10.9691	168.6226	30
31	4.5380	0.2204	70.7608	0.0141	15.5928	0.0641	11.2381	175.2333	31
32	4.7649	0.2099	75.2988	0.0133	15.8027	0.0633	11.5005	181.7392	32
33	5.0032	0.1999	80.0638	0.0125	16.0025	0.0625	11.7566	188.1351	33
34	5.2533	0.1904	85.0670	0.0118	16.1929	0.0618	12.0063	194.4168	34
35	5.5160	0.1813	90.3203	0.0111	16.3742	0.0611	12.2498	200.5807	35
40	7.0400	0.1420	120.7998	0.0083	17.1591	0.0583	13.3775	229.5452	40
45	8.9850	0.1113	159.7002	0.0063	17.7741	0.0563	14.3644	255.3145	45
50	11.4674	0.0872	209.3480	0.0048	18.2559	0.0548	15.2233	277.9148	50
55	14.6356	0.0683	272.7126	0.0037	18.6335	0.0537	15.9664	297.5104	55
60	18.6792	0.0535	353.5837	0.0028	18.9293	0.0528	16.6062	314.3432	60
65	23.8399	0.0419	456.7980	0.0022	19.1611	0.0522	17.1541	328.6910	65
70	30.4264	0.0329	588.5285	0.0017	19.3427	0.0517	17.6212	340.8409	70
75	38.8327	0.0258	756.6537	0.0013	19.4850	0.0513	18.0176	351.0721	75
80	49.5614	0.0202	971.2288	0.0010	19.5965	0.0510	18.3526	359.6460	80
85	63.2544	0.0158	1245.0871	0.0008	19.6838	0.0508	18.6346	366.8007	85
90	80.7304	0.0124	1594.6073	0.0006	19.7523	0.0506	18.8712	372.7488	90
95	103.0347	0.0097	2040.6935	0.0005	19.8059	0.0505	19.0689	377.6774	95
100	131.5013	0.0076	2610.0252	0.0004	19.8479	0.0504	19.2337	381.7492	100

Table C.12 Interest Rate Factors (6.0%)

	Single Payment		Equal Payment Series				Gradient Series		
N	Compound Amount Factor (F/P,i,N)	Present Worth Factor (P/F,i,N)	Compound Amount Factor (F/A,i,N)	Sinking Fund Factor (A/F,i,N)	Present Worth Factor (P/A,i,N)	Capital Recovery Factor (A/P,i,N)	Gradient Uniform Series (A/G,i,N)	Gradient Present Worth (P/G.i.N)	N
1	1.0600	0.9434	1.0000	1.0000	0.9434	1.0600	0.0000	0.0000	1
2	1.1236	0.8900	2.0600	0.4854	1.8334	0.5454	0.4854	0.8900	2
3	1.1910	0.8396	3.1836	0.3141	2.6730	0.3741	0.9612	2.5692	3
4	1.2625	0.7921	4.3746	0.2286	3.4651	0.2886	1.4272	4.9455	4
5	1.3382	0.7473	5.6371	0.1774	4.2124	0.2374	1.8836	7.9345	5
6	1.4185	0.7050	6.9753	0.1434	4.9173	0.2034	2.3304	11.4594	6
7	1.5036	0.6651	8.3938	0.1191	5.5824	0.1791	2.7676	15.4497	7
8	1.5938	0.6274	9.8975	0.1010	6.2098	0.1610	3.1952	19.8416	8
9	1.6895	0.5919	11.4913	0.0870	6.8017	0.1470	3.6133	24.5768	9
10	1.7908	0.5584	13.1808	0.0759	7.3601	0.1359	4.0220	29.6023	10
11	1.8983	0.5268	14.9716	0.0668	7.8869	0.1268	4.4213	34.8702	11
12	2.0122	0.4970	16.8699	0.0593	8.3838	0.1193	4.8113	40.3369	12
13	2.1329	0.4688	18.8821	0.0530	8.8527	0.1130	5.1920	45.9629	13
14	2.2609	0.4423	21.0151	0.0476	9.2950	0.1076	5.5635	51.7128	14
15	2.3966	0.4173	23.2760	0.0430	9.7122	0.1030	5.9260	57.5546	15
16	2.5404	0.3936	25.6725	0.0390	10.1059	0.0990	6.2794	63.4592	16
17	2.6928	0.3714	28.2129	0.0354	10.4773	0.0954	6.6240	69.4011	17
18	2.8543	0.3503	30.9057	0.0324	10.8276	0.0924	6.9597	75.3569	18
19	3.0256	0.3305	33.7600	0.0296	11.1581	0.0896	7.2867	81.3062	19
20	3.2071	0.3118	36.7856	0.0272	11.4699	0.0872	7.6051	87.2304	20
21	3.3996	0.2942	39.9927	0.0250	11.7641	0.0850	7.9151	93.1136	21
22	3.6035	0.2775	43.3923	0.0230	12.0416	0.0830	8.2166	98.9412	22
23	3.8197	0.2618	46.9958	0.0213	12.3034	0.0813	8.5099	104.7007	23
24	4.0489	0.2470	50.8156	0.0197	12.5504	0.0797	8.7951	110.3812	24
25	4.2919	0.2330	54.8645	0.0182	12.7834	0.0782	9.0722	115.9732	25
26	4.5494	0.2198	59.1564	0.0169	13.0032	0.0769	9.3414	121.4684	26
27	4.8223	0.2074	63.7058	0.0157	13.2105	0.0757	9.6029	126.8600	27
28	5.1117	0.1956	68.5281	0.0146	13.4062	0.0746	9.8568	132.1420	28
29	5.4184	0.1846	73.6398	0.0136	13.5907	0.0736	10.1032	137.3096	29
30	5.7435	0.1741	79.0582	0.0126	13.7648	0.0726	10.3422	142.3588	30
31	6.0881	0.1643	84.8017	0.0118	13.9291	0.0718	10.5740	147.2864	31
32	6.4534	0.1550	90.8898	0.0110	14.0840	0.0710	10.7988	152.0901	32
33	6.8406	0.1462	97.3432	0.0103	14.2302	0.0703	11.0166	156.7681	33
34	7.2510	0.1379	104.1838	0.0096	14.3681	0.0696	11.2276	161.3192	34
35	7.6861	0.1301	111.4348	0.0090	14.4982	0.0690	11.4319	165.7427	35
40	10.2857	0.0972	154.7620	0.0065	15.0463	0.0665	12.3590	185.9568	40
45	13.7646	0.0727	212.7435	0.0047	15.4558	0.0647	13.1413	203.1096	45
50	18.4202	0.0543	290.3359	0.0034	15.7619	0.0634	13.7964	217.4574	50
55	24.6503	0.0406	394.1720	0.0025	15.9905	0.0625	14.3411	229.3222	55
60	32.9877	0.0303	533.1282	0.0019	16.1614	0.0619	14.7909	239.0428	60
65	44.1450	0.0227	719.0829	0.0014	16.2891	0.0614	15.1601	246.9450	65
70	59.0759	0.0169	967.9322	0.0010	16.3845	0.0610	15.4613	253.3271	70
75	79.0569	0.0126	1300.9487	0.0008	16.4558	0.0608	15.7058	258.4527	75
80	105.7960	0.0095	1746.5999	0.0006	16.5091	0.0606	15.9033	262.5493	80
85	141.5789	0.0071	2342.9817	0.0004	16.5489	0.0604	16.0620	265.8096	85
90	189.4645	0.0053	3141.0752	0.0003	16.5787	0.0603	16.1891	268.3946	90
95	253.5463	0.0039	4209.1042	0.0002	16.6009	0.0602	16.2905	270.4375	95
100	339.3021	0.0029	5638.3681	0.0002	16.6175	0.0602	16.3711	272.0471	100

Table C.13 Interest Rate Factors (7.0%)

	Single Payment		Equal Payment Series				Gradient Series		
N	Compound Amount Factor (F/P,i,N)	Present Worth Factor (P/F,i,N)	Compound Amount Factor (F/A,i,N)	Sinking Fund Factor (A/F,i,N)	Present Worth Factor (P/A,i,N)	Capital Recovery Factor (A/P,i,N)	Gradient Uniform Series (A/G,i,N)	Gradient Present Worth (P/G.i.N)	N
1	1.0700	0.9346	1.0000	1.0000	0.9346	1.0700	0.0000	0.0000	1
2	1.1449	0.8734	2.0700	0.4831	1.8080	0.5531	0.4831	0.8734	2
3	1.2250	0.8163	3.2149	0.3111	2.6243	0.3811	0.9549	2.5060	3
4	1.3108	0.7629	4.4399	0.2252	3.3872	0.2952	1.4155	4.7947	4
5	1.4026	0.7130	5.7507	0.1739	4.1002	0.2439	1.8650	7.6467	5
6	1.5007	0.6663	7.1533	0.1398	4.7665	0.2098	2.3032	10.9784	6
7	1.6058	0.6227	8.6540	0.1156	5.3893	0.1856	2.7304	14.7149	7
8	1.7182	0.5820	10.2598	0.0975	5.9713	0.1675	3.1465	18.7889	8
9	1.8385	0.5439	11.9780	0.0835	6.5152	0.1535	3.5517	23.1404	9
10	1.9672	0.5083	13.8164	0.0724	7.0236	0.1424	3.9461	27.7156	10
11	2.1049	0.4751	15.7836	0.0634	7.4987	0.1334	4.3296	32.4665	11
12	2.2522	0.4440	17.8885	0.0559	7.9427	0.1259	4.7025	37.3506	12
13	2.4098	0.4150	20.1406	0.0497	8.3577	0.1197	5.0648	42.3302	13
14	2.5785	0.3878	22.5505	0.0443	8.7455	0.1143	5.4167	47.3718	14
15	2.7590	0.3624	25.1290	0.0398	9.1079	0.1098	5.7583	52.4461	15
16	2.9522	0.3387	27.8881	0.0359	9.4466	0.1059	6.0897	57.5271	16
17	3.1588	0.3166	30.8402	0.0324	9.7632	0.1024	6.4110	62.5923	17
18	3.3799	0.2959	33.9990	0.0294	10.0591	0.0994	6.7225	67.6219	18
19	3.6165	0.2765	37.3790	0.0268	10.3356	0.0968	7.0242	72.5991	19
20	3.8697	0.2584	40.9955	0.0244	10.5940	0.0944	7.3163	77.5091	20
21	4.1406	0.2415	44.8652	0.0223	10.8355	0.0923	7.5990	82.3393	21
22	4.4304	0.2257	49.0057	0.0204	11.0612	0.0904	7.8725	87.0793	22
23	4.7405	0.2109	53.4361	0.0187	11.2722	0.0887	8.1369	91.7201	23
24	5.0724	0.1971	58.1767	0.0172	11.4693	0.0872	8.3923	96.2545	24
25	5.4274	0.1842	63.2490	0.0158	11.6536	0.0858	8.6391	100.6765	25
26	5.8074	0.1722	68.6765	0.0146	11.8258	0.0846	8.8773	104.9814	26
27	6.2139	0.1609	74.4838	0.0134	11.9867	0.0834	9.1072	109.1656	27
28	6.6488	0.1504	80.6977	0.0124	12.1371	0.0824	9.3289	113.2264	28
29	7.1143	0.1406	87.3465	0.0114	12.2777	0.0814	9.5427	117.1622	29
30	7.6123	0.1314	94.4608	0.0106	12.4090	0.0806	9.7487	120.9718	30
31	8.1451	0.1228	102.0730	0.0098	12.5318	0.0798	9.9471	124.6550	31
32	8.7153	0.1147	110.2182	0.0091	12.6466	0.0791	10.1381	128.2120	32
33	9.3253	0.1072	118.9334	0.0084	12.7538	0.0784	10.3219	131.6435	33
34	9.9781	0.1002	128.2588	0.0078	12.8540	0.0778	10.4987	134.9507	34
35	10.6766	0.0937	138.2369	0.0072	12.9477	0.0772	10.6687	138.1353	35
40	14.9745	0.0668	199.6351	0.0050	13.3317	0.0750	11.4233	152.2928	40
45	21.0025	0.0476	285.7493	0.0035	13.6055	0.0735	12.0360	163.7559	45
50	29.4570	0.0339	406.5289	0.0025	13.8007	0.0725	12.5287	172.9051	50
55	41.3150	0.0242	575.9286	0.0017	13.9399	0.0717	12.9215	180.1243	55
60	57.9464	0.0173	813.5204	0.0012	14.0392	0.0712	13.2321	185.7677	60
65	81.2729	0.0123	1146.7552	0.0009	14.1099	0.0709	13.4760	190.1452	65
70	113.9894	0.0088	1614.1342	0.0006	14.1604	0.0706	13.6662	193.5185	70
75	159.8760	0.0063	2269.6574	0.0004	14.1964	0.0704	13.8136	196.1035	75
80	224.2344	0.0045	3189.0627	0.0003	14.2220	0.0703	13.9273	198.0748	80
85	314.5003	0.0032	4478.5761	0.0002	14.2403	0.0702	14.0146	199.5717	85
90	441.1030	0.0023	6287.1854	0.0002	14.2533	0.0702	14.0812	200.7042	90
95	618.6697	0.0016	8823.8535	0.0001	14.2626	0.0701	14.1319	201.5581	95
100	867.7163	0.0012	12381.6618	0.0001	14.2693	0.0701	14.1703	202.2001	100

Table C.14 Interest Rate Factors (8.0%)

	Single Payment		Equal Payment Series				Gradient Series		
N	Compound Amount Factor (F/P,i,N)	Present Worth Factor (P/F,i,N)	Compound Amount Factor (F/A,i,N)	Sinking Fund Factor (A/F,i,N)	Present Worth Factor (P/A,i,N)	Capital Recovery Factor (A/P,i,N)	Gradient Uniform Series (A/G,i,N)	Gradient Present Worth (P/G.i,N)	N
1	1.0800	0.9259	1.0000	1.0000	0.9259	1.0800	0.0000	0.0000	1
2	1.1664	0.8573	2.0800	0.4808	1.7833	0.5608	0.4808	0.8573	2
3	1.2597	0.7938	3.2464	0.3080	2.5771	0.3880	0.9487	2.4450	3
4	1.3605	0.7350	4.5061	0.2219	3.3121	0.3019	1.4040	4.6501	4
5	1.4693	0.6806	5.8666	0.1705	3.9927	0.2505	1.8465	7.3724	5
6	1.5869	0.6302	7.3359	0.1363	4.6229	0.2163	2.2763	10.5233	6
7	1.7138	0.5835	8.9228	0.1121	5.2064	0.1921	2.6937	14.0242	7
8	1.8509	0.5403	10.6366	0.0940	5.7466	0.1740	3.0985	17.8061	8
9	1.9990	0.5002	12.4876	0.0801	6.2469	0.1601	3.4910	21.8081	9
10	2.1589	0.4632	14.4866	0.0690	6.7101	0.1490	3.8713	25.9768	10
11	2.3316	0.4289	16.6455	0.0601	7.1390	0.1401	4.2395	30.2657	11
12	2.5182	0.3971	18.9771	0.0527	7.5361	0.1327	4.5957	34.6339	12
13	2.7196	0.3677	21.4953	0.0465	7.9038	0.1265	4.9402	39.0463	13
14	2.9372	0.3405	24.2149	0.0413	8.2442	0.1213	5.2731	43.4723	14
15	3.1722	0.3152	27.1521	0.0368	8.5595	0.1168	5.5945	47.8857	15
16	3.4259	0.2919	30.3243	0.0330	8.8514	0.1130	5.9046	52.2640	16
17	3.7000	0.2703	33.7502	0.0296	9.1216	0.1096	6.2037	56.5883	17
18	3.9960	0.2502	37.4502	0.0267	9.3719	0.1067	6.4920	60.8426	18
19	4.3157	0.2317	41.4463	0.0241	9.6036	0.1041	6.7697	65.0134	19
20	4.6610	0.2145	45.7620	0.0219	9.8181	0.1019	7.0369	69.0898	20
21	5.0338	0.1987	50.4229	0.0198	10.0168	0.0998	7.2940	73.0629	21
22	5.4365	0.1839	55.4568	0.0180	10.2007	0.0980	7.5412	76.9257	22
23	5.8715	0.1703	60.8933	0.0164	10.3711	0.0964	7.7786	80.6726	23
24	6.3412	0.1577	66.7648	0.0150	10.5288	0.0950	8.0066	84.2997	24
25	6.8485	0.1460	73.1059	0.0137	10.6748	0.0937	8.2254	87.8041	25
26	7.3964	0.1352	79.9544	0.0125	10.8100	0.0925	8.4352	91.1842	26
27	7.9881	0.1252	87.3508	0.0114	10.9352	0.0914	8.6363	94.4390	27
28	8.6271	0.1159	95.3388	0.0105	11.0511	0.0905	8.8289	97.5687	28
29	9.3173	0.1073	103.9659	0.0096	11.1584	0.0896	9.0133	100.5738	29
30	10.0627	0.0994	113.2832	0.0088	11.2578	0.0888	9.1897	103.4558	30
31	10.8677	0.0920	123.3459	0.0081	11.3498	0.0881	9.3584	106.2163	31
32	11.7371	0.0852	134.2135	0.0075	11.4350	0.0875	9.5197	108.8575	32
33	12.6760	0.0789	145.9506	0.0069	11.5139	0.0869	9.6737	111.3819	33
34	13.6901	0.0730	158.6267	0.0063	11.5869	0.0863	9.8208	113.7924	34
35	14.7853	0.0676	172.3168	0.0058	11.6546	0.0858	9.9611	116.0920	35
40	21.7245	0.0460	259.0565	0.0039	11.9246	0.0839	10.5699	126.0422	40
45	31.9204	0.0313	386.5056	0.0026	12.1084	0.0826	11.0447	133.7331	45
50	46.9016	0.0213	573.7702	0.0017	12.2335	0.0817	11.4107	139.5928	50
55	68.9139	0.0145	848.9232	0.0012	12.3186	0.0812	11.6902	144.0065	55
60	101.2571	0.0099	1253.2133	0.0008	12.3766	0.0808	11.9015	147.3000	60
65	148.7798	0.0067	1847.2481	0.0005	12.4160	0.0805	12.0602	149.7387	65
70	218.6064	0.0046	2720.0801	0.0004	12.4428	0.0804	12.1783	151.5326	70
75	321.2045	0.0031	4002.5566	0.0002	12.4611	0.0802	12.2658	152.8448	75
80	471.9548	0.0021	5886.9354	0.0002	12.4735	0.0802	12.3301	153.8001	80
85	693.4565	0.0014	8655.7061	0.0001	12.4820	0.0801	12.3772	154.4925	85
90	1018.9151	0.0010	12723.9386	0.0001	12.4877	0.0801	12.4116	154.9925	90
95	1497.1205	0.0007	18701.5069	0.0001	12.4917	0.0801	12.4365	155.3524	95
100	2199.7613	0.0005	27484.5157	0.0000	12.4943	0.0800	12.4545	155.6107	100

Table C.15 Interest Rate Factors (9.0%)

	Single Payment		Equal Payment Series				Gradient Series		
N	Compound Amount Factor (F/P,i,N)	Present Worth Factor (P/F,i,N)	Compound Amount Factor (F/A,i,N)	Sinking Fund Factor (A/F,i,N)	Present Worth Factor (P/A,i,N)	Capital Recovery Factor (A/P,i,N)	Gradient Uniform Series (A/G,i,N)	Gradient Present Worth (P/G.i.N)	N
1	1.0900	0.9174	1.0000	1.0000	0.9174	1.0900	0.0000	0.0000	1
2	1.1881	0.8417	2.0900	0.4785	1.7591	0.5685	0.4785	0.8417	2
3	1.2950	0.7722	3.2781	0.3051	2.5313	0.3951	0.9426	2.3860	3
4	1.4116	0.7084	4.5731	0.2187	3.2397	0.3087	1.3925	4.5113	4
5	1.5386	0.6499	5.9847	0.1671	3.8897	0.2571	1.8282	7.1110	5
6	1.6771	0.5963	7.5233	0.1329	4.4859	0.2229	2.2498	10.0924	6
7	1.8280	0.5470	9.2004	0.1087	5.0330	0.1987	2.6574	13.3746	7
8	1.9926	0.5019	11.0285	0.0907	5.5348	0.1807	3.0512	16.8877	8
9	2.1719	0.4604	13.0210	0.0768	5.9952	0.1668	3.4312	20.5711	9
10	2.3674	0.4224	15.1929	0.0658	6.4177	0.1558	3.7978	24.3728	10
11	2.5804	0.3875	17.5603	0.0569	6.8052	0.1469	4.1510	28.2481	11
12	2.8127	0.3555	20.1407	0.0497	7.1607	0.1397	4.4910	32.1590	12
13	3.0658	0.3262	22.9534	0.0436	7.4869	0.1336	4.8182	36.0731	13
14	3.3417	0.2992	26.0192	0.0384	7.7862	0.1284	5.1326	39.9633	14
15	3.6425	0.2745	29.3609	0.0341	8.0607	0.1241	5.4346	43.8069	15
16	3.9703	0.2519	33.0034	0.0303	8.3126	0.1203	5.7245	47.5849	16
17	4.3276	0.2311	36.9737	0.0270	8.5436	0.1170	6.0024	51.2821	17
18	4.7171	0.2120	41.3013	0.0242	8.7556	0.1142	6.2687	54.8860	18
19	5.1417	0.1945	46.0185	0.0217	8.9501	0.1117	6.5236	58.3868	19
20	5.6044	0.1784	51.1601	0.0195	9.1285	0.1095	6.7674	61.7770	20
21	6.1088	0.1637	56.7645	0.0176	9.2922	0.1076	7.0006	65.0509	21
22	6.6586	0.1502	62.8733	0.0159	9.4424	0.1059	7.2232	68.2048	22
23	7.2579	0.1378	69.5319	0.0144	9.5802	0.1044	7.4357	71.2359	23
24	7.9111	0.1264	76.7898	0.0130	9.7066	0.1030	7.6384	74.1433	24
25	8.6231	0.1160	84.7009	0.0118	9.8226	0.1018	7.8316	76.9265	25
26	9.3992	0.1064	93.3240	0.0107	9.9290	0.1007	8.0156	79.5863	26
27	10.2451	0.0976	102.7231	0.0097	10.0266	0.0997	8.1906	82.1241	27
28	11.1671	0.0895	112.9682	0.0089	10.1161	0.0989	8.3571	84.5419	28
29	12.1722	0.0822	124.1354	0.0081	10.1983	0.0981	8.5154	86.8422	29
30	13.2677	0.0754	136.3075	0.0073	10.2737	0.0973	8.6657	89.0280	30
31	14.4618	0.0691	149.5752	0.0067	10.3428	0.0967	8.8083	91.1024	31
32	15.7633	0.0634	164.0370	0.0061	10.4062	0.0961	8.9436	93.0690	32
33	17.1820	0.0582	179.8003	0.0056	10.4644	0.0956	9.0718	94.9314	33
34	18.7284	0.0534	196.9823	0.0051	10.5178	0.0951	9.1933	96.6935	34
35	20.4140	0.0490	215.7108	0.0046	10.5668	0.0946	9.3083	98.3590	35
40	31.4094	0.0318	337.8824	0.0030	10.7574	0.0930	9.7957	105.3762	40
45	48.3273	0.0207	525.8587	0.0019	10.8812	0.0919	10.1603	110.5561	45
50	74.3575	0.0134	815.0836	0.0012	10.9617	0.0912	10.4295	114.3251	50
55	114.4083	0.0087	1260.0918	0.0008	11.0140	0.0908	10.6261	117.0362	55
60	176.0313	0.0057	1944.7921	0.0005	11.0480	0.0905	10.7683	118.9683	60
65	270.8460	0.0037	2998.2885	0.0003	11.0701	0.0903	10.8702	120.3344	65
70	416.7301	0.0024	4619.2232	0.0002	11.0844	0.0902	10.9427	121.2942	70
75	641.1909	0.0016	7113.2321	0.0001	11.0938	0.0901	10.9940	121.9646	75
80	986.5517	0.0010	10950.5741	0.0001	11.0998	0.0901	11.0299	122.4306	80
85	1517.9320	0.0007	16854.8003	0.0001	11.1038	0.0901	11.0551	122.7533	85
90	2335.5266	0.0004	25939.1842	0.0000	11.1064	0.0900	11.0726	122.9758	90
95	3593.4971	0.0003	39916.6350	0.0000	11.1080	0.0900	11.0847	123.1287	95
100	5529.0408	0.0002	61422.6755	0.0000	11.1091	0.0900	11.0930	123.2335	100

Table C.16 Interest Rate Factors (10.0%)

	Single Payment		Equal Payment Series				Gradient Series		
N	Compound Amount Factor (F/P,i,N)	Present Worth Factor (P/F,i,N)	Compound Amount Factor (F/A,i,N)	Sinking Fund Factor (A/F,i,N)	Present Worth Factor (P/A,i,N)	Capital Recovery Factor (A/P,i,N)	Gradient Uniform Series (A/G,i,N)	Gradient Present Worth (P/G.i,N)	N
1	1.1000	0.9091	1.0000	1.0000	0.9091	1.1000	0.0000	0.0000	1
2	1.2100	0.8264	2.1000	0.4762	1.7355	0.5762	0.4762	0.8264	2
3	1.3310	0.7513	3.3100	0.3021	2.4869	0.4021	0.9366	2.3291	3
4	1.4641	0.6830	4.6410	0.2155	3.1699	0.3155	1.3812	4.3781	4
5	1.6105	0.6209	6.1051	0.1638	3.7908	0.2638	1.8101	6.8618	5
6	1.7716	0.5645	7.7156	0.1296	4.3553	0.2296	2.2236	9.6842	6
7	1.9487	0.5132	9.4872	0.1054	4.8684	0.2054	2.6216	12.7631	7
8	2.1436	0.4665	11.4359	0.0874	5.3349	0.1874	3.0045	16.0287	8
9	2.3579	0.4241	13.5795	0.0736	5.7590	0.1736	3.3724	19.4215	9
10	2.5937	0.3855	15.9374	0.0627	6.1446	0.1627	3.7255	22.8913	10
11	2.8531	0.3505	18.5312	0.0540	6.4951	0.1540	4.0641	26.3963	11
12	3.1384	0.3186	21.3843	0.0468	6.8137	0.1468	4.3884	29.9012	12
13	3.4523	0.2897	24.5227	0.0408	7.1034	0.1408	4.6988	33.3772	13
14	3.7975	0.2633	27.9750	0.0357	7.3667	0.1357	4.9955	36.8005	14
15	4.1772	0.2394	31.7725	0.0315	7.6061	0.1315	5.2789	40.1520	15
16	4.5950	0.2176	35.9497	0.0278	7.8237	0.1278	5.5493	43.4164	16
17	5.0545	0.1978	40.5447	0.0247	8.0216	0.1247	5.8071	46.5819	17
18	5.5599	0.1799	45.5992	0.0219	8.2014	0.1219	6.0526	49.6395	18
19	6.1159	0.1635	51.1591	0.0195	8.3649	0.1195	6.2861	52.5827	19
20	6.7275	0.1486	57.2750	0.0175	8.5136	0.1175	6.5081	55.4069	20
21	7.4002	0.1351	64.0025	0.0156	8.6487	0.1156	6.7189	58.1095	21
22	8.1403	0.1228	71.4027	0.0140	8.7715	0.1140	6.9189	60.6893	22
23	8.9543	0.1117	79.5430	0.0126	8.8832	0.1126	7.1085	63.1462	23
24	9.8497	0.1015	88.4973	0.0113	8.9847	0.1113	7.2881	65.4813	24
25	10.8347	0.0923	98.3471	0.0102	9.0770	0.1102	7.4580	67.6964	25
26	11.9182	0.0839	109.1818	0.0092	9.1609	0.1092	7.6186	69.7940	26
27	13.1100	0.0763	121.0999	0.0083	9.2372	0.1083	7.7704	71.7773	27
28	14.4210	0.0693	134.2099	0.0075	9.3066	0.1075	7.9137	73.6495	28
29	15.8631	0.0630	148.6309	0.0067	9.3696	0.1067	8.0489	75.4146	29
30	17.4494	0.0573	164.4940	0.0061	9.4269	0.1061	8.1762	77.0766	30
31	19.1943	0.0521	181.9434	0.0055	9.4790	0.1055	8.2962	78.6395	31
32	21.1138	0.0474	201.1378	0.0050	9.5264	0.1050	8.4091	80.1078	32
33	23.2252	0.0431	222.2515	0.0045	9.5694	0.1045	8.5152	81.4856	33
34	25.5477	0.0391	245.4767	0.0041	9.6086	0.1041	8.6149	82.7773	34
35	28.1024	0.0356	271.0244	0.0037	9.6442	0.1037	8.7086	83.9872	35
40	45.2593	0.0221	442.5926	0.0023	9.7791	0.1023	9.0962	88.9525	40
45	72.8905	0.0137	718.9048	0.0014	9.8628	0.1014	9.3740	92.4544	45
50	117.3909	0.0085	1163.9085	0.0009	9.9148	0.1009	9.5704	94.8889	50
55	189.0591	0.0053	1880.5914	0.0005	9.9471	0.1005	9.7075	96.5619	55
60	304.4816	0.0033	3034.8164	0.0003	9.9672	0.1003	9.8023	97.7010	60
65	490.3707	0.0020	4893.7073	0.0002	9.9796	0.1002	9.8672	98.4705	65
70	789.7470	0.0013	7887.4696	0.0001	9.9873	0.1001	9.9113	98.9870	70
75	1271.8954	0.0008	12708.9537	0.0001	9.9921	0.1001	9.9410	99.3317	75
80	2048.4002	0.0005	20474.0021	0.0000	9.9951	0.1000	9.9609	99.5606	80
85	3298.9690	0.0003	32979.6903	0.0000	9.9970	0.1000	9.9742	99.7120	85
90	5313.0226	0.0002	53120.2261	0.0000	9.9981	0.1000	9.9831	99.8118	90
95	8556.6760	0.0001	85556.7605	0.0000	9.9988	0.1000	9.9889	99.8773	95
100	13780.6123	0.0001	137796.1234	0.0000	9.9993	0.1000	9.9927	99.9202	100

Table C.17 Interest Rate Factors (11.0%)

	Single Payment		Equal Payment Series				Gradient Series		
N	Compound Amount Factor (F/P,i,N)	Present Worth Factor (P/F,i,N)	Compound Amount Factor (F/A,i,N)	Sinking Fund Factor (A/F,i,N)	Present Worth Factor (P/A,i,N)	Capital Recovery Factor (A/P,i,N)	Gradient Uniform Series (A/G,i,N)	Gradient Present Worth (P/G.i.N)	N
1	1.1100	0.9009	1.0000	1.0000	0.9009	1.1100	0.0000	0.0000	1
2	1.2321	0.8116	2.1100	0.4739	1.7125	0.5839	0.4739	0.8116	2
3	1.3676	0.7312	3.3421	0.2992	2.4437	0.4092	0.9306	2.2740	3
4	1.5181	0.6587	4.7097	0.2123	3.1024	0.3223	1.3700	4.2502	4
5	1.6851	0.5935	6.2278	0.1606	3.6959	0.2706	1.7923	6.6240	5
6	1.8704	0.5346	7.9129	0.1264	4.2305	0.2364	2.1976	9.2972	6
7	2.0762	0.4817	9.7833	0.1022	4.7122	0.2122	2.5863	12.1872	7
8	2.3045	0.4339	11.8594	0.0843	5.1461	0.1943	2.9585	15.2246	8
9	2.5580	0.3909	14.1640	0.0706	5.5370	0.1806	3.3144	18.3520	9
10	2.8394	0.3522	16.7220	0.0598	5.8892	0.1698	3.6544	21.5217	10
11	3.1518	0.3173	19.5614	0.0511	6.2065	0.1611	3.9788	24.6945	11
12	3.4985	0.2858	22.7132	0.0440	6.4924	0.1540	4.2879	27.8388	12
13	3.8833	0.2575	26.2116	0.0382	6.7499	0.1482	4.5822	30.9290	13
14	4.3104	0.2320	30.0949	0.0332	6.9819	0.1432	4.8619	33.9449	14
15	4.7846	0.2090	34.4054	0.0291	7.1909	0.1391	5.1275	36.8709	15
16	5.3109	0.1883	39.1899	0.0255	7.3792	0.1355	5.3794	39.6953	16
17	5.8951	0.1696	44.5008	0.0225	7.5488	0.1325	5.6180	42.4095	17
18	6.5436	0.1528	50.3959	0.0198	7.7016	0.1298	5.8439	45.0074	18
19	7.2633	0.1377	56.9395	0.0176	7.8393	0.1276	6.0574	47.4856	19
20	8.0623	0.1240	64.2028	0.0156	7.9633	0.1256	6.2590	49.8423	20
21	8.9492	0.1117	72.2651	0.0138	8.0751	0.1238	6.4491	52.0771	21
22	9.9336	0.1007	81.2143	0.0123	8.1757	0.1223	6.6283	54.1912	22
23	11.0263	0.0907	91.1479	0.0110	8.2664	0.1210	6.7969	56.1864	23
24	12.2392	0.0817	102.1742	0.0098	8.3481	0.1198	6.9555	58.0656	24
25	13.5855	0.0736	114.4133	0.0087	8.4217	0.1187	7.1045	59.8322	25
26	15.0799	0.0663	127.9988	0.0078	8.4881	0.1178	7.2443	61.4900	26
27	16.7386	0.0597	143.0786	0.0070	8.5478	0.1170	7.3754	63.0433	27
28	18.5799	0.0538	159.8173	0.0063	8.6016	0.1163	7.4982	64.4965	28
29	20.6237	0.0485	178.3972	0.0056	8.6501	0.1156	7.6131	65.8542	29
30	22.8923	0.0437	199.0209	0.0050	8.6938	0.1150	7.7206	67.1210	30
31	25.4104	0.0394	221.9132	0.0045	8.7331	0.1145	7.8210	68.3016	31
32	28.2056	0.0355	247.3236	0.0040	8.7686	0.1140	7.9147	69.4007	32
33	31.3082	0.0319	275.5292	0.0036	8.8005	0.1136	8.0021	70.4228	33
34	34.7521	0.0288	306.8374	0.0033	8.8293	0.1133	8.0836	71.3724	34
35	38.5749	0.0259	341.5896	0.0029	8.8552	0.1129	8.1594	72.2538	35
40	65.0009	0.0154	581.8261	0.0017	8.9511	0.1117	8.4659	75.7789	40
45	109.5302	0.0091	986.6386	0.0010	9.0079	0.1110	8.6763	78.1551	45
50	184.5648	0.0054	1668.7712	0.0006	9.0417	0.1106	8.8185	79.7341	50
55	311.0025	0.0032	2818.2042	0.0004	9.0617	0.1104	8.9135	80.7712	55
60	524.0572	0.0019	4755.0658	0.0002	9.0736	0.1102	8.9762	81.4461	60

Table C.18 Interest Rate Factors (12.0%)

	Single Payment		Equal Payment Series				Gradient Series		
N	Compound Amount Factor (F/P,i,N)	Present Worth Factor (P/F,i,N)	Compound Amount Factor (F/A,i,N)	Sinking Fund Factor (A/F,i,N)	Present Worth Factor (P/A,i,N)	Capital Recovery Factor (A/P,i,N)	Gradient Uniform Series (A/G,i,N)	Gradient Present Worth (P/G.i.N)	N
1	1.1200	0.8929	1.0000	1.0000	0.8929	1.1200	0.0000	0.0000	1
2	1.2544	0.7972	2.1200	0.4717	1.6901	0.5917	0.4717	0.7972	2
3	1.4049	0.7118	3.3744	0.2963	2.4018	0.4163	0.9246	2.2208	3
4	1.5735	0.6355	4.7793	0.2092	3.0373	0.3292	1.3589	4.1273	4
5	1.7623	0.5674	6.3528	0.1574	3.6048	0.2774	1.7746	6.3970	5
6	1.9738	0.5066	8.1152	0.1232	4.1114	0.2432	2.1720	8.9302	6
7	2.2107	0.4523	10.0890	0.0991	4.5638	0.2191	2.5515	11.6443	7
8	2.4760	0.4039	12.2997	0.0813	4.9676	0.2013	2.9131	14.4714	8
9	2.7731	0.3606	14.7757	0.0677	5.3282	0.1877	3.2574	17.3563	9
10	3.1058	0.3220	17.5487	0.0570	5.6502	0.1770	3.5847	20.2541	10
11	3.4785	0.2875	20.6546	0.0484	5.9377	0.1684	3.8953	23.1288	11
12	3.8960	0.2567	24.1331	0.0414	6.1944	0.1614	4.1897	25.9523	12
13	4.3635	0.2292	28.0291	0.0357	6.4235	0.1557	4.4683	28.7024	13
14	4.8871	0.2046	32.3926	0.0309	6.6282	0.1509	4.7317	31.3624	14
15	5.4736	0.1827	37.2797	0.0268	6.8109	0.1468	4.9803	33.9202	15
16	6.1304	0.1631	42.7533	0.0234	6.9740	0.1434	5.2147	36.3670	16
17	6.8660	0.1456	48.8837	0.0205	7.1196	0.1405	5.4353	38.6973	17
18	7.6900	0.1300	55.7497	0.0179	7.2497	0.1379	5.6427	40.9080	18
19	8.6128	0.1161	63.4397	0.0158	7.3658	0.1358	5.8375	42.9979	19
20	9.6463	0.1037	72.0524	0.0139	7.4694	0.1339	6.0202	44.9676	20
21	10.8038	0.0926	81.6987	0.0122	7.5620	0.1322	6.1913	46.8188	21
22	12.1003	0.0826	92.5026	0.0108	7.6446	0.1308	6.3514	48.5543	22
23	13.5523	0.0738	104.6029	0.0096	7.7184	0.1296	6.5010	50.1776	23
24	15.1786	0.0659	118.1552	0.0085	7.7843	0.1285	6.6406	51.6929	24
25	17.0001	0.0588	133.3339	0.0075	7.8431	0.1275	6.7708	53.1046	25
26	19.0401	0.0525	150.3339	0.0067	7.8957	0.1267	6.8921	54.4177	26
27	21.3249	0.0469	169.3740	0.0059	7.9426	0.1259	7.0049	55.6369	27
28	23.8839	0.0419	190.6989	0.0052	7.9844	0.1252	7.1098	56.7674	28
29	26.7499	0.0374	214.5828	0.0047	8.0218	0.1247	7.2071	57.8141	29
30	29.9599	0.0334	241.3327	0.0041	8.0552	0.1241	7.2974	58.7821	30
31	33.5551	0.0298	271.2926	0.0037	8.0850	0.1237	7.3811	59.6761	31
32	37.5817	0.0266	304.8477	0.0033	8.1116	0.1233	7.4586	60.5010	32
33	42.0915	0.0238	342.4294	0.0029	8.1354	0.1229	7.5302	61.2612	33
34	47.1425	0.0212	384.5210	0.0026	8.1566	0.1226	7.5965	61.9612	34
35	52.7996	0.0189	431.6635	0.0023	8.1755	0.1223	7.6577	62.6052	35
40	93.0510	0.0107	767.0914	0.0013	8.2438	0.1213	7.8988	65.1159	40
45	163.9876	0.0061	1358.2300	0.0007	8.2825	0.1207	8.0572	66.7342	45
50	289.0022	0.0035	2400.0182	0.0004	8.3045	0.1204	8.1597	67.7624	50
55	509.3206	0.0020	4236.0050	0.0002	8.3170	0.1202	8.2251	68.4082	55
60	897.5969	0.0011	7471.6411	0.0001	8.3240	0.1201	8.2664	68.8100	60

Table C.19 Interest Rate Factors (13.0%)

	Single Payment		Equal Payment Series				Gradient Series		
N	Compound Amount Factor (F/P,i,N)	Present Worth Factor (P/F,i,N)	Compound Amount Factor (F/A,i,N)	Sinking Fund Factor (A/F,i,N)	Present Worth Factor (P/A,i,N)	Capital Recovery Factor (A/P,i,N)	Gradient Uniform Series (A/G,i,N)	Gradient Present Worth (P/G.i.N)	N
1	1.1300	0.8850	1.0000	1.0000	0.8850	1.1300	0.0000	0.0000	1
2	1.2769	0.7831	2.1300	0.4695	1.6681	0.5995	0.4695	0.7831	2
3	1.4429	0.6931	3.4069	0.2935	2.3612	0.4235	0.9187	2.1692	3
4	1.6305	0.6133	4.8498	0.2062	2.9745	0.3362	1.3479	4.0092	4
5	1.8424	0.5428	6.4803	0.1543	3.5172	0.2843	1.7571	6.1802	5
6	2.0820	0.4803	8.3227	0.1202	3.9975	0.2502	2.1468	8.5818	6
7	2.3526	0.4251	10.4047	0.0961	4.4226	0.2261	2.5171	11.1322	7
8	2.6584	0.3762	12.7573	0.0784	4.7988	0.2084	2.8685	13.7653	8
9	3.0040	0.3329	15.4157	0.0649	5.1317	0.1949	3.2014	16.4284	9
10	3.3946	0.2946	18.4197	0.0543	5.4262	0.1843	3.5162	19.0797	10
11	3.8359	0.2607	21.8143	0.0458	5.6869	0.1758	3.8134	21.6867	11
12	4.3345	0.2307	25.6502	0.0390	5.9176	0.1690	4.0936	24.2244	12
13	4.8980	0.2042	29.9847	0.0334	6.1218	0.1634	4.3573	26.6744	13
14	5.5348	0.1807	34.8827	0.0287	6.3025	0.1587	4.6050	29.0232	14
15	6.2543	0.1599	40.4175	0.0247	6.4624	0.1547	4.8375	31.2617	15
16	7.0673	0.1415	46.6717	0.0214	6.6039	0.1514	5.0552	33.3841	16
17	7.9861	0.1252	53.7391	0.0186	6.7291	0.1486	5.2589	35.3876	17
18	9.0243	0.1108	61.7251	0.0162	6.8399	0.1462	5.4491	37.2714	18
19	10.1974	0.0981	70.7494	0.0141	6.9380	0.1441	5.6265	39.0366	19
20	11.5231	0.0868	80.9468	0.0124	7.0248	0.1424	5.7917	40.6854	20
21	13.0211	0.0768	92.4699	0.0108	7.1016	0.1408	5.9454	42.2214	21
22	14.7138	0.0680	105.4910	0.0095	7.1695	0.1395	6.0881	43.6486	22
23	16.6266	0.0601	120.2048	0.0083	7.2297	0.1383	6.2205	44.9718	23
24	18.7881	0.0532	136.8315	0.0073	7.2829	0.1373	6.3431	46.1960	24
25	21.2305	0.0471	155.6196	0.0064	7.3300	0.1364	6.4566	47.3264	25
26	23.9905	0.0417	176.8501	0.0057	7.3717	0.1357	6.5614	48.3685	26
27	27.1093	0.0369	200.8406	0.0050	7.4086	0.1350	6.6582	49.3276	27
28	30.6335	0.0326	227.9499	0.0044	7.4412	0.1344	6.7474	50.2090	28
29	34.6158	0.0289	258.5834	0.0039	7.4701	0.1339	6.8296	51.0179	29
30	39.1159	0.0256	293.1992	0.0034	7.4957	0.1334	6.9052	51.7592	30
31	44.2010	0.0226	332.3151	0.0030	7.5183	0.1330	6.9747	52.4380	31
32	49.9471	0.0200	376.5161	0.0027	7.5383	0.1327	7.0385	53.0586	32
33	56.4402	0.0177	426.4632	0.0023	7.5560	0.1323	7.0971	53.6256	33
34	63.7774	0.0157	482.9034	0.0021	7.5717	0.1321	7.1507	54.1430	34
35	72.0685	0.0139	546.6808	0.0018	7.5856	0.1318	7.1998	54.6148	35
40	132.7816	0.0075	1013.7042	0.0010	7.6344	0.1310	7.3888	56.4087	40
45	244.6414	0.0041	1874.1646	0.0005	7.6609	0.1305	7.5076	57.5148	45
50	450.7359	0.0022	3459.5071	0.0003	7.6752	0.1303	7.5811	58.1870	50
55	830.4517	0.0012	6380.3979	0.0002	7.6830	0.1302	7.6260	58.5909	55
60	1530.0535	0.0007	11761.9498	0.0001	7.6873	0.1301	7.6531	58.8313	60

Table C.20 Interest Rate Factors (14.0%)

	Single Payment		Equal Payment Series				Gradient Series		
N	Compound Amount Factor (F/P,i,N)	Present Worth Factor (P/F,i,N)	Compound Amount Factor (F/A,i,N)	Sinking Fund Factor (A/F,i,N)	Present Worth Factor (P/A,i,N)	Capital Recovery Factor (A/P,i,N)	Gradient Uniform Series (A/G,i,N)	Gradient Present Worth (P/G.i.N)	N
1	1.1400	0.8772	1.0000	1.0000	0.8772	1.1400	0.0000	-0.0000	1
2	1.2996	0.7695	2.1400	0.4673	1.6467	0.6073	0.4673	0.7695	2
3	1.4815	0.6750	3.4396	0.2907	2.3216	0.4307	0.9129	2.1194	3
4	1.6890	0.5921	4.9211	0.2032	2.9137	0.3432	1.3370	3.8957	4
5	1.9254	0.5194	6.6101	0.1513	3.4331	0.2913	1.7399	5.9731	5
6	2.1950	0.4556	8.5355	0.1172	3.8887	0.2572	2.1218	8.2511	6
7	2.5023	0.3996	10.7305	0.0932	4.2883	0.2332	2.4832	10.6489	7
8	2.8526	0.3506	13.2328	0.0756	4.6389	0.2156	2.8246	13.1028	8
9	3.2519	0.3075	16.0853	0.0622	4.9464	0.2022	3.1463	15.5629	9
10	3.7072	0.2697	19.3373	0.0517	5.2161	0.1917	3.4490	17.9906	10
11	4.2262	0.2366	23.0445	0.0434	5.4527	0.1834	3.7333	20.3567	11
12	4.8179	0.2076	27.2707	0.0367	5.6603	0.1767	3.9998	22.6399	12
13	5.4924	0.1821	32.0887	0.0312	5.8424	0.1712	4.2491	24.8247	13
14	6.2613	0.1597	37.5811	0.0266	6.0021	0.1666	4.4819	26.9009	14
15	7.1379	0.1401	43.8424	0.0228	6.1422	0.1628	4.6990	28.8623	15
16	8.1372	0.1229	50.9804	0.0196	6.2651	0.1596	4.9011	30.7057	16
17	9.2765	0.1078	59.1176	0.0169	6.3729	0.1569	5.0888	32.4305	17
18	10.5752	0.0946	68.3941	0.0146	6.4674	0.1546	5.2630	34.0380	18
19	12.0557	0.0829	78.9692	0.0127	6.5504	0.1527	5.4243	35.5311	19
20	13.7435	0.0728	91.0249	0.0110	6.6231	0.1510	5.5734	36.9135	20
21	15.6676	0.0638	104.7684	0.0095	6.6870	0.1495	5.7111	38.1901	21
22	17.8610	0.0560	120.4360	0.0083	6.7429	0.1483	5.8381	39.3658	22
23	20.3616	0.0491	138.2970	0.0072	6.7921	0.1472	5.9549	40.4463	23
24	23.2122	0.0431	158.6586	0.0063	6.8351	0.1463	6.0624	41.4371	24
25	26.4619	0.0378	181.8708	0.0055	6.8729	0.1455	6.1610	42.3441	25
26	30.1666	0.0331	208.3327	0.0048	6.9061	0.1448	6.2514	43.1728	26
27	34.3899	0.0291	238.4993	0.0042	6.9352	0.1442	6.3342	43.9289	27
28	39.2045	0.0255	272.8892	0.0037	6.9607	0.1437	6.4100	44.6176	28
29	44.6931	0.0224	312.0937	0.0032	6.9830	0.1432	6.4791	45.2441	29
30	50.9502	0.0196	356.7868	0.0028	7.0027	0.1428	6.5423	45.8132	30
31	58.0832	0.0172	407.7370	0.0025	7.0199	0.1425	6.5998	46.3297	31
32	66.2148	0.0151	465.8202	0.0021	7.0350	0.1421	6.6522	46.7979	32
33	75.4849	0.0132	532.0350	0.0019	7.0482	0.1419	6.6998	47.2218	33
34	86.0528	0.0116	607.5199	0.0016	7.0599	0.1416	6.7431	47.6053	34
35	98.1002	0.0102	693.5727	0.0014	7.0700	0.1414	6.7824	47.9519	35
40	188.8835	0.0053	1342.0251	0.0007	7.1050	0.1407	6.9300	49.2376	40
45	363.6791	0.0027	2590.5648	0.0004	7.1232	0.1404	7.0188	49.9963	45
50	700.2330	0.0014	4994.5213	0.0002	7.1327	0.1402	7.0714	50.4375	50

Table C.21 Interest Rate Factors (15.0%)

	Single Payment		Equal Payment Series				Gradient Series		
N	Compound Amount Factor (F/P,i,N)	Present Worth Factor (P/F,i,N)	Compound Amount Factor (F/A,i,N)	Sinking Fund Factor (A/F,i,N)	Present Worth Factor (P/A,i,N)	Capital Recovery Factor (A/P,i,N)	Gradient Uniform Series (A/G,i,N)	Gradient Present Worth (P/G.i.N)	N
1	1.1500	0.8696	1.0000	1.0000	0.8696	1.1500	-0.0000	-0.0000	1
2	1.3225	0.7561	2.1500	0.4651	1.6257	0.6151	0.4651	0.7561	2
3	1.5209	0.6575	3.4725	0.2880	2.2832	0.4380	0.9071	2.0712	3
4	1.7490	0.5718	4.9934	0.2003	2.8550	0.3503	1.3263	3.7864	4
5	2.0114	0.4972	6.7424	0.1483	3.3522	0.2983	1.7228	5.7751	5
6	2.3131	0.4323	8.7537	0.1142	3.7845	0.2642	2.0972	7.9368	6
7	2.6600	0.3759	11.0668	0.0904	4.1604	0.2404	2.4498	10.1924	7
8	3.0590	0.3269	13.7268	0.0729	4.4873	0.2229	2.7813	12.4807	8
9	3.5179	0.2843	16.7858	0.0596	4.7716	0.2096	3.0922	14.7548	9
10	4.0456	0.2472	20.3037	0.0493	5.0188	0.1993	3.3832	16.9795	10
11	4.6524	0.2149	24.3493	0.0411	5.2337	0.1911	3.6549	19.1289	11
12	5.3503	0.1869	29.0017	0.0345	5.4206	0.1845	3.9082	21.1849	12
13	6.1528	0.1625	34.3519	0.0291	5.5831	0.1791	4.1438	23.1352	13
14	7.0757	0.1413	40.5047	0.0247	5.7245	0.1747	4.3624	24.9725	14
15	8.1371	0.1229	47.5804	0.0210	5.8474	0.1710	4.5650	26.6930	15
16	9.3576	0.1069	55.7175	0.0179	5.9542	0.1679	4.7522	28.2960	16
17	10.7613	0.0929	65.0751	0.0154	6.0472	0.1654	4.9251	29.7828	17
18	12.3755	0.0808	75.8364	0.0132	6.1280	0.1632	5.0843	31.1565	18
19	14.2318	0.0703	88.2118	0.0113	6.1982	0.1613	5.2307	32.4213	19
20	16.3665	0.0611	102.4436	0.0098	6.2593	0.1598	5.3651	33.5822	20
21	18.8215	0.0531	118.8101	0.0084	6.3125	0.1584	5.4883	34.6448	21
22	21.6447	0.0462	137.6316	0.0073	6.3587	0.1573	5.6010	35.6150	22
23	24.8915	0.0402	159.2764	0.0063	6.3988	0.1563	5.7040	36.4988	23
24	28.6252	0.0349	184.1678	0.0054	6.4338	0.1554	5.7979	37.3023	24
25	32.9190	0.0304	212.7930	0.0047	6.4641	0.1547	5.8834	38.0314	25
26	37.8568	0.0264	245.7120	0.0041	6.4906	0.1541	5.9612	38.6918	26
27	43.5353	0.0230	283.5688	0.0035	6.5135	0.1535	6.0319	39.2890	27
28	50.0656	0.0200	327.1041	0.0031	6.5335	0.1531	6.0960	39.8283	28
29	57.5755	0.0174	377.1697	0.0027	6.5509	0.1527	6.1541	40.3146	29
30	66.2118	0.0151	434.7451	0.0023	6.5660	0.1523	6.2066	40.7526	30
31	76.1435	0.0131	500.9569	0.0020	6.5791	0.1520	6.2541	41.1466	31
32	87.5651	0.0114	577.1005	0.0017	6.5905	0.1517	6.2970	41.5006	32
33	100.6998	0.0099	664.6655	0.0015	6.6005	0.1515	6.3357	41.8184	33
34	115.8048	0.0086	765.3654	0.0013	6.6091	0.1513	6.3705	42.1033	34
35	133.1755	0.0075	881.1702	0.0011	6.6166	0.1511	6.4019	42.3586	35
40	267.8635	0.0037	1779.0903	0.0006	6.6418	0.1506	6.5168	43.2830	40
45	538.7693	0.0019	3585.1285	0.0003	6.6543	0.1503	6.5830	43.8051	45
50	1083.6574	0.0009	7217.7163	0.0001	6.6605	0.1501	6.6205	44.0958	50

Table C.22 Interest Rate Factors (16.0%)

	Single Payment		Equal Payment Series				Gradient Series		
N	Compound Amount Factor (F/P,i,N)	Present Worth Factor (P/F,i,N)	Compound Amount Factor (F/A,i,N)	Sinking Fund Factor (A/F,i,N)	Present Worth Factor (P/A,i,N)	Capital Recovery Factor (A/P,i,N)	Gradient Uniform Series (A/G,i,N)	Gradient Present Worth (P/G.i.N)	N
1	1.1600	0.8621	1.0000	1.0000	0.8621	1.1600	0.0000	0.0000	1
2	1.3456	0.7432	2.1600	0.4630	1.6052	0.6230	0.4630	0.7432	2
3	1.5609	0.6407	3.5056	0.2853	2.2459	0.4453	0.9014	2.0245	3
4	1.8106	0.5523	5.0665	0.1974	2.7982	0.3574	1.3156	3.6814	4
5	2.1003	0.4761	6.8771	0.1454	3.2743	0.3054	1.7060	5.5858	5
6	2.4364	0.4104	8.9775	0.1114	3.6847	0.2714	2.0729	7.6380	6
7	2.8262	0.3538	11.4139	0.0876	4.0386	0.2476	2.4169	9.7610	7
8	3.2784	0.3050	14.2401	0.0702	4.3436	0.2302	2.7388	11.8962	8
9	3.8030	0.2630	17.5185	0.0571	4.6065	0.2171	3.0391	13.9998	9
10	4.4114	0.2267	21.3215	0.0469	4.8332	0.2069	3.3187	16.0399	10
11	5.1173	0.1954	25.7329	0.0389	5.0286	0.1989	3.5783	17.9941	11
12	5.9360	0.1685	30.8502	0.0324	5.1971	0.1924	3.8189	19.8472	12
13	6.8858	0.1452	36.7862	0.0272	5.3423	0.1872	4.0413	21.5899	13
14	7.9875	0.1252	43.6720	0.0229	5.4675	0.1829	4.2464	23.2175	14
15	9.2655	0.1079	51.6595	0.0194	5.5755	0.1794	4.4352	24.7284	15
16	10.7480	0.0930	60.9250	0.0164	5.6685	0.1764	4.6086	26.1241	16
17	12.4677	0.0802	71.6730	0.0140	5.7487	0.1740	4.7676	27.4074	17
18	14.4625	0.0691	84.1407	0.0119	5.8178	0.1719	4.9130	28.5828	18
19	16.7765	0.0596	98.6032	0.0101	5.8775	0.1701	5.0457	29.6557	19
20	19.4608	0.0514	115.3797	0.0087	5.9288	0.1687	5.1666	30.6321	20
21	22.5745	0.0443	134.8405	0.0074	5.9731	0.1674	5.2766	31.5180	21
22	26.1864	0.0382	157.4150	0.0064	6.0113	0.1664	5.3765	32.3200	22
23	30.3762	0.0329	183.6014	0.0054	6.0442	0.1654	5.4671	33.0442	23
24	35.2364	0.0284	213.9776	0.0047	6.0726	0.1647	5.5490	33.6970	24
25	40.8742	0.0245	249.2140	0.0040	6.0971	0.1640	5.6230	34.2841	25
26	47.4141	0.0211	290.0883	0.0034	6.1182	0.1634	5.6898	34.8114	26
27	55.0004	0.0182	337.5024	0.0030	6.1364	0.1630	5.7500	35.2841	27
28	63.8004	0.0157	392.5028	0.0025	6.1520	0.1625	5.8041	35.7073	28
29	74.0085	0.0135	456.3032	0.0022	6.1656	0.1622	5.8528	36.0856	29
30	85.8499	0.0116	530.3117	0.0019	6.1772	0.1619	5.8964	36.4234	30
31	99.5859	0.0100	616.1616	0.0016	6.1872	0.1616	5.9356	36.7247	31
32	115.5196	0.0087	715.7475	0.0014	6.1959	0.1614	5.9706	36.9930	32
33	134.0027	0.0075	831.2671	0.0012	6.2034	0.1612	6.0019	37.2318	33
34	155.4432	0.0064	965.2698	0.0010	6.2098	0.1610	6.0299	37.4441	34
35	180.3141	0.0055	1120.7130	0.0009	6.2153	0.1609	6.0548	37.6327	35
40	378.7212	0.0026	2360.7572	0.0004	6.2335	0.1604	6.1441	38.2992	40
45	795.4438	0.0013	4965.2739	0.0002	6.2421	0.1602	6.1934	38.6598	45
50	1670.7038	0.0006	10435.6488	0.0001	6.2463	0.1601	6.2201	38.8521	50

Table C.23 Interest Rate Factors (18.0%)

	Single Payment		Equal Payment Series				Gradient Series		
N	Compound Amount Factor (F/P,i,N)	Present Worth Factor (P/F,i,N)	Compound Amount Factor (F/A,i,N)	Sinking Fund Factor (A/F,i,N)	Present Worth Factor (P/A,i,N)	Capital Recovery Factor (A/P,i,N)	Gradient Uniform Series (A/G,i,N)	Gradient Present Worth (P/G.i.N)	N
1	1.1800	0.8475	1.0000	1.0000	0.8475	1.1800	0.0000	0.0000	1
2	1.3924	0.7182	2.1800	0.4587	1.5656	0.6387	0.4587	0.7182	2
3	1.6430	0.6086	3.5724	0.2799	2.1743	0.4599	0.8902	1.9354	3
4	1.9388	0.5158	5.2154	0.1917	2.6901	0.3717	1.2947	3.4828	4
5	2.2878	0.4371	7.1542	0.1398	3.1272	0.3198	1.6728	5.2312	5
6	2.6996	0.3704	9.4420	0.1059	3.4976	0.2859	2.0252	7.0834	6
7	3.1855	0.3139	12.1415	0.0824	3.8115	0.2624	2.3526	8.9670	7
8	3.7589	0.2660	15.3270	0.0652	4.0776	0.2452	2.6558	10.8292	8
9	4.4355	0.2255	19.0859	0.0524	4.3030	0.2324	2.9358	12.6329	9
10	5.2338	0.1911	23.5213	0.0425	4.4941	0.2225	3.1936	14.3525	10
11	6.1759	0.1619	28.7551	0.0348	4.6560	0.2148	3.4303	15.9716	11
12	7.2876	0.1372	34.9311	0.0286	4.7932	0.2086	3.6470	17.4811	12
13	8.5994	0.1163	42.2187	0.0237	4.9095	0.2037	3.8449	18.8765	13
14	10.1472	0.0985	50.8180	0.0197	5.0081	0.1997	4.0250	20.1576	14
15	11.9737	0.0835	60.9653	0.0164	5.0916	0.1964	4.1887	21.3269	15
16	14.1290	0.0708	72.9390	0.0137	5.1624	0.1937	4.3369	22.3885	16
17	16.6722	0.0600	87.0680	0.0115	5.2223	0.1915	4.4708	23.3482	17
18	19.6733	0.0508	103.7403	0.0096	5.2732	0.1896	4.5916	24.2123	18
19	23.2144	0.0431	123.4135	0.0081	5.3162	0.1881	4.7003	24.9877	19
20	27.3930	0.0365	146.6280	0.0068	5.3527	0.1868	4.7978	25.6813	20
21	32.3238	0.0309	174.0210	0.0057	5.3837	0.1857	4.8851	26.3000	21
22	38.1421	0.0262	206.3448	0.0048	5.4099	0.1848	4.9632	26.8506	22
23	45.0076	0.0222	244.4868	0.0041	5.4321	0.1841	5.0329	27.3394	23
24	53.1090	0.0188	289.4945	0.0035	5.4509	0.1835	5.0950	27.7725	24
25	62.6686	0.0160	342.6035	0.0029	5.4669	0.1829	5.1502	28.1555	25
26	73.9490	0.0135	405.2721	0.0025	5.4804	0.1825	5.1991	28.4935	26
27	87.2598	0.0115	479.2211	0.0021	5.4919	0.1821	5.2425	28.7915	27
28	102.9666	0.0097	566.4809	0.0018	5.5016	0.1818	5.2810	29.0537	28
29	121.5005	0.0082	669.4475	0.0015	5.5098	0.1815	5.3149	29.2842	29
30	143.3706	0.0070	790.9480	0.0013	5.5168	0.1813	5.3448	29.4864	30
31	169.1774	0.0059	934.3186	0.0011	5.5227	0.1811	5.3712	29.6638	31
32	199.6293	0.0050	1103.4960	0.0009˙	5.5277	0.1809	5.3945	29.8191	32
33	235.5625	0.0042	1303.1253	0.0008	5.5320	0.1808	5.4149	29.9549	33
34	277.9638	0.0036	1538.6878	0.0006	5.5356	0.1806	5.4328	30.0736	34
35	327.9973	0.0030	1816.6516	0.0006	5.5386	0.1806	5.4485	30.1773	35
40	750.3783	0.0013	4163.2130	0.0002	5.5482	0.1802	5.5022	30.5269	40
45	1716.6839	0.0006	9531.5771	0.0001	5.5523	0.1801	5.5293	30.7006	45
50	3927.3569	0.0003	21813.0937	0.0000	5.5541	0.1800	5.5428	30.7856	50

Table C.24 Interest Rate Factors (20.0%)

| | Single Payment | | Equal Payment Series | | | | Gradient Series | | |
	Compound Amount Factor (F/P,i,N)	Present Worth Factor (P/F,i,N)	Compound Amount Factor (F/A,i,N)	Sinking Fund Factor (A/F,i,N)	Present Worth Factor (P/A,i,N)	Capital Recovery Factor (A/P,i,N)	Gradient Uniform Series (A/G,i,N)	Gradient Present Worth (P/G.i.N)	N
1	1.2000	0.8333	1.0000	1.0000	0.8333	1.2000	0.0000	0.0000	1
2	1.4400	0.6944	2.2000	0.4545	1.5278	0.6545	0.4545	0.6944	2
3	1.7280	0.5787	3.6400	0.2747	2.1065	0.4747	0.8791	1.8519	3
4	2.0736	0.4823	5.3680	0.1863	2.5887	0.3863	1.2742	3.2986	4
5	2.4883	0.4019	7.4416	0.1344	2.9906	0.3344	1.6405	4.9061	5
6	2.9860	0.3349	9.9299	0.1007	3.3255	0.3007	1.9788	6.5806	6
7	3.5832	0.2791	12.9159	0.0774	3.6046	0.2774	2.2902	8.2551	7
8	4.2998	0.2326	16.4991	0.0606	3.8372	0.2606	2.5756	9.8831	8
9	5.1598	0.1938	20.7989	0.0481	4.0310	0.2481	2.8364	11.4335	9
10	6.1917	0.1615	25.9587	0.0385	4.1925	0.2385	3.0739	12.8871	10
11	7.4301	0.1346	32.1504	0.0311	4.3271	0.2311	3.2893	14.2330	11
12	8.9161	0.1122	39.5805	0.0253	4.4392	0.2253	3.4841	15.4667	12
13	10.6993	0.0935	48.4966	0.0206	4.5327	0.2206	3.6597	16.5883	13
14	12.8392	0.0779	59.1959	0.0169	4.6106	0.2169	3.8175	17.6008	14
15	15.4070	0.0649	72.0351	0.0139	4.6755	0.2139	3.9588	18.5095	15
16	18.4884	0.0541	87.4421	0.0114	4.7296	0.2114	4.0851	19.3208	16
17	22.1861	0.0451	105.9306	0.0094	4.7746	0.2094	4.1976	20.0419	17
18	26.6233	0.0376	128.1167	0.0078	4.8122	0.2078	4.2975	20.6805	18
19	31.9480	0.0313	154.7400	0.0065	4.8435	0.2065	4.3861	21.2439	19
20	38.3376	0.0261	186.6880	0.0054	4.8696	0.2054	4.4643	21.7395	20
21	46.0051	0.0217	225.0256	0.0044	4.8913	0.2044	4.5334	22.1742	21
22	55.2061	0.0181	271.0307	0.0037	4.9094	0.2037	4.5941	22.5546	22
23	66.2474	0.0151	326.2369	0.0031	4.9245	0.2031	4.6475	22.8867	23
24	79.4968	0.0126	392.4842	0.0025	4.9371	0.2025	4.6943	23.1760	24
25	95.3962	0.0105	471.9811	0.0021	4.9476	0.2021	4.7352	23.4276	25
26	114.4755	0.0087	567.3773	0.0018	4.9563	0.2018	4.7709	23.6460	26
27	137.3706	0.0073	681.8528	0.0015	4.9636	0.2015	4.8020	23.8353	27
28	164.8447	0.0061	819.2233	0.0012	4.9697	0.2012	4.8291	23.9991	28
29	197.8136	0.0051	984.0680	0.0010	4.9747	0.2010	4.8527	24.1406	29
30	237.3763	0.0042	1181.8816	0.0008	4.9789	0.2008	4.8731	24.2628	30
31	284.8516	0.0035	1419.2579	0.0007	4.9824	0.2007	4.8908	24.3681	31
32	341.8219	0.0029	1704.1095	0.0006	4.9854	0.2006	4.9061	24.4588	32
33	410.1863	0.0024	2045.9314	0.0005	4.9878	0.2005	4.9194	24.5368	33
34	492.2235	0.0020	2456.1176	0.0004	4.9898	0.2004	4.9308	24.6038	34
35	590.6682	0.0017	2948.3411	0.0003	4.9915	0.2003	4.9406	24.6614	35
40	1469.7716	0.0007	7343.8578	0.0001	4.9966	0.2001	4.9728	24.8469	40
45	3657.2620	0.0003	18281.3099	0.0001	4.9986	0.2001	4.9877	24.9316	45

Table C.25 Interest Rate Factors (25.0%)

	Single Payment		Equal Payment Series				Gradient Series		
N	Compound Amount Factor (F/P,i,N)	Present Worth Factor (P/F,i,N)	Compound Amount Factor (F/A,i,N)	Sinking Fund Factor (A/F,i,N)	Present Worth Factor (P/A,i,N)	Capital Recovery Factor (A/P,i,N)	Gradient Uniform Series (A/G,i,N)	Gradient Present Worth (P/G.i.N)	N
1	1.2500	0.8000	1.0000	1.0000	0.8000	1.2500	0.0000	0.0000	1
2	1.5625	0.6400	2.2500	0.4444	1.4400	0.6944	0.4444	0.6400	2
3	1.9531	0.5120	3.8125	0.2623	1.9520	0.5123	0.8525	1.6640	3
4	2.4414	0.4096	5.7656	0.1734	2.3616	0.4234	1.2249	2.8928	4
5	3.0518	0.3277	8.2070	0.1218	2.6893	0.3718	1.5631	4.2035	5
6	3.8147	0.2621	11.2588	0.0888	2.9514	0.3388	1.8683	5.5142	6
7	4.7684	0.2097	15.0735	0.0663	3.1611	0.3163	2.1424	6.7725	7
8	5.9605	0.1678	19.8419	0.0504	3.3289	0.3004	2.3872	7.9469	8
9	7.4506	0.1342	25.8023	0.0388	3.4631	0.2888	2.6048	9.0207	9
10	9.3132	0.1074	33.2529	0.0301	3.5705	0.2801	2.7971	9.9870	10
11	11.6415	0.0859	42.5661	0.0235	3.6564	0.2735	2.9663	10.8460	11
12	14.5519	0.0687	54.2077	0.0184	3.7251	0.2684	3.1145	11.6020	12
13	18.1899	0.0550	68.7596	0.0145	3.7801	0.2645	3.2437	12.2617	13
14	22.7374	0.0440	86.9495	0.0115	3.8241	0.2615	3.3559	12.8334	14
15	28.4217	0.0352	109.6868	0.0091	3.8593	0.2591	3.4530	13.3260	15
16	35.5271	0.0281	138.1085	0.0072	3.8874	0.2572	3.5366	13.7482	16
17	44.4089	0.0225	173.6357	0.0058	3.9099	0.2558	3.6084	14.1085	17
18	55.5112	0.0180	218.0446	0.0046	3.9279	0.2546	3.6698	14.4147	18
19	69.3889	0.0144	273.5558	0.0037	3.9424	0.2537	3.7222	14.6741	19
20	86.7362	0.0115	342.9447	0.0029	3.9539	0.2529	3.7667	14.8932	20
21	108.4202	0.0092	429.6809	0.0023	3.9631	0.2523	3.8045	15.0777	21
22	135.5253	0.0074	538.1011	0.0019	3.9705	0.2519	3.8365	15.2326	22
23	169.4066	0.0059	673.6264	0.0015	3.9764	0.2515	3.8634	15.3625	23
24	211.7582	0.0047	843.0329	0.0012	3.9811	0.2512	3.8861	15.4711	24
25	264.6978	0.0038	1054.7912	0.0009	3.9849	0.2509	3.9052	15.5618	25
26	330.8722	0.0030	1319.4890	0.0008	3.9879	0.2508	3.9212	15.6373	26
27	413.5903	0.0024	1650.3612	0.0006	3.9903	0.2506	3.9346	15.7002	27
28	516.9879	0.0019	2063.9515	0.0005	3.9923	0.2505	3.9457	15.7524	28
29	646.2349	0.0015	2580.9394	0.0004	3.9938	0.2504	3.9551	15.7957	29
30	807.7936	0.0012	3227.1743	0.0003	3.9950	0.2503	3.9628	15.8316	30
31	1009.7420	0.0010	4034.9678	0.0002	3.9960	0.2502	3.9693	15.8614	31
32	1262.1774	0.0008	5044.7098	0.0002	3.9968	0.2502	3.9746	15.8859	32
33	1577.7218	0.0006	6306.8872	0.0002	3.9975	0.2502	3.9791	15.9062	33
34	1972.1523	0.0005	7884.6091	0.0001	3.9980	0.2501	3.9828	15.9229	34
35	2465.1903	0.0004	9856.7613	0.0001	3.9984	0.2501	3.9858	15.9367	35
40	7523.1638	0.0001	30088.6554	0.0000	3.9995	0.2500	3.9947	15.9766	40

Table C.26 Interest Rate Factors (30.0%)

	Single Payment		Equal Payment Series				Gradient Series		
N	Compound Amount Factor (F/P,i,N)	Present Worth Factor (P/F,i,N)	Compound Amount Factor (F/A,i,N)	Sinking Fund Factor (A/F,i,N)	Present Worth Factor (P/A,i,N)	Capital Recovery Factor (A/P,i,N)	Gradient Uniform Series (A/G,i,N)	Gradient Present Worth (P/G.i.N)	N
1	1.3000	0.7692	1.0000	1.0000	0.7692	1.3000	0.0000	0.0000	1
2	1.6900	0.5917	2.3000	0.4348	1.3609	0.7348	0.4348	0.5917	2
3	2.1970	0.4552	3.9900	0.2506	1.8161	0.5506	0.8271	1.5020	3
4	2.8561	0.3501	6.1870	0.1616	2.1662	0.4616	1.1783	2.5524	4
5	3.7129	0.2693	9.0431	0.1106	2.4356	0.4106	1.4903	3.6297	5
6	4.8268	0.2072	12.7560	0.0784	2.6427	0.3784	1.7654	4.6656	6
7	6.2749	0.1594	17.5828	0.0569	2.8021	0.3569	2.0063	5.6218	7
8	8.1573	0.1226	23.8577	0.0419	2.9247	0.3419	2.2156	6.4800	8
9	10.6045	0.0943	32.0150	0.0312	3.0190	0.3312	2.3963	7.2343	9
10	13.7858	0.0725	42.6195	0.0235	3.0915	0.3235	2.5512	7.8872	10
11	17.9216	0.0558	56.4053	0.0177	3.1473	0.3177	2.6833	8.4452	11
12	23.2981	0.0429	74.3270	0.0135	3.1903	0.3135	2.7952	8.9173	12
13	30.2875	0.0330	97.6250	0.0102	3.2233	0.3102	2.8895	9.3135	13
14	39.3738	0.0254	127.9125	0.0078	3.2487	0.3078	2.9685	9.6437	14
15	51.1859	0.0195	167.2863	0.0060	3.2682	0.3060	3.0344	9.9172	15
16	66.5417	0.0150	218.4722	0.0046	3.2832	0.3046	3.0892	10.1426	16
17	86.5042	0.0116	285.0139	0.0035	3.2948	0.3035	3.1345	10.3276	17
18	112.4554	0.0089	371.5180	0.0027	3.3037	0.3027	3.1718	10.4788	18
19	146.1920	0.0068	483.9734	0.0021	3.3105	0.3021	3.2025	10.6019	19
20	190.0496	0.0053	630.1655	0.0016	3.3158	0.3016	3.2275	10.7019	20
21	247.0645	0.0040	820.2151	0.0012	3.3198	0.3012	3.2480	10.7828	21
22	321.1839	0.0031	1067.2796	0.0009	3.3230	0.3009	3.2646	10.8482	22
23	417.5391	0.0024	1388.4635	0.0007	3.3254	0.3007	3.2781	10.9009	23
24	542.8008	0.0018	1806.0026	0.0006	3.3272	0.3006	3.2890	10.9433	24
25	705.6410	0.0014	2348.8033	0.0004	3.3286	0.3004	3.2979	10.9773	25
26	917.3333	0.0011	3054.4443	0.0003	3.3297	0.3003	3.3050	11.0045	26
27	1192.5333	0.0008	3971.7776	0.0003	3.3305	0.3003	3.3107	11.0263	27
28	1550.2933	0.0006	5164.3109	0.0002	3.3312	0.3002	3.3153	11.0437	28
29	2015.3813	0.0005	6714.6042	0.0001	3.3317	0.3001	3.3189	11.0576	29
30	2619.9956	0.0004	8729.9855	0.0001	3.3321	0.3001	3.3219	11.0687	30
31	3405.9943	0.0003	11349.9811	0.0001	3.3324	0.3001	3.3242	11.0775	31
32	4427.7926	0.0002	14755.9755	0.0001	3.3326	0.3001	3.3261	11.0845	32
33	5756.1304	0.0002	19183.7681	0.0001	3.3328	0.3001	3.3276	11.0901	33
34	7482.9696	0.0001	24939.8985	0.0000	3.3329	0.3000	3.3288	11.0945	34
35	9727.8604	0.0001	32422.8681	0.0000	3.3330	0.3000	3.3297	11.0980	35

Table C.27 Interest Rate Factors (35.0%)

	Single Payment		Equal Payment Series				Gradient Series		
N	Compound Amount Factor (F/P,i,N)	Present Worth Factor (P/F,i,N)	Compound Amount Factor (F/A,i,N)	Sinking Fund Factor (A/F,i,N)	Present Worth Factor (P/A,i,N)	Capital Recovery Factor (A/P,i,N)	Gradient Uniform Series (A/G,i,N)	Gradient Present Worth (P/G.i.N)	N
1	1.3500	0.7407	1.0000	1.0000	0.7407	1.3500	0.0000	0.0000	1
2	1.8225	0.5487	2.3500	0.4255	1.2894	0.7755	0.4255	0.5487	2
3	2.4604	0.4064	4.1725	0.2397	1.6959	0.5897	0.8029	1.3616	3
4	3.3215	0.3011	6.6329	0.1508	1.9969	0.5008	1.1341	2.2648	4
5	4.4840	0.2230	9.9544	0.1005	2.2200	0.4505	1.4220	3.1568	5
6	6.0534	0.1652	14.4384	0.0693	2.3852	0.4193	1.6698	3.9828	6
7	8.1722	0.1224	20.4919	0.0488	2.5075	0.3988	1.8811	4.7170	7
8	11.0324	0.0906	28.6640	0.0349	2.5982	0.3849	2.0597	5.3515	8
9	14.8937	0.0671	39.6964	0.0252	2.6653	0.3752	2.2094	5.8886	9
10	20.1066	0.0497	54.5902	0.0183	2.7150	0.3683	2.3338	6.3363	10
11	27.1439	0.0368	74.6967	0.0134	2.7519	0.3634	2.4364	6.7047	11
12	36.6442	0.0273	101.8406	0.0098	2.7792	0.3598	2.5205	7.0049	12
13	49.4697	0.0202	138.4848	0.0072	2.7994	0.3572	2.5889	7.2474	13
14	66.7841	0.0150	187.9544	0.0053	2.8144	0.3553	2.6443	7.4421	14
15	90.1585	0.0111	254.7385	0.0039	2.8255	0.3539	2.6889	7.5974	15
16	121.7139	0.0082	344.8970	0.0029	2.8337	0.3529	2.7246	7.7206	16
17	164.3138	0.0061	466.6109	0.0021	2.8398	0.3521	2.7530	7.8180	17
18	221.8236	0.0045	630.9247	0.0016	2.8443	0.3516	2.7756	7.8946	18
19	299.4619	0.0033	852.7483	0.0012	2.8476	0.3512	2.7935	7.9547	19
20	404.2736	0.0025	1152.2103	0.0009	2.8501	0.3509	2.8075	8.0017	20
21	545.7693	0.0018	1556.4838	0.0006	2.8519	0.3506	2.8186	8.0384	21
22	736.7886	0.0014	2102.2532	0.0005	2.8533	0.3505	2.8272	8.0669	22
23	994.6646	0.0010	2839.0418	0.0004	2.8543	0.3504	2.8340	8.0890	23
24	1342.7973	0.0007	3833.7064	0.0003	2.8550	0.3503	2.8393	8.1061	24
25	1812.7763	0.0006	5176.5037	0.0002	2.8556	0.3502	2.8433	8.1194	25
26	2447.2480	0.0004	6989.2800	0.0001	2.8560	0.3501	2.8465	8.1296	26
27	3303.7848	0.0003	9436.5280	0.0001	2.8563	0.3501	2.8490	8.1374	27
28	4460.1095	0.0002	12740.3128	0.0001	2.8565	0.3501	2.8509	8.1435	28
29	6021.1478	0.0002	17200.4222	0.0001	2.8567	0.3501	2.8523	8.1481	29
30	8128.5495	0.0001	23221.5700	0.0000	2.8568	0.3500	2.8535	8.1517	30

Table C.28 Interest Rate Factors (40.0%)

	Single Payment		Equal Payment Series				Gradient Series		
N	Compound Amount Factor (F/P,i,N)	Present Worth Factor (P/F,i,N)	Compound Amount Factor (F/A,i,N)	Sinking Fund Factor (A/F,i,N)	Present Worth Factor (P/A,i,N)	Capital Recovery Factor (A/P,i,N)	Gradient Uniform Series (A/G,i,N)	Gradient Present Worth (P/G.i.N)	N
1	1.4000	0.7143	1.0000	1.0000	0.7143	1.4000	0.0000	0.0000	1
2	1.9600	0.5102	2.4000	0.4167	1.2245	0.8167	0.4167	0.5102	2
3	2.7440	0.3644	4.3600	0.2294	1.5889	0.6294	0.7798	1.2391	3
4	3.8416	0.2603	7.1040	0.1408	1.8492	0.5408	1.0923	2.0200	4
5	5.3782	0.1859	10.9456	0.0914	2.0352	0.4914	1.3580	2.7637	5
6	7.5295	0.1328	16.3238	0.0613	2.1680	0.4613	1.5811	3.4278	6
7	10.5414	0.0949	23.8534	0.0419	2.2628	0.4419	1.7664	3.9970	7
8	14.7579	0.0678	34.3947	0.0291	2.3306	0.4291	1.9185	4.4713	8
9	20.6610	0.0484	49.1526	0.0203	2.3790	0.4203	2.0422	4.8585	9
10	28.9255	0.0346	69.8137	0.0143	2.4136	0.4143	2.1419	5.1696	10
11	40.4957	0.0247	98.7391	0.0101	2.4383	0.4101	2.2215	5.4166	11
12	56.6939	0.0176	139.2348	0.0072	2.4559	0.4072	2.2845	5.6106	12
13	79.3715	0.0126	195.9287	0.0051	2.4685	0.4051	2.3341	5.7618	13
14	111.1201	0.0090	275.3002	0.0036	2.4775	0.4036	2.3729	5.8788	14
15	155.5681	0.0064	386.4202	0.0026	2.4839	0.4026	2.4030	5.9688	15
16	217.7953	0.0046	541.9883	0.0018	2.4885	0.4018	2.4262	6.0376	16
17	304.9135	0.0033	759.7837	0.0013	2.4918	0.4013	2.4441	6.0901	17
18	426.8789	0.0023	1064.6971	0.0009	2.4941	0.4009	2.4577	6.1299	18
19	597.6304	0.0017	1491.5760	0.0007	2.4958	0.4007	2.4682	6.1601	19
20	836.6826	0.0012	2089.2064	0.0005	2.4970	0.4005	2.4761	6.1828	20
21	1171.3556	0.0009	2925.8889	0.0003	2.4979	0.4003	2.4821	6.1998	21
22	1639.8978	0.0006	4097.2445	0.0002	2.4985	0.4002	2.4866	6.2127	22
23	2295.8569	0.0004	5737.1423	0.0002	2.4989	0.4002	2.4900	6.2222	23
24	3214.1997	0.0003	8032.9993	0.0001	2.4992	0.4001	2.4925	6.2294	24
25	4499.8796	0.0002	11247.1990	0.0001	2.4994	0.4001	2.4944	6.2347	25
26	6299.8314	0.0002	15747.0785	0.0001	2.4996	0.4001	2.4959	6.2387	26
27	8819.7640	0.0001	22046.9099	0.0000	2.4997	0.4000	2.4969	6.2416	27
28	12347.6696	0.0001	30866.6739	0.0000	2.4998	0.4000	2.4977	6.2438	28
29	17286.7374	0.0001	43214.3435	0.0000	2.4999	0.4000	2.4983	6.2454	29
30	24201.4324	0.0000	60501.0809	0.0000	2.4999	0.4000	2.4988	6.2466	30

Table C.29 Interest Rate Factors (50.0%)

	Single Payment		Equal Payment Series				Gradient Series		
N	Compound Amount Factor (F/P,i,N)	Present Worth Factor (P/F,i,N)	Compound Amount Factor (F/A,i,N)	Sinking Fund Factor (A/F,i,N)	Present Worth Factor (P/A,i,N)	Capital Recovery Factor (A/P,i,N)	Gradient Uniform Series (A/G,i,N)	Gradient Present Worth (P/G.i.N)	N
1	1.5000	0.6667	1.0000	1.0000	0.6667	1.5000	0.0000	0.0000	1
2	2.2500	0.4444	2.5000	0.4000	1.1111	0.9000	0.4000	0.4444	2
3	3.3750	0.2963	4.7500	0.2105	1.4074	0.7105	0.7368	1.0370	3
4	5.0625	0.1975	8.1250	0.1231	1.6049	0.6231	1.0154	1.6296	4
5	7.5938	0.1317	13.1875	0.0758	1.7366	0.5758	1.2417	2.1564	5
6	11.3906	0.0878	20.7813	0.0481	1.8244	0.5481	1.4226	2.5953	6
7	17.0859	0.0585	32.1719	0.0311	1.8829	0.5311	1.5648	2.9465	7
8	25.6289	0.0390	49.2578	0.0203	1.9220	0.5203	1.6752	3.2196	8
9	38.4434	0.0260	74.8867	0.0134	1.9480	0.5134	1.7596	3.4277	9
10	57.6650	0.0173	113.3301	0.0088	1.9653	0.5088	1.8235	3.5838	10
11	86.4976	0.0116	170.9951	0.0058	1.9769	0.5058	1.8713	3.6994	11
12	129.7463	0.0077	257.4927	0.0039	1.9846	0.5039	1.9068	3.7842	12
13	194.6195	0.0051	387.2390	0.0026	1.9897	0.5026	1.9329	3.8459	13
14	291.9293	0.0034	581.8585	0.0017	1.9931	0.5017	1.9519	3.8904	14
15	437.8939	0.0023	873.7878	0.0011	1.9954	0.5011	1.9657	3.9224	15
16	656.8408	0.0015	1311.6817	0.0008	1.9970	0.5008	1.9756	3.9452	16
17	985.2613	0.0010	1968.5225	0.0005	1.9980	0.5005	1.9827	3.9614	17
18	1477.8919	0.0007	2953.7838	0.0003	1.9986	0.5003	1.9878	3.9729	18
19	2216.8378	0.0005	4431.6756	0.0002	1.9991	0.5002	1.9914	3.9811	19
20	3325.2567	0.0003	6648.5135	0.0002	1.9994	0.5002	1.9940	3.9868	20

Capital Tax Factors

Table D.1 Half Year Rule Capital Tax Factors

Tax Rate = 20%

Interest Rate - %	Declining Balance CCA Rate - %						
	4	8	10	20	25	30	40
1	0.8408	0.8231	0.8191	0.8105	0.8086	0.8074	0.8058
2	0.8680	0.8416	0.8350	0.8200	0.8166	0.8143	0.8114
3	0.8874	0.8567	0.8484	0.8286	0.8240	0.8208	0.8167
4	0.9019	0.8692	0.8599	0.8365	0.8309	0.8269	0.8217
5	0.9132	0.8799	0.8698	0.8438	0.8373	0.8327	0.8265
6	0.9223	0.8889	0.8785	0.8505	0.8433	0.8381	0.8310
7	0.9297	0.8968	0.8862	0.8567	0.8489	0.8431	0.8354
8	0.9358	0.9037	0.8930	0.8624	0.8541	0.8480	0.8395
9	0.9410	0.9098	0.8991	0.8678	0.8590	0.8525	0.8435
10	0.9455	0.9152	0.9045	0.8727	0.8636	0.8568	0.8473
11	0.9493	0.9200	0.9095	0.8774	0.8680	0.8609	0.8509
12	0.9527	0.9243	0.9140	0.8817	0.8721	0.8648	0.8544
13	0.9556	0.9282	0.9180	0.8858	0.8760	0.8685	0.8577
14	0.9583	0.9317	0.9218	0.8896	0.8797	0.8720	0.8609
15	0.9606	0.9350	0.9252	0.8932	0.8832	0.8754	0.8640
16	0.9628	0.9379	0.9284	0.8966	0.8865	0.8786	0.8670
17	0.9647	0.9406	0.9313	0.8997	0.8896	0.8816	0.8698
18	0.9664	0.9432	0.9340	0.9028	0.8926	0.8845	0.8726
19	0.9680	0.9455	0.9365	0.9056	0.8954	0.8873	0.8752
20	0.9694	0.9476	0.9389	0.9083	0.8981	0.8900	0.8778
21	0.9708	0.9496	0.9411	0.9109	0.9007	0.8926	0.8802
22	0.9720	0.9515	0.9431	0.9133	0.9032	0.8950	0.8826
23	0.9731	0.9532	0.9451	0.9157	0.9056	0.8974	0.8849
24	0.9742	0.9548	0.9469	0.9179	0.9078	0.8996	0.8871
25	0.9752	0.9564	0.9486	0.9200	0.9100	0.9018	0.8892
30	0.9792	0.9628	0.9558	0.9292	0.9196	0.9115	0.8989
35	0.9821	0.9676	0.9613	0.9367	0.9275	0.9197	0.9072
40	0.9844	0.9714	0.9657	0.9429	0.9341	0.9265	0.9143
45	0.9862	0.9745	0.9693	0.9480	0.9397	0.9324	0.9205
50	0.9877	0.9770	0.9722	0.9524	0.9444	0.9375	0.9259

Table D.2 Full Year Capital Tax Factors

Tax Rate = 20%

Interest Rate - %	Declining Balance CCA Rate - %						
	4	8	10	20	25	30	40
1	0.8400	0.8222	0.8182	0.8095	0.8077	0.8065	0.8049
2	0.8667	0.8400	0.8333	0.8182	0.8148	0.8125	0.8095
3	0.8857	0.8545	0.8462	0.8261	0.8214	0.8182	0.8140
4	0.9000	0.8667	0.8571	0.8333	0.8276	0.8235	0.8182
5	0.9111	0.8769	0.8667	0.8400	0.8333	0.8286	0.8222
6	0.9200	0.8857	0.8750	0.8462	0.8387	0.8333	0.8261
7	0.9273	0.8933	0.8824	0.8519	0.8438	0.8378	0.8298
8	0.9333	0.9000	0.8889	0.8571	0.8485	0.8421	0.8333
9	0.9385	0.9059	0.8947	0.8621	0.8529	0.8462	0.8367
10	0.9429	0.9111	0.9000	0.8667	0.8571	0.8500	0.8400
11	0.9467	0.9158	0.9048	0.8710	0.8611	0.8537	0.8431
12	0.9500	0.9200	0.9091	0.8750	0.8649	0.8571	0.8462
13	0.9529	0.9238	0.9130	0.8788	0.8684	0.8605	0.8491
14	0.9556	0.9273	0.9167	0.8824	0.8718	0.8636	0.8519
15	0.9579	0.9304	0.9200	0.8857	0.8750	0.8667	0.8545
16	0.9600	0.9333	0.9231	0.8889	0.8780	0.8696	0.8571
17	0.9619	0.9360	0.9259	0.8919	0.8810	0.8723	0.8596
18	0.9636	0.9385	0.9286	0.8947	0.8837	0.8750	0.8621
19	0.9652	0.9407	0.9310	0.8974	0.8864	0.8776	0.8644
20	0.9667	0.9429	0.9333	0.9000	0.8889	0.8800	0.8667
21	0.9680	0.9448	0.9355	0.9024	0.8913	0.8824	0.8689
22	0.9692	0.9467	0.9375	0.9048	0.8936	0.8846	0.8710
23	0.9704	0.9484	0.9394	0.9070	0.8958	0.8868	0.8730
24	0.9714	0.9500	0.9412	0.9091	0.8980	0.8889	0.8750
25	0.9724	0.9515	0.9429	0.9111	0.9000	0.8909	0.8769
30	0.9765	0.9579	0.9500	0.9200	0.9091	0.9000	0.8857
35	0.9795	0.9628	0.9556	0.9273	0.9167	0.9077	0.8933
40	0.9818	0.9667	0.9600	0.9333	0.9231	0.9143	0.9000
45	0.9837	0.9698	0.9636	0.9385	0.9286	0.9200	0.9059
50	0.9852	0.9724	0.9667	0.9429	0.9333	0.9250	0.9111

Table D.3 Half Year Rule Capital Tax Factors

Tax Rate = 25%

Interest Rate - %	Declining Balance CCA Rate - %						
	4	8	10	20	25	30	40
1	0.8010	0.7789	0.7739	0.7631	0.7608	0.7593	0.7573
2	0.8350	0.8020	0.7937	0.7750	0.7708	0.7679	0.7642
3	0.8592	0.8208	0.8105	0.7858	0.7800	0.7760	0.7708
4	0.8774	0.8365	0.8249	0.7957	0.7886	0.7837	0.7771
5	0.8915	0.8498	0.8373	0.8048	0.7966	0.7908	0.7831
6	0.9028	0.8612	0.8482	0.8131	0.8041	0.7976	0.7888
7	0.9121	0.8710	0.8578	0.8209	0.8111	0.8039	0.7942
8	0.9198	0.8796	0.8663	0.8280	0.8176	0.8099	0.7994
9	0.9263	0.8872	0.8739	0.8347	0.8238	0.8156	0.8043
10	0.9318	0.8939	0.8807	0.8409	0.8295	0.8210	0.8091
11	0.9366	0.9000	0.8869	0.8467	0.8350	0.8261	0.8136
12	0.9408	0.9054	0.8925	0.8521	0.8401	0.8310	0.8180
13	0.9446	0.9102	0.8976	0.8572	0.8450	0.8356	0.8222
14	0.9479	0.9147	0.9022	0.8620	0.8496	0.8400	0.8262
15	0.9508	0.9187	0.9065	0.8665	0.8539	0.8442	0.8300
16	0.9534	0.9224	0.9105	0.8707	0.8581	0.8482	0.8337
17	0.9558	0.9258	0.9141	0.8747	0.8620	0.8520	0.8373
18	0.9580	0.9289	0.9175	0.8785	0.8657	0.8557	0.8407
19	0.9600	0.9318	0.9207	0.8820	0.8693	0.8592	0.8440
20	0.9618	0.9345	0.9236	0.8854	0.8727	0.8625	0.8472
21	0.9635	0.9370	0.9264	0.8886	0.8759	0.8657	0.8503
22	0.9650	0.9393	0.9289	0.8917	0.8790	0.8688	0.8533
23	0.9664	0.9415	0.9313	0.8946	0.8820	0.8717	0.8561
24	0.9677	0.9435	0.9336	0.8974	0.8848	0.8746	0.8589
25	0.9690	0.9455	0.9357	0.9000	0.8875	0.8773	0.8615
30	0.9740	0.9534	0.9447	0.9115	0.8995	0.8894	0.8736
35	0.9777	0.9595	0.9516	0.9209	0.9093	0.8996	0.8840
40	0.9805	0.9643	0.9571	0.9286	0.9176	0.9082	0.8929
45	0.9828	0.9681	0.9616	0.9350	0.9246	0.9155	0.9006
50	0.9846	0.9713	0.9653	0.9405	0.9306	0.9219	0.9074

Table D.4 Full Year Capital Tax Factors

Tax Rate = 25%

Interest Rate - %	Declining Balance CCA Rate - %						
	4	8	10	20	25	30	40
1	0.8000	0.7778	0.7727	0.7619	0.7596	0.7581	0.7561
2	0.8333	0.8000	0.7917	0.7727	0.7685	0.7656	0.7619
3	0.8571	0.8182	0.8077	0.7826	0.7768	0.7727	0.7674
4	0.8750	0.8333	0.8214	0.7917	0.7845	0.7794	0.7727
5	0.8889	0.8462	0.8333	0.8000	0.7917	0.7857	0.7778
6	0.9000	0.8571	0.8438	0.8077	0.7984	0.7917	0.7826
7	0.9091	0.8667	0.8529	0.8148	0.8047	0.7973	0.7872
8	0.9167	0.8750	0.8611	0.8214	0.8106	0.8026	0.7917
9	0.9231	0.8824	0.8684	0.8276	0.8162	0.8077	0.7959
10	0.9286	0.8889	0.8750	0.8333	0.8214	0.8125	0.8000
11	0.9333	0.8947	0.8810	0.8387	0.8264	0.8171	0.8039
12	0.9375	0.9000	0.8864	0.8438	0.8311	0.8214	0.8077
13	0.9412	0.9048	0.8913	0.8485	0.8355	0.8256	0.8113
14	0.9444	0.9091	0.8958	0.8529	0.8397	0.8295	0.8148
15	0.9474	0.9130	0.9000	0.8571	0.8438	0.8333	0.8182
16	0.9500	0.9167	0.9038	0.8611	0.8476	0.8370	0.8214
17	0.9524	0.9200	0.9074	0.8649	0.8512	0.8404	0.8246
18	0.9545	0.9231	0.9107	0.8684	0.8547	0.8438	0.8276
19	0.9565	0.9259	0.9138	0.8718	0.8580	0.8469	0.8305
20	0.9583	0.9286	0.9167	0.8750	0.8611	0.8500	0.8333
21	0.9600	0.9310	0.9194	0.8780	0.8641	0.8529	0.8361
22	0.9615	0.9333	0.9219	0.8810	0.8670	0.8558	0.8387
23	0.9630	0.9355	0.9242	0.8837	0.8698	0.8585	0.8413
24	0.9643	0.9375	0.9265	0.8864	0.8724	0.8611	0.8438
25	0.9655	0.9394	0.9286	0.8889	0.8750	0.8636	0.8462
30	0.9706	0.9474	0.9375	0.9000	0.8864	0.8750	0.8571
35	0.9744	0.9535	0.9444	0.9091	0.8958	0.8846	0.8667
40	0.9773	0.9583	0.9500	0.9167	0.9038	0.8929	0.8750
45	0.9796	0.9623	0.9545	0.9231	0.9107	0.9000	0.8824
50	0.9815	0.9655	0.9583	0.9286	0.9167	0.9063	0.8889

Table D5 Half Year Rule Capital Tax Factors
Tax Rate = 30%

Interest Rate - %	Declining Balance CCA Rate - %						
	4	8	10	20	25	30	40
1	0.7612	0.7347	0.7286	0.7157	0.7130	0.7111	0.7088
2	0.8020	0.7624	0.7525	0.7299	0.7249	0.7215	0.7171
3	0.8311	0.7850	0.7726	0.7429	0.7360	0.7312	0.7250
4	0.8529	0.8038	0.7898	0.7548	0.7464	0.7404	0.7325
5	0.8698	0.8198	0.8048	0.7657	0.7560	0.7490	0.7397
6	0.8834	0.8334	0.8178	0.7758	0.7649	0.7571	0.7465
7	0.8945	0.8452	0.8293	0.7850	0.7733	0.7647	0.7530
8	0.9037	0.8556	0.8395	0.7937	0.7811	0.7719	0.7593
9	0.9115	0.8647	0.8486	0.8016	0.7885	0.7788	0.7652
10	0.9182	0.8727	0.8568	0.8091	0.7955	0.7852	0.7709
11	0.9240	0.8799	0.8642	0.8160	0.8020	0.7914	0.7764
12	0.9290	0.8864	0.8709	0.8225	0.8082	0.7972	0.7816
13	0.9335	0.8923	0.8771	0.8286	0.8140	0.8027	0.7866
14	0.9374	0.8976	0.8827	0.8344	0.8195	0.8080	0.7914
15	0.9410	0.9025	0.8878	0.8398	0.8247	0.8130	0.7960
16	0.9441	0.9069	0.8926	0.8448	0.8297	0.8178	0.8005
17	0.9470	0.9110	0.8970	0.8496	0.8344	0.8224	0.8048
18	0.9496	0.9147	0.9010	0.8541	0.8389	0.8268	0.8089
19	0.9520	0.9182	0.9048	0.8584	0.8432	0.8310	0.8128
20	0.9542	0.9214	0.9083	0.8625	0.8472	0.8350	0.8167
21	0.9562	0.9244	0.9116	0.8664	0.8511	0.8388	0.8203
22	0.9580	0.9272	0.9147	0.8700	0.8548	0.8425	0.8239
23	0.9597	0.9298	0.9176	0.8735	0.8584	0.8461	0.8273
24	0.9613	0.9323	0.9203	0.8768	0.8618	0.8495	0.8306
25	0.9628	0.9345	0.9229	0.8800	0.8650	0.8527	0.8338
30	0.9688	0.9441	0.9337	0.8938	0.8794	0.8673	0.8484
35	0.9732	0.9514	0.9420	0.9051	0.8912	0.8795	0.8607
40	0.9766	0.9571	0.9486	0.9143	0.9011	0.8898	0.8714
45	0.9793	0.9617	0.9539	0.9220	0.9095	0.8986	0.8807
50	0.9815	0.9655	0.9583	0.9286	0.9167	0.9063	0.8889

Table D6 Full Year Capital Tax Factors
Tax Rate = 30%

Interest Rate - %	Declining Balance CCA Rate - %						
	4	8	10	20	25	30	40
1	0.7600	0.7333	0.7273	0.7143	0.7115	0.7097	0.7073
2	0.8000	0.7600	0.7500	0.7273	0.7222	0.7188	0.7143
3	0.8286	0.7818	0.7692	0.7391	0.7321	0.7273	0.7209
4	0.8500	0.8000	0.7857	0.7500	0.7414	0.7353	0.7273
5	0.8667	0.8154	0.8000	0.7600	0.7500	0.7429	0.7333
6	0.8800	0.8286	0.8125	0.7692	0.7581	0.7500	0.7391
7	0.8909	0.8400	0.8235	0.7778	0.7656	0.7568	0.7447
8	0.9000	0.8500	0.8333	0.7857	0.7727	0.7632	0.7500
9	0.9077	0.8588	0.8421	0.7931	0.7794	0.7692	0.7551
10	0.9143	0.8667	0.8500	0.8000	0.7857	0.7750	0.7600
11	0.9200	0.8737	0.8571	0.8065	0.7917	0.7805	0.7647
12	0.9250	0.8800	0.8636	0.8125	0.7973	0.7857	0.7692
13	0.9294	0.8857	0.8696	0.8182	0.8026	0.7907	0.7736
14	0.9333	0.8909	0.8750	0.8235	0.8077	0.7955	0.7778
15	0.9368	0.8957	0.8800	0.8286	0.8125	0.8000	0.7818
16	0.9400	0.9000	0.8846	0.8333	0.8171	0.8043	0.7857
17	0.9429	0.9040	0.8889	0.8378	0.8214	0.8085	0.7895
18	0.9455	0.9077	0.8929	0.8421	0.8256	0.8125	0.7931
19	0.9478	0.9111	0.8966	0.8462	0.8295	0.8163	0.7966
20	0.9500	0.9143	0.9000	0.8500	0.8333	0.8200	0.8000
21	0.9520	0.9172	0.9032	0.8537	0.8370	0.8235	0.8033
22	0.9538	0.9200	0.9063	0.8571	0.8404	0.8269	0.8065
23	0.9556	0.9226	0.9091	0.8605	0.8438	0.8302	0.8095
24	0.9571	0.9250	0.9118	0.8636	0.8469	0.8333	0.8125
25	0.9586	0.9273	0.9143	0.8667	0.8500	0.8364	0.8154
30	0.9647	0.9368	0.9250	0.8800	0.8636	0.8500	0.8286
35	0.9692	0.9442	0.9333	0.8909	0.8750	0.8615	0.8400
40	0.9727	0.9500	0.9400	0.9000	0.8846	0.8714	0.8500
45	0.9755	0.9547	0.9455	0.9077	0.8929	0.8800	0.8588
50	0.9778	0.9586	0.9500	0.9143	0.9000	0.8875	0.8667

Table D7 Half Year Rule Capital Tax Factors

Tax Rate = 35%

Interest Rate - %	Declining Balance CCA Rate - %						
	4	8	10	20	25	30	40
1	0.7214	0.6904	0.6834	0.6683	0.6651	0.6630	0.6602
2	0.7690	0.7227	0.7112	0.6849	0.6791	0.6751	0.6699
3	0.8029	0.7492	0.7347	0.7001	0.6921	0.6865	0.6792
4	0.8284	0.7712	0.7548	0.7139	0.7041	0.6971	0.6879
5	0.8481	0.7897	0.7722	0.7267	0.7153	0.7071	0.6963
6	0.8640	0.8057	0.7874	0.7384	0.7257	0.7166	0.7043
7	0.8769	0.8194	0.8009	0.7492	0.7355	0.7255	0.7119
8	0.8877	0.8315	0.8128	0.7593	0.7447	0.7339	0.7191
9	0.8968	0.8421	0.8234	0.7686	0.7533	0.7419	0.7261
10	0.9045	0.8515	0.8330	0.7773	0.7614	0.7494	0.7327
11	0.9113	0.8599	0.8416	0.7854	0.7690	0.7566	0.7391
12	0.9172	0.8675	0.8494	0.7930	0.7762	0.7634	0.7452
13	0.9224	0.8743	0.8566	0.8001	0.7830	0.7699	0.7510
14	0.9270	0.8805	0.8631	0.8068	0.7894	0.7760	0.7567
15	0.9311	0.8862	0.8691	0.8130	0.7955	0.7819	0.7621
16	0.9348	0.8914	0.8747	0.8190	0.8013	0.7875	0.7672
17	0.9382	0.8961	0.8798	0.8246	0.8068	0.7928	0.7722
18	0.9412	0.9005	0.8845	0.8298	0.8120	0.7979	0.7770
19	0.9440	0.9046	0.8889	0.8348	0.8170	0.8028	0.7817
20	0.9465	0.9083	0.8931	0.8396	0.8218	0.8075	0.7861
21	0.9489	0.9118	0.8969	0.8441	0.8263	0.8120	0.7904
22	0.9510	0.9151	0.9005	0.8484	0.8306	0.8163	0.7946
23	0.9530	0.9181	0.9039	0.8524	0.8348	0.8204	0.7986
24	0.9548	0.9210	0.9070	0.8563	0.8387	0.8244	0.8024
25	0.9566	0.9236	0.9100	0.8600	0.8425	0.8282	0.8062
30	0.9636	0.9348	0.9226	0.8762	0.8593	0.8452	0.8231
35	0.9688	0.9433	0.9323	0.8892	0.8731	0.8594	0.8375
40	0.9727	0.9500	0.9400	0.9000	0.8846	0.8714	0.8500
45	0.9759	0.9554	0.9462	0.9090	0.8944	0.8817	0.8609
50	0.9784	0.9598	0.9514	0.9167	0.9028	0.8906	0.8704

Table D8 Full Year Capital Tax Factors

Tax Rate = 35%

Interest Rate - %	Declining Balance CCA Rate - %						
	4	8	10	20	25	30	40
1	0.7200	0.6889	0.6818	0.6667	0.6635	0.6613	0.6585
2	0.7667	0.7200	0.7083	0.6818	0.6759	0.6719	0.6667
3	0.8000	0.7455	0.7308	0.6957	0.6875	0.6818	0.6744
4	0.8250	0.7667	0.7500	0.7083	0.6983	0.6912	0.6818
5	0.8444	0.7846	0.7667	0.7200	0.7083	0.7000	0.6889
6	0.8600	0.8000	0.7813	0.7308	0.7177	0.7162	0.6957
7	0.8727	0.8133	0.7941	0.7407	0.7266	0.7237	0.7021
8	0.8833	0.8250	0.8056	0.7500	0.7348	0.7308	0.7083
9	0.8923	0.8353	0.8158	0.7586	0.7426	0.7375	0.7143
10	0.9000	0.8444	0.8250	0.7667	0.7500	0.7439	0.7200
11	0.9067	0.8526	0.8333	0.7742	0.7569	0.7500	0.7255
12	0.9125	0.8600	0.8409	0.7813	0.7635	0.7558	0.7308
13	0.9176	0.8667	0.8478	0.7879	0.7697	0.7614	0.7358
14	0.9222	0.8727	0.8542	0.7941	0.7756	0.7667	0.7407
15	0.9263	0.8783	0.8600	0.8000	0.7813	0.7717	0.7455
16	0.9300	0.8833	0.8654	0.8056	0.7866	0.7766	0.7500
17	0.9333	0.8880	0.8704	0.8108	0.7917	0.7813	0.7544
18	0.9364	0.8923	0.8750	0.8158	0.7965	0.7857	0.7586
19	0.9391	0.8963	0.8793	0.8205	0.8011	0.7900	0.7627
20	0.9417	0.9000	0.8833	0.8250	0.8056	0.7941	0.7667
21	0.9440	0.9034	0.8871	0.8293	0.8098	0.7981	0.7705
22	0.9462	0.9067	0.8906	0.8333	0.8138	0.8019	0.7742
23	0.9481	0.9097	0.8939	0.8372	0.8177	0.8056	0.7778
24	0.9500	0.9125	0.8971	0.8409	0.8214	0.8091	0.7813
25	0.9517	0.9152	0.9000	0.8444	0.8250	0.8250	0.7846
30	0.9588	0.9263	0.9125	0.8600	0.8409	0.8385	0.8000
35	0.9641	0.9349	0.9222	0.8727	0.8542	0.8500	0.8133
40	0.9682	0.9417	0.9300	0.8833	0.8654	0.8600	0.8250
45	0.9714	0.9472	0.9364	0.8923	0.8750	0.8688	0.8353
50	0.9741	0.9517	0.9417	0.9000	0.8833		0.8444

Table D9 Half Year Rule Capital Tax Factors

Tax Rate = 40%

Interest Rate - %	Declining Balance CCA Rate - %						
	4	8	10	20	25	30	40
1	0.6816	0.6462	0.6382	0.6209	0.6173	0.6148	0.6117
2	0.7359	0.6831	0.6699	0.6399	0.6333	0.6287	0.6228
3	0.7748	0.7133	0.6968	0.6572	0.6481	0.6417	0.6333
4	0.8038	0.7385	0.7198	0.6731	0.6618	0.6538	0.6434
5	0.8265	0.7597	0.7397	0.6876	0.6746	0.6653	0.6529
6	0.8445	0.7779	0.7571	0.7010	0.6865	0.6761	0.6620
7	0.8593	0.7936	0.7724	0.7134	0.6977	0.6863	0.6707
8	0.8716	0.8074	0.7860	0.7249	0.7082	0.6959	0.6790
9	0.8820	0.8195	0.7982	0.7355	0.7180	0.7050	0.6870
10	0.8909	0.8303	0.8091	0.7455	0.7273	0.7136	0.6945
11	0.8986	0.8399	0.8190	0.7547	0.7360	0.7218	0.7018
12	0.9054	0.8486	0.8279	0.7634	0.7442	0.7296	0.7088
13	0.9113	0.8564	0.8361	0.7715	0.7520	0.7370	0.7155
14	0.9166	0.8635	0.8436	0.7792	0.7593	0.7440	0.7219
15	0.9213	0.8699	0.8504	0.7863	0.7663	0.7507	0.7281
16	0.9255	0.8759	0.8568	0.7931	0.7729	0.7571	0.7340
17	0.9293	0.8813	0.8626	0.7995	0.7792	0.7632	0.7397
18	0.9328	0.8863	0.8680	0.8055	0.7852	0.7691	0.7452
19	0.9360	0.8909	0.8731	0.8112	0.7909	0.7747	0.7505
20	0.9389	0.8952	0.8778	0.8167	0.7963	0.7800	0.7556
21	0.9416	0.8992	0.8822	0.8218	0.8015	0.7851	0.7605
22	0.9440	0.9030	0.8863	0.8267	0.8064	0.7900	0.7652
23	0.9463	0.9064	0.8901	0.8313	0.8111	0.7948	0.7698
24	0.9484	0.9097	0.8937	0.8358	0.8157	0.7993	0.7742
25	0.9503	0.9127	0.8971	0.8400	0.8200	0.8036	0.7785
30	0.9584	0.9255	0.9115	0.8585	0.8392	0.8231	0.7978
35	0.9643	0.9352	0.9226	0.8734	0.8549	0.8393	0.8143
40	0.9688	0.9429	0.9314	0.8857	0.8681	0.8531	0.8286
45	0.9724	0.9490	0.9386	0.8960	0.8793	0.8648	0.8410
50	0.9753	0.9540	0.9444	0.9048	0.8889	0.8750	0.8519

Table D10 Full Year Capital Tax Factors

Tax Rate = 40%

Interest Rate - %	Declining Balance CCA Rate - %						
	4	8	10	20	25	30	40
1	0.6800	0.6444	0.6364	0.6190	0.6154	0.6129	0.6098
2	0.7333	0.6800	0.6667	0.6364	0.6296	0.6250	0.6190
3	0.7714	0.7091	0.6923	0.6522	0.6429	0.6364	0.6279
4	0.8000	0.7333	0.7143	0.6667	0.6552	0.6471	0.6364
5	0.8222	0.7538	0.7333	0.6800	0.6667	0.6571	0.6444
6	0.8400	0.7714	0.7500	0.6923	0.6774	0.6667	0.6522
7	0.8545	0.7867	0.7647	0.7037	0.6875	0.6757	0.6596
8	0.8667	0.8000	0.7778	0.7143	0.6970	0.6842	0.6667
9	0.8769	0.8118	0.7895	0.7241	0.7059	0.6923	0.6735
10	0.8857	0.8222	0.8000	0.7333	0.7143	0.7000	0.6800
11	0.8933	0.8316	0.8095	0.7419	0.7222	0.7073	0.6863
12	0.9000	0.8400	0.8182	0.7500	0.7297	0.7143	0.6923
13	0.9059	0.8476	0.8261	0.7576	0.7368	0.7209	0.6981
14	0.9111	0.8545	0.8333	0.7647	0.7436	0.7273	0.7037
15	0.9158	0.8609	0.8400	0.7714	0.7500	0.7333	0.7091
16	0.9200	0.8667	0.8462	0.7778	0.7561	0.7391	0.7143
17	0.9238	0.8720	0.8519	0.7838	0.7619	0.7447	0.7193
18	0.9273	0.8769	0.8571	0.7895	0.7674	0.7500	0.7241
19	0.9304	0.8815	0.8621	0.7949	0.7727	0.7551	0.7288
20	0.9333	0.8857	0.8667	0.8000	0.7778	0.7600	0.7333
21	0.9360	0.8897	0.8710	0.8049	0.7826	0.7647	0.7377
22	0.9385	0.8933	0.8750	0.8095	0.7872	0.7692	0.7419
23	0.9407	0.8968	0.8788	0.8140	0.7917	0.7736	0.7460
24	0.9429	0.9000	0.8824	0.8182	0.7959	0.7778	0.7500
25	0.9448	0.9030	0.8857	0.8222	0.8000	0.7818	0.7538
30	0.9529	0.9158	0.9000	0.8400	0.8182	0.8000	0.7714
35	0.9590	0.9256	0.9111	0.8545	0.8333	0.8154	0.7867
40	0.9636	0.9333	0.9200	0.8667	0.8462	0.8286	0.8000
45	0.9673	0.9396	0.9273	0.8769	0.8571	0.8400	0.8118
50	0.9704	0.9448	0.9333	0.8857	0.8667	0.8500	0.8222

Table D11 Half Year Rule Capital Tax Factors

Tax Rate = 45%

Interest Rate - %	Declining Balance CCA Rate - %						
	4	8	10	20	25	30	40
1	0.6418	0.6020	0.5929	0.5736	0.5694	0.5667	0.5631
2	0.7029	0.6435	0.6287	0.5949	0.5874	0.5823	0.5756
3	0.7466	0.6775	0.6589	0.6144	0.6041	0.5969	0.5875
4	0.7793	0.7058	0.6848	0.6322	0.6195	0.6106	0.5988
5	0.8048	0.7297	0.7071	0.6486	0.6339	0.6235	0.6095
6	0.8251	0.7501	0.7267	0.6636	0.6474	0.6356	0.6198
7	0.8417	0.7679	0.7440	0.6776	0.6599	0.6471	0.6295
8	0.8556	0.7833	0.7593	0.6905	0.6717	0.6579	0.6389
9	0.8673	0.7970	0.7729	0.7025	0.6828	0.6681	0.6478
10	0.8773	0.8091	0.7852	0.7136	0.6932	0.6778	0.6564
11	0.8859	0.8199	0.7963	0.7241	0.7030	0.6870	0.6645
12	0.8935	0.8296	0.8064	0.7338	0.7122	0.6958	0.6724
13	0.9002	0.8384	0.8156	0.7430	0.7210	0.7041	0.6799
14	0.9061	0.8464	0.8240	0.7515	0.7293	0.7120	0.6871
15	0.9114	0.8537	0.8317	0.7596	0.7371	0.7196	0.6941
16	0.9162	0.8603	0.8389	0.7672	0.7445	0.7268	0.7007
17	0.9205	0.8665	0.8454	0.7744	0.7516	0.7336	0.7072
18	0.9244	0.8721	0.8515	0.7812	0.7583	0.7402	0.7133
19	0.9280	0.8773	0.8572	0.7877	0.7647	0.7465	0.7193
20	0.9313	0.8821	0.8625	0.7938	0.7708	0.7525	0.7250
21	0.9342	0.8866	0.8674	0.7995	0.7767	0.7583	0.7305
22	0.9370	0.8908	0.8721	0.8050	0.7822	0.7638	0.7359
23	0.9396	0.8947	0.8764	0.8103	0.7875	0.7691	0.7410
24	0.9419	0.8984	0.8805	0.8152	0.7926	0.7742	0.7460
25	0.9441	0.9018	0.8843	0.8200	0.7975	0.7791	0.7508
30	0.9532	0.9162	0.9005	0.8408	0.8191	0.8010	0.7725
35	0.9598	0.9271	0.9130	0.8576	0.8368	0.8192	0.7911
40	0.9649	0.9357	0.9229	0.8714	0.8516	0.8347	0.8071
45	0.9690	0.9426	0.9309	0.8830	0.8642	0.8479	0.8211
50	0.9722	0.9483	0.9375	0.8929	0.8750	0.8594	0.8333

Table D12 Full Year Capital Tax Factors

Tax Rate = 45%

Interest Rate - %	Declining Balance CCA Rate - %						
	4	8	10	20	25	30	40
1	0.6400	0.6000	0.5909	0.5714	0.5673	0.5645	0.5610
2	0.7000	0.6400	0.6250	0.5909	0.5833	0.5781	0.5714
3	0.7429	0.6727	0.6538	0.6087	0.5982	0.5909	0.5814
4	0.7750	0.7000	0.6786	0.6250	0.6121	0.6029	0.5909
5	0.8000	0.7231	0.7000	0.6400	0.6250	0.6143	0.6000
6	0.8200	0.7429	0.7188	0.6538	0.6371	0.6250	0.6087
7	0.8364	0.7600	0.7353	0.6667	0.6484	0.6351	0.6170
8	0.8500	0.7750	0.7500	0.6786	0.6591	0.6447	0.6250
9	0.8615	0.7882	0.7632	0.6897	0.6691	0.6538	0.6327
10	0.8714	0.8000	0.7750	0.7000	0.6786	0.6625	0.6400
11	0.8800	0.8105	0.7857	0.7097	0.6875	0.6707	0.6471
12	0.8875	0.8200	0.7955	0.7188	0.6959	0.6786	0.6538
13	0.8941	0.8286	0.8043	0.7273	0.7039	0.6860	0.6604
14	0.9000	0.8364	0.8125	0.7353	0.7115	0.6932	0.6667
15	0.9053	0.8435	0.8200	0.7429	0.7188	0.7000	0.6727
16	0.9100	0.8500	0.8269	0.7500	0.7256	0.7065	0.6786
17	0.9143	0.8560	0.8333	0.7568	0.7321	0.7128	0.6842
18	0.9182	0.8615	0.8393	0.7632	0.7384	0.7188	0.6897
19	0.9217	0.8667	0.8448	0.7692	0.7443	0.7245	0.6949
20	0.9250	0.8714	0.8500	0.7750	0.7500	0.7300	0.7000
21	0.9280	0.8759	0.8548	0.7805	0.7554	0.7353	0.7049
22	0.9308	0.8800	0.8594	0.7857	0.7606	0.7404	0.7097
23	0.9333	0.8839	0.8636	0.7907	0.7656	0.7453	0.7143
24	0.9357	0.8875	0.8676	0.7955	0.7704	0.7500	0.7188
25	0.9379	0.8909	0.8714	0.8000	0.7750	0.7545	0.7231
30	0.9471	0.9053	0.8875	0.8200	0.7955	0.7750	0.7429
35	0.9538	0.9163	0.9000	0.8364	0.8125	0.7923	0.7600
40	0.9591	0.9250	0.9100	0.8500	0.8269	0.8071	0.7750
45	0.9633	0.9321	0.9182	0.8615	0.8393	0.8200	0.7882
50	0.9667	0.9379	0.9250	0.8714	0.8500	0.8313	0.8000

Table D13 Half Year Rule Capital Tax Factors
Tax Rate = 50%

Interest Rate - %	Declining Balance CCA Rate - %						
	4	8	10	20	25	30	40
1	0.6020	0.5578	0.5477	0.5262	0.5216	0.5185	0.5146
2	0.6699	0.6039	0.5874	0.5499	0.5416	0.5358	0.5285
3	0.7184	0.6417	0.6210	0.5715	0.5601	0.5521	0.5417
4	0.7548	0.6731	0.6497	0.5913	0.5773	0.5673	0.5542
5	0.7831	0.6996	0.6746	0.6095	0.5933	0.5816	0.5661
6	0.8057	0.7224	0.6963	0.6263	0.6082	0.5951	0.5775
7	0.8241	0.7421	0.7155	0.6417	0.6222	0.6079	0.5884
8	0.8395	0.7593	0.7325	0.6561	0.6352	0.6199	0.5988
9	0.8525	0.7744	0.7477	0.6694	0.6475	0.6313	0.6087
10	0.8636	0.7879	0.7614	0.6818	0.6591	0.6420	0.6182
11	0.8733	0.7999	0.7737	0.6934	0.6700	0.6523	0.6273
12	0.8817	0.8107	0.7849	0.7042	0.6803	0.6620	0.6360
13	0.8891	0.8205	0.7951	0.7144	0.6900	0.6712	0.6443
14	0.8957	0.8293	0.8045	0.7239	0.6992	0.6800	0.6524
15	0.9016	0.8374	0.8130	0.7329	0.7079	0.6884	0.6601
16	0.9069	0.8448	0.8210	0.7414	0.7161	0.6964	0.6675
17	0.9117	0.8516	0.8283	0.7494	0.7240	0.7040	0.6746
18	0.9160	0.8579	0.8350	0.7569	0.7315	0.7113	0.6815
19	0.9200	0.8637	0.8414	0.7641	0.7386	0.7183	0.6881
20	0.9236	0.8690	0.8472	0.7708	0.7454	0.7250	0.6944
21	0.9269	0.8740	0.8527	0.7773	0.7518	0.7314	0.7006
22	0.9300	0.8787	0.8578	0.7834	0.7580	0.7375	0.7065
23	0.9329	0.8830	0.8627	0.7892	0.7639	0.7434	0.7122
24	0.9355	0.8871	0.8672	0.7947	0.7696	0.7491	0.7177
25	0.9379	0.8909	0.8714	0.8000	0.7750	0.7545	0.7231
30	0.9480	0.9069	0.8894	0.8231	0.7990	0.7788	0.7473
35	0.9554	0.9190	0.9033	0.8418	0.8187	0.7991	0.7679
40	0.9610	0.9286	0.9143	0.8571	0.8352	0.8163	0.7857
45	0.9655	0.9362	0.9232	0.8700	0.8491	0.8310	0.8012
50	0.9691	0.9425	0.9306	0.8810	0.8611	0.8438	0.8148

Table D14 Full Year Capital Tax Factors
Tax Rate = 50%

Interest Rate - %	Declining Balance CCA Rate - %						
	4	8	10	20	25	30	40
1	0.6000	0.5556	0.5455	0.5238	0.5192	0.5161	0.5122
2	0.6667	0.6000	0.5833	0.5455	0.5370	0.5313	0.5238
3	0.7143	0.6364	0.6154	0.5652	0.5536	0.5455	0.5349
4	0.7500	0.6667	0.6429	0.5833	0.5690	0.5588	0.5455
5	0.7778	0.6923	0.6667	0.6000	0.5833	0.5714	0.5556
6	0.8000	0.7143	0.6875	0.6154	0.5968	0.5833	0.5652
7	0.8182	0.7333	0.7059	0.6296	0.6094	0.5946	0.5745
8	0.8333	0.7500	0.7222	0.6429	0.6212	0.6053	0.5833
9	0.8462	0.7647	0.7368	0.6552	0.6324	0.6154	0.5918
10	0.8571	0.7778	0.7500	0.6667	0.6429	0.6250	0.6000
11	0.8667	0.7895	0.7619	0.6774	0.6528	0.6341	0.6078
12	0.8750	0.8000	0.7727	0.6875	0.6622	0.6429	0.6154
13	0.8824	0.8095	0.7826	0.6970	0.6711	0.6512	0.6226
14	0.8889	0.8182	0.7917	0.7059	0.6795	0.6591	0.6296
15	0.8947	0.8261	0.8000	0.7143	0.6875	0.6667	0.6364
16	0.9000	0.8333	0.8077	0.7222	0.6951	0.6739	0.6429
17	0.9048	0.8400	0.8148	0.7297	0.7024	0.6809	0.6491
18	0.9091	0.8462	0.8214	0.7368	0.7093	0.6875	0.6552
19	0.9130	0.8519	0.8276	0.7436	0.7159	0.6939	0.6610
20	0.9167	0.8571	0.8333	0.7500	0.7222	0.7000	0.6667
21	0.9200	0.8621	0.8387	0.7561	0.7283	0.7059	0.6721
22	0.9231	0.8667	0.8438	0.7619	0.7340	0.7115	0.6774
23	0.9259	0.8710	0.8485	0.7674	0.7396	0.7170	0.6825
24	0.9286	0.8750	0.8529	0.7727	0.7449	0.7222	0.6875
25	0.9310	0.8788	0.8571	0.7778	0.7500	0.7273	0.6923
30	0.9412	0.8947	0.8750	0.8000	0.7727	0.7500	0.7143
35	0.9487	0.9070	0.8889	0.8182	0.7917	0.7692	0.7333
40	0.9545	0.9167	0.9000	0.8333	0.8077	0.7857	0.7500
45	0.9592	0.9245	0.9091	0.8462	0.8214	0.8000	0.7647
50	0.9630	0.9310	0.9167	0.8571	0.8333	0.8125	0.7778

ANSWERS TO SELECTED PROBLEMS

Chapter 2

2.1 8.33 years, 7.27 years   **2.3** Compound interest = $2382.03, simple interest = $2420   **2.5** $2518.86
2.8 (a) $2563.95   **2.10** 14.27 years   **2.13** $10,501.09   **2.16** $4606.13   **2.19** (a) $485.35
2.22 (a) $3154.70   **2.24** (a) $6204.28   **2.26** $2752.95   **2.28** $635.89   **2.31** $68,690,412   **2.35** $758.93
2.38 $315.95   **2.42** $397.45   **2.45** (2), (4)   **2.46** (2), (4), (5)   **2.48** (a)   **2.53** $37,696,550

Chapter 3

3.2 $r = 18\%$, $i_a = 19.56\%$ **3.4** $r = 38.68\%$, $i_a = 47.02\%$ **3.8** (a) $F = \$1044$ (b) $30,084   (c) $19,222
3.11 $F = \$6034.69$ **3.13** (d)   **3.16** $A = \$583.11$ **3.18** (b)   **3.21** (a) $A = \$356$ (b) $22.82
(c) $A = \$136.10$ **3.24** $r = 20.66\%$ **3.26** (a) $6231   (b) $16,351   (c) $141,476   **3.28** $F_1 = \$11,751.78$,
$F_2 = \$12,496.20$ **3.33** $A = \$627.50$ **3.35** $4055.28   **3.39** $64,429.09   **3.43** $P = \$546.45$
3.45 (a) $A = 10,000(\text{A/P}, 0.75\%, 24)$ (b) $B_{12} = A(\text{P/A}, 0.75\%, 12)$ **3.47** (a) $A = \$368.68$
(b) $i = 1.25\%$ per month   **3.50** $A = \$1654.38$ **3.52** $5135.82   **3.58** (a) $r = 21.996\%$ (b) $P = \$2771.39$
3.60 (a) $A = \$303.95$ (b) $293.97   **3.64** (a) $A = \$482.77$ (b) $522.74   **3.68** (c)   **3.70** $i_a = 10.84\%$
3.73 (a) $1227.20   (b) $938.04   (c) $i_a = 15.9\%$

Chapter 4

4.2 (a) 1 year   (b) 1.17 year   **4.5** (d)   **4.7** (a) $A_1 = 100$, $A_2 = 520$, $A_4 = 600$ (b) 20%   (c) $FW(15\%) =$
$219.78 > 0$, accept   **4.9** (a) $X = \$29.96$, $Y = \$49.35$ (b) 0   (c) ⓐ $= \$49.96$, ⓑ $= \$59.35$, ⓒ $= 17.91\%$
4.12 $1,599,319   **4.14** $FW(15\%)_A = -\$85.04$, $FW(15\%)_B = \$18.88$, $FW(15\%)_C = \$49.04$, $FW(15\%)_D = \$57.45$,
$FW(15\%)_E = \$0.12$ **4.18** (a) $267,785   (b) $29,456.40   **4.23** $PW(10\%)_A = \$860.46$, $PW(10\%)_B = \$886.28$
4.25 (a) $PW(15\%)_A = \$311.15$, $PW(15\%)_B = \$439.70$ (b) $FW(15\%)_A = \$411.50$, $FW(15\%)_B = \$581.50$
4.27 $CE(12\%)_A = -\$27,215$, $CE(12\%)_B = -\$45,226$, Select A   **4.30** (a) $PW(8\%)_A = -\$5993.50$,
$PW(8\%)_B = -\$6069.75$, Select A   (b) $PW(8\%)_A = -\$11,250$, $PW(8\%)_B = -\$11,034$ Select B
4.31 $PW(10\%)_A = \$58.55$, $PW(10\%)_B = \$82.96$ over 6-year analysis period   **4.34** (a) Infinite series (Model A
is replaced by Model B at the end of year 3) $PW(15\%)_{ABB\_B} = \$961.13$, $PW(15\%)_{BBB\_B} = \$1031.10$, Select
Model B   (b) $S = \$827.49$ **4.37** (a) $PW(15\%) = \$2557.74$ (b) $X = \$40,000$ (c) $3890   (d) Select A2
4.43 $PW(15\%) = \$6,895,810 > 0$, accept

Chapter 5

5.2 $AE(10\%) = -\$5957.13$ **5.4** $AE(10\%) = -\$64.68$ **5.6** $AE(12\%)_A = -\$77.35 < 0$, reject;
$AE(12\%)_B = \$69.95 > 0$, accept; $AE(12\%)_C = -\$655.24 < 0$, reject; $AE(12\%)_D = \$623.32 > 0$, accept
5.9 $AE(10\%) = \$28.10$ **5.12** $62,852   **5.15** $84,142   **5.18** $46,970.63   **5.20** 1017.8 hours/year
5.23 $0.52 per km   **5.25** 1298 hours   **5.27** (a) $AE(18\%)_A = \$2450.70$, $AE(18\%)_B = \$2632.60$
(b) $1.23/hour, $1.32/hour   (c) B is better   **5.29** bond, $AE(6.17\%) = \$79.59$; stock, $AE(6.17\%) = \$107.17$;
loan, $AE(6.17\%) = -\$300.57$ **5.32** $AE(12\%)_A = -\$7763.50$, $AE(12\%)_B = -\$6910.80$, Select B
5.36 7 passengers (6.56)   **5.40** tank/tower option: $AE(12\%) = \$19,894$, tank/hill option: $AE(12\%) = \$16,150$,
Select the "tank/hill" option.   **5.44** $0.06

Chapter 6

6.1 44.90% per year   **6.3** 17.68%   **6.4** **(a)** Project B   **(b)** 74.23% (Project A), 111.11% (Project B)
6.6 72.45%   **6.8** **(a)** Simple projects: A, B, E; nonsimple projects: C, D   **(b)** 28.08%   **(c)** $i°_A = 28.08\%$,
$i°_B = 33.97\%$, $i°_C = 4.59\%$, $i°_D = 4.59\%$, $i°_E = 8.78\%$ **6.10** **(b)** 10%, 40%   **6.13** 10%   **6.15** 13.94%
6.17 **(a)** $i°_A = 11.71\%$, $i°_B = 19.15\%$ **(b)** Select B   **6.19** **(a)** $i°_1 = 20\%$, $i°_2 = 18\%$, $i°_3 = 32.45\%$, -92.45%
(b) $IRR_1 = 20\%$, $IRR_2 = 18\%$, $IRR_3 = 31.07\%$ **(c)** Accept   **6.23** **(a)** $i°_1 = 20\%$, 40%, $PB_0(20\%) = -100$,
$PB_1(20\%) = 140$, $PB_2(20\%) = 0$ $(-, +, 0)$, mixed investment   **(b)** IRR = 10%   **(c)** Reject, IRR < 12%
6.27 $IRR_{bond-CD} = 10\% > 9\%$, Select bond   **6.29** **(a)** $IRR_{A2-A1} = 29.92\%$ **(b)** Select A2   **6.33** **(a)** $i°_{B-A} =$
0%, 30%, mixed investment, IRR = 16.96% > 15%, accept   **6.34** $IRR_{FMS-CMS} = 14.7\% < 15\%$, prefer
CMS   **6.37** **(a)** $i° = 10\%$, 20%   **(b)** IRR = 15.22%   **(c)** Select 1 ($IRR_{2-1} = 11.96\%$, $IRR_{3-1} = 9.16\%$)
6.39 Model C   **6.41** A_6 **6.42** $IRR_{B-A} = 15.98\%$, Select B   **6.45** **(a)** $IRR_{B-A} = 15.98\%$, Select B
(b) $IRR_{C-D} = 7.03\%$, Select D   **(c)** No IRR (Dominance), Select E   **6.47** **(a)** IRR = -23.18%, loss
(b) 98,905 copies

Chapter 7

7.1 **(a)** $56,700   **(b)** $60,700   **7.3** $1,500   **7.5** $1,591,500   **7.7** $31,000   **7.9** $D_1 = \$4,000$, $D_2 = \$2,400$,
$D_3 = \$1,440$, $D_4 = \$1,080$, $D_5 = \$1,080$,   **7.14** $30,000   **7.19** **(a)** $215,000   **(b)** $CCA_1 = \$21,500$,
$CCA_2 = \$38,700$, $CCA_3 = \$30,960$, $CCA_4 = \$24,768$, $CCA_5 = \$19,814$, $CCA_6 = \$15,852$, $CCA_7 = \$12,681$,
$CCA_8 = \$10,145$, $CCA_9 = \$8,116$, $CCA_{10} = \$6,493$, $CCA_{11} = \$5,194$, $CCA_{12} = \$4,155$ **7.23** $CCA_1 = \$10,000$,
$CCA_2 = \$19,000$, $CCA_3 = \$17,100$, $CCA_4 = \$15,390$, $CCA_5 = \$13,851$, $CCA_6 = \$12,466$, $CCA_7 = \$11,219$,
$CCA_8 = \$10,097$, $CCA_9 = \$9,088$, $CCA_{10} = \$8,179$,   **7.29** **(a)** Revised book depreciation = $8,833   **(b)** Total
tax depreciation = $11,274   **7.33** **(a)** $12,000   **(b)** $11,256   **(c)** $9,800   **(d)** $21,818   **(e)** $29,400

Chapter 8

8.1 Average taxation rate = 21.54%, Marginal taxation rate = 38.61%   **8.3** **(a)** Average taxation rate = 31.17%,
Marginal taxation rate = 45.58%   **(b)** Average taxation rate = 34.03%, Marginal taxation rate = 49.6%
8.5 **(a)** Ontario tax = $3,059, Newfoundland tax = $3,268   **(b)** Ontario tax = $10,041, Newfoundland
tax = $10,728   **(c)** Ontario tax = $18,649, Newfoundland tax = $19,715   **(d)** Ontario tax = $39,427,
Newfoundland tax = $40,195   **8.7** Tax payable = $15,983   **8.9** Year 1 = $39,095, Year 2 = $44,680,
Year 3 = $65,299   **8.13** **(a)** $2,432   **(b)** -204 **(c)** $-2,450$ **8.17** **(a)** $17,929   **(b)** $429   **(c)** $-\$34,571$
8.22 **(a)** Marginal Federal tax rate = 28.84%, Marginal Alberta tax rate = 15.50%   **(b)** Combined average tax
rate = 20.05%   **8.25** **(a)** Marginal tax rate = 44.34%   **(b)** Year 1 Average tax rate = 31.47%, Year 2 Average
tax rate = 31.24%   **(c)** Disposal tax effect = $1,009, Net salvage value = $33,991

Chapter 9

9.1 $A_0 = -\$54,000$, $A_1 = \$10,280$, $A_2 = \$17,423$, $A_3 = \$16,628$, $A_4 = \$21,807$, $A_5 = \$22,977$, $A_6 = \$20,588$
9.3 $A_0 = -\$1,100,000$, $A_1 = \$766,000$, $A_2 = \$790,840$, $A_3 = \$775,226$, $A_4 = \$764,215$, $A_5 = \$1,456,719$
9.5 $A_0 = -\$75,000$, $A_1 = \$13,500$, $A_2 = \$28,650$, $A_3 = \$26,355$, $A_4 = \$24,748$, $A_5 = \$23,624$, $A_6 = \$33,087$
9.7 **(a)** $A_0 = -\$26,000$, $A_1 = \$4,800$, $A_2 = \$5,892$, $A_3 = \$5,096$, $A_4 = \$4,540$, $A_5 = \$7,832$ **(b)** IRR = 2.58%
(c) Since IRR < 12%, not acceptable.   **9.9** $A_0 = -\$200,000$, $A_1 = \$132,000$, $A_2 = \$140,400$, $A_3 = \$134,280$,
$A_4 = \$129,996$, $A_5 = \$146,324$ **(b)** IRR = 61.46% (MARR 22%)   **9.11** $A_0 = 0$, $A_1 = \$8,000$, $A_2 = \$21,400$,
$A_3 = \$18,280$, $A_4 = \$16,996$, $A_5 = \$66,997$, $A_6 = \$64,898$, $A_7 = \$89,429$ **9.16** $A_0 = -\$50,000$, $A_1 = \$11,000$,
$A_2 = \$12,600$, $A_3 = \$11,880$, $A_4 = \$11,304$, $A_5 = \$18,216$, IRR = 8.8%   **9.22** $A_0 = -\$200,000$, $A_1 = \$132,000$,
$A_2 = \$140,000$, $A_3 = \$134,280$, $A_4 = \$129,996$, $A_5 = \$165,651$ **9.29** $A_0 = -\$15,000$, $A_1 = -\$1,860$,
$A_2 = -\$1,230$, $A_3 = -\$1,689$, $A_4 = -\$2,010$, $A_5 = -\$1,010$, $PW(20\%) = -\$19,756$, $AE(20\%) = -\$6,607$
9.32 $A_0 = -\$62,000$, $A_1 = \$15,720$, $A_2 = \$18,324$, $A_3 = \$16,427$, $A_4 = \$34,329$, $PW(8\%) = \$6,536$

Chapter 10

10.1 $f = 1.95\%$ **10.3** $f = 7.995\%$ **10.5** $P = \$15,056.31$ **10.7** $A'_0 = \$1000$, $A'_4 = \$1710$, $A'_5 = \$2466$, $A'_7 = \$3040$ **10.9** (a) $\$26,876$ (b) $\$26,876$, **10.11** $\$600$ (Actual dollars), $\$376.45$ (Constant dollars) **10.13** $\$6,494,215$ **10.16** (a) $A = \$4517.24$ (b) $A_1 = \$8244$ **10.19** (a) $A_0 = -\$10,000$, $A_1 = \$10,490$, $A_2 = \$53,123$ (b) $PW(18\%) = \$37,042$ **10.24** (a) $A_0 = -\$55,000$, $A_1 = \$5,171$, $A_2 = \$5,488$, $A_3 = \$14,388$, $A_4 = \$19,071$, $A_5 = \$29,631$ **10.29** (a) $PW(20\%)_A = -\$352,494$, $PW(20\%)_B = -\$329,058$, select Engine B (b) $AE(20\%)_A = -\$117,876$, $AE(20\%)_B = -\$110,030$, select Engine B (c) $AE(20\%)_A = -\$877,118$, $AE(20\%)_B = -\$818,802$, select Engine B

Chapter 11

11.1 (a) Flotation cost $= \$638,298$, number of shares $= 425,532$ (b) Flotation cost $= \$193,680$, number of $\$1000$ bonds $= 10,194$, interest payment $= \$1,223,280$ per year **11.3** $A_0 = \$0$, $A_1 = \$14,824$, $A_2 = \$19,130$, $A_3 = \$14,247$, $A_4 = \$10,385$, $A_5 = \$57,183 - \frac{\$2939}{i+0.3}$ **11.7** (a) $PW(15\%) = -\$21,670$ (b) $PW(15\%) = -\$21,969$ (c) Lease **11.9** (a) $PW(15\%)_{leasing} = -\$119,909$ (b) $PW(15\%)_{owing} = -\$129,732$ **11.11** (a) Buy option: $A_0 = \$0$, $A_1 = -\$5,032$, $A_2 = -\$4,659$, $A_3 = \$457 - \frac{\$167}{i+0.3}$; lease option: $A_0 = -\$500$, $A_1 = -\$3315$, $A_2 = -\$3315$, $A_3 = -\$2815$ (b) $PW(13\%)_{Buy} = -\$7,627$, $PW(13\%)_{Lease} = -\$7981$ **11.14** $\$17,591$

Chapter 12

12.1 (a) sunk cost $= \$118$ (b) depreciation base $= \$15,000$ (c) disposal tax effect $= \$47$ (d) opportunity cost $= \$2,547$ (e) Replace the old machine. $AE_{old} = -\$6,718$, $AE_{new} = -\$5,917$ **12.3** (a) cost base $= \$120,000$ (b) & (c) $PW(15\%) = -\$13,175$, $AE(15\%) = -\$3,930$ (b) & (d) $PW(15\%) = \$39,042$, $AE(15\%) = \$9,386$. Purchase the new machine. **12.5** Do not replace the defender now. $AE_{old} = \$13,374$, $AE_{new} = \$10,090$ **12.9** Economic service life $= 2$ years **12.13** Replace the defender now. $AE_{old} = -\$3,495$, $AE_{new} = \$3,241$ **12.17** (j_0,3), (j,3), (j,2) $PW(12\%) = \$17,850$ **12.22** Do not replace the defender now. $AE_{old}(15\%) = -\$7,188$, $AE_{new}(15\%) = -\$13,231$ **2.26** Replace the defender now. Challenger $AE(12\%) = -\$1,340$, Defender $AE_1 = -\$1,707$, $AE_2 = -\$1,789$, $AE_3 = -\$1,953$, $AE_4 = -\$2,084$, $AE_5 = -\$2,132$

Chapter 13

13.1 (c) **13.3** (a) 12 alternatives with "do-nothing" alternative (b) A, B, C, F: $PW(10\%) = \$288$ **13.5** (a) Project C (b) Project C **13.7** (a) 13 alternatives including "do-nothing" alternative (b) B, C, E, F with $PW(10\%) = \$3080$ **13.9** (a) $IRR_E = 4.37\%$, $IRR_F = 46.14\%$ (b) Project F **13.11** Select VI project. $PW(12\%)_{VI} = -\$3743$ **13.13** 1, 2, 3, 4, 7 net 90 million ($I = \$99$ M) **13.16** C_1D_4 $\$25,000$ **13.19** (a) AD (b) $\$6170$ from AD and $2000(F/P, 8\%, 3) = \$2519$ from left-over **13.21** (a) 9% **13.23** 14.56% **13.25** $k = 12.6\%$ **13.28** (a) IRR $= 50.35\% > 20\%$, accept $i_e = 20\%$ (b) IRR $= 28.45\% > 14.3\%$, accept $k = 14.3\%$

Chapter 14

14.5 (a) Select design A $B'C(i)_{A-B} = 2.56 > 1$ (b) No $B'C(i)_{A-C} = 2.57 > 1$ **14.7** Either A_1 or A_2 **14.10** Assuming no do-nothing alternative, select option 1 **14.13** (a) $\$18.67$ per tonne assuming no capital cost (b) Site 1 $= \$5.96$ per tonne, Site 2 $= \$7.42$ per tonne, Site 3 $= \$7.13$ per tonne, Site 4 $= \$7.55$ per tonne, Site 1 is the most economical choice

Chapter 15

15.1 (a) $PW(15\%) = \$3,185 > 0$, accept (b) $PW(15\%) = \$2,328 > 0$, accept (c) $\$35,865$ **15.3** (a) If $i < 20\%$, select 5-floor plan; If $i > 20\%$, select 2-floor plan **15.6** (a) Model A: $AE(10\%) = -\$1,340$, Model B: $AE(10\%) = -\$1,405$, select Model A (b) $\$793$ (c) Model A **15.8** 52.01% **15.10** $\$58,410$ **15.12** $\$15,335$ **15.15** $E[PW(12\%)] = \$521,460$ **15.18** (a) $PW(15.5\%) = -4,952 + 1.0413 X$ (b) $\$15,874$ (c) $65,058,341$ **15.20** (a) Contract A: $E[NPW] = \$40,000$, $Var[NPW] = 1,400,000,000$; Contract B: $E[NPW] = \$13,000$, $Var[NPW] = 381,000,000$; select A based on $E[NPW]$ (b) 0.72 **15.24** Cash flow: $\$1,184$, $\$2,165$; $PW(10\%) = \$784$ **15.26** Mean $= \$16.27$, Variance $= (4.54)^2$, $P\{NPW \le \$30\} = 100\%$

INDEX